MW00843469

Grundlehren der mathematischen Wissenschaften 258

A Series of Comprehensive Studies in Mathematics

Grundlehren der mathematischen Wissenschaften

A Series of Comprehensive Studies in Mathematics

A Selection

continued after index

Joel Smoller

Shock Waves and Reaction–Diffusion Equations

Second Edition

With 165 Illustrations

Springer-Verlag
New York Berlin Heidelberg London Paris
Tokyo Hong Kong Barcelona Budapest

Joel Smoller
Department of Mathematics
University of Michigan
Ann Arbor, MI 48109
USA

Mathematics Subject Classifications (1991): 35-02, 35L67, 35Q30, 35Q35, 73D05, 76L05

Library of Congress Cataloging in Publication Data
Smoller, Joel
    Shock waves and reaction–diffusion equations / Joel Smoller, —
2nd ed.
        p.   cm. — (Grundlehren der mathematischen Wissenschaften :
258)
    Includes bibliographical references and index.
    ISBN 0-387-94259-9 (New York). — ISBN 3-540-94259-9 (Berlin)
DM168.00
    1. Shock waves.   2. Reaction–diffusion equations.   I. Title.
II. Series.
QA927.S57   1994
515′.353—dc20   93-50712

Printed on acid-free paper.

Production coordinated by Brian Howe and managed by Henry Krell; manufacturing super-
vised by Vincent Scelta.
Typeset by Asco Trade Typesetting Ltd., Hong Kong
Printed and bound by R.R. Donnelley and Sons, Harrisonburg, VA.
Printed in the United States of America.

9 8 7 6 5 4 3 2 1

ISBN 0-387-94259-9 Springer-Verlag New York Berlin Heidelberg
ISBN 3-540-94259-9 Springer-Verlag Berlin Heidelberg New York

To Debbie, Alex, and Sally

Acknowledgment

> I get by with a little help from my friends.
>
> LENNON AND MCCARTNEY

Throughout my career, I have had the good fortune to be closely associated with many very gifted mathematicians. I learned a great deal from each of them; this book is really not much more than a compilation of what they taught me.

First, I want to thank Louis Nirenberg for initially inspiring me and for giving me an early opportunity to visit the Courant Institute. It was there that I met Edward Conway, my friend and collaborator, who showed me the mathematics of shock waves, and convinced me (with difficulty!) of the value of difference schemes. Years later, he introduced me to the equations of mathematical ecology, the study of which led to much fruitful joint research.

I must also thank Peter Lax, whose work has always interested me, for inviting me to visit the Courant Institute once again. The second time at the "Courant" I met my other good friend and collaborator, Charles Conley. It was he who enthusiastically taught me so much deep and exciting mathematics. Among N ($\geqslant 1$) other things, I learned from him the power and beauty of topological methods in analysis. He has also generously shared with me many of his brilliant insights. His ideas continue to influence my research to the present day.

To Takaaki Nishida, I owe thanks for sharing his ideas with me, and for actively collaborating with me on so many occasions.

Claude Bardos also deserves thanks for his willingness to discuss many of his ideas, his constant encouragement, and for introducing me to the Paris school of mathematics.

This book could not have been written were it not for the many able students who attended my courses on differential equations at Michigan. Their enthusiasm and hard work constantly served as a stimulation to me. In particular, special thanks too are due to Bob Gardner, David Hoff, Jeffrey Johnson, and Tai-Ping Liu, each of whom very quickly became my collaborator and teacher.

To my colleagues at Michigan: Paul Federbush, Carl Simon, Al Taylor, and especially to Jeff Rauch, Joe Ullman, and Ricky Wasserman, I owe many thanks for so often giving me their time in order to help me get over some rough spots.

Portions of this book are an outgrowth of lectures which I gave at the following institutions: Universidad Autonoma de Madrid, Université de Paris, Orsay, University of Warwick, and the University of Wisconsin. I am grateful to the mathematicians at these places for giving me the opportunity to lecture.

Charles Conley, Edward Conway, Bob Gardner, David Hoff, Tai-Ping Liu, Takaaki Nishida, Jeffrey Rauch, Blake Temple, and Peter Wolfe, each read portions of the manuscript. I am grateful to all of them for their many helpful suggestions.

Finally, I want to thank James Glimm for being a constant source of inspiration and encouragement over the years.

I should like to take this opportunity to thank both the U.S. Airforce Office of Scientific Research, and its mathematics director, Robert Buchel, as well as the National Science Foundation and its applied mathematics directors, Ettore Infante, and James Greenberg, for supporting my research.

Preface to the Second Edition

For this edition, a number of typographical errors and minor slip-ups have been corrected. In addition, following the persistent encouragement of Olga Oleinik, I have added a new chapter, Chapter 25, which I titled "Recent Results." This chapter is divided into four sections, and in these I have discussed what I consider to be some of the important developments which have come about since the writing of the first edition. Section I deals with reaction–diffusion equations, and in it are described both the work of C. Jones, on the stability of the travelling wave for the Fitz-Hugh–Nagumo equations, and symmetry-breaking bifurcations. Section II deals with some recent results in shock-wave theory. The main topics considered are L. Tartar's notion of compensated compactness, together with its application to pairs of conservation laws, and T.-P. Liu's work on the stability of viscous profiles for shock waves. In the next section, Conley's connection index and connection matrix are described; these general notions are useful in constructing travelling waves for systems of nonlinear equations. The final section, Section IV, is devoted to the very recent results of C. Jones and R. Gardner, whereby they construct a general theory enabling them to locate the point spectrum of a wide class of linear operators which arise in stability problems for travelling waves. Their theory is general enough to be applicable to many interesting reaction–diffusion systems.

Many people have been kind in assisting me with this revision; in particular, special thanks are due to Bob Gardner, Xabier Garaizer, David Hoff, Chen Jing, Tai-Ping Liu, Blake Temple, Rick Wasserman, Zhou-Ping Xin, and especially to Vijay Pant for his careful reading of the manuscript. In addition, I received many letters from mathematicians all over the world, who either corrected some of my errors, or showed me "different proofs." I wish to thank all of them for their interest.

In closing, I want to dedicate this edition to the memory of both Charles Conley and Edward Conway.

Preface to the First Edition

> ... the progress of physics will to a large extent depend on the progress of nonlinear mathematics, of methods to solve nonlinear equations ... and therefore we can learn by comparing different nonlinear problems.
>
> WERNER HEISENBERG

I undertook to write this book for two reasons. First, I wanted to make easily available the basics of both the theory of hyperbolic conservation laws and the theory of systems of reaction–diffusion equations, including the generalized Morse theory as developed by C. Conley. These important subjects seem difficult to learn since the results are scattered throughout the research journals.[1] Second, I feel that there is a need to present the modern methods and ideas in these fields to a wider audience than just mathematicians. Thus, the book has some rather sophisticated aspects to it, as well as certain textbook aspects. The latter serve to explain, somewhat, the reason that a book with the title *Shock Waves and Reaction–Diffusion Equations* has the first nine chapters devoted to linear partial differential equations. More precisely, I have found from my classroom experience that it is far easier to grasp the subtleties of nonlinear partial differential equations *after* one has an understanding of the basic notions in the linear theory.

This book is divided into four main parts: linear theory, reaction-diffusion equations, shock wave theory, and the Conley index, in that order. Thus, the text begins with a discussion of ill-posed problems. The aim here was to show that partial differential equations are not divorced from side conditions; indeed specific side conditions are required for specific equations. And in view of Lewy's example, which is presented in its entirety, no side conditions can force solutions on some equations. We discuss an example of a nonlinear scalar conservation law which has no global classical solution, thereby foreshadowing the notion of "weak" solution. In Chapter 2 we consider characteristics, an important notion which comes up widely in nonlinear contexts. Chapter 3 deals with the simple one-dimensional wave equation. Here is where we introduce the reader to the important ideas of

---

[1] This is not quite true; there are some good survey articles on shock waves (e.g., [Lx 5]) but these do not contain many proofs. Also in the theory of reaction–diffusion equations, there are the books [Fi] and [Mu], but they both seem to me to be research monographs.

domains of dependence, energy integrals, and finite differences. The purpose of the following chapter is to demonstrate the power, generality, and elegance of energy integral methods. In the course of the development we present several basic techniques for obtaining inequalities.

The next chapter is devoted to Holmgren's uniqueness theorem. We view it in a modern context, where we can use it later to motivate Oleinik's uniqueness theorems for conservation laws. In Chapter 6 we consider general hyperbolic operators and show how energy integrals, together with Fourier transform methods, are used to prove global existence theorems. The uniqueness of these solutions is obtained via Holmgren's theorem. Chapter 7 is devoted to the theory of distributions. The importance of this subject for linear operators is, of course, well known. This author firmly believes that the great advances in *nonlinear* partial differential equations over the last twenty years could not have been made were it not for distribution theory. The ideas of this discipline provided the conceptual framework for studying partial differential equations in the context of weak solutions. This "philosophy" carried over, rather easily, to many important nonlinear equations. In Chapters 8 and 9 we study linear elliptic and parabolic equations, respectively, and we prove the basic maximum principles. We also describe the estimates of Schauder, as well as those of Agmon, Douglis, and Nirenberg, which we need in later chapters. The proofs of these important estimates are (happily) omitted since it is difficult to improve upon the exposition given in Gilbarg-Trudinger [GT]. (We point out here that the material in Chapters 1–9 can serve as an introductory course in partial differential equations.)

A quick glance at the contents serves to explain the flavor of those topics which form the major portion of the book. I have made a deliberate effort to explain the main ideas in a coherent, readable manner, and in particular I have avoided excess generality. To be specific, Chapter 10 contains a discussion of how far one can go with the maximum principle for a scalar nonlinear parabolic (or elliptic) equation. It is used to prove the basic comparison and existence theorems; the latter done via the method of upper and lower solutions. The text contains several carefully chosen examples which are used both to illustrate the theorems and to prepare the way for some later topics; e.g., bifurcation theory. The next chapter begins with a development of the variational properties of the eigenvalues for a linear second-order elliptic operator on a bounded domain in $\mathbf{R}^n$. There follows a careful discussion of linearized stability for a class of evolution equations broad enough to include systems of reaction–diffusion equations. In Chapter 12, we give a complete development of degree theory in Banach spaces for operators of the form (Id. + compact). The discussion begins with the finite-dimensional case, culminating with Brouwer's fixed point theorem. This is applied to flows on manifolds; specifically, we give two applications, one to flows on balls and one to flows on tori. The Leray–Schauder degree is then developed, and we illustrate its use in nonlinear elliptic equations. The second half of this chapter is devoted to Morse theory. Our aim is to reinterpret the Morse index in an intrinsic topological way (using the stable manifold theorem), as the homotopy type of a quotient space. This is done in preparation for Chapters 22 and 23, where we consider Conley's extension of the Morse index. We give a proof of Reeb's theorem on the characterization of spheres in terms of Morse functions. The chapter ends with an appendix on algebraic topology where homotopy theory, homology theory, and cohomology theory are discussed. The goal was to make these important ideas accessible to nonspecialists.

In Chapter 13, some of the standard bifurcation theorems are proved; namely, those which come under the heading "bifurcation from a simple eigenvalue." We then use degree theory to prove the bifurcation theorems of both Krasnoselski and Rabinowitz. Again, these theorems are illustrated by applications to specific differential equations. In the final section we discuss, with an example, another more global type of bifurcation which we term "spontaneous" bifurcation. This is related back to earlier examples, and it is also made use of in Chapter 24.

Chapter 14 may be considered the "high point" in this group. It is here where the notion of an invariant region is defined, and all of the basic theorems concerning it are proved. As a first application, we prove a comparison theorem which allows us to obtain rather precise (but somewhat coarse) qualitative statements on solutions. We then give a general theorem on the asymptotic behavior of solutions. Thus, we isolate a parameter which, when positive, implies that for large time, every solution gets close to a spatially independent one; in particular, no bifurcation to nonconstant steady-state solutions can occur. There follows a section which makes quantitative the notion of an invariant region; the statement is that the flow is gradient-like near the boundary of this region. This means that attracting regions for the kinetic equations are also attracting regions for the full system of reaction–diffusion equations, provided that the geometry of the region under consideration is compatible with the diffusion matrix. In the final section, these results are applied to the general Kolmogorov form of the equations which describe the classical two-species ecological interactions, where now diffusion and spatial dependence are taken into account One sees here how the standard ecological assumptions lead in a fairly direct way to the mathematical conditions which we have considered.

In Chapter 15, we begin to discuss the theory of shock waves. This is a notoriously difficult subject due to the many subtleties not usually encountered in other areas of mathematics. The very fact that the entire subject is concerned with *discontinuous* functions, means that many of the modern mathematical techniques are virtually inapplicable. I have given much effort in order to overcome these difficulties, by leading the reader gently along, step by step. It is here where I have leaned most upon my classroom experience. Thus, the development begins with a chapter describing the basic phenomena: the formation of shock waves, the notion of a weak solution and its consequences, the loss of uniqueness, the entropy conditions, etc. These things are all explained with the aid of examples. There follows next a chapter which gives a rather complete description of the theory of a single conservation law: existence, uniqueness, and asymptotic behavior of solutions. The existence proof follows Oleinik and is done via the Lax–Friedrichs difference scheme. The reasons why I have chosen this method over the several other ones available are discussed at the beginning of the chapter; suffice it to say that it requires no sophisticated background, and that the method of finite differences is, in principle, capable of generalization to systems. The entrance into systems of conservation laws, is made via a discussion of the Riemann problem for the "p-system." Here it is possible to explain things geometrically, by actually drawing the shock- and rarefaction-wave curves. We then develop the basic properties of these waves, and following Lax, we solve the Riemann problem for general systems. These ideas are applied in the next chapter to the equations of gas dynamics, where we solve the Riemann problem for arbitrary data, both analytically and geometrically. We prove Weyl's entropy theorem, as well as von Neumann's shock-interaction theorem. The next chapter, the Glimm Difference Scheme, is one of the most difficult ones in the book (the others being Chapters 22 and 23 on the Conley index). Glimm's theorem continues to be the most important result in conservation laws, and it must be mastered by anyone seriously interested in this field. I feel that the proof is not nearly as difficult as is commonly believed, and I have tried hard to make it readable for the beginner.

The final chapter in this group is designed to give the reader a flavor of some of the general results that are known for systems, the emphasis being on systems of two equations. I have also included a proof of Oleinik's uniqueness theorem for the p-system; her paper is available only in the original Russian. Having been sufficiently "turned on" by the superb lectures of T. Nishida at Michigan (in academic year 1981/82), I was unable to resist including a chapter on quasi-linear parabolic systems. The main result here is Kanel's existence proof for the isentropic gas dynamics equations with viscosity.

With Chapter 22, I begin Part Four of the book. These last three chapters deal mainly with the Conley index, together with its applications. Thus, the first chapter opens with a long descriptive discussion in which the basic ideas of the theory are explained; namely the concept of an isolated invariant set and its index, together with their main properties. These are illustrated by an easily understood example, in which things are worked out in detail and the connections with the classical Morse index are noted. I have also included a discussion of the so-called "Hopf bifurcation," from this point of view. Although the sections which follow are independent of this one, I strongly recommend that the reader not skim over it, but rather that he give it serious thought. The remaining sections in this chapter contain all of the basic definitions, together with proofs of the existence of an isolating block, and the theorem that the index of the isolated invariant set is independent of the particular isolating block containing it. This is all done for flows, where the reader can "see" the geometrical and topological constructions. I have also given some applications to differential equations in $\mathbf{R}^n$, as well as a proof of the "connecting orbit" theorem. In Chapter 23, the theory is developed from a more general, more abstract point of view, in a form suitable for applications to partial differential equations. We define the notions of index pairs, and Morse decompositions of an isolated invariant set. The concept of local flow is also introduced, again with an eye towards the applications. We prove both the existence of index pairs for Morse decompositions, as well as the well-definedness of the Conley index. That is, we show that the index $h(S)$ of an isolated invariant set S, depends only on the homotopy class of the space N_1/N_0, where (N_1, N_0) is any index pair for S. This result immediately puts at our disposal the algebraic invariants associated with the cohomology groups which form exact sequences on a Morse decomposition of S. These are powerful tools for computing indices, in addition to being of theoretical use. They lead, for example, to an easy proof of the "generalized" Morse inequalities. We then prove the continuation property of the Conley index, in a rather general setting. The final section serves both to illustrate some of the theorems, as well as to derive additional results which will be used in the applications. We point out that these two chapters monotonically increase in difficulty as one proceeds. This is done by design in order to meet the needs of readers having assorted degrees of mathematical maturity—one can proceed along as far as his background will take him (and further, if he is willing to work hard!).

The last chapter contains a sample of the applications to travelling waves. We first study the shock structure problem of the existence of an orbit connecting two rest points, and in particular, we solve the shock structure problem for magnetohydrodynamic shock waves having arbitrary strength. We then prove the existence of a periodic travelling wave solution for the Nagumo equations. An isolating neighborhood is constructed, and the Conley index is explicitly computed, in order to demonstrate the different topological techniques which are involved. We also show how to obtain the desired information a different way by using an exact sequence of cohomology groups in order to determine the nontriviality of the index. Next follows a long section, where we apply the theory to reaction–diffusion equations, and we use the Conley index together with some previously obtained (global) bifurcation diagrams, to study the stability of steady-state solutions, and to determine in some cases, the entire global picture of the solution set. The chapter closes with a section in which we give some instability theorems for nonconstant stationary solutions of the Neumann problem.

Each of the four sections in this book (in any order) is suitable for a one-semester graduate course. In particular, as we have remarked earlier, the first section can be used for an introductory graduate-level course in partial differential equations. The prerequisite for this is one year of graduate-level mathematics as given in the average American university.

Contents

PART IV
The Conley Index

List of Frequently Used Symbols

$\mathbf{Z}$	the integers
$\mathbf{Z}_+$	the nonnegative integers
$\mathbf{R}_+$	the nonnegative real numbers
$\mathbf{R}$	the real numbers
$\mathbf{R}^n$	n-dimensional Euclidean space
$\mathbf{C}$	the complex numbers
L_p	the usual space of measurable functions whose pth power is Lebesgue integrable
L_∞	the space of measurable functions which are bounded almost everywhere
L_p^{loc}	the space of functions which are locally in L_p
C^k	the space of functions having k continuous derivatives, $k \in \mathbf{Z}_+$
C	the space C^0
spt f	the support of the function f
$\equiv$	denotes definition of a symbol
∂	boundary
$f \vert A$	the restriction of the function f to the set A
$f \vert_A$	$f \vert A$
$A \backslash B$	the elements in A which are not in B
$B(X, Y)$	the bounded operators from X into Y
$C([0, T]; X)$	the space of continuous curves defined on $0 \le t \le T$ with values in X
Σ^k	the pointed k-sphere
S^k	the k-sphere
A^t	the transpose of A
∇	the gradient operator
Δ	the Laplace operator
det	determinant
$[A]$	the equivalence class of A
$\sim$	homotopically equivalent to
$\approx$	isomorphic to
$I(N)$	the maximal invariant set in N
id.	identity

$\mathrm{int}(N)$	the interior of the set N
$\mathrm{cl}(N)$	the closure of the set N
$\bar{N}$	$\mathrm{cl}(N)$
$[v_1, \ldots, v_n]$	the linear space spanned by the vectors $\{v_1, \ldots, v_n\}$
$[v_1, \ldots, \hat{v}_k, \ldots, v_n]$	the linear space spanned by the vectors v_i, $1 \le i \le n, i \ne k$
$h(S)$	the Conley index of S
$\int^y f(s)$	the indefinite integral of f
δ_{ij}	the Kronecker delta; $\delta_{ij} = 0$ if $i \ne j$, and $\delta_{ii} = 1$
$\|u(\cdot, t)\|_B$	the B-norm of the function $u(\cdot, t)$, where t is fixed
$A \amalg B$	the disjoint union of A and B; i.e., $A \cup B$ where $A \cap B = \varnothing$
$C_0(\Omega)$	the space of continuous functions having compact support in Ω
$C_0^k(\Omega)$	$C_0(\Omega) \cap C^k(\Omega)$
b^+	the exit set on ∂B
b^-	the entrance set on ∂B
$\bar{O}$	the one-point pointed space
$\bar{1}$	the pointed zero sphere
du/dv	$\nabla u \cdot v$
$\alpha(\gamma)$	α-limit set of γ
$w(\gamma)$	w-limit set of γ
$C^{k,\alpha}$	the set of C^k functions whose kth derivative is Hölder continuous with component α

Our present analytical methods seem unsuitable for the solution of the important problems arising in connection with nonlinear partial differential equations and, in fact, with virtually all types of nonlinear problems in pure mathematics. The truth of this statement is particularly striking in the field of fluid dynamics. Only the most elementary problems have been solved analytically in this field . . . .

The advance of analysis is, at this moment, stagnant along the entire front of nonlinear problems. That this phenomenon is not of a transient nature but that we are up against an important conceptual difficulty . . . . yet no decisive progress has been made against them . . . which could be rated as important by the criteria that are applied in other, more successful (linear!) parts of mathematical physics.

It is important to avoid a misunderstanding at this point. One may be tempted to qualify these (shock wave and turbulence) problems as problems in physics, rather than in applied mathematics, or even pure mathematics. We wish to emphasize that it is our conviction that such an interpretation is wholly erroneous.

JOHN VON NEUMANN, 1946

Part I

Basic Linear Theory

Chapter 1

Ill-Posed Problems

Problems involving differential equations usually come in the following form: we are given an equation for the unknown function u, $P(u) = f$, on a domain Ω together with some "side" conditions on u. For example, we may require that u assumes certain preassigned values on $\partial\Omega$, or that u is in $L^2(\Omega)$, or that u is in class C^k in Ω. At first glance, it would seem that any of these extra conditions are quite reasonable, and that one is as good as the other. However, we shall see that this is far from being true, and that whichever additional supplementary conditions one assigns is intimately connected with the form of equation.

In general, the equations come from the sciences: physics, chemistry, and biology, and the "physical" equations come together with quite specific "side" conditions. At least, this is the way the theory of partial differential equations began. It is the purpose of this chapter to illustrate these ideas by some examples. The chapter ends with the remarkable example of H. Lewy [Le].

§A. Some Examples

1. Let Ω be the region in $\mathbf{R}^2$ defined by

$$\Omega = \{(x, y): x^2 + y^2 < 1, y > 0\},$$

and consider the "Cauchy problem"[1] in Ω for Laplace's equation:

$$\Delta u \equiv \frac{\partial^2 u}{\partial x^2} + \frac{\partial^2 u}{\partial y^2} = 0, \qquad (x, y) \in \Omega, \tag{1.1}$$

together with the "initial" conditions

$$u(x, 0) = 0, \qquad u_y(x, 0) = f(x), \qquad -1 < x < 1. \tag{1.2}$$

---

[1] This is often called an "initial-value" problem, for reasons to be made clear later.

Suppose that $u(x, y)$ is a C^2 solution of (1.1), (1.2) in Ω. We extend u to be a C^2 function in the unit disk by setting $u(x, y) = -u(x, -y)$, in the region $y < 0$. Since the unit disk is simply connected, the function

$$v(x, y) = \int_{(0,0)}^{(x,y)} u_y \, dx - u_x \, dy$$

is a harmonic conjugate of u (because $u + iv$ satisfies the Cauchy–Riemann equations). Thus $u + iv$ is an analytic function, so the same is true of u, and in particular, $u_y(x, 0) = f(x)$ must be a real analytic function. Thus the "data" $f(x)$, assigned along $y = 0$ cannot be arbitrary; it must be a real analytic function.

2. Consider the set of "initial-value" problems in the upper half-plane in $\mathbf{R}^2$, for $n = 1, 2, \ldots$,

$$\Delta u = 0, \qquad y > 0,$$

$$u(x, 0) = 0, \qquad u_y(x, 0) = \frac{\sin nx}{n}, \qquad x \in \mathbf{R}, \tag{P_n}$$

and

$$\Delta u = 0, \qquad y > 0,$$

$$u(x, 0) = 0, \qquad u_y(x, 0) = 0, \qquad x \in \mathbf{R}. \tag{P_0}$$

The problems (P_n) and (P_0) have the solutions

$$u^n(x, y) = \frac{1}{n^2} \sinh ny \sin nx,$$

and

$$u^0(x, y) = 0,$$

respectively. Observe that as $n \to \infty$, the data for (P_n) tends uniformly to zero, the data of (P_0). However, we have

$$\varlimsup_{n \to \infty} |u^n(x, y) - u^0(x, y)| = +\infty, \quad y > 0,$$

for each point (x, y). In fact, the functions u^n do not converge to u^0 in any reasonable topology. Thus arbitrarily small changes in the data lead to large changes in the solution; the mapping from the "data space" to "solution space" is not continuous—there is a lack of stability. If this mathematical problem were obtained from a physical problem, we would have that small changes in the *measurement* of the data would induce large changes in the predicted phenomena. Obviously such a mathematical theory would be useless.

3. Consider the equation in two independent variables, $u_{xy} = 0$. Since u_x is a pure function of x, say $u_x = \phi'(x)$, $u - \phi$ is independent of x, so that $u - \phi = \psi(y)$, for some function ψ. Thus all solutions of our partial differential equation are of the form $\phi(x) + \psi(y)$, where ϕ and ψ are (at least) differentiable functions.

Now consider the "boundary-value" problem in the unit square $Q = \{(x, y) : 0 < x, y < 1\}$:

$$u_{xy} = 0 \quad \text{in } Q,$$
$$u = u_0 \quad \text{on } \partial Q. \tag{1.3}$$

If $O = (0, 0)$, $P = (1, 0)$, $Q = (1, 1)$ and $R = (0, 1)$, then we see that u_0 must satisfy the relation

$$u_0(O) + u_0(Q) = u_0(P) + u_0(R). \tag{1.4}$$

It follows that (1.3) has no solution if (1.4) fails to hold.

We shall show in Chapter 3, that $u_{xy} = 0$ is actually equivalent to the wave equation, $u_{xx} - u_{yy} = 0$, via a change of variables. Thus the boundary value problem is, generally speaking, not correct for the wave equation.

These very simple examples show that one must be extremely careful in selecting the "data" for problems involving partial differential equations— if not, the corresponding problems may not be "well-posed." Fortunately, the problems obtained from the "physical world" are usually well-posed, i.e., they have unique solutions, which depend continuously on the "data," in some (reasonable) topology. However, as the next example shows, it may not be apparent how to show these things.

4. Consider the simple nonlinear equation

$$u_t + uu_x = 0, \tag{1.5}$$

defined in $(x - t)$ space, where $t > 0$. This equation arises in the study of nonlinear wave phenomena; e.g., in gas dynamics. Here t denotes the "time," and x the "space." It is required to obtain a solution of (1.5), defined for all $t > 0$ which satisfies the "initial" condition

$$u(x, 0) = u_0(x), \qquad x \in \mathbf{R}. \tag{1.6}$$

The equation (1.5) implies that along the curves

$$\frac{dt}{ds} = 1, \qquad \frac{dx}{ds} = u(x, t), \qquad (1.7)$$

u must be constant. Indeed, along such a curve

$$\frac{du}{ds} = u_t \frac{dt}{ds} + u_x \frac{dx}{ds} = u_t + uu_x = 0.$$

The curves defined by (1.7) obviously have slopes

$$\frac{dt}{dx} = \frac{1}{u(x, t)}.$$

Since u is constant along these curves, it is easy to see that these curves must all be straight lines, along which u is constant. That is, the value of u along these curves must equal the value of u at the point on the line $t = 0$ at which the given curve starts. Thus the solution is completely determined by its values at $t = 0$; i.e., by $u_0(x)$. This is illustrated in Figure 1.1.

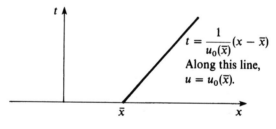

Figure 1.1

Now suppose that we take the following function for $u_0(x)$:

$$u_0(x) = \begin{cases} 1, & x < 0, \\ 1 - x, & 0 \le x \le 1, \\ 0, & x > 1. \end{cases} \qquad (1.8)$$

If we make the above computations for this data we find that

$$u(x, t) = \begin{cases} 1, & x < t, \\ (1 - x)/(1 - t), & t < x < 1, \\ 0, & x > 1, \end{cases}$$

and this function "blows up" at $t = 1$. Hence, no global solution (i.e., solution defined for all $t > 0$) exists! On the other hand, we have remarked that the problem (1.5), (1.6) arises in physical contexts. How do we get out of this dilemma? The answer is to weaken the notion of solution, and not demand that u be even continuous. Indeed, we have an example here of a "shock-wave" forming at time $t = 1$. This physical phenomenon occurs quite naturally (and generally!). We shall study such equations in Chapter 15.

A final word about this example. It is not too difficult to show that the lack of existence of a global solution is a distinctly nonlinear phenomenon —such things do not occur for linear equations. The following celebrated example of Hans Lewy [Le] shows that still worse things can happen, even for linear equations.

§B. Lewy's Example

We shall give an example here of a linear partial differential operator in which no solution exists, in any open set of the domain space. The operator in question is

$$A(u) = u_x + iu_y - 2i(x + iy)u_t, \qquad (x, y, t) \in \mathbf{R}^3, \tag{1.9}$$

where $i^2 = -1$. Notice that the coefficients are all analytic functions. Thus, if $f(x, y, t)$ were an analytic function, then from the classical Cauchy–Kowalewski theorem (see [Ga]), the equation $A(u) = f$ would admit analytic solutions. The point here is that we shall show that there are C^∞ functions f for which $A(u) = f$ has no solutions. This will be done in a sequence of lemmas, the most important of which is the first one. We begin by stating the theorem.

Theorem 1.1. *There exist C^∞ functions $f(x, y, t)$ such that the equation $A(u) = f$ has no $C^{1,\alpha}$ solution anywhere in $\mathbf{R}^3$.*

In fact, we shall show that the set of such f forms a set of the second category in the space of C^∞ functions.

Lemma 1.2. *Let $A(u)$ be defined by (1.9). Then the equation $A(u) = \phi'(t)$ has no C^1 solution in any neighborhood of $0 \in \mathbf{R}^3$, if ϕ is not analytic at $t = 0$.*

Proof. Let u be a solution in $|z| < R, |t| < T$, where $z = x + iy$. In the region $|t| < T, 0 \le r < R$, let

$$g(t, r) = \int_{|z| = r} u \, dz = \int_{|z| = r} u \, dx + iu \, dy.$$

From Green's theorem, we have

$$g(t, r) = \iint_{|z| \le r} (iu_x - u_y) \, dx \, dy = i \iint_{|z| \le r} (u_x + iu_y) \, dx \, dy$$

$$= i \int_0^r \int_0^{2\pi} (u_x + iu_y)\rho \, d\rho \, d\theta.$$

Set $y = r^2$, $\psi(t, y) = g(t, r^2)$; then $\psi_y = g_r r_y = g_r/2r$, so that using our equation, we find

$$\psi_y = \frac{1}{2} \int_{|z|=r^2} (u_x + iu_y) \frac{dz}{z}$$

$$= \frac{1}{2} \int_{|z|=r^2} 2izu_t \frac{dz}{z} + \frac{1}{2} \int_{|z|=r^2} \phi'(t) \frac{dz}{z}$$

$$= i \int_{|z|=r^2} u_t \, dz + \tfrac{1}{2}(2\pi i)\phi'(t)$$

$$= i\psi_t + i\pi\phi'(t).$$

If we let $S(t, y) = \psi(t, y) + \pi\phi(t)$, then $S_y = \psi_y$, $S_t = \psi_t + \pi\phi'(t) = -i\psi_y$ $= -iS_y$. Thus S satisfies the Cauchy–Riemann equations, and since S is a C^1 function, S is an analytic function of $t + iy$. Since $S(t, 0) = \pi\phi(t)$, we see that S is real on $y = 0$. By the Schwarz reflection principle, S can be extended to an analytic function in the region, $-R^2 < y \le 0$, $|t| < T$, by defining $S(t + iy) = \overline{S(t - iy)}$ for $y \le 0$. Thus $\pi\phi(t) = S(t, 0)$ is analytic at $t = 0$, and the proof of the lemma is complete. $\square$

We remark that, in effect, we have reduced the equation $A(u) = \phi'(t)$ to an equation in two independent variables, by integrating with respect to θ over S^1, (the boundary of the unit disc); this allows the technique of separation of variables to be applied.

By a change of coordinates, we get the following lemma.

Lemma 1.3. *If u is C^1 near (x_0, y_0, t_0), then the equation $A(u) = \phi'(t - t_0 - 2y_0 x + 2x_0 y)$ implies that ϕ is analytic in a neighborhood of 0.*

Proof. Let $\bar{x} = x - x_0$, $\bar{y} = y - y_0$, $\bar{t} = t - t_0 - 2y_0 x + 2x_0 y$; then

$$A(u) = u_{\bar{x}} - 2y_0 u_{\bar{t}} + i(u_{\bar{y}} + 2x_0 u_{\bar{t}}) - 2i[\bar{x} + x_0 + i(\bar{y} + y_0)]u_{\bar{t}}$$

$$= u_{\bar{x}} + iu_{\bar{y}} - 2i(\bar{x} + i\bar{y})u_{\bar{t}} = \phi'(\bar{t}).$$

Hence $A(u) = \phi'(t - t_0 - 2y_0 x + 2x_0 y)$ has a solution in a neighborhood of $(\bar{x}, \bar{y}, \bar{t})$ if and only if $A(u) = \phi'(\bar{t})$ has a solution in a neighborhood of $(0, 0, 0)$. $\square$

Now we shall construct periodic C^∞ nowhere analytic functions F of the form

$$F(x, y, t) = \sum_{j=1}^{\infty} a_j \psi(\zeta_j),$$

where: $\zeta_j = t - t_j - 2y_j x + 2x_j y$, the set $\{(x_j, y_j, t_j) : j = 1, 2, \dots\}$ is dense in $\mathbf{R}^3$, and ψ is nowhere analytic. Of course, this is somewhat delicate since the a_j must be chosen with care in order to ensure that F not be analytic.

Before doing this we must show that there are functions which are C^∞ but nowhere analytic. To this end, let

$$K_m = \{x \in \mathbf{R}^n : |x| \le m\}.$$

Then K_n is compact, $K_{n+1} \supset K_n$ and $\cup K_n = \mathbf{R}^n$. Let p_n be the seminorm on $C^\infty(\mathbf{R}^n)$ defined by (cf. below)

$$p_n(\phi) = \sup\{|D^\alpha \phi(x)| : x \in K_n, |\alpha| \le n\};$$

then $p_{n+1} \ge p_n$. We make $C^\infty(\mathbf{R}^n)$ into a complete metric space by defining the standard metric,

$$d(\phi, \psi) = \sum_{n=1}^{\infty} 2^{-n} \frac{p_n(\phi - \psi)}{1 + p_n(\phi - \psi)}.$$

Now observe that $\phi \in C^\infty(\mathbf{R}^n)$ is (real) analytic at $P \in \Omega$ if there is an $R > 0$ such that for $|x - P| < R$, we have

$$\phi(x) = \sum \frac{D^\alpha \phi(P)}{\alpha!} (x - P)^\alpha,$$

where $\alpha = (\alpha_1, \dots, \alpha_n)$, $\alpha_i \in \mathbf{Z}$, $D^\alpha = D_1^{\alpha_1} \cdots D_n^{\alpha_n}$, $\alpha! = \prod \alpha_i!$, and $(x - P)^\alpha = \prod (x_i - P_i)^{\alpha_i}$. Also, if ϕ is analytic at P then

$$\sup_\alpha \frac{D^\alpha \phi(P)}{\alpha!} r^{|\alpha|} < \infty \quad \text{for all } r, \quad 0 < r < R,$$

where $|\alpha| = \alpha_1 + \cdots + \alpha_n$. We define the set Φ by

$$\Phi = \{\phi \in C^\infty(\mathbf{R}^n) : \phi \text{ is real analytic at some } P \in \Omega\}.$$

We can now state the following result.

Proposition 1.4. *The set Φ is of the first category in $C^\infty(\mathbf{R}^n)$.*

We shall carry out the proof in several steps. First we define the sets $M(P, r)$ by

$$M(P, r) = \left\{ \phi \in \mathring{C}^\infty(\mathbf{R}^n): \sup_\alpha \frac{|D^\alpha\phi(P)|}{\alpha!} r^{|\alpha|} < \infty \right\},$$

where $\mathring{C}^\infty$ denotes the class of periodic C^∞ functions.

Lemma 1.5. *$M(P, r)$ is of the first category in $\mathring{C}^\infty(\mathbf{R}^n)$.*

Proof. Consider the linear functional L_α on $\mathring{C}^\infty(\mathbf{R}^n)$ defined by

$$L_\alpha(\phi) = \frac{D^\alpha\phi(P)}{\alpha!} r^{|\alpha|}.$$

Now it is evident that there do not exist $n \in \mathbf{Z}_+$ and constants C in $\mathbf{R}$ such that for all ϕ and α,

$$|L_\alpha(\phi)| \le C p_n(\phi).$$

Thus from the uniform boundedness principle[2] the set $M(P, r)$ is of the first category.

Now we fix $P \in \mathbf{R}^n$, and let

$$M(P) = \{\phi \in \mathring{C}^\infty(\mathbf{R}^n): \phi \text{ is analytic at } P\}.$$

Since $M(P) \subset \bigcup_{n=1}^\infty M(P, 1/n)$, we see from Lemma 1.5 that $M(P)$ is a set of the first category.

We can now complete the proof of Proposition 1.4. To this end, we choose $\{P_i\}$ to be a countable and dense set in Ω. Then $\bigcup_i M(P_i)$ is a set of the first category. If ϕ is real analytic at some $P \in \Omega$, then ϕ is real analytic at some P_i so $\phi \in M(P_i)$; hence $\Phi \subset \bigcup_i M(P_i)$, and so Φ is a set of the first category. The proof is complete. $\square$

We can now turn to the proof of the theorem. Choose $\psi(t) \in \mathring{C}^\infty(\mathbf{R}^1)$ such that ψ is nowhere analytic. Let $P_j = (x_j, y_j, t_j)$ be a countable dense set in $\mathbf{R}^3$, let $c_j = \max(j, |x_j|, |y_j|)$, and define $U_j = \{P \in \mathbf{R}^3: |P - P_j| < 1/j\}$, $j = 1, 2, \ldots$. Obviously every open set in $\mathbf{R}^3$ contains some U_j.

For a pair of positive integers m and n, and $U^{\text{open}} \subset \mathbf{R}^3$, let $H_{m,n}(U)$ be the class of Hölder continuous functions on U having constant m and exponent $1/n$, and let $H(U)$ be the class of Hölder continuous functions on

---

[2] Recall that this states that if X is a Banach space and $\{F_n\}$ is a countable family of semi-norms such that if $\sup_n F_n(x) < \infty$ for x in a set of second category, then there is a constant c with $F_n(x) \le c\|x\|$ for all n and x.

U. It is obvious that $H(U)$ is the union of the $H_{m,n}(U)$. Note that if $f \in C^2$ on some bounded open set containing $\bar{U}$, then $f \in H(U)$. Finally, set

$$H^j_{m,n} = H_{m,n}(U_j),$$

and let l_∞ denote the complete metric space of bounded real sequences $\sigma = (\sigma_0, \sigma_1, \sigma_2, \ldots)$, with norm $\|\sigma\| = \sup\{|\sigma_j| : j = 0, 1, 2, \ldots\}$. For $\sigma \in l_\infty$, define a real-valued function by

$$F_\sigma(x, y, t) = \sum_{j \geq 0} \sigma_j c_j^{-c_j} \psi'(\zeta_j), \qquad \zeta_j = t - t_j - 2y_j x + 2x_j y.$$

Claim 1. $F_\sigma \in C^\infty(\mathbf{R}^3)$.

If $v = (v_1, v_2, v_3)$, where each v_i is an integer ≥ 0, and $|v| = v_1 + v_2 + v_3$, then consider the formal series (in a given compact set $K \subseteq \mathbf{R}^3$),

$$D^v F_\sigma = \sum_{j \geq 0} \sigma_j c_j^{-c_j} \psi^{|v|+1}(\zeta_j)(-y_j)^{v_1}(x_j)^{v_2} 2^{v_1 + v_2}.$$

Since $|\sigma_j| \leq \|\sigma\|$, and $\|\psi^{|v|+1}\|_{L_\infty(K)} \leq M_{|v|+1}$ (for some finite constants $M_{|v|+1}$), and both $|x_j| \leq c_j$ and $|y_j| \leq c_j$, the formal series is majorized by

$$\|\sigma\| M_{|v|+1} 2^{|v|} \sum_{j \geq 0} c_j^{|v|-c_j}.$$

But for j large,

$$c_j^{|v|-c_j} \leq c_j^{c_j/2 - c_j} \leq c_j^{-j/2} \leq j^{-j/2} \leq j^{-2},$$

so that the formal series $D^v F_\sigma$ converges absolutely and uniformly in K. Since v and K were arbitrary we see that $F_\sigma \in C^\infty(\mathbf{R}^3)$; this proves Claim 1.

For positive integers m, n, l and j, set

$$B(m, n, l, j) = \{f \in H^j_{m,n} : \|D_i f\|_{L_\infty(\mathbf{R}^3)} \leq l, i = 1, 2, 3, \text{ and } f(P_j) = 0\}.$$

It is easy to see that $B(m, n, l, j)$ is a compact subset of $H^j_{m,n}$, under the topology of pointwise convergence of the function together with its first derivatives. Let

$$E(m, n, l, j) = \{\sigma \in l_\infty : A(u) = F_\sigma \text{ has a solution } u \in B(m, n, l, j)\}.$$

Fix m, n, l, j, and let $E = E(m, n, l, j)$.

Claim 2. *E is nowhere dense in* l_∞

We postpone the proof of the claim and complete the proof of the theorem. By the Baire category theorem l_∞ cannot be the union of the $E(m, n, l, j)$; thus let $\sigma_* \in l_\infty \setminus \bigcup E(m, n, l, j)$. Then $A(u) = F_{\sigma*}$ has no solution in any of the classes $B(m, n, l, j)$. Suppose that $A(u) = F_{\sigma*}$ had a Hölder continuous (on derivatives) solution $u(x, y, t)$ in $H(U)$ for some open subset $U \subset \mathbf{R}^3$. Pick j such that $U_j \subset U$. Note that we may assume that $u(P_j) = 0$; otherwise replace u by $v = u - u(P_j)$. Now $u \in H_{m,n}^j$ for some m, and n, and if $P \in U_j$,

$$\left| D_i u(P) \right| \leq \left| D_i u(P) - D_i u(P_j) \right| + \left| D_i u(P_j) \right|$$
$$\leq m + \max_{i=1,2,3} \left| D_i u(P_j) \right|.$$

Thus, if $l \geq m + \max_{i=1,2,3} |D_i u(p_j)|$, we see that $u \in B(m, n, l, j)$. This is the desired contradiction.

We complete the proof of the theorem by proving Claim 2. We first show that E is closed. Thus, let $\sigma_k \in E$, $\sigma_k \to \sigma$ in l_∞. We have $u_k \in B \equiv B(m, n, l, j)$ such that $A(u_k) = F_{\sigma_k}$. Since B is compact, we may assume that u_k converges in B; i.e., there exists $u \in B$ such that $u_k \to u$, $D_i u_k \to D_i u$ in U_j, $i = 1, 2, 3$. Then $A(u_k) \to A(u)$ in U_j and $F_{\sigma_k} \to F_\sigma$ in $L_\infty(\mathbf{R}^3)$. Thus, $A(u) = F(\sigma)$ so $\sigma \in E$ and E is closed.

We show that E is nowhere dense by showing that E has empty interior. Suppose, on the contrary that

$$S \equiv S(\sigma_0, \varepsilon) = \{\sigma \in l_\infty : \|\sigma - \sigma_0\| < \varepsilon\} \subseteq E.$$

Let $\delta = (0, \ldots, 0, \varepsilon/2, 0, \ldots)$; i.e., δ has a nonzero coordinate only in the jth slot, and let $\sigma = \sigma_0 + \delta$. Then $\|\sigma - \sigma_0\| = \varepsilon/2$ so $\sigma \in S$. Thus there is $u_1 \in H(N_j)$ such that $A(u_1) = F_\sigma = F_{\sigma_0} + F_\delta$. Since $\sigma_0 \in E$, there is a v_1 in $H(N_j)$ such that $A(v_1) = F_{\sigma_0}$. Therefore $A(u_1 - v_1) = F_\delta$, and $u_1 - v_1$ is in $H(N_j)$. But since

$$F_\delta = \frac{\varepsilon c_j^{-c_j} \psi'(\zeta_j)}{2}, \quad \text{where } \zeta_j = t - t_j - 2y_j x + 2yx_j,$$

we have $A(2c_j^{c_j}(u_1 - v_1)/\varepsilon) = \psi'(\zeta_j)$, where $2c_j^{c_j}(u_1 - v_1)/\varepsilon$ is in $C^1(\mathbf{R}^3)$. This contradicts Lemma 1.3, and completes the proof of the theorem. $\square$

NOTES

The results in §A of this chapter are classical, and are found in most standard books on the subject; see [Ga]. The equation (1.5) is called the Burgers equation (without viscosity) and its study by Hopf [Hf 2] was the starting point for the mathematical theory of shock waves; see Chapters 15 ff. The interesting example in §B is due to H. Lewy [Le]. It too was the starting point of a flurry of research on such equations; see [Ho 2].

Chapter 2

Characteristics and Initial-Value Problems

Roughly speaking, characteristics are curves which carry information. They are particularly relevant in the study of "initial-value" problems; that is, in solving partial differential equations, in which the solution surface is required to assume prescribed values "initially." Such a problem presupposes the existence of a distinguished coordinate, ξ, where the equation $\xi = 0$ defines the "initial" surface. Of course, as we have seen in the last chapter, one needs some kind of compatibility between the equation and the initial surface. The notion of characteristic serves to classify and make more precise these intuitive ideas.

We begin by considering a rather general linear partial differential operator of order m:

$$P(x, D) = \sum_{|\alpha| = m} a_\alpha(x)D^\alpha + \sum_{|\alpha| < m} a_\alpha(x)D^\alpha + a(x). \tag{2.1}$$

Here $x = (x_0, x_1, \ldots, x_n) \in \mathbf{R}^{n+1}$, α is a multi-index, $\alpha = (\alpha_0, \alpha_1, \ldots, \alpha_n)$, where each α_i is a nonnegative integer. $|\alpha| \equiv \sum \alpha_i$, is the length of α, $D = (D_0, D_1, \ldots, D_n)$ is a differential operator where $D_i = \partial/\partial x_i$, and D^α is the partial differential operator defined by $D^\alpha = \partial^{|\alpha|}/\partial x_0^{\alpha_0} \cdots \partial x_n^{\alpha_n}$. We also set $x^\alpha = x_0^{\alpha_0} \cdots x_n^{\alpha_n}$. Finally, we call

$$\sum_{|\alpha| = m} a_\alpha(x)D,$$

the *principal part* of the operator $P(x, D)$.

We consider the equation

$$P(x, D)u = 0, \tag{2.2}$$

and we assume that this equation is defined in a neighborhood of a smooth n-dimensional surface S given by $f(x) = 0$. The *initial-value* problem for (2.2) consists of assigning u and its derivatives of order $< m$ on S, and it is required to solve (2.2) in a neighborhood of S.

For example, if $P(x, D) = \partial^2/\partial t^2 - \partial^2/\partial x^2$ ($x_0 = t$), the one-dimensional wave operator, and $S = \{t = 0\}$, we assign $u(x, 0)$ and $u_t(x, 0)$, and try to solve the equation $u_{tt} - u_{xx} = 0$ in $t > 0$.

We proceed to solve the general problem as follows. First we change coordinates in a neighbourhood of S, by introducing the new independent variables $\xi_0, \xi_1, \ldots, \xi_n$, where $\xi_1, \ldots, \xi_n$ are (independent) coordinates on S, and $\xi_0 = f$; i.e., S corresponds to $\xi_0 = 0$.

In the above example, we can also put $\xi_0 = t$, $\xi_1 = x$. A less trivial example would be obtained if we set $S = \{(x, t): x = t\}$; then $\xi_0 = x - t$, $\xi_1 = x + t$.

Note that since $\xi_1, \ldots, \xi_n$ are coordinates in S, and u is given on S, all the derivatives of u with respect to $\xi_1, \ldots, \xi_n$ (interior derivatives) are known on S.

By repeated use of the chain rule, we obtain, for $\alpha = (\alpha_0, \ldots, \alpha_n)$, $|\alpha| = m$,

$$D^\alpha u = \frac{\partial^m u}{\partial x_0^{\alpha_0} \cdots \partial x_n^{\alpha_n}} = \frac{\partial^m u}{\partial \xi_0^m} \left(\frac{\partial \xi_0}{\partial x_0}\right)^{\alpha_0} \cdots \left(\frac{\partial \xi_0}{\partial x_n}\right)^{\alpha_n} + \cdots,$$

where the last dots represent derivatives of u with respect to ξ_0 of orders $< m$, together with derivatives of u with respect to ξ_i, $i \neq 0$, and are thus all *known quantities*. The equation (2.2) becomes

$$\sum_{|\alpha| = m} a_\alpha(x) \left(\frac{\partial \xi_0}{\partial x_0}\right)^{\alpha_0} \cdots \left(\frac{\partial \xi_0}{\partial x_n}\right)^{\alpha_n} \frac{\partial^m u}{\partial \xi_0^m} + \cdots = 0, \qquad (2.3)$$

where the dots represent quantities known on S. Now in order to be able to solve this equation for $\partial^m u/\partial \xi_0^m$, and to thereby obtain information on u not previously known, it is necessary and sufficient that

$$\sum_{|\alpha| = m} a_\alpha(x) \left(\frac{\partial \xi_0}{\partial x_0}\right)^{\alpha_0} \cdots \left(\frac{\partial \xi_0}{\partial x_n}\right)^{\alpha_n} \neq 0,$$

or equivalently, that

$$\sum_{|\alpha| = m} a_\alpha(x)(\nabla f(x))^\alpha \equiv \sum_{|\alpha| = m} a_\alpha(x) \left(\frac{\partial f(x)}{\partial x_0}\right)^{\alpha_0} \cdots \left(\frac{\partial f(x)}{\partial x_n}\right)^{\alpha_n} \qquad (2.4)$$

$$\neq 0, \quad \text{for all } x \in S.$$

This is the desired compatibility condition between the initial surface S and the operator $P(x, D)$. When (2.4) fails to hold, the initial-value problem would be unreasonable, and in this case we would say that S is a characteristic surface. Formally, we have the following definitions.

Definition 2.1. The surface $S = \{x: \phi(x) = 0\}$ is said to be *characteristic* at $p \in S$, for the operator (2.1) if

$$\sum_{|\alpha|=m} a_\alpha(x)\nabla\phi(x)^\alpha|_{x=p} = 0.$$

S is a *characteristic surface* for the operator (2.1) if it is characteristic at each point.

We call the equation

$$\sum_{|\alpha|=m} a_\alpha(x)\sigma^\alpha = 0, \tag{2.5}$$

$\sigma = (\sigma_0, \ldots, \sigma_n)$, the *characteristic equation* for the operator (2.1).

In these terms, we see that a surface S is characteristic at $p \in S$, for the operator (2.1), provided that the normal vector to S at p satisfies the characteristic equation.

We shall now discuss a few examples. In what follows, let $\sigma = (\sigma_0, \ldots, \sigma_n)$ denote a unit normal vector at a given point on S, i.e., σ_k is the cosine of the angle between the normal to S and the x_k-axis, and

$$\sum_0^n \sigma_k^2 = 1. \tag{2.6}$$

EXAMPLE 1. The n-dimensional wave equation:

$$u_{x_0 x_0} - \sum_{k=1}^n u_{x_k x_k} = 0.$$

(Here one usually denotes x_0 as t, the "time.") The characteristic equation is

$$\sigma_0^2 - \sum_1^n \sigma_k^2 = 0,$$

which, together with (2.6) yields $2\sigma_0^2 = 1$ and $\sigma_0 = \pm 1/\sqrt{2}$. Thus, a surface is characteristic for the wave equation if and only if its normal makes an angle of $\pi/4$ with respect to the x_0-axis. In particular, for the one-dimensional wave equation, $u_{tt} - u_{xx} = 0$, the characteristic surfaces are the lines which make $45°$ angles with the t-axis.

EXAMPLE 2. Laplace's equation: $\sum u_{x_i x_i} = 0$. Here the characteristic equation is $\sum \sigma_k^2 = 0$ which is incompatible with (2.6); thus there are no (real) characteristics.

EXAMPLE 3. The n-dimensional heat equation:

$$u_{x_0} - \sum_1^n u_{x_i x_i} = 0.$$

The characteristic equation is $\sum_1^n \sigma_i^2 = 0$ and thus from (2.6), $\sigma_0^2 = 1$. It follows that a surface is characteristic if and only if its normals are parallel to the x_0 axis; thus they are hyperplanes $x_0 = $ const.

EXAMPLE 4. First-order linear equation:

$$a(x, y)u_x + b(x, y)u_y = c(x, y)u + d(x, y).$$

The characteristic equation is $a\sigma_0 + b\sigma_1 = 0$; solving this together with (2.6) gives

$$\sigma = \pm \frac{1}{\sqrt{a^2 + b^2}} (b, -a).$$

It follows that the characteristic curves are solutions of the system of autonomous ordinary differential equations $\dot{x} = a(x, y)$, $\dot{y} = b(x, y)$.

Observe now, that if $\phi(x_0, \ldots, x_n) = 0$ is a characteristic surface for the operator (2.1), (2.3) shows that the differential equation imposes an additional restriction on the data; namely, the "known quantities," denoted by the dots in (2.3) must vanish. For example, consider the equation

$$u_{xx} - u_{tt} + a(x, t)u_x + b(x, t)u_t + c(x, t)u + d(x, t) = 0.$$

The line $S: \{x = t\}$ is a characteristic surface. Suppose that on S we assign data $u(x, x) = \phi(x)$ and $u_\theta(x, x) = \psi(x)$, where θ is the direction of the normal to S. If we change variables $\xi = x + t$, $\eta = x - t$, then S is given by $\eta = 0$. In these coordinates, the differential equation becomes

$$4u_{\xi\eta} + (a + b)u_\xi + (a - b)u_\eta + cu + d = 0. \tag{2.7}$$

Since $x = (\xi + \eta)/2$, the initial conditions become $u(\xi, 0) = \phi(\xi/2)$, and $u_\eta(\xi, 0) = \psi(\xi/2)$. Thus if we evaluate (2.7) on $\eta = 0$, we obtain

$$2\psi'(\xi/2) + \tfrac{1}{2}(a + b)\phi'(\xi/2) + \tfrac{1}{2}(a + b)\psi(\xi/2) + c\phi(\xi/2) + d = 0,$$

which is an additional restriction on the initial data.

NOTE

The results here are all classical; see [Ga].

Chapter 3

The One-Dimensional Wave Equation

In this chapter we study a simple but quite interesting equation for which the initial-value problem is well-posed. The ideas which we introduce here will be used in various places throughout the book, albeit at a "higher dialectical" level. The equation is derived from physical considerations, and in the case we consider here, the solution $u(x, t)$, may be thought of as describing the position of a vibrating string at a point x at a time t.

The equation we consider is

$$u_{tt} - c^2 u_{xx} = 0, \qquad (x, t) \in \mathbf{R} \times \mathbf{R}_+. \tag{3.1}$$

The constant c is usually called the wave-speed; it can be of either sign, and for definiteness, we choose $c > 0$.

Since x and t are both scalar variables, the characteristic surfaces S are really curves whose unit normal vectors $\bar{n} = (\phi, \psi)$ satisfy the equations

$$\psi^2 - c^2 \phi^2 = 0, \qquad \phi^2 + \psi^2 = 1,$$

from which it follows that

$$(\phi, \psi) = \frac{1}{\sqrt{1 + c^2}}(1, \pm c).$$

With the aid of Figure 3.1, we obtain $\tan \theta = \psi/\phi = \pm c$, $\tan \gamma = -\cot \theta = \pm c^{-1}$, so that the characteristics are the lines $t = \pm c^{-1}x + \text{const}$, or $x \pm ct = \text{const}$. That is, they are straight lines of slope $\pm c^{-1}$.

We change variables by setting $x + ct = \xi$, $x - ct = \eta$; then in these coordinates we find that (3.1) becomes $u_{\xi\eta} = 0$. We can immediately integrate this equation to get $u = f(\xi) + g(\eta)$, where f and g are arbitrary (but differentiable) functions. Thus, we find that the solution of (3.1) is given by

$$u(x, t) = f(x + ct) + g(x - ct), \tag{3.2}$$

where f and g are arbitrary. The functions $f(x + ct)$ and $g(x - ct)$ can be thought of as representing "waves" moving at speeds $\pm c$.

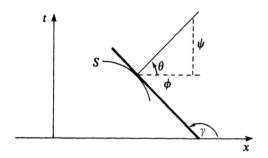

Figure 3.1

Now we consider the initial-value problem for (3.1); i.e., we seek a solution of (3.1) which at time $t = 0$ assumes the data

$$u(x, 0) = \alpha(x), \qquad u_t(x, 0) = \beta(x), \tag{3.3}$$

where $\alpha \in C^2$, $\beta \in C^1$. We find the functions f and g as follows. When $t = 0$,

$$\alpha(x) = f(x) + g(x), \qquad \beta(x) = c(f'(x) - g'(x)).$$

Thus

$$f(x) - g(x) = \frac{1}{c} \int_0^x \beta(s)\, ds,$$

and

$$f(x) = \tfrac{1}{2}\alpha(x) + \frac{1}{2c} \int_0^x \beta(s)\, ds,$$

so

$$g(x) = \tfrac{1}{2}\alpha(x) - \frac{1}{2c} \int_0^x \beta(s)\, ds.$$

This yields the following formula for the solution of the problem (3.1), (3.3):

$$u(x, t) = \tfrac{1}{2}[\alpha(x + ct) + \alpha(x - ct)] + \frac{1}{2c} \int_{x-ct}^{x+ct} \beta(s)\, ds. \tag{3.4}$$

Observe that this implies at once that the given problem is uniquely solvable. Next, note that if α_1 and β_1 are uniformly close to α and β, i.e.,

$$\|\alpha - \alpha_1\|_\infty < \varepsilon, \qquad \|\beta - \beta_1\|_\infty < \varepsilon, \qquad \varepsilon > 0,$$

and if $v(x, t)$ is the solution of (3.1) with data $v(x, 0) = \alpha_1(x)$, $v_t(x, 0) = \beta_1(x)$, then v is of the form (3.4), with α and β replaced by α_1 and β_1, respectively. Thus,

$$|u - v| \leq \frac{|\alpha(x + ct) - \alpha_1(x + ct)|}{2} + \frac{|\alpha(x - ct) - \alpha_1(x - ct)|}{2}$$

$$+ \frac{1}{2c} \int_{x-ct}^{x+ct} |\beta(s) - \beta_1(s)|\, ds,$$

so that for any $t > 0$,

$$\|u(\cdot, t) - v(\cdot, t)\|_{L_\infty(\mathbb{R})} \leq \varepsilon + \frac{1}{2c} \cdot \varepsilon \cdot 2ct = \varepsilon(1 + t).$$

Hence, for any $T > 0$, if $\varepsilon \to 0$, then $u - v \to 0$ uniformly in x on $[0, T]$. This shows that the solutions depend continuously on the data (in this topology!). The initial-value problem (3.1), (3.3) is thus well-posed.

We next examine more closely the formula (3.4) for the solution. Consider a point $P = (x_0, t_0)$, with $t_0 > 0$. We draw the characteristics of (3.1) through P, in the region $t < t_0$; these are called the *backward* characteristics, see Figure 3.2. The two characteristic lines meet the x-axis at the points $x_0 \pm ct_0$.

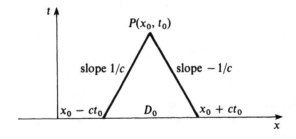

Figure 3.2

Now consider the region

$$D_0 = \{x : x_0 - ct_0 \leq x \leq x_0 + ct_0\}.$$

If we consider (3.4), we see that $u(x_0, t_0)$ depends only on the values of α and β in D_0; in fact, it depends only on $\alpha(x_0 \pm ct_0)$, the value of α at the endpoints of D_0, and the values of β on the complete interval D_0. In other words, if we change α and β outside of D_0, this will *not* affect the solution at P. We say that D_0 is the *domain of dependence* of $P(x_0, t_0)$. Of course, it is easy to see that if $P_1 = (x_1, t_1)$ is any point in the depicted triangle, then the domain of dependence of P_1 is also contained in D_0.

On the other hand, we may reverse things, and ask what points in $t > 0$ are influenced by the data in an interval on $t = 0$. Thus if $I = [a, b]$, we can draw the *forward* characteristics of slope $+1/c$ through b and of slope $-1/c$ through a; see Figure 3.3. The region D_I so determined is clearly (see (3.4)) the *domain* of *influence* of I, in the sense that the nonzero values of the initial functions α and β in I will affect the solution u only at points in D_I. We say that disturbances *propagate with speed c*. That is, suppose that α and β are supported in I. If an observer is at a point $\bar{x} \notin I$, say $\bar{x} > b$ (see Figure 3.3) he will not feel that disturbance for all times $t < (\bar{x} - b)/c$. However, once $t \geq (\bar{x} - b)/c$, the observer will (forever) be influenced by the data. We remark in passing that this is *not* the case in 3-space dimensions, (see [Ga]); the observer in $\mathbf{R}^3$ is not influenced by the data for all sufficiently large time. In this situation, a signal is felt at one particular time, and is not felt in all subsequent times—think of sound waves.

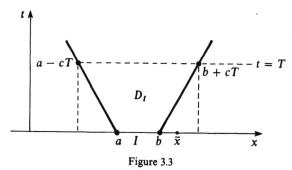

Figure 3.3

Next, let us consider the inhomogeneous wave equation

$$u_{tt} - c^2 u_{xx} = \phi(x, t), \qquad t > 0, \tag{3.5}$$

where ϕ is a given C^1 function. Suppose that $u \in C^2(G)$ solves (3.5), where G is a simply-connected region in $t > 0$ having a nice (= piecewise smooth) boundary. We can rewrite (3.5) as $(-c^2 u_x)_x + (u_t)_t = \phi$, integrate over G, and apply the divergence theorem to get

$$\int_{\partial G} -c^2 u_x \, dt - u_t \, dx = \iint_G \phi \, dx \, dt. \tag{3.6}$$

Now for G, we take the triangular region depicted in Figure 3.4.

Then along I, $dt = 0$, and $\int_I - c^2 u_x \, dt - u_t \, dx = \int_{x_1}^{x_2} - u_t \, dx$, while along C_2, $dx = -c \, dt$ and

$$\int_{C_2} - c^2 u_x \, dt - u_t \, dx = \int_{C_2} c u_x \, dx + c u_t \, dt$$

$$= \int_{C_2} d(cu) = cu(x_0, t_0) - cu(x_2, 0).$$

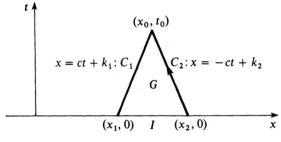

Figure 3.4

Similarly,

$$\int_{C_1} - c^2 u_x \, dt - u_t \, dx = cu(x_0, t_0) - cu(x_1, 0).$$

Hence (3.6) becomes

$$u(x_0, t_0) = \tfrac{1}{2}[u(x_1, 0) + u(x_2, 0)] + \frac{1}{2c} \int_{x_1}^{x_2} u_t(s, 0) \, ds + \frac{1}{2c} \iint_G \phi \, dx \, dt.$$

Now if we consider the initial-value problem for (3.5), with data (3.3), and use the fact that $x_1 = x_0 - ct_0$, and $x_2 = x_0 + ct_0$, we obtain the following formula for the solution:

$$u(x_0, t_0) = \tfrac{1}{2}[\alpha(x_0 + ct_0) + \alpha(x_0 - ct_0)] + \frac{1}{2c} \int_{x_0 - ct_0}^{x_0 + ct_0} \beta(s) \, ds$$

$$+ \frac{1}{2c} \iint_G \phi \, dx \, dt. \tag{3.7}$$

Thus, we have shown that if a solution to (3.5), (3.3) exists, then it is unique and is explicitly given by (3.7). Conversely, it is easy to check that (3.7) is indeed a solution of the problem. Note too that (3.7) reduces to (3.4) when $\phi = 0$, and that from the representation (3.7) it is easy to conclude that the solution depends continuously on the data, and on ϕ (with the same topology as before). Finally, we see that in this case, the *entire* region G is the domain of dependence for (x_0, t_0).

We turn now to a different problem for the wave equation; namely, a mixed (= initial-boundary-value) problem. Thus, in most physical situations, we are only concerned with a finite region of space, say an interval $a \le x \le b$, and we wish to solve (3.5) subject to prescribed initial values $u(x, 0)$ and $u_t(x, 0)$ on $a \le x \le b$, and also subject to "boundary" values prescribed on the lines $x = a$, $x = b$, in $t > 0$; see Figure 3.5. Physically one thinks of a

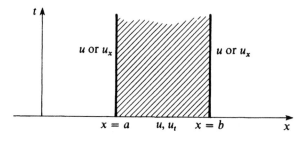

Figure 3.5

finite string, with a given initial displacement and velocity, which is being held by two observers at points a and b, who are now *moving* the string; it is required to find the position of the string at any time $t > 0$, when $a \leq x \leq b$.

Mathematically, it is required to solve the following mixed initial-boundary-value problem:

$$u_{tt} - c^2 u_{xx} = \phi(x, t), \qquad a < x < b, \quad t > 0, \tag{3.8}$$

with initial data

$$u(x, 0) = \alpha(x), \qquad u_t(x, 0) = \beta(x), \qquad a < x < b, \tag{3.9}$$

and with boundary data

$$\left.\begin{array}{l} u(a, t) = a(t) \quad \text{or} \quad u_x(a, t) = a(t) \\[2mm] u(b, t) = b(t) \quad \text{or} \quad u_x(b, t) = b(t) \end{array}\right\}, \qquad t > 0 \tag{3.10}$$

We shall first prove that the solution to (3.8)–(3.10) is unique (if it exists!). The technique is the simplest example of what is nowadays known as the *energy method*. We shall consider many generalizations of this technique in the subsequent chapters.

Theorem 3.1. *There is at most one (smooth) solution to any of the problems* (3.8)–(3.10).

Proof. Suppose that v and w are solutions; then $u = v - w$ is a solution to the problem

$$u_{tt} - c^2 u_{xx} = 0, \qquad a < x < b, \quad t > 0, \tag{3.11}$$

with initial data

$$u(x, 0) = u_t(x, 0) = 0, \qquad a < x < b, \tag{3.12}$$

and boundary data

$$u(a, t) = 0 \quad \text{or} \quad u_x(a, t) = 0$$
$$u(b, t) = 0 \quad \text{or} \quad u_x(b, t) = 0$$
$$\left.\right\}, \quad t > 0. \tag{3.13}$$

Our goal is to show that (3.11)–(3.13) imply that $u \equiv 0$. To this end, we consider the "energy integral"

$$I(t) = \frac{1}{2} \int_a^b (c^2 u_x^2 + u_t^2)\, dx.$$

If we differentiate I with respect to t, we find

$$\frac{dI}{dt} = \int_a^b (c^2 u_x u_{xt} + u_t u_{tt})\, dx = \int_a^b (c^2 u_x u_{xt} + c^2 u_t u_{xx})\, dx$$

$$= c^2 \int_a^b \frac{\partial}{\partial x}(u_x u_t)\, dx = c^2 u_x u_t \big|_{x=a}^{x=b} = 0,$$

in view of (3.11) and (3.13). Thus $I(t)$ is a constant, independent of t. Since (3.12) implies that $I(0) = 0$, we see that $I(t) = 0$ for all $t > 0$. It follows that $u_x(x, t) = u_t(x, t) \equiv 0$ for all $t \geq 0$, and all x, $a \leq x \leq b$. Hence u itself must be a constant, and in view of (3.12) again, $u \equiv 0$. This completes the proof. $\square$

We turn now to the existence problem; for simplicity, we assume that $c = 1$. We need a preliminary lemma, which reduces the problem to one involving finite differences.

Lemma 3.2. *The thrice differentiable function $u(x, t)$ is a solution of $u_{tt} - u_{xx} = 0$ if and only if u satisfies (cf. Figure 3.6) the difference equation*

$$u(x - k, t - h) + u(x + k, t + h) = u(x + h, t + k) + u(x - h, t - k) \tag{3.14}$$

for all $h, k \geq 0$.
(We refer to such a region as a characteristic rectangle.)

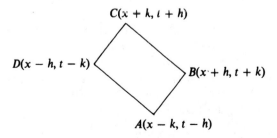

Figure 3.6

Proof. If u solves the differential equation, then $u(x, t)$ is of the form $u = f(x + t) + g(x - t)$ so that referring to Figure 3.6, we find $u(A) + u(C) = u(D) + u(B)$, since this equality holds for both f and g.

Conversely, if u solves the difference equation, then set $h = 0$ in (3.14), divide by k^2 and add $-2u(x, t)$ to both sides to obtain

$$\frac{u(x - k, t) - 2u(x, t) + u(x + k, t)}{k^2}$$
$$= \frac{u(x, t - k) - 2u(x, t) + u(x, t + k)}{k^2}, \quad (3.15)$$

By a Taylor expansion, we get the two equations

$$u(x \pm k, t) = u(x, t) \pm u_x(x, t)k + \tfrac{1}{2}u_{xx}(x, t)k^2 + k^2 o(k).$$

Putting these in (3.15) yields

$$u_{xx} - u_{tt} = o(k),$$

and thus if $k \to 0$, we find that u satisfies the differential equation. This completes the proof. $\square$

We shall use this lemma in order to construct a solution of the following mixed initial-boundary-value problem:

$$u_{tt} = c^2 u_{xx}, \qquad a < x < b, \quad t > 0, \tag{3.16}$$

$$u(x, 0) = \alpha(x), \qquad u_t(x, 0) = \beta(x), \qquad a < x < b, \tag{3.17}$$

$$u(a, t) = a(t), \qquad u(b, t) = b(t), \qquad t > 0, \tag{3.18}$$

where we assume the compatibility conditions $\alpha(a) = a(0)$ and $\beta(b) = b(0)$.

The idea is geometric and very simple. It relies on the observation that the boundary of the characteristic rectangle in Figure 3.6 is composed of characteristics. Thus, we divide the region $a < x < b$, $t > 0$ into four subregions I–IV, by drawing the characteristics from the points $(a, 0)$ and $(b, 0)$, see Figure 3.7 (where c is not necessarily equal to 1).

We observe first, that the solution in region I is completely determined by formula (3.4). To find the solution at a point P in region II, we construct the characteristic rectangle as depicted in Figure 3.7, and then use (3.14) to find $u(P)$. Similar remarks apply to region III. Thus the solution is now known in the closed regions I, II, and III. To find $u(Q)$ where Q lies in region IV, we again construct the characteristic rectangle as depicted in Figure 3.7, and then use (3.14). This yields u in the closed region IV. Now repeat this process to obtain u in the entire region $t > 0$, $a < x < b$.

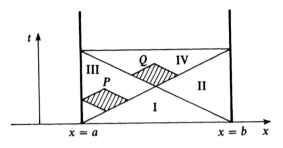

Figure 3.7

NOTES

These ideas are all classical. I learned of the difference scheme technique in John's notes [J 1].

Chapter 4

Uniqueness and Energy Integrals

We shall extend the method of energy integrals to more general second-order (hyperbolic) operators. This "energy" method is a basic technique in the modern theory of partial differential operators, and in the course of our development, we shall establish some interesting and important classical inequalities.

We begin by giving a somewhat different proof of the uniqueness theorem for the wave equation. It is this modification which we shall extend to more general equations. Thus, consider the wave equation, with $c = 1$ (for simplicity)

$$u_{tt} - u_{xx} = 0, \qquad (x, t) \in \mathbf{R} \times \mathbf{R}_+, \tag{4.1}$$

with vanishing initial data

$$u(x, 0) = u_t(x, 0) = 0, \qquad x \in \mathbf{R}. \tag{4.2}$$

We wish to prove that $u \equiv 0$.

Let $P = (x_0, t_0)$, with $t_0 > 0$, and let R be the triangular region APB in the $x - t$ plane, bounded by the line segment AB on $t = 0$, and the two characteristics AP and BP, see Figure 4.1. Let $0 < h < t_0$, and let Γ be the subregion of R cut off by the line $t = h$, as depicted in Figure 4.1. Multiplying (4.1) by $-2u_t$, gives

$$0 = -2u_t(u_{tt} - u_{xx}) = (u_x^2 + u_t^2)_t - 2(u_x u_t)_x.$$

If we integrate this over Γ and apply the divergence theorem, we find

$$0 = \int_{\partial \Gamma} \{(u_x^2 + u_t^2)t_v - 2u_x u_t x_v\} \, ds, \tag{4.3}$$

where t_v and x_v are the components of the outer unit normal to $\partial \Gamma$, and s denotes arc length. On the lines CA and DB, $x_v^2 = t_v^2 = \frac{1}{2}$, while on AB, $t_v = -1, x_v = 0$, and $t_v = 1, x_v = 0$ on CD. Thus, we obtain from (4.3)

$$0 = \int_{AB} -(u_x^2 + u_t^2) \, dx + \int_{CD} (u_x^2 + u_t^2) \, dx + \int_{AB+BD} \frac{1}{t_v} [u_x t_v - u_t x_v]^2 \, ds.$$

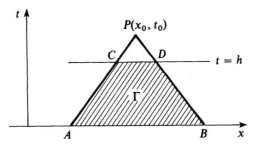

Figure 4.1

Using the nonnegativity of the last term, we find

$$\int_{AB} (u_x^2 + u_t^2)\, dx \ge \int_{CD} (u_x^2 + u_t^2)\, dx.$$

But since $u(0, x) = u_t(0, x) = 0$ along AB, we have

$$\int_{CD} (u_x^2 + u_t^2)\, dx = 0,$$

and thus $u_x^2 + u_t^2 = 0$ on CD. Hence $u_x = u_t = 0$ on CD. So since h was arbitrary, u is a constant in R. But on AB, $u = 0$; thus $u \equiv 0$ in R, so in particular $u(x_0, t_0) = 0$. Since $P = (x_0, t_0)$ was any point in $t > 0$, we have shown that $u \equiv 0$ in $t > 0$, $x \in \mathbf{R}$.

We shall show that this particular technique is quite general, and carries over for arbitrary second-order linear partial differential equations of the form

$$L(u) \equiv u_{tt} = \sum_{i,j=1}^{n} (a_{ij}(x)u_{x_i})_{x_j} - \sum_{i=1}^{n} a_i(x)u_{x_i} - a(x)u = 0, \qquad (4.4)$$

where we assume that a_{ij}, a_i, and a are continuous functions, with $a_{ij} \in C^2$, for each i and j. We also assume that $a_{ij} = a_{ji}$, and that there exist constants $c_m, c_M > 0$ such that

$$c_m^2 \sum_{i=1}^{n} \xi_i^2 \le \sum_{i,j=1}^{n} a_{ij}(x)\xi_i\xi_j \le c_M^2 \sum_{i=1}^{n} \xi_i^2, \qquad (4.5)$$

for each $\xi = (\xi_1, \ldots, \xi_n) \in \mathbf{R}^n$. This latter condition is called a *uniform ellipticity* condition on the square matrix $A(x) = (a_{ij}(x))$. It can also be written as

$$c_m^2 |\xi|^2 \le \langle A(x)\xi, \xi \rangle \le c_M^2 |\xi|^2,$$

where $\langle \, , \, \rangle$ denotes the usual Euclidean inner product. Note that for the wave equation $u_{tt} - \sum_{i=1}^n u_{x_i x_i} = 0$, $a_{ij} = \delta_{ij}$, the Kronecker delta, and $c_m = c_M = 1$.

Now let $P = (x_0, t_0)$, $t_0 > 0$, and through P we draw the backwards characteristic cone, generated by the characteristic surface S through P; i.e., the unit normal to S, $v = (v_t, v_{x_1}, \ldots, v_{x_n})$ satisfies the equation

$$v_t^2 - \sum a_{ij}(x) v_{x_i} v_{x_j} = 0. \tag{4.6}$$

We note that (4.5) implies that $v_t \neq 0$, so that the characteristic surface can be locally extended,[1] see Figure 4.2. We let $C(h)$ be the part of S generated by $AC \cup BD$ in Figure 4.2, and we let $R(h) = CD$ and $R(0) = AB$. Then the following theorem holds.

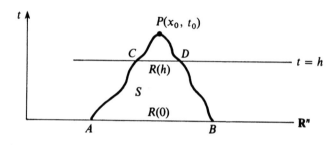

Figure 4.2

Theorem 4.1. *If $u = u_t = 0$ on $R(0)$, then $u = 0$ in S.*

Proof. Since S is compact, there exists an $M > 0$ such that $|a_i(x)| + |a(x)| < M$, if $x \in S$. Now we rewrite (4.4) using summation convention (repeated indices i and j are understood to be summed from 1 to n):

$$u_{tt} - (a_{ij}(x)u_{x_i})_{x_j} - a_i(x)u_{x_i} - a(x)u = 0. \tag{4.7}$$

Let $G = G_h$ be the region $ACDB$ in Figure 4.2, and let

$$E(h) = \int_{R(h)} (u_t^2 + a_{ij}u_{x_i}u_{x_j}) \, dx, \tag{4.8}$$

---

[1] The characteristic surface exists locally, but generally speaking, singularities (caustics) may develop in the large. One way out of this is to let Γ be the backward cone of speed c_M. This gives a less sharp result since the domain of dependence will be larger. We shall suppose that the characteristic surface is extendable to $t = 0$.

denote the "energy" along $t = h$, and $E(0)$ the energy along $t = 0$. Note that $E(h) \geq 0$ in view of (4.5). We shall show that there exists a constant $C = C(M, n, c_m, c_M)$ such that

$$E(h) \leq e^{Ch} E(0). \tag{4.9}$$

Since $E(0) = 0$ by hypothesis, we obtain from (4.9), that $E(h) = 0$, for all h, $0 \leq h < t_0$. Then $u_t^2 + a_{ij}u_{x_i}u_{x_j} = 0$, so that (4.5) implies $u_t^2 = u_{x_i}^2 = 0$, $i = 1, 2, \ldots, n$, in the entire region S, and thus u is a constant in S. Since $u = 0$ on $R(0)$, we have $u \equiv 0$ in S, and this is what we want to prove. We thus must show that (4.9) holds.

We multiply (4.4) by $2u_t$ to get

$$2u_t L(u) = 0, \tag{4.10}$$

and then compute. We have

$$2u_t u_{tt} = (u_t^2)_t,$$

$$-2u_t(a_{ij}u_{x_i})_{x_j} = (-2u_t \, a_{ij}u_{x_i})_{x_j} + 2u_{tx_j}a_{ij}u_{x_i}$$

$$= (-2u_t a_{ij}u_{x_i})_{x_j} + (a_{ij}u_{x_i}u_{x_j})_t,$$

so that

$$0 = 2u_t L(u) = (u_t^2 + a_{ij}u_{x_i}u_{x_j})_t - (2a_{ij}u_{x_i}u_t)_{x_j} - 2B, \tag{4.11}$$

where

$$B = a_i u_{x_i} u_t + auu_t. \tag{4.12}$$

We integrate (4.11) over G and apply the divergence theorem to get

$$\int_{\partial G} (u_t^2 + a_{ij}u_{x_i}u_{x_j})v_t - (2a_{ij}u_{x_i}u_t)v_{x_i} - 2\int_G B = 0,$$

so that with the obvious notation,

$$\int_{R(h)} + \int_{R(0)} + \int_{C(h)} - 2\int_G B = 0. \tag{4.13}$$

Now on $R(h)$, $v_t = 1$, $v_{x_i} = 0$, and thus

$$\int_{R(h)} = \int_{R(h)} (u_t^2 + a_{ij}u_{x_i}u_{x_j}) \, dx = E(h). \tag{4.14}$$

Similarly,

$$\int_{R(0)} = -E(0). \tag{4.15}$$

On $C(h)$, (4.6) implies that $v_t > 0$, and

$$
\begin{aligned}
\int_{C(h)} &= \int_{C(h)} \frac{1}{v_t} [u_t^2 v_t^2 + a_{ij} u_{x_i} u_{x_j} v_t^2 - 2a_{ij} u_{x_i} u_t v_{x_j} v_t] \\
&= \int_{C(h)} \frac{1}{v_t} [u_t^2 a_{ij} v_{x_i} v_{x_j} + a_{ij} u_{x_i} u_{x_j} v_t^2 - 2a_{ij} u_{x_i} u_t v_{x_i} v_t] \quad \text{(by (4.6))} \\
&= \int_{C(h)} \frac{a_{ij}}{v_t} [u_t^2 v_{x_i} v_{x_j} + u_{x_i} u_{x_j} v_t^2 - 2u_{x_i} v_{x_i} u_t v_t] \\
&= \int_{C(h)} \frac{a_{ij}}{v_t} [u_t v_{x_j} - u_{x_j} v_t][u_t v_{x_i} - u_{x_i} v_t] \\
&\geq \int_{C(h)} \frac{C_m^2}{v_t} (u_t v_{x_i} - u_{x_i} v_t)(u_t v_{x_i} - u_{x_i} v_t) \quad \text{(by (4.5))} \\
&\geq 0.
\end{aligned}
$$

This, together with (4.13)–(4.15) implies

$$E(h) \leq E(0) + 2 \int_G B. \tag{4.16}$$

We now estimate $\int_G B \equiv I$. To do this we need a lemma.

Lemma 4.2 (Poincaré's Inequality).[2]

$$\iint_{G_h} u^2 \, dx \, dt \leq h^2 \iint_{G_h} (u_t^2 + u_{x_i} u_{x_i}) \, dx \, dt. \tag{4.17}$$

Proof. Since $u(x, 0) = 0$, we can write

$$u(x, t) = \int_0^t u_\tau(x, \tau) \, d\tau,$$

and using the Schwarz inequality,

$$|u(x, t)| \leq \left(\int_0^t d\tau \right)^{1/2} \left(\int_0^t u_\tau^2 \, d\tau \right)^{1/2}$$

---

[2] See Chapter 11, Theorem 11.3 for a sharper version.

Squaring gives

$$u(x, t)^2 \leq t \int_0^t u_\tau^2 \, d\tau,$$

so that

$$\int_{R(t)} u^2 \, dx \leq t \int_{R(t)} \int_0^t u_\tau^2 \, d\tau \, dx = t \iint_{G_t} u_\tau^2 \, dx \, d\tau.$$

If we integrate this from $t = 0$ to $t = h$, we obtain

$$\iint_{G_h} u^2 \, dx \, dt = \int_0^h \int_{R(t)} u^2 \, dx \, dt \leq \int_0^h \left(t \iint_{G_\tau} u_\tau^2 \, dx \, d\tau \right) dt$$

$$\leq \int_0^h h \left(\iint_{G_h} u_\tau^2 \, dx \, d\tau \right) dt$$

$$\leq h^2 \iint_{G_h} u_\tau^2 \, dx \, d\tau,$$

and this implies the desired result, (4.17). $\quad\square$

Now we have

$$|I| \leq \int_0^h \int_{R(t)} |a_i u_{x_i} u_t + a u u_t| \, dx \, dt. \tag{4.18}$$

But, for each i, $|a_i u_{x_i} u_t| \leq M|u_{x_i} u_t| \leq M(u_{x_i}^2 + u_t^2)/2$; similarly $|a u u_t| \leq M(u^2 + u_t^2)/2$. Using these in (4.18) together with Poincaré's inequality yields

$$|I| \leq \int_0^h \int_{R(t)} \frac{M}{2}(u_{x_i} u_{x_i} + n u_t^2) + \frac{M}{2} h^2 (n u_t^2 + u_{x_i} u_{x_i}) + \frac{nM}{2} u_t^2.$$

But from (4.5), $u_{x_i} u_{x_i} \leq a_{ij} u_{x_i} u_{x_j}/c_m^2$, so that

$$|I| < \int_0^h \int_{R(t)} \left(\frac{M}{2c_m^2} + \frac{Mh^2}{2c_m^2} \right) a_{ij} u_{x_i} u_{x_j} + \left(M(n+1) + \frac{Mnh^2}{2} \right) u_t^2. \tag{4.19}$$

Let

$$\frac{C}{2} > \max \left(\frac{Mn}{2c_m^2} + \frac{Mnt_0^2}{2c_m^2}, M(n+1) + \frac{Mt_0^2}{2} \right);$$

then (4.19) implies

$$|I| \le \frac{C}{2} \int_0^h \int_{R(t)} u_t^2 + a_{ij} u_{x_i} u_{x_j} = \frac{C}{2} \int_0^h E(t)\, dt.$$

This together with (4.16) yields

$$E(h) \le E(0) + C \int_0^h E(t)\, dt, \qquad 0 \le h < t_0. \tag{4.20}$$

Now for $0 < t < t_0$, we have, from (4.20)

$$\frac{d}{dt}\left\{ e^{-Ct} \int_0^t E(\tau)\, d\tau \right\} = e^{-Ct}\left[E(t) - C \int_0^t E(\tau)\, d\tau \right] \le e^{-Ct} E(0).$$

Integrate this from $t = 0$ to $t = h$, to get

$$e^{-Ch} \int_0^h E(\tau)\, d\tau \le \frac{1 - e^{-Ch}}{C} E(0),$$

so

$$E(h) - E(0) \le C \int_0^h E(\tau)\, d\tau \le (e^{Ch} - 1)E(0),$$

from which (4.9) follows. This completes the proof. $\square$

We remark that the result that (4.20) yields (4.9) is commonly called *Gronwall's inequality*. It holds for any function $E(t) \ge 0$, and for any constant $C > 0$.

NOTES

The general theorem presented here was shown to me by E. C. Zachmanoglou.

Chapter 5

Holmgren's Uniqueness Theorem

There is a well-known theorem, called the Cauchy–Kowaleski theorem, which asserts that there exists a unique analytic solution of an analytic initial-value problem. Here, by an analytic initial-value problem, we mean a problem in which everything (the terms in the equation, the initial data, and the initial hypersurface), is analytic in a neighbourhood of a point (see [Ga]). The possibility is thereby left open as to whether there can exist a nonanalytic solution to this problem. Holmgren's uniqueness theorem denies this possibility. We shall also find this result useful in Chapter 6 where we shall apply it to determine qualitative information on domains of dependence. For this reason, we shall prove a rather general version of the theorem.

Before stating the theorem, we would like to recall a (nowadays) rather familiar technique used to prove uniqueness theorems. Thus, suppose that $\mathscr{H}$ is a Hilbert space, and A is an operator on $\mathscr{H}$. If $\mathscr{R}(A)$ and $\eta(A)$ denote the range and null-space of A, respectively, then we know that $\mathscr{R}(A^*) \subseteq \eta(A)^\perp$, where A^* is the adjoint of A, and $\perp$ denotes orthogonal complement. Thus, if $\mathscr{R}(A^*)$ is "large," $\eta(A)^\perp$ must be "large," so that $\eta(A)$ itself is "small." That is, the existence of sufficiently-many solutions of the adjoint equation implies that the null-space of A is zero; i.e., A has unique solutions. In symbols, if $Ax = Ay$, we choose w such that $A^*w = x - y$; then

$$\|x - y\|^2 = \langle x - y, x - y \rangle = \langle x - y, A^*w \rangle = \langle A(x - y), w \rangle = 0,$$

so that $x = y$. This is the idea behind Holmgren's method. It was a striking insight in Holmgren's time.

Let

$$L = \sum_{|\alpha| \leq m} A_\alpha(x)D^\alpha, \qquad x \in \mathbf{R}^n, \tag{5.1}$$

be an mth order linear differential operator with coefficients $A_\alpha(x)$ which are $N \times N$ matrices having coefficients in C^{m+2} (recall the notation defined in Chapter 2). For $\xi = (\xi_1, \ldots, \xi_n)$, we denote by

$$\mathscr{L}(x, \xi) = m! \sum_{|\alpha| = m} A_\alpha(x)\xi^\alpha, \qquad \xi^\alpha = \xi_1^{\alpha_1} \cdots \xi_n^{\alpha_n},$$

the *characteristic matrix* of L, and we let

$$P(x, \xi) = \det \mathscr{L}(x, \xi),$$

a form of degree mN, be the *characteristic form* of L.

We extend the notion of characteristic surface to vector-valued operators L as follows. A surface S given by $\phi(x) = 0$, is called *noncharacteristic* with respect to L at $x_0 \in S$ if

$$P(x_0, \nabla\phi(x_0)) \neq 0.$$

S is called noncharacteristic with respect to L if it is noncharacteristic at each point. For a scalar operator, it is easy to see that this condition reduces to the old one, namely

$$\sum_{|\alpha| = m} A_\alpha(x_0)\nabla\phi(x_0)^\alpha \neq 0.$$

We shall now motivate and explain the Holmgren approach. Suppose that G is a region in $\mathbf{R}^n$ with boundary, $\partial G = \Gamma_1 \cup \Gamma_2$. Let u be a column vector and v a row vector such that

$$D^\beta u(x) = 0 \quad \text{if } |\beta| \leq m - 1, \qquad x \in \Gamma_1. \tag{5.2}$$

$$D^\beta v(x) = 0 \quad \text{if } |\beta| \leq m - 1, \qquad x \in \Gamma_2. \tag{5.3}$$

We are interested in the uniqueness problem for

$$L(u) = 0 \quad \text{in } G \tag{5.4}$$

with data satisfying (5.2); i.e., we want to conclude that $u \equiv 0$. Here is the way of doing this via Holmgren's method.

If we integrate by parts, we get, from (5.2) and (5.3),

$$\int_G v\left(\sum_{|\alpha| \leq m} A_\alpha D^\alpha u\right) dx = \int_G \left(\sum_{|\alpha| \leq m} (-1)^{|\alpha|} D^\alpha(vA_\alpha)u\right) dx. \tag{5.5}$$

Let M be the (adjoint) operator defined by

$$M(v) = \sum_{|\alpha| \leq m} (-1)^{|\alpha|} D^\alpha(vA_\alpha). \tag{5.6}$$

Now suppose that the equation

$$M(v(x)) = w(x), \qquad x \in G, \tag{5.7}$$

with initial data (5.3), had a solution for each w. Then from (5.5) we obtain

$$\int_G wu \, dx = \int_G M(v)u \, dx = \int_G v(Lu) \, dx = 0, \qquad (5.8)$$

from which we conclude that $u \equiv 0$ (by setting $w = u^t$, the transpose of u). Thus, if the initial-value problem (5.7), (5.3) had a solution for all w, then the solution of (5.4), (5.2) is unique. The same conclusion holds if we can solve (5.7), (5.3) for all w in a dense set of functions; that is, if for $w \in C(\mathbf{R}^n)$, there was a sequence w_n converging uniformly to w, such that for each n, we can find v_n with

$$M(v_n) = w_n, \qquad D^\beta v_n = 0, \qquad |\beta| \le m - 1 \quad \text{on } \Gamma_2$$

Then

$$\int_G wu \, dx = \lim_{n \to \infty} \int_G w_n u \, dx = 0,$$

from which we could again conclude that $u \equiv 0$.

Thus, we have shown that the initial-value problem (5.4), (5.2) has at most one solution if the initial-value problem (5.7), (5.3) can be solved for a dense set of functions w. (We are leaving things a little vague here in that we have not defined a topology on the class of functions containing w; this will be made precise below.)

We now state and prove the theorem of this chapter.

Theorem 5.1 (Holmgren's Uniqueness Theorem). *The equation*

$$L(u) = \sum_{|\alpha| \le m} A_\alpha(x)D^\alpha u = f(x), \qquad x \in \mathbf{R}^n, \qquad (5.9)$$

with analytic coefficients A_α, with data

$$D^\beta u(x) = g_\beta(x), \qquad |\beta| \le m - 1, \qquad x \in S, \qquad (5.10)$$

given on an analytic noncharacteristic surface S, has at most one solution in a neighbourhood of S.

Before giving the proof, a few remarks are in order.

1. Note that neither g_β nor f is assumed to be analytic.

2. We can write (5.6) as

$$M^*(v^*) = \sum_{|\alpha| \le m} (-1)^{|\alpha|} D^\alpha(A_\alpha^* v^*), \qquad (5.6)$$

where the asterisk's denote adjoints. From this it follows that the principal part of M^* is

$$(-1)^m \sum_{|\alpha|=m} A_\alpha^*(x)D^\alpha,$$

and the characteristic equation of M^* is

$$(-1)^{mN} \det\left[\sum_{|\alpha|=m} A_\alpha \eta^\alpha\right] = \det\left[(-1)^m \sum_{|\alpha|=m} A_\alpha^* \eta^\alpha\right] = 0.$$

From this we see that the characteristic equations for M^* and L are the same. Hence, S is characteristic with respect to M^* if and only if S is characteristic with respect to L.

3. Since the coefficients of L are analytic functions, we are led by the Cauchy–Kowalewski theorem to take w to be in the class of analytic functions, these being dense in, e.g., $C(\mathbf{R}^n)$. The Cauchy–Kowalewski theorem guarantees the local existence of a solution v of the equation $M(v) = w$, with zero initial data on S, if S is analytic. But the region of existence depends on w. We must find a neighbourhood of S which is *independent* of w.

Proof of Theorem 5.1. Let $S = \{x : \phi(x) = 0\}$ be noncharacteristic with respect to L, and let S_0 be a compact subset of S. We deform S_0 by an analytic deformation (homotopy), keeping the boundary of S_0 fixed. To this end, we consider a family of surfaces $S_\lambda = \{x : \phi(x, \lambda) = 0\}$, $0 \le \lambda \le \bar{\lambda}$, each having the same boundary. The functions $\phi(x, \lambda)$ are analytic in both x and λ. If $0 \le \lambda_1 < \lambda_2 \le \bar{\lambda}$, we denote by $R_{\lambda_1 \lambda_2}$ the region determined by the S_λ, $\lambda_1 \le \lambda \le \lambda_2$; see Figure 5.1. We assume that $\partial S_0 = \{x : \phi(x) = 0\} \cap \{x : \alpha(x) = 0\}$ where α is an analytic function, and we set $\phi(x, \lambda) = (1 - \lambda)\phi(x) + \lambda\alpha(x)$. Note that for small λ, $P(x, \nabla\phi(x, \lambda)) \ne 0$ for x in the compact set S_λ, provided that $P(x, \nabla\phi(x)) \ne 0$ for x in the set $S_0 \cap \{x : \phi(x) = 0\}$. That is, we choose $\phi(x, \lambda)$ such that all the sets S_λ with $0 \le \lambda \le \bar{\lambda}$ are noncharacteristic with respect to L; this really defines $\bar{\lambda}$. Finally, let $R = R_{0\bar{\lambda}}$.

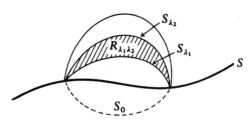

Figure 5.1

We shall show that a solution of $L(u) = f$ in R is determined uniquely by its data on S_0. We first need a lemma.

Lemma 5.2. *There exists an $\varepsilon > 0$ such that if $0 \leq \lambda < \mu \leq \bar{\lambda}$ and $\mu < \lambda + \varepsilon$, then the equation $M(v) = w$, in $R_{\lambda\mu}$, with data $D^\beta v = 0$ on S_μ, $|\beta| \leq m - 1$, has solutions v for w in a dense set in $C(R_{\mu\lambda})$.*

Proof. From the Cauchy–Kowalewski theorem, if K is a compact subset of a noncharacteristic analytic surface, and w is analytic in a neighborhood of K, there is a solution of $M(v) = w$ in a neighborhood of K, which has zero initial data on K.

Let w be an entire function of x and a parameter $\xi = (\xi_1, \ldots, \xi_n)$, and let the S_λ be noncharacteristic surfaces, as above. For each ξ with $|\xi| \leq 1$, and each $\mu, 0 \leq \mu \leq \bar{\lambda}$, we seek a solution $v(x, \xi, \mu)$ of the equation $M(v) = w(x, \xi)$, with zero initial data on S_μ.

We view the equations $M(v) = w$ as a system of equations for a function v of $2n + 1$ variables, $x_1, \ldots, x_n, \xi_1, \ldots, \xi_n, \mu$, for which no derivatives with respect to the last $(n + 1)$ variables enter the equations. We seek a solution in which all derivatives of v of order $\leq m - 1$, vanish on the analytic hypersurface $\phi(x, \mu) = 0$, in (x, ξ, μ)-space. Now the set

$$K = \{(x, \xi, \mu): x \in S_\mu, 0 \leq \mu \leq \bar{\lambda}, |\xi| \leq 1\},$$

is a compact analytic hypersurface in (x, ξ, μ) space. Furthermore, since the characteristic matrix of M is unchanged by the addition of more variables, K is noncharacteristic for M. Thus, again by the Cauchy–Kowalewski theorem, there exists a solution of our problem in some neighborhood of K.

Therefore there exists $\delta > 0$, such that for any ξ, μ with $0 \leq \mu \leq \bar{\lambda}$, and $|\xi| \leq 1$, the function $v(x, \xi, \mu)$ is defined for x in a δ-neighborhood of S_μ. It follows that by taking μ so close to λ such that the δ-neighborhoods of S_λ and S_μ meet, that we can find an $\varepsilon > 0$ such that $v(x, \xi, \mu)$ is defined in $R_{\lambda\mu}$ for every λ, μ with $0 < \mu - \lambda < \varepsilon$ and for all ξ, with $|\xi| \leq 1$. In total, if $0 < \mu - \lambda < \varepsilon$, then the problem $M(v) = w(x, \xi)$, in $R_{\lambda\mu}$ with zero data on S_μ has a solution for all ξ, with $|\xi| \leq 1$.

If we can find $w(x, \xi)$ such that their linear combinations are C^0-dense for $|\xi| \leq 1$, then the problem $M(v) = w$ in $R_{\lambda\mu}$, $D^\beta v = 0$ in S_μ, $|\beta| \leq m$, has solution v for a dense set of w in $C(R_{\lambda\mu})$, and this would complete the proof of the lemma. To this end we take $w(x, \xi) = \exp(x \cdot \xi)$. Then every monomial $x^\alpha = x_1^{\alpha_1} \cdots x_n^{\alpha_n}$ can be uniformly approximated in every bounded subset of x-space by linear combinations of the $w(x, \xi), |\xi| \leq 1$; namely,

$$x^\alpha = \lim_{\xi \to 0} \prod_{k=1}^{n} \left(\frac{e^{\xi_k x_k} - 1}{\xi_k} \right)^{\alpha_k}$$

Since the linear combinations of the monomials are dense, the same holds
true for linear combinations of the $w(x, \xi), |\xi| \leq 1$. Thus $M(v) = w$ has a
solution v in $R_{\lambda\mu}$ with zero data on S_μ, for a dense set of w if $\lambda - \mu < \varepsilon$. This
proves the lemma. $\square$

We can now complete the proof of the theorem. Thus, suppose that
$L(u) = 0$ in $R = R_{0\bar{\lambda}}$, and $D^\beta u = 0, |\beta| \leq m - 1$ on S_0. We find a finite set
$0 = \lambda_0 < \lambda_1 < \cdots < \lambda_k = \bar{\lambda}$ such that $\lambda_{j+1} - \lambda_j < \varepsilon, j = 0, 1, \ldots, k - 1$,
where the $\varepsilon > 0$ is obtained from our lemma. Using the lemma, we know
that for a dense set of w there are solutions of $M(v) = w$ in $R_{0\lambda_1}$, with zero
data on S_{λ_1}. Since $\partial(R_{0\lambda_1}) = S_0 \cup S_{\lambda_1}$, and u has zero data on S_0, and v
has zero data on S_{λ_1}, we obtain (as above) that $u \equiv 0$ in $R_{0\lambda_1}$. We repeat
this argument finitely many times to conclude $u \equiv 0$ in $R_{0\bar{\lambda}}$. This completes
the proof of the theorem. $\square$

We close this chapter with a small observation. We have really shown
that the solution of the initial-value problem is uniquely determined at all
points which can be reached by deforming a portion of the initial surface
analytically through noncharacteristic surfaces having the same boundary.
Thus the "region of uniqueness" depends only on the geometry of the
characteristic surface.

NOTES

The original theorem goes back to the beginning of the twentieth century;
the proof given here follows John [BJS], see also [Ga].

Chapter 6

An Initial-Value Problem for a Hyperbolic Equation

We consider the equation for the homogeneous operator P:

$$P(u) \equiv \sum_{|\alpha|=m} a_\alpha D^\alpha u = 0, \qquad x = (t, \xi) \in \mathbf{R}_+ \times \mathbf{R}^n, \tag{6.1}$$

with initial data

$$D_0^j u(0, \xi) = \psi_j(\xi), \qquad 0 \le j < m. \tag{6.2}$$

We assume that each a_α is constant, and that the hyperplane $t = 0$ is non-characteristic with respect to P. Since the normal vector to the surface $t = 0$ is $(1, 0, \dots, 0)$, this means $a_{m\dots0} \ne 0$; thus we assume $a_{m\dots0} = 1$.

We shall reduce this problem to the solvability of the equation (6.1) together with the data of the form

$$D_0^k u(0, \xi) = \begin{cases} 0, & 0 \le k < m - 1, \\ g(\xi), & k = m - 1. \end{cases} \tag{6.3}$$

Lemma 6.1. If $P(u) = 0$ with data (6.3) is solvable, then so is (6.1), (6.2).

Proof. If u solves (6.1), (6.2), we set

$$v(x) = u(x) - \sum_{k=0}^{m-1} \frac{1}{k!} \psi_k(\xi) t^k. \tag{6.4}$$

Then v solves a problem of the form

$$P(v) = \psi(x), \qquad D_0^k v(0, \xi) = 0, \qquad 0 \le k \le m - 1. \tag{6.5}$$

Conversely, if v solves (6.5), then u, defined by (6.4), solves (6.1) and (6.2). Now if $v(t, \xi, s)$ solves $P(v) = 0$, with data $D_0^k v(s, \xi, s) = 0$, $0 \le k < m - 1$ and $D_0^{m-1} v(s, \xi, s) = \psi(s, \xi)$, we set

$$u(t, \xi) = \int_0^t v(t, \xi, s)\, ds,$$

and we check easily that u solves (6.5). $\quad\square$

Thus, we consider the problem (6.1), (6.3). In order to solve this, we shall show that it is "formally" solvable by employing a separation of variables technique, together with a systematic use of the Fourier transform. If the "data" $g(y)$ decays sufficiently fast at infinity, we can show that the formal solution is actually a rigorous solution. We begin by recalling the definition and some standard properties of the Fourier transform (see [Ru 1]).

Definition 6.2. If $f \in L_1(\mathbf{R}^n)$, the Fourier transform $\hat{f}$, of f is defined by

$$\hat{f}(y) = (2\pi)^{-n/2} \int_{\mathbf{R}^n} e^{-i\xi \cdot y} f(\xi) \, d\xi, \qquad y \in \mathbf{R}^n,$$

where $\xi \cdot y = \sum \xi_i y_i$.

We recall that if $\hat{f}$ also is in $L_1(\mathbf{R}^n)$, then inversion holds; i.e.,

$$f(\xi) = (2\pi)^{-n/2} \int_{\mathbf{R}^n} e^{i\xi \cdot y} \, \hat{f}(y) \, dy.$$

Moreover, if $f \in C_0^s(\mathbf{R}^n)$, then $\sup_{y \in \mathbf{R}^n} |y^s \hat{f}(y)|$ is finite, so if $s > n$, this implies that $\hat{f} \in L_1(\mathbf{R}^n)$ and thus inversion holds. Finally, the mapping $f \to \hat{f}$ is an isometry from $L_2(\mathbf{R}^n)$ onto $L_2(\mathbf{R}^n)$.

We now return to the problem (6.1), (6.3). Before solving it, we shall first give some idea of the method. Thus, suppose that we can solve $P(u) = 0$, (6.3) for functions $g(y) = e^{iy \cdot \xi}$, where $\xi \in \mathbf{R}^n$; i.e., suppose that we can solve the equation $P(v) = 0$ with data

$$D_0^j v(0, \xi, y) = \begin{cases} 0, & 0 \leq k < m - 1, \\ e^{i\xi \cdot y}, & k = m - 1. \end{cases} \tag{6.6}$$

Then a formal solution to (6.1), (6.3) is given by

$$u(x) = (2\pi)^{-n/2} \int_{\mathbf{R}^n} \hat{f}(y) v(x, y) \, dy. \tag{6.7}$$

To see this, note that

$$P(u) = (2\pi)^{-n/2} \int_{\mathbf{R}^n} \hat{f}(y) P(v) \, dy = 0,$$

$$D_0^j u(0, \xi) = (2\pi)^{-n/2} \int_{\mathbf{R}^n} \hat{f}(y) D_0^k v(0, \xi, y) \, dy = 0,$$

if $0 \le k < m - 1$, while the inversion formula gives

$$D_0^{m-1} u(0, x) = (2\pi)^{-n/2} \int_{\mathbf{R}^n} \hat{f}(y) e^{ix \cdot y} \, dy = f(x).$$

In particular, this formal procedure will be valid provided that:

(i) equation (6.7) is well defined,
(ii) f satisfies the inversion formula,
(iii) the differential operators can be taken under the integral signs, and
finally,
(iv) $P(v) = 0$ with data, (6.6) has a solution.

We shall show that this procedure actually works if $f \in C_0^s(\mathbf{R}^n)$ where $s \ge m + n + 1$, provided that the operator is "hyperbolic," in the sense of the following definition.

Definition 6.3. The operator P is *hyperbolic* if the (algebraic) equation $P(t, y) = 0$ has only real roots t for each $y \in \mathbf{R}^n$.

We proceed with the details. We first solve $P(v) = 0$ with data, (6.6). To do this we use separation of variables, and seek a solution in the form

$$v(t, x, y) = e^{ix \cdot y} w(t, y).$$

Substituting this in the equation gives

$$0 = P(e^{i\xi \cdot y} w(t, y)) = \sum_{|\alpha| = m} a_\alpha (D_1^{\alpha_1} \cdots D_n^{\alpha_n} e^{i\xi \cdot y}) D_0^{\alpha_0} w(t, y)$$

$$= e^{i\xi \cdot y} \sum_{|\alpha| = m} a_\alpha [(iy_1)^{\alpha_1} \cdots (iy_n)^{\alpha_n} D_0^{\alpha_0} w]$$

$$= e^{i\xi \cdot y} \sum_{|\alpha| = m} a_\alpha (D_0^{\alpha_0} w)(iy)^\alpha = e^{i\xi \cdot y} P(D_0, iy) w.$$

Also, since

$$D_0^k v(0, \xi, y) = e^{i\xi \cdot y} D_0^k w(0, y), \qquad 0 \le k \le m - 1,$$

we see that w must solve the *ordinary* differential equation

$$P(D_0, iy) w(t, y) = 0, \tag{6.8}$$

with initial data

$$D_0^k w(0, y) = \begin{cases} 0, & 0 \le k < m - 1, \\ 1, & k = m - 1. \end{cases} \tag{6.9}$$

Conversely, a solution w of (6.8), (6.9) gives us the desired solution v. We thus proceed to solve (6.8), (6.9). It turns out that we can solve this problem explicitly, namely,

$$w(t, y) = \frac{1}{2\pi} \int_\Gamma \frac{e^{itz}}{P(iz, iy)} \, dz, \tag{6.10}$$

where Γ is any simple oriented contour surrounding the roots r of the polynomial equation $P(r, y) = 0$.

In this case, a formal solution of $P(u) = 0$, with data, (6.3) is given by

$$u(t, \xi) = (2\pi)^{-n/2} \int_{\mathbf{R}^n} e^{i\xi \cdot y} w(t, y) \hat{f}(y) \, dy, \tag{6.11}$$

where w is defined by (6.10). This formula would define an actual solution provided that $f \in C_0^s(\mathbf{R}^n)$ for sufficiently large s, and that w has polynomial growth in y, since if this were so, then we could pass the differential operators through the integral in (6.11). We shall show that the assumption of hyperbolicity ensures that w has polynomial growth in y.

Theorem 6.4. *If P is a hyperbolic operator, then (6.11) defines a solution $u \in C^m(\mathbf{R}^n)$ of $P(u) = 0$, with data, (6.3) if $f \in C_0^s(\mathbf{R}^n)$, $s = m + n + 1$.*

Proof. Let $R = \{r_k(y): 0 \le k \le m\}$ denote the (real) roots t of $P(t, y) = 0$. Let $\Gamma = \Gamma(y)$ be a (not necessarily connected) contour in $\mathbf{C}$ containing these roots, where $\Gamma = \partial\Omega$, and Ω is the union of squares of side 2, with centers at r_k. Notice that $\text{dist}(\Gamma, R) \ge 1$, and $l(\Gamma)$, the length of Γ, is equal to c, c independent of y.

Lemma 6.5. *There is an $M > 0$, depending only on P such that for each $k = 1, 2, \ldots, m$*

$$|r_k(y)| \le M|y|. \tag{6.12}$$

Proof. If θ is a root of the monic polynomial $p(x) = \sum_{i=0}^m b_i x^i$, and $\theta \ge 1$ then

$$-\theta^m = \sum_{i=0}^{m-1} b_i \theta^i, \quad \text{so} \quad |\theta| \le \sum_{i=0}^{m-1} |b_i||\theta|^{i-m} \le \sum_{i=0}^{m-1} |b_i|;$$

thus for any root θ, $|\theta| \le \max\{1, \sum_{i=0}^{m-1} |b_i|\}$. Applying this to the polynomial equation $P(r_k/|y|, y/|y|) = 0$, we find that the lemma holds with

$$M = \max\left\{1, \sum_{|\alpha|=m} \left|a_\alpha\left(\frac{y}{|y|}\right)^\alpha\right|\right\} = \max\left\{1, \sum_{|\alpha|=m} |a_\alpha|\right\}. \quad \square$$

Now if $r \in \Gamma$, $|r - r_k| \leq \sqrt{2}$ for some k and thus by the lemma $|r| \leq M|y|$ $+ \sqrt{2}$. Hence for $t \geq 0$,

$$
\begin{aligned}
|D_0^k w(t, y)| &= \left| \frac{i^k}{2\pi i^m} \int_\Gamma \frac{z^k e^{izt}}{P(z, y)} dz \right| \\
&\leq \frac{(M|y| + \sqrt{2})^k}{2\pi} \int_\Gamma \frac{|e^{izt}|}{|P(z, y)|} dz \\
&\leq \frac{(M|y| + \sqrt{2})^k}{2\pi} e^t c,
\end{aligned}
$$

since $z \in \Gamma$ implies both $|P(z, y)| \geq 1$ and $|e^{izt}| \leq e^t$. Thus for each $t > 0$, and each k, $0 \leq k \leq m$

$$
|e^{i\xi \cdot y} \hat{f}(y) D_0^k w(t, y)| = e^t O(|y^k \hat{f}(y)|) = e^t O(|y|^{-n-1}),
$$

as $|y| \to \infty$. This justifies all of the differentiations under the integral sign in (6.11), and completes the proof. $\square$

Corollary 6.6. *The equation* (6.1) *for the hyperbolic operator P with initial data* (6.2) *on the noncharacteristic hypersurface $t = 0$ is solvable if each $\psi_k(y)$ is in $C_0^s(\mathbf{R}^n)$, $s = m + n + 1$.*

We next show that the solution which we have just obtained is unique and has a bounded domain of dependence. This will be done with the aid of Holmgren's theorem. We recall that this theorem states that the equation $P(u) = 0$, with initial data given on an analytic noncharacteristic hypersurface, is uniquely determined in a region generated by analytic deformations of the original hypersurface.

Now by assumption, the hyperplanes $t = $ const. are noncharacteristic with respect to P. Hence the same holds true for an analytic hypersurface whose normal vector makes a sufficiently small angle with the t-axis. Thus, consider a hypersurface S of the form $t - f(\xi) = 0$. The characteristic equation for P with respect to S is $1 + A(-\nabla f) = 0$, where

$$
A(\xi) = \sum_{\substack{|\alpha| = m \\ \alpha_0 < m}} a_\alpha \xi_1^{\alpha_1} \cdots \xi_n^{\alpha_n}.
$$

Since $A(0) = 0$, there is a constant $\varepsilon > 0$, such that if $|\xi| < \varepsilon$, then $1 + A(\xi) \neq 0$. It follows that S is noncharacteristic provided $|\nabla f|^2 < \varepsilon^2$, at each point.

It is easy to write down a family of *analytic* hypersurfaces which are everywhere noncharacteristic. For example, for $(t_0, y_0) \in \mathbf{R} \times \mathbf{R}^n$, and $0 \leq \lambda$, let S_λ be the analytic, noncharacteristic hypersurface consisting of those points (t, y) in $\mathbf{R} \times \mathbf{R}^n$ for which

$$t - t_0 = -\left[\frac{\varepsilon t_0^2 + \varepsilon\lambda|y - y_0|^2}{\varepsilon + \lambda}\right]^{1/2}, \qquad |y - y_0| \le \frac{t_0}{\varepsilon}.$$

One checks easily that these generate the conical region $R = \{0 \le t < t_0 - \varepsilon|y - y_0|\} \cap \{|y - y_0| < t_0/\varepsilon\}$, having vertex at (t_0, y_0), and having the common boundary $\{t = 0\} \cap \{|y - y_0| = t_0/\varepsilon\}$.

Thus by Holmgren's theorem, u is determined uniquely in R by the values of $P(u)$ in R and by the data on $S_0 = \{t = 0\} \cap \{|y - y_0| < t_0\varepsilon\}$; that is, R is the domain of dependence of (t_0, y_0). We have thus proved the following theorem.

Theorem 6.7. *The solution of* (6.1), (6.2) *is unique and has a bounded domain of dependence. In particular, if each* $\psi_j(y)$ *has compact support, then the same is true for u on any line* $t = \text{const.} > 0$.

NOTES

The original ideas are due to Petrovsky and Gårding [Gå]; the proof given here is adopted from [BJS]. There is an extension to the case where P is not a homogeneous operator. Here one requires that the roots of $P_m(t, y) = 0$ are real *and distinct*. The initial-value problem need not be well-posed if this condition fails to be true [BJS]. For a nice geometric interpretation of hyperbolicity, see [BJS].

Chapter 7

Distribution Theory

§A. A Cursory View

Consider the partial differential equation (in $\mathbf{R}^2$), $u_{xy} = f$. If $\phi \in C_0^2(\mathbf{R}^2)$, then multiplying both sides of this equation by ϕ, and integrating by parts, gives $\int u \phi_{xy} = \int f \phi$. Now suppose that u is not necessarily smooth, but that this last equation holds for each $\phi \in C_0^\infty(\mathbf{R}^2)$. We could then say that f is the "weak" mixed derivative of u. We could actually go one step further and consider the linear functional on $C_0^\infty(\mathbf{R}^2)$ defined by

$$\phi \to \int_{\mathbf{R}^2} u \phi_{xy}.$$

Then we could say $u_{xy} = f$ "weakly" provided that the above linear functional agrees with the linear functional on $C_0^\infty(\mathbf{R}^n)$ given by $\phi \to \int_{\mathbf{R}^2} f \phi$. This is the key idea in the theory of distributions—replace functions by functionals.

We let Ω be an open set in $\mathbf{R}^n$, and define the differential operator D_j by $D_j = -i \, \partial/\partial x_j$. Note that this definition differs from that in Chapter 2; we include the $-i$ here in order to simplify things later where we employ Fourier transforms.

Definition 7.1. A *distribution* T on Ω is a linear functional on $C_0^\infty(\Omega)$ such that $T\phi_j \to 0$ for every sequence $\{\phi_j\} \subset C_0^\infty(\Omega)$ satisfying the following two properties:

(i) spt $\phi_j \subset K \subset \Omega$, where K is compact and independent of j,
(ii) $\| D^\alpha \phi_j \|_\infty \to 0$ as $j \to \infty$ for all α with $|\alpha| \geq 0$.

In this case we write $\phi_j \to 0$ in $C_0^\infty(\Omega)$.

We denote by $\mathscr{D}'(\Omega)$, the space of distributions on Ω. We pause to give a few examples.

EXAMPLE 1. Let f be a locally integrable function on $\mathbf{R}^n$; then the formula $T_f \cdot \phi = \int_{\mathbf{R}^n} f\phi$ defines a distribution on $\mathbf{R}^n$ since if $\phi \in C_0^\infty(\mathbf{R}^n)$,

$$|T_f \cdot \phi| \le \|\phi\|_\infty \int_{\text{spt }\phi} |f|.$$

Thus if $\phi_j \to 0$ in $C_0^\infty(\mathbf{R}^n)$, $T_f \phi_j \to 0$.

EXAMPLE 2. Let $a \in \mathbf{R}$, and define, for k a nonnegative integer,

$$\delta_a^k(\phi) = (-1)^k D^k \phi(a).$$

One checks easily that δ_a^k is a distribution. If $k = a = 0$, this distribution is usually called the "Dirac delta function," and is written simply as δ. We can obviously extend the definitions to functions on $\mathbf{R}^n$, $n > 1$.

EXAMPLE 3. Let μ be a Radon measure on $\mathbf{R}^n$; then μ defines a distribution by the formula $\mu(\phi) = \int_{\mathbf{R}^n} \phi \, d\mu$.

If $\{T_j\} \subset \mathscr{D}'(\Omega)$, we say T_j *converges to zero* in $\mathscr{D}'(\Omega)$, and write $T_j \to 0$, provided that $T_j\phi \to 0$ for each $\phi \in C_0^\infty(\Omega)$ (that is, T_j converges weakly to zero). For $T \in \mathscr{D}'(\Omega)$, we say that T is *zero* on an open set $\omega \subset \Omega$ if $T\phi = 0$ for each $\phi \in C_0^\infty(\omega)$. The *support* of $T \in \mathscr{D}'(\Omega)$ is the set of points in Ω which have no neighborhood on which T is zero. $\mathscr{E}'(\Omega)$ denotes the space of distributions having compact support in Ω.

EXAMPLE 4. If δ is the Dirac delta function, then spt $\delta = \{0\}$. To see this, observe that if $x \ne 0$, then there is an open set ω about x such that $0 \notin \omega$. So if spt $\phi \subset \omega$, then $\delta(\phi) = \phi(0) = 0$.

We now turn to the "raison d'être" for studying distribution theory. We begin with an example.

EXAMPLE 5 (Weak-Derivatives). Let u be a locally integrable function in Ω, and let α be any multi-index. A locally integrable function v on Ω is called the *α-weak derivative of u* if

$$\int_\Omega v\phi \, dx = (-1)^{|\alpha|} \int_\Omega u D^\alpha \phi \, dx,$$

for every $\phi \in C_0^\infty(\Omega)$. We write $v = D^\alpha u$. It is easy to check (Example 1) that $v \in \mathscr{D}'(\Omega)$. A function u is k-times weakly differentiable if all of its α-weak derivatives exist for $|\alpha| \le k$. The space of k-times weakly differentiable functions is denoted by $W^k(\Omega)$; clearly $C^k(\Omega) \subset W^k(\Omega) \subset \mathscr{D}'(\Omega)$. The notion of weak derivative is thus an extension of the usual concept of derivative, in which the above "integration by parts" formula is valid.

We can go one step further with this example and define the $W_p^k(\Omega)$ spaces by

$$W_p^k(\Omega) = \{u \in W^k(\Omega) : D^\alpha u \in L_p(\Omega) \text{ if } |\alpha| \leq k\}.$$

$W_p^k(\Omega)$ is a Banach space with the norm defined by

$$\|u\|_{p,k} = \left(\int \sum_{|\alpha| \leq k} (D^\alpha u)^p \, dx\right)^{1/p}.$$

$\mathring{W}_p^k(\Omega)$ is the closure of $C_0^k(\Omega)$ in $W_p^k(\Omega)$. (Thus one can take "derivatives"; i.e., weak derivatives on these spaces, and elements in these spaces can serve as candidates for solutions of differential equations.) These spaces are of importance in studying elliptic and parabolic equations, see Chapters 8, 9, and 10.

Definition 7.2. Let $T \in \mathscr{D}'(\Omega)$, and let α be any multi-index of nonnegative integers. The *derivative* $D^\alpha T$ is defined by

$$D^\alpha T \cdot \phi = (-1)^{|\alpha|} T \cdot D^\alpha \phi, \qquad \phi \in C_0^\infty(\Omega).$$

It is easy to see that $D^\alpha T$ again is in $\mathscr{D}'(\Omega)$. For, if $\phi_j \to 0$ in $C_0^\infty(\Omega)$, $D^\alpha \phi_j \to 0$ in $C_0^\infty(\Omega)$, and so $T \cdot D^\alpha \phi_j \to 0$. Thus, every distribution has a derivative which itself is a distribution; in short, distributions are "infinitely differentiable." In fact, all continuous functions have derivatives of all orders (in the above sense!). Therefore we may consider differential operators acting on distributions. In this way we can extend the class of objects which can serve as solutions to differential equations, from smooth functions to distributions. This increases our chances of finding a solution to a given differential equation. Of course, there remains the problem of "regularity"; i.e., the problem of showing that the distribution "solution" is actually a smooth function.

EXAMPLE 6 (The Heaviside Function). We define the function $H(x)$ by

$$H(x) = \begin{cases} 0, & x < 0, \\ 1, & x > 0. \end{cases}$$

$H \in \mathscr{D}'(\mathbf{R})$ since H is locally integrable. We show $H' = \delta$. To see this, let $\phi \in C_0(\mathbf{R})$, then

$$H' \cdot \phi = (-1)H \cdot \phi' = -\int_0^\infty \phi'(x) \, dx = \phi(0) = \delta \cdot \phi.$$

EXAMPLE 7. Let K be defined on $\mathbf{R}$ by

$$K(x) = \begin{cases} 0, & x < 0, \\ x, & x > 0. \end{cases}$$

Then $K' = H$; more generally, if

$$K_n(x) = \begin{cases} 0, & x < 0, \\ x^n/n!, & x > 0, \end{cases}$$

then $K'_n = K_{n-1}, n \geq 1$

EXAMPLE 8. Obviously $\delta' \cdot \phi = -\phi'(0)$ if $\phi \in C_0^\infty(\mathbf{R})$. Thus δ' is *not* a measure.

If $f \in C^\infty(\Omega)$, and $T \in \mathscr{D}'(\Omega)$, then we can define the *product* fT by

$$fT \cdot \phi = T(f\phi), \qquad \phi \in C_0^\infty(\Omega).$$

It is easy to see that $fT \in \mathscr{D}'(\Omega)$.

Next, we shall define the convolution of distributions. In order to motivate the definition, let's first consider again the situation for functions. Thus, if f and g are in $L_1(\mathbf{R}^n)$, the *convolution* of f and g, $f * g$ is the function defined by

$$(f * g)(x) = \int_{\mathbf{R}^n} f(y)g(x-y)\, dy.$$

The product is obviously commutative, and an application of Fubini's theorem shows $f * g \in L_1(\mathbf{R}^n)$; in fact $\| f * g \| \leq \| f \| \| g \|$, where the norms are L_1-norms. Thus $f * g \in \mathscr{D}'(\mathbf{R}^n)$. Hence, for $\phi \in C_0(\mathbf{R}^n)$

$$(f * g) \cdot \phi = \int (f * g)(x)\phi(x)\, dx = \int \left(\int f(y)g(x-y)\, dy \right)\phi(x)\, dx$$

$$= \int f(y)\left(\int g(x-y)\phi(x)\, dx \right) dy$$

$$= \int f(y)\left(\int g(z)\phi(y+z)\, dz \right) dy = f_y \cdot (g_z \cdot \phi(y+z)).$$

This motivates us to *define* the convolution of $T \in \mathscr{D}'(\mathbf{R}^n)$, $S \in \mathscr{E}'(\mathbf{R}^n)$ by

$$(T * S)\phi = T_x \cdot [S_y \cdot \phi(x+y)];$$

the same expression is taken as the definition if $T \in \mathscr{E}'(\mathbf{R}^n)$, and $S \in \mathscr{D}'(\mathbf{R}^n)$. Then it is not very hard to show the following properties are valid (cf. [Ho 2])

(A) $(T * S) \in \mathscr{D}'(\mathbf{R}^n)$.
(B) $T * S = S * T$.
(C) If $U \in \mathscr{E}'(\mathbf{R}^n)$, then $(T * S) * U = T * (S * U)$.

(D) $\text{spt}(T * S) \subset \text{spt } T + \text{spt } S$.

(E) If $f \in C_0^\infty(\Omega)$, and $T \in \mathscr{D}'(\Omega)$, then $T * f \in C^\infty(\mathbf{R}^n)$, and $(T * f)(x) = T_y \cdot f(x - y)$.

Proposition 7.3. *If* $T \in \mathscr{D}'(\mathbf{R}^n)$, *and* $S \in \mathscr{E}'(\mathbf{R}^n)$, *then*

$$D^\alpha(T * S) = (D^\alpha T) * S = T * (D^\alpha S).$$

Proof. If $\phi \in C_0(\mathbf{R}^n)$, then

$$\begin{aligned}
D^\alpha(T * S) \cdot \phi &= (-1)^{|\alpha|}(T * S)(D^\alpha\phi) = (-1)^{|\alpha|}T_x \cdot [S_y \cdot D^\alpha\phi(x + y)] \\
&= T_x[(-1)^{|\alpha|}S_y \cdot D^\alpha\phi(x + y)] = T_x[D^\alpha S_y \cdot \phi(x + y)] \\
&= (T * D^\alpha S) \cdot \phi.
\end{aligned}$$

Thus $D^\alpha(T * S) = T * D^\alpha S$; the rest follows from property (B) above. $\square$

EXAMPLE 9. Let $T \in \mathscr{D}'(\mathbf{R}^n)$, then $T * \delta = T$. For, if $\phi \in C_0^\infty(\mathbf{R}^n)$,

$$(T * \delta)\phi = T_x\delta_y\phi(x + y) = T_x\phi(x) = T \cdot \phi.$$

More generally, if α is any multi-index, $T * D^\alpha\delta = D^\alpha T$, since $T * D^\alpha\delta = D^\alpha T * \delta = D^\alpha T$.

We next turn to the Fourier transform of distributions. Just as not all functions have Fourier transforms, we must find a subclass of $\mathscr{D}'$ which have Fourier transforms. These functionals are defined on a space $\mathscr{S} \supset C_0^\infty$, so that $\mathscr{S}' \subset (C_0^\infty)' = \mathscr{D}'$.

We define the space $\mathscr{S}$ of *rapidly decreasing functions* by

$$\mathscr{S} = \left\{ f \in C^\infty(\mathbf{R}^n) : \lim_{|x| \to \infty} |x^\alpha D^\beta f(x)| = 0 \text{ for all multi-indices } \alpha \text{ and } \beta \right\}$$

A sequence $\{\phi_n\} \subset \mathscr{S}$ is said to *converge to zero in* $\mathscr{S}$, $\phi_n \to 0$ in $\mathscr{S}$, if

$$\lim_{n \to \infty} \sup_{x \in \mathbf{R}^n} |x^\alpha D^\beta\phi_n(x)| = 0, \tag{7.1}$$

for all multi-indices α and β. Note that $C_0^\infty \subset \mathscr{S}$, and that $\mathscr{S} \supsetneq C_0^\infty$, since, e.g., $e^{-|x|^2}$ is in $\mathscr{S} \backslash C_0^\infty$.

It follows easily from (7.1) that $\mathscr{S} \subset L_1(\mathbf{R}^n)$ so that each function in $\mathscr{S}$ has a Fourier transform. Moreover, a straightforward integration by parts shows that $\hat{f}$ is also in $L_1(\mathbf{R}^n)$ whenever $f \in \mathcal{S}$. Thus the inversion formula holds (see p. 40), for functions in $\mathcal{S}$.

Concerning the functions in the class $\mathcal{S}$, we have the following proposition:

Proposition 7.4. *Let $\phi, \psi \in \mathcal{S}$; then the following hold:*

(i) *The map $\phi \to \hat{\phi}$ is injective and bicontinuous from $\mathcal{S}$ onto $\mathcal{S}$.*

(ii) $(\hat{\phi})^{\hat{}} = \phi_-$, *where $\phi_-(x) = \phi(-x)$.*

(iii) $\widehat{D_j\phi} = -\xi_j\hat{\phi}$, *where we recall, $D_j = -i \, \partial/\partial x_j$.*

(iv) $\widehat{x_j\phi(x)} = D_j\hat{\phi}$.

(v) $\widehat{\phi * \psi} = (2\pi)^{+n/2}\hat{\phi}\hat{\psi}$.

(vi) $\widehat{\phi\psi} = (2\pi)^{-n/2}\hat{\phi} * \hat{\psi}$.

Proof. If $\phi \in \mathcal{S}$, then ϕ is in L_2 and by Plancherel's formula, $\|\hat{\phi}\|_2 = \|\phi\|_2$, so $\phi \to \hat{\phi}$ is 1–1. Since inversion holds, the map is also onto. In view of the equation

$$D^\alpha\hat{\phi}(y) = (2\pi)^{-n/2} \int e^{-ix\cdot y} x^\alpha \phi(x) \, dx,$$

we see that (iv) holds. Also,

$$y^\beta D^\alpha\hat{\phi}(y) = (2\pi)^{-n/2} \int (e^{-ix\cdot y}y^\beta)(-x^\alpha\phi(x)) \, dx$$

$$= (2\pi)^{-n/2} \int (D^\beta e^{-ix\cdot y})(x^\alpha\phi(x)) \, dx,$$

so

$$y^\beta D^\alpha\hat{\phi}(y) = (2\pi)^{-n/2} \int e^{-ix\cdot y} D^\beta((-1)^{|\beta|}x^\alpha\phi(x)) \, dx. \tag{7.2}$$

If we set $\alpha = 0$, we see that (iii) holds. (7.2) also shows that $y^\beta D^\alpha\hat{\phi}(y) \to 0$ as $|y| \to \infty$, so that

$$|y^\beta D^\alpha\hat{\phi}(y)| \le \|(1 + |x|)^{n+1}D^\beta(x^\alpha\phi(x))\|_\infty \int (1 + |x|)^{-(n+1)} \, dx.$$

Thus the map $\phi \to \hat{\phi}$ is continuous on $\mathscr{S}$. Using (ii) (which is obvious from the inversion formula), we see that $\hat{\phi} \to \phi$ is also continuous. Thus (i) holds. To prove (v), we compute

$$\widehat{\phi * \psi}(y) = (2\pi)^{-n/2} \int e^{-ix \cdot y}(\phi * \psi)(x)\,dx$$

$$= (2\pi)^{-n/2} \int e^{-ix \cdot y}\left(\int \phi(t)\psi(x-t)\,dt\right)dx$$

$$= (2\pi)^{-n/2} \int \phi(t)\left(\int e^{-ix \cdot y}\psi(x-t)\,dx\right)dt$$

$$= (2\pi)^{-n/2} \int \phi(t)\left(\int e^{-i(z+t)\cdot y}\psi(z)\,dz\right)dt$$

$$= (2\pi)^{-n/2} \int e^{-it \cdot y}\phi(t)\,dt \int e^{-iz \cdot y}\psi(z)\,dz$$

$$= (2\pi)^{n/2}\hat{\phi}(y)\hat{\psi}(y).$$

Finally, from (v),

$$\widehat{\hat{\phi} * \hat{\psi}} = (2\pi)^{n/2}\hat{\hat{\phi}}\hat{\hat{\psi}} = (2\pi)^{n/2}\phi_-\psi_-$$

so that

$$\hat{\phi} * \hat{\psi} = (2\pi)^{n/2}(\widehat{\phi_-\psi_-})_- = (2\pi)^{n/2}\widehat{\phi\psi}.$$

This proves (vi) and completes the proof. $\square$

Definition 7.5. The space of *tempered distributions*, $\mathscr{S}' \subset \mathscr{D}'$, consists of those functionals on $\mathscr{S}$ such that if $\phi_n \to 0$ in $\mathscr{S}$, then $T\phi_n \to 0$. If $T \in \mathscr{S}'$, the Fourier transform $\hat{T}$ of T is defined by $\hat{T} \cdot \phi = T \cdot \hat{\phi}$ for all $\phi \in \mathscr{S}$. $T_n \to 0$ in $\mathscr{S}'$ if $T_n \cdot \phi \to 0$ for each $\phi \in \mathscr{S}$.

Note that if $\phi_n \to 0$ in $\mathscr{S}$, $\hat{T}\phi_n = T\hat{\phi}_n \to 0$ in view of (i) above. Thus $\hat{T} \in \mathscr{S}'$. We now give some examples.

EXAMPLE 10. If $\phi \in \mathscr{S}$, then $\hat{\delta} \cdot \phi = \delta \cdot \hat{\phi} = (2\pi)^{-n/2} \int \phi(x)\,dx$; whence $\hat{\delta} = (2\pi)^{-n/2}1$.

EXAMPLE 11. If $a \in \mathbf{R}$, and $\phi \in \mathscr{S}$,

$$\widehat{e^{iax}} \cdot \phi = e^{iax} \cdot \hat{\phi} = \int e^{iax}\hat{\phi}(x)\,dx = (2\pi)^{n/2}\phi(a) = (2\pi)^{n/2}\delta_a \cdot \phi,$$

so that $(e^{iax})\hat{} = (2\pi)^{n/2}\delta_a$.

Theorem 7.6. *The map $T \to \hat{T}$ from $\mathscr{S}'$ into $\mathscr{S}'$ is injective and bicontinuous. Also, the following properties hold:*

(i)   *If $S \in \mathscr{E}'$, $T \in \mathscr{S}'$, then $\hat{S} \in C^{\infty}$ and $T * S \in \mathscr{S}'$.*

(ii) $\widehat{S * T} = \hat{S}\hat{T}$ *if $S \in \mathscr{E}'$ and $T \in \mathscr{S}'$.*

(iii) *If $P(\xi)$ is a polynomial, then $\widehat{P(D)u} = P(-\xi)\hat{u}$ for all $u \in \mathscr{S}'$.*

(iv) $\widehat{\xi_j S} = -D_j \hat{S}$ *if $S \in \mathscr{S}'$.*

The proofs of (i) and (ii) can be found in [Ho 2]. For (iii), we use (iv) of Proposition 7.4. Thus

$$\widehat{P(D)u} \cdot \phi = P(D)u \cdot \hat{\phi} = u \cdot P(-D)\hat{\phi} = u \cdot P(-\xi)\phi = \hat{u} \cdot P(-\xi)\phi$$
$$= P(-\xi)\hat{u} \cdot \phi.$$

The proof of (iv) is similar.

§B. Fundamental Solutions

We shall now show one way in which distributions are useful in studying partial differential equations. To this end, let P be a differential operator of order m with constant coefficients. We consider the equation

$$P(D)u = f, \qquad f \in C_0^{\infty}. \tag{7.3}$$

In order to solve this, it suffices to find a *fundamental solution*, i.e., a solution to the equation

$$P(D)v = \delta. \tag{7.4}$$

If one can solve (7.4), then if $u = v * f$, we have $P(D)u = P(D)v * f = \delta * f = f$. Thus it is only necessary to solve (7.4). We turn our attention to this problem.

We note that for certain differential operators it is easy to find fundamental solutions; e.g., if $\Delta \equiv \sum_{j=1}^{n} D_j^2$, we have the following lemma.

Lemma 7.7. *Let N be an integer such that $4N \geq n + 1$; then there exists $H \in L_2(\mathbf{R}^n)$ such that $(1 + \Delta)^N H = \delta$.*

Proof. Let $G(\xi) = (1 + |\xi|^2)^{-N}$; since $4N \geq n + 1$, $G \in L_2(\mathbf{R}^n)$. Thus (cf. Chapter 6) there is an $F \in L_2(\mathbf{R}^n)$ such that $\hat{F} = G$. Since $(1 + |\xi|^2)^{-N}G(\xi) = 1$, we have $(1 + |\xi|^2)^N(2\pi)^{-n/2}\hat{F}(\xi) = 1$ and from Theorem 7.6(iv), $(1 + \Delta)^N(2\pi)^{-n/2}F = \delta$. $\square$

The next lemma gives us a strategy for solving (7.3).

Lemma 7.8. *If there exists a subspace H such that $L_2(\mathbf{R}^n) \subseteq H \subseteq \mathscr{D}'(\mathbf{R}^n)$, and $P(D)H \supseteq H$, then a fundamental solution of P exists.*

Proof. Let F be as in Lemma 7.7. Then $F \in P(D)H$ so there is a G in H such that $P(D)G = F$. If $v = (1 - \Delta)^N G$, then

$$P(D)v = P(D)(1 - \Delta)^N G = (1 - \Delta)^N P(D)G = (1 - \Delta)^N F = \delta. \quad \square$$

In view of this lemma, our problem is to find such a space H. To this end, let $E(x) = \exp(|x|^2/2)$, and let $H_\pm$ be the spaces defined by

$$H_+ = \{f : Ef \in L_2(\mathbf{R}^n)\}, \qquad H_- = \{f : E^{-1}f \in L_2(\mathbf{R}^n)\}.$$

We define inner products on these spaces by

$$\langle f, g \rangle_+ = \int_{\mathbf{R}^n} E^2(x) f(x) \bar{g}(x)\, dx$$

$$\langle f, g \rangle_- = \int_{\mathbf{R}^n} E^{-2}(x) f(x) \bar{g}(x)\, dx.$$

It is an easy exercise to show that $H_\pm$ are both Hilbert spaces and that $C_0^\infty(\mathbf{R}^n)$ is dense in $H_\pm$. It will turn out that H_- is the space that we seek.

We first verify the easy part of the hypotheses in Lemma 7.8. Namely, we have the following lemma.

Lemma 7.9. $L_2(\mathbf{R}^n) \subseteq H_- \subseteq \mathscr{D}'(\mathbf{R}^n)$.

Proof. If $f \in L_2(\mathbf{R}^n)$ then

$$\|f\|_-^2 = \langle f, f \rangle_- = \int_{\mathbf{R}^n} E^{-2}(x) |f(x)|^2\, dx \le \int_{\mathbf{R}^n} |f(x)|^2\, dx < \infty,$$

so $f \in H_-$. On the other hand, if $\phi_n \to 0$ in $C_0^\infty(\mathbf{R}^n)$, and $f \in H_-$, then there is a compact set $K \subseteq \mathbf{R}^n$ such that each ϕ_n is supported in K. We have

$$f \cdot \phi_n = \int_K f \cdot \phi_n = \int_K (fE^{-1})(E\phi_n) \le \|f\|_- \left(\int_K E^2 \phi_n^2 \right)^{1/2}$$

$$\le \|f\|_- \|\phi_n\|_\infty \left(\int_K E^2 \right)^{1/2} \to 0 \quad \text{as } n \to \infty.$$

Thus $H_- \subseteq \mathscr{D}'(\mathbf{R}^n)$, and the proof is complete. $\quad \square$

We now obtain some properties of the spaces $H_\pm$.

Proposition 7.10.

(a) *The mapping $f \to E^2 f$ is an isometry from H_+ into H_- and takes $C_0^\infty(\mathbf{R}^n)$ onto itself.*

(b) *The dual of H_+ is H_-; here the duality pairing is with respect to the L_2-inner product.*

Proof. If $f \in H_+$, then

$$\| fE^2 \|_-^2 = \int (f^2 E^4) E^{-2} = \int f^2 E^2 = \| f \|_+^2,$$

so that the first part of (a) holds. The last sentence in (a) is obvious. To prove (b) let $\phi \in H_-$ and define a functional on H_+ by

$$\phi \cdot f = \int \phi f, \qquad f \in H_+.$$

Since

$$|\phi \cdot f| = \left| \int (\phi E^{-1})(E f) \right| \leq \| \phi \|_- \cdot \| f \|_+,$$

we see that ϕ defines a continuous linear functional on H_+. Conversely, if Φ is a continuous linear functional on H_+, then by the Riesz representation theorem, there exists $\phi_1 \in H_+$ such that

$$\Phi(f) = \langle \phi_1, f \rangle_+, \qquad f \in H_+.$$

Let $\phi_1 E^2 = \phi$; then

$$\Phi(f) = \int f(\phi_1 E^{+2}) = \int f\phi,$$

and

$$\| \phi \|_-^2 = \int |\phi|^2 E^{-2} = \int |\phi_1|^2 E^2 = \| \phi_1 \|_+^2 < \infty.$$

Thus $\phi \in H_-$ and Φ is represented by an element in H_-. This completes the proof. $\square$

We next prove the main lemma in this section.

Lemma 7.11. *If the inequality*

$$\|\phi\|_+^2 \le K\|P(D)\phi\|_+^2, \qquad \phi \in C_0^\infty(\mathbf{R}^n), \tag{7.5}$$

holds for all $\phi \in C_0^\infty(\mathbf{R}^n)$, where K is independent of ϕ, then P has a fundamental solution.

Proof. From Lemmas 7.8 and 7.9, it suffices to show that $H_- \subseteq P(D)H_-$. To do this, first note that the mapping $P(D)\phi \to \phi$, where $\phi \in C_0^\infty$, is well defined and continuous in the H_+ norm, in view of (7.5). We can thus extend this mapping to a mapping G, defined by continuity to the closure of $P(D)C_0^\infty$ in H_+, and setting G equal to zero on the orthogonal complement of this closure. This defines G on all of H_+, i.e.,

$$G: H_+ \to H_+, \qquad GP(D)\phi = \phi \quad \text{if } \phi \in C_0^\infty.$$

Now G induces a mapping $G^*: H_- \to H_-$ defined by $(G^*g, h) = (g, Gh)$, $g \in H_-, h \in H_+$, where the brackets denote the L_2 inner product. This mapping is obviously continuous since if $g_n \to 0$ in H_-, then $(g_n, Gh) \to 0$ so $(G^*g_n, h) \to 0$.

We can now show $H_- \subseteq P(D)H_-$. Thus, if $g \in H_-$, and $\phi \in C_0^\infty$,

$$g \cdot \phi = (g, \phi) = (g, GP(D)\phi) = (G^*g, P(D)\phi) = G^*g \cdot P(D)\phi,$$

so that $g \cdot \phi = P(-D)G^*g \cdot \phi$. It follows that $g = P(-D)G^*g$, and since $G^*g \in H_-$, we have proved that $P(-D)H_- \supseteq H_-$. If we now replace the polynomial $P(x)$ by $P(-x)$, the same argument shows that $P(D)H_- \supseteq H_-$ and the proof is complete. □

Thus, the existence of a fundamental solution for $P(D)$ reduces to proving the estimate (7.5). The heart of this inequality is basically algebraic, as we shall now show.

Let $\mathscr{A}$ be an algebra with unity 1 over the complex numbers $\mathbf{C}$, and let a and b be elements in $\mathscr{A}$ such that their commutator $[a, b] \equiv ab - ba = 1$. Let $p(x)$ and $q(x)$ be polynomials in the scalar x, with coefficients in $\mathbf{C}$.

Lemma 7.12.

$$q(b)p(a) = \sum_{k=0}^\infty \frac{(-1)^k}{k!} p^{(k)}(a)q^{(k)}(b),$$

where $p^{(k)}(x) = d^k p(x)/dx^k$.

Proof. We claim first that it suffices to prove the lemma for monomials, i.e., $p(x) = x^n$, $q(x) = x^m$, where m and n are nonnegative integers. To see this, let $\tilde{p}(x) = \sum \alpha_n x^n$, $\tilde{q}(x) = \sum \beta_m x^m$; then assuming the validity for monomials, we have

$$\tilde{q}(b)\tilde{p}(a) = \sum_{n,m} \beta_m \alpha_n b^m a^n = \sum_{m,n} \beta_m \alpha_n \sum_{k=0}^{\infty} \frac{(-1)^k}{k!} D^k(a^n)D^k(b^m)$$

$$= \sum_{k=0}^{\infty} \frac{(-1)^k}{k!} \sum_{m,n} \beta_m a_n D^k(a^n)D^k(b^m) = \sum_{k=0}^{\infty} \frac{(-1)^k}{k!} \tilde{p}^{(k)}(a)\tilde{q}^{(k)}(b),$$

which is of the desired form.

Thus, let $p(x) = x^n$, $q(x) = x^m$, we want to show

$$q(b)p(a) = \sum_{k=0}^{\infty} \frac{(-1)^k}{k!} n(n-1)\cdots(n-k+1)m(m-1)$$
$$\cdots(m-k+1)a^{n-k}b^{m-k}. \tag{7.6}$$

We do this by induction. Observe first that if $m = 0$ or $n = 0$, the result is trivial using Taylor's formula. Let $m = 1$; we induct on n. If $n \geq 1$,

$$q(b)p(a) = ba^n = (ba)a^{n-1} = (ab - 1)a^{n-1} = a(ba^{n-1}) - a^{n-1}.$$

By our induction hypothesis, $ba^{n-1} = a^{n-1}b - (n-1)a^{n-2}$, so

$$q(b)p(a) = a(a^{n-1}b - (n-1)a^{n-2}) - a^{n-1} = a^n b - (n-1)a^{n-1} - a^{n-1}$$
$$= a^n b - na^{n-1},$$

and the result holds for $m = 1$ and all n.

Now let n be arbitrary (and fixed), and we shall induct on m. Thus we assume

$$b^{n-1}a^m = \sum_{k=0}^{\infty} \frac{(-1)^k}{k!} c_{m,k} c_{n-1,k} a^{m-k} b^{n-1-k},$$

where $c_{\alpha,\beta} = \alpha(\alpha-1)\cdots(\alpha-\beta+1)$, $c_{\alpha,0} = 1$. We have

$$q(b)p(a) = b^n a^m = b(b^{n-1}a^m) = b \sum_{k=0}^{\infty} \frac{(-1)^k}{k!} c_{m,k} c_{n-1,k} a^{m-k} b^{n-1-k}$$

$$= \sum_{k=0}^{\infty} \frac{(-1)^k}{k!} c_{m,k} c_{n-1,k} (ba^{m-k}) b^{n-1-k}.$$

Since the result holds for $n = 1$ and any n, we have $ba^{m-k} = a^{m-k}b$ $-(m-k)a^{m-k-1}$, so that

$$q(b)p(a) = \sum_{k=0}^{\infty} \frac{(-1)^k}{k!} c_{m,k} c_{n-1,k} [a^{m-k}b - (m-k)a^{m-k-1}]b^{n-1-k}$$

$$= \sum_{k=0}^{\infty} \frac{(-1)^k}{k!} c_{m,k} c_{n-1,k} a^{m-k} b^{n-k}$$

$$- \sum_{k=0}^{\infty} \frac{(-1)^k}{k!} c_{m,k} c_{n-1,k} (m-k) a^{m-k-1} b^{n-1-k}.$$

In the first sum put $k = p$; in the second put $k = p - 1$ to get

$$q(b)p(a) = \sum_{p=0}^{\infty} \frac{(-1)^p}{p!} c_{m,p} c_{n-1,p} a^{m-p} b^{n-p}$$

$$- \sum_{p=1}^{\infty} \frac{(-1)^{p-1}}{(p-1)!} c_{m,p-1} c_{n-1,p-1} (m-p+1) a^{m-p} b^{n-p}$$

$$= c_{m,0} c_{n-1,0} a^m b^n$$

$$+ \sum_{p=1}^{\infty} \frac{(-1)^p}{(p-1)!} \left[\frac{c_{m,p} c_{n-1,p}}{p} + c_{m,p-1} c_{n-1,p-1} (m-p+1) \right]$$

$$\times a^{m-p} b^{n-p}.$$

Thus, we will be done if we can show that the last expression in brackets is $c_{m,p} c_{n,p}/p$. We have

$$\frac{c_{m,p} c_{n-1,p}}{p} + c_{m,p-1} c_{n-1,p-1} (m-p-1) = \frac{c_{m,p} c_{n-1,p} + p c_{m,p} c_{n-1,p-1}}{p}$$

$$= \frac{c_{m,p}}{p} [c_{n-1,p} + p c_{n-1,p-1}]$$

$$= \frac{c_{m,p}}{p} \cdot c_{n,p}.$$

The proof of the lemma is complete. □

We extend this result to the case of polynomials in several variables.

Lemma 7.13. *Let $a_1, \ldots, a_n, b_1, \ldots, b_n$ be elements of $\mathscr{A}$ such that the following commutation relations hold:*

(i) $[a_j, a_k] = [b_j, b_k] = [a_j, b_k] = 0$ if $j \neq k$.
(ii) $[a_j, b_j] = 1$.

If $p(x)$ and $q(x)$ are polynomials in $x = (x_1, \ldots, x_n)$ and $D^\alpha = (\partial/\partial x_1)^{\alpha_1} \cdots (\partial/\partial x_n)^{\alpha_n}$ for $\alpha = (\alpha_1, \ldots, \alpha_n)$, then

$$q(b)p(a) = \sum_\alpha \frac{(-1)^{|\alpha|}}{\alpha!} D^\alpha p(a) D^\alpha q(b).$$

Proof. As before, it suffices to prove the lemma for monomials, $p(x) = x_1^{\sigma_1} \cdots x_n^{\sigma_n}$, $q(x) = x_1^{\tau_1} \cdots x_n^{\tau_n}$. We have, in view of (i),

$$q(b)p(a) = b_1^{\tau_1} \cdots b_n^{\tau_n} a_1^{\sigma_1} \cdots a_n^{\sigma_n} = b_1^{\tau_1} a_1^{\sigma_1} \cdots b_n^{\tau_n} a_n^{\sigma_n}.$$

From Lemma 7.12,

$$b_i^{\tau_i} a_i^{\sigma_i} = \sum_{k_i = 0}^\infty \frac{(-1)^{k_i}}{k_i!} \left(\frac{\partial}{\partial x_i}\right)^{k_i} a_i^{\sigma_i} \left(\frac{\partial}{\partial x_i}\right)^{k_i} b_i^{\tau_i},$$

and so we are done. $\square$

We now consider $C_0^\infty(\mathbf{R}^n)$ as a subspace of $L_2(\mathbf{R}^n)$, and let $\mathscr{A}$ be the algebra of endomorphisms on $C_0^\infty(\mathbf{R}^n)$.

Let $A_1, \ldots, A_n$ be in $\mathscr{A}$ and set $B_j = -A_j^*$, where $*$ denotes the adjoint operation, taken with respect to the L_2-inner product. We assume that hypotheses (i) and (ii) of Lemma 7.13 hold. This implies that for any polynomial $R(x)$ with real coefficients, if $A = (A_1, \ldots, A_n)$, and $[A_i, A_j] = 0$ for $i \neq j$, $R(A^*) = R(A)^*$, since

$$R(A)^* = \left(\sum_\alpha a_\alpha A^\alpha\right)^* = \sum_\alpha a_\alpha (A^\alpha)^* = \sum_\alpha a_\alpha (A_1^{\alpha_1} \cdots A_n^{\alpha_n})^*$$

$$= \sum_\alpha a_\alpha (A_n^{\alpha_n} \cdots A_1^{\alpha_1})^* = \sum_\alpha a_\alpha (A_1^*)^{\alpha_1} \cdots (A_n^*)^{\alpha_n}$$

$$= R(A^*).$$

We shall now complete the proof of the existence of a fundamental solution; i.e., we have the following theorem.

Theorem 7.14. *If $P(D)$ is a differential operator with (real) constant coefficients, then P has a fundamental solution; i.e., the equation (7.4) is solvable in $\mathscr{D}'$.*

Proof. Define $Q(x) = P(-x)$, and let

$$A_i = \frac{1}{\sqrt{2}}\left(\frac{\partial}{\partial x_i} - x_i\right), \qquad A_i^* = \frac{1}{\sqrt{2}}\left(\frac{\partial}{\partial x_i} + x_i\right) = B_i.$$

An easy calculation shows that hypotheses (i) and (ii) of Lemma 7.13 hold for these A_i's and B_j's. Let $A = (A_1, \ldots, A_n)$, $B = -A^*$. Then

$$Q(B) = Q(-A^*) = P(A^*) = P(A)^*,$$

$$D^\alpha Q(B) = D^\alpha Q(-A^*) = D^\alpha P(A^*) = D^\alpha P(A)^* = [(-1)^{|\alpha|} D^\alpha P(A)]^*$$

$$= (-1)^{|\alpha|} [D^\alpha P(A)]^*$$

Thus, if we apply Lemma 7.13 to $P(x)$ and $Q(x)$ we obtain

$$P(A)^* P(A) = \sum \frac{1}{\alpha!} D^\alpha P(A) D^\alpha P(A)^*.$$

So if $\phi \in C_0^\infty$,

$$(P(A)^* P(A)\phi, \phi) = \sum_\alpha \frac{1}{\alpha!} (D^\alpha P(A) D^\alpha P(A)^*\phi, \phi),$$

and

$$\| P(A)\phi \|^2 = \sum_\alpha \frac{1}{\alpha!} \| D^\alpha P(A)^*\phi \|^2, \qquad \phi \in C_0^\infty, \tag{7.7}$$

where the norms are L_2 norms. Now if we apply (7.7) to $D^\beta P$ instead of P, where $|\beta| \le m = $ order of P, we get

$$\| D^\beta P(A)\phi \|^2 = \sum_\alpha \frac{1}{\alpha!} \| D^{\alpha+\beta} P(A)^*\phi \|^2$$

$$\le c_1 \sum_\alpha \frac{1}{\alpha!} \| D^\alpha P(A)^*\phi \|^2,$$

$$= c_1 \| P(A)\phi \|^2,$$

in view of (7.7), where c_1 depends only on P. Now choose β such that $D^\beta P(x) = \text{const.} \ne 0$. For this β, we obtain

$$\| \phi \|^2 \le c \| P(A)\phi \|^2, \qquad \phi \in C_0^\infty, \tag{7.8}$$

where c depends only on P. Applying (7.8) to the polynomial $P(\sqrt{2}x_1, \ldots, \sqrt{2}x_n)$, gives

$$\| \phi \|^2 \le c' \left\| P\left(\frac{\partial}{\partial x_1} - x_1, \ldots, \frac{\partial}{\partial x_n} - x_n\right)\phi \right\|^2, \tag{7.9}$$

where c' depends only P.

If now $\phi = E\psi$, $\psi \in C_0^\infty$, then

$$\left(\frac{\partial}{\partial x_j} - x_j\right)\phi = \left(\frac{\partial}{\partial x_j} - x_j\right)E\psi = \left(\frac{\partial}{\partial x_j} - x_j\right)\exp(\tfrac{1}{2}(x_1^2 + \cdots + x_n^2))\psi(x)$$

$$= \frac{\partial}{\partial x_j}\exp(\tfrac{1}{2}|x|^2)\psi - x_j\exp(\tfrac{1}{2}|x|^2)\psi = E\frac{\partial\psi}{\partial x_j}.$$

Thus $P(\partial/\partial x_1 - x_1, \ldots, \partial/\partial x_n - x_n)\phi = EP(D)\psi$, so (7.9) yields

$$\|\psi\|_+^2 = \|E\psi\|^2 \le C'\|EP(D)\psi\|^2 = C'\|P(D)\psi\|_+^2.$$

This completes the proof of the theorem. $\square$

There are other proofs of this theorem, e.g., see [Ho 2, 3], which are not as elementary, but which give more information on the fundamental solution; e.g., they give both global and local behavior of the fundamental solution. Another alternate method is based on taking the Fourier transform of (7.4), to obtain the equation $P(\xi)\hat{v} = 1$. The idea is to show that this "problem in division" has a solution $\hat{v}$ in $\mathscr{S}'$. One way of doing this is to show that the mapping $\phi \to P\phi$ from $\mathscr{S}$ into $\mathscr{S}$ has a *continuous* inverse. Once this is known, then for any $T \in \mathscr{S}'$, there exists an $S \in \mathscr{S}'$ such that $P(\xi)S = T$. To see this, observe that the linear functional $F: P(\xi)\phi(\xi) \to T(\phi)$ on $P\mathscr{S}$ is continuous so it can be extended, by the Hahn–Banach theorem, to a continuous linear functional on $\mathscr{S}$. Thus, there is an $S \in \mathscr{S}'$ such that S restricted to $P\mathscr{S}$ is T. If now $\phi \in \mathscr{S}$, $S(P\phi) = T\phi$; i.e., $PS = T$.

Thus the division problem is solvable in $\mathscr{S}'$ provided that the above mapping $\phi \to P\phi$ has a continuous inverse. This requires a careful study of the zero set of the polynomial $P(\xi)$. The main lemma is an interesting inequality [Ho 3]; namely, if Z is the real zero set of P, then there are constants $c > 0$ and $\mu > 0$ such that

$$|P(x)| \ge c \operatorname{dist}(x, Z)^\mu \quad \text{if } |x| \le 1, \qquad x \in \mathbf{R}^n.$$

Somewhat surprisingly, this in turn depends on an algebraic-logic theorem, called the Seidenberg–Tarski theorem [Ho]. Before stating it, we need a definition. Let P_i and Q_j be polynomials in n-variables, with real coefficients. A subset $F \subset \mathbf{R}^n$ is called *semialgebraic*, if it is the finite union of sets of the form

$$\{x \in \mathbf{R}^n : P_i(x) = 0, 1 \le i \le r, Q_j(x) > 0, 1 \le j \le s\}.$$

A *polynomial mapping* $\phi: \mathbf{R}^n \to \mathbf{R}^m$ is mapping of the form $x \to (\phi_1(x), \ldots, \phi_m(x))$, where each ϕ_j is a polynomial in n-variables, with real coefficients.

The Seidenberg–Tarski theorem states that the image of a semialgebraic set under a polynomial mapping is semialgebraic. (This is false for algebraic sets; e.g., let $F = \{(x, y) \in \mathbf{R}^2 : xy - 1 = 0\}$, and let $\phi(x, y) = y$.)

§C. Appendix

A. We shall give a proof of the existence of a "partition of unity" subordinate to a given covering of a compact set. Before doing this however, we shall have to construct C_0^∞-functions having certain properties. We shall refer to a C_0^∞-function as a *test function*.

It is easy to construct a test function ϕ supported in the unit ball in $\mathbf{R}^n$; namely, define $f : \mathbf{R} \to \mathbf{R}$ by

$$f(t) = e^{1/t} \quad \text{if } t < 0, \qquad f(t) = 0 \quad \text{if } t \geq 0,$$

and set $\phi(x) = f(|x|^2 - 1), x \in \mathbf{R}^n$. We can also construct test functions supported in arbitrary compact sets, as the following lemma shows.

Lemma A1. *Let Ω be an open set in $\mathbf{R}^n$ and let K be a compact subset of Ω. Then there exists $\psi \in C_0^\infty(\Omega)$ with $0 \leq \psi \leq 1$ and $\psi = 1$ in a neighborhood of K.*

Proof. Let $\delta = \text{dist}(K, \partial\Omega); \delta > 0$. Choose numbers $0 < \varepsilon < \mu < \varepsilon + \mu < \delta$. Let $K_\mu = \{x : \text{dist}(x, K) \leq \mu\}$, and let ϕ be the above function. Let w be the characteristic function of K_μ and set

$$w_\varepsilon(x) = \int_{\mathbf{R}^n} w(x - \varepsilon y)\bar{\phi}(y)\, dy = \varepsilon^{-n} \int_{\mathbf{R}^n} w(y)\bar{\phi}\left(\frac{x - y}{\varepsilon}\right) dy,$$

where $\bar{\phi}$ is a multiple of ϕ chosen so that $\int \bar{\phi} = 1$. Then spt $w_\varepsilon \subset K_{\mu + \varepsilon}$, w_ε is a C^∞-function, and $w_\varepsilon = 1$ in a neighborhood of K. $\square$

Lemma A2 (Partition of Unity). *Let $\Omega_1, \Omega_2, \ldots, \Omega_k$ be an open covering of the compact set $K \subset \Omega$. Then there are functions $\phi_j \in C_0^\infty(\Omega_j)$ such that $\phi_j \geq 0$, $\sum \phi_j \leq 1$, and $\sum \phi_j = 1$ in a neighborhood of K.*

Proof. For each j, choose a compact set $K_j \subset \Omega_j$ with $K \subset \cup K_j$. By Lemma A1, we can find $\psi_j \in C_0^\infty(\Omega_j)$ with $0 \leq \psi_j \leq 1$ and $\psi_j = 1$ in a neighborhood of K_j. Define

$$\phi_1 = \psi_1, \qquad \phi_j = \psi_j(1 - \psi_1) \cdots (1 - \psi_{j-1}), \qquad j > 1.$$

Then

$$\sum_1^k \phi_j = 1 - (1 - \psi_1) \cdots (1 - \psi_k),$$

and we see at once that all the statements in the theorem hold. $\square$

B. We prove that C_0^∞ is dense in L_p and C_0 is dense in L_p and C. To this end, let $\phi \in C_0^\infty(\mathbf{R}^n)$ with

$$\phi \geq 0, \qquad \|\phi\|_{L_1} = 1, \qquad \text{spt } \phi \subset \{|x| \leq 1\}.$$

If u is integrable, and $\varepsilon > 0$, we can "regularize" u by convolution; namely, define u_ε by

$$u_\varepsilon(x) = \int u(x - \varepsilon y)\phi(y)\, dy = \varepsilon^{-n} \int u(y)\phi\left(\frac{x - y}{\varepsilon}\right) dy.$$

Lemma B1. *If u is integrable and has compact support in $K \subset \Omega$, then $u_\varepsilon \in C_0^\infty(\Omega)$ if $\varepsilon < \text{dist}(K, \partial\Omega)$. If $u \in L_p$, $1 \leq p < \infty$, then $\|u_\varepsilon - u\|_p \to 0$ as $\varepsilon \to 0$. If u is continuous, then $\|u_\varepsilon - u\|_\infty \to 0$ as $\varepsilon \to 0$.*

Proof. It is easy to see that u_ε is continuous, and the fact that u_ε is in C^∞ comes from differentiating the expression for u_ε under the integral sign.

Let $d = \text{dist}(K, \partial\Omega)$. If $u_\varepsilon(x) \neq 0$, then there is a $y, |y| \leq 1$ such that $x - \varepsilon y \in K$; hence spt $u_\varepsilon \subset \{x : \text{dist}(x, K) < \varepsilon\}$ and this last set is compact for $\varepsilon < d$. Thus $u_\varepsilon \in C_0^\infty(\Omega)$.

Let u be continuous; then

$$u_\varepsilon(x) - u(x) = \int [u(x - \varepsilon y) - u(x)]\phi(y)\, dy,$$

and the uniform continuity of u on the compact set spt ϕ, shows that $\|u_\varepsilon - u\|_\infty \to 0$ as $\varepsilon \to 0$.

Let $u \in L_p$. From Minkowski's inequality, $\|u_\varepsilon\|_p \leq \|u\|_p$, since $\|\phi\|_{L_1} = 1$ and $\phi \geq 0$. Given $\eta > 0$, we can find a continuous function $\tilde{u}$ with compact support such that $\|u - \tilde{u}\|_p < \eta$, and thus $\|u_\varepsilon - \tilde{u}_\varepsilon\|_p < \eta$. Then using the result just proved,

$$\overline{\lim_{\varepsilon \to 0}} \|u_\varepsilon - u\|_p \leq \overline{\lim_{\varepsilon \to 0}} \|u_\varepsilon - u\|_p + \|u - \tilde{u}\|_p + \overline{\lim_{\varepsilon \to 0}} \|\tilde{u}_\varepsilon - \tilde{u}\|_p$$

$$< 2\eta. \quad \square$$

NOTES

The roots of the subject go back to the works of Bochner [Bo], Friedrichs [Fr 1, 2], and Sobolev [So], in addition to the use of "δ-functions" and the Heaviside function, in nonrigorous ways (see [Sz] for a nice historical discussion). The subject was put on a firm foundation, once and for all, by Schwartz [Sz]. The appearance of this book gave birth to a flood of research activity in partial differential equations. The notion of "weak derivative" is due to Friedrichs [Fr 2]; the spaces W_p^k go back to the work of Sobolev [So], and the space $\mathscr{S}$ is (more or less) discussed by Bochner [Bo]. The important idea of taking the Fourier transform of distributions, is first found in [Sz]; it is the key to the applications of distributions to partial differential equations. The notion of a fundamental solution is classical; the earliest existence theorems for fundamental solutions are due to Ehrenpreis [Ep] and Malgrange [Ma]; see also Hörmander [Ho 2]. The proof given here is due to Treves, and is taken from his Paris thesis [Tr 1]. The operators A_i and A_i^* in the proof of Theorem 7.14 arise in quantum mechanics, where they are known as "creation" and "annihilation" operators.

For a good exposition of the use of the Seidenberg Tarski theorem in differential equations, see [Ho 2]; and also [Go].

Chapter 8

Second-Order Linear Elliptic Equations

Solutions of elliptic equations represent *steady-state solutions*; i.e., solutions which do not vary with time. They often describe the asymptotic states achieved by solutions of time-dependent problems, as $t \to \infty$. Physically speaking, all the "rough spots" smooth out by the time this steady state is achieved.

There are three basic principles which are obeyed by solutions of elliptic equations. The first principle is, if solutions of elliptic equations have a minimum amount of smoothness, they in fact are "exceedingly" smooth, provided that their coefficients too are very smooth. Second, solutions of elliptic equations are determined, in bounded sets, by "their values" on the boundary of the set. Finally, solutions of elliptic equations obey some sort of maximum principle; that is, their values in a bounded set are majorized by their values on the boundary of the set.

These ideas can be easily illustrated by considering the simplest example of an elliptic equation, Laplace's equation in $\mathbf{R}^2$: $\Delta u \equiv u_{xx} + u_{yy} = 0$. One knows from elementary function theory, that all continuous solutions of $\Delta u = 0$; i.e., all harmonic functions, are smooth and in fact are analytic. This is not true for *all* solutions of $\Delta u = 0$; namely, just consider the function (which doesn't even define a distribution near the origin):

$$u(x, y) = \begin{cases} \text{Re } \exp(z^{-4}), & z \neq 0, z = x + iy, \\ 0, & z = 0. \end{cases}$$

This function satisfies $\Delta u = 0$, but it is not continuous at $(0, 0)$. To get an idea of what ought to be a "correct" problem for elliptic equations, suppose that u and v are C^2 functions in a domain Ω, where $\partial\Omega$ is sufficiently smooth. Since $v\Delta u + v_x u_x + v_y u_y = (vu_x)_x + (vu_y)_y$, the divergence theorem can be applied to yield

$$\iint_\Omega (v\Delta u + \nabla v \cdot \nabla u) = \int_{\partial\Omega} v \frac{du}{dn},$$

where du/dn is the derivative of u in the direction of the outward normal to Ω. If now $v = u$, and u is harmonic, we obtain

$$\iint_\Omega |\nabla u|^2 = \int_{\partial\Omega} u \frac{du}{dn}$$

This shows that the values of u on $\partial\Omega$, in some sense determine u in Ω. Thus, if there were two harmonic functions u_1 and u_2 which assumed the same values on $\partial\Omega$, then $u = u_1 - u_2$ would satisfy

$$\iint_\Omega |\nabla u|^2 = 0.$$

This means $u_x = u_y = 0$ in Ω so that u is constant in Ω. But $u = 0$ on $\partial\Omega$ forces $u \equiv 0$ in Ω; i.e., $u_1 = u_2$ in Ω. Observe that a similar argument works if u_1 and u_2 were two harmonic functions which agreed at each point $D \subset \partial\Omega$ and whose normal derivatives agreed at each point $N \subset \partial\Omega$, where $D \cup N = \partial\Omega$ and D has positive measure. (If meas $D = 0$, we could only conclude that u_1 and u_2 differed by a constant.) It seems reasonable then, to consider boundary-value problems for Laplace's equation.

Finally, again from function theory, one knows that harmonic functions obey a maximum principle; namely, if u is harmonic in Ω and continuous in $\bar\Omega$, then for any (x, y) in Ω,

$$\min_{\partial\Omega} u \le u(x, y) \le \max_{\partial\Omega} u.$$

We shall see that these properties all carry over, in some form or another, to solutions of second-order elliptic equations.

§A. The Strong Maximum Principle

We shall prove a general maximum principle for second-order elliptic equations. Thus, consider the linear partial differential equation

$$P(u) \equiv A(u) + au = f, \tag{8.1}$$

where

$$A(u) = \sum_{i,k=1}^n a_{ik}(x)u_{x_i x_k} + \sum_{i=1}^n a_i(x)u_{x_i}, \tag{8.2}$$

in a bounded domain $\Omega \subset \mathbf{R}^n$, with $\partial\Omega$, the boundary of Ω, being sufficiently smooth. We assume that $a_{ik}(x) = a_{ki}(x)$, $x \in \Omega$, and that the functions a_{ik}, a_i, a, and f are all continuous in $\overline{\Omega}$, $i, k = 1, 2, \dots, n$. We assume too that P is *elliptic*; by this we mean that for any $x \in \Omega$, and any $\xi = (\xi_1, \dots, \xi_n) \neq 0$

$$\sum_{i,k=1}^{n} a_{ik}(x)\xi_i\xi_k > 0. \tag{8.3}$$

Thus, if $\mathscr{A} = (a_{ik}(x))$ is the $n \times n$ matrix defined by the principal part of $P(x, D)$, (8.3) implies that the bilinear form determined by $\mathscr{A}$ is nonnegative; i.e., $(\mathscr{A}\xi, \xi) > 0$ for all $\xi \in \mathbf{R}^n$, $\xi \neq 0$.

By a *solution* of (8.1) we mean a function $u \in C(\overline{\Omega}) \cap C^2(\Omega)$ which satisfies (8.1) in Ω. Here is the statement of the strong maximum principle for solutions of (8.1):

Theorem 8.1. *Suppose that $a(x) \leq 0$ in $\overline{\Omega}$. If $f(x) \leq 0$ (resp. $f(x) \geq 0$) in $\overline{\Omega}$, then every nonconstant solution of (8.1) attains its negative minimum (resp. positive maximum), if it exists, on $\partial\Omega$ and not on Ω.*

Before we give the proof, we remark that the theorem is false if $a > 0$. To see this consider the equation $u_{xx} + u_{yy} + 2u = 0$, in the rectangle $\Omega = \{(x, y): 0 \leq x, y \leq \pi\}$; then $u(x, y) = \sin x \sin y$ takes its maximum value $+1$ at $(\pi/2, \pi/2)$.

The next corollary shows that the solution depends continuously on the boundary data, in the sup-norm.

Corollary 8.2. *Let u_1 and u_2 be solutions of $Pu = f$ in Ω, with $u_i = \phi_i$ on $\partial\Omega$, $i = 1, 2$. Then if $a \leq 0$ in $\overline{\Omega}$, $\max_{x \in \overline{\Omega}} |u_1(x) - u_2(x)| \leq \max_{x \in \partial\Omega} |\phi_1(x) - \phi_2(x)|$.*

Proof of Corollary. If $u = u_1 - u_2$, then $Pu = 0$ in Ω, and $u = \phi_1 - \phi_2$ on $\partial\Omega$. Thus the result follows at once from the theorem. □

Corollary 8.3. *The boundary-value problem $Pu = f$ in Ω, $u = \phi$ on $\partial\Omega$ has at most one solution if $a \leq 0$ in $\overline{\Omega}$.*

We are now ready to prove the theorem. This will follow easily from the following lemma.

Lemma 8.4. *If $Au \geq 0$ (resp. $Au \leq 0$) in Ω and $\exists x_0 \in \Omega$ such that $u(x) \leq u(x_0)$ (resp. $u(x) \geq u(x_0)$) for all $x \in \Omega$, then $u(x) \equiv u(x_0)$ in $\overline{\Omega}$.*

Proof of Theorem 8.1. Suppose that $f \geq 0$ in Ω. If u has a positive maximum at $x_0 \in \Omega$, and $u(x_0) = m$, let $M = \{x \in \Omega: u(x) = m\}$. Then M is closed and nonvoid. Also, if $x_0 \in M$, $u(x) \leq u(x_0)$ in an open ball S centered at x_0, and

$u(x) > 0$ in S. Since $A(u) = -au + f \geq 0$ in S, the lemma shows that $u(x) \equiv m$ in S, whence S is open. Since Ω is connected, $u(x) \equiv m$ in $\bar{\Omega}$. A similar proof works if $f \leq 0$ in Ω. $\square$

It remains to prove the lemma.

Proof of Lemma 8.4. We shall assume that $Au \geq 0$; a similar proof will hold if $Au \leq 0$. Suppose that $u(x_0) = m$ and let $M = \{x \in \Omega : u(x) = m\}$. Assuming that the theorem is false, M would be a proper nonvoid subset of Ω.

If $x_1 \in \Omega \backslash M$, we connect x_1 to x_0 by an arc γ in Ω. Since γ is compact, we can find $\delta > 0$ such that if $p \in \gamma$, $\text{dist}(p, \partial\Omega) \geq \delta > 0$. Since $u(x_0) > u(x_1)$, $u(x_0) > u(x)$ in some ball centered at x_1 of radius at most $\delta/2$. If x_1 moves along γ towards x_0, the boundary of this ball eventually contains a point in M. Let $\bar{x}$ be the center of the first ball whose boundary meets M. Thus, there exists a ball S whose closure is contained in Ω for which $\partial S \cap M \neq \phi$, but $S \cap M = \phi$. Let us again call x_0 the point where $\partial S \cap M \neq \phi$; see Figure 8.1.

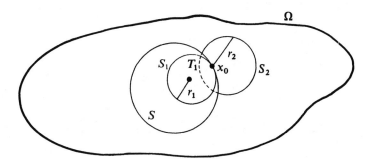

Figure 8.1

Let $S_1 \subset \bar{S}$ be a smaller ball of radius r_1 such that $x_0 \in \partial S_1$. Then $u < m$ in $\bar{S}_1 \backslash \{x_0\}$. Let $S_2 \subset \Omega$ be a third ball centered at x_0 having radius $r_2 < r_1$. If $\partial S_2 = T_1 \cup T_2$ where $T_1 = \partial S_2 \cap \bar{S}_1$, then T_1 is compact, so since $u < m$ on T_1, $u \leq m - \varepsilon$ on T_1 for some $\varepsilon > 0$.

We translate coordinates in $\mathbf{R}^n$ to put the origin at the center of S_1; this does not change the form of Au. Then we can consider the comparison function

$$h(x) = e^{-\alpha|x|^2} - e^{-\alpha r_1^2}, \tag{8.4}$$

where $\alpha > 0$, will be chosen shortly. A direct computation yields

$$e^{\alpha|x|^2} A(h) = 4\alpha^2 \sum_{i,k=1}^{n} a_{ik} x_i x_k - 2\alpha \sum_{i=1}^{n} (a_{ii} + a_i x_i). \tag{8.5}$$

Now since $r_2 < r_1$, $0 \notin \bar{S}_2$, and by the ellipticity hypothesis, $\sum_{i,k=1}^{n} a_{ik}x_ix_i \geq \sigma > 0$ in $\bar{S}_2$. Thus, (8.5) shows that $A(h) > 0$ in $\bar{S}_2$ if α is sufficiently large. Let $v(x) = u(x) + \varepsilon_1 h(x)$ and let $k = \max\{h(x) : x \in T_1\}$. Then on T_1,

$$v(x) \leq m - \varepsilon + \varepsilon_1 h(x) \leq m - \varepsilon + \varepsilon_1 k < m,$$

if $\varepsilon_1 < \varepsilon/k$. Having chosen $\varepsilon_1 < \varepsilon/k$, we see that on T_2, $h(x) < 0$ since $|x| > r_1$. Therefore

$$v(x) = u(x) + \varepsilon_1 h(x) < u(x) \leq m.$$

Thus, $v(x) < m$ on $T_1 \cup T_2 = \partial S_2$. Since $v(x_0) = u(x_0) = m$, we see that v has a maximum at a point $\tilde{x}$ in S_2. Hence $\nabla v(\tilde{x}) = 0$, and for any $\xi \in \mathbf{R}^n$,

$$\sum_{i,k=1}^{n} v_{x_i x_k}(\tilde{x})\xi_i\xi_k \leq 0.$$

But by ellipticity, we have

$$\sum_{i,k=1}^{n} a_{ik}(\tilde{x})\xi_i\xi_k \geq 0.$$

Thus, the elementary result from matrix theory (see the lemma below), yields

$$0 \geq \sum_{i,k=1}^{n} a_{ik}(\tilde{x})v_{x_i x_k}(\tilde{x}) = (Av)(\tilde{x}) = (Au)(\tilde{x}) + \varepsilon_1(Ah)(\tilde{x}) > (Au)(\tilde{x}),$$

which contradicts the assumption $(Au)(x) \geq 0$ in Ω. $\square$

Thus the proof will be complete if we prove the following elementary lemma. Let $\mathrm{tr}(A)$ denote the trace of the matrix A.

Lemma 8.5. *Suppose that A and B are symmetric $n \times n$ matrices with $A \geq 0$ and $B \leq 0$. Then $\mathrm{tr}(AB) \leq 0$.*

Proof. There exists an orthogonal matrix P such that

$$P^t BP = \Delta = \mathrm{diag}(b_1, \ldots, b_n), \qquad b_i \leq 0, \quad 1 \leq i \leq n.$$

Let $P^t AP = C = (c_{ij})$; C is symmetric and $C \geq 0$, so $c_{ii} \geq 0$, $1 \leq i \leq n$. Then

$$\mathrm{tr}(AB) = \mathrm{tr}(P^t ABP) = \mathrm{tr}(C\Delta) = \sum b_i c_{ii} \leq 0. \quad \square$$

It is often useful to know the behavior of u on $\partial\Omega$; in particular, we are interested in the signs of the directional derivatives of u on $\partial\Omega$. Recall that if

n is a unit outward-pointing normal vector at a point $p \in \partial\Omega$, the vector **v** is said to be *outward-pointing* from Ω at p if $\mathbf{n} \cdot v > 0$; we use the notation du/dv to represent the directional derivative of u at the boundary in the direction v.

Now suppose we are given equation (8.1) in Ω, where $f \geq 0$ and $a \leq 0$ in Ω, and $\partial\Omega$ is smooth, say C^1. From Theorem 8.1, we know that the maximum of u must be assumed at a point $p \in \partial\Omega$. It is clear that $du/dv \geq 0$ at p, where v is any outward-pointing vector. But in fact, a stronger result is true; namely, $du/dv > 0$ at p, unless u is constant in Ω. This is the content of the next theorem.

Theorem 8.6. *Suppose that $a(x) \leq 0$ in $\bar{\Omega}$. Let u be a solution of (8.1). Then if $f(x) \leq 0$ (resp. $f(x) \geq 0$) in $\bar{\Omega}$, and u achieves its negative minimum (resp. positive maximum) at $p \in \partial\Omega$, then every outward-pointing directional derivative of u at p is negative (resp. positive), unless u is identically equal to a constant in Ω.*

The proof will follow from a lemma.

Lemma 8.7. *Suppose that u is continuous in $\bar{\Omega}$, $Au \geq 0$ (resp. $Au \leq 0$) in $\bar{\Omega}$, and u achieves its maximum (resp. minimum) at $p \in \partial\Omega$. Then every outward-pointing directional derivative of u at p is positive (resp. negative) unless u is identically equal to a constant in Ω.*

Proof of Theorem 8.6. We suppose $f \geq 0$ in $\bar{\Omega}$ and u achieves its positive maximum at p; the proof of the other statement is similar. Since $\partial\Omega$ is smooth, we can find a ball $B \subset \bar{\Omega}$ with $\bar{B} \cap \bar{\Omega} = \{p\}$ and $u \geq 0$ in $\bar{B}$. If $du/dv \leq 0$ at p, then since $Au = -au + f \geq 0$ in B, our lemma implies that $u \equiv u(p)$ in $\bar{\Omega}$. $\square$

It remains to give the proof of Lemma 8.7. As above, we can find a ball $B \subset \bar{\Omega}$ with $\partial B \subset \Omega \cup \{p\}$. Let r be the radius of B, and let q be the center of B. We may assume, as in the proof of Lemma 8.4, that $q = 0$. Let K be a ball centered at p of radius $r/2$; see Figure 8.2.

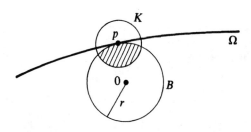

Figure 8.2

Define the auxiliary function $h(x)$ by

$$h(x) = e^{-\alpha|x|^2} - e^{-\alpha r^2},$$

where $\alpha > 0$ is chosen so large as to make $A(h) > 0$ in K (cf. the proof of Lemma 8.1). Again define $v(x) = u(x) + \varepsilon h(x)$. If $u \not\equiv u(p)$, then $u < u(p)$ in $(B \cup \partial B)\backslash\{p\}$. Choose ε so small as to make $v(x) \le v(p)$ on $\partial K \cap B$. Then by Theorem 8.1, $v(x) \le v(p)$ on $K \cap B$. Thus, at p, $du/dv + \varepsilon\,dh/dv = dv/dv \ge 0$. We shall show that $dh/dv < 0$ at p; this will imply that $du/dv > 0$ at p. Thus, we compute

$$\frac{\partial h}{\partial x_i} = -2\alpha x_i\, e^{-\alpha|x|^2},$$

so that if $\eta = (\eta_1,\dots,\eta_n)$ is the unit outward-pointing normal at p, then $\eta_i = x_i/|x|$. Thus, at p

$$\frac{dh}{dv} = -2\alpha|x|e^{-\alpha|r|^2}\sum v_i\eta_i < 0.$$

This completes the proof of Theorem 8.6. □

§B. A-Priori Estimates

An a-priori estimate for solutions of a partial differential equation is simply an inequality which is valid for all solutions whose data and coefficients obey certain restrictions. Such an estimate can be used to "continue" the solution as certain parameters vary. That is, one often perturbs the equation $P(u) = f$ by considering the family of equations $A_\varepsilon(u) \equiv (1 - \varepsilon)Q(u) + \varepsilon P(u) = f$, where $0 \le \varepsilon \le 1$, and the problem $Q(u) = f$ is known to be solvable (e.g., say $Q(u) = \Delta u$). If one considers the set $S = \{\varepsilon : A_\varepsilon(u) = f$ is solvable$\}$, then $S \ne \phi$, and the a-priori estimates are used to show that S is open and/or closed. If one can show that S is both open and closed, then $S = [0, 1]$; whence $\varepsilon = 1$ is in S and the original problem is solved. In Chapter 10, we shall see how a-priori estimates are employed in a different way (but however, for the same conclusion).

We begin with a result which follows in a fairly direct manner from the maximum principle. Consider the following boundary-value problem in a bounded domain Ω, where $\partial\Omega$ is smooth:

$$P(u) \equiv Au + au = f \quad \text{in } \Omega, \qquad u = \phi \quad \text{on } \partial\Omega, \tag{8.6}$$

where, in what follows, we always take $a \le 0$ in Ω, and Au to be defined by (8.2). We assume that $(a_{ik}(x))$ is a symmetric matrix and that the coefficients of

P, as well as f are continuous in $\overline{\Omega}$, and that ϕ is continuous on $\partial\Omega$. We assume that P is *uniformly elliptic*, which means that there is a constant $m > 0$ such that

$$\sum_{i,k=1}^{n} a_{ik}(x)\xi_i\xi_k \geq m|\xi|^2, \tag{8.7}$$

for all $\xi \in \mathbf{R}^n$. Let k be a bound for the quantities $|a_{ik}(x)|$, $|a_i(x)|$, and $|a(x)|$ in $\overline{\Omega}$, i, $k = 1, 1, \ldots, n$. Here is the first a-priori estimate.

Theorem 8.8. *If* $u \in C(\overline{\Omega}) \cap C^2(\Omega)$ *is a solution of* (8.6), *then there exists a constant* $M = M(m, \Omega, K)$ *such that*

$$\|u\|_{L_\infty(\overline{\Omega})} \leq \|\phi\|_{L_\infty(\partial\Omega)} + M\|f\|_{L_\infty(\overline{\Omega})} \tag{8.8}$$

We note that (8.8) shows that the problem (8.6) depends continuously (in the sup-norm), on the data ϕ and on the "right-hand side," f. Moreover, it yields still another proof of the uniqueness of the solution to the problem (8.6).

Proof. If we change coordinates by a linear transformation, the problem in the new coordinates stays of the same form, and (8.7) is invariant. Thus, there is no loss in generality if we assume that $x_1 \geq 0$ in $\overline{\Omega}$. Let

$$h(x) = \|\phi\|_{L_\infty(\partial\Omega)} + (e^{\alpha\xi} - e^{\alpha x_1})\|f\|_{L_\infty(\overline{\Omega})}, \tag{8.9}$$

where $\xi > \max\{x_1 : x \in \overline{\Omega}\}$, and $\alpha > 0$ is chosen so large that both

$$m\alpha^2 - k(\alpha + 1) \geq 1 \quad \text{and} \quad e^{\alpha\xi} > 2 \max_{x \in \overline{\Omega}} e^{\alpha x_1} \tag{8.10}$$

are valid. Note that $h(x) \geq \|\phi\|_{L_\infty(\partial\Omega)}$ if $x \in \partial\Omega$. Also, if $\xi = (1, 0, \ldots, 0)$, then (8.7) shows $a_{11} \geq m$ so

$$\begin{aligned}
-Ph &= -a\|\phi\|_\infty + [-ae^{\alpha\xi} + e^{\alpha x_1}(a_{11}\alpha^2 + a_1\alpha + a)]\|f\|_{L_\infty(\Omega)} \\
&\geq [-ae^{\alpha\xi} + e^{\alpha x_1}(a_{11}\alpha^2 + a_1\alpha + a)]\|f\|_{L_\infty(\Omega)} \\
&\geq [m\alpha^2 e^{\alpha x_1} - a(e^{\alpha\xi} - e^{\alpha x_1}) + a_1\alpha e^{\alpha x_1}]\|f\|_{L_\infty(\Omega)} \\
&\geq [m\alpha^2 - a + a_1\alpha]e^{\alpha x_1}\|f\|_{L_\infty(\Omega)} \\
&\geq [m\alpha^2 - k - k\alpha]e^{\alpha x_1}\|f\|_{L_\infty(\Omega)} \\
&= [m\alpha^2 - k(\alpha + 1)]e^{\alpha x_1}\|f\|_{L_\infty(\Omega)} \\
&\geq e^{\alpha x_1}\|f\|_{L_\infty(\Omega)} \geq \|f\|_{L_\infty(\Omega)}.
\end{aligned}$$

We shall use this to show

$$|u(x)| \leq h(x), \qquad x \in \overline{\Omega}. \tag{8.11}$$

This will imply (8.8) since if $x \in \overline{\Omega}$,

$$|u(x)| \leq h(x) \leq \|\phi\|_{L_\infty(\partial\Omega)} + M\|f\|_{L_\infty(\Omega)},$$

where $M = \max_{x\in\overline{\Omega}}(e^{\alpha\xi} - e^{\alpha x_1}) \leq e^{\alpha\xi} - 1$.

We put $v = u - h$; then on $\partial\Omega$, $v = \phi - h \leq 0$ and $Pv = Pu - Ph \geq f + \|f\|_{L_\infty(\Omega)} \geq 0$. Hence Theorem 8.1 shows $v \leq 0$ in $\overline{\Omega}$; i.e., $u \leq h$ in $\overline{\Omega}$. Similarly, if $v = u + h$, then on $\partial\Omega$, $v = \phi + h \geq 0$ and $Pv \leq f - \|f\|_{L_\infty(\Omega)} \leq 0$, so again from Theorem 8.1, $u + h \geq 0$ in $\overline{\Omega}$, and $u \geq -h$ in $\overline{\Omega}$. Thus (8.11) holds and the proof is complete. $\square$

We remark that if we do not assume that $a \leq 0$ in $\overline{\Omega}$, we can still obtain an estimate of the form

$$\|u\|_{L_\infty(\overline{\Omega})} \leq c(\|\phi\|_{L_\infty(\partial\Omega)} + \|f\|_{L_\infty(\overline{\Omega})}), \tag{8.12}$$

where c depends on k, m, and Ω, provided that Ω is sufficiently narrow in one direction, say the x_1-direction. More precisely, (8.12) will hold if

$$(e^{\alpha\xi} - 1)\|a\|_{L_\infty(\overline{\Omega})} < 1, \tag{8.13}$$

where ξ and α are as in the proof of the theorem. To see this, we let $b(x) = \min(a(x), 0)$, and write the equation as

$$A(u) + bu = (b - a)u + f = g.$$

We can now apply (8.8) to this equation to get

$$\|u\|_{L_\infty(\overline{\Omega})} \leq \|\phi\|_{L_\infty(\partial\Omega)} + (e^{\alpha\xi} - 1)\|g\|_{L_\infty(\overline{\Omega})}$$
$$\leq \|\phi\|_{L_\infty(\partial\Omega)} + (e^{\alpha\xi} - 1)(\|f\|_{L_\infty(\overline{\Omega})} + \|u\|_{L_\infty(\overline{\Omega})}\|a\|_{L_\infty(\overline{\Omega})}),$$

so that

$$\|u\|_{L_\infty(\overline{\Omega})} \leq \frac{\|\phi\|_{L_\infty(\partial\Omega)} + (e^{\alpha\xi} - 1)\|f\|_{L_\infty(\overline{\Omega})}}{[1 - (e^{\alpha\xi} - 1)]\|a\|_{L_\infty(\overline{\Omega})}}.$$

We note that this too implies uniqueness of solutions of (8.6).

§C. Existence of Solutions

In order to solve the problem (8.6), we require a sharper estimate than (8.8). Before stating this estimate we shall need definitions of certain Banach spaces of functions.

We consider the space $C^\alpha(\overline{\Omega})$, of Hölder continuous functions on $\overline{\Omega}$, having Hölder exponent α, with the norm

$$\|u\|_\alpha = \sup_{\substack{x,y\in\Omega \\ x\neq y}} \frac{|u(x) - u(y)|}{|x - y|^\alpha} < \infty.$$

We put a norm on $C^2(\Omega)$, by defining

$$\|u\|_2 = \|u\|_{L_\infty(\Omega)} + \max_{1\leq j\leq n} \|D_j u\|_{L_\infty(\Omega)} + \max_{1\leq i,j\leq n} \|D_i D_u u\|_{L_\infty(\Omega)}.$$

We also define $C^{2+\alpha}(\overline{\Omega})$ to be the class of C^2 functions on Ω whose second derivatives are in $C^\alpha(\overline{\Omega})$; this space is made into a Banach space by defining

$$\|u\|_{2+\alpha} = \|u\|_2 + \max_{1\leq i,j\leq n} \|D_i D_j u\|_\alpha.$$

The existence of a solution to (8.6), where A is uniformly elliptic depends on establishing the sharp a-priori estimate (the "Schauder estimate,")

$$\|u\|_{2+\alpha} \leq c\|f\|_\alpha, \tag{8.14}$$

for solutions u in class $C^{2+\alpha}$ of the problem

$$P(u) = f \quad \text{in } \Omega, \qquad u = 0 \quad \text{on } \partial\Omega. \tag{8.15}$$

Here $c = c(k, m, \Omega)$, where m is the ellipticity constant and K is a bound for the coefficients of P. We shall not prove this estimate here; see [GT] for a proof.

It can be shown (see [GT]) that $u \in C^{2+\alpha}(\overline{\Omega})$ if $f \in C^\alpha(\overline{\Omega})$ and $\phi \in C^{2+\alpha}(\Omega)$; i.e., the solution is considerably smoother than the right-hand side; in short, "the solution gains derivatives."

Theorem 8.9. *The problem* (8.15) *is uniquely solvable, for each* $f \in C^\alpha(\partial\Omega)$, *provided that A is uniformly elliptic in Ω, and the coefficients of P are in* $C^\alpha(\Omega)$.

Proof. We assume that (8.14) holds for all solutions $u \in C^{2+\alpha}$. The idea is to embed our problem in a family of problems

$$P_t(u) = tP(u) + (1 - t)\Delta u = f \quad \text{in } \Omega, \qquad u = 0 \quad \text{on } \partial\Omega, \tag{8.16}$$

where $t \in I = [0, 1]$. If $t = 0$, the problem is $\Delta u = f$ in Ω, $u = 0$ on $\partial\Omega$, which we consider to be solved (see [GT]). Let

$$T = \{t \in I : f \in C^\alpha(\overline{\Omega}) \Rightarrow \text{there is a solution } u \in C^{2+\alpha}(\overline{\Omega}) \text{ of (8.16)}\}.$$

We shall show that T is both open and closed; then $T = I$, and in particular, $1 \in T$. We first show that T is open. Thus, let $t_0 \in T$; we shall find an $\varepsilon > 0$ such that $|t - t_0| < \varepsilon$ implies that $t \in T$. To this end, let $t \in I$ and suppose $u \in C^{2+\alpha}(\overline{\Omega})$. We define a mapping Φ_t from $C^{2+\alpha}(\overline{\Omega})$ into itself given by $\Phi_t(u) = v$, where v is the unique solution of

$$P_{t_0} v = (t - t_0)[\Delta u - P(u)] + f \quad \text{in } \Omega, \qquad v = 0 \quad \text{on } \partial\Omega. \quad (8.17)$$

If Φ_t had a fixed point, i.e., if $\Phi_t(w) = w$, then $w = 0$ on $\partial\Omega$ and from (8.17)

$$P_{t_0} w = (t - t_0)[\Delta w - P(w)] + f \quad \text{in } \Omega,$$

so that $P_t w = f$. That is, fixed points of Φ_t correspond to solutions of (8.16). Thus, to show that T is open, we shall find an $\varepsilon > 0$ such that if $|t - t_0| < \varepsilon$, then Φ_t has a fixed point. This latter statement will be proved by showing that Φ_t is a contraction mapping for t sufficiently near t_0.

Thus, if u_1 and u_2 are in $C^{2+\alpha}(\overline{\Omega})$, let $v_1 = \Phi_t(u_1)$, $v_2 = \Phi_t(u_2)$, so that

$$P_{t_0} v_i = (t - t_0)[\Delta u_i - P(u_i)] + f \quad \text{in } \Omega, \qquad v_i = 0 \quad \text{on } \partial\Omega,$$

for $i = 1, 2$. Subtracting gives

$$P_{t_0}(v_1 - v_2) = (t - t_0)[\Delta(u_1 - u_2) - P(u_1 - u_2)].$$

Using (8.14) on this equation, we obtain

$$\|\Phi_t(u_1) - \Phi_t(u_2)\|_{2+\alpha} = \|v_1 - v_2\|_{2+\alpha} \le c|t - t_0| \|\Delta(u_1 - u_2)$$
$$- P(u_1 - u_2)\|_\alpha \le cc_1|t - t_0| \|u_1 - u_2\|_{2+\alpha}$$

for some constant c_1 independent of u_1, u_2, c, and t. If $\varepsilon = (2cc_1)^{-1}$, then for $|t - t_0| < \varepsilon$,

$$\|\Phi_t(u_1) - \Phi_t(u_2)\|_{2+\alpha} < \tfrac{1}{2}\|u_1 - u_2\|_{2+\alpha}.$$

Thus Banach's theorem implies that Φ_t has a fixed point. Since this is true for each t with $|t - t_0| < \varepsilon$, we see that T is open.

Since ε is independent of c, c_1, u_1, u_2, we can cover $[0, 1]$ by $[2/\varepsilon]$ intervals, each of length $\varepsilon/2$, so $0 \in T$ implies $1 \in T$. $\square$

We shall extend the last theorem to problems of the form

$$Pu = f \quad \text{in } \Omega, \qquad u = \phi \quad \text{on } \partial\Omega, \tag{8.18}$$

where $f \in C^\alpha(\overline{\Omega})$, $\phi \in C(\partial\Omega)$, and where the operator A is uniformly elliptic in Ω (cf. (8.7)), $a \leq 0$ in Ω, and the coefficients of P are in $C^\alpha(\overline{\Omega})$. Let K be a bound for $|a_{ik}|$, $|a_i|$, and $|a|$ in $\overline{\Omega}$, $i, k = 1, 2, \ldots, n$.

Theorem 8.10. *The problem $Pu = f$ in Ω, $u = \phi$ on $\partial\Omega$, where $f \in C^\alpha$ and $\phi \in C^0$, has a unique solution $u \in C^{2+\alpha}(\Omega)$.*

Proof. We first assume $\phi \in C^{2+\alpha}(\partial\Omega)$. Let h be the solution of

$$Ph = 0 \quad \text{in } \Omega, \qquad h = \phi \quad \text{on } \partial\Omega.$$

Using Theorem 8.9, let v be the solution of

$$Pv = f \quad \text{in } \Omega, \qquad v = 0 \quad \text{on } \partial\Omega.$$

Then if $u = v + h$, we have $Pu = f$ in Ω, and $u = \phi$ on $\partial\Omega$.

Now assume that $\phi \in C^0(\partial\Omega)$. Let ϕ_j be a sequence of functions in $C^{2+\alpha}(\partial\Omega)$ such that $\phi_j \to \phi$ uniformly on $\partial\Omega$ (for example, the ϕ_j can be polynomials obtained from the Weierstrass approximation theorem). Let u_j be the solution of $Pu_j = f$ in Ω, $u_j = \phi_j$ on $\partial\Omega$. By the maximum principle, we see that u_j converges uniformly in Ω to a function $u \in C^0(\overline{\Omega})$ and that $u = \phi$ in $\partial\Omega$.

Now if Q is any compact subset of Ω, one has the "interior estimate" for the solutions of $Pu = f$; namely,

$$\|u\|_{C^{2+\alpha}(\overline{Q})} \leq c(\|f\|_{C^\alpha(\overline{\Omega})} + \|u\|_{L^\infty(\Omega)}),$$

where $c = c(m, k, \Omega, Q)$. The proof of this is similar to (8.14); see [Ag]. It follows that

$$\|u_j - u\|_{C^{2+\alpha}(\overline{Q})} \leq c\|u_j - u\|_{L^\infty(\Omega)},$$

so that $u_j \to u$ in $C^{2+\alpha}(Q)$ for every compact $Q \subset \Omega$. Thus $u \in C^{2+\alpha}(\Omega)$ and $f = \lim_{j \to \infty} Pu_j = Pu$, at each point in Ω. This finishes the proof. $\square$

§D. Elliptic Regularity

In this short section, we shall describe some other methods of obtaining solutions for elliptic boundary-value problems; in particular, the notions of weak solution are important here. We shall only give the statement of the various results; the proofs can be found in [Ag] or [GT].

We consider a *weak* solution of (8.6); i.e., a function $u \in W_2^1(\Omega)$ which satisfies, for all $\psi \in C_0^1(\Omega)$,

$$\int_\Omega [A^* \psi)u + au\psi] \, dx = \int f\psi, \qquad (8.19)$$

and $(u - \phi) \in \mathring{W}_2^1(\Omega)$. One shows that if $f \in L^2(\Omega)$, and each a_i, a_{ij} is in $L_2(\Omega) \cap L_\infty(\Omega)$, then if $\phi \in W_2^1(\Omega)$, this problem is uniquely solvable. The proof is via functional analysis and Hilbert space techniques ([GT]). The remarkable thing is that the solution is actually a classical solution; this follows from the celebrated estimates of Agmon, Douglas, and Nirenberg, which we shall describe below.

Thus, we consider problem (8.6), where $f \in C^\alpha(\overline{\Omega})$, and ϕ is assumed to have a $C^{2+\alpha}$ extension into $\overline{\Omega}$. Then Agmon, Douglas, and Nirenberg [ADN] prove the inequality

$$\|u\|_{p,2} \le c_p(\|f\|_p + \|\phi\|_{p,2}),$$

where c is independent of u. Moreover, for classical solutions, one has in addition the Schauder estimate

$$\|u\|_{2+\alpha} \le c_\alpha(\|f\|_\alpha + \|\phi\|_{2+\alpha}),$$

where again c doesn't depend on u.

These estimates are used to show that the operator $T: f \to u$ given by $P(u) = f$ in Ω, $u = \phi$ on $\partial\Omega$ takes $C^\alpha(\overline{\Omega})$ continuously into $C^{2+\alpha}(\overline{\Omega})$, provided that ϕ can be extended to $W_2^2(\Omega)$, and $\partial\Omega$ is smooth, say of class C^2.

These results are all valid for more general boundary conditions. For example, we could take boundary conditions of the form

$$\frac{du}{dv} + p(x)u = \phi,$$

where v is any outward derivative on $\partial\Omega$, and $p(x) \ge 0$ on $\partial\Omega$. Again, for a proof see [GT].

NOTES

The strong maximum principle was proved by Hopf in [Hf 1]. The technique in the proof of Theorem 8.9 is due to Schauder, as are the estimates appearing in the text; see [GT] for modern statements, as well as references. The elliptic regularity theorems are due to Agmon, Douglis, and Nirenberg [ADN]; see also Agmon's book [Ag] as well as the Gilbarg–Trudinger book [GT].

Chapter 9

Second-Order Linear Parabolic Equations

Parabolic equations arise in diffusion processes, and more generally in "irreversible" time-dependent processes. Mathematically, this is reflected in the fact that the equations are not invariant under the reversal of time; i.e., under the transformation $t \to -t$. This means that knowledge about the "past" is lost as time increases. For example, there may be dissipation effects which lead to an increase in entropy and a consequent loss of information.

The simplest example of a parabolic equation is the equation of "heat conduction"

$$Hu \equiv u_t - k\Delta u = 0, \tag{9.1}$$

where Δ is the Laplace operator, and k is a positive constant. It is quite clear that this equation changes form when t is replaced by $-t$. Observe however, that (9.1) is preserved under the transformation $(x, t) \to (\alpha x, \alpha^2 t)$. This transformation also leaves invariant the expression $|x|^2/t$; indeed we shall see that this latter expression plays an important role in the study of (9.1).

We shall show that solutions of parabolic equations obey a maximum principle, and that parabolic operators are "smoothing," in the sense that the solutions are more smooth than the data.

§A. The Heat Equation

We consider the equation (9.1) in a cylindrical region of $x - t$ space of the form

$$\mathscr{D} = R \times (0, T), \qquad T < \infty,$$

where R is a bounded region in $\mathbf{R}^n$. We let

$$\mathscr{D}' = \mathrm{cl}(R) \times \{t = 0\} \cup \partial R \times [0, T];$$

see Figure 9.1.

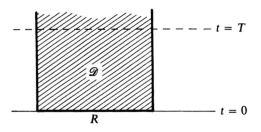

Figure 9.1

Theorem 9.1. *Suppose* $u \in C^0(\bar{\mathscr{D}}) \cap C^2(\mathscr{D})$ *is a solution of* (9.1). *Then both* max u *and* min u *in* $\bar{\mathscr{D}}$ *are taken on at a point in* $\mathscr{D}'$.

Proof. We prove the theorem for max u only. Thus, let $M = \max u$ in $\mathscr{D}'$. For $\varepsilon > 0$, define the auxiliary function

$$v(x, t) = u(x, t) + \varepsilon |x|^2 ;$$

then $Hv = -2nk\varepsilon < 0$. Let $\bar{t}$ be any number $0 < \bar{t} < T$ and let $K = \bar{R} \times [0, \bar{t}]$. The maximum of v in K cannot be taken on at a point in $R \times (0, \bar{t}]$, for at these points, $v_t \geq 0$ and $\Delta v \leq 0$, so $Hv \geq 0$. Thus max v on K is taken on at a point in $K \cap \mathscr{D}'$; i.e., at a point where $u \leq M$. Hence, in K, $u \leq v \leq M + \varepsilon c$, where c is an upper bound for $|x|^2$ in $\bar{R}$. Since ε is arbitrary, we see $u \leq M$ in K. Since $\bar{t}$ was arbitrary, the result follows.   □

Corollary 9.2. *The boundary-value problem* $Hu = \phi$ *in* $\mathscr{D}$ *and* $u = f$ *on* $\mathscr{D}'$, *has at most one solution in* $C^0(\bar{\mathscr{D}}) \cap C^2(\mathscr{D})$.

We also have a maximum principle for infinite regions of the form $\mathscr{D} = \mathbf{R}^n \times (0, T)$, $0 < T < \infty$. We also require u not to grow too fast at ∞.

Theorem 9.3. *Let* $u \in C(\bar{\mathscr{D}}) \cap C^2(\mathscr{D})$ *be a solution of* (9.1) *with* $u(x, 0) = f(x)$, $x \in \mathbf{R}^n$. *If* $|u(x, t)| \leq Me^{c|x|^2}$ *in* $\mathscr{D}$, *then* $|u(x, t)| \leq \|f\|_\infty$.

Proof. It suffices to prove that $u(x, t) \leq \sup\{f(x) : x \in \mathbf{R}^n\}$. Also, if we prove this inequality under the assumption $4cT < 1$, then the result follows upon dividing the interval $[0, T]$ into equal parts of lengths $l < \frac{1}{4}c$. Finally, for simplicity, we assume $k = 1$ in (9.1).

Now there is an $\varepsilon > 0$ such that $4c(T + \varepsilon) < 1$. Fix y, and for θ a positive constant, and $0 \leq t \leq T$, consider the function

$$u_\theta(x, t) = u(x, t) - \theta[4\pi(T + \varepsilon - t)]^{-n/2} \exp[|x - y|^2/4(T + \varepsilon - t)]$$

$$\equiv u(x, t) - \theta G(T + \varepsilon - t, (x - y)).$$

Since $G_t = \Delta G$, we have $(u_\theta)_t - \Delta u_\theta = 0$. If $\Omega = \{(x, t): |x - y| < \rho, 0 < t < T\}$, then by Theorem 9.1, $u_\theta(x, t) \le \max\{u(x, t): (x, t) \in \Omega'\}$. On $\{t = 0\} \cap \partial\Omega$, $u_\theta(x, 0) \le u(x, 0) \le \sup_x f(x)$, since $\theta G > 0$. On the remaining part of Ω',

$$u_\theta(x, t) \le Me^{c|x|^2} - \theta[4\pi(T + \varepsilon - t)]^{-n/2} \exp[\rho^2/4(T + \varepsilon - t)]$$

$$\le Me^{c(|y|+\rho)^2} - \theta[4\pi(T + \varepsilon)]^{-n/2} \exp[\rho^2/4(T + \varepsilon)] \le \sup_x f(x),$$

if ρ is sufficiently large. Thus $\max_{\Omega'} u_\theta \le \sup_x f(x)$, and thus

$$u_\theta(y, t) = u(y, t) - \theta[4\pi(T + \dot\varepsilon - t)]^{-n/2} \le \sup_x f(x).$$

If now we let $\theta \to 0$, we get the desired inequality. $\square$

Corollary 9.4. *The initial-value problem $Hu = \phi$ in $\mathscr{D}$, $u(x, 0) = f(x)$, $x \in \mathbf{R}^n$, has at most one solution $u \in C^2(\mathscr{D}) \cap C^0(\overline{\mathscr{D}})$, satisfying the growth condition $|u(x, t)| \le Me^{c|x|^2}$.*

We remark that this corollary is false if u grows too fast at infinity; see [Fn 3]. For example, if $n = 1$, the function

$$u(x, t) = \sum_{k=0}^{\infty} \frac{1}{2k!} x^{2k} \frac{d^k}{dt^k} e^{-1/t^2}$$

satisfies $Hu = 0$ and $u(x, 0) = 0$.

We now turn to the construction of a solution of the initial-value problem (9.1) with data

$$u(x, 0) = \phi(x). \tag{9.2}$$

We shall motivate our construction by giving a formal derivation of the solution. For simplicity, let $n = 1$ and $k = 1$. Then if we take the Fourier transform of (9.1) with respect to x, we get the equation

$$\frac{\partial}{\partial t} \hat{u}(\xi, t) = -\xi^2 \hat{u},$$

whose solution is $\hat{u}(\xi, t) = ce^{-\xi^2 t}$, where c is an arbitrary function of ξ. To evaluate c, we make the reasonable assumption that

$$\lim_{t \to 0+} \hat{u}(\xi, t) = (\lim_{t \to 0+} u(x, t))^\wedge.$$

This yields $c = \hat{\phi}(\xi)$, so that we have

$$\hat{u}(\xi, t) = \hat{\phi}(\xi)e^{-\xi^2 t}. \tag{9.3}$$

Now define the function

$$f(\lambda) = \int_{-\infty}^{\infty} e^{-a\xi^2} e^{-i\lambda\xi} \, d\xi = \int_{-\infty}^{\infty} e^{-a\xi^2} \cos \lambda\xi \, d\xi,$$

where a is a constant, $a > 0$. Then

$$f'(\lambda) = \frac{1}{2a} \int_{-\infty}^{\infty} [e^{-a\xi^2}(-2a\xi)] \sin \lambda\xi \, d\xi = -\frac{\lambda}{2a} \int_{-\infty}^{\infty} e^{-a\xi^2} \cos \lambda\xi \, d\xi.$$

Thus f satisfies the equation $f' + (\lambda/2a) f = 0$. It follows that $f = ke^{-\lambda^2/4a}$, where k is constant. To evaluate k, notice that $f(0) = k$ so that

$$k = \int_{-\infty}^{\infty} e^{-a\xi^2} \, d\xi = \sqrt{\pi/a}. \tag{9.4}$$

Thus

$$(e^{-a\xi^2})^\wedge = f(\lambda) = \sqrt{\frac{\pi}{a}} e^{-\lambda^2/4a},$$

and

$$e^{-\lambda^2/4a} = \sqrt{\frac{a}{\pi}} (e^{-a\xi^2})^\wedge.$$

If now $a = 1/4t$, we get

$$e^{-\lambda^2 t} = \sqrt{\frac{1}{4\pi t}} (e^{-\xi^2/4t})^\wedge.$$

Using this in (9.3) gives

$$\hat{u}(\xi, t) = \hat{\phi}\left(\frac{1}{\sqrt{4\pi t}} e^{-\xi^2/4t}\right)^\wedge = \left(\phi * \frac{1}{\sqrt{4\pi t}} e^{-x^2/4t}\right)^\wedge,$$

and finally

$$u(x, t) = \phi * \frac{1}{\sqrt{4\pi t}} e^{-x^2/4t} = \int_{-\infty}^{\infty} \phi(y) \frac{e^{(x-y)^2/4t}}{\sqrt{4\pi t}} \, dy.$$

We shall show now that this formula actually defines a solution of the initial-value problem (9.1), (9.2).

Theorem 9.5. *If $\phi(x)$ is continuous and uniformly bounded on $\mathbf{R}^n$, then*

$$u(x, t) = \int_{\mathbf{R}^n} (4k\pi t)^{-n/2} e^{-|z-x|2/4kt} \phi(z) \, dz$$

is the unique bounded solution of (9.1), (9.2). *This function is analytic for all*
$x \in \mathbf{C}^n$ *and for all* $t \in \mathbf{C}^1$ *with* Re $t > 0$.

Proof. Let $G(t, z - x) = (4\pi k t)^{-n/2} \exp(-|z - x|^2/4kt)$. Then since G and
all of its derivatives decay exponentially as $|z - x| \to \infty$ for each fixed $t > 0$,
we see that u defines an analytic function for any $x \in \mathbf{C}^n$ and $t \in \mathbf{C}^1$ with
Re $t > 0$. A straightforward calculation shows that $H(G) = 0$, for any $t > 0$
and any z; thus $Hu = 0$. It remains to show that (9.2) holds. To this end, we
shall first show that the following two properties hold:

(i) $\displaystyle \int_{\mathbf{R}^n} G(t, z - x) \, dz = 1$ for all $x \in \mathbf{R}^n$ and $t > 0$.

(ii) $\displaystyle \lim_{t \to 0_+} \int_{|z - x| > \sigma} G(t, z - x) \, dz = 0$ for every $\sigma > 0$.

For (i), note that (9.4) implies

$$\int_{-\infty}^{\infty} e^{-y_i^2/4kt} \, dy_i = \sqrt{4\pi kt},$$

so that

$$\int_{\mathbf{R}^n} G(t, z - x) \, dz = \int_{\mathbf{R}^n} G(t, y) \, dy = \prod_{i=1}^{n} \int_{\mathbf{R}} (4\pi kt)^{-1/2} e^{-y_i^2/4kt} \, dy = 1.$$

To show (ii), we write $z - x = r\xi, |\xi| = 1$, and

$$\int_{|z - x| > \sigma} G(t, z - x) \, dz = \int_{|\xi| = 1} \left(\int_{\sigma}^{\infty} r^{n-1} G(t, r\xi) \, dr \right) d\xi$$

$$= \int_{|\xi| = 1} \left(\int_{\sigma}^{\infty} r^{n-1} (4\pi kt)^{-n/2} e^{-r^2/4kt} \, dr \right) d\xi$$

$$= c \int_{\sigma/\sqrt{nkt}}^{\infty} \sigma^n e^{-\sigma^2} \, d\sigma,$$

for some constant c. Since this latter integral tends to zero as $t \to 0_+$, (ii)
follows.

We can now show (9.2). Thus, if $|\phi| < M$ on $\mathbf{R}^n$, and $\varepsilon > 0$ is given, let
$\sigma = \sigma(x) > 0$ be such that

$$|\phi(x) - \phi(z)| < \frac{\varepsilon}{3} \quad \text{if } r < \sigma,$$

$r = |z - x|$. From (ii), there exists $T = T(x, \varepsilon) > 0$ such that

$$\int_{r>\sigma} G(t, z - x)\, dz < \frac{\varepsilon}{3M}, \qquad 0 < t < T.$$

Thus, using (i), we have, for $0 < t < T$,

$$|u(x, t) - \phi(x)| = \left| \int_{\mathbf{R}^n} G(t, z - x)(\phi(x) - \phi(z))\, dz \right|$$

$$\leq \int_{\mathbf{R}^n} G(t, z - x)|\phi(x) - \phi(z)|\, dz$$

$$= \int_{r<\sigma} G(t, z - x)|\phi(x) - \phi(z)|\, dz$$

$$+ \int_{r>\sigma} G(t, z - x)|\phi(x) - \phi(z)|\, dz$$

$$< \frac{\varepsilon}{3} \int_{r<\sigma} G(t, z - x)\, dx + 2M \int_{r>\sigma} G(t, z - x)\, dz$$

$$< \frac{\varepsilon}{3} + 2M \cdot \frac{\varepsilon}{3M} = \varepsilon.$$

This shows that

$$\lim_{t \to 0+} u(x, t) = \phi(x), \tag{9.5}$$

so that (9.2) holds. To see that u is continuous at $t = 0$, note that on any compact set, σ can be chosen independently of x. Thus (9.5) holds uniformly for bounded x, and so u is continuous in $t \geq 0$. Finally, since

$$|u(x, t)| \leq M \int_{\mathbf{R}^n} G(t, z - x)\, dz = M,$$

we see that u is bounded. Thus, in view of Corollary 9.4, u must be the unique bounded solution of (9.1), (9.2). The proof is complete. $\square$

§B. Strong Maximum Principles

In this section, we shall prove an extension of Theorem 9.1 to general second-order parabolic equations. These theorems are the analogues of Theorems 8.1 and 8.6 which were valid for general second-order elliptic equations. Indeed, the proofs which we shall give are actually modifications of the "elliptic" proofs. The theorems we prove will be of use in studying nonlinear equations, in that they will enable us to obtain "comparison" theorems. These will be considered in Chapter 10.

The general linear parabolic second-order equation in n-space variables can be written in the form

$$Pu \equiv Au + au = f, \tag{9.6}$$

where $u = u(x, t)$, and $(x, t) \in \mathcal{D} \subset \mathbf{R}^n \times \mathbf{R}_+$, with $\mathcal{D}$ being a domain (i.e., an open, connected set). The operator A is defined by

$$Au = \sum_{i,j=1}^{n} a_{ij}(x, t) D_i D_j u + \sum_{i=1}^{n} a_i(x, t) D_i u - u_t, \tag{9.7}$$

where all of the coefficients a_{ij}, a_i, as well as the function $a = a(x, t)$, are bounded in $\mathcal{D}$. A is called *uniformly parabolic* in $\mathcal{D}$ if there is a constant $\mu > 0$ such that for any $\xi = (\xi_1, \ldots, \xi_n)$ in $\mathbf{R}^n$,

$$\sum_{i,j=1}^{n} a_{ij}(x, t) \xi_i \xi_j \geq \mu |\xi|^2, \tag{9.8}$$

for each $(x, t) \in \mathcal{D}$. We shall assume that the above conditions on A and a are valid, and (without loss of generality) that $a_{ij} = a_{ji}$, throughout this chapter.

In order to state the strong maximum theorems for (9.6), we need a simple definition. Namely, let (x_1, t_1) and (x_2, t_2) be any two points in $\mathcal{D}$. We say (x_1, t_1) is connected in $\mathcal{D}$ to (x_2, t_2) by a *horizontal* segment if $t_1 = t_2$ and the points can be joined by a line segment lying in $(t = t_1) \cap \mathcal{D}$. Similarly, we say that these points can be joined by an *upward vertical* segment, if $x_1 = x_2$, $t_1 < t_2$ and the line segment joining them is contained in $\mathcal{D}$. We can now state the first theorem; the strong maximum principle.

Theorem 9.6. *Suppose that A is uniformly parabolic in a domain $\mathcal{D}$, where $a \leq 0$ and $f \geq 0$ (resp. ≤ 0) in $\mathcal{D}$. Let $\sup_{\mathcal{D}} u = M \geq 0$ (resp. $\inf_{\mathcal{D}} u = M \leq 0$), and suppose that $u(\bar{x}, \bar{t}) < M$ for some $(\bar{x}, \bar{t}) \in \mathcal{D}$. Then $u(x, t) < M$ at all points in $\mathcal{D}$ which can be connected to $(\bar{x}, \bar{t})$ by an arc in $\mathcal{D}$ consisting of a finite number of horizontal and upward vertical segments.*

Corollary 9.7. *If $\mathcal{D}$ is a cylinder in $\mathbf{R}^n \times \mathbf{R}$ (i.e., $\mathcal{D} = \Omega \times [t_1, t_2]$, with $-\infty < t_1 < t_2 \leq \infty$), and the hypotheses of Theorem 9.6 hold, then $u = M$ everywhere in $(t \geq \bar{t}) \cap \mathcal{D}$.*

The proof of Theorem 9.6 follows from the next proposition, just as in the elliptic case; see Lemma 8.4.

Proposition 9.8. *Under the hypotheses of Theorem 9.6, assume $Au \geq 0$ (resp. ≤ 0) in $\mathcal{D}$, $a \leq 0$ in $\mathcal{D}$ and $\sup_{\mathcal{D}} u = M$. If $u < M$ at a point in $\mathcal{D}$, then the conclusion of Theorem 9.6 holds.*

In order to prove Proposition 9.8, we need three lemmas. In these we always assume $Au \geq 0$ in $\mathcal{D}$; the other case has a similar proof.

Lemma 9.9. *Let K be a ball with $\bar{K} \subset \mathscr{D}$, and suppose $u < M$ in K, and $u(x_1, t_1) = M$, with $(x_1, t_1) \in \partial K$. Then t_1 is either the largest or smallest t-value in K; that is, (x_1, t_1) is either at the "top" or "bottom" of K.*

Proof. Let $(\bar{x}, \bar{t})$ be the center of K, and let r be the radius of K. We shall assume $p \equiv (x_1, t_1)$, with $x_1 \neq \bar{x}$, and arrive at a contradiction. We may assume that p is the only boundary point on K with $u = M$; otherwise we can replace K by a smaller ball whose boundary is interior to K except at p.

Let K_1 be a ball centered at p having radius $r_1 < |x_1 - \bar{x}|$, with $K_1 \subset \mathscr{D}$; see Figure 9.2. (Note that if $p = (\bar{x}, t_1)$, then we cannot find such an r_1.) We write $\partial K_1 = \Gamma_1 \cup \Gamma_2$, where $\Gamma_1 = \partial K_1 \cap \bar{K}$, and Γ_2 is the complement. Since Γ_1 is compact and $u < M$ on Γ_1, there is a $\delta > 0$ with $u \leq M - \delta$ on Γ_1. Note too that $u \leq M$ on Γ_2.

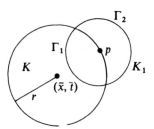

Figure 9.2

Now define the auxiliary function

$$h(x, t) = \exp\{-\alpha[|x - \bar{x}|^2 + (t - \bar{t})^2]\} - e^{-\alpha r^2},$$

where $\alpha > 0$ will be chosen below. Then $h > 0$ in K, $h = 0$ on ∂K, and $h < 0$ outside of $\bar{K}$. From (9.8), we have

$$Ah = \exp\{-\alpha[|x - \bar{x}|^2 + (t - \bar{t})^2]\} \left\{ 4\alpha^2 \sum_{i,j=1}^{n} a_{ij}(x_i - \bar{x}_i)(x_j - \bar{x}_j) \right.$$
$$\left. - 2\alpha \sum_{i=1}^{n} [a_{ii} + a_i(x_i - \bar{x}_i)] + (t - \bar{t}) \right\}$$
$$\geq \exp\{-\alpha[|x - \bar{x}|^2 + (t - \bar{t})^2]\}$$
$$\times \left\{ 4\alpha^2 \mu |x - \bar{x}|^2 - 2\alpha \sum_{i=1}^{n} [a_{ii} + a_i(x_i - \bar{x}_i)] + (t - \bar{t}) \right\}, \qquad (9.9)$$

and since $|x - \bar{x}| \geq |x_1 - \bar{x}| - r_1 > 0$ in $\bar{K}_1$, we can choose α so large as to make $Ah > 0$ in $\bar{K}_1$.

Let $v(x, t) = u(x, t) + \varepsilon h(x, t)$, where $\varepsilon > 0$ will be chosen below. Note that $Av > 0$ in K_1. Since $u \le M - \delta$ on Γ_1, we can choose ε so small that $v < M$ on Γ_1. Since $h < 0$ on Γ_2, and $u \le M$, $v < M$ on Γ_2. Thus $v < M$ on ∂K_1. Now on ∂K, $h = 0$ so $v(x_1, t_1) = M$. Thus v achieves its maximum on $\overline{K}_1$ at an interior point $\tilde{p}$. This implies that at $\tilde{p}$, each $v_{x_i} = 0$, $v_t = 0$, and $(D_i D_j v(\tilde{p}))$ is negative definite. Since $(a_{ij}(p))$ is positive definite, Lemma 8.5 implies that $(Av)(\tilde{p}) \le 0$ which contradicts $Av > 0$ in K_1. The proof is complete. $\square$

Lemma 9.10. *Let $\mathcal{D}$ be a domain in $x - t$ space and suppose $Au \ge 0$ in $\mathcal{D}$. Let $u \le M$ in $\mathcal{D}$ and $u(x_0, t_0) < M$ for some $(x_0, t_0) \in \mathcal{D}$. Let Γ be the component of $\{t = t_0\} \cap \mathcal{D}$ which contains (x_0, t_0). Then $u < M$ on Γ.*

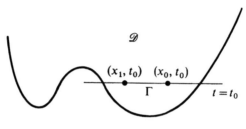

Figure 9.3

Note that this lemma gives part of the conclusion of Proposition 9.8; namely, it yields the result pertaining to horizontal segments.

Proof. Assume that $u(x_1, t_0) = M$ where $(x_1, t_0) \in \Gamma$. By moving (x_1, t_0) nearer to (x_0, t_0), if necessary, we can assume that

$$u(x, t_0) < M \quad \text{if } |x - x_0| < |x_1 - x_0|. \tag{9.10}$$

We may assume that (x_1, t_0) and (x_0, t_0) are connected by a line segment in $\mathcal{D}$; otherwise we can connect these points by a finite number of line segments in $\mathcal{D}$, and we can work with each such line segment to obtain a contradiction. Let L be the line segment from x_0 to x_1, and let $L(t_0) = \{(x, t_0) : x \in L\}$. Finally, define

$$\delta_0 = \min[|x_0 - x_1|, \text{dist}(L(t_0), \partial \mathcal{D})].$$

For x satisfying $0 < |x - x_0| < \delta_0$, define

$$d(x) = \text{dist}[(x, t_0), \mathcal{D} \cap \{(x, t) : u(x, t) = M\}].$$

Since $u(x_1, t_0) = M, d(x) \le |x - x_1|$.

By the previous lemma, the point in $\mathcal{D}$ nearest to (x, t_0), in which $u = M$, is of the form (x, t); hence $d(x) = |t_0 - t_1|$ so that either $u(x, t_0 + d(x)) = M$

or $u(x, t_0 - d(x)) = M$. If $\varepsilon > 0$ is a given small number, and $|\eta| = 1$,

$$d(x + \varepsilon\eta) \le \sqrt{\varepsilon^2 + d(x)^2} < d(x) + \frac{\varepsilon^2}{2d(x)} ; \qquad (9.11)$$

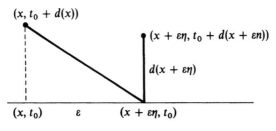

Figure 9.4

see Figure 9.4. Replacing x by $x + \varepsilon\eta$ and ε by $-\varepsilon$ gives

$$d(x) \le \sqrt{\varepsilon^2 + d(x + \varepsilon\eta)^2}$$

so that

$$d(x + \varepsilon\eta) \ge \sqrt{d(x)^2 - \varepsilon^2}. \qquad (9.12)$$

We shall now show that $d(x) \equiv 0$ in $|x - x_0| < \delta_0$. If this were so, then $u(x, t_0) \equiv M$ if $|x - x_1| < \delta_0 \le |x_0 - x_1|$, and this would contradict (9.10). Thus, assume that $d(x) > 0$, for some x, $|x - x_0| < \delta_0$ and let $0 < \varepsilon < d(x)$. If $|\eta| = 1$, we subdivide the line from (x, t_0) to $(x + \varepsilon\eta, t_0)$ into k equal parts. Then (9.11) and (9.12) give, for $0 \le i \le k - 1$,

$$d\left(x + \frac{i+1}{k}\varepsilon\eta\right) - d\left(x + \frac{i}{k}\varepsilon\eta\right) \le \frac{\varepsilon^2}{2k^2 d(x + (i/k)\varepsilon\eta)} \le \frac{\varepsilon^2}{2k^2 \sqrt{d(x)^2 - \varepsilon^2}}.$$

If we sum these from $i = 0$ to $i = k - 1$, we get

$$d(x + \varepsilon\eta) - d(x) \le \frac{\varepsilon^2}{2k\sqrt{d(x)^2 - \varepsilon^2}}.$$

Letting $k \to \infty$ shows $d(x + \varepsilon\eta) \le d(x)$ for each $\varepsilon > 0, |\eta| = 1$. Thus d is nonincreasing along the line from (x, t_0) to $(x + \varepsilon\eta, t_0)$. Since $d(x) \le |x - x_1|$ and $|x - x_1|$ can be made arbitrarily small for x near x_1, we see that $d(x) \equiv 0$ if $|x - x_1| < \delta_0$. This is a contradiction, and the proof of the lemma is complete. $\square$

Our last lemma pertains to upward vertical segments in $\mathscr{D}$.

Lemma 9.11. *Suppose that $Au \geq 0$ in $\mathcal{D}$ and that $u < M$ in $\mathcal{D} \cap \{t_0 < t < t_1\}$, for some $t_0 < t_1$. Then $u < M$ on $\mathcal{D} \cap \{t = t_1\}$.*

Proof. Suppose that $u(x_1, t_1) = M$ for some $(x_1, t_1) \in \mathcal{D}$. Let K be a ball of radius r centered at (x_1, t_1) contained in $t > t_0$. Define the auxiliary function

$$h(x, t) = \exp(-|x - x_1|^2 - \alpha|t - t_1|) - 1, \qquad \alpha > 0,$$

and compute

$$Ah \geq \exp(-|x - x_1|^2 - \alpha|t - t_1|)$$
$$\times \{4\mu|x - x_1|^2 - 2\sum[a_{ii} + a_i(x_i - (x_1)_i)] + \alpha\}.$$

Thus we can choose α so large as to make $Ah > 0$ in K for $t \leq t_1$.
    The paraboloid

$$|x - x_1|^2 + \alpha(t - t_1) = 0$$

is tangent to the hyperplane $t = t_1$ at (x_1, t_1) (cf. Figure 9.5). If Φ is the open region determined by this paraboloid, let $\Gamma_1 = \partial K \cap \Phi$, $\Gamma_2 = \partial \Phi \cap K$, and let D denote the open region determined by Γ_1 and Γ_2; cf. Figure 9.4 Since $u < M$ on the compact set Γ_1, we can find $\delta > 0$ such that $u \leq M - \delta$ on Γ_1.

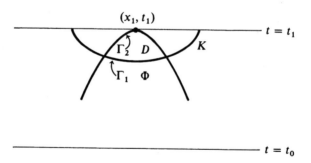

Figure 9.5

Let $v(x, t) = u(x, t) + \varepsilon h(x, t)$, where $\varepsilon > 0$ is chosen so small that

(i) $Av > 0$ in $\mathcal{D}$, and
(ii) $v \leq M$ on Γ_2.

Note that (ii) can be achieved since $h < 0$ on Γ_2. Since $v < M$ on Γ_1, we see that $v \leq M$ on ∂D. In view of (i), v cannot attain its maximum over $\bar{D}$ in D (cf. the last part of the proof of Lemma 9.9). Thus v attains its maximum on ∂D. Therefore M is the maximum of v in $\bar{D}$, and it is attained at $p = (x_1, t_1)$. It follows that $\partial v/\partial t \geq 0$ at p. But also $\partial h/\partial t = -\alpha < 0$ at p, so that $\partial u/\partial t > 0$ at p. Since the maximum of u on $t = t_1$ occurs at the interior point p, we have

$\nabla_x u(p) = 0$, and $(D_i D_j u(p))$ is negative definite. This implies, by Lemma 8.5, that $(Au)(p) < 0$, and gives the desired contradiction. □

We can now give the proof of Proposition 9.8. Let $p = (\bar{x}, \bar{t})$ and let q be any point which can be connected to p by an arc in $\mathscr{D}$ consisting of a finite number of horizontal and vertical upward segments. Thus there are points $Q_0 = p, Q_1, Q_2, \ldots, Q_k = q$ in $\mathscr{D}$ where Q_i is connected to Q_{i+1} by either a horizontal or upward vertical segment contained in $\mathscr{D}$. Lemmas 9.10 and 9.11 show that $u(q) < M$. This completes the proof. □

We shall now consider the behavior of the outward directional derivatives at those points on the boundary of $\mathscr{D}$ in which the maximum of u in $\overline{\mathscr{D}}$ is achieved. The results we obtain are analogous to the similar ones for elliptic equations; namely, these derivatives are nonzero.

We assume that A given in (9.7), is uniformly parabolic in the domain $\mathscr{D} \subset \mathbf{R}^n \times \mathbf{R}_+$, and that each coefficient a_{ij}, and a_i of A, as well as a, is bounded in $\mathscr{D}$.

Theorem 9.12. *Suppose that u is a solution of (9.6) in $\mathscr{D}$ and that $a \leq 0$ in $\mathscr{D}$. Suppose $f \geq 0$ in $\mathscr{D}$ and $\max_{\overline{\mathscr{D}}} u = M$ is attained at $p \in \partial\mathscr{D}$. Assume that $\partial\mathscr{D}$ is so regular at p that a ball S can be constructed through p with $\mathrm{int}(S) \subset \mathscr{D}$ and $u < M$ on $\mathrm{int}(S)$. Suppose too that the radial direction from the center of S to p is not parallel to the t-axis. Then $du(p)/dv > 0$ for every outward direction v. (A similar statement holds in the case $f \leq 0$ in $\mathscr{D}$, where $M = \min_{\overline{\mathscr{D}}} u$ and we conclude $du(p)/dv < 0$.)*

The condition that the radius vector to p is not parallel to the t-axis is necessary, as Theorem 9.1 shows.

The proof of the theorem follows easily from the next proposition.

Proposition 9.13. *Suppose that $Au \geq 0$, $M = \max u$ on $\overline{\mathscr{D}}$ is attained at $p \in \partial\mathscr{D}$, and that a ball S can be constructed through p which satisfies the conditions of Theorem 9.12. Then $du(p)/dv > 0$ for every outward direction v.*

Proof. We construct a ball S of radius r centered at (x_1, t_1) which is tangent to $\partial\mathscr{D}$ at $p \equiv (x_0, t_0)$ and a ball S_1 centered at p having radius $\rho < |x_1 - x_0|$; see Figure 9.6. Let $\Gamma_1 = \partial S_1 \cap \bar{S}, \Gamma_2 = \partial S \cap S_1$.

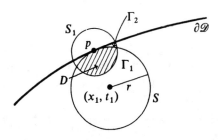

Figure 9.6

and let D be the region enclosed by Γ_1 and Γ_2. Since $u < M$ on Γ_1, there is a $\delta > 0$ such that (i) $u \le M - \delta$ on Γ_1. Also (ii) $u < M$ on $\Gamma_2 \backslash \{p\}$, and (iii) $u(p) = M$.

Let $h(x, t)$ be the auxiliary function defined by

$$h(x, t) = \exp\{-\alpha[|x - x_1|^2 + (t - t_1)^2]\} - e^{-\alpha r^2}, \quad \alpha > 0.$$

Clearly $h = 0$ on ∂S, and choosing α sufficiently large, we can achieve $Ah > 0$ on $\bar{D}$. Let $v = u + \varepsilon h, \varepsilon > 0$. Then $Av = Au + \varepsilon Ah > 0$ in D. In view of (i), we can choose ε so small that $v < M$ on Γ_1. From (ii), together with the fact that $h = 0$ on ∂S, we have $v < M$ on $\Gamma_2 \backslash \{p\}$, and $v(p) = M$. Thus we can apply Theorem 9.6 to conclude that the maximum of v in $\bar{D}$ is taken on only at p. Hence, at p, $dv/dv = du/dv + \varepsilon dh/dv \ge 0$. But also, at p, $dh/dv = -2v \cdot n\alpha r e^{-\alpha r^2}$, where n is the outward normal at p. Hence $dh(p)/dv < 0$ so that $du(p)/dv > 0$. The proof is complete. $\square$

The following (method of proof in the) corollary is important in applying the results of this section to nonlinear equations.

Corollary 9.14. *Theorems 9.6 and 9.12 hold if $M = 0$ without the restriction $a \le 0$.*

Proof. Let $w(x, t) = u(x, t)e^{-kt}$; then

$$0 \le (A + a)u = e^{kt}(A + a - k)w.$$

Since a is bounded from above, we can choose k so large that $a - k \le 0$. Thus, the hypotheses of the theorems are valid for w. It follows that the conclusions hold for w, and since $M = 0$, they also hold for u. $\square$

We end this chapter by noting that an existence theorem for the problem

$$Pu = f \quad \text{in } \mathscr{D}, \qquad u(x, 0) = \phi(x) \quad \text{in } \Omega, \qquad u(x, t) = h(x, t) \quad \text{on } \partial\Omega \times (0, T)$$

can be given along the same general lines as for the elliptic problem discussed at the end of the last chapter; see [Fn 3].

NOTES

The results on the heat equation are classical; see [Fn 3]. The example following Corollary 9.4 is due to Tychanov [Tv]. The results in §B are due to Nirenberg [Ni 1], and Friedman [Fn 1, 2]; a nice reference book is [Fn 3]; see also the Protter–Weinberger book, [PW].

Part II

Reaction–Diffusion Equations

Chapter 10

Comparison Theorems and Monotonicity Methods

In this chapter we begin to study nonlinear partial differential equations. The results which we obtain here all follow from the maximum principles which were obtained in Chapters 8 and 9. We shall show how they apply to nonlinear elliptic and parabolic partial differential equations. As a first application, we will use the strong maximum principles to prove comparison theorems; i.e., pointwise inequalities between different solutions. These say, roughly, that if u and v are two solutions, and if $u \leq v$ on $\partial \mathcal{D}$, it follows that $u \leq v$ on $\mathcal{D}$. Such theorems can be quite useful in obtaining qualitative information about solutions. For example, comparison theorems are often used to obtain information about the asymptotic behavior of solutions of parabolic equations as $t \to +\infty$. As a second application of the maximum principle, we shall show how it can be used to prove existence theorems. This is the method of "upper" and "lower" solutions, the solution being the limit of a monotone iteration scheme, where the monotonicity is a consequence of the maximum principle.

§A. Comparison Theorems for Nonlinear Equations

We consider first nonlinear parabolic equations

$$Pu = f(x, t, u), \qquad (x, t) \in \mathcal{D}, \tag{10.1}$$

where $\mathcal{D} = \Omega \times (0, T)$ is a bounded domain in $\mathbf{R}^n \times \mathbf{R}_+$, with $\partial \Omega$ smooth. Here

$$-Pu = u_t - Au \equiv u_t - \sum_{i,j=1}^{n} (a_{ij}(x, t)u_{x_i})_{x_j}, \tag{10.2}$$

where (a_{ij}) is a symmetric matrix, and each a_{ij} is bounded in $\mathcal{D}$. Furthermore, we assume that P is uniformly parabolic in $\mathcal{D}$, in the sense of Chapter 9, §B. Finally, we assume that f is C^1 in u, and Hölder continuous in x and t.

Let u and v each be C^2 functions of x in Ω, C^1 functions of t on $[0, T]$, and consider the following three conditions:

$$Pu - f(x, t, u) \leq Pv - f(x, t, v), \qquad (x, t) \in \mathcal{D}, \tag{10.3}$$

$$u(x, 0) \geq v(x, 0), \qquad x \in \Omega, \tag{10.4}$$

$$\frac{du}{dv} + \beta u \geq \frac{dv}{dv} + \beta v, \qquad (x, t) \in \partial\Omega \times (0, T), \tag{10.5}$$

where $\beta = \beta(x, t) \geq 0$ on $\partial\Omega \times (0, T)$, and v is any outward direction. Assuming these conditions, we have the following "comparison" theorem.

Theorem 10.1. *Under the above conditions on P and f, if (10.3)–(10.5) hold, then $u(x, t) \geq v(x, t)$ for all $(x, t) \in \bar{\mathcal{D}}$. Moreover, if an addition $u(x, 0) > v(x, 0)$ for each x in an open subset $\Omega_1 \subset \Omega$, then $u(x, t) > v(x, t)$ in $\bar{\Omega}_1 \times [0, T]$.*

Proof. We let $w(x, t) = u(x, t) - v(x, t)$, and note that $w(x, 0) \geq 0$ if $x \in \Omega$, and $dw/dv + \beta w \geq 0$ in $\partial\Omega \times (0, T)$. Furthermore, from (10.3), we have

$$Pw - f_u(x, t, \xi)w \leq 0, \qquad (x, t) \in \mathcal{D},$$

where $\xi = \theta u + (1 - \theta)v, 0 < \theta < 1$. Assume that $w < 0$ at some point in $\bar{\mathcal{D}}$. Then $e^{-kt}w < 0$ at this point, where k is chosen so large that $-f_u - k < 0$ in $\bar{\mathcal{D}}$. Let $z(x, t) = e^{-kt}w(x, t)$; then $Aw = e^{kt}Az$, $w_t = ke^{kt}z + e^{kt}z_t$, so in $\mathcal{D}$,

$$0 \geq P_w - f_u(x, t, \xi)w = e^{kt}[Pz - kz - f_u z].$$

Thus $Pz + (-k - f_u)z \leq 0$ in $\mathcal{D}$, $z(x, 0) \geq 0$ in Ω, and $(d/dv)z + \beta z \geq 0$ in $\partial\Omega \times (0, T)$. Since z is negative at some point in $\bar{\mathcal{D}}$, $0 > \min_{\bar{\mathcal{D}}} z(x, t) \equiv m$; hence there is a point $(\bar{x}, \bar{t}) \in \bar{\mathcal{D}}$ with $z(\bar{x}, \bar{t}) = m$. We consider three cases: First, if $(\bar{x}, \bar{t}) \in \mathcal{D}^{\text{int}}$ (or $\bar{t} = T$, and $\bar{x} \in \Omega$), then since $-f_u - k < 0$ in $\bar{\mathcal{D}}$ and $P_z + (-f_u - k)z \leq 0$ in $\mathcal{D}$, we may apply Lemma 9.11 to conclude $z(\bar{x}, T) = m < 0$. Now considering $z(x, T)$ as a function of x, this function has a minimum at $\bar{x}$, $z_{x_i}(\bar{x}, T) = 0$, $1 \leq i \leq n$, and the hessian matrix $(z_{x_i x_j}(\bar{x}, T))$ is positive definite. It follows that $Az(\bar{x}, T) > 0$ since A is strongly parabolic. Also $z_t(\bar{x}, T) \leq 0$ since $z(\bar{x}, t) \geq z(\bar{x}, T)$ for all $t < T$. We thus obtain the contradiction $Pz(\bar{x}, T) - (k + f_u)z(\bar{x}, T) > 0$. Finally, if $(\bar{x}, \bar{t}) = p \in \partial\Omega \times (0, T)$, then from Theorem 9.12, $\partial z(p)/\partial v < 0$. If $v = (v_1, \ldots, v_{n-1}, 0)$ is a direction perpendicular to the t-axis, then $\partial z(p)/\partial v = e^{-kt} \partial w(p)/\partial v$, and since $\beta z(p) \leq 0$, we get the contradiction

$$0 \leq e^{-kt}[\partial w(p)/\partial v + \beta w(p)] = \partial z(p)/\partial v + \beta z(p) < 0.$$

Thus $w \geq 0$ in $\bar{\mathcal{D}}$ so $u \geq v$ in $\mathcal{D}$. The proof of the second statement is similar. $\square$

In a similar way, we can consider the uniformly elliptic equation

$$Au = f(x, u), \qquad x \in \Omega, \tag{10.6}$$

where Ω and A are defined as above (a_{ij} and a_i being functions of x), and f is a C^1 function in x, and in u. Assume $f(x, u) = \alpha(x)u + g(x, u)$, where $g_u(x, 0) = 0$ for all $x \in \Omega$, and $\alpha(x) \geq 0$ in Ω.

Theorem 10.2. *Suppose that u and v are solutions of* (10.6) *and that on $\partial\Omega$,*

$$\frac{\partial u}{\partial v} + \beta u \geq \frac{\partial v}{\partial v} + \beta v,$$

where v is any outward direction, and $\beta = \beta(x)$. Then $u \geq v$ in $\bar{\Omega}$, if $g(x, u) \leq g(x, v)$ for all $x \in \Omega$.

Proof. Let $w = u - v$; then $\partial w/\partial v + \beta w \geq 0$ in $\partial\Omega$, and in Ω

$$Au - \alpha u = g(x, u) \leq g(x, v) = Av - \alpha v;$$

thus $Aw - \alpha w \leq 0$ in Ω. Now since $\alpha \geq 0$ in Ω, Theorems 8.1 and 8.6 together imply that $w \geq 0$ in Ω. This completes the proof.   □

As a simple application of these theorems, consider the parabolic differential equation

$$u_t = \Delta u - u^3, \qquad x \in \Omega \subset \mathbf{R}^n, \qquad t > 0, \tag{10.7}$$

where Δ is the n-dimensional Laplace operator on Ω. Suppose too that u satisfies the initial conditions

$$u(x, 0) = u_0(x), \qquad x \in \Omega, \tag{10.8}$$

and the boundary conditions

$$\frac{du}{dn} = 0 \quad \text{on } \partial\Omega \times (0, \infty), \tag{10.9}$$

where n is the outward normal derivative.
   Consider now the ordinary differential equations

$$\frac{dv}{dt} = -v^3, \quad v(0) = M; \qquad \frac{dw}{dt} = -w^3, \quad w(0) = m,$$

where $m \leq u_0(x) \leq M, x \in \Omega$.

Since u, v, and w satisfy (10.7) and (10.9), we have, from Theorem 10.1

$$w(t) \leq u(x, t) \leq v(t), \qquad t > 0, \tag{10.10}$$

uniformly in $x \in \Omega$. But it is trivial to see that all solutions of $dz/dt = -z^3$ tend to zero, as $t \to \infty$. Thus, from (10.10) we conclude that $u(x, t) \to 0$ as $t \to +\infty$, uniformly in $x \in \Omega$. This shows that all solutions of (10.7)–(10.9), having bounded initial values, tend to zero, uniformly in x, as $t \to +\infty$. Of course, we have not yet shown that (10.7)–(10.9) has a solution defined for all $t > 0$; this will follow from the results in Chapter 14.

§B. Upper and Lower Solutions

In this section we shall show how to solve nonlinear elliptic equations by iteration schemes. The convergence of these schemes is a consequence of the strong maximum principle. We shall also consider the problem of stability of our constructed solution, the solution now being considered as a "steady-state" solution of the associated time-dependent parabolic problem.

We begin by considering the nonlinear elliptic equation

$$Au + f(x, u) = 0, \qquad x \in \Omega, \tag{10.11}$$

together with Dirichlet boundary conditions

$$u = h \quad \text{on } \partial\Omega. \tag{10.12}$$

Here Ω is a bounded domain in $\mathbf{R}^n$ having smooth boundary, f is C^1 in u and C^α in x, $0 < \alpha < 1$, A is a uniformly elliptic operator defined by (10.2), and h is a function in class $C^{2+\alpha}$.

A C^2-function u_0 is called an *upper solution* of (10.11), (10.12) if u_0 satisfies the following two conditions:

$$Au_0 + f(x, u_0) \leq 0 \quad \text{in } \Omega \tag{10.13}$$

and

$$u_0 \geq h \quad \text{on } \partial\Omega. \tag{10.14}$$

We also assume that u_0 is *not* an actual solution. Similarly, we define a *lower solution* by reversing the inequalities in (10.13) and (10.14).

Theorem 10.3. *Suppose u_0 and v_0 are, respectively, upper and lower solutions of (10.11), (10.12), with $u_0 \geq v_0$ on Ω. Then there exists a solution u of (10.11), (10.12) such that $u_0 \geq u \geq v_0$, pointwise on Ω.*

We call attention to the fact that our proof will actually give a constructive, and computationally feasible method for obtaining the solution.

Proof. Let m and M denote, respectively, the minimum of v_0 and maximum of u_0 on $\bar{\Omega}$. Let $\kappa > 0$ be a constant which we choose so large that

$$\frac{\partial f}{\partial u} + \kappa > 0 \quad \text{if } (x, u) \in \bar{\Omega} \times [m, M]. \tag{10.15}$$

We define a mapping T from the space $C^2(\Omega)$ into itself, by $\beta = T\alpha$, where β is the solution of

$$\begin{cases} A\beta - \kappa\beta = -[f(x, \alpha) + \kappa\alpha], & x \in \Omega, \\ \beta = h \quad \text{on } \partial\Omega. \end{cases} \tag{10.16}$$

We note that the differential equation in (10.16) is a linear equation for β, and thus the solution exists from Theorem 8.10. We shall show that T is monotone in the sense that if $\alpha_1 \le \alpha_2$, then $T\alpha_1 \le T\alpha_2$, provided that $m \le \alpha_1, \alpha_2 \le M$. To see this, let $\beta_i = T\alpha_i$, $i = 1, 2$; then

$$(A - \kappa)\beta_i = -[f(x, \alpha_i) + \kappa\alpha_i] \quad \text{on } \Omega.$$

Thus

$$(A - \kappa)(\beta_2 - \beta_1) = -[f(x, \alpha_2) - f(x, \alpha_1) + \kappa(\alpha_2 - \alpha_1)], \quad x \in \Omega, \tag{10.17}$$

and

$$\beta_2 - \beta_1 = 0 \quad \text{on } \partial\Omega. \tag{10.18}$$

If now $\alpha_2 \ge \alpha_1$, then by choice of κ,

$$(A - \kappa)(\beta_2 - \beta_1) \le 0 \quad \text{in } \Omega,$$

so that (using (10.18)), the strong maximum principle (Theorem 8.1) implies that $\beta_2 > \beta_1$ in Ω, unless $\beta_1 = \beta_2$. But if $\beta_1 \equiv \beta_2$, then the right-hand side of (10.17) is identically zero, and $\alpha_1 \equiv \alpha_2$ in view of our choice of κ. Thus T is monotone in a somewhat stronger sense; namely, $\alpha_2 > \alpha_1$ implies that $T\alpha_2 > T\alpha_1$.

Next, we shall show that $\alpha > T\alpha$ provided that α is an upper solution. Thus, if $\beta = T\alpha$, then

$$(A - \kappa)(\beta - \alpha) = -[A\alpha + f(x, \alpha)] \ge 0 \quad \text{in } \Omega,$$

and on $\partial\Omega$, $(\beta - \alpha) = h - \alpha \le 0$. It follows again from the strong maximum principle for elliptic operators (Theorem 8.1), that $\beta < \alpha$, unless $\beta \equiv \alpha$. But this latter possibility is excluded since α is *not* a solution of (10.11), (10.12).

These remarks allow us to define inductively two sequences $\{u_n\}$ and $\{v_n\}$, by setting $(u_1, v_1) = (Tu_0, Tv_0)$, and defining for $n > 1$, $(u_n, v_n) = (Tu_{n-1}, Tv_{n-1})$. Since u_0 is an upper solution, $u_1 = Tu_0 < u_0$, and by the monotonicity of T, $Tu_1 < Tu_0 = u_1$. Thus $u_{n-1} > u_n$ for each n and similarly $v_n > v_{n-1}$. Again, since $u_0 \geq v_0$, it follows by induction, that $u_n \geq v_n$ for all n. We conclude therefore, that $\{u_n\}$ is a pointwise decreasing sequence, bounded from below by v_0. Thus the pointwise limit

$$\bar{u} = \lim_n u_n$$

exists at each point in Ω and $u_0 \geq \bar{u} \geq v_0$ in Ω.

Now let's take a closer look at the operator T. T is actually a composition of two operators, $T_2^{-1} \circ T_1$; namely, the nonlinear operator $T_1 : \alpha \to -f(x, \alpha) - \kappa\alpha$, followed by the inverse of the operator $T_2 : \beta \to \sigma$, where $(A - \kappa)\beta = \sigma$ in Ω, and $\beta = h$ on $\partial\Omega$. It is easy to see that T_1 maps the set of bounded pointwise convergent sequences into itself. Thus $\{T_1 u_n\}$ converges pointwise in Ω, and therefore it converges in $L^p(\Omega)$ for each $p \geq 1$. The elliptic regularity theorem (Chapter 8, §D) then implies that $\{Tu_n\} = \{T_2^{-1} \circ T_1 u_n\}$ converges in[1] $W_p^2(\Omega)$, and thus in $W_p^1(\Omega)$, $p > 1$. Since $W_p^1(\Omega)$ is embedded continuously in $C^\alpha(\bar{\Omega})$ if $\alpha = 1 - n/p$, $(p > n)$,[2] this sequence converges in $C^\alpha(\bar{\Omega})$. Then the Schauder elliptic estimates, and regularity theorems (Chapter 8, §D) imply that $\{Tu_n\} = \{u_n\}$ converges in $C^{2+\alpha}(\bar{\Omega})$. Since T takes $C^\alpha(\bar{\Omega})$ continuously into $C^{2+\alpha}(\bar{\Omega})$, we have

$$\bar{u} = \lim_n u_n = \lim_n Tu_{n-1} = T \lim_n u_n = T\bar{u}.$$

Thus $\bar{u}$ is a solution of (10.11), (10.12), and the proof is complete. □

We remark that either v_0 or u_0 could be an actual solution, and the proof would still go through. All that we really require is that not *both* u_0 and v_0 are solutions.

Note that we could obtain another solution $\bar{v}$ by defining $\bar{v} = \lim_n v_n$; then $\bar{v} \leq \bar{u}$ in Ω. However, we have no guarantee that $\bar{v}$ is different from $\bar{u}$. In any case, concerning these solutions, we have the following corollary.

Corollary 10.4. $\bar{u}$ *and* $\bar{v}$ *are, respectively, maximal and minimal solutions, in the sense that if u is any solution of (10.11), (10.12), with $v_0 \leq u \leq u_0$ in Ω, then $\bar{v} \leq u \leq \bar{u}$ in Ω.*

---

[1] $W_p^2(\Omega)$ is defined in Example 5 in Chapter 7.

[2] Strictly speaking this is not quite correct. $\mathring{W}_p^1(\Omega)$ is embedded continuously in $C^\alpha(\bar{\Omega})$ if $\alpha = 1 - n/p$, where $\mathring{W}_p^1(\Omega)$ is the closure of $C_0^1(\bar{\Omega})$ in the $W_p^1(\Omega)$ norm. But if we make the (reasonable) additional assumption that h has a smooth (C^2) extension to $\bar{\Omega}$, then we may assume $h = 0$, since we put $u = v + h$, where $Av + f(x, v) + Ah = 0$ in Ω, $v = 0$ on $\partial\Omega$. Then the argument is correct for f, since T_{u_n} now converges $\mathring{W}_p^1(\Omega)$; see [GT].

Proof. $u \le u_0$ implies $u = Tu < Tu_0 = u_1$, so by induction, $u < u_n$ for every n. Thus $u \le \bar{u}$; similarly $\bar{v} \le u$. $\square$

We remark that Theorem 10.3 has an analogue for parabolic problems of the form

$$\left(A - \frac{\partial}{\partial t}\right)u = f(x, t, u) \quad \text{in } \Omega \times (0, T), \qquad u(x, 0) = \phi(x), \qquad x \in \Omega,$$

$$u = h(x, t) \quad \text{on } \partial\Omega \times (0, T).$$

Namely, one defines the notions of upper and lower solutions in a straightforward analogous way, and one uses the strong maximal principle for parabolic equations (Theorems 9.6 and 9.12). We omit the details since in Chapter 14 we shall prove general existence theorems for parabolic systems. Next, we want to point out that these theorems also hold for more general boundary conditions; for example, (10.12) can be replaced by the condition

$$\frac{\partial u}{\partial v} + \beta(x)u = h(x) \quad \text{on } \partial\Omega,$$

where $\beta \ge 0$. For the details, see [Sa 2].

§C. Applications

In this section, we shall give a few applications of Theorem 10.3. We begin with the following example. Consider the equation in the bounded open set $\Omega \subset \mathbf{R}^2$,

$$\Delta u + au(1 - u) = 0, \qquad x \in \Omega, \qquad a > 0, \tag{10.19}$$

with homogeneous Dirichlet boundary conditions

$$u = 0 \quad \text{on } \partial\Omega. \tag{10.20}$$

Such equations arise in population dynamics, where the nonlinear function $au(1 - u)$ is referred to as a "logistic" nonlinearity; cf. Chapter 14. We are interested in constructing nonconstant positive solutions. We take for Ω the square $[-L, L] \times [-L, L]$. Let φ be the principal eigenfunction of $-\Delta$ on Ω with homogeneous Dirichlet boundary conditions (so that $-\Delta\phi = \lambda\phi$, $\lambda > 0$, $\phi > 0$ in Ω, $\phi = 0$ on $\partial\Omega$; see §A in Chapter 11). If[3] $a > \lambda$, and $\omega = \delta\phi$, $\delta > 0$, then for small δ,

$$\Delta\omega + a\omega(1 - \omega) = \delta\phi(-\lambda + a - a\delta\phi) \ge 0 \quad \text{in } \Omega,$$

---

[3] This will always be the case for large enough L; see §A, Chapter 11.

so that ω is a lower solution. If $z(x)$ is a nonconstant positive solution of the *ordinary* differential equation $z'' + az(1 - z) = 0$ on $-L < x < L$, with $z(\pm L) = 0$,[4] then z is an upper solution since $z \geq 0$ on $\partial\Omega$. It follows from our theorem that there is a solution u of (10.19), (10.20) with $\delta\phi(x, y) \leq u(x, y) \leq z(x)$ on Ω. Note that $u \not\equiv z$ since z does not satisfy (10.20).

As a second application of the theorem, we shall investigate the stability of the solutions which we have obtained from Theorem 10.3. Here these solutions will be considered as *steady-state* (i.e., equilibrium or time-independent) solutions of the associated parabolic equation

$$u_t = Au + f(x, u), \qquad (x, t) \in \Omega \times \mathbf{R}_+ \tag{10.21}$$

with Dirichlet boundary conditions

$$u = h \quad \text{on } \partial\Omega \times \mathbf{R}_+. \tag{10.22}$$

It is remarkable that the solutions obtained from the monotone iteration scheme in Theorem 10.3 are *always* stable. Before we prove this fact, we must define precisely our notion of stability.

We consider (10.21), (10.22) together with the initial condition

$$u(x, 0) = \phi(x), \qquad x \in \Omega. \tag{10.23}$$

We say that a solution $v = v(x)$ of (10.11), (10.12) is a *nunstable*[5] solution of (10.21), (10.22), if for every $\varepsilon > 0$, there is a $\delta > 0$ such that if $\| \phi - v \|_{L_\infty(\Omega)} < \delta$, then $\| u(\cdot, t) - v(\cdot) \|_{L_\infty(\Omega)} < \varepsilon$ for all $t > 0$, where $u(x, t)$ is the solution of (10.21)–(10.23). If it is true that $\lim_{t\to\infty} \| u(\cdot, t) - v(\cdot) \|_{L_\infty(\Omega)} = 0$, then we say that v is a *stable* solution of (10.21)–(10.23). Thus, our stability notions are taken in the sup-norm topology, (but other topologies are also useful (cf. [He])). We can now prove that our previously constructed solutions are stable.

Theorem 10.5. *Let v be a solution of (10.11), (10.12), with $v_0 \leq v \leq u_0$ in Ω, where u_0 and v_0 are upper and lower solutions of (10.11), (10.12), respectively.*

   (i)   *If*

$$v_0 \leq \phi \leq u_0 \quad \text{in } \Omega, \tag{10.24}$$

   the corresponding solution u of (10.21)–(10.23) satisfies $V_0(x, t) \leq u(x, t) \leq U_0(x, t)$, in $\Omega \times (0, T)$, where V_0 and U_0 are the solutions of (10.21)–(10.23) corresponding to $\phi = v_0$ and $\phi = u_0$, respectively.

   (ii)   *If $T^t v_0 \nearrow v$ and $T^t u_0 \searrow v$, then v is a stable solution of (10.21)–(10.23).*

---

[4] Again such a solution will exist if L is sufficiently large.
[5] nunstable = *not unstable*.

Before giving the proof, a few remarks are in order. First, when we write $T^n u_0 \searrow v$, we mean that v is the limit of the monotonically decreasing sequence of functions $\{T^n u_0\}$ $(\equiv \{u_n\} \equiv \{T_{u_n}\})$, which was constructed from the upper solution u_0, where T is the operator defined in the proof of Theorem 10.3 (cf. (10.16)). Similar meaning is given to $T^n v_0 \nearrow v$. Next, we call attention to the fact that (ii) actually gives information on the "domain of attraction" of the "rest point" (= rest state) v; namely, if $v_0(x) \le u(x, 0) \le u_0(x)$ for all $x \in \Omega$, then $u(x, t)$ tends to $v(x)$ as $t \to +\infty$, uniformly in $x \in \Omega$.

The proof will follow from two lemmas, which themselves are of independent interest.

Lemma 10.6. *Let $u_0(x)$ be an upper solution in Ω of the problem $Au + f(x, u)$ $= 0$, $u = 0$ on $\partial\Omega$. Let $u(x, t)$ be the*[6] *solution in $\mathcal{D} = \Omega \times (0, T)$ of the problem $u_t = Au + f(x, u), u(x, 0) = u_0(x), x \in \Omega$, where $u = 0$ on $\partial\Omega \times (0, T)$. Then $\partial u / \partial t \le 0$ in $\mathcal{D}$.*

Proof. In a manner analogous to the proof of Theorem 10.3, we define a sequence of functions $\{u_n\}$ in $\mathcal{D}$ by $u_0(x, t) = u_0(x)$, and for $n \ge 1$,

$$A u_n - \kappa u_n - \frac{\partial u_n}{\partial t} = -[f(x, u_{n-1}) + \kappa u_{n-1}],$$

$$(10.25)$$

$$u_n = 0 \quad \text{on } \partial\Omega \times (0, T), u_n(x, 0) = u_0(x).$$

Then as in the elliptic case, if $(x, t) \in \mathcal{D}$,

$$u_0(x) \ge u_1(x, t) \ge \cdots \ge u_{n-1}(x, t) \ge u_n(x, t) \ge \cdots, \qquad (10.26)$$

and $u_n \searrow v$, where $v_t = Av + f(x, v)$ in $\mathcal{D}$, $v(x, 0) = u_0(x)$ in Ω, and $v = 0$ on $\partial\Omega \times (0, T)$. By uniqueness,[6] we have $v = u$ in $\bar{\mathcal{D}}$.

Now for each n, we have by differentiating (10.25) with respect to t,

$$A\left(\frac{\partial u_n}{\partial t}\right) - \kappa\left(\frac{\partial u_n}{\partial t}\right) - \frac{\partial}{\partial t}\left(\frac{\partial u_n}{\partial t}\right) = Q(x, t) \qquad \frac{\partial u_n}{\partial t} = 0 \quad \text{on } \partial\Omega \times (0, T),$$

where Q is a bounded function on $\mathcal{D}$. Also, for each n, if $h > 0$ and $x \in \Omega$,

$$\frac{u_n(x, h) - u_n(x, 0)}{h} = \frac{u_n(x, h) - u_0(x)}{h} \le 0,$$

from (10.26). Hence $\partial u_n(x, 0) / \partial t \le 0$, $x \in \Omega$. The strong maximum theorem for parabolic equations (Theorem 9.6) implies that $\partial u_n / \partial t \le 0$ in $\mathcal{D}$. But as

---

[6] The uniqueness of the solution follows from Theorem 10.1.

in the proof of Theorem 10.3, it can be shown that u_n tends to u in $C^{1+\alpha}$ in the t-variable on $\mathscr{D}$. Hence $\partial u/\partial t \leq 0$ in $\mathscr{D}$. This proves the lemma. $\square$

We shall apply Lemma 10.6 to obtain the following result.

Lemma 10.7. *Let u_0 and v_0 be upper and lower solutions, respectively, of (10.11), (10.12), (with $h = 0$), $u_0 \geq v_0$ in Ω. Let $u(x, t)$ and $v(x, t)$ be the corresponding solutions of (10.21), (10.22) (for $h = 0$) with initial data u_0 and v_0, respectively. Then the pointwise limit, $\lim_{t \to \infty}(u(x, t), v(x, t)) = (\bar{u}(x), \bar{v}(x))$, exists for each $x \in \Omega$, and both $\bar{u}$ and $\bar{v}$ are solutions of (10.11), (10.12) (for $h = 0$).*

Proof. As described in the proof of the last lemma, $u_0 \geq u$, $v_0 \geq v$ and from Corollary 9.14, $u \geq v$ in $\mathscr{D} = \Omega \times (0, T)$. Using Lemma 10.6, $\partial u/\partial t \leq 0$, $\partial v/\partial t \geq 0$ in $\mathscr{D}$, so that

$$\lim_{t \to \infty}(u(x, t), v(x, t)) = (\bar{u}(x), \bar{v}(x)), \qquad x \in \Omega,$$

exists. Since $u_0 \geq v_0$, Theorem 10.1 gives $u \geq v$ in $\mathscr{D}$, and hence $\bar{u} \geq \bar{v}$ in Ω. It remains to show that $\bar{u}$ and $\bar{v}$ are solutions of (10.11), (10.12), with $h = 0$.

Now we know that u satisfies (10.21), $u = 0$ on $\partial\Omega \times \mathbf{R}_+$, and $u(x, 0) = u_0(x)$, $x \in \Omega$. If we multiply (10.21) by $\phi \in C_0^\infty(\Omega)$ and integrate over Ω, we get

$$\int_\Omega u_t \phi = \int_\Omega (Au)\phi + f\phi.$$

Since ϕ has compact support in Ω, we can integrate by parts to obtain

$$\int_\Omega u_t \phi = \int_\Omega u(A^*\phi) + f\phi.$$

Multiplying this by $1/T$, and integrating from 0 to T gives

$$\int_\Omega \frac{u(x, T) - u(x, 0)}{T} \phi \, dx = \int_\Omega \left\{ A^*\phi \frac{1}{T} \int_0^T u(x, t) \, dt + \frac{\phi}{T} \int_0^T f(x, u) \, dt \right\} dx.$$

$$(10.27)$$

Observe now that since we have the two estimates

$$|u(x, T)| \leq |u(x, T) - \bar{u}(x)| + |\bar{u}(x)|,$$

and

$$|f(x, u(x, t)) - f(x, \bar{u}(x))| \leq |f_u(x, \xi)| |u(x, t) - \bar{u}(x)|$$
$$\leq \bar{\kappa} |u(x, t) - \bar{u}(x)|,$$

where $\bar{\kappa}$ doesn't depend on t, we may apply the Lebesgue dominated convergence theorem to (10.27) as $T \to \infty$ to obtain

$$\int_\Omega [(A^*\phi)\bar{u} + f(x, \bar{u})\phi] \, dx = 0. \qquad (10.28)$$

Thus $\bar{u}$ is a *weak* solution of (10.11), (10.12) (with $h = 0$).

We shall show now that $\bar{u}$ is a *classical* solution of this boundary-value problem. The argument is similar to that of the proof of the regularity of the constructed solution, in Theorem 10.3. It goes as follows. First we observe that both A and A^* are invertible; this is just the strong maximum principle; namely, if $Au = 0$ in Ω and $u = 0$ on $\partial\Omega$, then u could not achieve a nonzero value in the interior of Ω. Now in the usual notation for the $L_2(\Omega)$ inner product, we have from (10.28),

$$(A^{-1}f, A^*\phi) = (f, (A^*)^{-1}A^*\phi) = (f, \phi) = -(\bar{u}, A^*\phi),$$

where f is evaluated on $\bar{u}$. Thus, since

$$(A^{-1}f + \bar{u}, A^*\phi) = 0$$

for all $\phi \in C_0^\infty(\Omega)$, and A^* is invertible, it follows that

$$A^{-1}f(x, \bar{u}(x)) + \bar{u}(x) = 0, \qquad x \in \Omega. \qquad (10.29)$$

Now the argument to show that $\bar{u}$ is regular proceeds just as before; namely $\bar{u} \in L_p(\Omega)$ for all $p \geq 1$ so $\bar{u} \in W_p^2(\Omega)$ by the elliptic regularity theorem [ADN]. If $p > n$, the embedding lemma [GT] implies $\bar{u} \in C^{1+\alpha}(\bar{\mathscr{D}})$, and the usual Schauder estimates show $\bar{u} \in C^{2+\alpha}(\bar{\Omega})$. Thus we may operate on both sides of (10.29) by A to conclude that $A\bar{u} + f(x, \bar{u}) = 0$ in Ω. Since $\bar{u} = 0$ on $\partial\Omega$, $\bar{u}$ is a classical solution. The same proof shows that $\bar{v}$ too is a classical solution. This completes the proof of Lemma 10.7. $\square$

We can now give the proof of Theorem 10.5. First of all, we can assume that for the problem (10.11), (10.12), $h = 0$; i.e., we have homogeneous boundary conditions. This is no loss in generality since we can always replace (10.11), (10.12) by the problem $Au + f(x, u + v) = 0$ in Ω, with $u + v = 0$ on $\partial\Omega$, where v is a solution of $Av = 0$ in Ω, $v = h$ on $\partial\Omega$. Then if u_0 is an upper solution for the original problem, $u_0 - v$ is an upper solution for the new problem. (Similar remarks apply for the associated parabolic equation.)

Now (i) is obviously a consequence of the comparison theorem, Theorem 10.1. For (ii), we have, from (i), $\bar{v}(x) \leq \lim_{t \to \infty} u(x, t) \leq \bar{u}(x)$, where $(\bar{u}(x), \bar{v}(x)) = \lim_{t \to \infty} (U_0(x, t), V_0(x, t))$, and all these limits exist by virtue of Lemma 10.6. Furthermore, each of these limiting functions are solutions of the elliptic boundary-value problem. Now by Corollary 10.4, $\lim_{n \to \infty} T^n v_0 \leq \lim_{t \to \infty}$

$u(x, t) \leq \lim_{n \to \infty} T^n u_0$; i.e.,

$$v(x) = \lim_{t \to \infty} u(x, t), \quad \text{uniformly in } x \in \Omega.$$

Hence v is stable, and the proof of the theorem is complete. $\square$

We shall apply the stability theorem to the following problem:

$$\Delta u + \mu u - u^3 = 0, \qquad x \in \Omega \subset \mathbf{R}^n, \tag{10.30}$$

where $u = 0$ on $\partial\Omega$. Here μ is a small positive parameter. It is obvious that $u \equiv 0$ solves this problem. If μ is sufficiently small, we claim that this is the only solution. To see this, multiply (10.30) by u and integrate over Ω to get

$$-\int_\Omega |\nabla u|^2 + \int_\Omega (\mu u^2 - u^4) = 0.$$

But as we shall show in the next chapter (in §A),

$$-\int_\Omega |\nabla u|^2 \leq -\lambda \int_\Omega |u|^2,$$

where $\lambda > 0$ in the largest eigenvalue of $-\Delta$ on Ω with homogenous Dirichlet boundary conditions. Therefore,

$$\int (\mu u^2 - u^4) \geq \lambda \int u^2,$$

so

$$0 \geq \int_\Omega -u^4 \geq (\lambda - \mu) \int_\Omega u^2 > 0,$$

if $\lambda > \mu$ and $u \not\equiv 0$. Thus if $\lambda > \mu$, $u \equiv 0$. This proves our claim.

Now let's suppose that $\lambda > \mu$, and κ is any positive number. If ω is the principal eigenfunction of $-\Delta$ on Ω with homogeneous Dirichlet boundary conditions, then $\omega \geq 0$[7] and $-\Delta\omega = \lambda\omega$; hence, if $\phi = \kappa\omega$,

$$(\Delta + \mu)\phi - \phi^3 = \kappa\omega[(\mu - \lambda) - \kappa^2\omega^2] \leq 0.$$

Thus $\kappa\omega$ is an upper solution, and similarly, $-\kappa\omega$ is a lower solution. Since the iterates $T^n(\pm\kappa\phi)$ both coverage to $u \equiv 0$ (by uniqueness), we see that the zero solution is stable, if $\lambda > \mu$. (In fact, it is not too hard to show that

---

[7] See Chapter 11, §A.

all solutions of the associated parabolic problem converge to 0 as $t \to \infty$, uniformly in $x \in \Omega$; see the proof of Theorem 10.5.)

Suppose now that $\lambda < \mu$; we shall show that there are two non-constant solutions of (10.30), with $u = 0$ on $\partial\Omega$. Again, if we put $u = \kappa\omega$, then u is a lower solution if $\kappa > 0$ is sufficiently small, and u is an upper solution if $\kappa < 0$ and κ is sufficiently small. If $\tilde{\Omega}$ is a domain such that $\tilde{\Omega} \supset \Omega$, and let $\tilde{\lambda}$ be the corresponding principal eigenvalue of $-\Delta$ on $\tilde{\Omega}$, with homogeneous Dirichlet boundary conditions.

Thus, if $\tilde{\phi} \geq 0$ is the corresponding eigenfunction and $u = C\tilde{\phi}$, then

$$\Delta u + \mu u - u^3 = u[(\mu - \tilde{\lambda}) - C^2\tilde{\phi}^2],$$

and since $\tilde{\phi} > 0$ on $\tilde{\Omega}$, we have $\Delta u + \mu u - u^3 < 0$ if $C > 0, |C|$ large, while $\Delta u + \mu u - u^3 > 0$ if $C < 0$, and $|C|$ is large. Thus, $C\tilde{\phi}$ is an upper solution if $C > 0$ and $|C|$ large, and it is a lower solution if $C < 0$ and $|C|$ large. Hence by Theorem 10.3, there are solutions u_1 and u_2 with

$$C\tilde{\phi} \geq u_1 \geq \kappa\phi \geq 0, \qquad \kappa, C > 0,$$
$$\kappa\phi \geq u_2 \geq C\tilde{\phi}, \qquad \kappa, C < 0.$$

As a final example, consider the problem

$$\Delta u + u^2 = 0, \qquad x \in \Omega, \quad u = 0 \quad \text{on } \partial\Omega. \qquad (10.31)$$

If ϕ is any solution, then $\Delta\phi = -\phi^2 \leq 0$ so by the maximum principle (Theorem 8.1), $\phi \geq 0$ on Ω. Suppose that $\phi \not\equiv 0$ on Ω, and let $\psi = \varepsilon\phi$, $\varepsilon > 0$. Then

$$\Delta\psi + \psi^2 = -\varepsilon\psi^2 + \varepsilon^2\psi^2 = \varepsilon(\varepsilon - 1)\psi^2.$$

Thus ψ is an upper solution if $0 < \varepsilon < 1$ and a lower solution if $\varepsilon > 1$. If u_ε is the corresponding solution of the parabolic problem $u_t = \Delta u + u^2, u = 0$ on $\partial\Omega$, $u(x, 0) = \psi$, $x \in \Omega$, then by Lemma 10.6, $\partial u_\varepsilon/\partial t > 0$ if $\varepsilon > 1$, and $\partial u_\varepsilon/\partial t < 0$ if $0 < \varepsilon < 1$. Since $\partial u_\varepsilon/\partial t > 0$ for every $\varepsilon > 1$ (Lemma 10.6), we see that $\phi(x)$ is not stable. Thus we have shown that any nonconstant solution of (10.31) *must be unstable*.

NOTES

The notion of upper and lower solutions was used by H. Keller and D. Cohen [KC], in special cases. The general results are due to Amann [Am 1] and Sattinger [Sa 2]. Our approach follows that of Sattinger. The stability theorem is due to Sattinger, as are most of the examples given in the text.

Chapter 11

Linearization

There is a well-known theorem in ordinary differential equations, going back to Poincaré, which states that the stability of a rest point can be inferred from "linearization." More precisely, if one considers the ordinary differential equation in $\mathbf{R}^n$, $u' = f(u)$, and $\bar{u}$ is a rest point (so that $f(\bar{u}) = 0$), then if the differential (matrix) $df(\bar{u})$ has all of its eigenvalues in the left-half plane, $\mathrm{Re}\, z < 0$, it follows that $\bar{u}$ is asymptotically stable; i.e., if u_0 is near $\bar{u}$, then the solution of the equation through u_0 tends to $\bar{u}$ as $t \to +\infty$. It is the main purpose of this chapter to prove an analogous theorem for partial differential equations, one which is sufficiently general to include systems of reaction–diffusion (parabolic) equations. In this context, the equation $u' = f(u)$ is replaced by an abstract equation of the form $u_t = Au + f(u)$, where u takes values in a Banach space B; i.e., for each t, $u(t)$ is in B, and A is a linear operator. The main example is the case where A is a linear elliptic operator. The "rest points" $\bar{u}$, in this setting now are solutions of the equation $Au + f(u) = 0$, and the linearized operator becomes $A + df(\bar{u})$. We shall show that if the spectrum of this operator lies in the left-half plane, then again one can conclude that $\bar{u}$ is asymptotically stable. This leads us quite naturally to a study of the spectrum of such linear operators. We shall undertake this study in §A, and in §B we shall prove a linearized stability theorem. The techniques which we develop here will be applied to specific problems in later chapters. In §C we shall give a useful extension of a result obtained in §A; this is the celebrated Krein–Rutman theorem.

§A. Spectral Theory for Self-Adjoint Operators

We consider a linear second-order elliptic operator

$$Pu \equiv Au + au \equiv \sum_{i,j=1}^{n} (a_{ij}(x)u_{x_i})_{x_j} + a(x)u \qquad (11.1)$$

defined in a bounded domain $\Omega \subset \mathbf{R}^n$, where $\partial\Omega$ is smooth. We assume that the coefficients of P are smooth, and that $a_{ij} = a_{ji}$ in Ω. We further assume that A is *strongly elliptic* in Ω, in the sense that there is an $\alpha > 0$ such that for all ξ,

$$\sum_{i,j} a_{ij}(x)\xi_i\xi_j \geq \alpha|\xi|^2, \qquad x \in \Omega. \tag{11.2}$$

We consider the operator (11.1) together with homogeneous boundary conditions of the form

$$\alpha(x)\frac{du}{dn} + \beta(x)u = 0 \quad \text{on } \partial\Omega, \tag{11.3}$$

where $\alpha(x) \geq 0$, $\beta(x) \geq 0$, $\alpha^2(x) + \beta^2(x) \neq 0$, and du/dn is the outward normal derivative of u on $\partial\Omega$.

We are interested in the "spectrum" of the operator P, where we regard P as an operator acting on functions in $W^2(\Omega)$ which satisfy (11.3) on $\partial\Omega$, into $L_2(\Omega)$.

Definition 11.1. The *spectrum* of P is the set of $\lambda \in \mathbf{C}$ for which $(P - \lambda I)$ is not invertible. We denote by $\sigma(P)$ the spectrum of P.

Since P is self-adjoint the spectrum of P is a subset of the real line. In order to analyze the spectrum more carefully, we need a few preliminary results.

Recall that a *compact* operator C on a Hilbert space H is one which takes bounded sets into precompact sets; i.e., into sets whose closure is compact. It is easy to see that compact operators are *not* invertible, so that $0 \in \sigma(P)$. Moreover, it is a classical result, going back to F. Riesz, that if C is compact, and $\lambda \in \mathbf{C}$, then $K = C - \lambda I$ is a Fredholm operator, i.e., dim ker K and dim coker K are both finite; for a proof see [BJS].

Lemma 11.2. *Let C be a compact operator on H. Then $\sigma(C)$ is discrete, and 0 is the only possible limit point of $\sigma(C)$.*

Proof. Let $\lambda \in \sigma(C)$, $\lambda \neq 0$. Then $C - \lambda I$ is a Fredholm operator so it has a finite-dimensional kernel [BJS]; i.e., the set $\{x \in H : (C - \lambda I)x = 0\}$ is finite dimensional. Let $E_0 = \ker(C - \lambda I)$, and let E_1 be the orthogonal complement of E_0. Then $H = E_0 \oplus E_1$, and E_1 is invariant under $C - \lambda I$. Take $\varepsilon > 0$ so small that $C - \lambda I - \varepsilon I$ is invertible on E_1 (the set of invertible operators is open), and since $C - \lambda I - \varepsilon I$ is invertible on E_0, we see that $C - \lambda I - \varepsilon I$ is invertible on H. Thus λ is isolated and the proof is complete. $\square$

Now since A is a second-order elliptic operator, A maps $W_2^2(\Omega)$ into $L_2(\Omega)$. Furthermore, A satisfies the celebrated Gårding's inequality; namely, (see [BJS]):

$$\langle A\phi, \phi \rangle_{L_2} \geq c_1\|\phi\|_{W_2^1} - c_2\|\phi\|_{L_2}, \qquad c_1 > 0, \quad c_2 > 0 \tag{11.4}$$

(in our context, the proof of (11.4) is much simpler than the general case). This inequality is used to show that A is a Fredholm operator with closed range; see [BJS].

If now k is a constant such that $k > c_2$, then (11.4) shows that $(A + kI)$ is injective. Since $A + kI$ has closed and dense range, we see that $A + kI$ is surjective and is thus an isomorphism. Let $\tilde{C} = (A + kI)^{-1}$ and let i be the inclusion $i: W_2^2(\Omega) \to L_2(\Omega)$. Then i is compact, so that $C = i\tilde{C}$ is also compact. Moreover, $\sigma(A + kI) = \{\lambda: \lambda^{-1} \in \sigma(\tilde{C})\} = \{\lambda: \lambda^{-1} \in \sigma(C)\}$. Thus the spectrum of $A + kI$ is discrete, and ∞ is its only possible accumulation point; the same conclusion is therefore valid for A. If $\lambda \in \sigma(A)$, then $Au = \lambda u, u \not\equiv 0$ so that

$$\int_\Omega (Au)u = \lambda \int_\Omega u^2.$$

If we integrate the left side by parts and use (11.2), we see that

$$-\alpha \int_\Omega u^2 \geq \lambda \int_\Omega u^2,$$

and this shows that $\lambda < 0$. Choose $k_1 < 0$ such that $k_1 + \sup\{|a(x)|: x \in \Omega\} < 0$. Then

$$\int_\Omega u(P + k_1)u \leq (k_1 - \alpha) \int_\Omega u^2 + \int_\Omega a(x)u^2 < 0.$$

It follows that the spectrum of $P + k_1$ lies in $\{\lambda < 0\}$. Thus, P can have at most a finite number of positive eigenvalues. We have thus proved the following theorem.

Theorem 11.3. *The strongly elliptic operator P, defined by (11.1) in the bounded set Ω, with boundary conditions (11.3), has discrete spectrum consisting only of real eigenvalues. If $a(x)$ is bounded in Ω, P can have at most a finite number of positive eigenvalues.*

We shall now show that the eigenvalues can be characterized as being extremals of certain variational problems. For simplicity, we shall replace Pu, as defined in (11.1), by the operator

$$Lu = \Delta u + a(x)u, \qquad x \in \Omega, \tag{11.5}$$

where $a \in L_\infty$ and Δ is the Laplacian on Ω. For the applications of this theory which we have in mind (in later chapters), this simplification is quite sufficient.

We shall study the equation $Lu = 0$, together with *one* of the following homogeneous boundary conditions on $\partial\Omega$,

$$B_1(u) \equiv \frac{du}{dn} + b(x)u = 0, \quad \text{or}$$

$$B_2(u) \equiv u = 0, \tag{11.6}$$

where b is a piecewise continuous function on $\partial\Omega$.

In order to study the eigenvalues and eigenfunction of the operator L, we define the following quadratic functionals on $W_2^1(\Omega)$:

$$Q_1(\phi) = \frac{\int_\Omega [-|\nabla\phi|^2 + a\phi^2] + \int_{\partial\Omega} b\phi^2}{\|\phi\|^2} \tag{11.7}$$

and

$$Q_2(\phi) = \frac{\int_\Omega [-|\nabla\phi|^2 + a\phi^2]}{\|\phi\|^2}, \tag{11.8}$$

where $\|\phi\|$ denotes the usual L_2-norm on Ω. Note that since a and b are bounded, both $Q_1(\phi)$ and $Q_2(\phi)$ are bounded from above, independently of ϕ.

Theorem 11.4. (1) *The supremum of Q_1 is assumed by a function ϕ_1, where ϕ_1 is an eigenfunction of L for the eigenvalue $Q_1(\phi_1)$, and $B_1(\phi_1) = 0$ on $\partial\Omega$.*

(2) *The supremum of Q_2, subject to the side condition $B_2(\phi) = 0$ on $\partial\Omega$, is assumed by a function ϕ_2, where ϕ_2 is an eigenfunction of L for the eigenvalue $Q_2(\phi_2)$.*

Proof. We shall not show that these variational problems have C^2 solutions; for a proof see, e.g., [Ag].

In order to prove that solutions of these problems are eigenfunctions, we introduce the bilinear forms,

$$F_1(\phi, \psi) = \int_G (-\nabla\phi \cdot \nabla\psi + a\phi\psi) + \int_{\partial G} b\phi\psi$$

and

$$F_2(\phi, \psi) = \int_G (-\nabla\phi \cdot \nabla\psi + a\phi\psi).$$

We consider the variational problem only for Q_1; the proof is similar for Q_2. Let ϕ_1 be a solution of this variational problem; since Q_1 is homogeneous of degree zero in ϕ^2, we may assume that $\|\phi_1\| = 1$. Let $\lambda_1 = Q_1(\phi_1)$, and let $\phi \in W_2^1(\Omega)$. If c is an arbitrary constant, then

$$Q_1(\phi_1 + c\phi) \le \lambda_1,$$

so that since $Q_1(\phi_1) = \lambda_1$, we have

$$2c\left\{F_1(\phi_1, \phi) - \lambda_1\langle\phi, \phi_1\rangle + \frac{c}{2}\|\phi\|^2[Q_1(\phi) - \lambda_1]\right\} \le 0,$$

where $\langle \ , \ \rangle$ denotes the L_2-inner product.

Since c is arbitrary, and $Q_1(\phi) \le \lambda_1$, we may conclude that

$$F_1(\phi_1, \phi) = \lambda_1\langle\phi_1, \phi\rangle. \qquad (11.9)$$

or, using the divergence theorem

$$\int_\Omega \phi L(\phi_1) + \int_{\partial\Omega} b\phi\phi_1 + \int_{\partial\Omega} \phi\frac{d\phi_1}{dn} = \lambda_1\int_\Omega \phi\phi_1.$$

In view of the arbitrariness of ϕ, this equation shows that $L(\phi_1) = \lambda_1\phi_1$ and that $B_1(\phi_1) = 0$ on $\partial\Omega$. Finally, if we put $\phi = \phi_1$ in (11.9), we see that $\lambda_1 = F_1(\phi_1, \phi_1) = Q_1(\phi_1)$. The proof is complete. $\square$

We shall refer to ϕ_1 and ϕ_2 as *principal eigenfunctions* for L, with the corresponding boundary values, and we call λ_1 and λ_2 the corresponding *principal eigenvalues*.

The variational character of the principal eigenvalues allows us to compare principal eigenvalues corresponding to different boundary conditions. This is the content of the next theorem.

Theorem 11.5. $\lambda_2 \le \lambda_1$.

Proof. The proof is obvious since in obtaining λ_2 we put on ϕ the additional condition that $\phi = 0$ on $\partial\Omega$; thus the maximum is taken over a smaller class of admissible functions. $\square$

In a similar way, we can prove the following theorem. (The notation below is the obvious one is this context.)

Theorem 11.6. (1) *If* $b(x) \ge \tilde{b}(x)$ *for all* $x \in \partial\Omega$, *then* $\lambda_1(b) \ge \lambda_1(b)$.
(2) *If* $a(x) \ge \tilde{a}(x)$ *for all* $x \in \Omega$, *then* $\lambda_1(a) \ge \lambda_1(\tilde{a})$ *and* $\lambda_2(a) \ge \lambda_2(\tilde{a})$.

Proof. These both follow at once from Theorem 11.4; for example, if $b(x)$ $\geq \tilde{b}(x)$, then $Q_1^b(\phi) \geq Q_1^{\tilde{b}}(\phi)$ for all admissible ϕ; whence the "sups" must also satisfy this inequality.

Let us now show that the principal eigenvalue depends continuously on a and b. To this end, let ϕ be any admissible function, $\|\phi\| = 1$. Then since

$$Q_1^{(a,b)}(\phi) - Q_1^{(\tilde{a},\tilde{b})}(\phi) \leq \|a - \tilde{a}\|_{L_\infty(\Omega)} + \|b - \tilde{b}\|_{L_\infty(\partial\Omega)} \equiv \sigma,$$

we see that $\lambda_{a,b} \leq \sigma + \lambda_{\tilde{a}\tilde{b}}$; interchanging the roles of $Q_1^{(a,b)}$ and $Q_1^{(\tilde{a},\tilde{b})}$ gives the result. We state this formally in the next theorem.

Theorem 11.7. *The principal eigenvalues, λ_2 and λ_1, depend continuously on a, and on a and b, respectively, in the sup norm topology.*

Next we shall show that the principal eigenvalues depend continuously on Ω. This requires that we make precise the notion of "closeness" of two domains Ω and Ω'.

Definition 11.8. Let $T = \text{Id.} + f$ be a C^1 injective transformation, mapping the domain Ω onto the domain Ω'. We say Ω is ε-close to Ω' if $\|f\|_\infty + \|df\|_\infty < \varepsilon$.

With this definition of closeness, we can prove the following theorem.

Theorem 11.9. *The principal eigenvalue of L, together with either boundary condition (11.6), varies continuously with Ω in the above C^1 topology. It also depends monotonically on Ω in the sense that if $\Omega \supset \tilde{\Omega}$, then $\lambda_\Omega \geq \lambda_{\tilde{\Omega}}$.*

Proof. Let $\varepsilon > 0$ be given. Then take Ω' to be ε-close to Ω. Now consider, e.g., the boundary condition $B_1 u = 0$. We write $Tx = x + f(x)$, and use the notation $x' = Tx$, $\phi(x) = \phi'(x')$, $x \in \bar{\Omega}$, $b(s) = b'(s')$, $s \in \partial\Omega$, and so on.

An easy computation shows that the functional Q_1 gets transformed into

$$\frac{\int_{\Omega'} |\nabla_{x'}\phi'|^2 + a(\phi')^2 + \int_{\partial\Omega'} b(\phi')^2 + o(\varepsilon)}{\|\phi'\|^2(1 + o(\varepsilon))^{-1}}.$$

It follows that

$$Q_1(\phi) = Q_1'(\phi') + o(\varepsilon),$$

from which the first result easily follows. The second statement is an easy consequence of the variational characterization of λ. □

We shall next prove an interesting fact concerning the principal eigenfunction; namely, we shall show that we can find one which doesn't change sign. This eigenfunction is very useful for comparison theorems, as we have seen in Chapter 10 (and as we shall see in subsequent chapters).

Theorem 11.10. *There always exists a principal eigenfunction which is of one sign in* Ω.

Proof. Suppose that $\bar{u}$ is a principal eigenfunction of L corresponding to one of the boundary conditions (11.6). Notice that $|\bar{u}|$ satisfies the same boundary condition as $\bar{u}$, since $|\bar{u}| = \pm \bar{u}$. Now we know that $Q_i(\bar{u})$ is a maximum for both $i = 1, 2$. If we can show that $Q_i(|\bar{u}|) = Q_i(\bar{u})$, then the theorem will obviously be proved. The trouble is that $\nabla(|u|)$ is not defined at points where $u = 0$. However, it is still true, for functions in $W_2^1(\Omega)$, that if f is a uniformly Lipschitzian function in $\mathbf{R}$, then the transformation $u \to f(u)$ takes $W_2^1(\Omega)$ into itself, and moreover $(\partial f(u)/\partial x_i) = f'(u) \, \partial u/\partial x_i$ (where we agree that the right-hand side vanishes if either one of its factors does); for a proof, see [Tr 2, p. 261]. Applying this fact to the function $f(u) = |u|$ gives the desired result. $\quad\square$

The variational method for the principal eigenvalue and eigenfunction of L, can easily be extended (see, e.g., [CH 1]) in order to obtain all of the eigenvalues $\{\lambda_n\}$, $\lambda_1 \geq \lambda_2 \geq \cdots$, and eigenfunctions $\{\phi_n\}$ of L, with either one of the boundary conditions (11.6). One then shows that $\lambda_n \to -\infty$, and that the set $\{\phi_n\}$ is a complete orthonormal system in $L_2(\Omega)$. Using this basis, we can prove some very important inequalities, relating L_2 norms of functions to their derivatives.

Theorem 11.11. *Let* $u \in W_2^1(\Omega)$; *then if* μ_1 *is the smallest positive eigenvalue of* $-\Delta$ *on* Ω (*with the appropriate boundary conditions*), *the following Poincaré inequalities hold*:

$$\|\nabla u\|^2 \geq \mu_1 \|u\|^2 \qquad \text{if } u = 0 \quad \text{on } \partial\Omega; \tag{11.10}$$

$$\|\nabla u\|^2 \geq \mu_1 \|u - \bar{u}\|^2 \quad \text{if } \frac{du}{dn} = 0 \quad \text{on } \partial\Omega, \tag{11.11}$$

where $\bar{u} = (\text{meas. } \Omega)^{-1} \int_\Omega u$; *and*

$$\|\Delta u\|^2 \geq \mu_1 \|\nabla u\|^2 \quad \text{if } u \in W_2^2(\Omega) \quad \text{and} \quad \frac{du}{dn} = 0 \quad \text{on } \partial\Omega. \tag{11.12}$$

If $\partial\Omega = \Gamma_1 \cup \Gamma_2$, *where* Γ_2 *has positive* $(n-1)$-*dimensional measure, then*

$$\|\nabla u\|^2 \geq \mu \|u\|^2, \tag{11.13}$$

for every $u \in W_2^1(\Omega)$, *with* $du/dn = 0$ *on* Γ_1, $a(x) \, du/dn + b(x)u = 0$ *on* Γ_2, $b > 0$. *Here* $\mu > 0$ *is independent of* u.

Proof. It suffices to prove these results for functions in $C^2(\Omega)$, and then to use standard approximation-type arguments.

Suppose first that $u \in C^2(\Omega)$, and $u = 0$ on $\partial\Omega$. Let $\{\psi_k\}$ be a complete orthonormal system of eigenfunctions of $-\Delta$ on Ω with homogeneous Dirichlet boundary conditions ($\psi_k = 0$ on $\partial\Omega$ for all k), and let $\{\mu_k\}$ be their corresponding eigenvalues. Then

$$u = \sum_{k=1}^{\infty} a_k \psi_k, \qquad -\Delta u = \sum_{k=1}^{\infty} a_k \mu_k \psi_k,$$

and thus

$$\|\nabla u\|^2 = \int_\Omega \langle \nabla u, \nabla u \rangle = -\int_\Omega \langle u, \Delta u \rangle$$

$$= \sum_{k=1}^{\infty} \mu_k a_k^2 \geq \mu_1 \sum_{k=1}^{\infty} a_k^2 = \mu_1 \|u\|^2.$$

This proves (11.10). If now we consider $-\Delta$ with homogeneous Neumann boundary conditions, i.e., $du/dn = 0$ on $\partial\Omega$, then $\psi_1 = 1$ is a principal eigenfunction corresponding to the eigenvalue $\mu_1 = 0$. Thus we have

$$u = \sum_{k=1}^{\infty} a_k \psi_k, \qquad \Delta u = -\sum_{k=2}^{\infty} a_k \mu_k \psi_k, \quad \text{so that}$$

$$\|\nabla u\|^2 = -\int_\Omega \langle u, \Delta u \rangle = \sum_{k=2}^{\infty} \mu_k a_k^2 \geq \mu_2 \sum_{k=2}^{\infty} a_k^2 = \mu_2 \|u - \bar{u}\|^2,$$

since $\bar{u} = \int_\Omega \langle u, \psi_1 \rangle = a_1$. To prove (11.12), we write $\Delta u = \sum_{k=2}^{\infty} a_k \psi_k$, where $a_k = \int_\Omega \langle \Delta u, \psi_k \rangle$, $\psi_1 = 1$ and $a_1 = \psi_1 \int_{\partial\Omega} \Delta u = \int_{\partial\Omega} du/dn = 0$. Since $u = \sum_{k=2}^{\infty} -a_k \mu_k^{-1} \psi_k + a_1 \psi_1$, we have

$$\|\nabla u\|^2 = \int_{\partial\Omega} \left\langle u, \frac{du}{dn} \right\rangle - \int_\Omega \langle u, \Delta u \rangle = -\int_\Omega \langle u, \Delta u \rangle$$

$$= \sum_{k=2}^{\infty} a_k^2 \mu_k^{-1} \leq \mu_2^{-1} \sum_{k=2}^{\infty} a_k^2 = \mu_2^{-1} \|\Delta u\|^2;$$

thus (11.12) holds. For the proof of (11.13), we refer the reader to [Mc]. □

§B. Linearized Stability

We consider Cauchy problems for semilinear equations, of the form

$$u_t = Au + f(u), \qquad u(t_0) = u_0, \tag{11.14}$$

where $u(t)$ takes values in a Banach space $(X, \|\cdot\|)$, and A generates a *continuous semigroup* on X. By this we mean that the operator[1] $T(t) = e^{tA}$ ($t \geq 0$), on X satisfies the following conditions:

- (i) $T(0) = I$, the identity.
- (ii) $T(t)T(s) = T(t + s)$ for all $s, t \geq 0$.
- (iii) $\lim_{t \to 0^+} T(t)x = x$ for each $x \in X$.
- (iv) The mapping $t \to T(t)x$ is continuous on $t > 0$ for each $x \in X$.

The reader can easily check that if $Au = -\Delta u$ on a bounded domain $\Omega \subset \mathbf{R}^n$, with, say, homogeneous Dirichlet boundary conditions, then A generates a continuous semigroup.

Now returning to the equation (11.14), we make the following assumptions:

1. The mapping $u \to f(u)$ is locally Lipschitzian; i.e., for all $u, v \in X$,

$$\| f(u) - f(v) \| \leq k(\|u\|, \|v\|) \|u - v\|,$$

where k is a continuous nonnegative real-valued function, which is increasing in each variable.

2. f is Fréchet differentiable, with Fréchet derivative df, (see Chapter 13, §A) and the mapping $u \to df_u$ is continuous from X to Hom X. (Hom X is the space of linear maps on X onto itself, with the usual norm-topology.) We also require that for any bounded set $B_1 \subset B$, there is a constant $c > 0$ such that

$$\| f(u) - f(v) - df_v(u - v) \| \leq c \|u - v\|^2, \qquad \forall u, v \in B_1. \tag{11.15}$$

We define $C([t_1, t_2]; X)$ to be the set of continuous functions $u(t)$ defined on $t_1 \leq t \leq t_2$ taking values in X. With this notation, we say that $u \in C([t_1, t_2]; X)$ is a *solution* of (11.14) provided that for every $t \in [t_1, t_2]$, we have

$$u(t) = e^{(t - t_1)A}u(t_1) + \int_{t_1}^{t} e^{(t - s)A} f(u(s)) \, ds. \tag{11.16}$$

Thus we see that the equation (11.14) is satisfied in a generalized sense; however, for "parabolic" problems, one shows that solutions of (11.16) are

---

[1] $e^{tA} = \lim_{n \to \infty} (1 - tA/n)^{-n}$ in the strong operator topology.

"weak" solutions of (11.14), and then one uses the parabolic regularity theorems to conclude that the solutions are classical solutions; see, e.g., [Re], In our context, for u_t to exist we want $Au(t)$ to make sense; i.e., we require that $u \in C([0, T]; D(A))$, and to achieve this, the natural hypotheses is that $f: D(A) \to D(A)$ is a locally Lipschitz function. If $u(0) \in D(A)$, then one expects to have $u \in C^1([0, T]; X) \cap C([0, T]; D(A))$.

We shall now prove the existence of a local solution by a standard contraction-mapping argument.

Theorem 11.12. *Let $u_0 \in X$ and let $t_0 > 0$. Then there exists a $\delta > 0$, depending only on $\|u_0\|$, such that (11.14) has a solution $u \in C([t_0, t_0 + \delta]; X)$.*

Proof. If $t_1 > t_0$, define a mapping K from $C([t_0, t_1]; X)$ into itself by

$$(Ku)(t) = e^{(t - t_0)}u_0 + \int_{t_0}^t e^{(t - s)A}f(u(s)) \, ds, \qquad t_0 \le t \le t_0 + \delta.$$

We seek a fixed point of K. For $0 < \delta \le 1$, we define

$$S = \{u \in C([t_0, t_0 + \delta]; X): \|u(t) - e^{(t - t_0)A}u_0\| \le 1, \qquad t_0 \le t \le t_0 + \delta.$$

Let

$$M_1 = 1 + \sup\{\|e^{\tau A}\| : 0 \le \tau \le 1\}(1 + \|u_0\|),$$

and

$$M = \sup_{\|u\| \le M_1} \|f(u)\|.$$

Then if $u, v \in S$, $\|u(s)\|, \|v(s)\| \le M_1, t_0 \le s \le t_0 + \delta$, and if $t_0 \le t \le t_0 + \delta$, and $2\delta M_1 M < 1$,

$$\|(Ku)(t) - e^{(t - t_0)A}u_0\| = \left\| \int_{t_0}^t e^{(t - s)A}f(u(s)) \, ds \right\|$$

$$\le (t - t_0)M_1 M \le \delta M_1 M < \tfrac{1}{2}.$$

Thus K maps S into S. Furthermore,

$$\|(Ku)(t) - (Kv)(t)\| \le 2(t - t_0)MM_1 < 1,$$

so that K is a contraction on S. Therefore by Banach's theorem, K has a fixed point in S. $\square$

We next show that the solution of (11.14) is unique.

Theorem 11.13. *There is a unique solution of (11.14) in $C([t_1, t_2]; X)$.*

Proof. If u and v were two such solutions, then

$$u(t) - v(t) = \int_{t_1}^t e^{(t-s)A}(f(u(s)) - f(v(s)))\, ds.$$

Let

$$M = \sup_{0 \le \tau \le t_2 - t_1} \|e^{\tau A}\| + \sup_{t_1 \le t \le t_2} k(\|u(t)\|, \|v(t)\|);$$

then

$$\|u(t) - v(t)\| \le M^2 \int_{t_1}^t \|u(s) - v(s)\|\, ds,$$

so that by Gronwall's inequality (see Chapter 4), we have $u(t) = v(t)$ on $t_1 \le t \le t_2$. This completes the proof. $\square$

We shall next show how we can "glue" local solutions together.

Theorem 11.14. *Let u_1 and u_2 be solutions of (11.14) for $t_0 \le t \le t_1$ and $t_1 \le t \le t_2$, respectively, where $u_1(t_1) = u_2(t_1)$. Then the function $u(t)$ defined by*

$$u(t) = \begin{cases} u_1(t), & t_0 \le t \le t_1, \\ u_2(t), & t_1 \le t \le t_2, \end{cases}$$

is a solution of (11.14) in $t_0 \le t \le t_2$.

Proof. We need to show that (11.16) holds for $t_0 \le t \le t_2$. This is obvious on $t_0 \le t \le t_1$; if $t_1 \le t \le t_2$, then

$$u(t) = u_2(t) = e^{(t-t_1)A}u_1(t_1) + \int_{t_1}^t e^{(t-s)A} f(u_2(s))\, ds$$

$$= e^{(t-t_1)A}\left[e^{(t_1-t_0)A}u_1(t_0) + \int_{t_0}^t e^{(t_1-s)A} f(u_1(s))\, ds \right]$$

$$+ \int_{t_1}^t e^{(t-s)A} f(u_2(s))\, ds$$

$$= e^{(t-t_0)A}u(t_0) + \int_{t_0}^{t_1} e^{(t-s)A} f(u(s))\, ds$$

$$+ \int_{t_1}^t e^{(t-s)A} f(u(s))\, ds,$$

$$= e^{(t-t_0)A}u(t_0) + \int_{t_0}^t e^{(t-s)} f(u(s))\, ds.$$

This completes the proof. $\square$

We now prove an important result. Namely, if we have a solution $u(t)$ defined for $0 \leq t \leq T$, then by taking initial data sufficiently close to $u(0)$, the corresponding solutions exist at least for time T.

Theorem 11.15. *Let $u \in C([0, T]; X)$ be a solution of (11.14). Then there exists a neighborhood N of $u(0)$ such that if $\phi_0 \in N$, there is a solution $\phi \in C([0, T]; X)$ of (11.14), with $\phi(0) = \phi_0$. Moreover, there is a constant $c > 0$ such that for all such ϕ_0 in N,*

$$\| u(t) - \phi(t) \| \leq c \| u_0 - \phi_0 \|. \tag{11.17}$$

Proof. Suppose $u_i \in C([0, T]; X)$, $i = 1, 2$. Then if $t \leq T$,

$$u_i = e^{tA} u_i(0) + \int_0^t e^{(t-s)A} f(u_i(s)) \, ds,$$

so that if $\delta(t) = u_1(t) - u_2(t)$, we have

$$\delta(t) = e^{tA} \delta(0) + \int_0^t e^{(t-s)A} [f(u_1(s)) - f(u_2(s))] \, ds.$$

If $M > 1 + \sup\{\|e^{tA}\| + k(\|u_1(t)\| + 1, \|u_2(t)\|) : 0 \leq t \leq T\}$, then

$$\| \delta(t) \| \leq M^2 \| \delta(0) \| + M^2 \int_0^t \| \delta(s) \| \, ds.$$

Thus Gronwall's inequality gives

$$\| \delta(t) \| \leq M^2 \| \delta(0) \| e^{M^2 t} \leq M^2 \| \delta(0) \| e^{M^2 T}. \tag{11.18}$$

It follows that if

$$\| \delta(0) \| < M^{-2} e^{-M^2 T}. \tag{11.19}$$

then $\| \delta(t) \| < 1$, so that $\| u_2(t) \| \leq 1 + \sup\{\|u_1(t)\| : 0 \leq t \leq T\}$. In particular, if $u_1 = u$, we see that solutions with data $u_2(0)$ satisfying (11.19), where $u_1 = u$, exist for at least time T. (It is a standard result, just as in the case of ordinary differential equations, that solutions continue to exist, provided that their norms stay bounded; see Theorem 14.4, Chapter 14). Finally, (11.17) follows from (11.18). This completes the proof. $\square$

Now let $u \in C([0, T]; X)$, and let N be as in the last theorem. Suppose that $\phi_0 \in N$, and let $\phi \in C([0, T]); X)$, be the solution of (11.14) with $\phi(0) = \phi_0$. We define the *solution operator* $S : N \to X$ by

$$S(t)(\phi(0)) = \phi(t), \qquad 0 \leq t \leq T. \tag{11.20}$$

Our immediate aim is to show that $S(t)$ is Fréchet differentiable, and that the derivative

$$\gamma(t) = dS_{\phi(0)}\gamma(0)$$

solves the linearized equations

$$\gamma_t = A\gamma + df_{\phi(t)}\gamma, \qquad \gamma(0) = \gamma_0. \tag{11.21}$$

Recall that we assumed that the mapping $t \to df_{\langle\phi(t)\rangle}$ is continuous with values in Hom X. This will be used to show that linear equations are solvable.

Theorem 11.16. *Let $D(t)$ be a continuous linear operator on X, defined for $0 \le t \le T$. Then for each $\gamma_0 \in X$, there is a unique solution $\gamma \in C([0, T]; X)$ of the linear equation $\gamma_t = A\gamma + D(t)\gamma$, with $\gamma(0) = \gamma_0$.*

Proof. We must show that the equation

$$\gamma(t) = e^{tA}\gamma_0 + \int_0^t e^{(t-s)A}D(s)\gamma(s)\,ds \tag{11.22}$$

is solvable for $0 \le t \le T$. The local existence of solutions follows as in Theorem 11.12. To obtain the existence on $0 \le t \le T$, we again need an a-priori estimate. But this is easy; in fact, if

$$M > \sup\{\|e^{tA}\| + \|D(t)\| : 0 \le t \le T\},$$

then from (11.22) we get

$$\|\gamma(t)\| \le M\|\gamma_0\| + M^2 \int_0^t \|\gamma(s)\|\,ds,$$

and so by Gronwall's inequality, $\|\gamma(t)\| \le M\|\gamma(0)\|e^{M^2 t}$. □

Now if $u(t)$ is a solution of (11.14), $0 \le t \le T$, and $u(0) \in N$, we define the *linearized solution operator* $S_L : X \to X$ by

$$S_L(t)\gamma_0 = \gamma(t), \qquad 0 \le t \le T, \tag{11.23}$$

where $\gamma(t)$ is the solution of (11.21) with $\gamma(0) = \gamma_0$. We can now prove that $S(t)$ is Fréchet differentiable.

Theorem 11.17. *Let N be as in Theorem 11.15, and let $u \in C([0, T]; X)$ be a solution of (11.14). Then $S(t)$, $0 \le t \le T$, is Fréchet differentiable, and if $\phi_0 \in N$, $dS(t)_{\phi_0} = S_L(t)$.*

Proof. We must show that

$$\| S(t)(\phi_0 + h) - S(t)\phi_0 - S_L(t)h \| = o(h) \quad \text{as } h \to o.$$

To this end, let $\gamma(t) = S_L(t)h$, $\phi(t) = S(t)\phi_0$, $\bar{\phi}(t) = S(t)(\phi_0 + h)$, $\delta(t) = \bar{\phi}(t) - \phi(t)$. Then as in the proof of Theorem 11.15,

$$\delta(t) = e^{tA}h + \int_0^t e^{(t-s)A}[f(\bar{\phi}(s)) - f(\phi(s))] \, ds.$$

Using (11.15), we can write

$$\delta(t) = e^{tA}h + \int_0^t e^{(t-s)A}[df_{\phi(s)} \cdot \delta(s) + \rho(s)] \, ds,$$

where $\| \rho(s) \| \leq c_1 \| \delta(s) \|^2, 0 \leq t \leq T$, and c_1 is independent of h if $\| h \| \leq 1$. From (11.17),

$$\left\| \int_0^t e^{(t-s)A}\rho(s) \, ds \right\| \leq c \| \delta(0) \|^2 = c \| h \|^2,$$

for some constant c. If $\sigma(t) = \delta(t) - \gamma(t)$, then

$$\sigma(t) = \int_0^t e^{(t-s)A} df_{\phi(s)} \sigma(s) \, ds + O(\| h \|^2).$$

Thus Gronwall's inequality gives $\| \sigma(t) \| \leq c \| h \|^2 = o(\| h \|)$. If we unravel this inequality, we find that it is precisely what we wanted to show. $\square$

We shall end this chain of ideas with the following theorem.

Theorem 11.18. *$S(t)$ is continuously Fréchet differentiable on N (see Theorem 11.15).*

Proof. We must show that if $u_1(0)$ and $u_2(0)$ are in N, and are sufficiently close, and if S_L^1 and S_L^2 are the associated linearized solution operators, then $\| S_L^1(t)h - S_L^2(t)h \|$ is small, uniformly for $\| h \| \leq 1$.
  Set $\gamma_i(t) = S_L^i(t)h, i = 1, 2$, and $\delta = \gamma_1 - \gamma_2$; then

$$\delta(t) = \int_0^t e^{(t-s)A}[df_{u_1(s)}\gamma_1(s) - df_{u_2(s)}\gamma_2(s)] \, ds. \tag{11.24}$$

Using (11.17) and the smoothness of f, we can find $\eta > 0$ such that if $\|u_1(0) - u_2(0)\| < \eta$, then $u_2(0) \in N (= N(u_1(0)))$, and

$$\|df_{u_1(s)} - df_{u_2(s)}\| \leq c\|u_1(0) - u_2(0)\|, \qquad 0 \leq s \leq T.$$

We also have

$$df_{u_1(s)}\gamma_1(s) - df_{u_2(s)}\gamma_2(s) = df_{u_2(s)}\delta + (df_{u_1(s)} - df_{u_2(s)})\gamma_1(s),$$

so that

$$\|\delta(t)\| \leq c_1 \int_0^t \|\delta(s)\| + c_2\|u_1(0) - u_2(0)\|,$$

where c_1 and c_2 depend only on η if $\|u_1(0) - u_2(0)\| \leq \eta$ and $\|h\| \leq 1$. Again using Gronwall's inequality, we find

$$\|\delta(t)\| \leq c_3\|u_1(0) - u_2(0)\|, \qquad 0 \leq t \leq T.$$

This estimate gives the desired result. $\square$

We have enough machinery now to take up the problem of linearized stability.

Definition 11.19. Let $\bar{u} \in X$ be an equilibrium solution of (11.14); i.e., $\bar{u}$ is a solution independent of t, so that $A\bar{u} + f(\bar{u}) = 0$. Then $\bar{u}$ is said to be *linearly stable* if there are numbers $\bar{t} > 0$ and $\sigma > 0$, such that $\|S_L(t)\| \leq e^{-\sigma t}$ if $t \geq t$, where S_L is the linearized operator at $\bar{u}$.

We shall give some equivalent formulations.

Theorem 11.20. *Let $\bar{u}$ and S_L be as in the previous definition. Then the following are equivalent*:

(i)   *$\bar{u}$ is linearly stable.*
(ii) $\|S_L(t)\| \to 0$ *as $t \to +\infty$.*
(iii) *For some $t > 0$, the spectrum of $S_L(t)$ lies in $\{|z| < 1\}$.*

If in addition, $S_L(t)$ is a compact operator for $t \geq t_0 > 0$, then these conditions are equivalent to

(iv) *There exists an $\alpha < 0$ such that the spectrum of $A + df_{\bar{u}}$ lies in the half-space Re $z \leq \alpha$.*

Proof. It is clear that (i) implies (ii) and that (ii) implies (iii). To see that (iii) implies (i), we use the spectral radius formula (see [Wi]),

$$\lim_{n \to \infty} \|S_L(nt)\|^{1/n} = \text{spect. radius } (S_L(t)) < 1.$$

Thus we can find $\sigma \in \mathbf{R}$ such that $1 > \sigma >$ [spect. radius $(S_L(t))$]. Then

$$\| S_L(nt) \| \leq \sigma^n \quad \text{if } n \text{ is large; say } n \geq n_0.$$

Since $S_L(t) = S_L(t - [t/n_0]n_0)S_L([t/n_0]n_0)$, where $[\cdot]$ denotes the greatest integer symbol, we have

$$\| S_L(t) \| \leq \sup_{0 \leq s \leq n_0} \| S_L(s) \| \cdot \sigma^{[t/n_0]},$$

and this implies linear stability of $\bar{u}$; since $\sigma < 1$. Thus (i), (ii) and (iii) are equivalent.

Suppose now that $S_L(t)$ is compact for $t \geq t_0$. Since

$$S_L(t) = \exp t(A + df_{\bar{u}})$$

we may invoke the spectral mapping theorem (see [Ru 3]) to conclude

$$\sigma(S_L(t)) = \{e^{tz} : z \in \sigma(A + df_{\bar{u}})\}, \qquad t \geq t_0.$$

Thus (iii) and (iv) are equivalent. This completes the proof of the theorem. $\square$

We remark that if we consider equilibria of parabolic systems; e.g., $A = D\Delta$, where D is a diagonal matrix with positive entries, then $S_L(t)$ is compact for all $t > 0$, since it is "smoothing" (see, e.g., [Fn 3]).

We are now ready to prove the main result of this section; namely, that linearized stability implies stability. We first must define precisely what we mean by stability.

Definition 11.21. Let $\bar{u}$ be an equilibrium solution of (11.14). Then $\bar{u}$ is *stable* if there is a neighborhood N of $\bar{u}$ and positive numbers c, α such that if u is a solution of (11.14) with $u(0) \in N$, then u exists for all $t > 0$ and

$$\| u(t) - \bar{u} \| \leq c e^{-\alpha t} \| u(0) - \bar{u} \|, \qquad t > 0.$$

Theorem 11.22. *If $\bar{u}$ is a linearly stable solution of* (11.14), *then $\bar{u}$ is stable.*

Proof. Let $S_L(t)$ denote the Fréchet derivative of $S(t)$ at $\bar{u}$. Choose $\bar{t} > 0$ so large that $\| S_L(t) \| \leq e^{-\sigma t}$ if $t \geq \bar{t}$. Let $0 < \beta < \alpha$. We can find an open set $N = \{u \in X : \| u - \bar{u} \| < r\}$, with the property that if $u(t)$ solves (11.14) and $u(0) \in N$, then (by Theorem 11.15), $u \in C([0, \bar{t}]; X)$, and moreover, $\| \tilde{S}_L(\bar{t}) \| \leq e^{-\beta \bar{t}}$, where $\tilde{S}_L(\bar{t})$ is the Fréchet derivative of $S(\bar{t})$ at $u(0)$. The latter assertion follows from Theorem 11.18.

Now using the mean-value theorem [Di], we can write, for $u, v \in N$,

$$\| S(\bar{t})u - S(\bar{t})v \| \leq e^{-\beta t} \| u - v \|. \tag{11.25}$$

Moreover, since $S(\bar{t})\bar{u} = \bar{u}$, (11.25) shows that $S(\bar{t})$ maps N into itself. Thus, if $\phi \in N$, Theorem 11.14 shows that $S(t)\phi$ exists for all $t > 0$.

Now if k is any positive integer,

$$
\begin{aligned}
\| S(k\bar{t})u(0) - \bar{u} \| &= \| S(k\bar{t})u(0) - S(k\bar{t})\bar{u} \| \\
&= \| S(\bar{t})^k u(0) - S(\bar{t})^k \bar{u} \| \\
&= \| S(\bar{t})^k(u(0) - \bar{u}) \|.
\end{aligned}
$$

Therefore, from (11.25) we find

$$
\| S(k\bar{t})u(0) - \bar{u} \| \le e^{-k\beta\bar{t}} \| u(0) - \bar{u} \|. \tag{11.26}
$$

If now $t > 0$ is given, let $k = [t/\bar{t}]$, and $t_0 = t - k\bar{t}$. Then $0 \le t_0 \le \bar{t}$, and since $\bar{u}(t) = S(t_0)\bar{u}(0)$, (11.17) implies that

$$
\begin{aligned}
\| u(t) - \bar{u} \| &= \| S(t)u(0) - \bar{u} \| \\
&= \| S(t_0)S(k\bar{t})u(0) - S(t_0)\bar{u} \| \\
&\le c \| S(k\bar{t})u(0) - \bar{u} \|,
\end{aligned}
$$

where the constant is independent of $u(0) \in N$, and $t \in \mathbf{R}_+$. This, together with (11.26) implies that

$$
\| u(t) - \bar{u} \| \le ce^{-k\beta t} \| u(0) - \bar{u} \| \le ce^{-\delta t} \| u(0) - \bar{u} \|,
$$

for some $\delta > 0$. This proves stability, and completes the proof of the theorem. $\square$

§C. Appendix: The Krein–Rutman Theorem

The Krein–Rutman theorem gives conditions under which a not necessarily self-adjoint operator possesses a principal eigenvalue with corresponding eigenfunction of one sign. These things are of use in constructing comparison functions for nonlinear equations; cf. Chapter 10, §C. The proof of the main result is done via a degree-theoretic argument (see Chapter 12, §A and §B); however the statement of the main result is close in spirit to the material in §A and we thus include it here.

Let E be a Banach space. A subset $K \subset E$ is called a *cone* if K is closed under addition and multiplication by nonnegative scalars. If K is a closed cone in E, with int $K \ne \phi$, and L is a linear map, $L: K\backslash\{0\} \to$ int K, then L is called *strongly positive* with respect to K. A map $T: E \to E$ is called *compact* if T maps bounded sets into precompact sets; i.e., into sets with compact closure. Here is the main result.

Theorem A1 (Krein–Rutman). *Let K be a closed cone in E with int $K \neq \phi$ and suppose that if $u \in K \backslash \{0\}$, then $-u \notin K$. Let L be a compact linear operator on E which is strongly positive with respect to K. Then L has a unique eigenvector $v \in$ int K with $\|v\| = 1$, and the corresponding eigenvalue μ is positive. Moreover, v is the unique eigenvector of L in K, and $\mu \leq \sup\{|\lambda| : \lambda$ is in the spectrum of $L\}$.*

In the applications, E is usually a space of functions, and K consists of the nonnegative functions in E. Thus the theorem asserts the existence of a "positive" eigenfunction associated to the "principal" eigenvalue.

Before giving the proof, we need a few preliminary results.

Let $T : \mathbf{R} \times E \to E$ be compact, and consider the equation

$$u = T(\lambda, u). \tag{A1}$$

A solution is a pair (λ, u) satisfying (A1). Suppose that $T(0, u) = 0$ for all u in E. Then $(0, 0)$ is a solution of (A1), and it is the unique solution when $\lambda = 0$. If T is smooth near $(0, 0)$, then the Banach space version of the implicit function theorem (Theorem 13.3) implies that (A1) has a curve of solutions $(\lambda, u(\lambda))$ for λ near 0 with $u(0) = 0$. If, for example, $T(\lambda, u) \equiv \lambda K(u)$ with K compact, then the Schauder fixed point theorem (Theorem 12.15) gives a solution $(\lambda, u(\lambda))$ for each small λ. However, as we shall presently show, a much stronger result is true.

Theorem A2. *If T is compact on $\mathbf{R} \times E$, and $T(\lambda, 0) = 0$ for all $u \in E$, then (A1) possesses a component (i.e., a maximal connected set) of solutions $C = C^+ \cup C^-$ where $C^\pm \in \mathbf{R}_\pm \times E$, $C^\pm$ are both unbounded, and $C^+ \cap C^- = \{(0, 0)\}$.*

Proof. Define $S = \{(\lambda, u) \in \mathbf{R} \times E : \Phi(u) \equiv u - T(\lambda, u) = 0\}$, and let C be the component of S which contains $(0, 0)$. Let $C^\pm = C \cap (\mathbf{R}_\pm \times E)$; then $C^+ \cap C^- = \{(0, 0)\}$. It remains to show that $C^\pm$ are unbounded sets.

Suppose that C^+ is bounded. We claim that there exists a bounded open set U in $\mathbf{R}_+ \times E$ such that $C^+ \subset U$ and $S \cap \partial U = \phi$. To see this, let C_ε be a uniform ε-neighborhood of C^+ in $\mathbf{R}_+ \times E$. Then $S \cap \text{cl}(C_\varepsilon)$ is a compact metric space with the topology induced by $\mathbf{R}_+ \times E$, via the compactness of T. Set $K_1 = C^+$ and $K_2 = S \cap \partial C_\varepsilon$ (∂ in $\mathbf{R}_+ \times E$). By the usual standard separation lemma, there exist disjoint compact sets $\tilde{K}_i$, $i = 1, 2$, with $K_1 \subset \tilde{K}_1$, $K_2 \subset \tilde{K}_2$ and $\tilde{K}_1 \cup \tilde{K}_2 = S \cap \text{cl}(C_\varepsilon)$. Let $\delta = \min(\text{dist}(\tilde{K}_1, \tilde{K}_2), \text{dist}(\tilde{K}_1, \partial C_\varepsilon))$, and let U be a uniform $\delta/3$ neighborhood of $\tilde{K}_1$; then U has the desired properties.

For $\lambda \geq 0$, set $U_\lambda = \{u \in E : (\lambda, u) \in U\}$, and consider the degree $d(\Phi(\lambda, \cdot), 0_\lambda, 0)$. Using the homotopy invariance,

$$d(\Phi(\lambda, \cdot), 0_\lambda, 0) = d(\Phi(0, \cdot), 0_0, 0) = d(\text{Id.}, 0_0, 0) = 1.$$

But also, if λ is large, $0_\lambda = \phi$ so that $d(\Phi(\lambda, \cdot), 0_\lambda, 0) = 0$, again using the homotopy invariance. This contradiction shows C^+ is unbounded; similarly C^- is unbounded. The proof is complete. $\square$

We remark that in the applications it is sometimes the case that $T(\lambda, 0) = 0$ for all λ in $\mathbf{R}$ in which case we trivially get the unbounded sets $C^\pm$. In bifurcation problems, one can often get other interesting solutions in addition to these trivial ones; see Chapter 13.

Corollary A3. *Let K be a cone in E, and let $T: \mathbf{R} \times K \to K$ be compact with $T(0, u) = 0$ for all $u \in K$. Then the equation $T(\lambda, u) = u$ has a component of solutions $C = C^+ \cup C^-$ with $C^\pm \subset \mathbf{R}^\pm \times K$, $C^\pm$ both unbounded and $C^+ \cap C^- = \{(0, 0)\}$.*

Proof. Extend the function $T(0, \cdot)$ to all of E by defining $T(0, x) = 0$ if $x \in E \backslash K$, and then extend T to $\mathbf{R} \times E$ with values in K by the Dugundji extension theorem ([Sw, p. 119]). If we now use our theorem, and note that the range of T lies in K, so that the zeros of $\Phi(\lambda, u) \equiv u - T(\lambda, u)$ lie in $\mathbf{R} \times K$, we see that the result follows. $\square$

We can now prove the Krein–Rutman theorem. We shall only show that there is a $v \in \text{int } K$ and $\mu > 0$ such that $Lv = \mu v$. (For the other statements we refer the reader to the original paper [KR].)

If $u \in K$ we will write $u \geq 0$, and $u \geq v$ means $(u - v) \in K$. Let $w \in K \backslash \{0\}$. Then there is an $M > 0$ such that $MLw \geq w$. Indeed, if not, then $Lw - M^{-1}w \notin K$ for all $M > 0$ and so $Lw \notin \text{int } K$, contrary to the hypothesis.

Let $\varepsilon > 0$ be given and define a compact map $T_\varepsilon: \mathbf{R} \times E \to E$ by $T_\varepsilon(\lambda, u) = \lambda L(u + \varepsilon w)$. By the above corollary, there is an unbounded component C_ε of solutions of the equation $u = T_\varepsilon(\lambda, u)$ in $\mathbf{R} \times K$, which contains $(0, 0)$. We claim $C_\varepsilon \subset [0, M] \times K$. To see this, suppose $(\lambda, u) \in C_\varepsilon$; then $u = \lambda Lu + \lambda \varepsilon Lw \geq \lambda \varepsilon Lw \geq \lambda \varepsilon w/M$. Thus $Lu \geq (\lambda \varepsilon/M)Lw \geq (\lambda \varepsilon/M^2)w$. But $u \geq \lambda Lu$, so that $u \geq (\lambda/M)^2 \varepsilon w$. By induction we find $u \geq (\lambda/M)^n \varepsilon w$ for every $n \in \mathbf{Z}_+$. If $\lambda > M$, we conclude $w \leq 0$ and so $-w \in K$, a contradiction. Thus $\lambda \leq M$. Since C_ε is unbounded, there is a u_ε in K such that $(\lambda_\varepsilon, u_\varepsilon) \in C_\varepsilon$ and $\|u_\varepsilon\| = 1$. Let $\varepsilon \to 0$; we may assume that for some subsequence $\{\varepsilon_k\}$, $\lambda_{\varepsilon_k} \to \lambda \in [0, M]$. From the equation $u_\varepsilon = \lambda_\varepsilon Lu_\varepsilon + \lambda_\varepsilon \varepsilon Lw$, we see $u_{\varepsilon_k} \to v \in K$, $\|v\| = 1$, and $v = \lambda Lv$. Moreover, $\|v\| = 1$ implies $\lambda > 0$ and thus $v \in \text{int } K$. It follows that v is a positive eigenvector with corresponding eigenvalue $\mu = \lambda^{-1}$. $\square$

As an example of an E, K, and L where the above conditions hold, consider the following situation. Let D be a bounded domain in $\mathbf{R}^n$ having smooth boundary, and define

$$\mathscr{L}u = -\sum a_{ij}(x)u_{x_i x_j} + \sum b_i(x)u_{x_i} + c(x)u,$$

where a_{ij}, b_i, c are all in $C^\alpha(\bar{D})$, $\mathscr{L}$ is uniformly elliptic on $\bar{D}$, and $c \geq 0$ in $\bar{D}$. For $a \in C^\alpha(\bar{D})$, $a > 0$, set $E = C^{1+\alpha}(\bar{D})$ and $v = Lu$, where $u \in E$ and v satisfies

$$\mathscr{L}v = a(x)v, \qquad x \in D, \qquad v = 0 \quad \text{on } \partial D.$$

Then L is compact and linear. Let K be defined by

$$K = \text{cl}\left\{ u \in E : u > 0 \text{ in } \Omega \text{ and } \frac{\partial u}{\partial n} < 0 \text{ on } \partial\Omega \right\},$$

where n is the outward-pointing unit normal vector on ∂D. Then K is a closed cone in E, $u \in K\backslash\{0\}$ implies that $-u \notin K$, and $\text{int } K \neq \phi$. Moreover, $u \in K\backslash\{0\}$ implies that Lu is in $\text{int } K$, via the strong maximum principle (Theorem 8.1). Thus L is strongly positive with respect to K. It follows from the Krein–Rutman theorem that the smallest characteristic value of L is positive, and has a unique eigenvector in K.

NOTES

The results in §A are classical, and our proofs are adapted from Courant and Hilbert [CH 1]. The proof of Theorem 11.11 is taken from Conway Hoff and Smoller [CHS]. The material in §B is taken from Rauch [R]; more general results can be found in Henry [He] and Mora [Mx]. The proof of the Krein–Rutman Theorem, [KR], given here was shown to me by Paul Rabinowitz.

Chapter 12

Topological Methods

The invention of modern topology goes back to Poincaré, who was led to it in his study of the differential equations of celestial mechanics. Its development was taken over, for quite a while, by people who interestingly enough, seemed to have completely forgotten its origins. Perhaps this really was necessary in order that the subject develop rapidly. In any case, already in the twenties and thirties, people like Morse, Leray, Schauder, and others, were applying topological methods to differential equations. It is our purpose here to explain the relevance of some of these techniques to nonlinear differential equations.

We begin with a study of the (Brouwer) degree of a mapping in finite-dimensional spaces. This remarkable integer-valued function shows that the values of a continuous mapping from a bounded domain $U \subset \mathbf{R}^n$, into $\mathbf{R}^n$, are, in some sense, determined by its values on the boundary of U. We obtain the standard properties of the degree, and we use these to prove the Brouwer fixed-point theorem. This is then applied to two problems in ordinary differential equations. We then consider the extension of the degree to an infinite-dimensional setting. This is the celebrated Leray–Schauder degree for mappings of the form $I - K$, where K is a compact operator. We show how it is used to obtain solutions of partial differential equations. In §C we introduce the reader to Morse theory, and we describe some of the important basic results. This is in preparation for Chapters 22 and 23, in which we shall consider Conley's important and far-reaching generalization of Morse theory. In the final section, we give a very brief description of some standard results in algebraic topology that are needed in this chapter as well as in Chapters 22 and 23.

§A. Degree Theory in $\mathbf{R}^n$

Loosely speaking, the degree of a mapping[1] f from an open subset $U \subset \mathbf{R}^n$ into $\mathbf{R}^n$ is an integer determined solely by its values on ∂U, which when nontrivial, implies that f has a zero in U. We shall require of the degree

---

[1] In §§A, B, and C, all functions are assumed to be (at least) continuous. We will not always bother to state this.

that it be "stable under pertubations." By this we mean that if g is "close" to f, then to g is assigned the same integer as is assigned to f. Indeed, we shall do better, in that the degree will turn out to also be a homotopy invariant. This notion will be discussed in the subsequent development.

Let U be a bounded open connected subset of $\mathbf{R}^n$, let $y_0 \in \mathbf{R}^n$, and let f map U into $\mathbf{R}^n$. We are interested in finding solutions of the equation $f(x) = y_0$. We shall define an integer which tells when this equation has solutions. This integer is easily defined on a dense set of functions in $C^0(U)$, and on a dense set of points y in the range $R(f)$ of f. Then the definition will be extended to all continuous functions and all y_0 in $R(f)$ where $y_0 \notin f(\partial U)$. The dense set of functions will be $C^1(\bar{U}, \mathbf{R}^n)$, and the dense set of points will be the "regular values" of $f \in C^1(\bar{U}, \mathbf{R}^n)$ in the sense of the following definition. (Recall that $f \in C^k(\bar{U}, \mathbf{R}^n)$ if f is a C^k mapping from U into $\mathbf{R}^n$, and $f, Df, \ldots, D^k f$ are all continuous on $\bar{U}$; see Chapter 13, §A.)

Definition 12.1. (i) $x_0 \in U$ is a *regular point* of f if $df(x_0)$ is nonsingular; otherwise x_0 is called a *critical point* of f.

(ii) $y_0 \in \mathbf{R}^n$ is called a *regular value* of f if $f^{-1}(y_0)$ contains no critical points of f; otherwise y_0 is called a *critical value*.

The denseness of the set of regular values comes from the celebrated theorem of A. Sard:

Theorem 12.2. *If $f \in C^1(\bar{U}, \mathbf{R}^n)$, then the set of critical values of f has measure zero.*

Proof. Since $U = \bigcup_{k=1}^{\infty} C_k$ where each C_k is a hypercube, we may assume that U itself is a hypercube $C_k = C$ of side λ. We then divide each "edge" of C into N equal parts; this gives a partition of C into N^n congruent hypercubes, the face of each one having $(n-1)$-dimensional volume λ/N. Let Q denote any one of these smaller hypercubes. If x and x_0 are in Q, then by the mean value theorem we can write

$$f(x) = f(x_0) + df_{x_0}(x - x_0) + (x - x_0)o(\lambda/N),$$

since f has *uniformly* continuous derivative in Q. Now given $\varepsilon > 0$, we choose N so large that the $o(\lambda/N)$ term equals $\gamma_n \varepsilon$, where γ_n is independent of N. Thus

$$|f(x) - f(x_0) - df_{x_0}(x - x_0)| \le \tilde{C}\varepsilon\lambda/N, \tag{12.1}$$

where $\tilde{C}$ doesn't depend on N. Assume now that x_0 is a critical point; then $\det(df_{x_0}) = 0$ so (12.1) shows that Q gets mapped by f into a "cylinder" whose "base" lies in an $(n-1)$-dimensional plane whose "height" is at most $\tilde{C}\varepsilon\lambda/N$. Since $|f(x) - f(y)| \le M|x - y| \le M\lambda/N$, where M is an upper bound for $|df|$ in C, we see that the base of the cylinder has $(n-1)$-dimensional volume at most $M^{n-1}\lambda^{n-1}/N^{n-1}$. It follows that the image of Q has volume not exceeding $M^{n-1}\tilde{C}\varepsilon\lambda^n/N^n$. If we sum over all such Q, we find that the image of the critical point set has volume not exceeding

$$\frac{N^n \cdot M^{n-1}\tilde{C}\varepsilon\lambda^n}{N^n} = M^{n-1}\tilde{\tilde{C}}\lambda^n\varepsilon.$$

Since $\varepsilon > 0$ was arbitrary, this shows that the set of critical values in each C_k has measure zero. The same then holds for U. $\square$

In what follows, we shall frequently use the following condition:

$$f(x) \neq y_0 \quad \text{if } x \in \partial U. \tag{12.2}$$

Suppose that $f \in C^1(\bar{U}; \mathbf{R}^n)$. Let y_0 be a regular value of f. Then the set

$$f^{-1}(y_0) = \{x \in \bar{U} : f(x) = y_0\}$$

is finite. This follows from the inverse function theorem; namely, $f^{-1}(y_0)$ must be discrete, and thus cannot have a limit point in $\bar{U}$. We define the *degree* of f at y_0 as

$$d(y_0) = \sum_{x \in f^{-1}(y_0)} \text{sgn}[\det df_x]. \tag{12.3}$$

Note that $d(y_0)$ is an integer, positive, negative, or zero.

We want to extend this definition in two directions; namely

(i)   to functions $f \in C^1(\bar{U}, \mathbf{R}^n)$ at points y_0 satisfying (12.2), where df_{x_0} is singular; and

(ii) to continuous functions; i.e., to $f \in C(\bar{U}, \mathbf{R}^n)$.

The idea for (i) is to use Sard's theorem to approximate y_0 by a sequence $y_k \to y_0$, where the y_k are regular values of f, and then to define $d(f, y_0, U) = \lim_{k\to\infty} d(f, y_k, U)$. This requires that we show that the limit exists, is finite, and is independent of the approximating sequences. To show (ii), we can approximate f by $f_k \in C^1(\bar{U}, \mathbf{R}^n)$ and define $d(f, y_0, U) = \lim_{k\to\infty} d(f_k, y_0, U)$; again we must show that this limit exists, is finite, and doesn't depend on the approximating sequence. This program can be carried out (see [BB]), but it involves a good deal of work. Instead of doing this, we prefer to give an alternate, much quicker approach to degree theory, which is based on differential forms. For the reader who is not familiar with this notion, we have included a short appendix at the end of this section which gives the basic ideas and results.

Thus as above, let $f \in C^1(U, \mathbf{R}^n)$, where U is an open set with compact closure. Let $y_0 \in \mathbf{R}^n$ satisfy (12.2). Here is the basic definition:

Definition 12.3. Let $y_0 \in \mathbf{R}^n \backslash f(\partial U)$, and let[2] $\mu = \phi(y)\, dy$ be a C^∞ n-form on $\mathbf{R}^n$ having compact support $K \subset \mathbf{R}^n \backslash f(\partial U)$, such that $y_0 \in K$, and $\int \mu = 1$. We define the *degree of f at y_0* to be

$$d(f, U, y_0) = \int_U \mu \circ f. \tag{12.4}$$

Differential forms μ, which satisfy these conditions, are called *admissible* for y_0 and f.

We must show that $d(f, U, y_0)$ is well defined; i.e., that it is independent of μ. This is a consequence of the following lemma.

Lemma 12.4. *Let $\mu = \phi(y)\, dy$ be a C^∞ n-form on $\mathbf{R}^n$ having compact support K, satisfying $\int \mu = 0$. Then there exists an $(n-1)$-form ω with $\mu = d\omega$ and* $\operatorname{spt} \omega \subset K$.

To see now that the degree is well defined, suppose that μ and η are admissible for y_0 and f. Then $\mu - \eta$ satisfies the conditions of the lemma, so there is an $(n-1)$-form ω having compact support in K with $\mu - \eta = d\omega$. Then from the divergence theorem

$$\int_U \mu \circ f - \int_U \eta \circ f = \int_U (\mu - \eta) \circ f = \int_U d\omega \circ f$$

$$= \int_U d(\omega \circ f) = \int_{\partial U} \omega \circ f = 0,$$

since $(\operatorname{spt} \omega) \subset K$ implies that $\omega = 0$ on $f(\partial U)$; thus

$$\int_U \mu \circ f = \int_U \eta \circ f,$$

and $d(f, U, y_0)$ is well defined.

We now prove Lemma 12.4. We may assume that $\operatorname{spt} \mu$ is contained in a cube C in $\mathbf{R}^n$. We must show that we can write ϕ as

$$\phi(y) = \sum_1^n \frac{\partial \psi_i}{\partial y_i},$$

---

[2] We use the notation $dy = dy_1 \wedge dy_2 \wedge \cdots \wedge dy_n$ throughout this section.

where spt $\psi_i \subset C$ for each i. The proof proceeds by induction on the dimension n. The case $n = 1$ is easy; namely, put $\psi_1(y) = \int_{-\infty}^y \phi(t)\,dt$, and note that ψ_1 has compact support ($\int \mu = 0$), and that $d\psi_1 = \mu$. Thus, we assume that the lemma is true in n-dimensions, and we prove it for $(n + 1)$-dimensions.

Let $(y, t) = (y_1, \ldots, y_n, t)$, $y_{n+1} = t$, and set

$$\theta(y) = \int_{-\infty}^{\infty} \phi(y, t)\,dt.$$

Now $\int \theta\,dy = 0$, so by our induction hypothesis,

$$\theta(y) = \sum_1^n \frac{\partial \psi_i}{\partial y_i},$$

where spt $\psi_i \subset C_n$, the projection of C on y-space $1 \le i \le n$. Let $\rho(t)$ be a C_0^∞ function with support in the projection of C on the t-axis, and $\int \rho(t)\,dt = 1$. Consider the function $\phi(y, t) - \rho(t)\theta(y)$. We have

$$\int [\phi(y, t) - \rho(t)\theta(y)]\,dt = \int \phi(y, t)\,dt - \theta(y) \int \rho(t)\,dt$$

$$= \theta(y) - \theta(y) = 0,$$

so that if we put

$$\psi(y, t) = \int_{-\infty}^t [\phi(y, s) - \rho(s)\theta(y)]\,ds,$$

we see that $\partial \psi / \partial t = \phi(y, t) - \rho(t)\theta(y)$, and that $\psi(y, t)$ has support in C. Finally,

$$\phi(y, y_{n+1}) = \phi(y, t) = \frac{\partial \psi}{\partial t} + \rho(t)\theta(y)$$

$$= \frac{\partial \psi}{\partial y_{n+1}} + \sum_1^n \frac{\partial(\rho(y_{n+1})\psi_i(y))}{\partial y_i},$$

as desired. This completes the proof of Lemma 12.4. □

We shall now obtain some basic properties of the degree.

Property 1. If $|y_1 - y_0|$ is small, $d(f, U, y_1) = d(f, U, y_0)$.

Proof. If μ is admissible for y_0, then it is admissible for y_1, if $|y_0 - y_1|$ is sufficiently small. □

Property 2. *If y_0 is a regular value for f, then $d(f, U, y_0) = d(y_0)$. In particular, $d(f, U, y_0) = 0$ if $y_0 \notin f(\bar{U})$.*

Proof. Let $f^{-1}(y_0) = \{x_1, x_2, \ldots, x_m\}$. By the inverse function theorem, each x_i has a neighborhood N_i on which f is a homeomorphism. We may choose these so small that $N_i \cap N_j = \phi$ if $i \neq j$. Set $N = \cap f(N_i)$; then $y_0 \in N$. Let μ have support in N and let it be admissible for y_0 and f. Then

$$d(f, U, y_0) = \int_U \mu \circ f = \sum_{i=1}^m \int_{N_i} \mu \circ f$$

$$= \sum_{i=1}^m \int_{f(N_i)} \operatorname{sgn}[\det df_{x_i}]\mu \quad \text{(see (12.11) in the appendix)}$$

$$= \sum_{i=1}^m \operatorname{sgn}[\det df_{x_i}] \int_N \mu \quad (\int_{f(N_i)\setminus N} \mu = 0)$$

$$= \sum_{i=1}^m \operatorname{sgn}[\det df_{x_i}]$$

$$= d(y_0).$$

Finally, the last statement is immediate, since if $y_0 \notin f(\bar{U})$, then y_0 is a regular value and $d(y_0) = 0$. $\square$

Corollary 12.5. *If y lies in the same component of y_0 (in $\mathbf{R}^n \setminus f(\partial U)$), then $d(f, U, y) = d(f, U, y_0)$. In particular, the degree is an integer-valued function.*

For, we may connect y to y_0 by an arc Γ in the component. Property 1 and the compactness of Γ give the desired result. The second statement follows by Sard's theorem and Property 2. $\square$

Property 3 (Homotopy Invariance). *Let $\{f_t(\cdot)\}$, be a continuous (from $[0, 1]$ to $L_\infty(U)$) one-parameter family of mappings taking $\bar{U} \times [0, 1]$ into $\mathbf{R}^n$, which is C^1 on U, for each fixed t in $[0, 1]$. Assume that $y_0 \notin f_t(\partial U), 0 \leq t \leq 1$. Then $d(f_t, U, y_0)$ is independent of t.*

Proof. Let $Y = \{f_t(x): x \in \partial U, 0 \leq t \leq 1\}$; then $y_0 \notin Y$, and Y is compact. Let μ be admissible for y_0 and each f_t, where (spt μ) $\cap Y = \phi$. Then

$$d(f_t, U, y_0) = \int_U \mu \circ f_t,$$

and this function is easily seen to be continuous in t. Since the degree is integer valued, it must be constant in t. $\square$

Property 4 (Dependence Only on Boundary Values). *If $f\vert_{\partial U} = g\vert_{\partial U}$, and $y_0 \notin f(\partial U) = g(\partial U)$, then $d(f, U, y_0) = d(g, U, y_0)$.*

Proof. Apply Property 3 to the family $tf + (1 - t)g$, $0 \le t \le 1$. Since f and g agree on ∂U, the hypotheses of Property 3 are satisfied. $\square$

Property 5. *Let* $\{U_i\}$ *be a countable family of disjoint open sets contained in* U. *Let* $y_0 \notin f(\bar{U} \backslash \bigcup U_i)$. *Then* $d(f, U_i, y_0)$ *is zero for all but a finite number of* i, *and*

$$d(f, U, y_0) = \sum_i d(f, U_i, y_0).$$

Proof. $\bar{U} \backslash \bigcup U_i$ is closed and thus compact; hence $f(\bar{U} \backslash \bigcup U_i)$ is compact. Let N be a connected neighborhood of y_0 disjoint from this latter compact set, and let y be a regular value in N (Sard's theorem). Then by Corollary 12.5, $d(f, U, y_0) = d(f, U, y)$, and for each i, $d(f, U_i, y_0) = d(f, U_i, y)$. Since $f^{-1}(y)$ is finite, it must be contained in a finite number of the U_i's, say $U_1, \ldots, U_k$. Then if $f^{-1}(y) = \{x_1, \ldots, x_k\}$, where $x_i \in U_i$,

$$d(f, U, y_0) = d(f, U, y) = d(y)$$

$$= \sum_1^k \mathrm{sgn}[\det df_{x_i}]$$

$$= \sum_1^k d(f, U_i, y)$$

$$= \sum_1^k d(f, U_i, y_0). \quad \square$$

Property 6 (Excision). *Let* Q *be a closed set in* U, *and suppose* $y_0 \notin f(Q)$; *then*

$$d(f, U, y_0) = d(f, U \backslash Q, y_0)$$

Proof. If we set $U_1 = U \backslash Q$, then the result follows immediately from Property 5. $\square$

Property 7. *Let* U *and* $\tilde{U}$ *be bounded open subsets of* $\mathbf{R}^n$ *and* $\mathbf{R}^m$, *respectively, and suppose* $f \in C^1(clU, \mathbf{R}^n)$, *and* $\tilde{f} \in C^1(cl\tilde{U}, \mathbf{R}^m)$. *Then if* $y_0 \in \mathbf{R}^n \backslash f(\partial U)$, *and* $\tilde{y}_0 \in \mathbf{R}^m \backslash \tilde{f}(\partial \tilde{U})$,

$$d(f \times \tilde{f}, U \times \tilde{U}, (y_0, \tilde{y}_0)) = d(f, U, y_0) \, d(\tilde{f}, \tilde{U}, \tilde{y}_0). \tag{12.5}$$

Proof. $(y_0, \tilde{y}_0) \in \mathbf{R}^{n+m} \backslash (f \times \tilde{f})(\partial(U \times \tilde{U}))$, so that the left-hand side of (12.5) is defined. Let μ and $\tilde{\mu}$ be admissible for (y_0, f) and $(\tilde{y}_0, \tilde{f})$, respectively. Then $\mu \wedge \tilde{\mu}$ is an admissible $(n + m)$-form for $f \times \tilde{f}$ at $(y_0, \tilde{y}_0)$. Thus

$$\int_{U \times \tilde{U}} (\mu \wedge \tilde{\mu}) \circ (f \times \tilde{f}) = \int_U \mu \circ f \int_{\tilde{U}} \tilde{\mu} \circ \tilde{f},$$

from which the result follows. $\square$

Property 8. *If the vectors $f(x)$ and $g(x)$ never point in opposite directions for $x \in \partial U$ (i.e., $f(x) + \lambda g(x) \neq 0$ for all $\lambda \geq 0$, $x \in \partial U$), then $d(f, U, 0) = d(g, U, 0)$, provided that the right-hand side is defined; i.e., $0 \notin g(\partial U)$.*

Proof. This is an immediate consequence of the homotopy invariance, using the homotopy $t[f(x)] + (1 - t)[g(x)]$. $\square$

In particular, if U is a ball in $\mathbf{R}^n$ centered at the origin, and $f(x)$ never points opposite to x for $x \in \partial U$, then the equation $f(x) = 0$ has a solution inside U. This holds since

$$d(f, U, 0) = d(I, U, 0) = 1,$$

where I is the identity in $\mathbf{R}^n$.

We need one more property which will be used later.

Property 9 (Composition). *Let $f \in C^1(\overline{U}, V)$, $g \in C^1(\overline{V}, \mathbf{R}^n)$, where U and V are bounded open subsets of $\mathbf{R}^n$, and let $\{V_j\}$ be the set of open connected subsets of $V \backslash f(\partial U)$, (whose closures are disjoint compact subsets contained in V). Then if $z_0 \in \mathbf{R}^n \backslash (g \circ f)(\partial U)$,*

$$d(g \circ f, U, z_0) = \sum_j d(f, U, V_j)\, d(g, V_j, z_0),$$

and the sum on the right is finite.

(Here $d(f, U, v)$ is constant for all $v \in V_j$, by Corollary 12.5; thus, $d(f, U, V_j)$ is defined to be $d(f, U, v_j)$, $v_j \in V_j$.)

Proof. We may suppose that z_0 is a regular value of both g and $g \circ f$. Then

$$d(g \circ f, U, z_0) = \sum_{\substack{u \in U \\ (g \circ f)(u) = z_0}} \operatorname{sgn} \det d(g \circ f)_u$$

$$= \sum_{\substack{u \in U \\ (g \circ f)(u) = z_0}} \operatorname{sgn} \det dg_{f(u)} \operatorname{sgn} \det df_u$$

$$= \sum_{\substack{v \in V \\ g(v) = z_0}} \operatorname{sgn} \det dg_v \sum_{\substack{u \in U \\ f(u) = v}} \operatorname{sgn} \det df_u$$

$$= \sum_{\substack{v \in V \\ g(v) = z_0}} \operatorname{sgn} \det dg_v\, d(f, U, v).$$

If v is in a component of $V \setminus f(\partial U)$, then this component is disjoint from $f(\overline{U})$ so that $d(f, U, v) = 0$ (by Property 2 and Corollary 12.5). Thus by Property 5, $d(f, U, v) = \sum_j d(f, U, V_j)$, and

$$d(g \circ f, U, z_0) = \sum_j d(f, U, V_j) \sum_{\substack{v \in V_j \\ g(v) = z_0}} \text{sgn} \det dg_v$$

$$= \sum_j d(f, U, V_j) \, d(g, V_j, z_0). \quad \square$$

We have proved many of the important properties of degree theory, with one notable exception; namely, we have not shown that the degree is a topological notion. In other words, we have not extended the concept of degree to *continuous* mappings. We now turn our attention to this problem.

Let $f \in C(\overline{U}, \mathbf{R}^n)$, and let $\{f_n\}$ be a sequence of functions in $C^1(\overline{U}, \mathbf{R}^n)$ such that f_n converges uniformly to f on $\overline{U}$. If $y_0 \notin f(\partial U)$, then for n sufficiently large, $y_0 \notin f_n(\partial U)$ so that $d(f_n, U, y_0)$ is defined. We set

$$d(f, U, y_0) = \lim_{n \to \infty} d(f_n, U, y_0). \tag{12.6}$$

Lemma 12.6. *The above limit exists and is independent of $\{f_n\}$.*

Proof. Let $\delta = \text{dist}(y_0, f(\partial U))$; then $0 < \delta < \infty$ since $f(\partial U)$ is compact. Let $\{g_n\}$ be another sequence of functions in $C^1(\overline{U}, \mathbf{R}^n)$ with g_n converging uniformly to f in $\overline{U}$. We choose an integer N so large that $n \geq N$ implies that $\|f_n - f\| + \|g_n - f\| < \delta/2$, where $\|\cdot\| = \|\cdot\|_{L_\infty(U)}$.

Suppose that

$$y_0 = t f_n(x) + (1 - t) g_n(x) \quad \text{for some } x \in \partial U, n \geq N, \qquad 0 \leq t \leq 1.$$

Then

$$
\begin{aligned}
y_0 - f(x) &= y_0 - t f(x) - (1 - t) f(x) \\
&= t[f_n(x) - f(x)] + (1 - t)[g_n(x) - f(x)],
\end{aligned}
$$

so $|y_0 - f(x)| < \delta/2$. This is impossible. Hence, if $n \geq N$ we can apply the homotopy invariance property to the family $t f_n + (1 - t) g_n, 0 \leq t \leq 1$, and conclude that

$$d(f_n, U, y_0) = d(g_n, U, y_0). \tag{12.7}$$

Thus, if the limit exists, it is independent of $\{f_n\}$. To see that the limit indeed exists, we merely apply the same argument to f_n and f_m with $n, m \geq N$. Then, as in (12.7), we conclude that $d(f_n, U, y_0) = d(f_m, U, y_0)$, so the sequence $\{d(f_n, U, y_0)\}$ stabilizes (i.e., is constant), for $n \geq N$. This completes the proof. $\square$

Remark. With the aid of this lemma, we note that if $y_0 \notin f(\partial U)$, then there is a neighborhood N of f (L_∞-topology) such that all C^1 functions in N have the same degree.

Theorem 12.7. *The Properties 1–9 are valid for continuous functions. Moreover, if y_0 and y_1 are in the same component of $\mathbf{R}^n \backslash f(\partial U)$, then $d(f, U, y_0) = d(f, U, y_1)$.*

Proof. We shall prove these in a convenient order. Property 1 is immediate from our above remark. Property 3 then follows from Property 1 since $d(f_t, U, y_0)$ is continuous in t and has discrete range (the integers). Property 4 follows as before from Property 3. Property 2 must be stated as follows: if $y_0 \notin f(\bar{U})$ then $d(f, U, y_0) = 0$; this is obvious since it holds for each of the approximating C^1-mappings. All of the remaining properties, as well as the statement in the last sentence, are similarly obtainable from the C^1 approximations. The proof is considered complete. $\square$

We conclude this development with two important topological results. Suppose that ϕ is a continuous mapping $\phi: \partial U \to \mathbf{R}^n \backslash \{y_0\}$. If f is *any* continuous extension of ϕ to U, then $d(f, U, y_0)$ is, of course, defined. We claim however, that this degree is independent of the particular extension. Indeed, if g is any other extension, then if $f_t \equiv tf + (1 - t)g, 0 \le t \le 1$, the homotopy property shows that $d(f_t, U, y_0)$ is independent of t. We can thus give an unambiguous meaning to $d(\phi, U, y_0)$. This yields the following important properties.

Property 10. *Let $f, g \in C(U, \mathbf{R}^n)$ and suppose f and g can be continuously extended to $\bar{U}$, and $f|_{\partial U} = g|_{\partial U}$. If $y_0 \notin f(\partial U)$, then $d(f, U, y_0) = d(g, U, y_0)$; thus the degree depends only on $f|_{\partial U}$.*

Property 11. *If $\phi \in C(\partial U, \mathbf{R}^n)$, and $y_0 \notin \phi(\partial U)$, then $d(\phi, U, y_0)$ depends only on the homotopy class of ϕ.*

Proof. Let $\phi_t, 0 \le t \le 1$, be a homotopy deformation of ϕ, with $\phi_0 = \phi$, such that $y_0 \notin \phi_t(\partial U)$, if $0 \le t \le 1$. If f_0 and f_1 are, respectively, continuous extensions of ϕ_0 and ϕ_1 to all of U, then since $y_0 \notin \phi_t(\partial U)$, so we again see that $d(\phi_t, U, y_0)$ is independent of t. This proves the result. $\square$

We shall now give a few of the standard applications of the theory which we have developed. Here is one easy result. Suppose $f \in C(\mathbf{R}^n, \mathbf{R}^n)$ and for some $k > 0, k \in \mathbf{Z}$,

$$\lim_{|x| \to \infty} \frac{\langle f(x), x \rangle}{|x|^k} = +\infty.$$

Then for each $y_0 \in \mathbf{R}^n$, there is an x_0 in $\mathbf{R}^n$ with $f(x_0) = y_0$; i.e., f is *onto*.

To see this, note that we may assume $y_0 = 0$ since otherwise, we can replace $f(x)$ by $f(x) - y_0$ and the hypotheses still hold. Now there is an $r > 0$ such that

$$\langle f(x), x \rangle > 0 \quad \text{if } |x| = r.$$

If $f(x) \neq 0$ for $|x| = r$, then $\langle f(x), x \rangle > 0$ implies that $f(x)$ never points opposite to x if $|x| = r$. Thus $f(x) = 0$ has a solution $\bar{x}$ with $|\bar{x}| < r$, by Property 8.

As a next application, we have the following so-called "no retraction theorem."

Theorem 12.8. *Let D be the open unit ball in $\mathbf{R}^n$. Then there is no continuous mapping $f : \bar{D} \to \partial D$ such that $f|_{\partial D}$ is the identity.*

Proof. Were there such a mapping, then since $0 \in \mathbf{R}^n \backslash f(\partial D)$, Property 4 shows that

$$d(f, D, 0) = d(I, D, 0).$$

But $d(I, D, 0) = 1$, by Property 2 and (12.3). Thus $d(f, D, 0) = 1$, so again by Property 2, $0 \in f(D)$. This is impossible since $f(D) \subset \partial D$. $\quad\square$

Theorem 12.9 (Brouwer Fixed Point Theorem). *Let D be any set homeomorphic to an open ball in $\mathbf{R}^n$, and let $\phi : \bar{D} \to \bar{D}$ be continuous. Then ϕ has a fixed point in $\bar{D}$; i.e., there is an $\bar{x} \in \bar{D}$ with $\phi(\bar{x}) = \bar{x}$.*

Proof. First note that if D' is homeomorphic to D and if the theorem holds for D', then it holds for D. In fact, let ψ map D' homeomorphically onto D. Then $\psi^{-1}\phi\psi$ maps $\bar{D}'$ continuously into itself; hence it has a fixed point p by hypothesis; so $\psi^{-1}\phi\psi(p) = p$. But then $\phi(\psi(p)) = \psi(p)$ so $\psi(p)$ is a fixed point of ϕ. We may therefore assume that D is the unit ball centered at the origin.

If ϕ had no fixed points, then for each $x \in \bar{D}$, the points x and $\phi(x)$ define a line: $\lambda x + (1 - \lambda)\phi(x)$, $\lambda \in \mathbf{R}$. Let $f(x)$ be the unique point on this line having norm 1, where $\lambda \geq 1$. Then f maps $\bar{D}$ into ∂D continuously, and $f|_{\partial D} = I$. This contradicts the last result and thus the theorem is proved. $\quad\square$

We turn now to a few applications of Brouwer's theorem to differential equations. Recall that a closed trajectory (periodic orbit) of a plane autonomous system of ordinary differential equations must contain "within it" a rest point. We shall generalize this result to n-dimensional autonomous systems. Thus consider the system of ordinary differential equations

$$\dot{x} = f(x), \tag{12.8}$$

where $x \in \mathbf{R}^n$ and f is an n-vector which is locally Lipschitzian.

Theorem 12.10. *Suppose that D is homeomorphic to an n-ball and that every trajectory of (12.8) starting in $\bar{D}$ at $t = 0$ stays inside $\bar{D}$ for all $t > 0$. Then $\bar{D}$ contains at least one rest point of the system.*

Proof. Let $\{t_k\}$ be a sequence of positive real numbers which converges to zero, and consider the "time-t_k" mappings, which carry any $x_0 \in \bar{D}$ into $x_0 \cdot t_k$, the point on the trajectory through x_0 at "time" t_k. This is a continuous mapping of $\bar{D}$ into $\bar{D}$, so by Brouwer's theorem, there is a point x_k with $x_k \cdot t_k = x_k$. Using the compactness of $\bar{D}$, and passing to a subsequence, we may assume that the sequence $\{x_k\}$ converges to $\bar{x} \in \bar{D}$.

We shall show that $\bar{x}$ is a rest point of (12.8). To this end, note that for each k, and each t

$$x_k \cdot \left[\frac{t}{t_k} \right] t_k = x_k \tag{12.9}$$

where $[\ ]$ denotes the greatest integer. Since the left-hand side of (12.9) tends to $\bar{x} \cdot t$ as $k \to \infty$, we see that $\bar{x} \cdot t = \bar{x}$, for all t. Thus $\bar{x}$ is a rest point, as asserted. $\square$

We shall next prove a theorem on periodic orbits.

Theorem 12.11. *Let (12.8) be a three-dimensional system, and let T be a (solid) 3-torus such that each trajectory of (12.8) starting in T at $t = 0$ stays inside T for all $t > 0$. Suppose too that every trajectory in T doesn't[3] "turn around." Then T contains a periodic orbit.*

Proof. Referring to Figure 12.1, let the disk D be a cross-section of T. If $x \in D$, then the trajectory through x meets D again at some positive time.

---

[3] That is, every point x moving along a trajectory in T has positive angular velocity about the axis of T. If the x_3-axis is chosen as the axis of T, this means that $x_1 \dot{x}_2 - x_2 \dot{x}_1 = x_1 f_2(x) - x_2 f_1(x) > 0$.

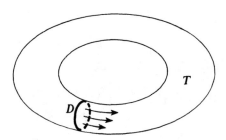

Figure 12.1

Thus the flow induces a continuous mapping $x \to \phi(x)$ of D into D. Brouwer's theorem implies the existence of a fixed point $\bar{x}$ of this mapping; i.e., $\bar{x} = \phi(\bar{x})$. But then the trajectory through $\bar{x}$ is clearly a periodic orbit.   □

Note. ϕ is often called the "first-return map," or the "Poincaré map."

We close this section with a short appendix on differential forms. For more details, the interested reader should consult [Fl].

Appendix: Differential Forms

Let P be a point in $\mathbf{R}^n$. A *one-form* at P is an expression of the form

$$\sum_1^n a_i \, dx_i, \quad a_i \text{ constant.}$$

These form an n-dimensional vector space. The *q-forms* at P are expressions of the form

$$\sum a_I \, dx_{i_1} \wedge \cdots \wedge dx_{i_q}, \quad a_I \text{ constants.}$$

If U is a domain in $\mathbf{R}^n$, a q-form ω on U is defined by choosing a q-form at each point p in U, and doing it smoothly. Thus

$$\omega = \sum a_I(x) \, dx_{i_1} \wedge \cdots \wedge dx_q,$$

where each a_I is smooth. These form a vector space over $\mathbf{R}$ which we denote as $F^{(q)}(U)$; q is called the *degree* of ω.

There is an operator d, called *exterior differentiation*, $d: F^{(q)}(U) \to F^{(q+1)}(U)$, which is characterized by the following axioms:

(i) $d(\omega + \lambda) = d\omega + d\lambda.$
(ii) $d(\omega \wedge \lambda) = d\omega \wedge \lambda + (-1)^{\deg \omega} \omega \wedge d\lambda.$
(iii) $d(d\omega) = 0$ *for every* ω.
(iv) *If f is a smooth real-valued function*

$$df = \sum_i \frac{\partial f}{\partial x_i} \, dx_i.$$

Thus, if ω is a smooth $(n-1)$-form, so that

$$\omega = \sum_{j=1}^n (-1)^{j-1} f_j(y) \, dy_1 \wedge \cdots \wedge dy_{j-1} \wedge dy_{j+1} \wedge \cdots \wedge dy_n,$$

then

$$d\omega = \sum_{j=1}^{n} \frac{\partial f_j}{\partial y_j} dy_1 \wedge \cdots \wedge dy_n.$$

We shall denote $dy_1 \wedge \cdots \wedge dy_n$ by dy. If μ is a smooth n-form, $\mu = f(y)dy$, then its *pull back* with respect to a mapping $\phi \in C^1(U, \mathbf{R}^n)$ is defined by

$$(\mu \circ \phi)(x) = f(\phi(x)) \det d\phi_x \, dx.$$

We may integrate n-forms; in particular, we have the *divergence theorem*: if ω is an $(n - 1)$-form

$$\int_{\partial U} \omega = \int_{U} d\omega.$$

Therefore, if ω has compact support in U, then

$$\int_{U} d\omega = 0.$$

Finally, we recall the "change of variables" formula from advanced calculus: If $y = \phi(x)$ is a smooth bijective transformation with $d\phi$ nowhere singular, then

$$\int_{\mathbf{R}^n} f(y) \, dy = \int_{\mathbf{R}^n} f(\phi(x)) |\det d\phi_x| \, dx, \tag{12.10}$$

so that for an n-form $\mu = f(y) \, dy$,

$$\int_{\mathbf{R}^n} \mu = \operatorname{sgn} \det d\phi \int_{\mathbf{R}^n} \mu \circ \phi. \tag{12.11}$$

Here the term "sgn" is needed because of orientation considerations. Finally, ϕ "commutes" with exterior differentiation; namely, $d(\omega \circ \phi) = (d\omega) \circ \phi$.

§B. The Leray–Schauder Degree

We want to extend the concept of degree to infinite-dimensional spaces so as to make it applicable to partial differential and integral equations. This is indeed possible, and not very difficult, once one isolates the "correct" class of mappings. That is, we must find a class of mappings to which the notion of degree makes sense. That continuity is not enough can be seen by the following example.

EXAMPLE.**12.12.** Let l_2 denote the space of infinite sequences $x = (x_1, x_2, \ldots)$ with $\|x\|^2 = \sum |x_i|^2 < \infty$. Let D be the closed unit ball in l_2 and let T be the transformation on l_2 defined by $Tx = (\sqrt{1 - \|x\|^2}, x_1, x_2, \ldots)$. T is clearly continuous, and if $\|x\| \le 1$, $\|Tx\|^2 = (1 - \|x\|^2) + \|x\|^2 = 1$. Thus T maps D into ∂D. This implies that T cannot have any fixed points. For, if x were a fixed point, then $\|x\| = \|Tx\| = 1$ so $Tx = x$ gives

$$(0, x_1, x_2, \ldots) = (x_1, x_2, \ldots).$$

But then $x = 0$, and this is impossible.

This example shows that the Brouwer fixed point theorem is not valid in Banach spaces with merely a continuity hypothesis. The correct class of mappings in the infinite-dimensional framework, is the so-called compact operators; see Chapter 11. We reformulate the definition of compact operators in Banach spaces, for mappings which are not necessarily linear.

Definition 12.13. A continuous mapping T of a subset U of a Banach space B into B is called *compact* if $\overline{T(K)} \equiv \mathrm{cl}(T(K))$ is compact for every closed and bounded subset $K \subset U$.

The property of compact mappings which enables us to extend Brouwer's theorem to infinite dimensions is given in the following lemma.

Lemma 12.14. *Let K be a closed and bounded subset of a Banach space B, and suppose that $T: K \to B$ is compact. Then T is a uniform limit (i.e., limit in the norm-topology on operators) of finite-dimensional mappings (i.e., mappings whose ranges are finite dimensional).*

Proof. Let $\varepsilon > 0$ be given. Since $\overline{T(K)}$ is compact, it can be covered by open balls $N_1, \ldots, N_{j(\varepsilon)}$ each of radius ε, with centers $x_1, \ldots, x_{j(\varepsilon)}$, respectively. Let $\{\psi_i(x): 1 \le i \le j(\varepsilon)\}$ be a partition of unity on $\overline{T(K)}$ subordinate to the cover $N_1, \ldots, N_{j(\varepsilon)}$; i.e., for each $i, \psi_i \ge 0$, $\mathrm{spt}\,\psi_i \subset N_i$, and if $x \in \overline{T(K)}$, $\sum \psi_i(x) = 1$ (see the appendix to Chapter 7). Set

$$T_\varepsilon(x) = \sum_1^{j(\varepsilon)} \psi_i(T(x))x_i, \qquad x \in K. \tag{12.12}$$

Then clearly $T_\varepsilon(x)$ has finite range, and

$$\|T(x) - T_\varepsilon(x)\| = \left\| \sum_1^{j(\varepsilon)} \psi_i(T(x))[x_i - T(x)] \right\|.$$

If $\psi_i(T(x)) \ne 0$, then $\psi_i(T(x)) > 0$ so $T(x) \in N_i$ and $\|x_i - T(x)\| < \varepsilon$. Therefore $\|T(x) - T_\varepsilon(x)\| < \varepsilon$, uniformly in x. This completes the proof. $\square$

We shall call T_ε, as defined in (12.12), an ε-*approximation to T.*

This last result enables us to prove the analogue of Brouwer's theorem for Banach spaces; in this context it is called the *Schauder fixed point theorem.*

Theorem 12.15. *Let D be a closed convex bounded subset of a Banach space B, and let $T: D \to D$ be compact. Then T has a fixed point.*

Proof. Let T_ε be an ε-approximation to T, and notice that $T_\varepsilon(x)$ lies in the convex hull of the x_i's. If V_ε is the linear space spanned by $x_1, x_2, \ldots, x_{j(\varepsilon)}$, then $T_\varepsilon(x)$ lies in the convex hull of V_ε and by convexity, this set lies in D. It follows that the range of T_ε lies in $D \cap V_\varepsilon$. Thus, T_ε maps the set $D \cap V_\varepsilon$ into itself. By the Brouwer fixed point theorem, there is a point x_ε in $D \cap V_\varepsilon$ such that $T_\varepsilon(x_\varepsilon) = x_\varepsilon$.

Since T is compact, $\{T(x_\varepsilon)\}$ has a convergent subsequence, which we again call $\{T(x_\varepsilon)\}$. We claim that $\{x_\varepsilon\}$ is a Cauchy sequence. To see this, note that for any $\varepsilon_1, \varepsilon_2 > 0$,

$$\| x_{\varepsilon_1} - x_{\varepsilon_2} \| \le \| x_{\varepsilon_1} - T(x_{\varepsilon_1}) \| + \| T(x_{\varepsilon_1}) - T(x_{\varepsilon_2}) \| + \| T(x_{\varepsilon_2}) - x_{\varepsilon_2} \|$$

$$= \| T_{\varepsilon_1}(x_{\varepsilon_1}) - T(x_{\varepsilon_1}) \| + \| T(x_{\varepsilon_1}) - T(x_{\varepsilon_2}) \|$$

$$+ \| T(x_{\varepsilon_2}) - T_{\varepsilon_2}(x_{\varepsilon_2}) \|$$

$$\le \varepsilon_1 + \varepsilon_2 + \| T(x_{\varepsilon_1}) - T(x_{\varepsilon_2}) \|,$$

and the claim follows. Thus $\{x_\varepsilon\}$ has a limit; i.e., $x_\varepsilon \to \bar{x}$, and $\bar{x} \in D$ since D is closed. Therefore by continuity, $T(x_\varepsilon) \to T(\bar{x})$. But as

$$\| T(x_\varepsilon) - \bar{x} \| \le \| T(x_\varepsilon) - T_\varepsilon(x_\varepsilon) \| + \| x_\varepsilon - \bar{x} \|$$

$$\le \varepsilon + \| x_\varepsilon - \bar{x} \|,$$

we see that $T(x_\varepsilon) \to \bar{x}$. Thus $T(\bar{x}) = \bar{x}$, and the proof is complete. $\square$

We are now going to extend the concept of degree to mappings of the form $T = I - K$, where I is the identity and K is compact. Thus, let U be a bounded open subset of a Banach space B and let T map $\bar{U}$ into B. Let $y_0 \in B \backslash T(\partial U)$. We claim that $T(\partial U)$ is a closed set. Indeed, if F is any closed bounded set, then $T(F)$ is closed. To see this, let $x_n \in F$, and suppose $x_n - K(x_n) = T(x_n) \to y$. By compactness of K, we may assume (passing to a subsequence, if necessary), that $K(x_n) \to k$. Thus, $x_n \to k + y = x$, and $x \in F$ since F is closed. But since K is continuous, $K(x) = k$; hence $y = x - K(x)$ and our assertion follows.

Since $T(\partial U)$ is closed, and $y_0 \notin T(\partial U)$, we let $\text{dist}(y_0, T(\partial U)) = \delta > 0$. Let $\varepsilon < \delta/2$, and let K_ε be an ε-approximation of K which has range in the finite-dimensional space V_ε containing y_0. Set $T_\varepsilon = I - K_\varepsilon$. Then, $T_\varepsilon(x) \ne$

y_0 if $x \in \partial U$ (otherwise $\exists x_0 \in \partial U$ with $T(x_0) = y_0$ and $\| y_0 - T(x_0) \| \leq$ $\| y_0 - T_\varepsilon(x_0) \| + \| T_\varepsilon(x_0) - T(x_0) \| < \varepsilon < \delta/2$). Thus, for the mapping

$$T_\varepsilon|_{V_\varepsilon \cap \bar{U}} : V_\varepsilon \cap \bar{U} \to V_\varepsilon,$$

the degree $d(T_\varepsilon, V_\varepsilon \cap U, y_0)$ is defined.

We now set

$$d(T, U, y_0) = d(T_\varepsilon, V_\varepsilon \cap U, y_0); \tag{12.13}$$

this is called the *Leray–Schauder degree*. We must show that it is well defined. To do this, we need a lemma.

Lemma 12.16. *Let U be a bounded open subset of $\mathbf{R}^n = \mathbf{R}^{n_1} \oplus \mathbf{R}^{n_2}, n_1 + n_2 = n$. Let f be a mapping from $\bar{U}$ into $\mathbf{R}^n$ of the form $I + \phi$, where $\phi : \bar{U} \to \mathbf{R}^{n_1} \times \{0\}$. Let $(y_0, 0) \in \mathbf{R}^{n_1} \times \{0\} \backslash f(\partial U)$; then $d(f, U, y_0) = d(f|_{U_1}, U_1, y_0)$, where U_1 is the projection of U on $\mathbf{R}^{n_1}$.*

Proof. It suffices to prove the lemma for $f \in C^1(U, \mathbf{R}^n)$ and then extend it to continuous f by the usual approximation arguments. Also, we can assume that $y_0 = 0$ (otherwise we translate coordinates).

We shall write $x \in \mathbf{R}^n$ uniquely as $x = x_1 + x_2$ where $x_i \in \mathbf{R}^n$, $i = 1, 2$. We choose functions $f_i(x_i)$ in $C_0^\infty(\mathbf{R}^{n_i})$ having small support about $0 \in \mathbf{R}^{n_i}$, $f_i(0) > 0$, $f_i \geq 0$, and such that $\int_{\mathbf{R}^{n_i}} f_i(x_i) \, dx_i = 1$. By (12.4),

$$d(f, U, 0) = \int_U (f_1 f_2) \circ f.$$

Thus from (12.10), since $x + \phi(x) = [x_1 + \phi(x_1 + x_2)] + x_2$,

$$d(f, U, 0) = \int_{\mathbf{R}^{n_1}} \int_{\mathbf{R}^{n_2}} f_1(x_1 + \phi(x_1 + x_2)) f_2(x_2) |\det(I + d\phi_{x_1})| \, dx_1 \, dx_2,$$

so that if $f_2 \to \delta$, the "delta function," we have

$$d(f, U, 0) = \int_{\mathbf{R}^{n_1}} f_1(x_1 + \phi(x_1)) |\det(I + d\phi_{x_1})| \, dx_1$$

$$= \int_{\mathbf{R}^{n_1}} f_1 \circ f = d(f|_{U_1}, U_1, 0).$$

This proves the lemma. $\qquad \square$

We can now show that the degree, as given by (12.13), doesn't depend on V_ε. Note first that if $V = V_\varepsilon \oplus N$, where N is finite dimensional, then it follows from Lemma 12.16 that $d(T_\varepsilon, V \cap U, y_0) = d(T_\varepsilon, V_\varepsilon \cap U, y_0)$. Thus, if K_η is another approximation of K such that $K_\eta: U \to V_\eta$, $\eta < \delta/2$, and $V = V_\varepsilon \oplus V_\eta$, then again by Lemma 12.16,

$$d(T_\varepsilon, V_\varepsilon \cap U, y_0) = d(T_\varepsilon, V \cap U, y_0), \quad \text{and}$$

$$d(T_\eta, V_\eta \cap U, y_0) = d(T_\eta, V \cap U, y_0).$$

If $T_t = tT_\eta + (1 - t)T_\varepsilon$, then $y_0 \notin T_t(\partial U)$ from the definition of δ, and by the homotopy invariance,

$$d(T_\varepsilon, V \cap U, y_0) = d(T_\eta, V \cap U, y_0),$$

so that

$$d(T_\varepsilon, V_\varepsilon \cap U, y_0) = d(T_\eta, V_\eta \cap U, y_0).$$

This shows that (12.13) doesn't depend on how we approximate K.

Now just as we have done before, we can extend all of the properties of degree that hold in $\mathbf{R}^n$, to Banach spaces B, where the mappings T that we consider are of the form $I - K$, where K is compact. In addition, the mapping T need only be defined on ∂U, with range in $B \backslash \{y_0\}$, where $K: \partial U \to B$ is compact. Then as before, $d(T, U, y_0)$ depends only on the homotopy class of mappings $T: \partial U \to B \backslash \{y_0\}$, where the homotopy consists of mappings of the form $I - K_t$, $0 \le t \le 1$, with $K_t(\cdot)$ compact, which send $\partial U \times [0, 1]$ into $B \backslash \{y_0\}$.

Finally, if $T = I - K: \bar{U} \to B$ and $d(T, U, y_0) \neq 0$, then $T(x) = y_0$, for some $x \in \bar{U}$. Indeed, if $y_0 \notin T(\bar{U})$, then since $T(\bar{U})$ is closed, $\text{dist}(T(\bar{U}), y_0) > 0$. Thus we can find an ε-approximation T_ε of T (actually, there is an ε-approximation K_ε of K, and $T_\varepsilon \equiv I - K_\varepsilon$), such that $T_\varepsilon: V_\varepsilon \cap U \to V_\varepsilon$ and $y_0 \notin T(\overline{V_\varepsilon \cap U})$. It follows that $d(T_\varepsilon, V_\varepsilon \cap U, y_0) = 0$ so that $d(T, U, y_0) = 0$ in view of (12.13).

We conclude the theoretical development of the Leray–Schauder degree with the notion of the "index" of an isolated solution.

Thus, let B be a Banach space, U an open subset of B, and let $f \in C^1(U, B)$. We assume that f is nonzero on ∂U and that $K = I - f$ is a compact mapping.

Now suppose that u_0 is a zero of f, and that $df_{u_0} = I - dK_{u_0}$ is an isomorphism. Let $S_\varepsilon(u_0)$ be a small ball of radius ε about u_0, containing no other solutions of $f = 0$. Since $f = I - K$, we can compute $d(f, S_\varepsilon(u_0), 0)$, and for small ε, this is independent of ε (Property 5); it is called the *index of f at u_0*, and written $i_f(u_0)$.

From Theorem 13.2, $A = dK_{u_0}$ is compact and consider the set of real eigenvalues of A greater than 1. If λ is any such eigenvalue, let η_λ denote its multiplicity, i.e.,

$$\eta_\lambda = \dim \bigcup_{k=1}^{\infty} N[(\lambda I - A)^k],$$

where $N[T]$ denotes the null-space of T. The standard theory of linear compact operators shows that $\eta_\lambda < \infty$. (In fact if $S_k = N((\lambda I - A)^k)$, and $s_k = \dim S_k$, then we shall show that $s_k < \infty$ for all k, and $s_k = s_{k+1}$ for large k. If $\{x_j\}$ is a sequence of unit vectors in S_k, then $x_j = Kx_j$ where K is compact. Thus a subsequence converges, so we have shown that the unit ball in S_k is compact; thus $s_k < \infty$. Next, if the S_k are strictly increasing, we can choose $x_k \in S_k$ such that $\|x_k\| = 1$ and $\operatorname{dist}(x_k, S_{k-1}) \geq \frac{1}{2}$. Then if $j > k$,

$$
\|Ax_j - Ax_k\| = \|(A - \lambda I)x_j + \lambda x_j - Ax_k\|
$$

$$
= |\lambda| \left\| x_k - \left[\frac{1}{\lambda}(\lambda I - A)x_k + \frac{1}{\lambda}Ax_j \right] \right\|
$$

$$
\geq \frac{|\lambda|}{2}.
$$

Thus $\{Ax_k\}$ has no convergent subsequence, and this violates the compactness of A.)

The following theorem shows that we can explicitly compute the index of f at u_0 in terms of the η_λ.

Theorem 12.17. $i_f(u_0) = (-1)^\sigma$, where $\sigma = \sum_{\lambda > 1} \eta_\lambda$.

Proof. Consider first the finite-dimensional case. If P is a nonsingular $n \times n$ matrix, and $Q = I - P$, then clearly $i_p(0) = d(P, S_\varepsilon, 0) = \operatorname{sgn} \det P$, where S_ε is a small ball about 0. We claim that the theorem holds in this case; i.e., that

$$
\operatorname{sgn} \det P = (-1)^\sigma, \qquad \sigma = \sum_{\lambda > 1} \eta_\lambda(Q).
$$

To see this, note that since P is real, if $\lambda_1, \ldots, \lambda_k$ are the real eigenvalues of P, having multiplicities $m_1, \ldots, m_k$, respectively, then

$$
\operatorname{sgn} \det P = \operatorname{sgn} \prod_1^k \lambda_i^{m_i} = \operatorname{sgn} \prod_{\substack{\lambda_i < 0 \\ m_i \text{ odd}}} \lambda_i^{m_i} = (-1)^\alpha,
$$

where $\alpha = \sum\{\text{odd } m_i\}$. On the other hand, if $\tau = \sum\{\text{odd } \eta_\lambda(Q)\}$, then $(-1)^\sigma = (-1)^\tau$. But if λ is a real eigenvalue of Q, $\lambda = 1 - \lambda_i$ for some i, $1 \leq i \leq k$ and $\lambda > 1$ if $\lambda_i < 0$. Hence

$$
(-1)^\sigma = (-1)^\tau = (-1)^\alpha = \operatorname{sgn} \det P.
$$

This proves our claim.

To continue the proof, we first see that we may assume that $u_0 = 0$. Using the homotopy $t^{-1}K(tx)$, $0 < t \le 1$ of K into A, we find that $d(I - K, S_\varepsilon, 0) = d(I - A, S_\varepsilon, 0)$, where $S_\varepsilon = S_\varepsilon(0)$. Now write $B = B_1 \oplus B_2$ where B_1 is spanned by $\bigcup \{N(\lambda I - A)^k : \lambda \ge 1, k = 0, 1, 2, \ldots\}$ and $AB_2 \subset B_2$. Using Property 7,

$$d(I - A, S_\varepsilon, 0) = d((I - A)\big|_{B_1}, S_\varepsilon \cap B_1, 0) \cdot d((I - A)\big|_{B_2}, S_\varepsilon \cap B_2, 0).$$

Now if $u_2 \in B_2$ and $(I - tA)u_2 = 0$ for some t, $0 \le t \le 1$, then $u_2 = 0$. Thus we may deform $I - A$ to the identity in $S_\varepsilon \cap B_2$; this gives

$$d(I - A, S_\varepsilon, 0) = d((I - A)\big|_{B_1}, S_\varepsilon \cap B_1, 0) = (-1)^\sigma,$$

since B_1 is a finite-dimensional space. The proof is complete.   □

Remark. Let A be a compact linear operator, let $(I - A)$ be invertible, and let λ_0 be an eigenvalue of A having multiplicity m, with corresponding eigenvector u_0. Then $d(\lambda I - A, S_\varepsilon(u_0), 0)$ changes by a factor $(-1)^m$ as λ crosses λ_0. To show this we observe that (as before) it suffices to consider the case where A is an $n \times n$ matrix. In this case, we note that $\det(\lambda I - A) = \Pi(\lambda - \lambda_j)^{m_j}$ where the λ_j are the eigenvalues of A with their respective multiplicities m_j. Then if $\varepsilon > 0$ is small,

$$\operatorname{sgn} \det(\lambda_0 + \varepsilon - A) = (-1)^m \operatorname{sgn} \det(\lambda_0 - \varepsilon - A),$$

and the result follows.

We shall now briefly show how the Leray–Schauder degree is applied to specific problems. (In the next chapter, we shall show how it is used to obtain a global bifurcation theorem.)

The idea is to show that all solutions must be in some ball U in an appropriate function space; i.e., to obtain an a-priori estimate. Then one proves that the degree $d(T, U, y_0) \ne 0$, by using the homotopy invariance. In other words, by a suitable homotopy one "connects" the original problem to a problem in which the degree is easily seen to be nonzero. Let us now illustrate this technique in somewhat more detail.

Consider the problem

$$\begin{cases} \Delta u + f(x, u, Du) = 0, & x \in \Omega, \\ u = 0 & \text{on } \partial\Omega, \end{cases} \tag{12.14}$$

where f is a C^∞ function. As in Chapter 10, §C (see (10.29)), we can recast this problem in the form $u - KF(u) = 0$, $K = \Delta^{-1}$, where $F(u) = f(x, u, Du)$, and since K is smoothing, it is a compact operator in an appropriate function space. That is, if $\|u\|_{L_\infty(\Omega)} \le 1$, then $\|Ku\|_1 \le C$ where C is independent of u and $\|\cdot\|_1$ denotes the norm in $C^1(\Omega)$. We then can apply degree theory in the Banach space $B = \{u \in C^1(\Omega): u = 0 \text{ on } \partial\bar\Omega\}$. This requires that we obtain an a-priori estimate. For f not growing too fast at infinity in all of its arguments, we can show that for any solution of $u - KF(u) = 0$ (see, e.g., [Ni 2, p. 49])

$$\|u\|_1 \le C. \tag{12.15}$$

Having this estimate enables us to prove the existence of a solution. Thus, let U be the ball $\|u\|_1 \le 1 + C$, contained in B, and let $T: U \to B$ be defined by

$$T(u) = u - KF(u). \tag{12.16}$$

We seek a solution to the equation $T(u) = 0$. In view of (12.15), if $u \in \partial U$, then $T(u) \ne 0$. Thus, $d(T, U, 0)$ is defined. In order to compute this degree, we consider the mappings

$$T_t(u) = u - tKF(u), \qquad 0 \le t \le 1.$$

By the homotopy property, $d(T, U, 0) = d(I, U, 0) = 1$. Thus (12.16) has a solution $\bar u$ in B. It is not too hard to show, for any solution of $Tu = 0$, that under reasonable growth conditions on f, $\bar u$ is smooth.

§C. An Introduction to Morse Theory

Morse theory studies the topological and analytical properties of gradient vector fields; that is, of functions which can be written as df, for some $f \in C^2(\Omega, \mathbf{R})$, where Ω is an open set in $\mathbf{R}^n$. It is a rich and varied discipline, which makes connections with many diverse areas of mathematics and its applications. For our purposes, it is the concept of the "Morse index" which we shall find most useful. Briefly, this is a nonnegative integer, assigned to each nondegenerate critical point of f, which measures the "degree of stability" of this point when considered as a rest point of the associated gradient dynamical system. Our development in this section will be rather brief, and perhaps even sketchy in places, since in Chapters 22 and 23 we shall consider a far-reaching extension of these ideas due to C. Conley.

Let $f \in C^2(\Omega, \mathbf{R})$; then $\bar x$ is called a *critical* point of f if $df(\bar x) = 0$; here df denotes the gradient of f with respect to the usual Euclidean metric. We are interested in the structure of f near the critical points. If $x = \phi(y)$ is a smooth bijective transformation, and $F(y) = f(\phi(y))$, then $dF(\bar y) = df(\phi(\bar y))d\phi(\bar y) = df(\bar x)\,d\phi(\bar y) = 0$. Thus critical points are preserved under nonsingular smooth maps; i.e., they have an invariant character. (This, in fact, allows the whole theory to actually be developed on any smooth manifold, but we will not pursue this generality.)

One of our goals is to investigate the existence of critical points of f. Simple examples show that there need not be any in general. This is due to a lack of compactness; in fact, if f were defined on a compact manifold M without boundary, then $\max f$ and $\min f$ are obviously critical values. One way to remedy the lack of compactness is to require f to be suitably restricted at infinity. For example, suppose that the following condition holds:

There is an $R > 0$, and a $\delta > 0$, such that

$$|u| > R \quad \text{implies} \quad |df(u)| > \delta. \tag{$*$}$$

Using $(*)$, we see that the set $S = \{u : |df(u)| \leq \delta\}$ is compact, and lies in the ball of radius R centered at 0. But then the maximum and minimum values of f on S are critical values.

It is not too hard to show that, (in $\mathbf{R}^1$), $(*)$ implies the following condition:

$$|f(u)| \to \infty \quad \text{as } |u| \to \infty. \tag{12.17}$$

(In fact, by Taylor's theorem, if $|u| > R$, $|f(u) - f(0)| \geq \delta|u|$.) Now if we strengthen (12.17) to

$$f(u) \to +\infty \quad \text{as } |u| \to \infty,$$

then $\inf f$ is attained, and at this point, $df = 0$. We remark that this condition implies much more; for example, if f has two distinct relative minima, it must have a third critical point. We shall give a very simple proof of this in Chapter 22, §C.

In order to be able to give a reasonable classification of the critical points of f, it is necessary that they be "isolated." Analytically, we require that at each critical point $\bar{x}$, the hessian matrix d^2f is nonsingular. Such critical points are called *nondegenerate*. If all the critical points of f are nondegenerate we call f a *Morse function*.

That nondegenerate critical points are isolated follows immediately from the inverse function theorem. Moreover, if as above, $F(y) = f(\phi(y))$ where ϕ is bijective, then if $df(\bar{x}) = 0$, and d^2f is nonsingular at $\bar{x}$, we have for any n-vector ξ,

$$d^2F(y)(\xi, \xi) = d^2f(\phi(y))(\xi, \xi) + df(\phi(y)) d^2\phi(y)(\xi, \xi), \tag{12.18}$$

and since $df(\phi(\bar{y})) = df(\bar{x}) = 0$, we see that $d^2F(\bar{y})(\xi, \xi) = d^2f(\bar{x})(\xi, \xi) \neq 0$. Thus the property of a critical point being nondegenerate is also preserved under smooth bijective changes of coordinates.

We shall now prove two important lemmas; the first shows that Morse functions are dense in the C^2-topology on compacta, and the second shows that each such f has a canonical form near each nondegenerate critical point.

Lemma 12.18. *For any bounded open set* $\Omega \subset \mathbf{R}^n$, *the Morse functions are generic in* $C^2(U, \mathbf{R})$; *i.e., they form an open and dense set.*

Proof. Let $f \in C^2(\Omega, \mathbf{R})$, $a \in \mathbf{R}^n$, and consider the function $f_a(x) = f(x) + \langle a, x \rangle$. We wll show that for almost all a, f_a is a Morse function. Thus, consider the mapping $g: \Omega \to \mathbf{R}^n$, given by $g: x \to df(x)$. The derivative of f_a at any $\bar{x}$ is given by

$$(df_a)(\bar{x}) = g(\bar{x}) + a.$$

Therefore $\bar{x}$ is a critical point of f_a iff $g(\bar{x}) = -a$. Also, since f_a and f have the same hessians, $d^2f(\bar{x}) = dg(\bar{x})$. If $-a$ is a regular value for g, then if $g(\bar{x}) = -a$, $dg(\bar{x})$ is nonsingular. Thus $\bar{x}$ is a nondegenerate critical point of f_a whenever $-a$ is a regular value of g. By Sard's theorem (Theorem 12.2), $-a$ is a regular value of g for almost all $a \in \mathbf{R}^n$. This shows that the Morse functions are dense.

Next if f is a Morse function and g is C^2-close to f, then obviously g is also a Morse function. Thus the Morse functions are also open. $\quad\square$

Lemma 12.19 (Morse Lemma). *Let* $f \in C^2(\Omega, \mathbf{R})$, *and let* $\bar{x}$ *be a nondegenerate critical point of* f. *Then there is a local coordinate system near* $\bar{x}$ *such that*

$$f(x) = f(\bar{x}) + \sum_1^n \varepsilon_i x_i^2, \tag{12.19}$$

where $\varepsilon_i = \pm 1$.

Before giving the proof, we need a lemma which gives a parametric form of the diagonalization of symmetric matrices. Let A^t denote the transpose of the matrix A, and let $\mathscr{S}_n$ denote the real symmetric $n \times n$ matrices.

Lemma. *Let* $D = \operatorname{diag}(d_1, \ldots, d_n)$ *be a diagonal* $n \times n$ *matrix, where* $d_i = \pm 1$. *Then there is a neighborhood* $U \subset \mathscr{S}_n$ *of* D, *and an analytic map* σ *from* U *into the real nonsingular* $n \times n$-*matrices such that* $\sigma(D) = I$, *and if* $\sigma(R) = S$, *then* $S^t R S = D$.

Proof. Let $R \in \mathscr{S}_n$ be so close to D that $r_{11} \neq 0$ and $\operatorname{sgn} r_{11} = \operatorname{sgn} d_1$. Let T be the linear transformation on $\mathbf{R}^n$, $x = Ty$, where

$$x_1 = \left[y_1 - \frac{r_{12}}{r_{11}} y_2 - \cdots - \frac{r_{1n}}{r_{11}} y_n \right] / \sqrt{|r_{11}|},$$

$$x_k = y_k / \sqrt{|r_{11}|}, \qquad k > 1.$$

Then

$$T^tRT = \begin{bmatrix} d_1 & 0 & 0 & \dots & 0 \\ 0 & & & & \\ 0 & & R_1 & & \\ \vdots & & & & \\ 0 & & & & \end{bmatrix},$$

where $R_1 \in \mathscr{S}_{n-1}$. If R is close to D, then R_1 is close to $D_1 = \mathrm{diag}(d_2, \dots, d_n)$; in particular, it will be invertible. Note that T and R_1 are analytic functions of R. By induction on n, we assume that there is an $(n - 1) \times (n - 1)$ non-singular matrix $S_1 = \sigma(R_1)$, depending analytically on R_1 such that $S_1^t R_1 S_1 = D_1$. Now we define $\sigma(R) = S$ where $S = TK$, and

$$K = \begin{bmatrix} 1 & 0 & 0 & \dots & 0 \\ 0 & & & & \\ 0 & & S_1 & & \\ \vdots & & & & \\ 0 & & & & \end{bmatrix};$$

then $S^t RS = D$. This proves the lemma. $\quad\square$

We can now prove the Morse lemma. We may assume that Ω is a convex open set in $\mathbf{R}^n$, $\bar{x} = 0$, and $f(\bar{x}) = 0$. Furthermore, by a linear coordinate change, we may assume that the Hessian matrix $D = (f_{x_i x_j}(0))$ is diagonal, with the first k diagonal entries equal to $+1$, and the remaining equal to -1.

Now there is a smooth function $x \to R_x$ from Ω to $\mathscr{S}_n$ such that if $x \in \Omega$ and $R_x = (r_{ij}(x))$, then

$$f(x) = \sum_{i,j=1}^{n} r_{ij}(x) x_i x_j,$$

and $R_0 = D$. To see this note that

$$f(x) = \int_0^1 \frac{df(tx)}{dt} dt = \sum_{i=1}^{n} x_i \int_0^1 \frac{\partial f(tx)}{\partial x_i} dt$$

Now since $df(0) = 0$, we have

$$\frac{\partial f(tx)}{\partial x_i} = \int_0^1 \frac{d}{ds} \left(\frac{\partial f(stx)}{\partial x_i} \right) ds = \sum_{j=1}^{n} x_j \int_0^1 \frac{\partial^2 f(stx)}{\partial x_i \partial x_j} ds,$$

so that

$$f(x) = \sum_{i,j=1}^{n} x_i x_j \int_0^1 \int_0^1 \frac{\partial^2 f(stx)}{\partial x_i \partial x_j} ds\, dt \equiv \sum_{i,j=1}^{n} r_{ij}(x) x_i x_j.$$

If $\sigma(R)$ is the matrix-valued function in the lemma, set $\sigma(R_x) = S_x$; S_x is an $n \times n$ nonsingular matrix. Define a map $\phi: U \to \mathbf{R}^n$, by $\phi(x) = S_x^{-1}x$, where $U \subset \Omega$ is a small neighborhood of 0. Since $d\phi(0) = I$, the inverse function theorem implies that ϕ is a smooth homeomorphism on U, if U is small. Set $y = \phi(x)$; then

$$f(x) = x^t R_x x = y^t(S_x^t R_x S_x)y = y^t D y = \sum_{i=1}^{n} d_i y_i^2,$$

as desired.   □

For a given bilinear functional H on $\mathbf{R}^n$, we define the *index* of H to be the maximal dimension of a (vector) subspace of $\mathbf{R}^n$ on which H is positive definite. We may therefore speak of the *Morse index of the critical point* $\bar{x}$ *of* f, and by this we mean the index of $d^2f(\bar{x})$. Now let k be the number of $\varepsilon_i > 0$ in (12.19). We shall show that k is the Morse index of the critical point $\bar{x}$.

Lemma 12.20. *If k is the number of positive ε_i in (12.19), then k is the index of $d^2f(\bar{x})$. Hence k is independent of the transformations used to obtain the representation (12.19).*

Proof. Let $y(x) = (y_1(x), \ldots, y_n(x))$ be any coordinate system, and let m be the index of $\bar{x}$. If in this coordinate system, f had the form

$$f(x) = f(\bar{x}) + \sum_{j=1}^{k} y_j(x)^2 - \sum_{j=k+1}^{n} y_j(x)^2,$$

then

$$\frac{\partial^2 f(\bar{x})}{\partial y_i \, \partial y_j} = \begin{cases} -2, & \text{if } i = j > k, \\ +2, & \text{if } i = j \le k, \\ 0, & \text{otherwise.} \end{cases}$$

Thus, the matrix representing d^2f with respect to the basis determined by these new coordinates is

$$\mathrm{diag}(\underbrace{2, 2, \ldots, 2}_{k}, -2, -2, \ldots, -2).$$

It follows that there is a subspace V of dimension $n - k$ on which d^2f is negative definite, and one of dimension k on which d^2f is positive definite. Hence $k \le m$. If there were a subspace of dimension greater than k on which d^2f was positive definite, this subspace would meet V, which clearly is impossible. Thus $k = m$, and we are done.   □

These last two results show that f has a unique canonical representation near each nondegenerate critical point, and hence such critical points can be classified according to their Morse indices k, $1 \leq k \leq n$. But we can go even further with this index.

Suppose that $\bar{x}$ is a nondegenerate critical point of f. If we consider the "gradient" system defined by

$$\dot{x} = \nabla f(x), \tag{12.20}$$

then clearly $\bar{x}$ is a rest point. If $H(\bar{x})$ denotes the hessian matrix of f at $\bar{x}$, then the Morse lemma shows that $H(\bar{x})$ is nonsingular. In this case we say that $\bar{x}$ is a *nondegenerate rest point* of (12.20).

Now let's consider the flow for the "linearized equations"

$$\dot{y} = H(\bar{x})y. \tag{12.21}$$

Clearly $y = 0$ is a rest point for the system (12.21). From the last two lemmas, we see that if the index of $\bar{x}$ is k, then $H(\bar{x})$ has k positive eigenvalues and $(n - k)$ negative eigenvalues. Let $\lambda_1 \geq \cdots \geq \lambda_k > 0 > \lambda_{k+1} \geq \cdots \geq \lambda_n$ denote the eigenvalues of $H(\bar{x})$, with $r_1, r_2, \ldots, r_n$ their corresponding eigenvectors. Then the following theorem is well known. (We use the notation $y_0 \cdot t$ to denote the point on the trajectory of (12.21) through y_0, after t units of time.)

Linear Stable Manifold Theorem. *There are manifolds M_k and M_{n-k} of dimensions k and $(n - k)$, respectively, with the following properties:*

(i)   *if $y_0 \in M_k$, then $y_0 \cdot t \to 0$ as $t \to -\infty$;*
(ii) *if $y_0 \in M_{n-k}$, $y_0 \cdot t \to 0$ as $t \to +\infty$; and*
(iii) *near $y = 0$, M_k is spanned by $r_1, \ldots, r_k$, and M_{n-k} is spanned by $r_{k+1}, \ldots, r_n$.*

Thus, $\mathbf{R}^n \backslash \{0\} = M_k \oplus M_{n-k}$, where $M_k \cap M_{n-k} = \phi$. We call M_{n-k} and M_k the *stable* (resp. *unstable*) *manifold* at the rest point 0 of (12.21). Now for our purposes, we are interested in the (nonlinear) "stable manifold" theorem. This states that *in a small neighborhood N of the nondegenerate rest point $\bar{x}$ of (12.20), the conclusions (i)–(iii) above, are valid*; thus the flow of (12.20) near $\bar{x}$, is topologically equivalent to the flow near 0 of (12.21). (For a proof, see [CL].) We again refer to M_{n-k} and M_k as the *stable* (resp. *unstable*) *manifold* at $\bar{x}$ of (12.20). Thus the Morse index is a measure of the "instability" of the critical point, now considered as a rest point of (12.20).

Now we shall show how this theorem enables us to give an invariant topological sense to the Morse index; i.e., to the equation (12.20). (The reader can consult §D for unfamiliar topological notions.) For this purpose, we consider Figure 12.2. The stable manifold theorem implies that we can

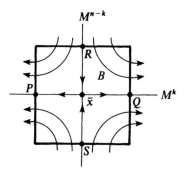

Figure 12.2

find an open set B containing $\bar{x}$ in which M_k meets B in a k-ball B^k having as boundary a $(k-1)$ sphere S^{k-1}, and M_{n-k} meets B in an $(n-k)$ ball B^{n-k}. (In Figure 12.2, P and Q denote S^{k-1}, while $\bar{x}P \cup \bar{x}Q$ denotes B_k, and $\bar{x}R \cup \bar{x}S$ denotes B^{n-k}.) If we denote the entrance and exit points of the flow on ∂B by

$$b^- = \{x \in \partial B : \exists \varepsilon > 0, \ni : x \cdot (-\varepsilon, 0) \cap B = \phi\},$$

and

$$b^+ = \{x \in \partial B : \exists \varepsilon > 0, \ni : x \cdot (0, \varepsilon) \cap B = \phi\},$$

respectively, then B can be represented as $B^k \times B^{n-k}$, while b^+ and b^- can be represented as $\partial B^k \times B^{n-k}$ and $B^k \times \partial B^{n-k}$, respectively. Thus we have the homotopy equivalence of the pairs $(B, b^+) \sim (B^k, \partial B^k) \sim (B^k, S^{k-1})$, and, on collapsing the boundary (i.e., going over to the quotient space), B/b^+ is a (pointed) k-sphere, $B/b^+ = \Sigma^k$. Similarly, $B/b^- = \Sigma^{n-k}$. We may thus say that the *topological Morse index of* $\bar{x}$ is a (pointed) k-sphere. Note that this shows that the Morse index of f at $\bar{x}$ depends only the values of f on ∂B.

Now suppose that we consider a function g which is C^2 close to f near $\bar{x}$; say $\|f - g\| \le \varepsilon$, where the norm is the C^2-norm in a small neighborhood of $\bar{x}$. We claim that for ε sufficiently small, g has a nondegenerate critical point near $\bar{x}$. That g has a critical point near $\bar{x}$ if ε is small, follows from degree theory. Namely, since $\bar{x}$ is a nondegenerate critical point of f, there is a neighborhood U of $\bar{x}$ in which $\bar{x}$ is the only zero of ∇f. Then if g is sufficiently close to f, the remark following the proof of Lemma 12.6, shows that $d(\nabla g, U, 0) \ne 0$. Thus ∇g is zero in U. The fact that this critical point is nondegenerate follows from Lemma 12.18.

Now if g is close enough to f, let y be the unique critical point of g in B, where B is as above, a small neighborhood of $\bar{x}$. Then we see that the Morse index of y is again a pointed k-sphere. It follows that the Morse index is "invariant under continuation"; i.e., invariant under small perturbations.

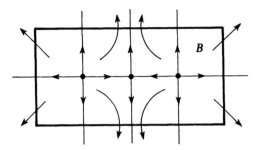

Figure 12.3

In Chapter 22, we shall give Conley's extension of the Morse index to more general invariant sets than nondegenerate rest points. In this more general context, the analogue of (12.19) is not available, and the index is stated in terms of the homotopy class of the space B/b^+. To get an indication of the more general situation, consider a flow (12.20) in $\mathbf{R}^2$ with three rest points, as depicted in Figure 12.3. The set B contains the rest points as depicted. If b^+ has the same meaning as before, we see that $b^+ = \partial B$, and hence B/b^+ is homotopically equivalent to Σ^2, the pointed 2-sphere; cf. Figure 12.4. On the other hand, if $\bar{x}$ is a nondegenerate critical point of f of index 2, (in $\mathbf{R}^2$), then $\bar{x}$ is a relative maximum of f, and thus it is a "repellor" for the gradient flow (12.20). Thus the topological Morse index of $\bar{x}$ is, as above, easily computed to be Σ^2. Therefore the invariant set in B (Figure 12.3) has the index of a "repellor." Clearly this is consistent with Figure 12.3, since B is a "repelling" region.

Figure 12.4

We shall now consider the question as to whether critical points of a given index exist. We may as well suppose that we are on an n-manifold M; i.e., $f \in C^2(M, \mathbf{R})$. We define

$$M^a = \{x \in M : f(x) \leq a\};$$

then we can state the first important theorem.[4]

---

[4] The reader not familiar with the terminology should consult §D.

Theorem 12.21. *Let $f \in C^2(M, \mathbf{R})$ and let $a < b$. If the set $f^{-1}([a, b])$ $= \{x \in M : a \leq f(x) \leq b\}$ is compact and contains no critical points of f, then M^a is diffeomorphic to M^b.*

The idea of the proof is to "push" M^b down to M^a along the trajectories of the gradient flow $\dot{x} = df(x)$; see Figure 12.5. See [Mr 1] for a proof.

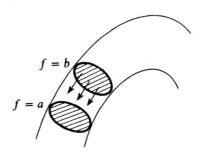

Figure 12.5

The second theorem describes the situation near a nondegenerate critical point.

Theorem 12.22. *Let $f \in C^2(M, \mathbf{R})$, and let $\bar{x}$ be a nondegenerate critical point of Morse index k. With $c = f(\bar{x})$, suppose that for some $\varepsilon > 0$, $f^{-1}([c - \varepsilon, c + \varepsilon])$ is compact and contains no other critical point of f. If ε is sufficiently small, the set $M^{c+\varepsilon}$ has the homotopy type of $M^{c-\varepsilon}$ with an $(n - k)$-cell attached.*

Thus we see that the index of the critical point $\bar{x}$ is intimately related to the topology of the manifold in the vicinity of $\bar{x}$. Again see [Mr 1] for a proof of Theorem 12.22.

As an application of these theorems, we can prove the following interesting result.

Theorem 12.23. *If M is a compact manifold, and there exists an $f \in C^1(M, \mathbf{R})$ having only two critical points, both of which are nondegenerate, then M is homeomorphic to a sphere.*

Proof. The critical points must be at max f and min f. Let $f(x_0) = 0$ be the minimum and $f(x_1) = 1$ the maximum values. For small ε, the sets M^ε $= f^{-1}([0, \varepsilon])$ and $f^{-1}([1 - \varepsilon, 1])$ are closed n-cells; this follows from Lemma 12.19. By Theorem 12.21, M^ε and $M^{1-\varepsilon}$ are homeomorphic. Thus M is the union of two n-cells $M^{1-\varepsilon}$ and $f^{-1}([1 - \varepsilon, 1])$ "glued" together at this common boundary. Hence M is homeomorphic to an n-sphere. $\square$

We remark that, in his celebrated paper [Mr 1], Milnor uses this theorem to find a 7-sphere with a nonstandard differentiable structure, which is homeomorphic to the usual 7-sphere; namely, he finds a function on it with exactly two nondegenerate critical points.

We shall conclude this section by showing how one can use Morse theory to describe the relationship between the topology of M and the critical points of a real-valued function on M. This will be done by giving a collection of inequalities, called the Morse inequalities. (Again, we refer the reader to the §D for the terminology and results.)

Suppose that k is an integer, and that we have a function S_k defined on pairs of spaces (X, A), $A \subset X$, satisfying the following properties:

   (i)   *Subadditivity*: If $X \supset Y \supset Z$, then $S_k(X, Z) \leq S_k(X, Y) + S_k(Y, Z)$.
   (ii) *Dimension*: If E_r is an r-cell, with boundary ∂E_r, then $S_k(E_r, E_r) = \delta_{kr}$, the Kronecker delta.
   (iii) *Homotopy Invariance*: If (X, A) and (Y, B) have the same homotopy type, then $S_k(X, A) = S_k(Y, B)$.
   (iv) *Excision*. If U is an open subset in the interior of A, then $S_k(X \backslash U, A \backslash U) = S_k(X, A)$.

As a consequence of (i), if $X_0 \subset X_1 \subset \cdots \subset X_m$, then $S_k(X_n, X_0) \leq \sum_1^m S_k(X_i, X_{i-1})$; this follows by induction. Therefore, writing $S_k(X) = S_k(X, \phi)$, we have if $X_0 = \phi$,

$$S_k(X_m) \leq \sum_{i=1}^m S_k(X_i, X_{i-1}). \tag{12.22}$$

Now let M be an n-manifold, and let f be a Morse function on M. Let $a_1 < a_2 < \cdots < a_m$ be such that M^{a_i} contains exactly i critical points of f, and let $M = M^{a_m}$. Then if $a_0 < a_1$, $\phi = M^{a_0} \subset M^{a_1} \subset \cdots \subset M^{a_m} = M$, and (12.22) gives the following theorem.

Theorem 12.24. *Let C_λ denote the number of critical points of f having index $n - \lambda$ on M. Then $R_\lambda(M) \leq C_\lambda$, where $R_\lambda(M)$ is the λth Betti number of M.*

Proof. Using (12.22),

$$S_\lambda(M) \leq \sum_{i=1}^m S_\lambda(M^{a_i}, M^{a_{i-1}})$$

$$\leq \sum_{i=1}^m S_\lambda(M^{a_{i-1}} \cup E_{\lambda_i}, M^{a_{i-1}})$$

$$= \sum_{i=1}^m S_\lambda(E_{\lambda_i}, \partial E_{\lambda_i}),$$

by Theorem 12.22, and the excision property above. We apply these inequalities to the Betti numbers $R_\lambda(X, A)$. That these satisfy the above properties (ii)–(iv) is a consequence of their definition as being the ranks of the homology groups $H_\lambda(X, A)$, and that they satisfy (i) follows from the exact sequence: $\ldots \to H_\lambda(Y, Z) \to H_\lambda(X, Z) \to H_\lambda(X, Y) \to \ldots$; see Axiom 4 in §D. Thus the above inequalities give

$$R_\lambda(M) \le \sum_{i=1}^{m} R_\lambda(E_{\lambda_i}, \partial E_{\lambda_i}) = C_\lambda,$$

and the proof is complete. $\square$

There are in fact more general relations between the Betti numbers R_λ of a compact differentiable manifold and the number of critical points of a Morse function f on M. These are the Morse inequalities:

$$R_0 \le C_0,$$

$$R_1 - R_0 \le C_1 - C_0,$$

$$R_2 - R_1 + R_0 \le C_2 - C_1 + C_0,$$
$$\vdots$$
$$R_{n-1} - R_{n-2} + \cdots \pm R_0 \le C_{n-1} - C_{n-2} + \cdots \pm C_0,$$

$$R_n - R_{n-1} + \cdots \pm R_0 = C_n - C_{n-1} + \cdots \pm C_0.$$

The proof is not very difficult, but we shall not give it here, since a more general statement will be given in Chapter 23, §B.

§D. A Rapid Course in Topology

1. Manifolds

An *n-dimensional manifold* M is a Hausdorff space M together with a countable open covering $\{U_i\}$ of M, and mappings $\{f_i\}$ such that f_i is a homeomorphism of U_i onto an open subset of $\mathbf{R}^n$, and if $U_i \cap U_j \ne \phi$, the mapping $f_i \circ f_j^{-1}$ is differentiable and has a nonsingular Jacobian; see Figure 12.6. Some examples of manifolds are *n*-spheres S^n, *n*-balls B^n, and *n*-tori T^n.

A *diffeomorphism* between two manifolds M and N in $\mathbf{R}^n$ is a homeomorphism $f : M \to N$ such that f and f^{-1} are differentiable.

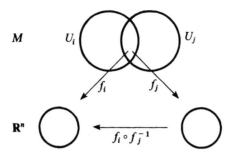

Figure 12.6

2. Homology Theory

We first state the axioms of homology theory.

Let $\mathscr{C}$ be a collection of pairs of topological spaces (X, A), where $A \subset X$. A *homology theory* on $\mathscr{C}$ is a set of three functions H, $*$, and ∂, on $\mathscr{C}$, defined as follows.

(i)   For each pair (X, A) in $\mathscr{C}$ and each integer $k \geq 0$, H assigns an abelian group $H_k(X, A)$, called the *k-dimensional homology group of the pair* (X, A).

(ii) For each continuous mapping $f : (X, A) \to (Y, B)$ on $\mathscr{C}$ onto $\mathscr{C}$ with $f(X) \subset Y, f(A) \subset B$, $*$ assigns a group homomorphism $f_* : H_k(X, A) \to H_k(Y, B)$, for each integer $k \geq 0$. f_* is called the homomorphism *induced* by f.

(iii) For each $(X, A) \in \mathscr{C}$ and each $q \geq 1$, ∂ is a group homomorphism $\partial : H_q(X, A) \to H_{q-1}(A, \phi) \equiv H_{q-1}(A)$.

The functions and groups are required to satisfy the following seven axioms: (all spaces are assumed to be in $\mathscr{C}$, and all functions are assumed to be continuous).

Axiom 1. *If i is the identity mapping $i : (X, A) \to (X, A)$, then $i_* : H_q(X, A) \to H_q(X, A)$ is the identity homomorphism.*

Axiom 2. *If $f : (X, A) \to (Y, B)$, and $g : (Y, B) \to (Z, C)$, then $(g \circ f)_* = g_* \circ f_*$.*

Axiom 3. $\partial \circ f_* = f_* \circ \partial$.

Axiom 4. *If $i : A \to X$, and $j : X \to (X, A)$ denote inclusion maps, then the following sequence is exact (the image of each map is the kernel of the next map):*

$$\cdots \to H_q(A) \xrightarrow{i_*} H_q(X) \xrightarrow{j_*} H_q(X, A) \xrightarrow{\partial} H_{q-1}(A) \to \cdots$$

Axiom 5. *If* $f : (X, A) \to (Y, B)$, *and* $g : (X, A) \to (Y, B)$ *are homotopic,*[5] *then* $f_* = g_*$.

Axiom 6. *If* U *is an open subset of* X *and* $\bar{U} \subset$ *int* A, *the inclusion* $i : (X\backslash U, A\backslash U)$ $\to (X, A)$ *induces an isomorphism* $i_* : H_q(X\backslash U, A\backslash U) \to H_q(X, A)$, *for each integer* $q \geq 0$.

Axiom 7. *If* $q \neq 0$, $H_q(\theta) = 0$, *where* θ *is any point.*

If (p, p) is a fixed point in $\mathscr{C}$, and $H_0(p) = G$, G is called the *coefficient group* of the homology theory. If $G = \mathbf{Z}$, the integers, then if

$$S^n = \{x \in \mathbf{R}^n : |x| = 1\} \quad \text{and} \quad B^n = \{x \in \mathbf{R}^n : |x| \leq 1\},$$

it is easy to show the following results:

If $n > 0$,
$$H_q(S^n) = \begin{cases} \mathbf{Z}, & \text{if } q = 0 \text{ or } q = n, \\ 0, & \text{otherwise,} \end{cases}$$

$$H_q(S^0) = \begin{cases} \mathbf{Z} + \mathbf{Z}, & \text{if } q = 0, \\ 0, & \text{otherwise,} \end{cases}$$

$$H_q(B^n, S^n) = \begin{cases} \mathbf{Z}, & \text{if } q = n, \\ 0, & \text{otherwise.} \end{cases}$$

Now suppose that M is an n-dimensional compact manifold. Then M can be thought of as an n-complex, i.e., a finite number of simplicies with the following two properties: (a) if A is a simplex of M, then every face of A is also in M, and (b) any two simplices of M have either empty intersection, or they intersect along a common face. [We recall the notion of a *simplex* in $\mathbf{R}^n$. First, we say that a system of points $a_0, a_1, \ldots, a_r$ in $\mathbf{R}^n$, $r \leq n$ is *independent*, if the set of vectors $(a_1 - a_0), \ldots, (a_r - a_0)$ is linearly independent; this is equivalent to: if $\lambda_i \in \mathbf{R}$ satisfy $\sum_0^r \lambda_i a_i = 0$, and $\sum_0^r \lambda_i = 0$, then $\lambda_0 = \lambda_1 = \cdots = 0$. Now let $a_0, a_1, \ldots, a_r$ be a set of independent points in $\mathbf{R}^n$. Then an *r-simplex* is the set A^r of points of the form $\sum_0^r \lambda_i a_i$, where the $\lambda_i \in \mathbf{R}_+$ satisfy $\sum_0^r \lambda_i = 1$. The points $a_0, \ldots, a_r$ are called the *vertices* of A^r. An n-simplex is sometimes called an *n-cell* or an *n-face*.]

If $A_1^r, \ldots, A_{\alpha(r)}^r$ is the set of all r-simplices of the complex M, an element of the form $\sum_i g_i A_i^r$ where $g_i \in G$, is called an *r-chain*. The set of all r-chains

---

[5] See the discussion in the next section on homotopy.

forms an abelian group $C^r(M)$, under componentwise addition. If A^r is an r-simplex of M, the *boundary*, ∂A^r, of A^r is defined by

$$\partial A^r = B_0^{r-1} + B_1^{r-1} + \cdots + B_r^{r-1}, \qquad \partial A^0 = 0,$$

where the $B_i^{r-1}, 0 \leq i \leq r$, denote the set of all $(r-1)$-faces of A^r. It is fairly straightforward to show that $\partial\partial A^r = 0$. An r-chain is called a *cycle* if $\partial A^r = 0$. The set of all r-cycles also forms an abelian group, which we denote by $Z_r = Z_r(M, G)$. We say that an r-cycle of M is *homologous to zero* if it is the boundary of an $(r-1)$-chain. These cycles form a group B_r; clearly $B_r \subset Z_r$. We can thus form the factor group $H_r(M) = Z_r/B_r$; this group is called the *r-dimensional homology group* of M over G.

Now if we consider pairs (X, A) with $A \subset X \subset M$ and A closed in X, then we can consider the chain spaces $C_r(X)$ and $C_r(A)$ and the groups $C_r(X)/C_r(A)$, as the chains on the *pair* (X, A). As above, this leads to the groups $H_r(X, A)$. If (Y, B) is another such pair, and $f:(X, A) \to (Y, B)$ is continuous with $f(X) \subset Y$, and $f(A) \subset B$, then f induces a mapping $f_*: H_r(X, A) \to H_r(Y, B)$ defined by $f_*: [x] \to [f(x)]$ (where $[\ ]$ denotes the equivalence class). It is not hard to show that f_* is a homomorphism. It remains to define the boundary homomorphisms on the homology groups; i.e., the operators $\partial: H_{n+1}(X, A) \to H_n(A)$. For this, we need a little "diagram chasing." Thus consider the diagram of maps and spaces where $\tilde\partial$ are the boundary operators on the chain complexes:

$$C_{n+1}(A) \hookrightarrow C_{n+1}(X) \overset{j}{\to} C_{n+1}(X, A) \equiv C_{n+1}(X)/C_{n+1}(A)$$
$$\downarrow{\scriptstyle\tilde\partial} \qquad \downarrow{\scriptstyle\tilde\partial} \qquad\qquad \downarrow{\scriptstyle\tilde\partial}$$
$$C_n(A) \quad \hookrightarrow C_n(X) \quad \to C_n(X, A) \quad \equiv C_n(X)/C_n(A).$$

To define ∂ on $H_{n+1}(X, A)$, we take an element $\bar z = z + C_{n+1}(A) \in C_{n+1}(X, A)$ where $\tilde\partial(\bar z) = C_n(A)$ (i.e., $\bar z \in Z_{n+1}(X, A)$). Then since the map j is surjective there is an element $z \in C_{n+1}(X)$ with $j(z) = z$. We then set $\partial(z) = \tilde\partial(z) \in Z_n(A)$. This defines the boundary operator ∂. We can then show that the collection of all such pairs of spaces (X, A), together with the groups $H_r(X, A)$ and the maps f_* and ∂, satisfy the above axioms and therefore give a homology theory.

Next, we note that the groups $H_r(X, A)$ are all finitely generated. In general, if G is a finitely generated abelian group, then G is a direct sum of cyclic groups $A_1, \ldots, A_p, B_1, \ldots, B_q$, where the A_i's are free abelian groups (and thus are isomorphic to the integers $\mathbf{Z}$), and the B_i's have finite order. The number p is called the *rank* of G. In our case, we call the rank of $H_r(M) \equiv H_r(M, \phi)$, the *r-th Betti number* of M, and denote it by β_r. The number

$$\chi(M) = \sum (-1)^r \beta_r$$

is a topological invariant of M, called the "Euler–Poincaré characteristic" of M. We could more generally also consider the numbers

$$R(X, A) = \operatorname{rank} H_\lambda(X/A), \qquad \lambda \in \mathbf{Z}, \lambda \geq 0;$$

these are called the *Betti numbers* of the pair (X, A).

We can now describe an alternate definition of degree of a continuous mapping (in finite-dimensional spaces). This description has the advantage that we do not have to first consider smooth mappings, and thus avoids all the approximation arguments which went through §A.

Thus, let U be a bounded open set in $\mathbf{R}^n$ whose boundary, we assume for simplicity, is homeomorphic to S^{n-1}. If $f \in C(U, \mathbf{R}^n)$, and $f(\partial U) \neq 0$, the function

$$\bar{f}(x) = \frac{f(x)}{|f(x)|}$$

maps ∂U into S^{n-1}. Hence $\bar{f}$ induces a homomorphism $\bar{f}_*: H_r(\partial U) \to H_r(S^{n-1})$ (integer coefficients are assumed). Since ∂U and S^{n-1} are homeomorphic, and $H_{n-1}(S^{n-1}) = \mathbf{Z}$, we have $H_{n-1}(\partial U) = \mathbf{Z}$. But since every group homomorphism of $\mathbf{Z}$ into $\mathbf{Z}$ is of the form $z \to kz$, for some $k \in \mathbf{Z}$ (easy exercise), we see that $\bar{f}_*(z) = kz$. The integer k is called the *degree* of f, with respect to the origin, and is denoted by $d(f, U, 0)$.

The degree $d(f, U, y_0)$ can be defined similarly, provided that $f(\partial U) \subset \mathbf{R}^n \backslash \{y_0\}$. We merely consider the mapping $(f - y_0)/|f - y_0| : \partial U \to S^{n-1}$.

It can be shown that this definition of degree agrees with the one which we gave in §A [Hu]. This is done by proving that continuous mappings f, as we have considered above, can be classified up to homotopy equivalence (see §3), by this integer k, i.e., the degree of a mapping is the *only* homotopy invariant associated with continuous mappings of ∂U into S^{n-1}. Since we have shown that the definition of degree given in §A is also a homotopy invariant, the result follows.

3. Homotopy Theory

A topological pair is an ordered pair (X, A) where X is a topological space, and A is a closed subspace of X. The *product* of pairs (X, A) and (Y, B) is the pair $(X \times Y, X \times B \cup Y \times A)$. If $A = \phi$, then we write (X, A) as X.

A *map* from (X, A) into (Y, B) is a continuous function $f: X \to Y$ such that $f(A) \subset B$. Suppose that f and g are maps from (X, A) into (Y, B). We say that f is *homotopic* to g, and write $f \sim g$, if there is a map $\phi: (X, A) \times [0, 1] \to (Y, B)$ such that $\phi(x, 0) = f(x)$, and $\phi(x, 1) = g(x)$ for all $x \in X$. It is easy to show that $\sim$ is an equivalence relation on the set of maps from (X, A) into (Y, B). We denote by $[f]$, the equivalence class of f; i.e., $[f]$ is the set of all maps $g: (X, A) \to (Y, B)$ such that $g \sim f$.

This notion allows us to define a homotopy between pairs of spaces; namely, we say that the pair (X, A) is *homotopically equivalent to*, or *of the same homotopy type as* the pair (Y, B) (in symbols, $(X, A) \sim (Y, B)$), if there are maps $f: (X, A) \to (Y, B)$ and $g: (Y, B) \to (X, A)$, such that $g \circ f \sim 1(X, A)$, and $f \circ g \sim 1(Y, B)$ (where, for any set S, $1(S)$ is the identity function from S to S). The maps are called *homotopy equivalences*, and each is the homotopy inverse of the other. Thus, pairs of spaces which are "homotopic" are continuously "deformable" to each other. Again, the notion of homotopic pairs of spaces is an equivalence relation, and $[X, A]$ denotes the homotopy class of (X, A); i.e., the set of pairs (Y, B) such that $(Y, B) \sim (X, A)$.

We pause to consider some examples. First consider the solid torus in $\mathbf{R}^3$, $T^2 = S^1 \times D^2$, where D^2 is the unit disk in $\mathbf{R}^2$, centered at $(1, 0)$. We claim that $T^2 \sim S^1$. Thus consider the maps $f: T^2 \to S^1$, and $g: S^1 \to T^2$, defined by $f(\theta, x) = \theta$, and $g(\theta) = (\theta, 0)$, where $\theta \in S^1$, $x \in D^2$, and 0 is the origin in $\mathbf{R}^2$. Clearly $(f \circ g)(\theta) = \theta$, so $f \circ g \sim 1(S^1)$. To show that $g \circ f \sim 1(T^2)$, note that $(g \circ f)(\theta, x) = (\theta, 0)$; thus if we define a map $\phi: T^2 \times [0, 1] \to T^2$, by $\phi(\theta, x, t) = (\theta, tx)$, we see that $\phi(\cdot, \cdot, 1) = 1(T^2)$, and $\phi(\cdot, \cdot, 0) = g \circ f$. This proves our claim.

Next observe that S^1 can be continuously deformed to a point; namely, just shrink the circle radially to its center. But S^1 is *not* of the same homotopy type as a point. To see this, note that the maps $f: S^1 \to \{\theta_1\}$, and $g: \{\theta_1\} \to S^1$, $\theta_1 \in S^1$, given by $f(\theta) = \theta_1$, and $g(\theta_1) = \bar{\theta}$, satisfy $(f \circ g)(\theta_1) = \theta_1$, (so $f \circ g \sim 1(\theta_1)$), and $(g \circ f)(\theta) = \bar{\theta}$. But one can show that there is no map $\phi: S^1 \times [0, 1] \to S^1$ which is the identity for $t = 1$, and is a constant map for $t = 0$. (This statement follows easily from the results in 4.c. below.)

If X is a compact Hausdorff space, and A is compact, we call (X, A) a *compact Hausdorff* pair.

It is easy to show that compositions of homotopic maps are homotopic, i.e., if $\bar{f}, \bar{g}: (X, A) \to (Y, B)$, $f, g: (Y, B) \to (Z, C)$, and if $\bar{f} \sim \bar{g}$ and $f \sim g$, then $f \circ \bar{f} \sim g \circ \bar{g}$.

A more subtle result is the following:

Proposition 12.25. *Suppose that we are given pairs and maps*

$$(X_1, A_1) \overset{f_{21}}{\to} (X_2, A_2) \overset{f_{32}}{\to} (X_3, A_3) \overset{f_{43}}{\to} (X_4, A_4),$$

such that $f_{32} \circ f_{21}$ and $f_{43} \circ f_{32}$ are homotopy equivalences. Then each of the maps f_{21}, f_{32}, and f_{43} are homotopy equivalences.

Proof. Let $f_{13}: (X_3, A_3) \to (X_1, A_1)$ and $f_{24}: (X_4, A_4) \to (X_2, A_2)$ be homotopy inverses for $f_{32} \circ f_{21}$ and $f_{43} \circ f_{32}$, respectively. Then

$$f_{13} \circ (f_{32} \circ f_{21}) \sim (f_{13} \circ f_{32}) \circ f_{21} \sim 1(X_1, A_1).$$

Thus $f_{13} \circ f_{32}$ is a "left homotopy inverse" for f_{21}. Now

$$
\begin{aligned}
f_{21} \circ (f_{13} \circ f_{32}) &\sim (f_{24} \circ f_{43} \circ f_{32}) \circ (f_{21} \circ f_{13} \circ f_{32}) \\
&\sim (f_{24} \circ f_{43}) \circ (f_{32} \circ f_{21} \circ f_{13}) \circ f_{32} \\
&\sim (f_{24} \circ f_{43}) \circ f_{32} \sim 1(X_2, A_2).
\end{aligned}
$$

The other cases are proved similarly. □

We now shall next turn to the notion of a "pointed space." This will be used in Chapters 22 and 23.

We first recall what is meant by the "quotient" topology. Given a pair (X, A), the space X/A is called the *quotient space* and it is defined as follows. Let $\bar{X}$ be those points $x \in X\backslash A$, together with the set A. We put a topology on $\bar{X}$ by calling a set open in $\bar{X}$ provided that the union of its members is open in X. Then $\bar{X}$ is called the *quotient space* X/A. Geometrically, this corresponds to "collapsing" the space A to a point.

We regard X/A as the pair $(\bar{X}, A)$. There is a natural map $X \to \bar{X}$ defined by $x \to [x]$, where $[x]$ denotes the (smallest) set containing x. We call this map the *quotient map*. It is continuous, and surjective except when $A = \phi$, in which case it misses the point A.

One can show fairly easily the following properties:

(i)   If (X, A) is a compact Hausdorff pair, then so is X/A.
(ii) If (X, A) and (Y, B) are compact Hausdorff pairs, and $f : (X, A) \to (Y, B)$ is continuous, then the induced map $\bar{f} : X/A \to Y/B$ defined by $\bar{f}([x]) = [f(x)]$ is well defined and continuous.

Now let (X, A) be a topological pair; if A consists of one point, we call the pair a *pointed space*. Thus, we may regard X/A as the pointed space $(\bar{X}, A)$.

Definition 12.26. If (X, x_0) and (Y, y_0) are pointed spaces their *sum* (or *wedge*) $(X, x_0) \vee (Y, y_0)$ is defined by

$$
(X, x_0) \vee (Y, y_0) = (X \cup Y)/\{x_0, y_0\},
$$

and their *product* $(X, x_0) \wedge (Y, y_0)$ is

$$
(X \times Y)/[(X \times \{y_0\}) \cup (\{x_0\} \times Y)]
$$

The reader should draw pictures to illustrate these constructions; in particular, the sum can be regarded as "glueing" together the spaces (X, x_0) and (Y, y_0) at their distinguished points.

It can be shown, [Sp], that both the sum and product are well defined on homotopy equivalence classes, and as operations on homotopy classes, they satisfy the associative, commutative and distributive laws.

The additive identity, $\bar{0}$, is the pointed one-point space; the multiplicative identity, $\bar{1}$ is the pointed two-point space. We denote by Σ^n, the homotopy type of the pointed n-sphere.

Proposition 12.27. (i) *if* $[(X, x_0)] \vee [(Y, y_0)] = \bar{0}$, *then* $([X, x_0]) = ([Y, y_0])$
$= \bar{0}$.

(ii) $\Sigma^n \wedge \Sigma^m = \Sigma^{n+m}$.

We omit the proof of (ii); see Lemma 22.27 for the proof of (i).

4. Cohomology Theory

In Chapters 22 and 23, we shall be interested in ascertaining whether or not two spaces are homotopic; for example, whether a pointed n-sphere is homotopic to a pointed m-sphere. One way to do this is to find algebraic invariants for homotopy classes, which can be computed reasonably easy. The definitions which we give below constitute a step in this direction. Most of the proofs are straightforward; see [Sp] for the details.

4.a. Let X be a compact metric space, and $X^n = X \times \cdots \times X$. Define $\Delta^{n+1}X = \{x = (x_0, \ldots, x_n) \in X^{n+1} : x_0 = x_1 = \cdots = x_n\}$; $\Delta^{n+1}X$ is called the *diagonal* in X^n. Let R be a given commutative ring; e.g., the integers, the reals, or $\mathbf{Z}_n$.

We let $C^n(X) = C^n(X, R)$ be the set of functions $\phi: X^{n+1} \to R$, and $C_0^n(X)$ the subset of $C^n(X)$ consisting of those ϕ which are zero in a neighborhood of $\Delta^{n+1}X$.

We define "co-boundary" maps $\delta^n: C^n(X) \to C^{n+1}(X)$, by

$$\delta^n\phi(x_0, \ldots, x_n) = \sum_{r=0}^{n} (-1)^r \phi(x_0, \ldots, \hat{x}_r, \ldots, x_{n+1}),$$

where the "hat" means that the corresponding x_r is omitted.

Let A be a compact subset of X, and define $C^n(X, A) \equiv C^n(X, A, R)$ to be the set of ϕ in $C^n(X)$ such that $\phi|_{A^{n+1}}$ belongs to $C_0^n(A)$.

For a given $f: X \to Y$, define $f^\#: C^n(Y) \to C^n(X)$ by

$$f^\#\phi(x_0, \ldots, x_n) = \phi(f(x_0), \ldots, f(x_n)).$$

Finally, for ϕ_1, ϕ_2 in $C^n(X)$, and r_1, r_2 in R, define $\phi = r_1\phi_1 + r_2\phi_2 \in C^n(X)$, by $\phi(x) = r_1\phi_1(x) + r_2\phi_2(x)$.

One can verify at once the following statements.

(i) If R is the reals or $\mathbf{Z}_2$, $C^n(X, R)$ is a vector space over R and both $C_0^n(X)$ and $C^n(X, A)$ are subspaces. (If R is not a field, the words "vector space" should be replaced by "R-module.")

(ii) δ^n is a homomorphism from $C^n(X)$ to $C^{n+1}(X)$, taking $C_0^n(X)$ into $C_0^{n+1}(X)$, and $C^n(X, A)$ into $C^{n+1}(X, A)$.

(iii) $\delta^{n+1} \circ \delta^n$ is the zero homomorphism.

(iv)   If (X, A) and (Y, B) are compact pairs, and $f : (X, A) \to (Y, B)$, then $f^* : C^n(Y) \to C^n(X)$ is a homomorphism.

(v)   If f is continuous, and $f : (X, A) \to (Y, B)$, then $f^*[C^n(Y, B)] \subset C^n(X, A)$, and $f^*[C_0^n(Y)] \subset C_0^n(X)$.

(vi)   If $f : (X, A) \to (Y, B)$ and $g : (Y, B) \to (Z, C)$ are continuous, then $(g \circ f)^* = f^* \circ g^*$.

(vii) $f^* \circ \delta^n = \delta^n \circ f^*$.

The reader should compare these statements with the axioms of homology given in Part 2, above.

4.b. Let $\bar{C}(X, A)(\equiv \bar{C}(X, A, R))$ be the group $C^n(X, A)/C_0^n(X)$, and let $\bar{\delta}^n : \bar{C}^n(X, A) \to \bar{C}^{n+1}(X, A)$ be the homomorphism induced by δ^n (4.a.(ii)). Similarly, if $f : (X, A) \to (Y, B)$, is continuous, let $\bar{f}^* : \bar{C}^n(Y, B) \to \bar{C}^n(X, A)$ be the map induced by f^* (4.a.(iv)). Then the following statements hold:

(i) $\bar{\delta}^{n+1} \circ \bar{\delta}^n = 0$.

(ii) $\bar{f}^* \circ \bar{g}^* = \overline{(g \circ f)}^*$.

(iii) $\bar{f}^* \circ \bar{\delta}^n = \bar{\delta}^n \circ \bar{f}^*$.

(iv)   If $B \subset A \subset X$ and i and j denote the inclusions, $i : (A, B) \subset (X, B)$ and $j : (X, B) \subset (X, A)$, then the sequence

$$0 \to \bar{C}^n(X, A) \xrightarrow{\bar{j}^*} \bar{C}^n(X, B) \xrightarrow{\bar{i}^*} \bar{C}^n(A, B) \to 0$$

is exact; i.e., the image of each homomorphism is the kernel of the next one.

(v)   For any X, let $i : X \to X/\phi$ (where X/ϕ is the disjoint union of X and a point) denote the inclusion map. Then $i^* : \bar{C}^n(X/\phi) \to \bar{C}^n(X)$ is an isomorphism.

4.c. Let (X, A) be a compact pair, let $Z^n = \ker \bar{\delta}^n$, and let $B^n = \operatorname{im} \bar{\delta}^{n-1}$ (where $B^0 = 0$). We define $H^n(X, A)(\equiv H^n(X, A, R))$ to be the quotient group

$$H^n(X, A) = Z^n/B^n.$$

$H^n(X, A)$ is called the nth $(Alexander)$ $cohomology$ $module$ of the pair (X, A). These are the algebraic invariants of homotopy classes.

If $f : (X, A) \to (Y, B)$ is continuous, then we write

$$f^r : H^r(Y, B) \to H^r(X, A) \quad \text{as the map induced by } \bar{f}^*.$$

Here are some more properties in this set-up:

(i) If X is a single point, then $H^0(X) = R$, and $H^r(X) = 0$ if $r > 0$. If $X = \bar{0}$ (the pointed one-point space; see Section 3, above), then $H^r(X) = 0$ for all r. If $X = \bar{1}$ (see Section 3), $H^r(X) = R$ if $r = 0$, and $H^r(X) = 0$ if $r > 0$.

(ii) For any compact triple (i.e., $B \subset A \subset X$), there is a "canonically defined" map $\delta^r: H^r(A, B) \to H^{r+1}(X, A)$ such that if $f: (X, A, B) \to (X', A', B')$ is continuous, then the rows of the following diagram are exact and the diagram commutes; i.e., $f^* \circ \delta^r = \delta^r \circ f^*$, for all r:

$$0 \to H^0(X', A') \to H^0(X', B') \to H^0(A', B') \xrightarrow{\delta^\circ} H^1(X', A') \to \dots$$
$$\quad\;\; \downarrow^{f^0} \qquad\qquad \downarrow^{f^0} \qquad\qquad \downarrow^{f^0} \qquad\qquad\quad \downarrow^{f^1}$$
$$0 \to H^0(X, A) \;\to H^0(X, B) \;\to\; H^0(A, B) \xrightarrow{\delta^\circ} H^1(X, A) \to \dots$$

(iii) If $C \subset A \subset X$, with C open and A closed in X, then the inclusion $i: (X \backslash C, A \backslash C) \to (X, A)$ induces isomorphisms $i^r: H^r(X \backslash C, A \backslash C) \to H^r(X, A)$, for every $r \geq 0$.

(iv) If f_0 is homotopic to f_1, then $f_0^r = f_1^r$, for all r. In particular, if f is a homotopy equivalence from (X, A) to (Y, B) then the maps f^r are all isomorphisms.

(v)
$$H^n(\Sigma^r) = \begin{cases} R, & \text{if } n = r, \\ 0, & \text{if } n \neq r. \end{cases}$$

In particular Σ^n is not homotopic to Σ^m, if $n \neq m$.

(vi) If A and B are disjoint spaces and $A \perp\!\!\!\perp B$ denotes their disjoint union, then for every $n \in \mathbf{Z}_+$

$$H^n(A \perp\!\!\!\perp B) = H^n(A) \oplus H^n(B).$$

For a discussion of Čech cohomology, see [Sp].

NOTES

In §A and §B, I have followed [BB, Ni 2, Sw], with a few minor modifications. Sard's theorem, Theorem 12.2, is valid in a more general context; see [Sw]. Theorems 12.10 and 12.11 are (by now) standard applications of Brouwer's fixed point theorem. The development of degree theory via differential forms stems from a paper of E. Heinz [Hn]. Theorem 12.15 is due to Schauder [Sc]; the extension of the notion of degree to Banach spaces is due to Leray and Schauder, see [Sc]. The basic ideas of Morse theory are, of course, due to M. Morse; see Milnor's book [Mr], for the original references. I have followed the approach in [BB], with certain modifications. The proofs of Theorems 12.21 and 12.22 can be found in [Mr]; Theorem 12.23 is also taken from there. The material in §D is rather standard, see, e.g., [Sp]. In the sections on homotopy and cohomology, I have followed Conley's approach fairly closely; see [Cy 2]. For an interesting survey of recent work in partial differential equations related to the material in this chapter, see Nirenberg's paper, [Ni 3].

Chapter 13

Bifurcation Theory

Many problems in mathematics, and its applications to theoretical physics, chemistry, and biology, lead to a problem of the form

$$f(\lambda, x) = 0, \tag{13.1}$$

where f is an operator on $\mathbf{R} \times B_1$ into B_2, with B_1, B_2 Banach spaces. For example, (13.1) could represent a system of differential or integral equations, depending on a parameter λ. We are interested in the structure of the solution set; namely, the set

$$f^{-1}(0) = \{(\lambda, x) \in \mathbf{R} \times B_1 : f(\lambda, x) = 0\}. \tag{13.2}$$

In particular, we seek conditions on f in order that we can determine when a solution $(\bar{\lambda}, \bar{x})$ of (13.1) lies on a "curve" of solutions $(\lambda, x(\lambda))$, at least locally; i.e., for $|\lambda - \bar{\lambda}| < \varepsilon$. We may also inquire as to when $(\bar{\lambda}, \bar{x})$ lies on several solution curves, $(\lambda, x_1(\lambda)), (\lambda, x_2(\lambda)), \ldots$.

This last question leads naturally to the concept of a bifurcation point. Thus suppose that $\Gamma : (\lambda, x(\lambda))$, is a curve of solutions of (13.1). Let $(\lambda_0, x_0) \equiv (\lambda_0, x(\lambda_0))$ be an interior point on this curve, with the property that every neighborhood of (λ_0, x_0) in $\mathbf{R} \times B_1$ contains solutions of (13.1) which are not on Γ (Figure 13.1). Then (λ_0, x_0) is called a *bifurcation point* with respect to Γ. Solutions of (13.1) near (λ_0, x_0) and not on Γ, are often loosely referred to as "bifurcating solutions," or the "bifurcation set." Note that the definition does not guarantee the existence of a continuous branch of "bifurcating solutions" emanating from (λ_0, x_0).

Several questions are of interest to us:

(i) Given a curve Γ of solutions of (13.1), what conditions guarantee that it contains a bifurcation point? Obviously, if the implicit function theorem is applicable at a point, bifurcation cannot occur there.

(ii) What is the structure of $f^{-1}(0)$ near a bifurcation point? How is this related to the spectrum of the linearized equations?

(iii) If the bifurcation set is, say, a curve Γ', can it be continued in the large? Does "secondary bifurcation" occur; i.e., does Γ' contain bifurcation points? (see Figure 13.1).

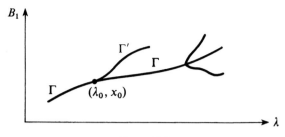

Figure 13.1

(iv) For problems in which solutions of (13.1) are equilibria of a given evolution equation, can one infer any information concerning the stability properties of the bifurcating solution?

We shall attack these problems in this chapter, and somewhat more generally in Chapters 22 and 23. Our results here, for the most part, will come from a rather systematic use of the implicit function theorem. Since our approach will take place in an abstract framework, we shall give a general version of the implicit function theorem. In §A and §B we shall prove some of the standard results in bifurcation theory which stem from the implicit function theorem. In §C we shall prove two general bifurcation theorems using degree theory, while §D is devoted to a description of some different aspects of bifurcation. This is done by giving a fairly complete picture of a special, but interesting, example.

§A. The Implicit Function Theorem

We shall begin by reviewing the definition of the derivative of mappings in finite-dimensional spaces.

Let Ω be an open subset of $\mathbf{R}^m$, and suppose that $f \in C^1(\Omega, \mathbf{R}^n)$. If $a \in \Omega$, then the *derivative* of f at a, written $f'(a)$ or df_a, is the $n \times m$ matrix

$$df_a = \left[\frac{\partial f_i(a)}{\partial x_j} \right],$$

where $f = (f_1, \ldots, f_n)$. Now df_a being a matrix, is thus a bounded linear map from $\mathbf{R}^m$ to $\mathbf{R}^n$; $df_a : x \to df_a x$. Hence $df_a \in B(\mathbf{R}^m, \mathbf{R}^n)$, where we are using the notation $B(X, Y)$ to denote the set of bounded linear maps $X \to Y$.

Since df_x exists for all $x \in \Omega$, we can consider the function $df : \Omega \to B(\mathbf{R}^m, \mathbf{R}^n)$, defined by $x \to df_x$. Observe that $f \in C^1(\Omega, \mathbf{R}^n)$ implies $df \in C(\Omega, B(\mathbf{R}^m, \mathbf{R}^n))$.

If $\xi, a \in \mathbf{R}^m$, and $\phi \in C^1(\Omega, \mathbf{R})$, the *differential* of ϕ at a, acting on ξ is

$$d\phi_a \cdot \xi = \nabla \phi(a) \cdot \xi,$$

where ∇ denotes the usual gradient operator. Applying this last equation to the components of f, we obtain (from Taylor's theorem), the estimate

$$|f(a + \xi) - f(a) - df_a \cdot \xi| = o(|\xi|) \quad \text{as } |\xi| \to 0. \tag{13.3}$$

It is this equation which we shall use to extend the concept of derivative to mappings on Banach spaces.

Thus, let X and Y be Banach spaces, and let $f \in C(\Omega, Y)$, where Ω is open in X.

Definition 13.1. f is (Fréchet) *differentiable* at $a \in \Omega$ if there exists $T \in B(X, Y)$ such that

$$\| f(a + \xi) - f(a) - T\xi \| = o(\| \xi \|) \quad \text{as } \| \xi \| \to 0.$$

T is called the (Fréchet) *derivative* of f at a, and we write $T = df_a$.

It is not too hard to show that if T exists, then it is unique; see, e.g., [LuS]. If f is Fréchet differentiable at each $a \in \Omega$, we write $f \in C^1(\Omega, Y)$.

Now we want to consider higher derivatives of mappings on Banach spaces. We define the *second derivative* d^2f_a of $f \in C(\Omega, Y)$, at a point $a \in \Omega$, to be the bilinear form $d^2f_a: X \times X \to Y$ which satisfies

$$\| f(a + \xi) - f(a) - df_a\xi - \tfrac{1}{2}d^2f_a(\xi, \xi) \| = o(\| \xi \|^2) \quad \text{as } \| \xi \| \to 0.$$

Continuing in this way, the kth order derivative of f at a is the k-linear mapping $d^kf_a: X^k \to Y$ satisfying

$$\left\| f(a + \xi) - f(a) - df_a\xi - \tfrac{1}{2}d^2f_a(\xi, \xi) - \cdots - \frac{1}{k!}d^kf_a(\xi, \ldots, \xi) \right\|$$

$$= o(\| \xi \|^k) \quad \text{as } \| \xi \| \to 0.$$

If f has a kth order derivative at each $a \in \Omega$, we write $f \in C^k(\Omega, Y)$. It is not very hard to show that $C^k(\Omega, Y)$ is closed under composition; for a proof, see [BP].

It is often useful to have the following theorem; it implies that if $T = I - K$ where K is compact, then dT has the same structure.

Theorem 13.2. *Let $f \in C^1(\Omega, Y)$ be compact near $a \in \Omega$. Then df_a is compact.*

Proof. Suppose the theorem is false. Then there is a $\delta > 0$ and a sequence $\{\xi_i\}$, with $\| \xi_i \| \leq 1$, such that $\| T\xi_i - T\xi_j \| \geq \delta > 0$ for all $i \neq j$. (We are denoting df_a by T.) We can find $\gamma > 0$ so small that whenever $\| \xi \| \leq \gamma$,

$$\| f(a + \xi) - f(a) - T\xi \| < \frac{\delta}{3}\| \xi \|.$$

Now $\| \gamma \xi_i \| \le \gamma$, so if $i \ne j$,

$$\| f(a + \gamma \xi_i) - f(a + \gamma \xi_j) \| \ge \gamma \| T\xi_i - T\xi_j \| - \| f(a + \gamma \xi_i) - f(a) - \gamma T\xi_i \|$$
$$- \| f(a + \gamma \xi_j) - f(a) - \gamma T\xi_j \|$$
$$\ge \frac{\delta \gamma}{3}.$$

Thus, although the sequence $\{a + \gamma \xi_i\}$ is bounded, the sequence $\{f(a + \gamma \xi_i)\}$ obviously has no convergent subsequence. This violates the compactness of f. $\square$

Now in preparation for the implicit function theorem, we consider functions defined on product spaces. Thus, let B_1, B_2 and B_3 be Banach spaces and let U be open in $B_1 \times B_2$. If $f: U \to B_3$, and $u = (u_1, u_2) \in U$, we let U_1 be the cross section, $U_1 = \{x_1 \in B_1 : (x_1, u_2) \in U\}$. We say that f is differentiable with respect to the x_1 variable at (u_1, u_2) if the function $g(x_1) = f(x_1, u_2)$ is differentiable at u_1. When this holds, we write $dg_u = D_1 f(u_1, u_2)$; dg_u is a linear mapping from U_1 into B_3. f is said to be differentiable with respect to x_1 on U, if it is differentiable with respect to x_1 at each $u \in U$. Of course, the usual properties for partial derivatives hold in this general context. In particular, if f is differentiable at $u = (u_1, u_2) \in U$, then f is differentiable with respect to both x_1 and x_2 at u, and for all $(\xi_1, \xi_2) \in B_1 \times B_2$,

$$df_u(\xi_1, \xi_2) = D_1 f(u) \cdot \xi_1 + D_2 f(u) \cdot \xi_2.$$

Furthermore, the mapping $(x_1, x_2) \to (D_1 f(x_1, x_2), D_2 f(x_1, x_2))$ is in $C(U, B(U_1, B_3) \times B(U_2, B_3))$. Finally, if $f \in C(U, B_3)$, and f is continuously differentiable with respect to both variables (i.e., $D_1 f: U \to B(U_1, B_3)$ and $D_2 f: U \to B(U_2, B_3)$, are continuous), then $f \in C^1(U, B_3)$. The proofs of these statements are easy; see, e.g., [BP].

We can now state the implicit function theorem in the Banach space context; we omit the proof (see, e.g., [Cr 2]).

Theorem 13.3 (Implicit Function Theorem). *Let $f \in C(U, B)$ where U is open in $\Lambda \times B_1$, and Λ, B_1 and B are Banach spaces. Assume that:*

(i) $f(\lambda_0, u_0) = 0$ *for some* $(\lambda_0, u_0) \in U$,
(ii) $D_2 f: (\lambda, u) \to D_2 f(\lambda, u)$ *is continuous in a neighborhood of* (λ_0, u_0), *and*
(iii) $D_2 f(\lambda_0, u_0)$ *is nonsingular (i.e., has a bounded inverse); equivalently, $D_2 f(\lambda_0, u_0)$ is a continuous bijective (linear) mapping.*

Then there exists a continuous "curve" $u = u(\lambda)$ defined in a neighborhood N of λ_0, such that $u(\lambda_0) = u_0$, and $f(\lambda, u(\lambda)) \equiv 0$ in N. These are the only solutions of $f(\lambda, u) = 0$ in N. Finally, if $f \in C^k(U, B)$, then $u \in C^k(N, B)$.

Thus, according to the implicit function theorem, we can only expect bifurcation at (λ_0, u_0) if $D_2 f(\lambda_0, u_0)$ is singular. On the other hand, even if $D_2 f(\lambda_0, u_0)$ is singular, (λ_0, u_0) need *not* be a bifurcation point. This of course, is all quite well known. However, let us observe that if $F(\lambda, x) = \lambda - x^3$, then $F(0, 0) = 0 = F_x(0, 0)$; yet the equation $F(\lambda, x) = 0$ uniquely determines x as a function of λ. On the other hand, if $G(\lambda, x) = \lambda - x^2$, then $G(0, 0) = 0 = G_x(0, 0)$ and $(0, 0)$ is a bifurcation point. The difference comes from a consideration of the derivatives, $F_x = -3x^2$ and $G_x = -2x$; namely, the latter changes sign at $x = 0$, while the former does not.

In order to get a better feeling about what is happening here, let's consider the case where f is a mapping on a finite-dimensional space, where for simplicity, we assume that $f(\lambda, 0) = 0$ for all $\lambda \in \mathbf{R}$. Using Taylor's theorem, we can write

$$f(\lambda, u) = L_0 u + (\lambda - \lambda_0)L_1 u + r(\lambda, u), \tag{13.4}$$

where $L_0 = D_2 f(\lambda_0, 0)$ and $L_1 = D_1 D_2 f(\lambda_0, 0)$ are $n \times n$ matrices, and $r \in C^2$ satisfies

$$r(\lambda, 0) \equiv 0, D_2 r(\lambda_0, 0) = D_1 D_2 r(\lambda_0, 0) = 0. \tag{13.5}$$

Now one can show, using degree theory, that if $\det(L_0 + (\lambda - \lambda_0)L_1)$ changes sign at λ_0, then $(\lambda_0, 0)$ is a bifurcation point; (see [Kr, p. 196]). We shall not prove this here since we are going to prove a more general theorem in §C. What we will prove is the so called "bifurcation from a simple eigenvalue" theorem. Here is a preliminary, finite-dimensional statement.

Theorem 13.4. *Let U be an open subset of $\mathbf{R} \times \mathbf{R}^n$ and let $f \in C^2(U, \mathbf{R}^n)$ be given by* (13.4), *where r satisfies* (13.5). *Assume that the null space of L_0 is spanned by u_0, and that $L_1 u_0$ is not in the range of L_0. Denote $\{u_0\}^\perp$ by Z. Then there is a $\delta > 0$ and a C^1-curve $(\lambda, \phi): (-\delta, \delta) \to \mathbf{R} \times Z$ such that* (i) $\lambda(0) = \lambda_0$, (ii) $\phi(0) = 0$, *and* (iii) $f(\lambda(s), s(u_0 + \phi(s))) = 0$ *for $|s| < \delta$. Furthermore, there is a neighborhood of $(\lambda_0, 0)$ such that any zero of f either lies on this curve or is of the form $(\lambda, 0)$.*

Thus $(\lambda_0, 0)$ is a bifurcation point for f. This situation is depicted in Figure 13.2. Note that we get some information about the "direction" of bifurcation; i.e., $du/ds = u_0$ at $s = 0$. Furthermore, if we use the notation $N(T)$ and $R(T)$ to denote, respectively, the null space and range of an operator T, then the conditions $N(L_0) = \text{span}\{u_0\}$, and $L_1 u_0 \notin R(L_0)$ are equivalent to the fact that $\mu = 0$ is a simple root of $\det(L_0 + \mu L_1)$; hence, in particular, $\det(L_0 + (\lambda - \lambda_0)L_1)$ changes sign at $\lambda = \lambda_0$.

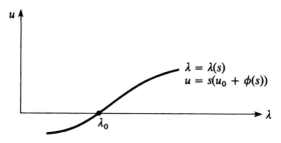

Figure 13.2

Proof of Theorem 13.4. The idea is to introduce a new parameter s which enables us to immediately apply the implicit function theorem on the "quotient" space Z. Thus, we define a new function $F(s, \lambda, z)$ by

$$F(s, \lambda, z) = \begin{cases} s^{-1}f(\lambda, s(u_0 + z)), & \text{if } a \neq 0, \\ D_2 f(\lambda, 0)(u_0 + z), & \text{if } s = 0. \end{cases}$$

Using (13.4) we can easily check that $F \in C^1(U \times Z, \mathbf{R}^n)$. Now since $F(0, \lambda_0, 0) = 0$, and[1] $D_2 D_3 F(0, \lambda_0, 0) \cdot (\tilde{\lambda}, \xi) = L_0 \xi + \tilde{\lambda} L_1 u_0$ is nonsingular, we can apply the implicit function theorem, and solve $F(s, \lambda, z) = 0$, thereby obtaining functions $\lambda = \lambda(s)$, $z = \phi(s)$ defined for $|s| < \delta$, near the point $(0, \lambda_0, 0)$. For the uniqueness, we observe that since $D_2 D_3 F(0, \lambda_0, 0)$ is an isomorphism, it follows by a continuity argument, that for every ε, $0 < \varepsilon < 1$, there is a neighborhood N of $(\lambda_0, 0)$ with the property that every solution $(\lambda, u) \equiv (\lambda, su_0 + w)$ of $f = 0$ (with $w \in Z$) in N satisfies the estimate $\|w\| < 2|s|\varepsilon$, if $|s| + |\lambda - \lambda_0|$ is small. Indeed, if $(\lambda, u) \in U$, $w \in Z$, then since the mapping $(\lambda, w) \to L_0 w + (\lambda - \lambda_0)L_1 u_0$ is nonsingular, there is a $k > 0$ such that

$$L_0 w + (\lambda - \lambda_0)L_1 u_0 = v \quad \text{implies} \quad \|w\| + |\lambda - \lambda_0| < k\|v\|. \quad (13.6)$$

Now given any $\varepsilon > 0$, the hypotheses (13.5) on r together with the fact that $r(\lambda, su_0 + w) - r(\lambda, 0) = r(\lambda, su_0 + w)$, imply that there exists a $\delta > 0$ such that

$$\|w\| + |s| + |\lambda - \lambda_0| < \delta \quad \text{implies} \quad \|r(\lambda, su_0 + w)\| < \varepsilon(|s| + \|w\|). \quad (13.7)$$

But noting that $f(\lambda, su_0 + w) = 0$ can be written in the form

$$L_0 w + s(\lambda - \lambda_0)L_1 u_0 = -(\lambda - \lambda_0)L_1 w + r(\lambda, su_0 + w),$$

---

[1] Namely, $F(0, \lambda, z) = D_2 f(\lambda, 0)(u_0 + z)$, so $D_3 F(0, \lambda, 0)\xi|_{\lambda=0} = D_2 f(0, 0)\xi = L_0 \xi$, and $D_2 F(0, 0, 0)\tilde{\lambda} = \tilde{\lambda} D_1 D_2 f(0, 0)u_0 = \tilde{\lambda} L_1 u_0$. Thus $d_{(\lambda, z)}F(\xi, \tilde{\lambda}) = L_0 \xi + \tilde{\lambda} L_1 u_0$.

we find that (13.6) and (13.7) yield

$$\|w\| + |s||\lambda - \lambda_0| \leq k(|\lambda - \lambda_0|\,\|L_1\|\,\|w\|) + \varepsilon(|s| + \|w\|).$$

Thus, if $0 < \varepsilon < 1$,

$$\tfrac{1}{2}\|w\| < (1 - k|\lambda - \lambda_0|\,\|L\| - \varepsilon)\|w\| \leq \varepsilon|s|, \quad \text{if } |\lambda - \lambda_0| \text{ is small}.$$

From this estimate it follows that if $u \neq 0$, then $s \neq 0$ (otherwise $su_0 + w = 0$), and so $F(s, \lambda, s^{-1}w) = 0$. Thus, by the uniqueness part of the implicit function theorem, if ε is sufficiently small, $(s, \lambda, s^{-1}w) = (s, \lambda(s), \phi(s))$. Hence $\lambda = \lambda(s)$, and $u = su_0 + w = su_0 + s\phi(s)$. This completes the proof. $\square$

Now let us consider the analogue of this theorem in the infinite-dimensional context. We will often assume the condition

$$f(\lambda, 0) \equiv 0, \qquad \lambda \in S \subset \mathbf{R}, \tag{13.8}$$

where S will vary according to the specific context. Here is the infinite-dimensional version of Theorem 13.4.

Theorem 13.5 (Bifurcation from a Simple Eigenvalue). *Let $U = S \times V$ be an open subset of $\mathbf{R} \times X$ and let $f \in C^2(U, Y)$, where X and Y are Banach spaces. Suppose that (13.8) holds, and let $L_0 = D_2 f(\lambda_0, 0)$ and $L_1 = D_1 D_2 f(\lambda_0, 0)$. Assume that the following conditions hold:*

(i)   *$N(L_0)$ is one-dimensional, spanned by u_0.*
(ii) *$R(L_0)$ has co-dimension 1; i.e., $\dim[Y/R(L_0)] = 1$.*
(iii) *$L_1 u_0 \notin R(L_0)$.*

Let Z be any closed subspace of X such that $X = [\text{Span } u_0] \oplus Z$ (i.e., any $x \in X$ can be uniquely written as $x = \alpha u_0 + z$, $\alpha \in \mathbf{R}$, $z \in Z$). Then the conclusions of Theorem 13.4 hold.

The proof of this theorem is virtually identical to that of Theorem 13.4. We remark that in the finite-dimensional case, $\dim N(L_0) = 1$ is equivalent to $\operatorname{codim} R(L_0) = 1$; this is not necessarily true in infinite dimensions.

We now give two examples to illustrate this theorem.

EXAMPLE 1. Let $[0, 1]$ be the unit interval in $\mathbf{R}$, and let C^k, $k = 0$ or 2 be the Banach space $C^k([0, 1], \mathbf{R})$, with its usual norm. Denote by $C_0^{2+\alpha}$, the subspace $\{u \in C^{2+\alpha} : u'(0) = u'(1) = 0\}$. Let $f : \mathbf{R} \times C_0^{2+\alpha} \to C^\alpha$ be defined by

$$f(\lambda, u) = u'' + \lambda[u + u^2 h(u)],$$

where h is smooth. Then $f(\lambda, 0) = 0$ for all $\lambda \in \mathbf{R}$,

$$D_2 f(\lambda, 0)v = v'' + \lambda v \quad \text{and} \quad D_1 D_2 f(\lambda, 0)v = v, \qquad v \in C_0^2.$$

Thus $D_2 f(\lambda, 0)$ is nonsingular if and only if $\lambda \neq n^2\pi^2$, $n = 1, 2, \ldots$; otherwise

$$N(D_2 f(n^2\pi^2, 0)) = \text{span of } \{\cos n\pi x\}, \quad \text{and}$$

$$R(D_2 f(n^2\pi^2, 0)) = \left\{ u \in C^0 : \int_0^1 u(x) \cos n\pi x \, dx = 0 \right\}.$$

It follows that $\cos n\pi x \notin R(D_2 f(n^2\pi^2, 0))$. Now if we define Z as

$$Z = \left\{ u \in C_0^2 : \int_0^1 u(x) \cos n\pi x \, dx = 0 \right\},$$

then all the hypotheses of the preceding theorem are satisfied. Therefore bifurcation occurs at all points $(n^2\pi^2, 0) \subset \mathbf{R} \times C_0^2$, $n = 1, 2, \ldots$.

EXAMPLE 2. Let Ω be a bounded region in $\mathbf{R}^n$, and consider the problem

$$\Delta u - \lambda\phi(u) = 0 \quad \text{in } \Omega, \qquad au + b\frac{du}{dn} = 0 \quad \text{on } \partial\Omega,$$

where $a^2 + b^2 \neq 0$, and du/dn is the normal derivative of u on $\partial\Omega$. Here ϕ is smooth, $\phi(0) = 0$, and $\phi'(0) \neq 0$. We let $f(\lambda, u) = \Delta u - \lambda\phi(u)$, with the domain and range of f being the obvious analogues of those in Example 1. The linearized operator at $u = 0$ is

$$D_2 f(\lambda, 0)v = \Delta v - \lambda\phi'(0)v, \qquad av + b\frac{dv}{dn} = 0 \quad \text{on } \partial\Omega.$$

If λ_0 is an eigenvalue with one-dimensional null space spanned by u_0 (so, for example, if λ_0 is the principal eigenfunction of the linearized operator; see footnote 3, below), then since the linearized operator is self-adjoint, it follows that $R(D_2 f(\lambda_0, 0))$ is, as in the previous example, the set of v satisfying

$$\int_\Omega v u_0 \, dx = 0.$$

Since $D_1 D_2 f(\lambda_0, 0)v = -\phi'(0)v$, and $\phi'(0) \neq 0$, we see that $D_1 D_2 f(\lambda_0, 0) \notin R(D_2 f(\lambda_0, 0))$. Thus the hypotheses of Theorem 13.5 hold, and $(\lambda_0, 0)$ is a bifurcation point.

In the remainder of this section, we shall describe the very useful "method of Lyapunov–Schmidt." It is a procedure which provides a systematic way to reduce the dimension of the space in which one tries to solve an equation of the form $f(\lambda, u) = 0$ near a singular point. In fact, it often reduces infinite-dimensional problems to finite-dimensional ones.

The idea is to attempt to split the space into two sets and to "project" the equation into each one of them. Of the two equations that are obtained, one is solvable by the implicit function theorem, and the other is often one in a finite-dimensional space, and thus can be attacked by topological techniques, such as degree theory or Morse theory.

Let Λ, B_1, and B_2 be Banach spaces, and suppose $f \in C^k(U, B_2)$ where U is a neighborhood of the origin in $\Lambda \times B_1$. We assume that $f(0, 0) = 0$, and we are interested in solutions of $f(\lambda, x) = 0$ near $(0, 0)$. We make the important assumption that $D_2 f(0, 0)$ is a Fredholm operator (see Chapter 11, §A); i.e.,

    (i) $N(D_2 f(0, 0)) \equiv X_1$ is finite dimensional, and
    (ii) $R(D_2 f(0, 0)) \equiv X_2$ is a closed subspace of B_2 having finite codimension.

In view of (ii), we may write B_2 as a direct sum $B_2 = Z_2 \oplus X_2$, where Z_2 is a finite-dimensional space. Let Q be the projection operator onto X_2 (i.e., $Q(z, x) = x$). We can also write $B_1 = Z_1 \oplus X_1$. Then if we apply the operators Q and $I - Q$ to the equation $f(\lambda, x) = 0$, we see that

$$f(\lambda, x) = 0 \quad \text{iff} \quad Qf(\lambda, x) = 0 \quad \text{and} \quad (I - Q)f(\lambda, x) = 0.$$

Now the composition Qf maps $\Lambda \times Z_1 \times X_1$ into X_2. If we let $\tilde{X} = \Lambda \times X_1$ and apply the implicit function theorem to $g(\tilde{x}, z_1) \equiv f(\lambda, z_1 + x_1) = 0$, we see that since $D_2 g(0, z_1) = D_2 f(0, z_1)$ is an isomorphism, there is a unique solution $z_1 = \phi(\lambda, x_1)$ near $(0, 0)$, of $Qf = 0$; i.e.,

$$Qf(\lambda, x_1 + \phi(\lambda, x_1)) = 0.$$

Thus, $x_1 + \phi(\lambda, x_1)$ solves $f(\lambda, x) = 0$ iff

$$(I - Q)f(\lambda, x_1 + \phi(\lambda, x_1)) = 0. \tag{13.9}$$

Now $R(I - Q)$ being finite dimensional, means that (13.9) is a *finite* number of equations; it is called the bifurcation equation. If Λ is also a finite-dimensional parameter space, then (13.9) becomes a finite-dimensional system of equations.

EXAMPLE 3. Let $B_1 = \mathbf{R} \times B$, where B is a Banach space, and let S be a compact operator taking B into itself. Suppose further that $r: B_1 \to B$ satisfies (13.5). Let

$$f(\lambda, u) = u - \lambda S(u) + r(\lambda, u),$$

and recall by Theorem 13.2, $T = dS_u$ is a compact linear operator. Now the null space of

$$D_2 f(\lambda_0, 0)(\lambda, \xi) = (I - \lambda_0 T)\xi$$

is the set $\{(\bar{\lambda}, \xi) : \xi \in N(I - \lambda_0 T)$ and thus has dimension

$$n = 1 + \dim N(I - \lambda_0 T).$$

The standard Riesz–Schauder theory (see [Sr]) implies that $I - \lambda_0 T$ is a Fredholm operator of index zero; thus the codimension of $R(I - \lambda_0 T)$ is also n. The Lyapunov–Schmidt procedure then leads to a system of n equations in $(n + 1)$ unknowns.

§B. Stability of Bifurcating Solutions

In many contexts, a solution $\bar{x}$ of an equation $f(x) = 0$ corresponds to a "steady-state" solution of a dynamical system; i.e., of a time-dependent problem

$$x' = f(x), \qquad ' = \frac{d}{dt}.$$

A very important problem is to decide whether or not the steady-state solution is stable. Let us recall what is meant by this (cf. Chapter 11, §B). Thus, suppose that $\bar{x}$ is perturbed (slightly!) to the state $\bar{x} + \xi$ and we consider the initial-value problem

$$x' = f(x), \qquad x(0) = \bar{x} + \xi.$$

We want to determine whether this solution tends to $\bar{x}$ as $t \to +\infty$, or even if it stays close to $\bar{x}$ for all $t > 0$. If we consider the "linearized" problem

$$\xi' = df_{\bar{x}}\xi,$$

then if the spectrum of $df_{\bar{x}}$ lies in the left-half plane, the solution ξ decays exponentially to zero as $t \to +\infty$. In this case, it is natural to say that $\bar{x}$ is (linearly) stable. On the other hand, if the spectrum contains points in the right-half plane, we say that $\bar{x}$ is (linearly) unstable. We are interested here in the stability properties of bifurcating solutions.[2]

To be a little more specific, let us consider the equations

$$u' = f(\lambda, u), \quad \text{where } f(\lambda, 0) \equiv 0, \quad \lambda \in \mathbf{R}, \tag{13.10}$$

---

[2] In Chapters 22 and 23 this problem is considered from a generalized "Morse theoretic" point of view.

and suppose that we have the following situation. For $\lambda < \bar{\lambda}$, the spectrum $\sigma(D_2 f(\lambda, 0))$, lies in the left-half plane, Re $z < 0$, while for $\lambda > \bar{\lambda}$, $\sigma(D_2 f(\lambda, 0))$ intersects Re $z > 0$. Then at $\lambda = \bar{\lambda}$, at least one point of the spectrum crosses the imaginary axis; i.e., a point ir, with r real, is in $\sigma(D_2 f(\bar{\lambda}, 0))$. Now if $r = 0$, then, generally speaking, the trivial solution bifurcates into new equilibrium solutions, while if $r \neq 0$ it bifurcates into periodic solutions. This latter circumstance occurs with a so-called "Hopf bifurcation" (see, e.g., [HKW]). In this section we shall be concerned only with bifurcation into equilibria.

We shall consider now two examples which illustrate the theorem we are going to state.

EXAMPLE 1. Let Ω be a bounded domain in $\mathbf{R}^n$ with $\partial\Omega$ smooth and consider the Dirichlet problem

$$\Delta u - \mu u - u^2 = 0 \quad \text{in } \Omega, \qquad u = 0 \quad \text{on } \partial\Omega. \tag{13.11}$$

We recall from Chapter 11 §A, that if λ_0 is the principal eigenvalue of Δ, with homogeneous Dirichlet boundary conditions, then $\lambda_0 < 0$, and λ_0 has an associated eigenfunction u_0 which we may assume to be positive on the interior of Ω. Furthermore, it is not too hard to show, using the nodal properties of eigenfunctions, that λ_0 is a simple eigenvalue in the sense that the associated eigenspace is one-dimensional.[3]

Now from Example 2 in the last section $(\lambda_0, 0)$ is a bifurcation point, and the set of solutions near $(\lambda_0, 0)$ consists only of the trivial branch $(\mu, 0)$ and the smooth curve

$$(\mu(s), u(s)) = (\mu(s), su_0 + s\phi(s)), \qquad |s| < \delta, \tag{13.12}$$

where $(\mu(0), u(0)) = (\lambda_0, 0)$. Moreover, for each s with $|s| < \delta$, $\phi(s)$ is in the L_2-orthogonal complement of u_0.

We want to calculate the sign of $\mu'(0)$. To this end, substitute (13.12) into (13.11), divide by s, differentiate with respect to s, and set $s = 0$; this gives

$$\Delta\phi'(0) - \lambda_1\phi'(0) - \mu'(0)u_0 - u_0^2 = 0.$$

Now multiply by u_0 and integrate over Ω to get

$$\int u_0 \Delta\phi'(0) - \lambda_1 \int u_0 \phi'(0) - \mu'(0) \int u_0^2 - \int u_0^3 = 0.$$

---

[3] In fact, the nodes of the nth eigenfunction divide Ω into at most n subdomains; see [CH 1]. If u_2 were an eigenfunction of Δ corresponding to λ_0 and $u_2 \perp u_1$, then u_2 must change sign in Ω. Thus, $(u_2 - u_1)$ would be an eigenfunction corresponding to λ_0 having at least two nodes.

If we integrate the first term by parts, we find

$$\mu'(0) \int u_0^2 = -\int u_0^3,$$

and thus $\mu'(0) < 0$. If we write $\lambda(s) = \mu(s) - \lambda_0$, then this curve is as depicted in Figure 13.3(a); it corresponds to solution branches.

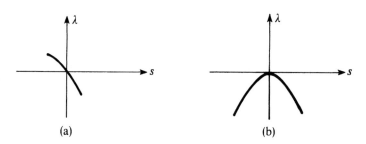

(a)   (b)

Figure 13.3

EXAMPLE 2. Consider the Dirichlet problem

$$\Delta u - \lambda u - u^3 = 0 \quad \text{in } \Omega, \qquad u = 0 \quad \text{on } \partial\Omega,$$

where Ω is as in Example 1. Here again $(\lambda_0, 0)$ is a bifurcation point, and we have the bifurcating solutions as in (13.11). Computations similar to those in Example 1 show $\lambda'(0) = 0$ and $\lambda''(0) < 0$; this yields Figure 13.3(b).

Definition 13.6. Let X and Y be Banach spaces and let L_0 and $K \in B(X, Y)$. We say that $\mu \in \mathbf{C}$ is a K-*simple eigenvalue* of L_0 (with eigenfunction u_0) if the following three conditions hold:

   (i) $\dim N(L_0 - \mu K) = \operatorname{codim} R(L_0 - \mu K) = 1$,
   (ii) u_0 spans $N(L_0 - \mu K)$,
   (iii) $Ku_0 \notin R(L_0 - \mu K)$.

The terminology comes from the case where $X = Y$, $K = I$, and L_0 is a compact operator. Under these circumstances, $\mu \neq 0$ is an I-simple eigenvalue of L_0 if and only if μ is a simple eigenvalue of L_0. Note too that we could weaken the definition by not requiring L_0 to be a bounded operator. In this case if $X = Y$, $K = I$, and μ is an isolated point in the spectrum of L_0, then μ is an I-simple eigenvalue for L_0 if and only if μ is a simple eigenvalue of L_0.

Let us observe that in this terminology, the hypotheses (i)–(iii) of Theorem (13.5) can be stated as: $L_0 \equiv D_2 f(\lambda_0, 0)$ has 0 as an $L_1 \equiv D_1 D_2 f(\lambda_0, 0)$

simple eigenvalue. The importance of this notion comes from the next lemma, which implies that K-simple eigenvalues "continue" along the bifurcating branches; i.e., for s near zero, $D_2 f(\lambda(s), u(s))$ has a unique (small) K-simple eigenvalue $\mu(s)$. We shall also see that sign $\mu(s) = $ sign $s\lambda'(s)$, and this equation will determine whether the bifurcating solutions are stable or unstable.

Lemma 13.7. *Let μ_0 be a K-simple eigenvalue of L_0, with eigenfunction u_0. Then there exists $\rho > 0$ such that if $\| L - L_0 \| < \rho$, L has a unique K-simple eigenvalue $\eta(L)$, with eigenfunction $w(L) = u_0 + z(L)$; here $X = \text{span}\{u_0\} \oplus Z$, and $z(L) \in Z$. Also $\eta(L_0) = \mu_0$, $w(L_0) = u_0$, and the mapping $L \to (\eta(L), w(L))$ is smooth.*

Now let the hypotheses of Theorem 13.5 hold, i.e., as we have seen above, $D_2 f(\lambda_0, 0)$ has 0 as a $D_1 D_2 f(\lambda_0, 0)$ simple eigenvalue. Let $\lambda(s)$ and $u(s) \equiv s(u_0 + \phi(s))$ be the bifurcating curve as provided by this theorem. We shall use the notation provided by this theorem. We shall also use the notation $f'(s) = D_2 f(\lambda(s), u(s))$. Now suppose that $X \subset Y$, the inclusion $i: X \to Y$ is continuous, and 0 is an i-simple eigenvalue of $D_2 f(\lambda_0, 0)$. Then by Lemma 13.7, there exist functions

$$\lambda \to (\gamma(\lambda), v(\lambda)), \qquad s \to (\eta(s), w(s))$$

defined on neighborhoods of λ_0 and 0, respectively, into $\mathbf{R} \times X$, such that $(\gamma(\lambda_0), v(\lambda_0)) = (0, u_0) = (\eta(0), w(0))$, $v(\lambda) - u_0 \in Z$, $w(s) - u_0 \in Z$, and on these neighborhoods,

$$D_2 f(\lambda, 0)v(\lambda) = \gamma(\lambda)v(\lambda), \tag{13.13}$$

$$f'(s)w(s) = \eta(s)w(s). \tag{13.14}$$

Namely, set $\eta(s) = \eta(f'(s))$, $w(s) = w(f'(s))$, where η and w are the functions provided by Lemma 13.7, which is now considered along the curve $(\lambda(s), u(s))$; this gives (13.14). Equation (13.13) comes from the same lemma, now considered along the curve $(\lambda, 0)$; i.e., $\eta(D_2 f(\lambda, 0)) = \gamma(\lambda)$, and $w(D_2 f(\lambda, 0)) = v(\lambda)$. Note that these functions, being compositions of smooth functions, are themselves smooth.

Observe that in the case where $f(\lambda, u) = 0$ is the equilibrium equation of (13.10) then the natural choice of K is the identity I, and equations (13.13) and (13.14) are the linearized equations. Namely, (13.13) is the linearization about $(\lambda, 0)$, and one would study its spectrum as λ crosses λ_0, while (13.14) is the linearization about the bifurcating solution; it would be studied near $s = 0$.

Now the following theorem is used to determine the stability of the bifurcating solution.

Theorem 13.8. *Let the hypotheses of Theorem 13.5 hold, and let γ and η be defined as above. Then $\gamma'(\lambda_0) \neq 0$, and if $\eta(s) \neq 0$ for s near 0,*

$$\lim_{s \to 0} \frac{s\lambda'(s)\gamma'(\lambda_0)}{\eta(s)} = -1. \tag{13.15}$$

We omit the proof of both Lemma 13.7 and Theorem 13.8; see [CRa 1].

We shall show how (13.15) is used to discuss the stability of the bifurcating solutions by returning to Examples 1 and 2 above. First, consider Example 1. Let $(u(s), \mu(s))$ be the bifurcating branch of solutions and as before, let $\lambda(s) = \mu(s) - \lambda_0$. The equation in Example 1 along the bifurcating curve becomes

$$f(\lambda, u) \equiv (\Delta - \lambda - \lambda_0)u - u^2 = 0.$$

Hence $D_2(\lambda, 0)u_0 = (\Delta - \lambda - \lambda_0)u_0 = -\lambda u_0$. Thus using (13.13), we get $-\lambda u_0 = \gamma(\lambda)u_0$, so that $\gamma(\lambda) = -\lambda$. Now we have seen that $\lambda'(0) = \mu'(0) < 0$. Thus, from (13.15) for s near zero,

$$\operatorname{sgn} \eta(s) = -\operatorname{sgn} s.$$

Hence $\eta(s) > 0$ if $s < 0$, and the solution is unstable. For $s > 0$, $\eta(s) < 0$ and the solution is stable. Note too, $\gamma'(\lambda_0) < 0$ so that $\gamma < 0$ if $\mu > \lambda_0$, i.e., $\lambda > 0$ and $(\lambda, 0)$ is stable, while for $\lambda < \lambda_0$, this trivial branch is unstable. These are depicted in Figure 13.4(a). Similarly, for Example 2, we get the situation depicted in Figure 13.4(b). (Of course we are considering the principal eigenvalues in these examples.)

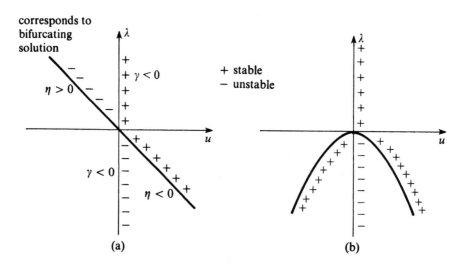

(a)   (b)

Figure 13.4

§C. Some General Bifurcation Theorems

In the last two sections we studied what may be called "local" bifurcation theory. Our results stemmed from the implicit function theorem, and the bifurcating solutions were only shown to exist near certain distinguished solutions. We shall now give some bifurcation results of a more global nature; the proofs rely on degree theory.

Let B be a Banach space, and let $f \in C(U, B)$ where U is an open subset of $\mathbf{R} \times B$. We assume that f is of the form

$$f(\lambda, u) = u - \lambda L u + h(\lambda, u), \tag{13.16}$$

where

(i) $1/\mu$ is an eigenvalue of L of odd multiplicity,
(ii) $L: B \to B$ is compact and linear,
(iii) $h: U \to B$ is compact,
(iv) $h(\lambda, u) = o(\|u\|)$ at $u = 0$, uniformly on bounded λ-intervals.

Our first theorem gives conditions for $(\mu, 0)$ to be a bifurcation point. An obvious necessary condition is that $I - \mu L$ is not invertible; hence a necessary condition for $(\mu, 0)$ to be a bifurcation point is that $1/\mu$ is an eigenvalue of L.

Theorem 13.9. *Suppose that* (i)–(iv) *hold. Then* $(\mu, 0)$ *is a bifurcation point of* $f(\lambda, x) = 0$.

Proof. If the result were false, then for $|\lambda - \mu|$ sufficiently small, if $\varepsilon > 0$ is sufficiently small, $d(f(\lambda, \cdot), S_\varepsilon, 0)$ is defined and is independent of λ; here S_ε is the ball of radius ε centered at 0. It follows from Theorem 12.17 that

$$d(f(\lambda, \cdot), S_\varepsilon, 0) = (-1)^{\beta_\lambda},$$

where $\beta_\lambda = \sum \{$multiplicity of eigenvalues of L greater than $1/\lambda\}$. If $\lambda_1 > \mu$, and $\lambda_2 < \mu$ then $\beta_{\lambda_1} - \beta_{\lambda_2}$, being the multiplicity of $1/\mu$ (see the remark after the proof of Theorem 12.17), is odd, by hypothesis. This violates the fact that $d(f(\lambda, \cdot), S_\varepsilon, 0)$ is independent of λ, and proves the theorem. $\square$

The second theorem is of a more global nature, in that it is a statement *in the large* about the bifurcation curve which branches out of $(\lambda_0, 0)$.

Theorem 13.10. *Suppose that the above hypotheses* (i)–(iv) *hold with* μ *replaced by* λ_0. *Let* S *denote the closure of the set of nontrivial zeros of* f. *Then* S *contains a component* C (*i.e., a maximal connected subset*) *which meets* $(\lambda_0, 0)$ *and either*

(i) C *is noncompact in* U (*if* $U = \mathbf{R} \times B$, *this means that* C *is unbounded*), *or*

(ii) C *meets* $u = 0$ *in a point* $(\bar{\lambda}, 0)$ *where* $\bar{\lambda} \neq \lambda_0$ *and* $1/\bar{\lambda}$ *is an eigenvalue of* L.

Proof. Suppose that the theorem is false. Since L is compact, the only possible limit point of the eigenvalues of L is zero (Lemma 11.2). Thus in any finite

λ-interval, there are only a finite number of *characteristic values* (i.e., reciprocals of eigenvalues) of L. Take $\delta > 0$ so small that L has no characteristic values in $I = (\lambda_0 - \delta, \lambda_0 + \delta)$ other than λ_0. Let C be the maximal connected subset of $S \cup \{(\lambda_0, 0)\}$ containing $(\lambda_0, 0)$; our assumption implies that C is compact (at this point it may be that $C = \{(\lambda_0, 0)\}$!). The proof now proceeds in two steps. First we shall find a bounded open subset $\mathcal{O} \subset U$ such that:

(a) $(\lambda_0, 0) \in \mathcal{O}$;
(b) $\partial\mathcal{O} \cap S = \phi$ (∂ taken in U); and
(c) $\mathcal{O}$ contains no nontrivial solutions other than $\{(\lambda, 0): \lambda_0 - \delta < \lambda < \lambda_0 + \delta\}$. Then we shall apply the homotopy invariance of degree in order to get a contradiction.

For step 1, let U_ε be an ε-neighborhood of C in U, where $0 < \varepsilon \ll 1$. Since (ii) holds, and $(\lambda, 0)$ is an isolated solution if λ is not a characteristic value, we may assume that U_ε contains no solution $(\lambda, 0)$ if $|\lambda - \lambda_0| > \varepsilon$. Now let $K = \overline{U}_\varepsilon \cap S$; then K is compact, and by construction, $\partial U_\varepsilon \cap C = \phi$. Set $A = \partial U_\varepsilon \cap S$, and $B = C$; then A and B are closed, nonvoid subsets of K, and $A \cap B = \phi$. At this point we may apply the following lemma [Wh, p. 15]. *Let K be a compact metric space, and let A and B be disjoint, closed, nonvoid subsets of K. Then either there exists a subcontinuum of K which meets both A and B, or there are disjoint closed subsets K_A, K_B containing A and B, respectively, with $K = K_A \cup K_B$.* Since $C \cap \partial U_\varepsilon = \phi$, and C is maximal, the first alternative is excluded, and we conclude that there are disjoint closed subsets K_A and K_B of K with $\partial \overline{U}_\varepsilon \cap S \subset K_A$, $C \subset K_B$, and $K = K_A \cup K_B$. Now let $\mathcal{O}$ be an ε-neighborhood of K_B where $\varepsilon < \text{dist}(K_A, K_B)$; then $\mathcal{O}$ satisfies conditions (a), (b), and (c) above.

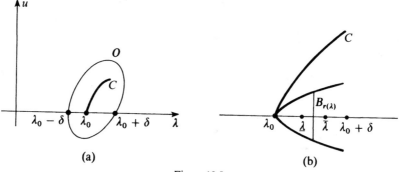

Figure 13.5

Let $O_\mu = \{u \in B: (\mu, u) \in O\}$. For each $\lambda \neq \lambda_0$, choose $r(\lambda) > 0$ such that $(\lambda, 0)$ is the only solution of $f = 0$ on $\{\lambda\} \times B_{r(\lambda)}$, where $B_{r(\lambda)}$ is the ball in B of radius $r(\lambda)$ about $u = 0$ (Figure 13.5(b)).

If $\lambda \neq \lambda_0$, $f = 0$ has no solution in $\{\lambda\} \times \partial(O_\lambda \backslash \overline{B}_{r(\lambda)})$, and thus $d(\lambda) \equiv d(f(\lambda, \cdot), O_\lambda \backslash B_{r(\lambda)}, 0)$ is well defined. Using the homotopy invariance of degree, $d(\lambda)$ is constant for $\lambda > \lambda_0$, $|\lambda - \lambda_0| < \delta$. Since C is compactly contained in O, there is a $\lambda > \lambda_0$ such that $O_\lambda \neq \phi$ and $O_\lambda \backslash \bar{B}_{r(\lambda)}$ contains no solution of $f = 0$. Thus for $\lambda > \lambda_0$

$$d(f(\lambda, \cdot), O_\lambda \backslash \bar{B}_{r(\lambda)}, 0) = 0. \tag{13.17}$$

By a similar argument, this equation holds for $\lambda < \lambda_0$, again for $|\lambda - \lambda_0| < \delta$. Then again using the homotopy invariance, we obtain

$$d(f(\lambda, \cdot), O_\lambda, 0) = \text{const.}, \quad \text{if } |\lambda - \lambda_0| < \delta. \tag{13.18}$$

Now let $\bar{\lambda}$ and $\underline{\lambda}$ be such that $\lambda_0 - \delta < \underline{\lambda} < \lambda_0 < \bar{\lambda} < \lambda_0 + \delta$. Then $O_{\bar{\lambda}} = B_{r(\bar{\lambda})} \cup (O_{\bar{\lambda}} \backslash B_{r(\bar{\lambda})})$, so by the additive property of degree (Chapter 12, §A, Property 5), which is also true in the general Banach space context; see Chapter 12, §B, we have

$$d(f(\bar{\lambda}, \cdot), O_{\bar{\lambda}}, 0) = d(f(\bar{\lambda}, \cdot), O_{\bar{\lambda}} \backslash B_{r(\bar{\lambda})}, 0) + d(f(\bar{\lambda}, \cdot), B_{r(\bar{\lambda})}, 0)$$
$$= d(f(\bar{\lambda}, \cdot), B_{r(\bar{\lambda})}, 0),$$

where we have used (13.17). Hence, from (13.18),

$$d(f(\underline{\lambda}, \cdot), \dot{B}_{r(\underline{\lambda})}, 0) = d(f(\bar{\lambda}, \cdot), B_{r(\bar{\lambda})}, 0). \tag{13.19}$$

Next, define the homotopy F, taking f to $I\lambda - L$ by

$$F(\lambda, u, \theta) = u - \lambda L u + \theta h(\lambda, u), \quad 0 \le \theta \le 1.$$

If $r(\bar{\lambda})$ is sufficiently small, we see that $F(\bar{\lambda}, u, \theta) \neq 0$ for every $(u, \theta) \in \partial B_{r(\bar{\lambda})} \times [0, 1]$, since (d) holds. Thus we obtain

$$0 \neq d(f(\bar{\lambda}, \cdot), B_{r(\bar{\lambda})}, 0) = d(I - \bar{\lambda}L, B_{r(\bar{\lambda})}, 0),$$

and similarly

$$d(f(\underline{\lambda}, \cdot), B_{r(\underline{\lambda})}, 0) = d(I - \underline{\lambda}L, B_{r(\underline{\lambda})}, 0).$$

These two equations, in conjunction with (13.19), give

$$d(I - \bar{\lambda}L, B_{r(\bar{\lambda})}, 0) = d(I - \underline{\lambda}L, B_{r(\underline{\lambda})}, 0). \tag{13.20}$$

But λ_0 is a characteristic value of L of odd multiplicity; hence by the remark after the proof of Theorem 12.17,

$$d(I - \bar{\lambda}L, B_{r(\bar{\lambda})}, 0) = -d(I - \underline{\lambda}L, B_{r(\underline{\lambda})}, 0) \neq 0.$$

This contradicts (13.20), and the proof is complete. $\square$

We now discuss an application of this theorem.

EXAMPLE. Consider the boundary-value problem

$$f(\lambda, u) \equiv u'' + \lambda\psi(u) = 0, \qquad 0 < x < 1, \tag{13.21}$$

with homogeneous Neumann boundary conditions,

$$u'(0) = u'(1) = 0.$$

Here we assume $\psi(u) = -u(u - a)(u - b)$, where $a < 0 < b$. We have seen, in Example 1 of §A, that the points $(\lambda, u) = (n^2\pi^2, 0)$, $n = 1, 2, \ldots$, are all bifurcation points for $f(\lambda, u) = 0$. We wish to consider the global behavior of these bifurcation branches. (Note that we can write our problem in the equivalent form $u + \lambda K\psi(u) = 0$, (cf. (12.16)), where K is compact, and this equation is easily seen to be of the form (13.16).

To this end, we first recall some standard properties of linear ordinary differential operators (see [CH 1]). Thus, let L be defined on $0 < x < 1$ by $Lv = v'' + a(x)v$, where $\|a\|_\infty < \infty$, with homogeneous Neumann boundary conditions at $x = 0$ and $x = 1$. Then L has a decreasing sequence of simple eigenvalues $\lambda_0 > \lambda_1 > \cdots > \lambda_n > \cdots$, where $\lambda_n \to -\infty$, and if v_n is the eigenfunction corresponding to λ_n, $n \geq 1$, then v_n has precisely n zeros on $(0, 1)$, where at each zero, $v_n' \neq 0$ (i.e., each is a nodal zero).

Now let u be a solution of (13.21) which bifurcates out of $(n^2\pi^2, 0)$; then from Theorem 13.5, we can write $u = u(s)$ where $u(s) = sv_n + s\phi(s)$, for s near zero. Thus for $|s|$ small, we see that u has exactly n-nodal zeros on $(0, 1)$. Since λ_n is a simple eigenvalue of the linearization of (13.21) (so λ_n has *odd* multiplicity equal to 1), we may apply Theorem 13.10. Now a nonzero solution u of (13.21) cannot have $u(\bar{x}) = u'(\bar{x}) = 0$ for some $\bar{x}$ [0, 1], since this would violate the uniqueness theorem for the initial-value problem of the first-order system $u' = w$, $w' = -f(u)$. We conclude that $u(s)$ has exactly n nodal zeros for *all* s for which it is defined. It follows that the branch of bifurcating solutions containing $(\lambda_n, 0)$ cannot come back to the set $u = 0$, and must therefore be unbounded.

In fact, more is true; namely, there cannot be any secondary bifurcation off of this curve; for a proof, see [SW]. The bifurcation diagram takes the form as in Figure 13.6.

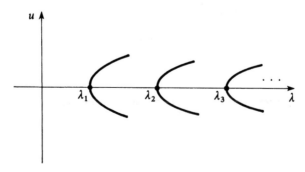

Figure 13.6

§D. Spontaneous Bifurcation; An Example

Up to now, we have considered bifurcation from a particular solution of the equation $f(\lambda, u) = 0$; the distinguished solution usually being one of the form $(\lambda, 0)$. This is what we might term "local" bifurcation. But there is another type of bifurcation which can occur; one of a distinctly global nature. It is what we could call "spontaneous" bifurcation, whereby the solution suddenly "appears," when λ crosses a certain critical value. We shall illustrate this phenomenon by means of a nontrivial important example.

We consider the ordinary differential equation

$$u'' + f(u) = 0, \qquad |x| < L, \tag{13.22}$$

with homogeneous Dirichlet boundary conditions

$$u(\pm L) = 0. \tag{13.23}$$

Here $f(u)$ is the cubic polynomial

$$f(u) = -u(u - a)(u - 1), \qquad 0 < a < \tfrac{1}{2}. \tag{13.24}$$

Notice that $u \equiv 0$ is always a solution; we are interested in nonconstant solutions.

In studying this problem, we shall consider the interval length L as the bifurcation parameter. However, note that if we make the change of variables $y = L^{-1}x$, then (13.22) and (13.23) become

$$L^2 u_{yy} + f(u) = 0, \qquad |y| < 1, \quad u(\pm 1) = 0,$$

so if $\lambda = 1/L^2$, we obtain a problem of the familiar type. We prefer, however, to consider the equation (13.22) because as we shall see presently, its solutions can be given a nice geometric interpretation.

We rewrite (13.22) as a first-order system

$$u' = v, \qquad v' = -f(u), \qquad |x| < L, \tag{13.22'}$$

and consider the phase plane for (13.22), as depicted in Figure 13.7. We let $F'(u) = f(u)$, $F(0) = 0$, and observe that the function $H(u, v) = v^2/2 + F(u)$, (the "total energy"), is constant along orbits of (13.22). We define A by $A^2/2 = F(1)$; see Figure 13.7. It is clear that solutions of (13.22)', (13.23) correspond to those orbits of (13.22)' which "begin" on the interval $(0, A)$ on the v-axis (i.e., the line $u = 0$), and "end" on the v-axis, and take "time" (parameter length) $2L$ to make the voyage. We have depicted one such an orbit as the curve joining p to q in Figure 13.7.

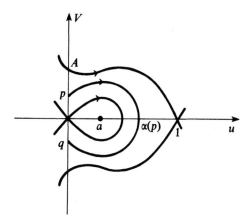

Figure 13.7

Now let $T(p)$ denote the "time" that an orbit starting at p takes to arrive at $\alpha(p)$, as depicted in the figure; here $H(\alpha(p), 0) = H(0, p)$ so that

$$p^2 = 2F(\alpha(p)). \tag{13.25}$$

Using the first equation in (13.22)′, together with the invariance of H along orbits, we find that

$$\sqrt{2}\,T(p) = \int_0^{\alpha(p)} \frac{du}{\sqrt{F(\alpha(p)) - F(u)}}. \tag{13.26}$$

Notice that solutions of (13.22)′, (13.23) correspond to curves for which $T(p) = L$. This leads us to investigate the shape of the graph of T.

To this end, we recall that the domain of T is the open interval $(0, A)$. Furthermore, if p is near 0 or A, then the orbit through p comes near the respective rest points $(0, 0)$ or $(1, 0)$, of (13.22)′, and hence $T(p)$ must be very large. Since T is a smooth function, we see that T must achieve its minimum on $(0, A)$; say at p_0. Obviously $T(p_0) > 0$; see Figure 13.8.

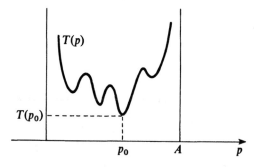

Figure 13.8

Now if $L < T(p_0)$, then it is clear that there are no nonconstant solutions of our problem. But precisely when $L = T(p_0)$, *a nonzero solution suddenly appears!* Viewed from the opposite side, we see that for $L \geq T(p_0)$, there are nonconstant solutions, but as soon as L gets below $T(p_0)$, these solutions disappear; i.e., they "cancel" each other.[4] This cancellation phenomenon will be studied in Chapter 22 in a generalized Morse-theoretic framework; see Theorem 22.33. For now, our interest in this phenomenon is that it is an example of a distinctly different type of bifurcation phenomenon than the one we have hitherto studied.

Actually, we can go much further with this example, and give a precise (qualitative) description of the graph of $T(p)$. By this we mean that we can count the exact number of critical points of T. This will be used in Chapter 24.

Theorem 13.11. *Let f be given by* (13.24) *and let* $T(p)$ *be defined by* (13.26). *Then T has exactly one critical point.*

Before giving the proof of this result, let us pause to interpret its significance. Thus, the theorem says that the graph of T is *not* as depicted in Figure 13.8, but rather, it is as in Figure 13.9. What this means, is that for

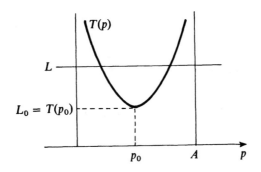

Figure 13.9

$L < L_0, u \equiv 0$ is the only solution; when $L = L_0$, a global bifurcation occurs and we obtain a nonconstant solution, while for every $L > L_0$, there are precisely two nonconstant solutions. (We have already remarked earlier that $u \equiv 0$ is a solution for every $L > 0$.)

If we now view (13.22) as the steady-state equation of the full time-dependent problem

$$u_t = u_{xx} + f(u), \quad |x| < L, \quad t > 0, \qquad u(\pm L, t) = 0, \quad t > 0, \qquad (13.27)$$

---

[4] We may say (*not* tongue in cheek, cf. Chapter 22), that the nonconstant solutions continue to the empty set, while as L increases to $T(p_0)$, a solution appears out of the empty set.

then we have an *exact* count of the number of "rest points" ($\equiv$ steady-state solutions) as a function of L. Using the Conley index we can give a global description of *all* solutions of (13.27); this will be carried out in Chapter 24, §D.

Proof of Theorem 13.11. It suffices to prove that T has at most one critical point (since we already know that T has *at least* one critical point). To achieve this, we define $S(\alpha)$ by

$$S(\alpha) = \int_0^\alpha \frac{du}{\sqrt{F(\alpha) - F(u)}}. \tag{13.28}$$

From (13.25) we see that $\alpha'(p) > 0$; thus the equation $\sqrt{2}\,T'(p) = S'(\alpha)\alpha'(p)$, shows that $T'(p) = 0$ if and only if $S'(\alpha) = 0$. Therefore it suffices to prove that S has at most one critical point. For this we will show

$$S''(\alpha) > 0 \quad \text{if } S'(\alpha) = 0. \tag{13.29}$$

To this end we change variables in the integral in (13.28) by writing $u = \alpha \sin \psi$; then

$$S(\alpha) = \int_0^{\pi/2} (F(\alpha) - F(\alpha \sin \psi))^{-1/2}\alpha \cos \psi \, d\psi.$$

This allows us to compute the derivatives; namely,

$$2S'(\alpha) = \int_0^{\pi/2} G^{-3/2}(2G - \alpha G') \cos \psi \, d\psi, \tag{13.30}$$

and

$$2S''(\alpha) = \int_0^{\pi/2} [G^{-3/2}(2G' - \alpha G'' - G') + (2G - \alpha G')(-3/2)G^{-5/2}G']$$
$$\times \cos \psi \, d\psi, \tag{13.31}$$

where $G(\alpha) = F(\alpha) - F(\alpha \sin \psi)$.
   Now if $S'(\alpha) = 0$, then for this α,

$$2S''(\alpha) = kS'(\alpha) + 2S''(\alpha), \tag{13.32}$$

where k is a constant, to be chosen in a moment. We have from (13.30), (13.31), and (13.32),

$$2S'' = \int_0^{\pi/2} \left[G^{-5/2}\left\{\frac{-3}{2}G' + kG\right\}(2G - \alpha G') + G^{-3/2}(G' - \alpha G'') \right] \cos \psi \, d\psi.$$

Now put $k = 3/\alpha$ in this equation to get

$$2S'' = \int_0^{\pi/2} \left[G^{-5/2} \left(\frac{3}{2\alpha}\right)(2G - \alpha G')^2 + G^{-3/2}(G' - \alpha G'') \right] \cos \psi \, d\psi$$

$$\geq \int_0^{\pi/2} G^{-3/2}(G' - \alpha G'') \cos \psi \, d\psi. \tag{13.33}$$

But

$$G'(\alpha) = f(\alpha) - f(\alpha \sin \psi) \sin \psi, \tag{13.34}$$

and

$$G''(\alpha) = f'(\alpha) - f'(\alpha \sin \psi) \sin^2 \psi.$$

Thus, from (13.33)

$$2S''(\alpha) \geq \int_0^{\pi/2} G^{-3/2}[f(\alpha) - f(\alpha \sin \psi)$$

$$\times \sin \psi - \alpha f'(\alpha) + \alpha f'(\alpha \sin \psi) \sin^2 \psi] \cos \psi \, d\psi$$

$$= \frac{1}{\alpha^2} \int_0^\alpha (F(\alpha) - F(u))^{-3/2}[\alpha f(\alpha) - u f(u) - \alpha^2 f'(\alpha) + u^2 f'(u)] \, du.$$

$$\tag{13.35}$$

Writing

$$\theta(x) = 2F(x) - xf(x),$$

gives

$$\theta'(x) = f(x) - xf'(x), \qquad x\theta'(x) = xf(x) - x^2 f'(x),$$

and from (13.35),

$$2S''(\alpha) \geq \frac{1}{\alpha^2} \int_0^\alpha (F(\alpha) - F(u))^{-3/2}(\alpha \theta'(\alpha) - u\theta'(u)) \, du. \tag{13.36}$$

Now for the cubic (13.24), we have

$$x\theta'(x) = 2x^3\left(x - \frac{a+1}{2}\right);$$

the graph of this function is depicted in Figure 13.10(a). Thus if $\alpha \geq (a + 1)/2$, then (13.36) shows $S''(\alpha) > 0$; that is, $\alpha \geq (a + 1)/2$ implies that (13.29) holds. On the other hand from (13.30),

$$2S'(\alpha) = \frac{1}{\alpha} \int_0^\alpha \frac{\theta(\alpha) - \theta(u)}{(F(\alpha) - F(u))^{3/2}} \, du, \tag{13.37}$$

and since

$$\theta(x) = \frac{x^3}{2} [x - \tfrac{2}{3}(a + 1)],$$

we see that $\theta(x)$ has the graph as in Figure 13.10(b). Thus $S'(\alpha) \neq 0$ if $\alpha < (a + 1)/2$.

We have therefore proved (13.29). From this it follows that S, and thus T has at most one critical point. The proof is complete. $\square$

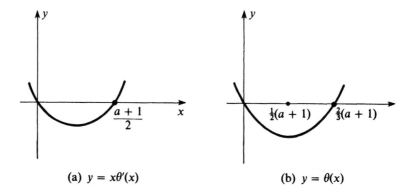

(a) $y = x\theta'(x)$ (b) $y = \theta(x)$

Figure 13.10

NOTES

The proof Theorem 13.3 can be found in Crandall [Cr 2]; see also [Di] and [Ni 2]. Theorem 13.5 is due to C–R [CR 1], following earlier work of Sattinger [Sa 1]. The method of Lyapunov–Schmidt has recently been used by Golubitsky and Schaeffer [GoS] to give a theory of "imperfect bifurcation" via singularity theory. The notion of a K-simple eigenvalue is also due to C–R, as is Lemma 13.7 and Theorem 13.8; see also Sattinger [Sa 1]. Theorem 13.9 due to Krasnoselski [Kr], was the first general bifurcation theorem, and Theorem 13.10 is Rabinowitz' celebrated global bifurcation theorem [Ra]. If one considers positive operators, Dancer [Da] and Turner

[Tu] have shown that one may drop the hypothesis of odd multiplicity in Rabinowitz' theorem. Theorem 13.11 is due to Smoller and Wasserman [SW].

It should here be mentioned again that there is a corresponding theory of bifurcation to periodic solutions, the so-called "Hopf Bifurcation." For a description of this, see the monographs [HKW] and [MM].

Chapter 14

Systems of Reaction–Diffusion Equations

In recent years, systems of reaction–diffusion equations have received a great deal of attention, motivated by both their widespread occurrence in models of chemical and biological phenomena, and by the richness of the structure of their solution sets. In the simplest models, the equations take the form

$$\frac{\partial u}{\partial t} = D\Delta u + f(u), \qquad x \in \Omega \subset \mathbf{R}^k, \quad t > 0, \tag{14.1}$$

where $u \in \mathbf{R}^n$, D is an $n \times n$ matrix, and $f(u)$ is a smooth function. The combination of diffusion terms together with the nonlinear interaction terms, produces mathematical features that are not predictable from the vantage point of either mechanism alone. Thus, the term $D\Delta u$ acts in such a way as to "dampen" u, while the nonlinear function $f(u)$ tends to produce large solutions, steep gradients, etc. This leads to the possibility of threshold phenomena, and indeed this is one of the interesting features of this class of equations.

It is possible to consider the system (14.1) as an ordinary differential equation, $\dot{u} = A(u)$ defined on a Banach space B. Thus, for each $t > 0$, $u(\cdot, t)$ is a point in B, and the equation (14.1) determines a curve in B which describes the evolution of u. With this viewpoint, the associated steady-state equation

$$D\Delta u + f(u) = 0,$$

determines the "rest points" of the equation. Indeed, if there is sufficient compactness available, this description is quite useful.

The compactness requirement can be fulfilled in a very straightforward and natural way; namely, by requiring that the system (14.1) admit bounded invariant regions; i.e., bounded regions Σ in phase space (i.e., u-space), with the property that if the "data" lies in Σ, then the solution $u(x, t)$ lies in Σ for all $x \in \Omega$ and all $t > 0$. Thus Σ provides an a-priori sup-norm bound on u, and it follows that if the datum lies in Σ, then the solution exists for all $t > 0$.

The region Σ can likewise be thought of as an "attracting" region for the equation $\dot{u} = A(u)$, and it is from this that the compactness is obtained.

In §A we shall show that (14.1) has a (local) solution which exists on an interval $0 \le t \le \tau$, where τ depends only on the sup-norm of the data. From this it follows easily that if we obtain an a-priori sup-norm bound, then (14.1) has solutions which are globally defined in time. This is done in §B where we find necessary and sufficient conditions for our equations to admit invariant regions. We show that these conditions hold for the standard systems; e.g., the Hodgkin–Huxley equations, and the ecology equations with diffusion.

In §C we use invariant regions to prove a comparison theorem for solutions of (14.1) on bounded domains Ω, where u satisfies homogenous Neumann boundary conditions. We construct vector fields related to f, and show that solutions of (14.1) can be estimated componentwise, by solutions of the ordinary differential equations associated with these vector fields. The theorem is illustrated by some examples from mathematical ecology. In §D we introduce the notion of a contracting rectangle, and we construct a (local) Lyapunov function for such sets. We point out how these are used to obtain qualitative information for solutions. The large time behavior of solutions satisfying homogeneous Neumann boundary conditions, is studied in §E. If there is an invariant region Σ, we isolate a parameter σ, depending on Σ, the spatial region Ω, and the diffusion matrix, which when positive implies that *all* solutions decay as $t \to \infty$ to spatially homogeneous functions; i.e., to solutions of the associated "kinetic" equations, $\dot{u} = f(u)$. The condition $\sigma > 0$ is valid, for example, when the diffusion is, in a certain precise sense, large with respect to the other parameters in the problem. It implies too that there cannot be any bifurcation of constant solutions into nonhomogeneous steady-state ones. In the final section we show how the general Kolmogorov form of the equations for two species population dynamics, with diffusion and spatial effects taken into account, fits into our theoretical framework in a very natural and satisfactory way. That is, the ecologically reasonable assumption, that every community must by its very nature be constained by resource limitations, implies that the equations admit arbitrarily large bounded invariant regions.

§A. Local Existence of Solutions

The development which we give here is closely related to the results in Chapter 11, §B. The difference is that we consider here a more narrow class of equations, and we are thus able to obtain more exact results. Namely, we obtain quantitative information on the time of existence of the local solution.

We consider, for simplicity, the system (14.1) in a single space variable, where D is diagonal:

$$u_t = Du_{xx} + f(u), \qquad x \in \mathbf{R}, \quad t > 0. \tag{14.2}$$

(For the more general boundary-value problems in $\mathbf{R}^k$, see [Fn 3].) Here $u = (u_1, \ldots, u_n)$, $D = \text{diag}(d_1, \ldots, d_n)$, with $d_i \geq 0$ for all i. Together with this equation we consider the initial data

$$u(x, 0) = u_0(x), \quad x \in \mathbf{R}. \tag{14.3}$$

We shall obtain solutions which are continuous functions of time having values in certain Banach spaces. We proceed to describe these spaces.

Assume that B is a Banach space of functions on $\mathbf{R}$ with values in $\mathbf{R}^n$ (i.e., B consists of vector-valued functions); $\|\cdot\|_B$ denotes the B-norm, and $\|\cdot\|_\infty$ the L_∞-norm.

Definition 14.1. B is *admissible* if the following four conditions hold:

(1) B is a subset of the bounded continuous functions on $\mathbf{R}$, and if $w \in B$, $\|w\|_B \geq \|w\|_\infty$.
(2) B is translation-invariant; i.e., $w \circ \tau \in B$ for every $w \in B$ and every translate $\tau : \mathbf{R} \to \mathbf{R}$. Also $\|w \circ \tau\|_B = \|w\|_B$.
(3) If $f : \mathbf{R}^n \to \mathbf{R}^n$ is smooth, and $f(0) = 0$, then $f(w) \in B$ for every $w \in B$, and for any $M > 0$, there is a constant $k(M)$ such that

$$\|f(w) - f(w')\|_B \leq k(M)\|w - w'\|_B, \tag{14.4}$$

for all w, w' in B with $\|w\|_\infty, \|w'\|_\infty \leq M$.
(4) If $\tau_h : \mathbf{R} \to \mathbf{R}$ denotes translation by h (i.e., $\tau_h(x) = x + h$, then for each $w \in B$,

$$\|w \circ \tau_h - w\|_B \to 0 \quad \text{as } h \to 0.$$

It is easy to check that the following Banach spaces are admissible:

(A) $BC = \{$bounded, uniformly continuous functions on $\mathbf{R}\}$.
(B) $BC \cap L_p$, $p \geq 1$.
(C) $BC_0 = \{w \in BC : |w(x)| \to 0 \text{ as } |x| \to \infty\}$.

Next, let $j \in \mathscr{S}$ (see Chapter 7, §A), $j \geq 0$, $\int j = 1$, and define $j_\varepsilon(x) = \varepsilon^{-1}j(\varepsilon^{-1}x)$. Then consider the condition

(5) For each $w \in B$, $\|j_\varepsilon * w - w\|_B \to 0$ as $\varepsilon \to 0$. (Here $*$ denotes convolution; see Chapter 7, §A.)

Then it is easy to see that conditions (4) and (5) are equivalent; for example, since

$$j_\varepsilon * w = \int_{\mathbf{R}} j_\varepsilon(h)(\tau_h w)\, dh \to w,$$

as $\varepsilon \to 0$ (see Chapter 7, Appendix), it follows that condition (4) implies condition (5); we do not prove the other implication as it will not be used.

Now let's consider the example of the scalar heat equation

$$u_t = au_{xx}, \qquad u(x, 0) = u_0(x), \qquad a > 0,$$

in $t > 0$. We assume that $u_0 \in B$, where B is admissible. The solution to this problem is (see Chapter 9, Theorem 9.5)

$$u(x, t) = \int_{\mathbb{R}} u_0(y) \frac{\exp[-(x - y)^2/4at]}{\sqrt{4\pi at}} \, dy.$$

If we define $j(x) = \exp[-x^2/4a]/\sqrt{4\pi a}$, and

$$j_{\sqrt{t}}(x) \equiv \frac{j(x/\sqrt{t})}{\sqrt{t}} = \frac{\exp[-x^2/4at]}{\sqrt{4\pi at}},$$

then we may write

$$u(x, t) = j_{\sqrt{t}} * u_0.$$

Hence condition (5) implies that u is a continuous function of t (it is only necessary to check this at $t = 0$), with values in B, for $0 \le t < \infty$.

We define $C([0, T]; B)$ to be the Banach space of continuous functions on $[0, T]$ with values in B, normed by

$$\|w\| = \sup_{0 \le t \le T} \|w(t)\|_B.$$

Then as in Chapter 11, the function $u \in C([0, T]; B)$ satisfies (14.2), and

$$u(x, 0) = u_0(x), \tag{14.5}$$

if and only if u satisfies

$$u(x, t) = \int_{\mathbb{R}} G(x - y, t)u_0(y) \, dy + \int_0^t \int_{\mathbb{R}} G(x - y, t - s)f(u(y, s)) \, dy \, ds$$

$$= G(t) * u_0 + \int_0^t G(t - s) * f(u(s)) \, ds, \tag{14.6}$$

where

$$G(t) = \mathrm{diag}(g_1(t), g_2(t), \ldots, g_n(t)), \quad \text{and}$$

$$g_i(t) = (4\pi d_i t)^{-1/2} \exp\left[\frac{-x^2}{4d_i t}\right], \qquad i = 1, 2, \ldots, n. \tag{14.7}$$

Concerning this representation, we make the following remarks:

(A)   If $u \in C([0, t]; B)$, then the integrand in (14.6) is a continuous func-
      tion of s with values in B, so the integral is really a Riemann integral.
(B)   If μ is a finite Borel measure, and $w \in B$, then $\mu * w \in B$, and $\|\mu * w\|_B$
      $\leq |\mu| \|w\|_B$, where $|\mu|$ denotes the total variation of μ. (This is true
      because the Riemann integral is a limit of approximate sums.) Thus,
      the function G defined by (14.7) can serve as a μ, and since the total
      variation of this measure is 1 (see Chapter 9, Theorem 9.5), we obtain
      the useful inequality

$$\|G(t) * u\|_B \leq \|u\|_B, \qquad u \in B, \quad t \geq 0. \tag{14.8}$$

With these preliminaries out of the way, we shall show how to solve (14.2),
(14.3) for a short time interval. We assume $f(0) = 0$.

Theorem 14.2. *Let $u_0 \in B$; then there exists $t_0 > 0$, where t_0 depends only on
f and $\|u_0\|_\infty$, such that (14.6) has a unique solution in $C([0, t_0]; B)$, and
$\|u\| \leq 2\|u_0\|_B$.*

Proof. For $t_0 > 0$, let

$$\Gamma = \{u \in C([0, t_0]; B): \|u(t) - G(t) * u_0\|$$
$$\leq \|u_0\|_B, \text{ and } \|u(t) - G(t) * u_0\|_\infty \leq \|u_0\|_\infty, 0 \leq t \leq t_0\}.$$

Then $\Gamma \neq \phi$ (since $0 \in \Gamma$), and Γ is closed.

If $u \in \Gamma$, then (14.8) shows that $\|u(t)\|_B \leq 2\|u_0\|_B$. Therefore, since B is
admissible, (14.3) shows that there is a constant k, depending only on $\|u_0\|_\infty$
(and not on t), such that if u and v are in Γ, and $0 \leq t \leq t_0$,

$$\|f(u(t)) - f(v(t))\|_B \leq k\|u(t) - v(t)\|_B \leq k\|u - v\|; \tag{14.9}$$

thus

$$\|f(u) - f(v)\| \leq k\|u - v\|. \tag{14.10}$$

Let $t_0 = 1/2k$; then t_0 depends only on f and $\|u_0\|_\infty$. Define a mapping Φ
on $C([0, t_0]; B)$ into itself by

$$\Phi u(t) = G(t) * u_0 + \int_0^t G(t - s) * f(u(s))\, ds.$$

Observe that a fixed point of Φ is a solution of (14.6).

We claim that Φ maps Γ into itself. To see this, let $u \in \Gamma$. Since $g_i(t), t \geq 0$,
is a positive measure having total mass 1, we have, for $0 \leq t \leq t_0$

$$\|\Phi u(t) - G(t) * u_0\|_B \leq \int_0^t \|G(t - s) * f(u(s))\|_B\, ds.$$

Now from (14.9) with $v = 0$, together with (14.8),

$$\| f(u(t)) \|_B \le k \| u(t) \|_B \le 2k \| u_0 \|_B, \qquad 0 \le t \le t_0.$$

Therefore

$$\| \Phi(u(t)) - G(t) * u_0 \|_B \le 2k \int_0^t \| u_0 \|_B \, dt$$

$$\le 2k t_0 \| u_0 \|_B$$

$$= \| u_0 \|_B.$$

Similarly
$$\| \Phi(u(t)) - G(t) * u_0 \|_\infty \le \| u_0 \|_\infty.$$

This proves our claim. Next, we show that Φ is a contraction on Γ. Thus, if $u, v \in \Gamma$,

$$\| \Phi(u(t)) - \Phi(v(t)) \|_B \le \int_0^t \| G(t - s) * [f(u(s)) - f(v(s))] \|_B \, ds$$

$$\le \int_0^t \| f(u(s)) - f(v(s)) \|_B \, ds \quad \text{(Remark B)}$$

$$\le k \int_0^t \| u(s) - v(s) \|_B \, ds$$

$$\le k t_0 \| u - v \|$$

$$= \tfrac{1}{2} \| u - v \|.$$

Therefore
$$\| \Phi u - \Phi v \| \le \tfrac{1}{2} \| u - v \|,$$

and so Φ is a contraction on Γ. We may apply Banach's fixed point theorem to conclude that Φ has a unique fixed point in Γ. $\square$

We must show that there cannot be a solution outside of Γ. This will follow from the next lemma.

Lemma 14.3. *Let $u, v \in C([0, T]; B)$ be solutions of (14.1) on $0 \le t \le T$, where $\| u \|_\infty, \| v \|_\infty \le M$. Then there is a constant $k = k(M)$ such that*

$$\| u(t) - v(t) \|_B \le e^{kt} \| u(0) - v(0) \|_B, \qquad 0 \le t \le T. \qquad (14.11)$$

Proof. If $t \in [0, T]$,

$$u(t) - v(t) = G(t) * (u(0) - v(0)) + \int_0^t G(t - s) [f(u(s)) - f(v(s))] \, ds.$$

Now since B is admissible, there is a $k = k(M)$, such that

$$\| f(u(s)) - f(v(s)) \|_B \le k \| u(s) - v(s) \|_B.$$

Thus

$$\| u(t) - v(t) \|_B \le \| u(0) - v(0) \|_B + k \int_0^t \| u(s) - v(s) \|_B \, ds,$$

so that Gronwall's inequality (Chapter 4) gives the desired result. This proves the lemma and completes the proof of Theorem 14.2. $\square$

Remark. If f is linear, i.e., $f(u) = Au$ where A is an $n \times n$ matrix, then the solution exists for all $t > 0$. This follows from (14.10) where k can be taken to depend only on A.

Suppose now that we can prove an a-priori bound for solutions of (14.2), (14.3), of the following form; namely, there is a constant $c > 0$, depending only on $\| u_0 \|_\infty$ such that if u is any solution of (14.2), (14.3) in $0 \le t \le T$, then $\| u(\cdot, t) \|_\infty \le c$. Then we claim that the solution of this problem exists for all time t, $0 \le t \le T$. To see this, we use Theorem 14.2 to obtain a solution in $0 \le t \le \tau$, $\tau > 0$. Then taking $u(\cdot, \tau)$ as the data on $t = \tau$, we use Theorem 14.2 to get a solution on $\tau \le t \le \tau + \sigma$, where $\sigma = \sigma(\| u(\cdot, \tau) \|_\infty)$. We repeat this process to find a solution on $\tau + \sigma \le t \le \tau + 2\sigma$, and eventually after a finite number of steps we obtain a solution on $0 \le t \le T$. We state this formally as the following theorem.

Theorem 14.4. *Let B be an admissible Banach space, and let $u_0 \in B$. If the solution is a-priori bounded in the L_∞-norm on $0 \le t \le T \le \infty$, then the solution of (14.2), (14.3) exists for all t, $0 \le t \le T$, and $u(\cdot, t) \in B$, $0 \le t \le T$.*

§B. Invariant Regions

The notion of an invariant region is the most important idea in this chapter. It allows us to prove global existence theorems, and thereby provides a suitable theoretical foundation and framework for studying the large time behavior of solutions. In the subsequent sections of this chapter, we shall use it to obtain some general theorems which can then be applied to specific systems of equations.

We consider the system[1]

$$\frac{\partial v}{\partial t} = \varepsilon D v_{xx} + M v_x + f(v, t), \qquad (x, t) \in \Omega \times \mathbf{R}_+, \qquad (14.12)$$

together with the initial data

$$v(x, 0) = v_0(x), \qquad x \in \Omega. \qquad (14.13)$$

Here $\varepsilon > 0$, Ω is an open interval in $\mathbf{R}$, $D = D(v, x)$, and $M = M(v, x)$, are matrix-valued functions defined on an open subset $U \times V \subset \mathbf{R}^n \times \Omega$, $D \geq 0$, $v = (v_1, v_2, \ldots, v_n)$ and f is a smooth mapping from $U \times \mathbf{R}_+$ into $\mathbf{R}^n$. If Ω is not all of $\mathbf{R}$, we will assume that v satisfies specific boundary conditions; e.g., Dirichlet and Neumann boundary conditions. We assume that this problem has a local (in time) solution on some set X of smooth functions from Ω to $\mathbf{R}^n$; i.e., given a function $v_0 \in X$, there is a $\delta > 0$ and a smooth solution $v(x, t)$ of (14.12), (14.13) defined for $x \in \Omega$ and $t \in [0, \delta)$, such that $v(\cdot, t) \in X$, $0 \leq t < \delta$. The topology on X should be at least as strong as the compact-open topology (uniform convergence on compact subsets of Ω).

Definition 14.5. A closed subset $\Sigma \subset \mathbf{R}^n$ is called a (positively) invariant region for the local solution defined by (14.12), (14.13), if any solution $v(x, t)$ having all of its boundary and initial values in Σ, satisfies $v(x, t) \in \Sigma$ for all $x \in \Omega$ and for all $t \in [0, \delta)$.

We always assume that if $u \in X$, there is a compact set $K \subset \Omega$ such that if $x \notin K$, then $u(x) \in \Sigma^{int}$. We call this condition K.

For example, let us consider the simple heat equation $u_t = u_{xx}$, where X is the space of C^2 functions which tend to zero as $|x| \to \infty$. Then if

$$\Sigma = \{u: -1 \leq u \leq 1\},$$

the maximum principle (see Theorem 9.1) shows that Σ is invariant, and condition K is certainly satisfied. Observe that condition K is always valid if we consider (14.12) on a bounded domain, with the standard boundary conditions lying in Σ^{int}; the condition is needed only to recover some measure of compactness when we are on unbounded domains.

---

[1] The ideas which we shall present in this chapter can be carried over, without any difficulty, to more general systems in several space variables, and also to equations on bounded domains in $m \geq 1$ space variables, satisfying general boundary conditions (see [CCS]). We consider (14.12) merely for simplicity in presentation.

The invariant regions Σ will be made up of the intersection of "half space"; i.e., we consider regions Σ of the form

$$\Sigma = \bigcap_{i=1}^{m} \{v \in U : G_i(v) \leq 0\}, \qquad (14.14)$$

where G_i are smooth real-valued functions defined on open subsets of U, and for each i, the gradient dG_i never vanishes.

Now if there is a solution v of (14.12), (14.13), with boundary data and initial data $v(x, 0)$ in Σ for all $x \in \Omega$, which is *not* in Σ for all $t > 0$, then there is a function G_i, a time t_0 such that for $t \leq t_0$ and $x \in \mathbf{R}$, $G_i \circ v(x, t) \leq 0$ and for any $\varepsilon > 0$, $\exists t'$ and $x' \in \mathbf{R}$ with $t_0 < t' < t_0 + \varepsilon$ such that $G_i \circ v(x', t') > 0$.

Thus, if the assumptions

$$G_i \circ v(x_0, t) < 0 \quad \text{for } 0 \leq t < t_0 \quad \text{and} \quad G_i \circ v(x_0, t_0) = 0, \qquad (14.15)$$

together imply that

$$\frac{\partial(G_i \circ v)}{\partial t} < 0 \quad \text{at } (x_0, t_0), \qquad (14.16)$$

then Σ must be invariant.[2] (Since

$$\frac{(G_i \circ v)(x_0, t) - (G_i \circ v)(x_0, t_0)}{t - t_0} > 0 \quad \text{if } t < t_0,$$

it follows that $\partial(G_i \circ v)/\partial t \geq 0$ at (x_0, t_0); this contradiction means that $(G_i \circ v)(x_0, t_0) < 0$.)

Before proving the first theorem, we shall need the following definition.

Definition 14.6. The smooth function $G : \mathbf{R}^n \to \mathbf{R}$ is called *quasi-convex* at v if whenever $dG_v(\eta) = 0$, then $d^2 G_v(\eta, \eta) \geq 0$.

Theorem 14.7. *Let Σ be defined by (14.14), and suppose that for all $t \in \mathbf{R}_+$ and for every $v_0 \in \partial\Sigma$ (so $G_i(v_0) = 0$ for some i), the following conditions hold:*

(1) *dG_i at v_0 is a left eigenvector of $D(v_0, x)$, and $M(v_0, x)$, for all $x \in \Omega$.*
(2) *If $dG_i D(v_0, x) = \mu\, dG_i$, with $\mu \neq 0$, then G_i is quasi-convex at v_0.*
(3) *$dG_i(f) < 0$ at v_0, for all $t \in \mathbf{R}_+$*

Then Σ is invariant for (14.12), for every $\varepsilon > 0$.

Proof. For simplicity in notation, let $G = G_i$. To show that Σ is invariant, we assume that (14.15) holds for $x_0 \in \Omega$, and we shall show (14.16). Thus, at (x_0, t_0),

$$\frac{\partial(G \circ v)}{\partial t} = dG(v_t) = dG(\varepsilon D v_{xx} + M v_x + f).$$

---

[2] It is here where condition K is used.

Now since dG is a left eigenvector of D and M, we have at $v_0 = v(x_0, t_0)$,

$$dGD = \mu\, dG \quad \text{and} \quad dGM = \lambda\, dG.$$

This implies that

$$\frac{\partial(G \circ v)}{\partial t} = \varepsilon\mu\, dG(v_{xx}) + \lambda\, dG(v_x) + dG(f). \tag{14.17}$$

Now we claim that at (x_0, t_0),

$$dG(v_x) = 0. \tag{14.18}$$

To see this, define $h(x) = G \circ v(x, t_0)$; then $h(x_0) = 0$, and $h'(x) = dG(v_x(x, t_0))$. If $h'(x_0) > 0$, then $h(x) > 0$ for $x > x_0$, if $|x - x_0|$ is small (see Figure 14.1).

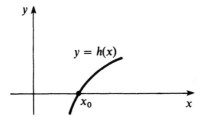

Figure 14.1

Thus $G \circ v(x, t_0) > 0$ for x close to x_0, and so $G \circ v(x, t) > 0$ for $|t - t_0| < \varepsilon$, for some $\varepsilon > 0$; in particular $G \circ v(x, t) > 0$ for some x and some $t < t_0$. This violates (14.15). Similarly, $h'(x_0) < 0$ is impossible. Thus $dG(v_x(x_0, t_0)) = h'(x_0) = 0$, and this proves the claim.

Observe too that with $h(x)$ as defined above, $h''(x_0) \le 0$; otherwise we would arrive at a contradiction similar to the one above. It follows that

$$0 \ge h''(x_0) = d^2G(v_x, v_x) + dG(v_{xx}). \tag{14.19}$$

Now suppose that $\mu \ne 0$; then $\mu > 0$, so from the second hypothesis, together with our claim, we find $d^2G(v_x, v_x) \ge 0$ at (x_0, t_0). Therefore from (14.19), $dG(v_{xx}) \le 0$ at (x_0, t_0). Thus (14.17) gives

$$\frac{\partial(G \circ v)}{\partial t} \le dG(f) < 0,$$

in view of the third hypothesis. This completes the proof. □

Remarks. (i) We could have replaced hypotheses (2) and (3) by (2′) and (3′), where D is positive definite, and

(2′) If $dG_i D(v_0, x) = \mu\, dG_i$, with $\mu \neq 0$, then G is *strongly convex* at v_0; i.e., if $dG_{i_0}(\eta) = 0$, $\eta \neq 0$, then $d^2 G_{iv_0}(\eta, \eta) > 0$, and

(3′) $dG_i(f) \leq 0$ at v_0.

(ii) If D and M are diagonal matrices, and $G_i = u_i - c_i$ for some constant c_i, then G_i is everywhere quasi-convex, and dG_i is a left eigenvector of both D and M. Therefore, the half space

$$\{u : u_i - c_i \leq 0\},$$

is invariant for (14.12), for every $\varepsilon > 0$, provided that $f_i(u_1, u_2, \ldots, u_{i-1}, c_i, u_{i+1}, \ldots, u_n) < 0$, where f_i is the ith component of f. This gives the following useful corollary.

Corollary 14.8. (a) *Suppose that D and M are diagonal matrices. Then any region of the form*

$$\Sigma = \bigcap_{i=1}^{n} \{u : a_i \leq u_i \leq b_i\} \tag{$*$}$$

is invariant for (14.12), *for all $\varepsilon > 0$, provided that f points strictly into Σ on $\partial\Sigma$; i.e., provided that hypothesis* (3) *of the theorem is valid.*

(b) *If $D = I$, the identity matrix, then any convex region Σ, in which f points into Σ on $\partial\Sigma$, is invariant for* (14.12), *if $M = 0$.*

We shall refer to such an invariant region $(*)$ as an *invariant rectangle.*

Corollary 14.9. *Consider the system* (14.2), *with data $u_0 \in BC_0$. If the system admits a bounded invariant region Σ, and $u_0(x) \in \Sigma$ for all $x \in \mathbf{R}$, then the solution exists for all $t > 0$.*

Proof. This is immediate from Theorem 14.4. □

In many applications, the vector field f satisfies the above weaker condition (3′); i.e., f is tangent to $\partial\Sigma$ at certain places. This is especially true for equations involving positive quantities, like population densities, or chemical concentrations. In these cases, one finds that the positive "orthant" is invariant for the vector field f, since f is of the form

$$f(u) = (u_1 M_1(u), u_2 M_2(u), \ldots, u_n M_n(u)).$$

However, in this case, f is tangent to the coordinate hyperplanes. It is therefore desirable to have an extension of Theorem 14.7 which covers this case. In order to obtain such a result, we must assume that solutions of (14.12) depend "continuously" on f. Accordingly, we make the following definition.

Definition 14.10. The system (14.12) is called *f-stable* if, whenever f is the limit of functions in f_n the C^1-topology on compacta, for all $t > 0$, then any solution of (14.12), (14.13) is the limit in the compact-open topology, of solutions of (14.12), (14.13), where f is replaced by f_n.

In these terms, we can extend Theorem 14.7 as follows.

Theorem 14.11. *If the system (14.12) is f-stable then we may replace condition (3) by*

(3') $dG_i(f) \leq 0$ *at* v_0 *for all* $t \in \mathbf{R}_+$,

and the same conclusion holds as in Theorem 14.7.

Proof. We let $h(v)$ be a smooth vector field which is bounded on $\partial\Sigma$, and which points into Σ on $\partial\Sigma$; i.e., $dG_i(h) < 0$ on $\partial\Sigma$, for each $i = 1, 2, \ldots, m$. If $\delta > 0$ is a small constant, we consider the system

$$\frac{\partial v}{\partial t} = \varepsilon D v_{xx} + M v_x + f + \delta h, \qquad (x, t) \in \Omega \times \mathbf{R}_+, \qquad (14.20)$$

together with the initial data (14.13). Now we may apply Theorem (14.7) to this system, thereby obtaining solutions $\{v_\delta\}$, $0 < \delta \leq \delta_0$, which satisfy $G_i \circ v_\delta(x, t) \leq 0$ for all i, for all $x \in \mathbf{R}$, and, in view of Corollary 14.9, for all $t > 0$. Since (14.12) is f-stable, it follows that $v(x, t) \in \Sigma$ for all $(x, t) \in \Omega \times \mathbf{R}_+$. $\square$

We shall now derive some necessary conditions which must hold if Σ is an invariant set. In order to exclude extraneous trivial cases, we assume that for each G_i, there is an $(n - 1)$-dimensional subset of $G_i = 0$ which meets Σ; in other words, we want to rule out the situation as is depicted in Figure 14.2, where $\Sigma \cap \{\phi \leq 0\} = \Sigma$, but $\phi = 0$ is obviously unecessary for defining Σ.

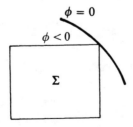

Figure 14.2

Theorem 14.12. *Let* Σ *be defined by* (14.14), *and suppose that* Σ *is an invariant region for* (14.12), *for fixed* $\varepsilon > 0$, *where* $f = f(v, t)$ *and* D *is a positive definite matrix. Then the following conditions hold at each point* v_0 *on* $\partial\Sigma$ *(say,* $G_i(v_0) = 0$*):*

    (1) dG_i *is a left eigenvector of* D *at* v_0 *for all* $x \in \Omega$.
    (2) G_i *is quasi-convex at* v_0,
    (3) $dG_i(f) \leq 0$ *for all* $t \geq 0$.

Proof. Let's again write $G_i = G$. If dG is not a left eigenvector of D, we can find $\zeta \in \mathbf{R}^n$, $\eta \in \mathbf{R}^n$, and $\lambda \in \mathbf{R}$, such that, at v_0 (cf. Figure 14.3):

    (a) $dG(\zeta) < 0, dG(D\zeta) > 0$.
    (b) $dG(\eta) = 0$.
    (c) $\lambda \varepsilon \, dG(D\zeta) + dG(M\eta) + dG(f) > 0$.
    (d) $\lambda \, dG(\zeta) + d^2G(\eta, \eta) < 0$.

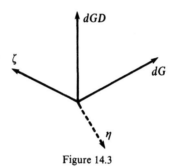

Figure 14.3

We consider two cases. First suppose that v_0 is not a "corner" point of Σ; i.e., there is an $(n - 1)$-dimensional neighborhood of v_0 which lies in $(G_i = 0) \cap \Sigma$. Let

$$U(x) = v_0 + x\eta + \tfrac{1}{2}\lambda x^2 \zeta, \qquad G \circ U(x) = h(x);$$

then

$$G \circ U(x) = h(x) = h(0) + xh'(0) + \frac{x^2}{2}h''(0) + o(x^2)$$

$$= G \circ U(0) + x \, dG(U'(0)) + \frac{x^2[d^2G(U'(0), U'(0)) + dG(U''(0))]}{2}$$
$$+ o(x^2)$$

$$= G(v_0) + x \, dG(\eta) + \frac{x^2[d^2G(\eta, \eta) + dG(\lambda\zeta)]}{2} + o(x^2)$$

$$= \frac{x^2}{2}[d^2G(\eta, \eta) + \lambda \, dG(\zeta)] + o(x^2).$$

It follows that for some $\delta > 0$, $G \circ U(x) \leq 0$ for $|x| \leq \delta$, and thus since we are in Case 1, $U(x) \in \Sigma$, if $|x| < \delta$. (Notice that this is not necessarily true if $G_j(v_0) = 0$ for some $j \neq i$, since v_0 could be a "corner" point; see Figure 14.4.)

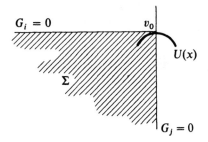

Figure 14.4

Now let $u_0(x)$ be a smooth function with values in Σ such that $u_0(x) = U(x)$, if $|x| < \delta$, and let $v(x, t)$ be the corresponding solution of (14.12) with this data. Then

$$dG(v_t) = \varepsilon \, dG(Dv_{xx}) + dG(Mv_x) + dG(f),$$

so that at $x = t = 0$,

$$dG(v_t) = [\varepsilon\lambda \, dG(D\zeta) + dGM(\eta + \lambda x \zeta) + dG(f)]|_{\substack{x=0 \\ t=0}}$$
$$= \varepsilon\lambda \, dG(D\zeta) + dG(M\eta) + dG(f)$$
$$> 0,$$

in view of (c). This contradicts the invariance of Σ.

We now consider the case where $G(v_0) = G_i(v_0) = 0$, and $G_j(v_0) = 0$, for some $j \neq i$. In this case, we choose a sequence $\{v_n\}$ such that $G(v_n) = 0$, $G_j(v_n) \neq 0$ for $j = 1, \ldots, m$ and $v_n \to v_0$. By what we have just proved, we have $dG_{v_n} D(v_n) = \lambda(v_n) \, dG_{v_n}$, and passing to the limit as $n \to \infty$, we obtain the desired result (1).

Next, suppose that G is not quasi-convex at v_0. Then there exists an $\eta \in \mathbf{R}^n$ with

$$dG(\eta) = 0 \quad \text{and} \quad d^2G(\eta, \eta) < 0 \quad \text{at } v_0.$$

Now from (1), $dGD = \theta \, dG$ at v_0, where $\theta > 0$. We choose $\zeta \in \mathbf{R}^n$ such that at v_0,

$$dG(\zeta) > 0 \quad \text{and} \quad dG(\zeta) + d^2G(\eta, \eta) < 0,$$

and then we choose $\lambda \in \mathbf{R}$ so that at v_0

$$\varepsilon\lambda^2\theta \, dG(\zeta) + \lambda \, dG(M\eta) + dG(f) > 0.$$

Let

$$U(x) = v_0 + \lambda x \eta + \tfrac{1}{2}\lambda^2 x^2 \zeta;$$

then

$$G \circ U(x) = \frac{\lambda^2 x^2 [dG(\zeta) + d^2G(\eta, \eta)]}{2} + o(x^2),$$

so that for small x, say $|x| < \delta$, $G \circ U(x) \in \Sigma$.[3]

---

[3] We are assuming that v_0 is not a "corner" point; as before, corner points are treated by a limiting argument.

Now as above, define $u_0(x)$ to be a smooth function with values in Σ such that $u_0(x) = U(x)$ if $|x| < \delta$. Let v be the corresponding solution of (14.12) with this data. Then

$$dG(v_t)\big|_{\substack{x=0 \\ t=0}} = \varepsilon \theta \lambda^2 \, dG(\zeta) + \lambda \, dG(M\eta) + dG(f) > 0,$$

so that Σ is not invariant.

Finally, let $v(t)$ be a solution of the problem

$$\frac{dv}{dt} = f(v, t), \qquad v(0) = v_0.$$

Then v is a solution of (14.12), so by the invariance of Σ, we know that $v(t) \in \Sigma$ for $t \geq 0$. It follows that $dG(f) \leq 0$ at v_0. This completes the proof of the theorem. $\square$

Our next result combines our previous results, and gives conditions, both necessary and sufficient in order that Σ be invariant for systems of reaction–diffusion equations; it is an immediate consequence of Theorems 14.11 and 14.12.

Theorem 14.13. *Let Σ defined by (14.14), and consider the system (14.12) with $M = 0$, D positive definite and $f = f(v, t)$. Suppose that this system is f-stable. Then Σ is a positively invariant region for (14.12) for fixed $\varepsilon > 0$ if and only if the following hold at each boundary point v_0 of Σ (so $G_i(v_0) = 0$):*

(a) *dG_i is a left eigenvector of D.*
(b) *G_i is a quasi-convex at v_0.*
(c) *$dG_i(f) \leq 0$.*

We next give necessary and sufficient conditions in order that Σ be invariant for all $\varepsilon > 0$. This will be useful in obtaining a-priori estimates independent of $\varepsilon > 0$. Thus, consider the system

$$v_t = \varepsilon D v_{xx} + M v_x + f(v, t). \tag{14.21}$$

Theorem 14.14. *Let Σ be defined as in (14.14), and let D be a positive definite matrix. Suppose that (14.21) is f-stable. Then Σ is invariant for (14.21) for every $\varepsilon > 0$ if and only if the following conditions hold at each $v_0 \in \partial\Sigma$ (so $G_i(v_0) = 0$):*

(i)   *dG_i is a left eigenvector of D and M.*
(ii)   *G_i is quasi-convex at v_0.*
(iii) *$dG_i(f) \leq 0$.*

Proof. If conditions (i)–(iii) hold, then Theorem 14.11 shows that Σ is invariant. Conversely, if Σ is invariant, then in view of Theorem 14.12, we need only show that dG_i is a left eigenvector of M.

If $G = G_i$, and dG is not a left eigenvector of M at v_0, then we can find $\eta \in \mathbf{R}^n$ such that at v_0,

$$dG(\eta) = 0 \quad \text{and} \quad dG(M\eta) > 0.$$

Since G is quasi-convex at v_0, $d^2G(\eta, \eta) \geq 0$. Now we choose $\zeta \in \mathbf{R}^n$ and $\lambda \in \mathbf{R}$ so that at v_0,

$$dG(\zeta) + d^2G(\eta, \eta) < 0 \quad \text{and} \quad dG(D\zeta) + \lambda\, dG(M\eta) + dG(f) > 0.$$

Let $\varepsilon = \lambda^{-2}$, and set

$$U(x) = v_0 + \lambda x \eta + \frac{\lambda^2 x^2 \zeta}{2};$$

then

$$G \circ U(x) = \frac{\lambda^2 x^2 [d^2G(\eta, \eta) + dG(\zeta)]}{2} + o(x^2),$$

so that[4] $U(x)$ is in Σ for $|x| < \delta$. Let u_0 be a smooth function with values in Σ, such that $u_0(x) = U(x)$, if $|x| < \delta$, and let $v(x, t)$ be the corresponding solution of (14.21) with this data. Then

$$dG(v_t)\big|_{\substack{x=0 \\ t=0}} = \{\varepsilon\, dG(D\lambda^2\zeta) + dG[M(\lambda\eta) + \lambda^2 x\zeta] + dG(f)\}\big|_{\substack{x=0 \\ t=0}}$$

$$= dG(D\zeta) + \lambda\, dG(M\eta) + dG(f)$$

$$> 0,$$

and so Σ is not invariant. This completes the proof of the theorem. $\square$

(We remark that the condition "for every $\varepsilon > 0$," cannot be omitted in the statement of Theorem 14.14. Indeed, consider the example

$$u_t = u_{xx} + Mu_x - cu,$$

where u is an n-vector, M is a constant matrix, and c is a positive constant. The claim is that if $|M|^2 \leq 4c$, then any region of the form $\{u : |u| \leq R\}$ is invariant for this system. Indeed, if we multiply the equation by u^t (the transpose of u), we obtain

$$\left(\frac{|u|^2}{2}\right)_t = u^t u_{xx} + u^t M u_x - c|u|^2$$

$$= (u^t u_x)_x - |u_x|^2 + u^t M u_x - c|u|^2$$

$$\leq \left(\frac{|u|^2}{2}\right)_{xx} - |u_x|^2 + \frac{\varepsilon}{2}|u|^2 + \frac{1}{2\varepsilon}|M|^2|u_x|^2 - c|u|^2$$

$$\leq \left(\frac{|u|^2}{2}\right)_{xx} + \left(\frac{|M|^2}{4} - c\right)|u|^2$$

$$\leq \left(\frac{|u|^2}{2}\right)_{xx},$$

where we have set $\varepsilon = |M|^2/2$. Thus $|u(x, t)|^2 \leq \|u_0\|^2$, for any matrix M, having small norm; i.e., u need not be an eigenvector of M^t.)

---

[4] As in the proof of Theorem 14.12, this holds if v_0 is not a "corner" point; otherwise we use a limiting procedure as before.

We pause now to give some examples which illustrate our results.

EXAMPLE 1 (Hodgkin–Huxley equations). The equations take the form

$$cu_t = R^{-1}u_{xx} + g(u, v, w, z),$$

$$v_t = \varepsilon_1 v_{xx} + g_1(u)(h_1(u) - v),$$

$$w_t = \varepsilon_2 w_{xx} + g_2(u)(h_2(u) - w),$$

$$z_t = \varepsilon_3 z_{xx} + g_3(u)(h_3(u) - z), \qquad (14.22)$$

where c and R are positive constants, the ε_i are nonnegative constants, and g is defined by

$$g(u, v, w, z) = k_1 v^3 w(c_1 - u) + k_2 z^4(c_2 - u)$$
$$+ k_3(c_3 - u), \qquad c_1 > c_3 > 0 > c_2,$$

where the k_i are positive constants. Furthermore, $g_i > 0$, $1 > h_i > 0$, $i = 1, 2, 3$. In this model, the variables v, w, and z represent chemical concentrations, and are thus nonnegative, while u denotes electric potential. The equations are a mathematical model for the physiological phenomenon of signal transmission across axons.

If we set $U = (u, v, w, z)$, $F(U) = [c^{-1}g, g_1(h_1 - v), g_2(h_2 - w), g_3(h_3 - z)]$, and $D = \text{diag}((Rc)^{-1}, \varepsilon_1, \varepsilon_2, \varepsilon_3)$, then we can write the system as

$$U_t = DU_{xx} + F(U). \qquad (14.23)$$

Furthermore, since D is a diagonal matrix, we may apply Corollary 14.8(a) and seek invariant "rectangles" for our system. Thus let

$$\Sigma = \left\{ (u, v, w, z) : 0 \leq \begin{matrix} v \\ w \\ z \end{matrix} \leq \begin{matrix} \alpha \\ \beta \\ \gamma \end{matrix}, \bar{c}_2 \leq u \leq \bar{c}_1 \right\},$$

where $\bar{c}_1 \geq c_1$, $\bar{c}_2 \leq c_2$, and $\alpha, \beta, \gamma \geq 1$. The claim is that Σ is invariant for (14.23). To see this, we first note that $(0, -1, 0, 0)$ is a left eigenvector of D; thus if we set $G(u, v, w, z) = -v$, then

$$dG(F)\big|_{v=0} = g_1(u)(v - h(u))\big|_{v=0} = -g(u)h_1(u) < 0.$$

It follows that $-v = G(u, v, w, z) \leq 0$; i.e., $v \geq 0$. Similarly, we can show that $w, z \geq 0$. Next, we set $G = v - \alpha$, and compute

$$dG(F)\big|_{v=\alpha} = g_1(u)(h_1(u) - \alpha) < g_1(u)(1 - \alpha) \leq 0.$$

Therefore $v \le \alpha$, and similarly $w \le \beta$, $z \le \gamma$. Also if we set $G = u - \bar{c}_1$, then

$$dG(F)|_{u=\bar{c}_1} = \frac{1}{c}[k_1 v^3(c_1 - \bar{c}_1) + k_2 z^4(c_2 - \bar{c}_1) + k_3(c_3 - \bar{c}_1)] < 0,$$

so that $u \le \bar{c}_1$. Finally, put $G = \bar{c}_2 - u$; then

$$dG(F)|_{u=\bar{c}_2} = \frac{1}{c}[k_1 v^3(\bar{c}_2 - c_1) + k_2 z^4(\bar{c}_2 - c_2) + k_3(\bar{c}_2 - c_3)] < 0,$$

and thus $u \ge \bar{c}_2$. These calculations show that Σ is an invariant region. Consequently, we have shown that (14.22) admits *arbitrarily large* invariant regions. It follows then, from Theorem 14.4 that if $u_0 \in BC_0$ (§A), then the Hodgkin–Huxley equations, with this initial data, are (uniquely) solvable for all $t > 0$.

EXAMPLE 2 (Fitz-Hugh–Nagumo equations). These equations are considered to be models for the Hodgkin–Huxley equations. They are given by

$$v_t = v_{xx} + f(v) - u, \qquad u_t = \varepsilon u_{xx} + \sigma v - \gamma u.$$

Here σ, γ, and ε are constants, with $\sigma, \gamma > 0$ and $\varepsilon \ge 0$. The function $f(v)$ has the qualitative form of a cubic polynomial; for definiteness, we may take $f(v) = -v(v - \beta)(v - 1)$, where $0 < \beta < \frac{1}{2}$. We shall give a geometric construction of the existence of arbitrarily large invariant rectangles (cf. Corollary 14.8(a)). For this we refer to Figure 14.5, where we have drawn the zero sets of the components of the vector field $F = (f(v) - u, \sigma v - \gamma u)$; the $+$ and $-$ signs refer to the signs of the respective functions, on each side of their zero sets. The region Σ is constructed as follows: The lines $u = \text{const.}$ are chosen so that the top one is above the zero set of $\sigma v - \gamma u$, and the bottom one is below it. The lines $v = \text{const.}$ are taken to be on both sides of the zero set of $f(v) - u$. It is obvious (from the picture!) that F points into Σ on $\partial \Sigma$. Since we can construct arbitrarily large invariant rectangles of this form, we see that these equations also admit global solutions for any data in BC_0.

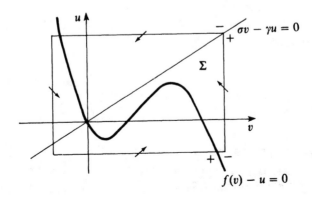

Figure 14.5

EXAMPLE 3. These equations arise in the study of superconductivity of liquids (see [CCS]). Let $u = (u_1, \ldots, u_n)$, and $D = \text{diag}(\alpha_1, \ldots, \alpha_n)$, $\alpha_i \geq 0$, where not all $\alpha_i = 0$. The system is given by

$$u_t = D\Delta u + (1 - |u|^2)u.$$

We consider two cases. First, suppose that $\alpha_i \neq \alpha_j$ for some $i \neq j$. Then it is obvious that any rectangle containing the unit disc $|u| = 1$ is invariant. Now let's consider the case where $\alpha_1 = \alpha_2 = \cdots = \alpha_n$. Take Σ to be any convex region which contains the unit disc $|u| = 1$ in its interior. Then Corollary 14.8(b) implies that Σ is invariant. In particular, if

$$\Sigma_\delta = \{u : |u|^2 \leq 1 + \delta\}, \qquad \delta > 0,$$

then Σ_δ is invariant. We want to show that the unit disc, Σ_0, is invariant. We cannot apply Corollary 14.8(b) directly since the vector field $(1 - |u|^2)u$ vanishes identically on $\partial\Sigma_0$. However, if $u(x, 0) = u_0(x) \in \Sigma_0$, for all x, then $u_0(x) \in \Sigma_\delta$ for every $\delta > 0$. Thus the solution $u(x, t) \in \Sigma_\delta$, for each $\delta > 0$, for all x, and all $t > 0$; whence $u(x, t) \in \Sigma_0$ for all x and all $t > 0$.

Note that this example illustrates an interesting (rather obvious) theorem; namely, that the *intersection of invariant regions is an invariant region*.

EXAMPLE 4 (Field–Noyes equations). These equations serve as models for the Belousov–Zhabotinsky reactions in chemical kinetics; see, e.g., ([HM] and [HK]) for a discussion of this very interesting chemical reaction. The equations take the form

$$u_t = \varepsilon_1 \Delta u + \alpha(v - uv + u - \beta u^2),$$

$$v_t = \varepsilon_2 \Delta v + \alpha^{-1}(\gamma w - v - uv),$$

$$w_t = \varepsilon_3 \Delta w + \delta(u - w).$$

Here u, v, and w denote chemical concentrations, $\varepsilon_i \geq 0$, and $\alpha, \beta, \gamma, \delta > 0$ are constants, and $\beta \approx 10^{-8}$. Let

$$a > \beta^{-1}, \qquad c > a, \qquad b > \gamma c, \quad \text{and}$$

$$\Sigma = \{(u, v, w) : 0 \leq u \leq a, 0 \leq v \leq b, 0 \leq w \leq c\}.$$

We claim that Σ is invariant. To see this, we set

$$V = [\alpha(v - uv - \beta u^2), \alpha^{-1}(\gamma w - v - uv), \delta(u - w)];$$

then successively, if

$$G = -u, \quad \nabla G \cdot V|_{u=0} = -\alpha v \leq 0 \quad \text{in } \Sigma, \qquad \text{so } u \geq 0;$$

$$G = -v, \quad \nabla G \cdot V|_{v=0} = -\alpha^{-1}\gamma w \leq 0 \quad \text{in } \Sigma, \quad \text{so } v \geq 0;$$

$$G = -w, \quad \nabla G \cdot V|_{w=0} = -\delta u \leq 0 \quad \text{in } \Sigma, \qquad \text{so } w \geq 0;$$

$$G = w - c, \quad \nabla G \cdot V|_{w=c} = \delta(u - c) < \delta(u - a) \leq 0, \quad \text{so } w \leq c;$$

$$G = v - b, \quad \nabla G \cdot V|_{v=b} = \alpha^{-1}(\gamma w - b - ub) \leq \alpha^{-1}(\gamma c - b - ub)$$
$$\leq \alpha^{-1}(\gamma c - b) < 0, \qquad \text{so } v \leq b;$$

finally if $G = u - a$, then

$$\nabla G \cdot V|_{u=a} = \alpha(v - av + a - \beta a^2) = \alpha[v(1 - a) + a(1 - a\beta)] < 0,$$

so $u \leq a$. This proves that Σ is invariant.

EXAMPLE 5. The "p-system," with viscosity (cf. [Lx 2], [K 1]).

$$v_t - u_x = \varepsilon v_{xx}, \qquad u_t + p(v)_x = \varepsilon u_{xx}, \qquad \varepsilon > 0. \qquad (14.24)$$

We assume that $p' < 0, p'' > 0$, and $p > 0$ in $v > 0$. The system can be written in the form

$$U_t = M(U)U_x + \varepsilon U_{xx},$$

where $U = (v, u)$, and

$$M(U) = \begin{bmatrix} 0 & 1 \\ -p' & 0 \end{bmatrix}. \qquad (14.25)$$

If

$$r = u - \int^v \sqrt{-p'(\xi)}\, d\xi, \qquad s = u + \int^v \sqrt{-p'(\xi)}\, d\xi,$$

denote the Riemann invariants (see Chapter 17, §B), then since ∇r and ∇s are left eigenvectors of M, the region

$$\Sigma = \{(v, u): r \leq r_0, s \geq s_0\}$$

is invariant. If $p(v) = v^{-\gamma}, \gamma \geq 1$, then Σ can be depicted as in Figure 14.6. This shows that if we denote the solution of (14.24) by $(v_\varepsilon, u_\varepsilon)$, then we have

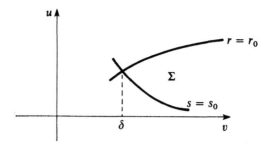

Figure 14.6

$v_\varepsilon(x, t) \geq \delta > 0$, for all x and $t > 0$, where δ *doesn't depend on* ε. If $\gamma > 1$, then one can show (see Chapter 17, §A) that the curves $r = $ const. and $s = $ const. are asymptotic to straight lines. Thus, in this case, $(\gamma > 1)$, we have an additional estimate of the form

$$\left| u_\varepsilon(x, t) \right| \leq c, \qquad (x, t) \in \mathbf{R} \times \mathbf{R}_+,$$

where c is *independent of* ε.

We can use this last example to illustrate the fact that the notion of being an invariant region is *not* an "open" condition. Thus, consider the system

$$v_t - u_x = \varepsilon v_{xx}, \qquad u_t + p(v)_x = \mu u_{xx}, \qquad \varepsilon, \mu > 0. \qquad (14.26)$$

If $\varepsilon = \mu$, then we have seen that Σ, as defined above, is invariant for this system. However, if $\varepsilon \neq \mu$, then Σ is not an invariant region. In fact, if $\varepsilon \neq \mu$, (14.26) *admits no invariant regions*. This is true because the diffusion matrix for (14.26), $D = \operatorname{diag}(\varepsilon, \mu)$, admits only $(1, 0)$ and $(0, 1)$ as left eigenvectors, and these are not left eigenvectors of (14.25); cf. Theorem 14.12.

§C. A Comparison Theorem

With this section we begin the study of the asymptotic behavior of solutions, as $t \to +\infty$. Our main result is a comparison theorem which estimates the solution of a system of reaction-diffusion equations in terms of solutions of an associated systems of *ordinary* differential equations. Comparison theorems usually are obtained as a consequence of a "maximum principle"; e.g., see Chapter 10. However, in the case of *systems* of equations, there are usually no general maximum principles available. The theorem which we obtain here by-passes any maximum principles; i.e., we shall give a direct proof of the comparison theorem. As the examples at the end of this section demonstrate, our result is general enough to catch the "gross" asymptotic behavior in that it locates the attracting regions in phase space.

We consider the system

$$u_t = D\nabla^2 u + f(u, t), \qquad (x, t) \in \Omega \times \mathbf{R}_+, \qquad (14.27)$$

where Ω is a bounded domain in $\mathbf{R}^m$, $\partial\Omega$ is smooth, $u \in \mathbf{R}^n$, and f is a smooth mapping from an open set $U \subset \mathbf{R}^n$ into $\mathbf{R}^n$ for each $t \geq 0$. Here we take D to be a constant diagonal matrix with nonnegative entries. Finally, we assume that (14.27) admits a bounded invariant region

$$\Sigma = \prod_1^n [a_i, b_i], \qquad (14.28)$$

where $-\infty < a_i < b_i < \infty$, $i = 1, 2, \ldots, n$.

Together with (14.27), we have the initial conditions

$$u(x, 0) = u_0(x) = (u_1^0, u_2^0, \ldots, u_n^0)(x), \qquad (14.29)$$

where $u_0(x)$ lies in Σ for all $x \in \Omega$. In addition, we assume that u satisfies homogeneous Neumann boundary conditions:

$$\frac{du}{dn} = 0 \quad \text{on } \partial\Omega \times \mathbf{R}_+, \qquad (14.30)$$

where du/dn denotes differentiation in the outward normal direction on $\partial\Omega$.

We partition the set $\mathbf{Z}_n = \{1, 2, \ldots, n\}$ into two disjoint sets σ_M and σ_m, and for each such partition, we define functions f_i^+ and f_i^-, $i = 1, 2, \ldots, n$, by

$$f_i^+(u, t) = \sup\{f_i(\xi_1, \ldots, \xi_{i-1}, u_i, \xi_{i+1}, \ldots, \xi_n, t):$$

$$a_j \leq \xi_j \leq u_j, \text{ if } j \in \sigma_M, j \neq i,$$

$$u_j \leq \xi_j \leq b_j, \text{ if } j \in \sigma_m, j \neq i\},$$

$$f_i^-(u, t) = \inf \{\text{same set as above}\}.$$

Note that f_i^+ is nondecreasing (resp. nonincreasing) in the jth argument if $j \neq i, j \in \sigma_M$ (resp. $j \in \sigma_m$), while f_i^- is nonincreasing (resp. nondecreasing) in the jth argument if $j \neq i, j \in \sigma_M$ (resp. $j \in \sigma_m$).

Now we set

$$h_i(v, t) = \begin{cases} f_i^+(v, t), & \text{if } i \in \sigma_M, \\ f_i^-(v, t), & \text{if } i \in \sigma_m, \end{cases}$$

and

$$H_{\sigma_M}(v, t) = (h_1(v, t), \ldots, h_n(v, t)).$$

Observe that the number of (vector) functions H_{σ_M} that we have constructed is equal to 2^n, the number of subsets of $\mathbf{Z}_n$.

We pause here to give an example. Thus take $n = 2$, and let

$$f = (f_1, f_2) \equiv (uM(u, v), vN(u, v)), \tag{14.31}$$

where

$$u_1 = u, \qquad u_2 = v, \qquad a_1 \le u \le b_1, \qquad a_2 \le v \le b_2.$$

The set $\mathbf{Z}_2 = \{1, 2\}$ has four subsets; namely,

(a) $\sigma_M = \{1\}$,   (b) $\sigma_M = \{2\}$,   (c) $\sigma_M = \{1, 2\}$,   (d) $\sigma_M = \phi$.

We shall construct H_{σ_M} for each of these cases.

(a) $\sigma_M = \{1\}$, $\sigma_m = \{2\}$,

$$f_1^+(u, v) = \sup_{v \le \xi \le b_2} f_1(u, \xi) = u \sup_{v \le \xi \le b_2} M(u, \xi),$$

$$f_2^+(u, v) = \sup_{a_1 \le \xi \le u} f_2(\xi, v) = v \sup_{a_1 \le \xi \le u} N(\xi, v) \equiv vN^+(u, v),$$

$$f_1^-(u, v) = \inf_{v \le \xi \le b_2} f_1(u, \xi) = u \inf_{v \le \xi \le b_2} M(u, \xi) \equiv uM^-(u, v),$$

$$f_2^-(u, v) = \inf_{a_1 \le \xi \le u} f_2(\xi, v) = v \inf_{a_1 \le \xi \le u} N(\xi, v),$$

$$H_{\sigma_M} \qquad = H_{\{1\}} = (f_1^+, f_2^-).$$

(b) $\sigma_M = \{2\}$; here one calculates as above,

$$H_{\{2\}} = (f_1^-, f_2^+).$$

(c) $\sigma_M = \{1, 2\}$; here

$$H_{\{1,2\}} = (f_1^+, f_2^+), \quad \text{where } f_1^+ = u \sup_{a_2 \le \xi \le v} M(u, \xi) \equiv uM^+(u, v),$$

and so $H_{\{1,2\}} = (uM^+, vN^+)$.

(d) $\sigma_M = \phi$; here

$$H_\phi = (f_1^-, f_2^-) = (uM^-, vN^-), \quad \text{where } N^-(u, v) = \inf_{u \le \xi \le b_1} N(\xi, v).$$

We shall refer to the vector fields

$$(uM^+, vN^+) \quad \text{and} \quad (uM^-, vN^-) \tag{14.32}$$

as the *maximal* and *minimal* vector fields, respectively, associated to the vector field (14.31), relative to Σ.

The functions H_{σ_M} will be considered as vector fields, and the orbits of the associated flows will define the "comparison" functions. In order that they determine well-defined flows, we need the following basic lemma.

Lemma 14.15. *Let Σ be defined by (14.28) and let $p = p(u, t)$ be Lipschitz continuous in Σ, for each $t \geq 0$. Let*

$$p^+(u, t) = \sup\{p(u_1, \xi_2, \ldots, \xi_n, t): a_i \leq \xi_i \leq u_i, i = 2, 3, \ldots, n\}.$$

Then p^+ is Lipschitz continuous in Σ, with the same Lipschitz constant as p.

Proof. We may assume that $a_i = 0$, $i = 1, 2, \ldots, n$, For $u \in \Sigma$, set

$$A_u = \{\xi \in \mathbf{R}^n: 0 \leq \xi_j \leq u_j, j > 1, u_1 = \xi_1\};$$

then $p^+(u, t) = \sup\{p(\xi, t): \xi \in A_u\}$. Let $u, v \in \Sigma$. Since p is continuous and A_u is compact, there exists ξ' in A_u such that

$$p^+(u, t) = p(\xi', t).$$

Let $\xi_0' = \max(0, \xi' + v - u)$; i.e., $(\xi_0')_j = \max(0, \xi_j' + v_j - u_j)$, $j = 1, 2, \ldots,$ n. Now it is clear that $\xi_0' \in A_v$ (see Figure 14.7), and that

$$|\xi_0' - \xi'| \leq |u - v|. \tag{14.33}$$

To see this, note that

$$|\xi_0' - \xi'|^2 = \Sigma[\xi_j' - \max(0, \xi_j' + v_j - u_j)]^2,$$

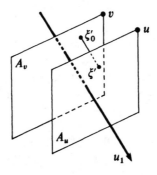

Figure 14.7

and if $\xi_j' + v_j - u_j \geq 0$, $\xi_j' - \max(0, \xi_j' + v_j - u_j) = u_j - v_j$, while if $\xi_j' + v_j - u_j < 0$, then $\xi_j - \max(0, \xi_j' + v_j - u_j) = \xi_j' < u_j - v_j$. Thus (14.33) is valid, and so suppressing the t's we have

$$p^+(u) - p^+(v) = \sup_{\xi \in A_u} p(\xi) - \sup_{\xi \in A_v} p(\xi)$$

$$\leq p(\xi') - p(\xi_0') \leq k|\xi' - \xi_0'| \leq k|u - v|.$$

The proof is completed by interchanging the roles of u and v. $\quad\square$

Now let $v(t)$ be the solution of the problem

$$\dot{v} = H_{\sigma_M}(v, t), \qquad v(0) = v^0, \tag{14.34}$$

where

$$v_i^0 = \begin{cases} \sup u_i^0(x), & x \in \bar{\Omega}, \quad \text{if } i \in \sigma_M, \\ \inf u_i^0(x), & x \in \bar{\Omega}, \quad \text{if } i \notin \sigma_M. \end{cases}$$

In view of Lemma 14.15, the problem (14.34) has a unique solution. Observe that $v(t)$ is the unique solution of the problem

$$v_t = D\nabla^2 v + H_{\sigma_M}(v, t), \qquad v(x, 0) = v^0, \qquad \frac{dv}{dn} = 0 \quad \text{on } \partial\Omega \times \mathbf{R}_+. \tag{14.35}$$

We can now state the principal result of this section, i.e., the following comparison theorem holds.

Theorem 14.16. *Under the above hypotheses, the following inequalities are valid for all* $(x, t) \in \Omega \times \mathbf{R}_+$:

$$v_i(t) \geq u_i(x, t), \quad \text{if } i \in \sigma_M,$$

$$v_i(t) \leq u_i(x, t), \quad \text{if } i \notin \sigma_M. \tag{14.36}$$

Before giving the proof we point out that (14.36) shows that every component u_i of u is bounded by solutions of ordinary differential equations for all $t > 0$, uniformly in $x \in \Omega$. As we have mentioned in the introduction, such generality cannot catch anything more than the "rough" asymptotic properties of solutions.

Proof. We first note that Σ is an invariant region for (14.34). Indeed, if e.g., $v = (1, 0, \ldots, 0)$, then $H_{\sigma_M} \cdot v = h_1$, and on, say $u = b_1$, we have (suppressing t)

$$H_{\sigma_M} \cdot v = h_1(b_1, u_2, \ldots, u_n) = f_1^+(b_1, u_2, \ldots, u_n),$$

if $1 \in \sigma_M$. But $f_1^+(b_1, u_2, \ldots, u_n) \leq 0$, since Σ is invariant for (14.27). Similarly, $H_{\sigma_M} \cdot v \leq 0$ if $1 \notin \sigma_M$. Thus $v(t) \in \Sigma$ for all $t \geq 0$.

Now let $w = (w_1, w_2, \ldots, w_n)$, where

$$w_i = \begin{cases} u_i - v_i, & \text{if } i \in \sigma_M, \\ v_i - u_i, & \text{if } i \notin \sigma_M, \end{cases}$$

and put $G(w, t) = (g_1(w, t), \ldots, g_n(w, t))$, where

$$g_i(w, t) = \begin{cases} f_i(u, t) - h_i(v, t) & \text{if } i \in \sigma_M, \\ h_i(v, t) - f_i(u, t), & \text{if } i \notin \sigma_M. \end{cases}$$

Observe that g_i is actually a function of w and t since $u_i = w_i + v_i$ if $i \in \sigma_M$, while $u_i = v_i - w_i$ if $i \notin \sigma_M$, and $v_i = v_i(t)$. Also w satisfies the equation

$$w_t = D\nabla^2 w + G(w, t), \qquad (x, t) \in \Omega \times \mathbf{R}_+,$$

the homogeneous Neumann boundary conditions, $dw/dn = 0$ on $\partial\Omega \times \mathbf{R}_+$, and by construction, the (componentwise) inequality $w(x, 0) \leq 0$, $x \in \Omega$.

We shall prove that $w(x, t) \leq 0$ in $\Omega \times \mathbf{R}_+$; this is equivalent to (14.36). To do this, we shall show that G doesn't point out of the "negative" orthant in $\mathbf{R}^n$ (i.e., the set $w_i \leq 0$, $i = 1, 2, \ldots, n$), and then apply Theorem 14.11. Thus, we must show that

$$g_i(w, t) \leq 0, \quad \text{whenever } w_i = 0 \quad \text{and} \quad w_j \leq 0, \quad j \neq i.$$

We assume that $i \in \sigma_M$; if $i \notin \sigma_M$, the proof is similar. Now since $w_i = 0$, we have $v_i = u_i$, and

$$g_i(w, t) = f_i(u, t) - h_i(v, t) = f_i(u, t) - f_i^+(v, t)$$
$$\leq f_i^+(u_1, \ldots, u_i, \ldots, u_n, t) - f_i^+(v_1, \ldots, v_{i-1}, u_i, v_{i+1}, \ldots, v_n, t).$$

Suppose that $1 \in \sigma_M$ and $1 \neq i$; then by definition of f_i^+,

$$f_i^+(u_1, u_2, \ldots, u_n, t) \leq f_i^+(v_1, u_2, \ldots, u_n, t), \tag{14.37}$$

since $u_1 \leq v_1$ and f_i^+ is nondecreasing in its first argument ($1 \in \sigma_M$). If now $1 \notin \sigma_M$, then $w_1 = v_1 - u_1 \leq 0$ and f_i^+ is again nonincreasing in its first argument, so (14.37) holds as before. Continuing in this way for every $j \neq i$, we find

$$f_i^+(u_1, \ldots, u_{j-1}, u_j, \ldots, u_i, \ldots, u_n, t)$$
$$\leq f_i^+(u_1, \ldots, u_{j-1}, v_j, \ldots, u_i, \ldots, u_n, t).$$

It follows easily from this that $g_i(w, t) \leq 0$. The proof is complete. $\square$

We shall now apply this theorem to some systems which arise in the mathematical theory of population dynamics, or, as it is often called, *mathematical ecology*. These equations will actually be studied in §E from a more general point of view. Our goal here is merely to illustrate the comparison theorem.

EXAMPLE 1 (Predator–Prey Equations). We investigate the two species predator–prey interaction, where we assume that both species are continuously distributed throughout a region Ω, and are undergoing diffusion. The equations we consider are of the form

$$u_t = \alpha\Delta u + uM(u, v),$$

$$v_t = \beta\Delta v + vN(u, v), \tag{14.38}$$

where $(x, t) \in \Omega \times \mathbf{R}_+$. Here u and v denote the population densities of the prey and predator, respectively, and the functions M and N are their corresponding growth rates. The predator–prey interaction is defined by the following conditions:

$$M_v < 0, \qquad N_u > 0,$$

which merely state that the prey growth rate M decreases as the predator population increases, and that an increase in prey is favorable for the growth rate N of the predator. As a specific example, we may take

$$M(u, v) = -(u - d)(u - 1) - cv \quad \text{and} \quad N(u, v) = -\mu - \alpha v + cu,$$

where $0 < d < 1$, and c, α, and μ are positive constants with $d < \mu/c < 1$. We depict the zero sets of M and N in Figure 14.8. Note that (14.38) admits arbitrarily large bounded invariant rectangles in the positive quadrant $u \geq 0, v \geq 0$.

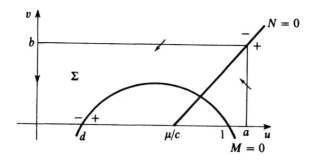

Figure 14.8

Together with initial conditions

$$u(x, 0) = u_0(x), \qquad v(x, 0) = v_0(x), \qquad x \in \Omega, \qquad (14.39)$$

we assume that the system satisfies homogeneous Neumann boundary conditions

$$\frac{du}{dn} = \frac{dv}{dn} = 0 \quad \text{on } \partial\Omega \times \mathbf{R}_+, \qquad (14.40)$$

where d/dn denotes differentiation in the outward normal direction to $\partial\Omega$. This "no flux" condition is quite reasonable from an ecological point of view; namely, one imagines that the species are, say, on an island, or in a valley surrounded by mountains, from which they cannot escape.

Now the results of §A can be easily extended to initial-boundary-value problems [CCS]. Thus if $u_0 \geq 0$ and $v_0 \geq 0$, in Ω, then the problem (14.38)–(14.40) has a (unique), globally defined solution (u, v) which satisfies $u(x, t) \geq 0$ and $v(x, t) \geq 0$ for $(x, t) \in \Omega \times \mathbf{R}_+$.

We let $\Sigma = \{(u, v) : 0 \leq u \leq a, 0 \leq v \leq b\}$ be an invariant rectangle, as depicted in Figure 14.8, and we compute the maximal vector field (uM^+, vN^+) relative to Σ (see (14.32)). Rather than write these functions out explicitly, we shall merely sketch the phase plane; see Figure 14.9. If we denote by (u^+, v^+) the solution of

$$\dot{u} = uM^+(u, v), \qquad \dot{v} = vN^+(u, v), \qquad (u(0), v(0)) = (U, V), \quad (14.41)$$

where

$$(U, V) = \left(\sup_{\bar{\Omega}} u_0(x), \sup_{\bar{\Omega}} v_0(x) \right), \qquad (14.42)$$

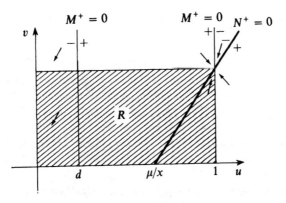

Figure 14.9

then from Theorem 14.16, we have the estimates

$$0 \le u(x, t) \le u^+(t), \qquad 0 \le v(x, t) \le v^+(t), \qquad (x, t) \in \Omega \times \mathbf{R}_+. \qquad (14.43)$$

These enable us to conclude qualitative information as regards the solution (u, v). For example, if $u_0(x) < d$ for all $x \in \Omega$, then we see from Figure 14.9, that $(u^+(t), v^+(t))$ tends to $(0, 0)$ as $t \to +\infty$, whence, the same is true of (u, v), uniformly in x.[5] Furthermore, it is easy to see that the shaded rectangle R, in Figure 14.9, is a *global* attractor for all solutions of (14.38)–(14.40) in the sense that given any neighborhood of R, there is a $T > 0$ such that if $t > T$, $(u(x, t), v(x, t))$ lies in this neighborhood for all $x \in \Omega$.

EXAMPLE 2 (Competing-Species Equations). We consider the problem (14.38)–(14.40) as representing two interacting species in competition with each other. In this case, we assume the conditions

$$M_v < 0 \quad \text{and} \quad N_u < 0.$$

Moreover, rather than give explicit forms for M and N, we shall only draw the phase plane of a vector field (uM, vN); see Figure 14.10(a); we have

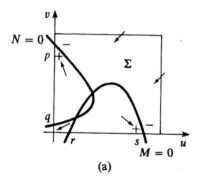
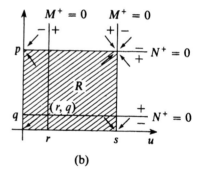

(a)   (b)

Figure 14.10

also sketched the maximal vector field (uM^+, vN^+) in Figure 14.10(b). As before we consider the equations (14.41), (14.42), whose solutions we denote by $(u^+(t), v^+(t))$. Then from Theorem 14.16, the estimates (14.43) are valid. Thus, we see that the shaded region R is a global attractor for all solutions of (14.38)–(14.40), and that if $u_0(x) \le r$ and $v_0(x) \le q$ for all $x \in \Omega$, then $(u, v) \to (0, 0)$ as $t \to \infty$, uniformly in x. Moreover, if $u_0(x) \le r$ for all $x \in \Omega$, then $u(x, t) \to 0$ as $t \to \infty$, uniformly in x; this holds since $(u^+(t), v^+(t))$ tends to

---

[5] This has an interesting ecological interpretation; namely if the initial *prey* density is too low, the prey, and thus the predator, must become extinct.

(0, 0) or (0, q), as $t \to \infty$. Similarly, if $v_0(x) \leq q$ for all $x \in \Omega$, then $v(x, t) \to 0$ as $t \to \infty$. These last two statements can be interpreted as a weak form of the "competitive exclusion principle" in mathematical ecology.

EXAMPLE 3 (Symbiosis). This is the last of the classical ecological interactions. In this case we have the inequalities

$$M_v > 0 \quad \text{and} \quad N_u > 0,$$

and we can take, as an example, the vector field whose phase plane is depicted in Figure 14.11. Here the zero sets of M, M^+, and M^- are all the same, as are those of N, N^+, and N^-. Thus if $(u^+(t), v^+(t))$ is as above, and $(u^-(t), v^-(t))$ is the solution of

$$\dot{u} = uM^-(u, v), \qquad \dot{v} = vN^-(u, v), \qquad (u(0), v(0)) = \left(\inf_{\bar{\Omega}} u_0(x), \inf_{\bar{\Omega}} v_0(x) \right),$$

and $u_0(x) > 0$, $v_0(x) > 0$ for all $x \in \bar{\Omega}$, then from Theorem 14.16, we have the estimates

$$u^-(t) \leq u(x, t) \leq u^+(t),$$

$$v^-(t) \leq v(x, t) \leq v^+(t),$$

uniformly in x for $t \geq 0$. It follows that (see Figure 14.11)

$$p \geq \varlimsup_{t \to \infty} u(x, t) \geq \varliminf_{t \to \infty} u(x, t) \geq p,$$

so that $u(x, t) \to p$, uniformly in x as $t \to \infty$. Similarly, $\|v(\cdot, t) - q\|_{L_\infty(\Omega)} \to 0$ as $t \to \infty$. Thus, this two-species symbiotic interaction eventually settles down to a uniform state, and the ratio of the two population densities becomes constant.

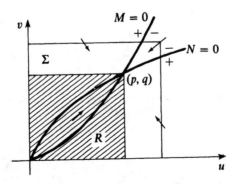

Figure 14.11

§D. Decay to Spatially Homogeneous Solutions

For a given system of reaction–diffusion equations, say (14.1), there is associated with it the system of ordinary differential equations

$$\frac{du}{dt} = f(u), \tag{14.44}$$

all of whose solutions are, of course, spatially homogeneous solutions of (14.1). If (14.1) is defined on a bounded domain Ω, and if solutions are required to satisfy homogeneous Neumann boundary conditions ($du/dn \equiv \nabla_x u \cdot \mathbf{n} = 0$, where $\mathbf{n}$ denotes the outer normal on $\partial\Omega$), then solutions of (14.44) even satisfy the boundary conditions. Under suitable circumstances, it might be reasonable to assume that u does not vary too much from point to point in space, and that transport processes can be ignored. In this case one might hope that the full set of equations could be replaced by (14.44). This is often referred to in the biological and chemical literature as the "lumped parameter assumption." Of course, this assumption cannot always be valid; for example, if (14.1) admits nonconstant steady-state solutions then (14.44) would certainly *not* be a good approximation to the full system. Our goal in this section is to give a clear understanding of *when*, and *in what sense* the lumped parameter assumption is valid. We shall be concerned with the questions: When do solutions of (14.1) decay as $t \to \infty$ to spatially homogeneous (independent of x) solutions? and, how are these solutions related to solutions of (14.44)? That we are able to give precise mathematical answers to these questions depends on the existence of a bounded invariant region.

Thus for simplicity, we consider the system (14.1), where we assume that D is a constant, positive definite matrix, together with homogeneous Neumann boundary conditions

$$\frac{du}{dn} = 0 \quad \text{on } \partial\Omega \times \mathbf{R}_+. \tag{14.45}$$

We assume that (14.1) admits a bounded invariant region Σ, and we define a parameter σ by

$$\sigma = \lambda d - M. \tag{14.46}$$

Here $\lambda > 0$, is the first nonzero eigenvalue of $-\Delta$ on Ω with homogeneous Neumann boundary conditions (see Chapter 11, §A), $d > 0$ is the smallest eigenvalue of D, and

$$M = \max\{|df_u| : u \in \Sigma\}.$$

Thus $M < \infty$, since Σ is compact.

Our main hypothesis is to assume that σ is positive. This will allow us to study the validity of the lumped parameter assumption, at least for large values of t. Before we get into the details, we would like to discuss the significance of the assumption $\sigma > 0$. For this, note that we can consider it in two ways; namely

$$\lambda > \frac{M}{d} \quad \text{or} \quad d > \frac{M}{\lambda}.$$

In the first case, we can view it as saying that λ is "large," and since λ is inversely proportional to the squared diameter of Ω (see [CH 1]), we may interpret $\sigma > 0$ as saying that the spatial region Ω is "small." In the second case, $\sigma > 0$ can be looked upon as saying that the diffusion is "strong" relative to the reaction terms f. In both of these cases, small Ω, and big diffusion, it is reasonable to expect that spatial inhomogeneities are insignificant and become quickly damped out. The following theorem shows this indeed to be the case.

Theorem 14.17. *Consider the problem* (14.1), (14.3) *in* Ω, *with boundary data* (14.45). *Assume that* (14.1) *admits a bounded invariant region* Σ, *and that* $\{u_0(x) : x \in \Omega\} \subset \Sigma$. *If σ is positive, then there exist constants* $c_i > 0, 1 \leq i \leq 4$, *such that the following hold for* $t > 0$:

(1) $\|\nabla_x u(\cdot, t)\|_{L_2(\Omega)} \leq c_1 e^{-\sigma t}$,

(2) $\|u(\cdot, t) - \bar{u}(t)\|_{L_2(\Omega)} \leq c_2 e^{-\sigma t}$, *where*

$$\bar{u}(t) = \frac{1}{|\Omega|} \int_\Omega u(x, t) \, dt \qquad (|\Omega| = \text{measure of } \Omega), \qquad (14.47)$$

and $\bar{u}$ satisfies

$$\frac{d\bar{u}}{dt} = f(\bar{u}) + g(t), \qquad \bar{u}(0) = \frac{1}{|\Omega|} \int_\Omega u_0(x) \, dx, \qquad (14.48)$$

with

(3) $|g(t)| \leq c_3 e^{-\sigma t}$.

If D is a diagonal matrix, then (2) *may be strengthened to*

$$\|u(\cdot, t) - \bar{u}(t)\|_{L_\infty(\Omega)} \leq c_4 e^{-\sigma t}. \qquad (14.49)$$

Before giving the details of the proof we shall make a few remarks. First, note that if $\sigma > 0$, the theorem shows that solutions u of (14.1) decay exponentially fast to their spatial average $\bar{u}$. Furthermore, $\bar{u}$ satisfies an equation, (14.48), which becomes a better and better approximation to (14.44) as $t \to \infty$.

Thus, if $\sigma > 0$, we can say that the lumped parameter assumption is valid. Next, we point out that the proof will show that $c_i = O(\|\nabla u_0\|_{L_2(\Omega)})$, $1 \leq i \leq 3$, and that $c_4 = O(\|\nabla u_0\|_{L_\infty(\Omega)})$, where ∇ (always) denotes the spatial gradient.

Proof. Let $\|\cdot\|$ denote the $L_2(\Omega)$ norm, and define

$$\phi(t) = \tfrac{1}{2}\|\nabla u(\cdot, t)\|^2 = \tfrac{1}{2}\int_\Omega \langle \nabla u, \nabla u \rangle \, dx.$$

Then integrating by parts, and using (14.45), we have

$$\dot\phi = \int_\Omega \langle \nabla u, \nabla u_t \rangle = \int_\Omega \langle \nabla u, \nabla D\Delta u \rangle + \int_\Omega \langle \nabla u, \nabla f \rangle$$

$$= -\int_\Omega \langle \Delta u, D\Delta u \rangle + \int_\Omega \langle \nabla u, df_u(\nabla u) \rangle$$

$$\leq -d\int_\Omega |\Delta u|^2 + M\int_\Omega |\nabla u|^2$$

$$\leq -\lambda d\int_\Omega |\nabla u|^2 + M\int |\nabla u|^2 \quad \text{(by Theorem 11.11)}$$

$$= 2(M - \lambda d)\phi(t) = -2\sigma\phi(t).$$

It follows easily from this that

$$\phi(t) \leq \phi(0)e^{-2\sigma t},$$

and thus (1) is proved. Now let $\bar u$ be the spatial average of u, defined by (14.47). Then from Theorem 11.11, we have

$$\lambda \|u(\cdot, t) - \bar u(t)\|^2 \leq \|\nabla u(\cdot, t)\|^2,$$

and this, in conjunction with (1) gives (2). Next, if $v = \bar u$, then from (14.45)

$$\dot v = \frac{1}{|\Omega|}\int u_t = \frac{1}{|\Omega|}\int_\Omega (D\Delta u) + \frac{1}{|\Omega|}\int_\Omega f(u)$$

$$= \frac{1}{|\Omega|}\int f(u) = f(v) + \frac{1}{|\Omega|}\int_\Omega (f(u) - f(v))$$

$$\equiv f(v) + g(t).$$

Now since f is smooth, and Σ is compact, there is an $M_1 > 0$ independent of t, such that

$$|g(t)| \leq \frac{1}{|\Omega|} \int_\Omega |f(u) - f(v)| \leq \frac{M_1}{|\Omega|} \int_\Omega |u - v|$$

$$\leq M_1 |\Omega|^{-1/2} \|u(\cdot, t) - v(t)\|_{L_2(\Omega)},$$

where we have used the Schwarz inequality. Thus $\bar{u}$ satisfies (14.48) and (3) is valid. Finally, if D is a diagonal matrix, $D = \text{diag}(d_1, \ldots, d_n)$, then (14.2) can be written as

$$\frac{\partial u_i}{\partial t} = d_i \frac{\partial^2 u_i}{\partial x^2} + f_i(u), \qquad i = 1, 2, \ldots, n.$$

Since Σ is a bounded invariant region, each $f_i(u)$ is bounded. From a theorem of Nash this implies that $u \in C^\alpha$ (α independent of t), and thus by a result of Schauder, $u \in C^{2+\alpha}$, so that $|\nabla u|$ is bounded independent of t (see [CHS] for details). Now take $p > \max(3, n)$; then Sobolev's inequality gives for $(x, t) \in \Omega \times \mathbf{R}_+$,

$$|u(x, t) - \bar{u}(t)|^p \leq \int_\Omega |u(\cdot, t) - \bar{u}(t)|^p + \int_\Omega |\nabla[u(\cdot, t) - \bar{u}(t)]|^p$$

$$\leq c \int_\Omega |u(\cdot, t) - \bar{u}(t)|^2 + k \int_\Omega |\nabla u(\cdot, t)|^2,$$

and this implies (14.49). The proof of the theorem is complete. □

There are several implications we can draw from this theorem. We have already noted that if $\sigma > 0$, the solution of the partial differential equations gets exponentially close to its average, as $t \to +\infty$. Furthermore, from a result of Markus [Ms], it follows that if $\sigma > 0$, the ω-limit sets[6] of solutions of the partial differential equations coincide with the ω-limit sets of the ordinary differential equation (14.44). Finally, if $\sigma > 0$, then the theorem shows that there cannot exist any nonconstant solutions of the (elliptic) system

$$D\Delta u + f(u) = 0, \qquad x \in \Omega, \tag{14.50}$$

with homogeneous Neumann boundary conditions $du/dn = 0$ on $\partial\Omega$. This follows since solutions of (14.50) depend only on x, and the theorem implies that they must tend to solutions independent of x.

---

[6] The ω-limit set of a solution $u(x, t)$ is the intersection over $t \geq 0$ of the closure of this "orbit" starting at the point (x, t), with $x \in \Omega$, for each $x \in \Omega$.

This last remark is in agreement with the examples we have considered in Chapter 13, §A. Thus, we can consider a scalar equation of the form

$$du_{xx} + f(u) = 0, \qquad |x| < L, \quad d > 0, \tag{14.51}$$

with homogeneous Neumann boundary data. If we make the change of variable $y = d^{-1/2}x$, (14.51) goes over into

$$\ddot{u} + f(u) = 0, \qquad |y| < \frac{L}{\sqrt{d}}, \tag{14.52}$$

with the same type of boundary conditions. Now (14.52) is the type of equation which we have considered in Chapter 13. We have seen there that if the length of the interval is small, then there are no nonconstant solutions. In other words, if d is large, or if L is small, there can be only constant solutions of (14.51), with the given boundary conditions. We now can see that this particular result is a consequence of our general theorem.

Remark. We close this section by pointing out that the technique which we used in the proof of (14.49) can be used to obtain derivative estimates for solutions of (14.2) in the case where D is diagonal, $D = (d_1, \ldots, d_n)$, which satisfy either homogeneous Neumann boundary conditions (14.2), or homogeneous Dirichlet boundary conditions

$$u = 0 \quad \text{on } \partial\Omega \times \mathbf{R}_+,$$

or more general boundary conditions. These estimates will hold provided we have an a-priori bound on solutions; in particular, they are valid if (14.2) admits a bounded invariant region.

Thus we consider the single equation, for simplicity in one space variable

$$w_t = dw_{xx} + g(u).$$

Here $w = u_i$, $d = d_i$, $g = f_i$, $1 \le i \le n$. We set $v^k = (\partial/\partial t)^k w$; then v^1 satisfies the equation

$$v_t^1 = dv_{xx}^1 + \nabla g \cdot v.$$

Since u is bounded, we have as in the proof of the theorem, $u \in C^{2+\alpha}$, with respect to x, where α and the constant are independent of t. Thus $w_t = v^1$ is in C^α so the last equation shows, as before, that if $g \in C^1$, then $v^1 \in C^{2+\alpha_1}$ with respect to x, where α_1 and the constant are independent of t. Repeating this argument, we find that if $f \in C^{k-1}$, then v^k is bounded and is in $C^{\alpha k}$. Going back to the equation for w, we see that since $v^1 = w_t$ is bounded, then dw_{xx} is bounded, so by the usual elliptic estimates, $w \in C^{2+\alpha}$ with respect to x, where α (not necessarily the same α as above) and the constant are independent of t. Now differentiating the w equation with respect to t, we find

$$w_{tt} = d^2 w_{xxxx} + dg(w)_{xx} + \nabla g \cdot v.$$

If $f \in C^1$, then $v^2 = w_{tt}$ is bounded, and we see that w_{xxxx} is bounded so $w_{xx} \in C^{2+\alpha}$ and so on. Thus, *if $f \in C^k$, then $(\partial/\partial t)^i w$ is in C^{α_i}, $1 \le i \le k$, and $u \in C^{2k+\alpha}$.* All of the Hölder exponents as well as the constants are independent of t. In fact, they only depend on the equation, and on Σ, the bounded invariant region.

§E. A Lyapunov Function for Contracting Rectangles

In many theorems concerning global existence and stability of solutions, as well as their asymptotic behavior as $t \to \infty$, a central role is played by regions which are contracting for the vector field $f(u)$ in the following sense.

Definition 14.18. A bounded rectangle $R \subset \mathbf{R}^n$ is *contracting* for the vector field $f(u)$, if for every point $u \in \partial R$, $f(u) \cdot \mathbf{n}(u) < 0$, where $\mathbf{n}(u)$ is the outward pointing normal at u.

Thus R is contracting for f, if and only if the orbits of $u' = f(u)$ cross ∂R transversally and enter R in positive time; i.e., R is an "attractor" for the vector field f. For example, for the Fitz-Hugh–Nagumo equations (Example 2 in §B), Figure 14.5 clearly shows that Σ is a contracting rectangle.

The importance of contracting rectangles is that they allow us to define functionals which decrease in time along solutions of the equations (14.2); i.e., the "flow" defined by (14.2) is "gradient-like" near the boundary of Σ.

We consider the system (14.2), where D is a diagonal matrix $D = \text{diag}(d_1, d_2, \ldots, d_n)$, with each $d_i \ge 0$. In view of Corollary 14.8, we may assume that (14.2) admits contracting rectangles. Thus, let R be a rectangle in $\mathbf{R}^n$ containing the origin in its interior, and let $|\cdot|_R$ be the usual norm on $\mathbf{R}^n$ defined by R; namely,

$$|u|_R = \inf\{t \ge 0 : u \in tR\}. \tag{14.53}$$

Thus, $|u|_R$ is the smallest multiple of R which contains the vector u. Next, we define a continuous functional F_R on the bounded continuous functions by

$$F_R(w) = \sup\{|w(x)|_R : x \in \mathbf{R}\}. \tag{14.54}$$

The basic result in this section is a quantitative version of the invariant region theorem, Theorem 14.7. It is given by the following. (Recall that for a function $g : \mathbf{R} \to \mathbf{R}$, the *upper Dini derivative* of g at t is defined by

$$\bar{D}g(t) = \overline{\lim_{h \to 0}} \frac{g(t + h) - g(t)}{h}.\Big)$$

Theorem 14.19. *Let $f(u)$ be a vector field on $\mathbf{R}^n$, and let R be a rectangle with $0 \in \operatorname{int}(R)$. Suppose that $u(\cdot, t) \in BC_0$ is a smooth solution of (14.2), for $|t - T| < \delta$. If R is a contracting rectangle for $f(u)$, then there is an $\eta > 0$ such that*

$$\bar{D}F_R(u(\cdot, T)) \leq \frac{-2\eta}{L}, \qquad (14.55)$$

where L is the length of the shortest side of R.

Thus (14.55) shows that $u(\cdot, t)$ must lie in a smaller rectangle for $T < t < T + \delta_1$, for some $\delta_1 > 0$.

Proof. Let $u = (u_1, \ldots, u_n)$, $f = (f_1, \ldots, f_n)$, and let R be defined by the inequalities $l_i \leq u_i \leq r_i$, $l_i < 0 < r_i$, $i = 1, 2, \ldots, n$. Multiplying R by a scalar if necessary,[7] we may assume that $F_R(u(\cdot, T)) = 1$. Thus $a_i \leq u_i(x, T) \leq b_i$, for all x. We say that $u(x, t)$ is in the jth right-hand face of R if $u_j(x, t) = r_j$, with a similar definition for the jth left-hand face.

If now $u(x, T) \in \partial R$, then there is a subset $J \subset \{1, 2, \ldots, n\}$ such that $u(x, T)$ is in one of the jth faces if and only if $j \in J$. Since R is contracting for f, there is an $\eta > 0$ such that for all $u \in \partial R$,

$$f(u) \cdot \bar{n}(u) < -\eta,$$

where $\bar{n}$ is the outward-pointing normal at u. If, e.g., $u(\bar{x}, T)$ is in the jth right-hand face, then $u_j(x, T) \leq r_j$ for all x, with equality at $x = \bar{x}$. Thus $d_j \partial^2 u_j(\bar{x}, T)/\partial x^2 \leq 0$, since $F_R(u(\cdot, T)) = 1$. Consequently,

$$\frac{\partial u_j}{\partial t} = \frac{a_j \partial^2 u_j}{\partial x^2} + f_j(u) < -\eta,$$

so that $u_j(\bar{x}, T + h) < r_j - \eta h$ for small h. By continuity, this holds for all x in a neighborhood of $\bar{x}$.

If $K = \{x : u(x, T) \in \partial R\}$, then K is compact, and by what we have just shown, there is an open set $U \supset K$ such that if $x \in U$, and h is small,

$$u(x, T + h) \in \left([1 - \eta h] \min_{1 \leq j \leq n} (r_j, l_j)^{-1} \right) R.$$

---

[7] If R is multiplied by the scalar s, then $F_R(u(\cdot, T))$ gets multiplied by s^{-1}, so the same is true of $\bar{D}F_R(u(\cdot, T))$, and L too gets multiplied by s.

If $x \notin U$, then $u(x, T)$ is interior to R, so there is a compact set $Q \subset \text{int } R$, and an $h_0 > 0$ such that $u(x, T + h) \in Q$ if $|h| < h_0$ for all x. Thus, for sufficiently small h, $u(x, T + h)$ is in $(1 - h\eta/L)R$ for all x, and so

$$F_R(u(x, T + h)) \le 1 - \frac{2h\eta}{L}.$$

Therefore, for all x

$$\frac{F_R(u(x, T + h)) - F_R(u(x, T))}{h} \le -2\eta/L,$$

and this gives (14.55). $\square$

The result just obtained allows us to prove via Lyapunov's method that certain solutions tend to constant states, and that under fairly general conditions all solutions tend to "attracting regions" in u-space. Theorem 14.19 can sometimes be used instead of the comparison theorem of §C. It is often easier to use, and one can conclude qualitative information merely from the phase portrait of $u' = f(u)$ (which one presumably already knows from the existence of the invariant regions). It is not necessary to calculate, say, the "maximal" and "minimal" vector fields. For example, consider the Fitz-Hugh–Nagumo equations, Example 2 in §B. Referring to Figure 14.5, we see that there are arbitrarily large contracting rectangles containing $(0, 0)$. Thus, we may conclude, from Theorem 14.19, that *all* solutions, tend, as $t \to \infty$, to the invariant region Σ as depicted in the figure. In fact, all solutions must enter the *interior* of Σ in finite time. So, in particular, we can conclude at once that Σ contains all steady-state solutions; i.e., all solutions of the (elliptic) system

$$v_{xx} + f(v) - u = 0, \qquad \varepsilon u_{xx} + \sigma v - \gamma u = 0.$$

As a second application, we again consider the Fitz-Hugh–Nagumo equations, and we assume that

$$-f'(0) > \frac{\sigma}{\gamma}.$$

Then it is not too hard to construct a family of contracting rectangles about $(0, 0)$ of the form τR, where R is a rectangle containing $(0, 0)$, and τ is small. If the data curve, $(v_0(x), u_0(x))$, lies in the union of these sets, then our theorem implies that the corresponding solution $(v(x, t), u(x, t))$ tends to $(0, 0)$ uniformly in x as $t \to \infty$.

§F. Applications to the Equations of Mathematical Ecology

We shall apply some of our results to a particularly interesting class of reaction–diffusion equations; namely, the Kolomogorov form of the equations which describe the classical two-species interaction, where now diffusion and spatial dependence are taken into account. Thus we consider the equations

$$u_t = \alpha \Delta u + u M(u, v),$$

$$v_t = \beta \Delta v + v N(u, v), \qquad (x, t) \in \Omega \times \mathbf{R}_+, \tag{14.56}$$

where Ω is a bounded region in $\mathbf{R}^m$, and $\alpha \geq 0$, $\beta \geq 0$ are constants. Here u and v are scalar functions of (x, t), and represent population densities; M and N are their respective growth rates. We assume M and N to be smooth.

Together with (14.56), we have the initial conditions

$$(u(x, 0), v(x, 0)) = (u_0(x), v_0(x)), \qquad x \in \Omega, \tag{14.57}$$

and the homogeneous Neumann boundary conditions

$$\frac{du}{dn} = \frac{dv}{dn} = 0 \quad \text{on } \partial\Omega \times \mathbf{R}_+. \tag{14.58}$$

It is assumed that both u_0 and v_0 are bounded nonnegative smooth functions.

The boundary conditions (14.58) are to be interpreted as "no flux" conditions; i.e., there is no migration of either species across $\partial\Omega$. Ω is here considered as the habitat of u and v.

We have had occasion to consider these equations before, in §C, but now our goal is to study them from a general point of view, and to show how nicely they fit into theoretical framework which we have developed.

Under minimal hypotheses, we are already able to prove the following (basic!) result.

Lemma 14.20. *For all* $(x, t) \in \bar{\Omega} \times \mathbf{R}_+$, *both* $u(x, t) \geq 0$, *and* $v(x, t) \geq 0$.

Proof. Let

$$\Sigma_0 = \{(u, v) : u \geq 0, v \geq 0\};$$

then $\partial\Sigma_0 = \{u = 0\} \cup \{v = 0\}$. It is clear that the vector field (uM, vN) does not point out of Σ_0, so our result follows from Theorem 14.11, provided that $u_0(x) > 0$ and $v_0(x) > 0$ for all $x \in \Omega$. If either function is allowed to be zero, then we approximate it by positive functions, and use (14.10) on $0 \leq t \leq T$, for any preassigned $T > 0$. $\square$

We now consider the three classical ecological interactions; they are determined by the signs of the partial derivatives M_v and N_u. In the *predator–prey* interaction, the derivatives are of opposite sign:

$$M_v < 0, \quad -N_u > 0, \quad \text{for } u > 0, v > 0, \tag{P}$$

where u denotes the prey density, and v the predator density.

Competition refers to the case when both derivatives are negative:

$$M_v < 0, \quad N_u < 0, \quad \text{for } u > 0, v > 0. \tag{C}$$

In *symbiosis*, both derivatives are positive:

$$M_v > 0, \quad N_u > 0, \quad \text{for } u > 0, v > 0. \tag{S}$$

These are the minimal assumptions which are imposed in virtually all discussions. However, even in the context of the ordinary differential equations, $(\dot{u}, \dot{v}) = (uM, vN)$, many studies have shown the importance of imposing further conditions which reflect an ultimate growth limit of each species. In other words, there are always present, in any environment, specific *resource limitations* which place a definite upper bound on the growth rates. We shall show that such limits to growth are intimately connected with pointwise bounds on u and v; i.e., they imply the existence of bounded invariant regions. Our specific assumptions are as follows. First, we require that both M and N change sign in $u > 0, v > 0$. Next, for each of the specific interactions, we require the following:

Predator–Prey
PL 1. There is a $k_0 > 0$ such that $M(u, 0) < 0$ for all $u > k_0$.
PL 2. There is a function l such that $N(u, v) < 0$ for all $u > 0$ and $v > l(u)$.

Competition
CL 1. There is a $k_0 > 0$ such that $M(u, 0) < 0$ for all $u > k_0$.
CL 2. There is a $l_0 > 0$ such that $N(0, v) < 0$ for all $v > l_0$.

Symbiosis
SL 1. There is a function k such that $M(u, v) < 0$ for $v > 0$ and $u > k(v)$.
SL 2. There is a function l such that $N(u, v) < 0$ for $u > 0$ and $v > l(u)$.
SL 3. $k(v) = o(v)$ and $l(u) = o(u)$ for large values of their arguments.

These conditions need a little explanation. In the case of PL 1 we are saying that even when there are *no* predators ($v = 0$); i.e., under the most favorable conditions for u, the environment does not allow growth once the density exceeds a critical value, k_0. Conditions CL 1 and CL 2 are to be interpreted similarly. In the case of PL 2 we notice that because of condition

(P) an increase of u represents an enrichment of the environment for v. But PL 2 insures that no matter what the value of u the growth rate for v becomes negative once v becomes large enough. Conditions SL 1 and SL 2 are to be interpreted similarly. Note that all of these assumptions hold for the examples we have considered in §C.

Now because of the smoothness of N, we see that in all of these cases, there is a function $v = l(u)$ such that $N(u, l(u)) = 0$ for all $u > 0$, $l(u) > 0$; hence, for P and S

$$N_u(u, l(u)) + N_v(u, l(u))l'(u) = 0.$$

Since $N_u > 0$ in conditions (P) and (S), and since $N_v(u, l(u)) \le 0$ because of PL 2 and SL 2, it follows that $l'(u) > 0$. In a similar way, we see that $k'(v) \ge 0$. This is merely a statement of the fact that in P and S, an increase of u is advantageous for the growth of v. Similarly, in symbiosis, an increase of v is advantageous for u. Condition SL 3 requires that this enhancement diminish for very large values of the densities.

We see then that in each interaction, the following condition is satisfied;

(L) *Limitation to Growth*: There are nondecreasing nonnegative functions $k = k(v)$ and $l = l(u)$, such that

   (i)   if $u > k(v)$, then $M(u, v) < 0$, and
   (ii) if $v > l(u)$, then $N(u, v) < 0$.

For example, in the predator–prey case, $k(v) = k_0$ and if $v > 0$, then $M(u, v) < M(u, 0) < 0$ if $u > k_0$. In the case of competition, again $M(u, v) < M(u, 0) < 0$ if $u > k_0$; similarly, $N(u, v) < 0$ if $v > l_0$. Of course (L) is obvious for symbiosis.

We assume that k and l are the smallest functions satisfying (L). Now it is easy to see that in each case, we have either $k(v) = o(v)$ as $v \to \infty$ or $l(u) = o(u)$ as $u \to \infty$. We make the additional assumption that in every case one or the other of the following hold for all sufficiently large values of their arguments:

$$k(v) = O(v) \quad \text{and} \quad l(u) = o(u), \quad \text{or}$$

$$k(v) = o(v) \quad \text{and} \quad l(u) = O(u). \tag{14.59}$$

Note again that these assumptions hold for the examples we have considered in §C.

Now we define the set B by

$$B = \{(u, v) : u \ge k(v) \text{ and } v \ge l(u)\}.$$

Our assumptions imply that B is an unbounded subset of the positive quadrant, having nonvoid interior. The set B need not be connected since the curves $u = k(v)$ and $v = l(u)$ can intersect more than once. However, in view of (14.59), the set of intersection points, $\{(u_\alpha, v_\alpha)\}$, of these curves must lie in a bounded, therefore compact set. Let $K = \max u_\alpha$ and $L = \max v_\alpha$. Since k and l are nondecreasing functions, (K, L) is an intersection point. Let

$$B_\infty = B \cap \{(u, v); u \geq K, v \geq L\}.$$

Then B_∞ is the unique unbounded component of B. In Figures 14.12(a) and (b) we have sketched examples of B_∞ for the predator–prey and symbiosis interactions, respectively.

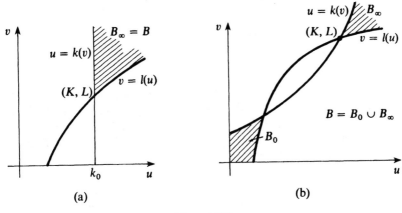

Figure 14.12

Now for $a > 0$ and $b > 0$, let

$$\Sigma(a, b) = \{(u, v): 0 \leq u \leq a, 0 \leq v \leq b\}.$$

Theorem 14.21. *If condition* (L) *holds, and* $(a, b) \in \text{int}(B)$ *then* $\Sigma(a, b)$ *is invariant for* (14.56), (14.58).

Proof. If $(a, b) \in \text{int}(B)$, then $a > k(b)$, so since $k' \geq 0$, $b \geq v$ implies $k(b) \geq k(v)$ and thus $a > k(v)$. Therefore, by (L), $M(a, v) < 0$. Thus $aM(a, v) < 0$ if $0 \leq v \leq b$. Similarly, $bN(u, b) < 0$ if $0 \leq u \leq a$. Then Lemma 14.20, together with Theorem 14.11 imply that $\Sigma(a, b)$ is invariant. $\square$

Now since u_0 and v_0 are bounded and B is unbounded, we can find $(a, b) \in \text{int } B$ such that $0 \leq u_0(x) < a$, and $0 \leq v_0(x) < b$ for all $x \in \Omega$. Thus the last theorem implies that $0 \leq u(x, t) \leq a$, and $0 \leq v(x, t) \leq b$ for all $(x, t) \in \Omega \times \mathbf{R}_+$. But we can make a much stronger statement (Theorem 14.25). Before doing this, however, we need a few lemmas. Recall that M^+ and N^+ are defined in equation (14.32), (and preceding).

Lemma 14.22. *If condition* (L) *holds, then the functions* M^+ *and* N^+ *are negative in the interior of* B.

Proof. Let $(u, v) \in$ int B; then $u > k(v)$, and since k is nondecreasing, $u > k(\theta)$ if $0 \le \theta \le v$. Hence $M(u, \theta) < 0$ if $0 \le \theta \le v$, so $M^+(u, v) = \{\sup M(u, \theta): 0 \le \theta \le v\} < 0$. Similarly, $N^+(u, v) < 0$. $\quad\square$

Lemma 14.23. B *is an invariant set for the maximal vector field* $(uM^+(u, v), vN^+(u, v))$.

Proof. Note that ∂B is made up of segments of one of the following types:

    (i) $u \equiv 0$,
    (ii) $v \equiv 0$,
    (iii) $v = l(u)$, $l' \ge 0$,
    (iv) $u = k(v)$, $k' \ge 0$.

Now trajectories of the maximal vector field clearly cannot cross segments of the type (i) or (ii). Also, in the interior of B, $v > l(u)$ and $N^+ < 0$ (by Lemma 14.22). Thus since $N^+(u, l(u)) = 0$ (by the minimality of l), trajectories of (uM^+, vN^+) starting in B, cannot cross segments of the type (iii). Similarly, they cannot cross segments of the type (iv). Thus B is invariant for the maximal vector field. $\quad\square$

As an immediate corollary of the last two lemmas, we have

Lemma 14.24. *Every solution curve of the maximal vector field, which for* $t = 0$ *is in* B_∞, *converges to the rest point* (K, L).

Now consider any solution of (u, v) of (14.56)–(14.58). As we have noted above, we can find a point (a, b) in B_∞ such that

$$0 \le u_0(x) < a, \qquad 0 \le v_0(x) < b, \qquad x \in \Omega.$$

Now let $(u^+(t), v^+(t))$ be the solution of

$$\dot{u} = uM^+(u, v), \qquad \dot{v} = vN^+(u, v), \qquad (u(0), v(0)) = (a, b).$$

From Lemma 14.24 it follows that

$$\lim_{t \to \infty}(u^+(t), v^+(t)) = (K, L),$$

and from Theorem 14.16, as applied to the maximal vector field, we have

$$u(x, t) \le u^+(t) \quad \text{and} \quad v(x, t) \le v^+(t).$$

Therefore we can conclude that

$$\overline{\lim_{t \to \infty}} \, u(x, t) \le K \quad \text{and} \quad \overline{\lim_{t \to \infty}} \, v(x, t) \le L, \tag{14.60}$$

uniformly for $x \in \Omega$. We have thus proved the following theorem.

Theorem 14.25. *Let Ω be a bounded domain with smooth boundary, and let M and N satisfy* (L) *and one of the conditions* (14.59). *Then every solution of* (14.56)–(14.58) *which is initially nonnegative remains so for all $t > 0$ and satisfies* (14.60).

Thus, in particular, we see that for every $\varepsilon > 0$, there is a $T > 0$ such that $(u(x, t), v(x, t))$ lies in $\Sigma(K + \varepsilon, L + \varepsilon)$ for all $x \in \Omega$ and all $t \ge T$. For the examples in §C, the $\Sigma(K, L)$ are the shaded regions in Figures 14.9, 14.10, and 14.11. We emphasize that K and L are determined *only* by the functions M and N. Thus our result shows clearly the limitation to growth, and in fact more. Namely, it implies the existence of a global attracting rectangle in phase space for all positive solutions of the partial differential equations.

NOTES

The local existence theorems are classical; see, e.g., [Fn]. We have followed the rather different approach given by Rauch and Smoller in [RS]. The notion of invariant region is due to Chueh, Conley and Smoller [CCS], and independently to Weinberger [Wi], (for scalar diffusion matrices). The necessary conditions for the existence of invariant regions is found in [CCS].

The theorems given here can be easily generalized to equations in several space variables of a more general form; e.g.,

$$u_t = D\Delta u + \sum_1^m A_j \frac{\partial u}{\partial x_j} + f$$

with general boundary conditions. Here it is necessary to also require that dG_i be a left eigenvector of each matrix A_j. The proofs are virtually the same as those which we have given here; see [CCS]. There are related papers on invariant regions due, e.g., to Alikakos, [Al 1, 2], Amann [Am 2], Bebernes, Chueh, and Fulks [BCF], and Kuiper [Kp]. Estimates on derivatives are obtained in these papers, as well as in [CCS].

The examples in §B are taken from [CCS]. For more detailed studies of Examples 2, 3, and 4, see [BDG], [RS], and [Ty], respectively. The observation concerning equation (14.26) was first made in [CCS].

Theorem 14.16 in §C is due to Gardner [Gd 3]; it extends earlier work of Conway and Smoller [CS 3]. See also [FT]. Lemma 14.15 is also found in [CS 3]; the elegant proof given here is due to D. Wagner. The theorem in

§D is due to Conway, Hoff, and Smoller [CHS]; see also Othmer [Ot] for earlier, less general, less precise results. The Lyapunov function construction in §E is due to Rauch and Smoller [RS]. The material in §F is taken from Conway and Smoller [CS 5]; see also [CS 4]. Related work has been done by Alikakos in [Al 1, 2]. These papers also contain references to the spatially homogeneous ecology equations; i.e., the Kolmogoroff form, [Ko], of the ordinary differential equations $\dot{u} = uM$, $\dot{v} = vN$. See also [SZ] for the important early work.

There are many papers devoted to a *single* reaction-diffusion equation; see, e.g., [AW], [FM], and the references therein. For systems of ordinary differential equations related to the equations studied here, see e.g., [HM], [Kop], [HK], [Ty]. For a discussion of the ecology equations from a biological viewpoint, see [My], and [Sm]. See also the books [Fi], [Mu], as well as the paper [AN].

There is a vast literature on pattern formation, stability, and bifurcation theory, as applied to chemical, biological, and ecological equations; see, e.g., [MiN], [MNY], [Mg 1–4], and [Mo].

The notion of an invariant region can be applied to finite difference approximations. D. Hoff, in [Hof 2], has shown that under the same conditions as we have imposed for differential equations; i.e., the conditions which imply the existence of invariant regions, the finite-difference approximations converge to the solution.

For reaction-diffusion equations with nonlinear boundary conditions, see the papers of Hernandez [Hz 1, 2]. Related work can be found in [CW], [Gb 4], and [Sk].

Part III

The Theory of Shock Waves

Chapter 15

Discontinuous Solutions of Conservation Laws

In this chapter we shall begin the study of quasi-linear systems of the form

$$u_t + f(u)_x = 0, \tag{15.1}$$

where $u = (u_1, \ldots, u_n) \in \mathbf{R}^n$, $n \geq 1$, and $(x, t) \in \mathbf{R} \times \mathbf{R}_+$. We assume that the vector-valued function f is C^2 in some open subset $\Omega \subset \mathbf{R}^n$. These equations are commonly called *conservation laws* in analogy to the examples of such systems which arise in physics; see the examples below.

Systems of the form (15.1) arise in the study of nonlinear wave phenomena, when dissipation effects, such as viscosity, are neglected. It is typical of such systems that they admit discontinuous solutions; i.e., shock waves, so that the equation must be understood in some generalized sense. It is our goal here to develop a rigorous mathematical theory for problems involving these equations. In this chapter we shall consider some examples and describe certain phenomena. In the next chapter we will consider the case of a scalar equation ($n = 1$), and in the succeeding chapters we will be concerned with the more interesting, and difficult, case of systems; i.e., $n > 1$.

We start by giving some examples of systems of the form (15.1).

EXAMPLE 1. The equations of gas dynamics for an inviscid, non-heat conducting gas in Lagrangian coordinates can be written in the form ([LL])

$$
\begin{aligned}
v_t - u_x &= 0 \quad \text{(conservation of mass)}, \\
u_t + p_x &= 0 \quad \text{(conservation of momentum)}, \\
E_t + (up)_x &= 0 \quad \text{(conservation of energy)},
\end{aligned}
\tag{15.2}
$$

where v is the specific volume; $v = 1/\rho$ and ρ is the density, u is the velocity, and E is the specific energy; $E = e + u^2/2$, with e the internal energy. The pressure p is a given function of e and v, which depends on the particular gas under consideration; this relation is often called the *equation of state*.

EXAMPLE 2. In Eulerian coordinates, the above equations take the form

$$\rho_t + (\rho u)_x = 0 \quad \text{(conservation of mass)},$$

$$(\rho u)_t + (\rho u^2 + p)_x = 0 \quad \text{(conservation of momentum)}, \quad (15.3)$$

$$\left[\rho\left(\frac{u^2}{2} + e\right)\right]_t + [\rho u(\tfrac{1}{2}u^2 + i)]_x = 0 \quad \text{(conservation of energy)}.$$

Here $i = e + p/\rho$ is the specific enthalpy, and the equation of state is a given function $e = e(v, s)$, where s is the specific entropy. The pressure p is obtained from the formula $p = e_v$. The change of coordinates is given by $(h, \tau) \to (x, t)$, where $t = \tau$ and $h = \int_{-\infty}^{x(h,\tau)} \rho(s, \tau)\,ds$; this change takes (15.3) into (15.2) if we set $x = h$, $t = \tau$ in (15.2). Choosing ρ, u, and s as dependent variables, we can easily reduce (15.3) to the form (15.1).

EXAMPLE 3 (The p-system). The equations written in Lagrangian coordinates are

$$v_t - u_x = 0, \qquad u_t + p(v)_x = 0. \tag{15.4}$$

When $p(v) = kv^{-\gamma}$ ($\gamma \geq 1$ and $k > 0$ are constants), the equations (15.4) are a model for isentropic (= constant entropy) gas dynamics. These equations are related to certain second-order equations. Thus, differentiating the first equation with respect to t and the second with respect to x gives

$$v_{tt} + p(v)_{xx} = 0.$$

On the other hand, in simply connected regions, the first equation in (15.4) implies the existence of a function ϕ such that $v = \phi_x$ and $u = \phi_t$. The second equation then becomes

$$\phi_{tt} + p(\phi_x)_x = \phi_{tt} + p'(\phi_x)\phi_{xx} = 0.$$

If $p' < 0$, this is a *nonlinear* wave equation where the speed of propagation, $\sqrt{-p'}$, depends on ϕ_x. This type of equation has been extensively studied numerically, beginning with the classical paper of Fermi, Pasta, and Ulam [FPU].

It is natural to consider the initial-value problem for (15.1); i.e., to find a solution of (15.1) defined in $t > 0$, which assumes the given data

$$u(x, 0) = u_0(x), \qquad x \in \mathbf{R}, \tag{15.5}$$

at time $t = 0$. If the function u_0 is smooth, then it is easy to construct a unique *local* solution; i.e., a solution defined only for $0 < t < T$. This limitation is real, and not due to the techniques used to construct the solution. Rather, it is a consequence of the nonlinearity of the equations; i.e., the dependence of f upon u. We shall study this problem in the next section.

§A. Discontinuous Solutions

We begin with perhaps the simplest example which leads to discontinuous solutions. Consider the single equation, called the *Burgers equation*,[1]

$$u_t + \left(\frac{u^2}{2}\right)_x = 0, \tag{15.6}$$

which we can write in the form

$$u_t + uu_x = 0. \tag{15.7}$$

This equation has the rather remarkable property that *the only C^1 functions which satisfy it in $t > 0$, are those which are monotonically nondecreasing in x, for each fixed $t > 0$.* To see this, suppose that u is a C^1 solution in $t > 0$. For any point $(x_0, t_0), t_0 > 0$, we consider the unique solution curve $x(t)$ which solves

$$\frac{dx}{dt} = u(x, t), \qquad x(t_0) = x_0. \tag{15.8}$$

Along this curve, which is a characteristic curve of (15.7) (cf. Chapter 2),

$$\frac{d}{dt} u(x(t), t) = u_x \frac{dx}{dt} + u_t = u_x u + u_t = 0.$$

Thus u is constant along characteristics, and from (15.8), we can conclude that the characteristics have constant slope. In other words, the characteristics are straight lines having *speed* (i.e., reciprocal of slope) equal to the value of u along these lines. This shows at once that for each $t \geq 0, u(x_1, t) \leq u(x_2, t)$ if $x_1 < x_2$; otherwise the characteristic lines would meet at some point in $t > 0$. But this cannot happen since u is supposedly C^1 in $t > 0$.

Thus, if $u(x, 0) = u_0(x)$ and $u_0'(\bar{x}) < 0$ for some $\bar{x}$, then (15.6) *cannot* have a solution defined in all of $t > 0$. To understand better the obstruction which prevents the solution from being "globally" defined; i.e., defined in all of $t > 0$, let's consider the C^∞ function u_0 whose graph is depicted in Figure 15.1. Here $u_0(x) = 1$ if $x \leq 0, u_0(x) = 0$ for $x \geq 1$ and $u_0' \leq 0$. We can sketch the characteristics, (15.8), in Figure 15.2. (For definiteness, suppose that $u_0(x) = 1 - x$ if $x \in [0, 1]$.)

---

[1] This is really what one might more properly call the "Burgers equation without viscosity"; we use the term "Burgers equation" only for brevity.

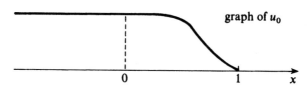

Figure 15.1

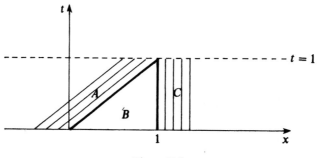

Figure 15.2

Consider the region $0 \le t \le 1$. In region A, the characteristics have slope 1 and the solution has the constant value 1; in region C the characteristics have infinite slope, and the solution has the constant value 0. But in region B, we see that the decrease of u from 1 to 0 takes place over an x-interval whose length tends to zero as t increases. At the point $(1, 1)$, a *continuous* solution is overdetermined, since different characteristics meet there, and they each carry different values of u. If $0 < t_0 < t_1 < t_2 < 1$, then the graphs of $u(\cdot, t_i)$ are as depicted in Figure 15.3; the graph of $u(\cdot, t)$ steepens as t increases to 1, max $u_x(\cdot, t)$ becomes unbounded, and the curve "breaks" at $t = 1$.

More generally, consider the initial-value problem for the scalar u,

$$u_t + f(u)_x = 0, \quad t > 0; \qquad u(x, 0) = u_0(x), \quad x \in \mathbf{R}. \tag{15.9}$$

We can rewrite the equation as $u_t + f'(u)u_x = 0$, and consider the characteristics

$$\frac{dt}{ds} = 1, \qquad \frac{dx}{ds} = f'(u).$$

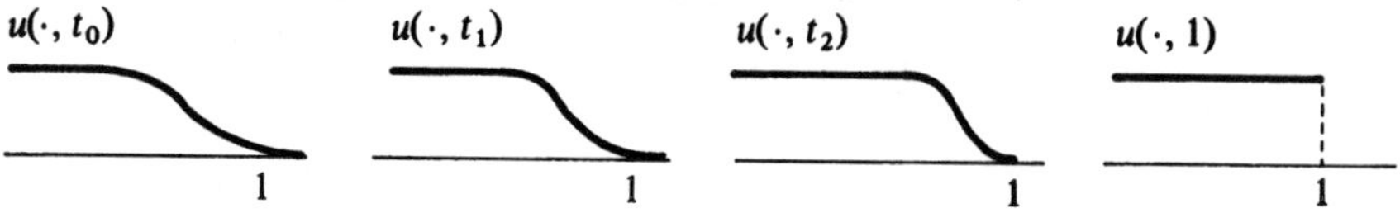

$u(\cdot, t_0)$ $\qquad\qquad$ $u(\cdot, t_1)$ $\qquad\qquad$ $u(\cdot, t_2)$ $\qquad\qquad$ $u(\cdot, 1)$

1 $\qquad\qquad\qquad$ 1 $\qquad\qquad\qquad$ 1 $\qquad\qquad\qquad$ 1

Figure 15.3

Along such a curve,

$$\frac{du}{ds} = u_t \frac{dt}{ds} + u_x \frac{dx}{ds} = u_t + f'(u)u_x = 0.$$

Thus, again u is constant along the characteristics. Since the slope of the characteristic is $1/f'(u)$, the characteristics are straight lines, having slope determined by their values at $t = 0$; i.e., by $u_0(x)$. So, if there are points $x_1 < x_2$ with

$$0 < m_1 = \frac{1}{f'(u_0(x_1))} < \frac{1}{f'(u_0(x_2))} = m_2,$$

then the characteristics starting at $(x_1, 0)$ and $(x_2, 0)$ will cross in $t > 0$; cf. Figure 15.4. Along l_i, $u(x, t) \equiv u_0(x_i)$, $i = 1, 2$. Thus at P the solution *must* be discontinuous. Note that this conclusion is independent of the smoothness properties of f and u_0; they can each be analytic, and still we cannot obtain a globally defined solution. The phenomenon is a purely nonlinear one.

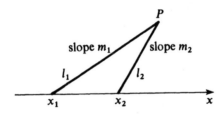

Figure 15.4

We can be a bit more explicit and see analytically that discontinuities must form if u_0' is negative at some point. Thus, consider (15.9), and assume that $f'' > 0$. Since the characteristics are straight lines, if (x, t) is any point with $t > 0$, we let $y(x, t)$ denote the unique point on the x-axis which lies on the characteristic through (x, t). Since u is constant along characteristics, and $tf'(u) = x - y$, we see that u must implicitly be given by

$$u(x, t) = u_0(x - tf'(u(x, t))).$$

Now if u_0 is a differentiable function, then we can invoke the implicit function theorem and solve this last equation for u, provided that t is sufficiently small. We find

$$u_t = - \frac{f'(u)u_0'}{1 + u_0' f''(u)t} \quad \text{and} \quad u_x = \frac{u_0'}{1 + u_0' f''(u)t}.$$

Now if $u_0'(x) \geq 0$ for all x, then these formulas show that grad u stays bounded for all $t > 0$, and the solution u exists for all time. On the other hand, if $u_0' < 0$ at some point, both u_x and u_t become unbounded when $1 + u_0' f''(u)t$ tends to zero.

Thus, the lesson is clear: if we adhere to the notion that a solution must be smooth, then we must content ourselves with solutions which exist for only a finite time. But in several instances in physics, equations of the above type arise naturally, and "discontinuous" solutions have been considered (with striking success!). We shall show how to overcome this problem in §B. But first, one more example.

EXAMPLE 4 (The Shock Tube and Riemann's Problem in Gas Dynamics). Consider a long, thin, cylindrical tube containing a gas separated by a thin membrane. We assume that the gas is at rest on both sides of the membrane, but that it is of different constant pressures and densities on each side. At time $t = 0$, the membrane is broken, and the problem is to determine the ensuing motion of the gas. This problem was studied by Riemann in his basic paper [Ri], and is now known by his name.

Let (u_l, ρ_l, p_l) and (u_r, ρ_r, p_r) denote the velocity, density and pressure on both sides of the membrane. We consider the case where $u_l = u_r = 0$, $\rho_l > \rho_r, p_l > p_r$, and all of these quantities are constant; see Figure 15.5.

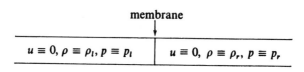

Figure 15.5

We are really looking therefore, at the initial-value problem for (15.2), with initial data

$$(u, \rho, p)(x, 0) = \begin{cases} (0, \rho_l, p_l), & x < 0, \\ (0, \rho_r, p_r), & x > 0. \end{cases}$$

(Note that we take u, ρ, and p as dependent variables; e and i are functions of ρ and p.) We shall consider this problem in detail in Chapters 17 and 18.

The solution can be depicted graphically as in Figure 15.6. We see that the initial discontinuity breaks up into two discontinuities, the shock wave and the contact discontinuity; these depend on the data. This is in sharp distinction from the case of linear equations in which discontinuities propagate along the characteristics. If we plot, e.g., ρ as a function of x, for a

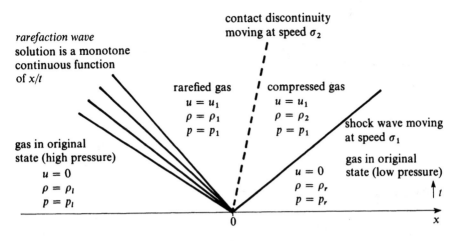

Figure 15.6

fixed $t > 0$, we find that ρ has the form given in Figure 15.7. Notice that in the "rarefaction-wave" region, the density decreases continuously from ρ_l to ρ_1. The contact discontinuity is due to the original discontinuity in the data; it is present in the linear approximations to the equations. The shock wave is the discontinuity due to the nonlinearities in the equations; it is analogous to the example which we have discussed for the scalar equations.

The reader will note that we have not yet said in what sense the above solution satisfies the original equations. This is a nontrivial problem since a discontinuous function cannot be differentiable. Fortunately, the equations are in "divergence" form; i.e., of the form (15.1), which can be written as $\text{div}[u, f(u)] = 0$. This allows us to extend the notion of "weak" solution (cf. Chapter 7, §A) to these nonlinear equations. This will be discussed in the next section.

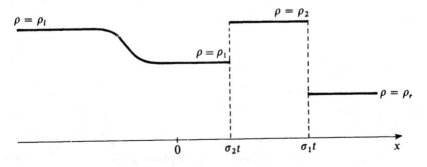

Figure 15.7

§B. Weak Solutions of Conservation Laws

We shall generalize the notion of solution for equations of the form (15.1).
Thus, consider the initial-value problem for (15.1) in $t > 0$:

$$u_t + f(u)_x = 0, \qquad u(x, 0) = u_0(x). \tag{15.9}$$

Let us suppose for the moment, that u is a classical solution of (15.9). Let C_0^1
be the class of C^1 functions ϕ which vanish outside of a compact subset in
$t \geq 0$, i.e., $(\text{spt } \phi) \cap (t \geq 0) \subseteq D$, where D is the rectangle $0 \leq t \leq T, a \leq x \leq b$,
so chosen that $\phi = 0$ outside of D, and on the lines $t = T, x = a$, and $x = b$;
see Figure 15.8. We multiply (15.9) by ϕ and integrate over $t > 0$, to get

$$\iint\limits_{t>0} (u_t + f_x)\phi \, dx \, dt = \iint\limits_{D} (u_t + f_x)\phi \, dx \, dt = \int_a^b \int_0^T (u_t + f_x)\phi \, dx \, dt = 0.$$

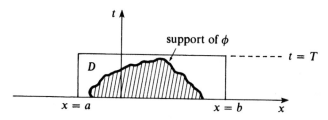

Figure 15.8

Now integrating by parts gives

$$\int_a^b \int_0^T u_t\phi = \int_a^b u\phi \Big|_{t=0}^{t=T} dx - \int_a^b \int_0^T u\phi_t \, dx \, dt$$

$$= \int_a^b - u_0(x)\phi(x, 0) \, dx - \int_a^b \int_0^T u\phi_t \, dx \, dt,$$

and

$$\int_0^T \int_a^b f_x\phi \, dx \, dt = \int_0^T f\phi \Big|_{x=a}^{x=b} dt - \int_0^T \int_a^b f\phi_x \, dx \, dt.$$

Thus we finally obtain

$$\iint\limits_{t\geq0} (u\phi_t + f(u)\phi_x) \, dx \, dt + \int_{t=0} u_0\phi \, dx = 0. \tag{15.10}$$

We have shown that if u is a classical solution of (15.9), then (15.10) holds for all $\phi \in C_0^1$. But (15.10) makes perfectly good sense if u and u_0 are merely bounded and measurable. We are thus led to the following definition of a solution of (15.9).

Definition 15.1. A bounded measurable function $u(x, t)$ is called a *weak solution* of the initial-value problem (15.9) with bounded and measurable initial data u_0, provided that (15.10) holds for all $\phi \in C_0^1$.

Henceforth, we shall drop the word "weak" when speaking of solutions, but we shall always mean solutions in the sense of Definition 15.1.

Note that if (15.10) holds for all $\phi \in C_0^1$, then if u happens to be C^1, u is a classical solution of (15.9). This is easy to see, since if ϕ has support in $t > 0$, then we may integrate (15.10) by parts and get for all such ϕ,

$$\iint_{t>0} (u_t + f_x)\phi = 0,$$

so that $u_t + f(u)_x = 0$. If now $\phi \in C_0^1$, we can multiply this last equation by ϕ and integrate by parts over D to obtain

$$\iint_{t>0} (u\phi_t + f(u)\phi_x)\, dx\, dt + \int_{t=0} u(x, 0)\phi(x, 0)\, dx = 0.$$

Comparing this with (15.10) gives

$$\int_{t=0} (u(x, 0) - u_0(x))\phi(x, 0)\, dx = 0,$$

and since u_0 is continuous, the arbitrariness of ϕ gives $u(x, 0) = u_0(x)$. Thus, we have shown that the concept of solution given by Definition 15.1 is a true generalization of the classical notion of solution.

We shall now show that not every discontinuity is permissible; in fact, the condition (15.10) places severe restrictions on the curves of discontinuity. To this end, let Γ be a smooth curve across which u has a jump discontinuity; i.e., u has well-defined limits on both sides of Γ, and u is smooth away from Γ. Let P be any point on Γ, and let D be a small ball centered at P. We assume that in D, Γ is given by $x = x(t)$. Let D_1 and D_2 be the components of D which are determined by Γ; see Figure 15.9. Let $\phi \in C_0^1(D)$; then from (15.10),

$$0 = \iint_D (u\phi_t + f\phi_x)\, dx\, dt = \iint_{D_1} (u\phi_t + f\phi_x)\, dx\, dt + \iint_{D_2} (u\phi_t + f\phi_x)\, dx\, dt.$$

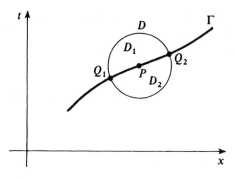

Figure 15.9

Now using the fact that u is C^1 in D_i, the divergence theorem gives

$$\iint_{D_i} (u\phi_t + f\phi_x)\, dx\, dt = \iint_{D_i} (u\phi)_t + (f\phi)_x = \int_{\partial D_i} \phi(-u\, dx + f\, dt).$$

Since $\phi = 0$ on ∂D, these line integrals are nonzero only along Γ. Thus, if $u_l = u(x(t) - 0, t)$, and $u_r = u(x(t)) + 0, t)$, then we have

$$\int_{\partial D_1} \phi(-u\, dx + f\, dt) = \int_{Q_1}^{Q_2} \phi(-u_l\, dx + f(u_l)\, dt),$$

$$\int_{\partial D_2} \phi(-u\, dx + f\, dt) = -\int_{Q_1}^{Q_2} \phi(-u_r\, dx + f(u_r)\, dt).$$

Therefore,

$$0 = \int_{\Gamma} \phi(-[u]\, dx + [f(u)]\, dt),$$

where $[u] = u_l - u_r$, the jump across Γ, and similarly, $[f(u)] = f(u_l) - f(u_r)$. Since ϕ was arbitrary, we conclude that

$$s[u] = [f(u)] \tag{15.11}$$

at each point on Γ, where $s = dx/dt$. We call s the *speed* of the discontinuity; it is the reciprocal of the slope. We have thus shown that (15.10) implies the relation (15.11), an equation which ties together the values of u on both sides of the curve of discontinuity with f and the speed of the discontinuity. Relation (15.11) is called the *jump condition*; in gas dynamics it is known as the *Rankine–Hugoniot condition*.

EXAMPLE 5. If we consider the Burgers equation (15.6), we obtain from (15.11)

$$s(u_l - u_r) = \tfrac{1}{2}(u_l^2 - u_r^2); \qquad (15.12)$$

whence, since $u_l \neq u_r$, $s = (u_l + u_r)/2$. Thus the speed of the discontinuity is the average of the limiting values on both sides of the discontinuity. We can use this relation to solve explicitly initial-value problems which are not classically solvable.

Thus consider the Burgers equation with data

$$u(x, 0) = \begin{cases} 1, & x < 0, \\ 1 - x, & 0 \le x \le 1, \\ 0, & x > 1. \end{cases}$$

Using the method of characteristics, we obtain the formula for u given by

$$u(x, t) = \begin{cases} 1, & x < t, \\ (1 - x)/(1 - t), & t \le x \le 1, \\ 0, & x > 1, \end{cases}$$

and we see that this is defined only for $t < 1$. For $t \ge 1$, we use (15.12) with $u_l = 1$, $u_r = 0$, and get $s = \tfrac{1}{2}$; thus if $t \ge 1$, we may define u by

$$u(x, t) = \begin{cases} 1, & x < 1 + \tfrac{1}{2}(t - 1), \\ 0, & x > 1 + \tfrac{1}{2}(t - 1). \end{cases}$$

We depict this solution in Figure 15.10.

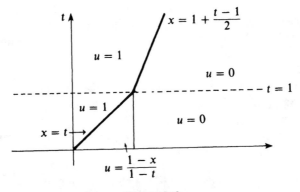

Figure 15.10

For a second example, consider the p-system given by equations (15.4). In this case (15.11) gives

$$s[v] = -[u], \qquad s[u] = [p(v)]. \tag{15.13}$$

Since not both $[u] = 0$ and $[v] = 0$, we can eliminate s from these equations to obtain

$$(u_l - u_r)^2 = (v_l - v_r)(p(v_r) - p(v_l)). \tag{15.14}$$

This shows that for fixed (v_l, u_l), the possible values (v_r, u_r) must lie along the curve given by (15.14). We shall study this curve in Chapter 17, §A.

In our effort to solve initial-value problems which were not solvable classically, we were led to extend the class of solutions. In doing this, we run the risk of admitting several solutions and thus of losing uniqueness. That this anxiety is well-founded follows from the next example.

EXAMPLE 6. Again consider the Burgers equation (15.6) with data

$$u_0(x) = \begin{cases} 0, & x < 0, \\ 1, & x > 0. \end{cases}$$

The method of characteristics determines the solution everywhere in $t > 0$ except in the sector $0 < x < t$. The following two functions differ precisely in this sector.

$$u_1(x, t) = \begin{cases} 0, & x < t/2, \\ 1, & x > t/2, \end{cases} \qquad u_2(x, t) = \begin{cases} 0, & x < 0, \\ x/t, & 0 < x < t, \\ 1, & x > t. \end{cases}$$

It is easy to check that these are both solutions of our problem. It is interesting to note that u_2 is a continuous function. The fact that a continuous solution can have discontinuous initial values is again a distinct feature of nonlinear equations; it is the "converse" of the appearance of discontinuities in solutions which were initially continuous.

A rather spectacular example of the loss of uniqueness is given by the following example.

EXAMPLE 7. Again consider the Burgers equation (15.6), but now with data

$$u_0(x) = \begin{cases} 1, & x < 0, \\ -1, & x > 0. \end{cases} \tag{15.15}$$

For each $\alpha \geq 1$, this problem has a solution u_α defined by (cf. Figure 15.11)

$$u_\alpha(x, t) = \begin{cases} +1, & 2x < (1 - \alpha)t, \\ -\alpha, & (1 - \alpha)t < 2x < 0, \\ \alpha, & 0 < 2x < (\alpha - 1)t, \\ -1, & (\alpha - 1)t < 2x. \end{cases} \qquad (15.16)$$

Thus the problem (15.6), (15.15) has a continuum of solutions!

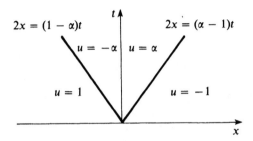

Figure 15.11

Now equations of the above form (or more generally, systems of con-servation laws), arise in the physical sciences and so we must have some mechanism to pick out the "physically relevant" solution. Thus, we are led to impose an a-priori condition on solutions which distinguishes the "correct" one from the others. In the case of a single equation $u_t + f(u)_x = 0$ ($n = 1$), where $f'' > 0$, we shall show below (in Chapter 16) that there is a unique solution which satisfies the "entropy" condition

$$\frac{u(x + a, t) - u(x, t)}{a} \leq \frac{E}{t}, \qquad a > 0, \quad t > 0, \qquad (15.17)$$

where E is independent of x, t, and a. This condition implies that if we fix $t > 0$, and we let x go from $-\infty$ to $+\infty$, then we can only jump down; i.e., in one direction across a discontinuity—hence the reason for the word "entropy." If we return to Example 7, then we see that (15.17) is satisfied only when $\alpha = 1$; i.e., u_1 is the distinguished solution; see (15.16).

The condition (15.17) is a bit strange when it is encountered for the first time. One may ask why should any solution satisfy that. A partial answer to this question can be given by the following argument, which, in a certain sense, shows that (15.17) is really quite natural. (A completely rigorous derivation of (15.17) will be given in the next chapter.)

Thus, consider the scalar equation $u_t + f(u)_x = 0$. In order to obtain (15.17), we note that we have already shown that for smooth solutions, $u_0' \geq 0$, and

$$u_x = \frac{u_0'}{1 + u_0' f'' t}.$$

Thus, if $u_0' = 0$, then $u_x = 0$, and if $u_0' > 0$, then

$$u_x \leq \frac{u_0'}{u_0' f'' t} = \frac{1}{f'' t} \leq \frac{E}{t},$$

where $E = 1/\mu$ and $\mu = \inf f''$; this is (15.17).

This loss of uniqueness has long been recognized in the context of gas dynamics. Thus, the Rankine–Hugoniot conditions admit "rarefaction" shocks; i.e., discontinuities across which the entropy decreases, and of course these solutions are physically unacceptable.

It is worth noting that as the above examples show, the loss of uniqueness occurs in the class of piecewise smooth functions. Uniqueness was lost not because we admitted as solutions functions which were unnecessarily "wild" from a regularity point of view; it is deeper than that. The "good" and "bad" shocks are indistinguishable from the point of view of regularity. Some condition of a qualitative or structural nature is needed to rule out the undesirable solutions.

If we consider (15.17) again, where $f'' > 0$, then as we have observed, the jump must always be downward as we increase in x; i.e., $u_l > u_r$. If we write the jump condition in the form

$$s = \frac{f(u_l) - f(u_r)}{u_l - u_r} = f'(\xi),$$

where $u_l > \xi > u_r$, then we obtain the *entropy inequality*

$$f'(u_l) > s > f'(u_r). \tag{15.18}$$

It states that the shock speed is intermediate to the characteristic speeds on both sides of the shock. It is this condition which we will later generalize to systems.

As we have just seen, our notion of solution forced us to radically alter our approach to the uniqueness question. We must expect other difficulties as well. This will be illustrated in the next two examples.

EXAMPLE 8. Observe that a smooth solution of $u_t + uu_x = 0$ also satisfies both of the following equations:

$$u_t + \left(\frac{u^2}{2}\right)_x = 0 \quad \text{and} \quad \left(\frac{u^2}{2}\right)_t + \left(\frac{u^3}{3}\right)_x = 0.$$

But this is not true for discontinuous solutions. To see this, note that the first one satisfies the jump relation $s(u_l - u_r) = (u_l^2 - u_r^2)/2$, or $s = (u_l + u_r)/2$. The second satisfies the equation $s = 2(u_l^2 + u_l u_r + u_r^2)/3(u_l + u_r)$. Since these are obviously different, the same is true of the class of discontinuous solutions.

Thus, one and the same differential equation can be written in two different divergence forms, each of which has a different set of (weak) solutions. The only possible reaction to this is to give up the idea of the differential equation being something basic, and to relegate it to a subsidiary role. The starting point of discussion must be a particular divergence expression; this choice will be decided upon by extraneous conditions. In the applications to physics, the choice will be unambiguous, since the particular set of equations will follow from integral conservation laws; e.g., the conservation of mass, energy, etc.

The next example shows that (weak) solutions are not preserved under smooth nonlinear transformations of the equations.

EXAMPLE 9. We consider the single equation $u_t + f(u)_x = 0$, where $f'' > 0$. The nonlinear transformation $u \to v = f'(u)$ maps smooth solutions of this equation into smooth solutions of the Burgers equation, $v_t + vv_x = 0$. But it does not map discontinuous solutions into themselves. This again follows from the jump condition. The original equation implies

$$s = \frac{f(u_l) - f(u_r)}{u_l - u_r},$$

while the transformed equation yields

$$s = \frac{v_l + v_r}{2} = \frac{f'(u_l) + f'(u_r)}{2}.$$

Since these two values are generally unequal, we see that if u satisfies the original equation, $v = f'(u)$ is generally *not* a solution of the Burgers equation.

Again this phenomenon has long been recognized in the context of gas dynamics. By elementary calculations, one can derive a new set of equations which has the same smooth solutions as does (15.2); see, e.g., equation (18.3)'.

However, one of these equations is $S_t = 0$, where S is the entropy. The jump relations would thus yield $S_l - S_r = 0$; i.e., the entropy would be preserved across a shock. However, it is well known that the entropy must *increase* across a shock, as follows from the original equations. This precise point was the flaw in Riemann's discussion of the shock-tube problem. It was not clarified until several years later.

§C. Evolutionary Systems

In studying time-dependent processes, we frequently consider the initial-value problem. We shall call a system of differential equations an *evolutionary system* if the initial-value problem is well-posed; i.e., if the solution in any compact subset of $t > 0$ depends continuously on its initial values in the L_∞-norm.

We begin by considering two classes of linear systems with constant coefficients

$$u_t = Au_x, \tag{15.19}$$

and

$$u_t = Au_x + Bu_{xx}, \tag{15.20}$$

where $u = u(x, t) \in \mathbf{R}^n$, $x \in \mathbf{R}$ and $t \geq 0$. Here A and B are constant $n \times n$ matrices. We are interested in studying the evolutionary character of these systems. To this end, we shall check the correctness of the initial-value problem in the class of plane wave solutions; i.e., solutions of the form

$$u(x, t) = \xi e^{i(\lambda t + \mu x)}, \tag{15.21}$$

where ξ is a constant n-vector. We shall assume that the initial values $u(x, 0) = \xi e^{i\mu x}$ are bounded; this forces μ to be real.

We shall first consider the system (15.19). Substituting (15.21) in (15.19) gives the equation $\mu A \xi = \lambda \xi$, so that λ/μ is an eigenvalue of A; say $\lambda/\mu = a + ib$. Then

$$u(x, t) = \xi e^{i\mu(at + x)} e^{-\mu bt},$$

and so $|u(x, t)| = |u(x, 0)| e^{-\mu bt}$. Thus if $b \neq 0$, we could choose $|\xi| = N^{-1}$ and $\mu = (-2N \log N)/b$, to get

$$\max |u(x, 0)| = \frac{1}{N}, \quad \text{while} \max \left| u\left(x, \frac{1}{N}\right) \right| = N.$$

Thus in order that the solution depend continuously on the data, we must take $b = 0$; this forces A to have real eigenvalues. In this case, we call the system (15.19) *weakly hyperbolic*. If in addition the eigenvalues of A are distinct, we say that (15.19) is a *hyperbolic* system.

Now let's turn our attention to (15.20). If we substitute (15.21) into (15.20), we find that

$$i\lambda\xi = i\mu A\xi - \mu^2 B\xi,$$

so that $i\lambda/\mu^2$ is an eigenvalue of $(i/\mu) A - B$. Since the eigenvalues of a matrix depend continuously upon the coefficients, we see that

$$-\frac{i\lambda}{\mu^2} = c + id + r(\mu) + is(\mu),$$

where $c + id$ is an eigenvalue of B, and $r^2 + s^2 \to 0$ as $|\mu| \to \infty$. This gives

$$|u(x, t)| = |\xi| e^{-[c + r(\mu)]\mu^2 t}.$$

Hence, if $c < 0$, we will again get magnification for large μ and the system will not be evolutionary. We thus assume that the eigenvalues of B have nonnegative real parts; if $c = 0$ then the correctness will be determined by the properties of A too, and not just those of B.

If we now consider the quasi-linear systems

$$u_t + f(u)_x = 0 \quad \text{and} \quad u_t + f(u)_x = (B(u)u_x)_x,$$

then we shall require that the matrix $df(u)$ has real and distinct eigenvalues for every u in the domain in question, and we shall call the first equation a *hyperbolic system*. Similarly, we shall require that $B(u)$ have eigenvalues with nonnegative real parts. In this case we shall call the second equation a *parabolic system*.

EXAMPLE 10. We consider the gas dynamics equations in Eulerian coordinates (15.3). The first two equations can easily be shown to be equivalent to

$$\rho_t + u\rho_x + \rho u_x = 0,$$

$$u_t + uu_x + \frac{1}{\rho} p_x = 0.$$

Using these equations in the third equation of (15.3) we obtain

$$s_t + us_x = 0.$$

(To do this, we use the thermodynamic relation $p = -e_v$, where $e = e(v, s)$ is the equation of state, and $\rho = v^{-1}$.) Then using $p_x = e_{vv}v^2\rho_x - e_{vs}s_x$, we can finally write our equations in the form $U_t + A(U)U_x = 0$, where $U = (\rho, u, s)$ and

$$A(U) = \begin{bmatrix} u & \rho & 0 \\ e_{vv}\rho^{-2} & u & -e_{vs} \\ 0 & 0 & u \end{bmatrix}$$

The characteristic polynomial for A is

$$\phi(\lambda) = (u - \lambda)[(u - \lambda)^2 - \rho^{-1}e_{vv}].$$

Now it is a thermodynamical requirement that $p_v < 0$. Accordingly, since $e_{vv} = -p_v > 0$, we see that ϕ has real and distinct roots, and thus the system is hyperbolic.

EXAMPLE 11. We consider the equations of gas dynamics for a viscous, heat conducting fluid in Eulerian coordinates. They take the form

$$\rho_t + (\rho u)_x = 0,$$

$$(\rho u)_t + (\rho u^2 + p)_x = \mu u_{xx}, \tag{15.22}$$

$$\left[\rho\left(\frac{u^2}{2} + e\right)\right]_t + \left[\rho u\left(\frac{u^2}{2} + i\right)\right]_x = \mu(uu_x)_x + \kappa T_{xx},$$

where T is the temperature. We are assuming that the *viscosity coefficient* μ, and the *thermal conductivity coefficient* κ are both positive constants. If we take as dependent variables ρ, u, and T, and note that both e and i are functions of ρ and T, then the matrix multiplying $(\rho, u, T)_{xx}$ is

$$B = \begin{bmatrix} 0 & 0 & 0 \\ 0 & \mu & 0 \\ 0 & \mu u & \kappa \end{bmatrix},$$

which has 0, μ and κ as eigenvalues. Thus the system (15.22) is parabolic according to our definition.

With these notions established in our minds, we shall now consider a different approach to the study of conservation laws; this approach being motivated by physical considerations. Thus, hyperbolic conservation laws often arise in models of physical processes which ignore the effects of various dissipative (and dispersive) mechanisms. In models at the next level of exactness, these mechanisms make their appearance felt by the presence of higher-order derivatives in the equations multiplied by small coefficients called *viscosity coefficients*, in analogy with gas dynamics. The consistency of the models would then demand that solutions of the two sets of equations be "close" in some sense, and in the limit, as the viscosity coefficients tend to zero, the solutions of the higher-order equations should converge to the solution of the system of first-order conservation laws. The protypical example is of course gas dynamics, where equations (15.3) are the hyperbolic system, and (15.22) is the higher-order parabolic system.

These ideas can be carried one step further; namely, we can try and obtain the existence of solutions for the hyperbolic equations as limits of solutions of "some" system of parabolic equations as the viscosity coefficients tend to zero. This technique is usually called the *viscosity method*. Presumably the parabolic system has smooth solutions[2] so that these should be easier to construct, but the more difficult problem is to obtain estimates independent of the viscosity coefficients, so as to allow the passage to the limit. There are some results on the existence of solutions of the parabolic system,[3] but the viscosity method has been shown to work only in the case of a single equation [O 1]; no such theorem has been proved for general systems at the date of this writing (Spring, 1982).[4]

But we can consider the viscosity method at a simple lower level, in what we may call a "first approximation." Thus, consider the nonlinear systems

$$u_t + f(u)_x = 0, \tag{15.23}$$

and

$$u_t + f(u)_x = \varepsilon B u_{xx}, \qquad \varepsilon > 0, \tag{15.24}$$

where in both cases, $u = u(x, t) \in \mathbf{R}^n$, and both systems are assumed to be evolutionary. Thus $df(u)$ has only real and distinct eigenvalues, and the *constant* matrix B has eigenvalues with positive real parts. We shall refer to B as a "*viscosity*" matrix. We want solutions of (15.23) to be obtainable as limits of solutions of (15.24) as $\varepsilon \to 0$. We shall find *necessary* conditions for this to be true.

Consider a small neighborhood N of a point p on a discontinuity $x = x(t)$ of (15.23). Let $s = x'(t_p)$, and note that if N is small, we can assume that as a first approximation, the solution of (15.23) is constant on rays of the form $x - st = \text{const.}$ It is thus natural to expect that solutions of (15.24) also

---

[2] However, if the initial data is not smooth, the solution may not be smooth; see [HS], where it is shown that for the equations of gas dynamics, initial discontinuities in the density persist for all positive time.

[3] See [NS 3].

[4] See however the interesting paper of DiPerna [6].

depend only on $x - st$ in N. Such solutions are called *travelling wave* solutions. Thus, if $\xi = x - st$ and $u = u(\xi)$ is to be a solution of (15.24), we substitute this in (15.24) to get the ordinary differential equation

$$-su' + f(u)' = \varepsilon Bu'' \qquad ' = \frac{d}{d\xi}, \qquad (15.25)$$

where the solutions of this equation depend on ε and B. Now we fix B and make the dependence on ε explicit by writing $u = u(\xi; \varepsilon)$. If we change variables in (15.25) by writing $\zeta = \xi/\varepsilon$, we get

$$-su' + f(u)' = Bu'', \qquad ' = \frac{d}{d\zeta}, \qquad (15.26)$$

and so $u(\xi; \varepsilon) = u(\xi/\varepsilon; 1) = u(\zeta; 1)$. Thus as ε changes, the solution curve remains invariant, and only the parametrization changes. Now we can integrate (15.26) once to obtain the equation

$$-su + f(u) + c = Bu', \qquad (15.27)$$

where c is a constant. But if we assume that solutions of this equation tend to the given discontinuity (u_l, u_r) (at p), we must have

$$\lim_{\zeta \to -\infty} u(\zeta; 1) = u_l \quad \text{and} \quad \lim_{\zeta \to +\infty} u(\zeta; 1) = u_r, \qquad (15.28)$$

since $|\zeta| \to \infty$ is equivalent to $\varepsilon \to 0$, and $\zeta < 0$ iff $x - st < 0$, while $\zeta > 0$ iff $x - st > 0$. Thus, from (15.27), we see that u_l and u_r must *both* be rest points[5] of (15.27). Since u_l is a rest point, $c = su_l - f(u_l)$ and (15.27) becomes

$$Bu' = -s(u - u_l) + f(u) - f(u_l). \qquad (15.29)$$

In view of the jump conditions (15.11), we see that the right-hand side of (15.29) also vanishes at u_r!

Thus we have both (15.28) and (15.29); our object is to see what this implies about the limiting solution. The interesting thing which we shall show is that not all solutions of (15.23) can be obtained this way; in other words, the "viscosity method" imposes an extra condition on solutions of (15.23). It is presumably a way of finding the "physically relevant" or unique solution. We shall examine this now in the context of a single equation; in Chapter 24, we shall study the general case of systems.

---

[5] If $x(t)$ is a solution of $\dot{x} = \phi(x)$, and $\lim_{t \to \infty} x(t) = x_0 \in \mathbf{R}^n$, then if $\phi(x_0) \neq 0$, the flow takes small neighborhoods U of x_0 into a neighborhood disjoint from U. If $x(t)$ ever got into U, then it must leave U in later time. Thus $\phi(x_0) = 0$.

Thus suppose $u(\zeta) \in \mathbf{R}$ is a solution of

$$u' = -s(u - u_l) + f(u) - f(u_l) \equiv \phi(u),$$

subject to the boundary conditions (15.28). We assume first that $u_l > u_r$. Since $\phi(u_l) = \phi(u_r) = 0$, the uniqueness theorem for ordinary differential equations shows that the graph of $u(\zeta)$ cannot cross the lines $u = u_l$ and $u = u_r$. Thus $u_l > u(\zeta) > u_r$ for all ζ. Uniqueness considerations again rule out $\phi(\bar{u}) = 0$ for any $\bar{u}$ between u_l and u_r, for in such a case, the line $u = \bar{u}$ would prevent $u(\zeta)$ from decreasing to u_r. Thus $\phi(u) < 0$ if $u_l > u > u_r$. It follows easily from this that $\phi'(u_l) \geq 0 \geq \phi'(u_r)$, and thus $f'(u_l) - s \geq 0 \geq f'(u_r) - s$. Hence, discontinuities which are limits of parabolic equations satisfy

$$f'(u_l) \geq s \geq f'(u_r). \tag{15.30}$$

(If $u_r > u_l$, the same conclusion would follow.) Notice that this is like the previously mentioned entropy inequality (15.18). Observe that (15.30) is slightly weaker than (15.18), but on the other hand we did not assume $f'' \neq 0$.

EXAMPLE 6 (revisited). We examine the solution u_1, and compute $f'(u_l) \equiv f'(0) < \frac{1}{2} = s < 1 = f'(1) \equiv f'(u_r)$. Thus (15.30) doesn't hold, and we conclude that u_1 is not the limit of travelling wave solutions of $u_t + (u^2/2)_x = \varepsilon u_{xx}$ as $\varepsilon \to 0$.

The situation hinted at by the above arguments is indeed valid, at least in the case of a single equation. Indeed, O. A. Oleinik [O 1] showed that for every bounded and measurable initial function, there is a unique smooth solution of the parabolic equation defined for all $t > 0$, and as $\varepsilon \to 0$, the solution converges to a solution of the scalar conservation law obeying the entropy condition (15.17). This program has not yet been carried out for systems.

§D. The Shock Inequalities

The purpose of this section is to derive an analogue of (15.18) for systems of conservation laws. In order to see how to do this, we begin with a scalar linear equation in a quarter plane

$$u_t + a u_x = 0, \qquad x > 0, \quad t > 0,$$

where a is a constant. We suppose that we are given the initial conditions $u(x, 0)$ in $x \geq 0$, and the boundary conditions $u(0, t)$ in $t \geq 0$. We ask to what extent do these values determine u in the full quarter plane? It is clear that u must be constant along the lines $x - at = $ const. Now if $a < 0$, u is determined along $x = 0$ by its initial values (see Figure 15.12(a)). Thus in this case, no boundary conditions can be given. On the other hand, if $a > 0$, we see from Figure 15.12(b), that the boundary conditions along $x = 0$ must be given in order to determine u in the entire quarter plane.

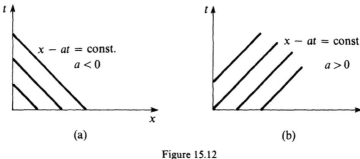

Figure 15.12

Next, suppose that we have a hyperbolic system in the quarter plane $x > 0, t > 0$:

$$u_t + Au_x = 0, \qquad u \in \mathbf{R}^n,$$

where A is a constant $n \times n$ matrix with eigenvalues $\lambda_1 < \cdots < \lambda_k < 0 < \lambda_{k+1} < \cdots < \lambda_n$. Let P be a nonsingular matrix such that $P^{-1}AP = \text{diag}(\lambda_1, \ldots, \lambda_n) \equiv \Lambda$. If $v = P^{-1}u$, then $v_t + \Lambda v_x = 0$, so that the system decouples into n scalar equations $v_t^i + \lambda_i v_x^i = 0$, $i = 1, 2, \ldots, n$. Thus if $i \leq k$, $\lambda_i < 0$, and $v^i(0, t)$ is determined by the initial data. If $i > k$, $\lambda_i > 0$ and we must specify $v^i(0, t)$, $i = k + 1, \ldots, n$. Now each v^i is a linear combination of the u_i's, so we see that we must specify $(n - k)$ conditions on the components of u on the boundary $x = 0$.

More generally, if we don't have a quarter plane problem, but instead we have a boundary which moves with speed s ($s = 0$ for the quarter plane problem) and if $\lambda_1 < \cdots < \lambda_k < s < \lambda_{k+1} < \cdots < \lambda_n$, then we must give $(n - k)$ boundary conditions, in order to specify the solution in the region $x - st > 0$, $t > 0$.

Now if instead of a boundary, we have a discontinuity of the hyperbolic system of conservation laws (15.1), these remarks can easily be extended. Thus let $\lambda_1(u) < \cdots < \lambda_n(u)$ be the eigenvalues of df, and let u_l and u_r, respectively, be the values of u on the left and right sides of the discontinuity which moves with speed s. Suppose that

$$\lambda_k(u_r) < s < \lambda_{k+1}(u_r).$$

Then from the above reasoning, we should specify $(n - k)$ conditions on the right boundary of the discontinuity. Similarly, looking at the left boundary, if

$$\lambda_j(u_l) < s < \lambda_{j+1}(u_l),$$

we must specify j conditions on the left boundary of the discontinuity. Now the jump conditions $s(u_l - u_r) = f(u_l) - f(u_r)$, (see (15.11)), are n equations connecting the values on both sides of the discontinuity with s. But since $u_l - u_r \neq 0$, we can eliminate s from these equations, to get $(n - 1)$ equations (or $(n - 1)$ conditions) between u_l and u_r. Thus we should require

$$(n - k) + j = n - 1 \quad \text{or} \quad j = k - 1.$$

In view of these considerations, we should admit a discontinuity $(u_l, u_r; s)$ provided that for some index k, $1 \leq k \leq n$, the following inequalities hold:

$$\lambda_k(u_r) < s < \lambda_{k+1}(u_r),$$
$$\lambda_{k-1}(u_l) < s < \lambda_k(u_l). \tag{15.31}$$

We shall call such a discontinuity a *k-shock wave*, or simply, a *k-shock*. The inequalities (15.31) will be called the *entropy inequalities*, or the *(Lax) shock conditions*.

Notice that if $n = 1$, then $\lambda(u) = f'(u)$ and the conditions (15.31) reduce to $f'(u_l) > s > f'(u_r)$, our previously obtained condition (15.18).

We can go one step further by rewriting (15.31) in the form

$$\lambda_k(u_r) < s < \lambda_k(u_l),$$
$$\lambda_{k-1}(u_l) < s < \lambda_{k+1}(u_r).$$

This shows that for one and only one index k is the shock speed s intermediate to the characteristic speeds λ_k on both sides of the shock. Actually (15.31) is a type of stability condition since it persists under small perturbations. In Chapter 17, we shall see that (15.31), when applied to the gas dynamics equations (15.2), is equivalent to the fact that the entropy S increases across the shock wave. The conditions (15.31) will also be useful in constructing a general theory of shock waves, as well as in obtaining shock waves as limits of travelling waves for systems. These will be discussed in later chapters.

§E. Irreversibility

Physical processes described by smooth solutions of hyperbolic equations are generally reversible in time: if we know the solution at one time, we can obtain the solution in the past as well as in the future. However, if the process is described by discontinuous solutions, then there is a high degree of irreversibility. This is not surprising if we believe that solutions of hyperbolic equations are obtainable as limits of solutions of parabolic equations, since it is known that for these latter equations, the "backwards" problem is not well-posed. Namely, the transformation $t \rightarrow -t$ does *not* leave the equations invariant.

Physically speaking, in the context of gas dynamics, this irreversibility has long been recognized; entropy increases across a shock, and hence information, the reciprocal of entropy, is lost. We cannot tell what initiated an explosion just from watching the particles blow up.

We shall illustrate this mathematically by means of a simple example. For $0 \le \varepsilon \le 1$, we define $u_\varepsilon(x, t)$ to be a solution of the equation

$$u_t + \left(\frac{u^2}{2}\right)_x = 0;$$

namely for $t \le \varepsilon$, we set

$$u_\varepsilon(x, t) = \begin{cases} 1, & x < t - \varepsilon/2, \\ \dfrac{x - \varepsilon/2}{t - \varepsilon}, & t - \varepsilon/2 < x < \varepsilon/2, \\ 0, & x > \varepsilon/2, \end{cases}$$

while for $t > \varepsilon$ we define

$$u_\varepsilon(x, t) = \begin{cases} 1, & x < t/2, \\ 0, & x > t/2. \end{cases}$$

The u_ε are depicted in Figure 15.13. It is easy to check that they are all solutions; moreover, each u_ε satisfies the entropy condition (15.18). Thus

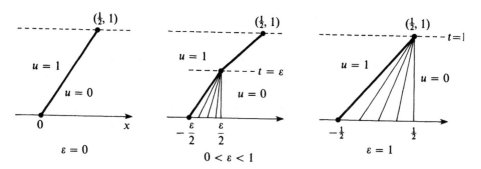

Figure 15.13

they are all "correct" solutions. The point that we wish to emphasize here
is that *all* of these solutions coincide at $t = 1$:

$$u_\varepsilon(x, 1) = \begin{cases} 1, & x < \frac{1}{2}, \\ 0, & x > \frac{1}{2}. \end{cases}$$

At $t = 1$, we know that a shock has formed. We don't know *when* it was
formed, nor can we even say *how* it was formed.

We emphasize again that all of these solutions are the "right" ones, in
that they belong to the class of solutions, which, as we shall prove in the next
chapter, are uniquely determined by their initial values. But this uniqueness
is only in the direction of increasing t. Two solutions in this class which
agree at $t = t_0$ must be equal for all $t > t_0$, but need not be equal for $t < t_0$.
It is in this strong sense that solutions of conservation laws are irreversible.

In the next chapter we shall give a more precise mathematical statement
of irreversibility. There we will show that the operator

$$T_t: L_\infty \to L_1^{\text{loc}}, \qquad t > 0,$$

defined by $u_0(\cdot) \to u(\cdot, t)$, where u is the unique solution having data u_0, is a
compact operator. Thus T_t^{-1} is not continuous: small changes in the solution
at $t_0 > 0$ can produce large changes for $t < t_0$. In other words, the backwards
problem is not well-posed.

NOTES

The rigorous mathematical theory of conservation laws dates from the 1950
paper of E. Hopf [Hf 2]. This was followed by a series of papers by O. Oleinik
(culminating in [O 1]), together with the paper of P. Lax [Lx 2]. The theory
of discontinuous solutions had to wait until the notions of weak solutions
and distributions became familiar. Of course, much important work had
been done prior to this by the fluid dynamicists beginning in the nineteenth
century. An excellent survey of the field up to 1948 is the classical book of
Courant and Friedrichs [CF]; see also [Nn 1].

During the Second World War, the invention of planes capable of exceed-
ing the speed of sound created a need for a better theoretical understanding
of shock phenomena and the gas dynamics equations. Many "pure" mathematicians become attracted to this field, among whom we mention, J. von
Neumann [Nn 1] and H. Weyl [Wy]. In the 1950s, nonlinear wave phenomena caught the attention of theoretical physicists; see, e.g., the classical paper
of Fermi, Pasta, and Ulam [FPU].

The problem of turbulence has always held an attraction for mathematicians. The equation (15.6) studied by Burgers in [Bu], was considered to be
a model for turbulence and was the one studied by Hopf in [Hf 2]. He considered the equation with viscosity, $u_t + u u_x = \theta u_{xx}$, and showed that the

transformation $u = -2\theta v_x/v$, takes this into the linear heat equation $v_t = \theta v_{xx}$; see also Cole, [Co]. This remarkable fact enabled Hopf to completely analyze the equation.

The shock tube problem was considered in the pioneering paper of Riemann [Ri]. The ideas in that paper laid the foundation for the theoretical work in gas dynamics. Riemann's solution of the problem contains a flaw; namely, he used the equation $S_t = 0$, where S denotes the entropy. As we have noted, this equation is formally derivable from the equations for the conservation of mass, momentum, and energy, but is not valid in the presence of shock waves. The notion of entropy was not fully understood in Riemann's time. The Rankine–Hugoniot conditions are classical, as is the concept of a rarefaction shock, i.e., a discontinuous solution across which the entropy *decreases*.

The viscosity method goes back to the earliest mathematical studies [Hf 2], [L]. A nice discussion is given in the important paper of Gelfand [Ge]. The equations of gas dynamics, e.g., (15.3), were considered by physicists as being models for the full equations (15.22) (with the dissipation mechanisms present), the reason being that almost nothing could be done (mathematically) with (15.22). There is reason to believe, however, that perhaps the equations (15.22) are really the "better" equations to consider, in view of the fact that one can (often) avoid discontinuous solutions.[2] The equations (15.22) have recently been considered with great success by the Japanese school, e.g., Itaya [It], Matsumura and Nishida [MN], following earlier work of Kanel' [K]; see also the recent paper of Nishida and Smoller [NS 3].

The problem (15.29), (15.28) has long been known to physicists [CF], and was brought to the attention of the mathematical community by Gelfand in [Ge]. For the gas dynamics equations, it was considered first by H. Weyl [Wy], and an elegant solution was given by Gilbarg in [Gi]. More general results were obtained by Conley–Smoller in [CS 1–5]. The analogous problem for the equations of magnetohydrodynamics; i.e., the equations for an ionized gas in a magnetic field, was studied by Germain [Gr], and Conley–Smoller in [CS 6,8]; see Chapter 24, §B.

The shock inequalities (15.31) are due to Lax [L x 2]; the corresponding statements for gas dynamics are classical [CF]. The notion of irreversibility was known for some time in the setting of gas dynamics; again see [CF].

A nice survey paper, which contains much of the mathematical results up to about 1962 is [Rz]; see also the book [RY].

Chapter 16

The Single Conservation Law

In this chapter we shall obtain precise mathematical results on the existence and uniqueness of solutions for a single conservation law. In addition we shall also study the asymptotic behavior of our constructed solution. The existence problem will be attacked via a finite-difference method. Thus we shall replace the given differential equation by a finite-difference approximation depending on mesh parameters Δx and Δt. For every such pair $(\Delta x, \Delta t)$ we shall construct a solution $U_{\Delta x, \Delta t}$ of the finite-difference equation, and we shall then obtain estimates which enable us to pass to the limit as the mesh parameters tend to zero in a certain definite way. The estimates which we obtain will be in the sup-norm and in the total variation-norm of the approximants, both sets of estimates being independent of the mesh parameters. It is worth noting that we are forced into obtaining bounds on the variation of the approximants, rather than (the usually encountered) bounds on derivatives, since the latter bounds would imply via the standard compactness criteria, that the limit would be *continuous*; we know that this is not generally true.

There are at least four other mathematically different approaches to the existence problem for a scalar conservation law. They are via: (i) calculus of variations and Hamilton–Jacobi theory; (ii) the viscosity method; (iii) non-linear semigroup theory;[1] and (iv) the method of characteristics. All of these methods are very interesting, and in particular, (iii) is quite elegant. We have decided to give a proof via finite differences for several reasons. Namely, first of all, the other methods require rather more sophisticated background than we have thus far presented. But more importantly, it seems to us that the finite difference methods are more capable of being generalized to the case of *systems* of conservation laws. Indeed, as of this writing (Spring, 1982), the only way of obtaining existence theorems for systems is via a finite-difference approach (albeit, a very nonstandard one; see Chapter 19). Finally, we believe that more mathematicians should familiarize themselves with finite-difference techniques, since it forms a powerful and elegant tool in nonlinear mathematics, in addition to being well-suited for computational purposes.

---

[1] In [Te 2], Temple has shown that this method cannot work for a broad class of systems in two dependent variables.

We shall show that our solution satisfies the entropy inequality (15.17), and furthermore, that it is the only solution with this property. Thus, the constructed solution is the "correct" one. The uniqueness proof is really a nonlinear version of Holmgren's technique (Chapter 5), which in principle is capable of generalization to systems (see §E, Chapter 20). The entropy inequality will also be used to study the asymptotic behavior of the solutions as t tends to infinity. It turns out that if the initial data has compact support, the solution decays in $L_1(\mathbf{R})$, at a specific rate, to a particular type of solution called an N-wave. We shall also study the decay of solutions having periodic data.

§A. Existence of an Entropy Solution

We consider the scalar conservation law in a single space variable

$$u_t + f(u)_x = 0, \qquad t > 0, \quad x \in \mathbf{R}, \tag{16.1}$$

with initial data

$$u(x, 0) = u_0(x), \qquad x \in \mathbf{R}. \tag{16.2}$$

We shall assume that $u_0 \in L_\infty(\mathbf{R})$, and that $f'' > 0$ on the (convex hull of the) range of u_0. We recall from the last chapter that by a solution of (16.1), (16.2) we mean a locally bounded, and measurable function $u(x, t)$, which satisfies

$$\iint_{t>0} (u\phi_t + f(u)\phi_x) \, dx \, dt + \int_{t=0} u_0 \phi \, dx = 0, \tag{16.3}$$

for every test function $\phi \in C_0^1$. Here is the main theorem of this section.

Theorem 16.1. Let $u_0 \in L_\infty(\mathbf{R})$, and let $f \in C^2(\mathbf{R})$ with $f'' > 0$ on $\{u : |u| \le \|u_0\|_\infty\}$. Then there exists a solution u of (16.1), (16.2) with the following properties:

(a) $|u(x, t)| \le \|u_0\|_\infty \equiv M, (x, t) \in \mathbf{R} \times \mathbf{R}_+$.
(b) There is a constant $E > 0$, depending only on M, $\mu = \min \{f''(u): |u| \le \|u_0\|_\infty\}$ and $A = \max\{|f'(u)| : |u| \le \|u_0\|_\infty\}$, such that for every $a > 0$, $t > 0$, and $x \in \mathbf{R}$,

$$\frac{u(x + a, t) - u(x, t)}{a} < \frac{E}{t}. \tag{16.4}$$

(c) u is stable and depends continuously on u_0 in the following sense: If u_0, $v_0 \in L_\infty(\mathbf{R}) \cap L_1(\mathbf{R})$ with $\|v_0\|_\infty \le \|u_0\|_\infty$, and v is the corresponding constructed solution of (16.1) with initial data v_0, then for every x_1 and

$x_2 \in \mathbf{R}$, with $x_1 < x_2$, and every $t > 0$,

$$\int_{x_1}^{x_2} |u(x, t) - v(x, t)| \, dx \le \int_{x_1 - At}^{x_2 + At} |u_0(x) - v_0(x)| \, dx. \qquad (16.5)$$

Before beginning the proof of this theorem, a few remarks are in order. First, note that (16.5) shows that our *constructed* solution is stable and unique, but leaves open the possibility of the existence of another solution not satisfying (16.5). However, as we have remarked earlier, the entropy condition (16.4) implies uniqueness, as we shall show in the next section. Next, we wish to point out that property (a) is not valid for systems; indeed, a major difficulty in the existence of solutions for systems is in obtaining sup-norm estimates. Next, we call attention to the fact that (16.4) can be viewed as a regularly theorem, in the sense that it implies that for any $t > 0$, the solution $u(\cdot, t)$ is of locally bounded total variation. To see this, let c_1 be a constant such that $c_1 > E/t$, and let $v(x, t) = u(x, t) - c_1 x$. Then, if $a > 0$, (16.4) implies that

$$v(x + a, t) - v(x, t) = u(x + a, t) - u(x, t) - c_1 a \le a\left(\frac{E}{t} - c_1\right) < 0,$$

so that v is a decreasing function, and thus v has locally bounded total variation. Since the same holds true for the function $c_1 x$, we see that our claim is valid. Thus, even though the data is merely in L_∞ on $t = 0$, the solution immediately becomes fairly regular in $t > 0$. So, for example, we can conclude that $u(\cdot, t)$ has most a countable number of jump discontinuities, and that it is differentiable almost everywhere, etc. This striking regularity property of the solution is, of course, a purely nonlinear phenomenon. On the other hand, if one accepts that solutions can be constructed via the viscosity method, then it is not too surprising that the solution is fairly regular since one knows that solutions of parabolic equations display a high degree of regularity.

Finally, we observe that (16.5) shows that the solution has a finite speed of propagation; this follows by setting $v_0 \equiv v \equiv 0$ in (16.5).

The proof of our theorem is not easy, and will follow from a series of lemmas. We begin by defining the difference scheme. Thus, let the upper-half plane $t \ge 0$ be covered by a grid $t = kh$, $x = nl$, $n = 0, \pm 1, \pm 2, \dots, k = 0, 1, 2, \dots$, where $h = \Delta t > 0$, $l = \Delta x > 0$.

We consider the following *difference approximation* to (16.1):

$$\frac{u_n^{k+1} - (u_{n+1}^k + u_{n-1}^k)/2}{h} + \frac{f(u_{n+1}^k) - f(u_{n-1}^k)}{2l} = 0, \qquad (16.6)$$

defined on our grid, where we are using the notation $u_\beta^\alpha = u(\beta l, \alpha h)$. We take a fixed representative of u_0, and define

$$u_n^0 = (1/l) \int_{(n-1)l/2}^{(n+1)l/2} u_0(x) \, dx.$$

Let $M = \|u_0\|_{L_\infty}$,

$$A = \max_{|u| \le M} |f'(u)|, \tag{16.7}$$

and

$$\mu = \min_{|u| \le M} f''(u). \tag{16.8}$$

Note that both $A > 0$ and $\mu > 0$. We choose the mesh parameters h and l so that the following stability condition[2] holds:

$$\frac{Ah}{l} \le 1. \tag{16.9}$$

The following lemma is the difference scheme analogue of (a).

Lemma 16.2. *For every $n \in \mathbf{Z}$, $k \in \mathbf{Z}_+$, $|u_n^k| \le M$.*

Proof. We write the difference equation (16.6) in the form

$$u_n^{k+1} = -\frac{h}{2l} [f(u_{n+1}^k) - f(u_{n-1}^k)] + \tfrac{1}{2}(u_{n+1}^k + u_{n-1}^k)$$

$$= -\frac{h}{2l} f'(\theta_n^k)(u_{n+1}^k - u_{n-1}^k) + \tfrac{1}{2}(u_{n+1}^k + u_{n-1}^k), \tag{16.10}$$

where θ_n^k is between u_{n+1}^k and u_{n-1}^k. Then

$$u_n^{k+1} = \left(\frac{1}{2} + \frac{h}{2l} f'(\theta_n^k)\right) u_{n-1}^k + \left(\frac{1}{2} - \frac{h}{2l} f'(\theta_n^k)\right) u_{n+1}^k.$$

If we inductively assume $|u_n^k| \le M$ for all $n \in \mathbf{Z}$, $k \in \mathbf{Z}_+$, then (16.9) shows that

$$\frac{1}{2} \pm \frac{h}{2l} f'(\theta_n^k) \ge 0.$$

Thus

$$|u_n^{k+1}| \leq M\left(\frac{1}{2} + \frac{h}{2l}f'(\theta_n^k)\right) + M\left(\frac{1}{2} - \frac{h}{2l}f'(\theta_n^k)\right) = M.$$

This proves the lemma. ☐

The next lemma is perhaps the most important one in this chapter. It is the discrete version of the entropy condition (16.4).

Lemma 16.3 (Entropy Condition). *If* $c = \min(\mu/2, A/4M)$, *then for* $n \in \mathbf{Z}$, $k \in \mathbf{Z}_+$,

$$\frac{u_n^k - u_{n-2}^k}{2l} \leq \frac{E}{kh}, \quad \text{where } E = \frac{1}{c}. \tag{16.11}$$

Proof. Let

$$z_n^k = \frac{u_n^k - u_{n-2}^k}{2l};$$

then from the difference scheme (16.6),

$$z_n^{k+1} = \tfrac{1}{2}(z_{n+1}^k + z_{n-1}^k) + 2f(u_{n-1}^k)\frac{h}{4l^2} - \frac{f(u_{n+1}^k) + f(u_{n-3}^k)}{4l^2}h$$

$$= \tfrac{1}{2}(z_{n+1}^k + z_{n-1}^k) - \frac{h}{4l^2}[f(u_{n-3}^k) - f(u_{n-1}^k)] - \frac{h}{4l^2}[f(u_{n+1}^k) - f(u_{n-1}^k)]$$

$$= \tfrac{1}{2}(z_{n+1}^k + z_{n-1}^k) - \frac{h}{4l^2}[f'(u_{n-1}^k)(u_{n-3}^k - u_{n-1}^k)$$

$$\qquad + \tfrac{1}{2}f''(\theta_1)(u_{n-1}^k - u_{n-3}^k)^2] - \frac{h}{4l^2}[f'(u_{n-1}^k)(u_{n+1}^k - u_{n-1}^k)$$

$$\qquad + \tfrac{1}{2}f''(\theta_2)(u_{n+1}^k - u_{n-1}^k)^2],$$

where θ_1 is between u_{n-3}^k and u_{n-1}^k, and θ_2 is between u_{n-1}^k and u_{n+1}^k. We thus have

$$z_n^{k+1} = \left[\frac{1}{2} + \frac{h}{2l}f'(u_{n-1}^k)\right]z_{n-1}^k + \left[\frac{1}{2} - \frac{h}{2l}f'(u_{n-1}^k)\right]z_{n+1}^k$$

$$\qquad - \frac{h}{2}[(z_{n-1}^k)^2 f''(\theta_1) + (z_{n+1}^k)^2 f''(\theta_2)].$$

Define

$$\tilde{z}_n^k = \max\{z_{n-1}^k, z_{n+1}^k, 0\};$$

then if $\tilde{z}_n^k = 0$ for all n, the estimate (16.11) surely holds since $z_n^k \leq \tilde{z}_{n-1}^k = 0 \leq E/kh$. We can thus assume that some $\tilde{z}_n^k \neq 0$. Suppose that $\tilde{z}_n^k = z_{n+1}^k$; the case where $\tilde{z}_n^k = z_{n-1}^k$ is treated similarly. Using (16.9), we have

$$z_n^{k+1} \leq \left[\frac{1}{2} + \frac{h}{2l} + f'(u_{n-1}^k)\right] z_{n-1}^k + \left[\frac{1}{2} - \frac{h}{2l} f'(u_{n+1}^k)\right] z_{n-1}^k$$
$$- ch[(z_{n+1}^k)^2 + (z_{n-1}^k)^2]$$
$$\leq \left[\frac{1}{2} + \frac{h}{2l} f'(u_{n-1}^k)\right] \tilde{z}_n^k + \left[\frac{1}{2} - \frac{h}{2l} f'(u_{n+1}^k)\right] \tilde{z}_n^k - ch(z_{n+1}^k)^2,$$

so that

$$z_n^{k+1} \leq \tilde{z}_n^k - ch(\tilde{z}_n^k)^2. \tag{16.12}$$

Now since $|u_n^k| \leq M$ (by Lemma 16.2), we have from (16.9), and the fact that $c \leq A/4M$,

$$z_n^k \leq |z_n^k| \leq \frac{M}{l} \leq \frac{M}{Ah} \leq \left(\frac{M}{h}\right)\left(\frac{1}{4Mc}\right) = \frac{1}{4ch};$$

thus

$$z_n^k \leq \frac{1}{4ch}. \tag{16.13}$$

Let

$$M^k = \max_n \{\tilde{z}_n^k\},$$

and notice that $M^k \geq 0$. Let $\phi(y) = y - chy^2$. Since $\phi' = 1 - 2chy$, ϕ is an increasing function if $y \leq 1/(2ch)$. But from (16.13),

$$\tilde{z}_n^k \leq M^k \leq \frac{1}{4ch} < \frac{1}{2ch}, \tag{16.14}$$

so that $\phi(\tilde{z}_n^k) \leq \phi(M^k)$, and this gives

$$\tilde{z}_n^k - ch(\tilde{z}_n^k)^2 \leq M^k - ch(M^k)^2.$$

Thus from (16.12), $z_n^{k+1} \leq M^k - ch(M^k)^2$ for all $n \in \mathbf{Z}$. It follows that

$$M^{k+1} \leq M^k - ch(M^k)^2. \tag{16.15}$$

We shall show that this implies

$$M^k \le \frac{1}{ckh + 1/M^0}.$$ \hfill (16.16)

Assuming this, we have

$$z_n^k \le M^k \le \frac{1}{ckh + 1/M^0} \le \frac{1}{ckh} = \frac{E}{kh},$$

and this is (16.11). In order to prove (16.16), we proceed by induction. The case $k = 0$ is trivial. Thus assume that (16.16) holds; we shall show that it holds when k is replaced by $k + 1$. Now from (16.16),

$$\frac{1}{M^k} \ge ckh + \frac{1}{M^0}$$

so $1 - chM^k \ge 1 - ckhM^k \ge 1/M^0 \ge 0$, and thus $1 - (chM^k)^2 \ge 0$. Now from (16.15), $M^{k+1} \le M^k(1 - chM^k)$ so that

$$\frac{M^{k+1}}{1 - chM^k} \le M^k \le \frac{M^k}{1 - (chM^k)^2},$$

and thus

$$M^{k+1} \le \frac{M^k}{1 + chM^k}$$

$$= \frac{1}{ch + 1/M^k}$$

$$\le \frac{1}{ch + ckh + 1/M^0}$$

$$= \frac{1}{c(k + 1)h + 1/M^0}.$$

Thus (16.16) holds for all k, and the proof is complete. $\quad\square$

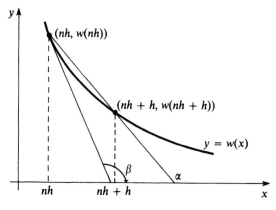

Figure 16.1

The next lemma shows that the variation of the difference approximants is locally bounded whenever $kh > 0$; i.e., in $t > 0$.

Lemma 16.4 (Space Estimate). *For any $X > 0$ and $kh \geq \alpha > 0$, there is a constant c depending on M, X and α, but independent of h and l, such that*

$$\sum_{|n| \leq X/l} \left| u_{n+2}^k - u_n^k \right| \leq c. \tag{16.17}$$

(Here, and in what follows, the summation is over all $n \in \mathbf{Z}$ for which $|n| \leq X/l$.)

Proof. Set $v_n^k = u_n^k - c_1 nl$, where c_1 is chosen so large that $E/\alpha < c_1$. Then using (16.11),

$$v_{n+2}^k - v_n^k = u_{n+2}^k - u_n^k - 2c_1 l$$

$$\leq \frac{2lE}{kh} - 2c_1 l$$

$$\leq 2l\left(\frac{E}{\alpha} - c_1\right) < 0.$$

Thus

$$\sum_{|n| \leq X/l} \left| u_{n+2}^k - u_n^k \right| \leq \sum_{|n| \leq X/l} \left| v_{n+2}^k - v_n^k \right| + \sum_{|n| \leq X/l} 2c_1 l$$

$$= - \sum_{|n| \le X/l} (v_{n+2}^k - v_n^k) + 2c_1 l \left(\frac{2X + l}{l} \right)$$

$$\le 2 \max_{|n| \le X/l} |v_n^k| + c_2 X$$

$$\le 2M + 2c_1 X + c_2 X.$$

This completes the proof. $\square$

In order to show that the difference approximants converge to a solution of (16.1), (16.2) in the upper-half plane $t > 0$ (and not merely on each line $t = \text{const.} > 0$), it is necessary to estimate the differences of each approximating solution on different time levels in some suitable norm. The following estimate does just this; namely it shows that the difference approximants are L_1 (locally) Lipschitz continuous in t. Before stating the result, note that from (16.6) the values of u_n^k where $n - k$ is even are computed independently from those where $n - k$ is odd.

Lemma 16.5 (Time Estimate). *If $h/l \ge \delta > 0$, and $l, h \le 1$, then there exists an $L > 0$, independent of h and l such that if $k > p$, where $k - p$ is even and $ph \ge \alpha > 0$,*

$$\sum_{|n| \le X/l} |u_n^k - u_n^p| l \le L(k - p)h. \tag{16.18}$$

A similar estimate holds if $k - p$ is odd.

Proof. We shall express u_n^k in terms of u_n^p, $k - p$ even. To do this we consider the difference scheme (16.6), which gives (cf. (16.10)),

$$u_n^k = -\frac{h}{2l} f'(\theta_n^{k-1})(u_{n+1}^{k-1} - u_{n-1}^{k-1}) + \tfrac{1}{2}(u_{n+1}^{k-1} + u_{n-1}^{k-1})$$

$$= \left(\frac{1}{2} + f'(\theta_n^{k-1}) \frac{h}{2l} \right) u_{n-1}^{k-1} + \left(\frac{1}{2} - f'(\theta_n^{k-1}) \frac{h}{2l} \right) u_{n+1}^{k-1}, \tag{16.19}$$

or

$$u_n^k = a_{n,n-1}^{k,k-1} u_{n-1}^{k-1} + a_{n,n+1}^{k,k-1} u_{n+1}^{k-1},$$

where $a_{n,n-1}^{k,k-1} + a_{n,n+1}^{k,k-1} = 1$, and $a_{n,n-1}^{k,k-1}, a_{n,n+1}^{k,k-1} \ge 0$.

Applying this to u_{n-1}^k and u_{n+1}^k gives a formula of the form

$$u_n^{k+1} = A u_{n-2}^{k-1} + B u_n^{k-1} + C u_{n+2}^{k-1},$$

where $A + B + C = 1$, and $A, B, C \geq 0$. Hence

$$\left|u_n^{k+1} - u_n^{k-1}\right| \leq A\left|u_{n-2}^{k-1} - u_n^{k-1}\right| + C\left|u_{n+2}^{k-1} - u_n^{k-1}\right|.$$

If we multiply this by Δx and sum we get

$$\sum_{|n| \leq X/l} \left|u_n^{k+1} - u_n^{k-1}\right| \Delta x \leq c \, \Delta x,$$

where c is the constant in (16.17). Now if $k - p$ is even, the triangle inequality gives

$$\sum_{|n| \leq X/l} \left|u_n^k - u_n^p\right| \Delta x \leq \sum_{i=p}^{k-2} \sum_{|n| \leq X/l} \left|u_n^{i+2} - u_n^i\right| \Delta x$$

$$\leq (k - p)c \, \Delta x$$

$$\leq \frac{c}{\delta}(k - p) \, \Delta t.$$

This proves the result, with $L = c/\delta$. $\quad\square$

We remark that the condition $h/l \geq \delta > 0$ means that the solutions of the difference equations have a finite speed of propagation, i.e., u_n^k is determined by the values of u_n^0 for a bounded set of n's. Such a condition is *not* true for difference approximations to parabolic equations.

We next prove the discrete analogue of (16.5).

Lemma 16.6 (Stability). *Let $\{u_n^k\}$ and $\{v_n^k\}$ be solutions of the finite-difference scheme* (16.6), *corresponding to the initial values $\{u_n^0\}$, and $\{v_n^0\}$, respectively, where $\sup_n |u_n^0| \leq M$ and $\sup_n |v_n^0| \leq M$. Then if $k > 0$,*

$$\sum_{|n| \leq N} \left|u_n^k - v_n^k\right| \cdot l \leq \sum_{|n| \leq N+k} \left|u_n^0 - v_n^0\right| \cdot l. \qquad (16.20)$$

Proof. Let $w_n^k = u_n^k - v_n^k$; then from (16.6), we have

$$w_n^{k+1} = u_n^{k+1} - v_n^{k+1} = \frac{u_{n+1}^k + u_{n-1}^k}{2} - \frac{h}{2l}\left[f(u_{n+1}^k) - f(u_{n-1}^k)\right]$$

$$- \frac{v_{n+1}^k + v_{n-1}^k}{2} + \frac{h}{2l}\left[f(v_{n+1}^k) - f(v_{n-1}^k)\right]$$

$$= \frac{u_{n+1}^k - v_{n+1}^k}{2} - \frac{h}{2l}\left[f(u_{n+1}^k) - f(v_{n+1}^k)\right]$$

$$+ \frac{u_{n-1}^k - v_{n-1}^k}{2} + \frac{h}{2l}\left[f(u_{n-1}^k) - f(v_{n-1}^k)\right]$$

$$= \tfrac{1}{2}w_{n+1}^k - \frac{h}{2l}w_{n+1}^k f'(\theta_{n+1}^k) + \tfrac{1}{2}w_{n-1}^k$$

$$+ \frac{h}{2l}w_{n-1}^k f'(\theta_{n-1}^k),$$

where $\theta^k_{n\pm1}$ is between $u^k_{n\pm1}$ and $v^k_{n\pm1}$. Thus

$$w^{k+1}_n = \left[\frac{1}{2} - \frac{h}{2l}f'(\theta^k_{n+1})\right]w^k_{n+1} + \left[\frac{1}{2} + \frac{h}{2l}f'(\theta^k_{n-1})\right]w^k_{n-1}.$$

Since (16.9) holds, the coefficients of w^k_{n+1} and w^k_{n-1} are both nonnegative; using this we have

$$\sum_{|n|\leq N}|w^{k+1}_n| \leq \sum_{|n|\leq N}\left[\frac{1}{2} - \frac{h}{2l}f'(\theta^k_{n+1})\right]|w^k_{n+1}|$$

$$+ \sum_{|n|\leq N}\left[\frac{1}{2} + \frac{h}{2l}f'(\theta^k_{n-1})\right]|w^k_{n-1}|$$

$$= \sum_{m=1-N}^{1+N}\left[\frac{1}{2} - \frac{h}{2l}f'(\theta^k_m)\right]|w^k_m| + \sum_{m=-1-N}^{N-1}\left[\frac{1}{2} + \frac{h}{2l}f'(\theta^k_m)\right]|w^k_m|$$

$$\leq \sum_{|m|\leq N+1}\left[\frac{1}{2} - \frac{h}{2l}f'(\theta^k_m)\right]|w^k_m| + \sum_{|m|\leq N+1}\left[\frac{1}{2} + \frac{h}{2l}f'(\theta^k_m)\right]|w^k_m|$$

$$= \sum_{|m|\leq N+1}|w^k_m|.$$

It follows from this by induction, that the lemma holds. □

We are now prepared to investigate the convergence of the difference approximations. Rather than consider the difference approximations to be defined only on the mesh points, we wish to consider them as functions defined in the upper-half plane, $t \geq 0$. To accomplish this, we construct a family of functions $\{U_{h,l}\}$ from the $\{u^k_n\}$ by defining

$$U_{h,l}(x, t) = u^k_n, \quad \text{if } nl \leq x < (n+2)l, \quad kh \leq t < (k+1)h. \quad (16.21)$$

Thus the value of $U_{h,l}$ in the rectangle $nl \leq x < (n+1)l$, $kh \leq t < (k+1)h$, is the value of the difference approximation at the point (nl, kh); see Figure 16.2.

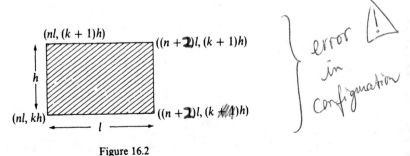

$(nl, (k+1)h)$ $((n+2)l, (k+1)h)$

h

(nl, kh) $((n+2)l, (k+4)h)$

l

error in configuration

Figure 16.2

We now show that the set of functions $\{U_{h,l}\}$ is compact in the topology of L_1 convergence on compacta.

Lemma 16.7. *There exists a sequence $\{U_{h_i,l_i}\} \subset \{U_{h,l}\}$ which converges to a measurable function $u(x\,t)$ in the sense that for any $X > 0$, $t > 0$, and $T > 0$, both*

$$\int_{|x| \leq X} |U_{h_i,l_i}(x, t) - u(x, t)|\, dx \to 0, \tag{16.22}$$

and

$$\int_{0 \leq t \leq T} \int_{|x| \leq X} |U_{h_i,l_i}(x.\ t) - u(x, t)|\, dx\, dt \to 0, \tag{16.23}$$

as $i \to \infty$ (i.e., $(h_i, l_i) \to (0, 0)$). Furthermore, the limit function satisfies $\sup_{\substack{x \in \mathbb{R} \\ t > 0}} |u(x, t)| \leq M$, and the stability inequality (16.5).

Proof. From Lemmas 16.2 and 16.4, the set of functions $\{U_{h,l}\}$, considered as functions of x, are uniformly bounded, and have uniformly bounded total variation on each bounded interval on any line $t = \text{const.} > 0$ (the uniformity being of course, with respect to h and l). By Helley's theorem ([Nt]), we can find a subsequence $\{U'_{h,l}\}$, which converges at each point on any bounded interval of this line, and by a standard diagonal process, we can construct a subsequence $\{U''_{h,l}\}$ from $\{U'_{h,l}\}$ which converges at *every* point of this line $t = \text{const.} > 0$, as h and l tend to zero.

Let $\{t_m\}$ be a countable and dense subset of the interval $(0, T)$. By a further diagonal process, we can select a subsequence $\{U_{h_i,l_i}\}$ from $\{U''_{h,l}\}$, which converges at every point on each line $t = t_m$, $m = 1, 2, \ldots$ as $i \to \infty$; i.e., as h_i and l_i tend to zero.

We set $U_i = U_{h_i,l_i}$; we shall show that this sequence of functions converges at each $t \in (0, T]$, so that in the limit we indeed obtain a function $u(x, t)$ which is defined in the strip $0 < t \leq T$. To do this, we first show that for every $t \in (0, T]$,

$$I_{i,j}(t) \equiv \int_{-X}^{X} |U_i(x, t) - U_j(x, t)|\, dx \to 0 \tag{16.24}$$

as $i, j \to \infty$; i.e., that $\{U_i(\,\cdot\,, t)\}$ is a Cauchy sequence in $L_1(|x| \leq X)$.

For $t \in (0, T]$, we find a subsequence $\{t_{m_s}\} \subset \{t_m\}$ such that $t_{m_s} \to t$ as $s \to \infty$; let $\tau_s = t_{m_s}$. Then

$$I_{ij}(t) \leq \int_{-X}^{X} |U_i(x, t) - U_i(x, \tau_s)| \, dx + \int_{-X}^{X} |U_i(x, \tau_s) - U_j(x, \tau_s)| \, dx$$

$$+ \int_{-X}^{X} |U_j(x, \tau_s) - U_j(x, t)| \, dx \equiv I_1 + I_2 + I_3. \tag{16.25}$$

In view of the choice of $\{t_m\}$, and the Lebesgue bounded convergence theorem, we see that $I_2 \to 0$ as $i, j \to \infty$. Next, if we use the notation $[\sigma]$ to denote the greatest integer in σ, then by definition of U_i, $U_i(x, t) = U_i(x, [t/h_i]h_i)$. Thus

$$I_1 = \int_{-X}^{X} \left| U_i\left(x, \left[\frac{t}{h_i}\right]h_i\right) - U_i\left(x, \left[\frac{\tau_s}{h_i}\right]h_i\right) \right| dx$$

$$\leq \sum_{|n| \leq X/l_i + 1} \int_{nl_i}^{(n+1)l_i} \left| U_i\left(x, \left[\frac{t}{h_i}\right]h_i\right) - U_i\left(x, \left[\frac{\tau_s}{h_i}\right]h_i\right) \right| dx$$

$$= \sum_{|n| \leq X/l_i + 1} \left| u_n^{[t/h_i]} - u_n^{[\tau_s/h_i]} \right| l_i$$

$$\leq L \left| \left[\frac{t}{h_i}\right] - \left[\frac{\tau_s}{h_i}\right] \right| h_i,$$

where we have used Lemma 16.6. Thus $I_1 \leq L[|t - \tau_s| + h_i]$, and similarly, $I_3 \leq L[|t - \tau_s| + h_i]$, so that

$$I_1 + I_3 \leq 2L[|t - \tau_s| + h_i]. \tag{16.26}$$

If now $\varepsilon > 0$ is given, we choose τ_s so that $4L|t - \tau_s| < \varepsilon$. For this fixed s, we choose i and j so large that $2I_2 < \varepsilon$. Then for these i and j, we have from (16.25) and (16.26), that $I_{ij}(t) < \varepsilon$; this proves (16.24).

It follows that the sequence $\{U_i(x, t)\}$ has a limit $u(x, t)$, for each fixed $t, 0 < t \leq T$. We now show that $I_{ij}(t) \to 0$ as $i, j \to \infty$, uniformly in t, $0 < \tau \leq t \leq T$. To see this, let $\varepsilon > 0$ be given, and choose a finite subset $\mathscr{F} \subset \{t_m\}$ with the property that if $0 \leq t \leq T$, there is a t_m in $\mathscr{F}$ such that $2L|t - t_m| < \varepsilon/2$. If we choose i and j so large that $2I_2 < \varepsilon$ for all $t_m \in \mathscr{F}$, then for these i, j, we have $I_{ij}(t) < \varepsilon$; this gives us the desired uniformity in t.

Using the uniform convergence, we have that for any $\tau, 0 < \tau \leq T$,

$$\int_{\tau}^{T} I_{ij}(t) \, dt \to 0 \tag{16.27}$$

as $i, j \to \infty$. Now write

$$\int_0^T \int_{-X}^X |U_i(x, t) - U_j(x, t)| \, dx \, dt = \int_0^\tau \int_{-X}^X |U_i(x, t) - U_j(x, t)| \, dx \, dt$$
$$+ \int_\tau^T \int_{-X}^X |U_i(x, t) - U_j(x, t)| \, dx \, dt.$$

$$(16.28)$$

If we choose τ, $0 < \tau \le T$, so small that $8MX\tau < \varepsilon$, then for this fixed τ, we take i, j so large that the second integral on the right side of (16.28) is at most $\varepsilon/2$ (this can be done in view of (16.27)). Then for these i and j, we have

$$\int_0^T I_{ij}(t) \, dt < \varepsilon.$$

It follows that $u(x, t)$ is measurable, and that (16.23) holds. Moreover, since local convergence in L_1 implies pointwise convergence almost everywhere of a subsequence, we see that since each $|U_i| \le M$, the limit function satisfies the same inequality, where, if necessary, we redefine u on a set of measure zero. Finally (16.5) holds in view of Lemma 16.6. This completes the proof. $\square$

We shall next show that the limit function $u(x, t)$ satisfies the entropy inequality (16.4); this, of course, will follow from Lemma 16.3, which is the "discrete" version of (16.4).

Lemma 16.8. *The function $u(x, t)$ constructed in Lemma 16.7 satisfies an entropy inequality of the form* (16.4).

Proof. It suffices to show that if $(x_1 - x_2) > 2l_i$ and $t > h_i$, that

$$\frac{U_i(x_1, t) - U_i(x_2, t)}{x_1 - x_2} < \frac{2E}{t - h_i},$$

$$(16.29)$$

where the U_i's are defined as in the proof of Lemma 16.7, and the constant E is defined in Lemma 16.3. This is so because if (16.29) holds, we get our desired result by passing to the limit as $i \to \infty$.

Thus, let $x_1 > x_2$, and note that

$$U_i(x_j, t) = U_i\left(x_j - \eta_j, \left[\frac{t}{h_i}\right]h_i\right), \qquad j = 1, 2,$$

for some η_j, where $0 \le \eta_j < l_i$. Thus

$$\frac{U_i(x_1, t) - U_i(x_2, t)}{x_1 - x_2} = \frac{1}{(x_1 - x_2)} \sum (u_n^k - u_{n-2}^k),$$

where $k = [t/h_i]$ and we sum over all integers n lying in the interval $[x_2 - \eta_2, x_1 - \eta_1]$. Using Lemma 16.3, we have

$$
\begin{aligned}
\frac{U_i(x_1, t) - U_i(x_2, t)}{x_1 - x_2} &\leq \frac{E(x_1 - \eta_1 - x_2 + \eta_2)}{[t/h_i]h_i(x_1 - x_2)} \\
&\leq \frac{E(x_1 - \eta_1 - x_2 + \eta_2)}{(t - h_i)(x_1 - x_2)} \\
&= \frac{E}{t - h_i} + \frac{(\eta_2 - \eta_1)E}{(t - h_i)(x_1 - x_2)} \\
&\leq \frac{2E}{t - h_i},
\end{aligned}
$$

since $\eta_2 - \eta_1 \leq l_i \leq 1$, for i large. This completes the proof. $\square$

An alternate proof can be given based on the fact that in the proof of Lemma 16.4 we showed that $v_n^k = u_n^k - c_1 nl$ is decreasing in n for each fixed k; the same is true of the limit; namely $u(x, t) - c_1 x$.

We shall now show that our function u is indeed a solution of (16.1), (16.2). The proof is via an interesting "summation by parts" technique.

Lemma 16.9. *Let $l_i \to 0$ as $i \to \infty$, and suppose that for $\phi \in C_0^3$*

$$
\lim_{i \to \infty} \int_{-\infty}^{\infty} [U_i(x, 0) - u_0(x)]\phi(x, 0)\, dx = 0. \tag{16.30}
$$

Then u satisfies (16.3); i.e., u is a solution of (16.1), (16.2).

Remark. Since ϕ is bounded, (16.30) holds if for every $X > 0$,

$$
\lim_{i \to \infty} \int_{-X}^{X} |U_i(x, 0) - u_0(x)|\, dx = 0. \tag{16.31}
$$

However, since u_0 is bounded and measurable, there exist step functions $U_i(x, 0)$, constant on intervals $nl \leq x < (n + 1)l, n \in \mathbf{Z}$, which converge locally in L_1 to u_0; i.e., which satisfy (16.31). We take these functions to define the initial values of our difference scheme (16.6).

Now the solutions u_n^k of our difference equation (16.6) can be written in the form

$$
\frac{u_n^{k+1} - u_n^k}{h} - \frac{u_{n+1}^k - 2u_n^k + u_{n-1}^k}{2l^2} \cdot \frac{l^2}{h} + \frac{f(u_{n+1}^k) - f(u_{n-1}^k)}{2l} = 0, \tag{16.32}
$$

as one can easily verify.

We multiply (16.32) by $\phi_n^k = \phi(nl, kh)$, and get

$$\frac{\phi_n^{k+1} u_n^{k+1} - \phi_n^k u_n^k}{h} - u_n^{k+1} \frac{\phi_n^{k+1} - \phi_n^k}{h} + \frac{l^2}{2h} \frac{2\phi_n^k - \phi_{n+1}^k - \phi_{n-1}^k}{l^2} u_n^k$$

$$+ \frac{\phi_{n+1}^k u_n^k - \phi_n^k u_{n-1}^k}{2h} + \frac{\phi_{n-1}^k u_n^k - \phi_n^k u_{n+1}^k}{2h}$$

$$+ \frac{\phi_{n+1}^k f(u_{n+1}^k) - \phi_{n-1}^k f(u_{n-1}^k)}{2l} - f(u_{n+1}^k) \frac{\phi_{n+1}^k - \phi_n^k}{2l}$$

$$- f(u_{n-1}^k) \frac{\phi_n^k - \phi_{n-1}^k}{2l} = 0. \tag{16.33}$$

Since ϕ has compact support, we may assume that $\phi_n^k = 0$, if $k \geq [T/h]$. Then we multiply (16.33) by hl and sum over all $n \in \mathbf{Z}$, $k \in \mathbf{Z}_+$. Since the first, fourth, fifth, and sixth terms are "telescoping," they cancel, except for the first term with $k = 0$. Thus we get

$$-l \sum_n u_n^0 \phi_n^0 + hl \left\{ \sum_{k,n} \left[-u_n^{k+1} \frac{\phi_n^{k+1} - \phi_n^k}{h} - \frac{l^2}{2h} \frac{\phi_{n+1}^k - 2\phi_n^k + \phi_{n-1}^k}{l^2} u_n^k \right] \right.$$

$$\left. - \sum_{k,n} f(u_{n+1}^k) \frac{\phi_{n+1}^k - \phi_n^k}{2l} - \sum_{k,n} f(u_{n-1}^k) \frac{\phi_n^k - \phi_{n-1}^k}{2l} \right\} = 0.$$

Since U_{hl} is piecewise constant, ϕ is smooth, and the integrals are limits of step functions, we can write

$$- \int_{t=0} U_{h,l} \phi + \delta_1 - \iint_{t \geq 0} U_{h,l} \phi_t + \delta_2 - \frac{l^2}{2h} \iint_{t \geq 0} U_{h,l} \phi_{xx}$$

$$+ \delta_3 - \iint_{t \geq 0} f(U_{h,l}) \phi_x + \delta_4 = 0,$$

where the $\delta_i \to 0$ uniformly, as $h, l \to 0$. We replace $U_{h,l}$ by U_i to get

$$\iint_{t \geq 0} U_i \phi_t + f(U_i)\phi_x + \frac{l_i^2}{2h_i} \iint_{t \geq 0} U_i \phi_{xx}$$

$$+ \int_{t=0} U_i \phi = \delta(l_i, h_i), \tag{16.34}$$

where $\delta(l_i, h_i) \to 0$ as $i \to \infty$. Now let $i \to \infty$; since $U_i \to u$ locally in L_1 (Lemma 16.8), and $l_i \to 0$, $l_i^2/h_i \to 0$ (since l_i/h_i is bounded), we see that

$$\iint_{t \geq 0} U_i \phi_t + \frac{l_i^2}{2h_i} \iint_{t \geq 0} U_i \phi_{xx} \to \iint_{t \geq 0} u \phi_t. \tag{16.35}$$

Next, by choice of the initial values,

$$\int_{t=0} U_i \phi \rightarrow \int_{t=0} u_0 \phi \quad \text{as } i \rightarrow \infty. \tag{16.36}$$

Also,

$$\left| \iint_{t \geq 0} [f(U_i) - f(u)]\phi_x \right| \leq \| \phi_x \|_{L_\infty} \iint_{\text{spt } \phi} |f(U_i) - f(u)|$$

$$\leq \| \phi_x \|_{L_\infty} \iint_{\text{spt } \phi} |f'(\xi)| |U_i - u|,$$

where ξ is some intermediate point. Thus there is a constant C independent of i such that

$$\left| \iint_{t \geq 0} [f(U_i) - f(u)]\phi_x \right| \leq C \iint_{\text{spt } \phi} |U_i - u|,$$

and this implies that

$$\iint_{t \geq 0} f(U_i)\phi_x \rightarrow \iint_{t \geq 0} f(u)\phi_x. \tag{16.37}$$

Then using (16.35), (16.36), and (16.37), and passing to the limit as $i \rightarrow \infty$ in (16.34) completes the proof of the lemma.   □

We can now complete the proof of the theorem. First, from our remark before the proof of this last lemma, we see that (16.31) holds, and thus, so does (16.30). But since C_0^3 is dense in C_0^1 (in the topology of uniform convergence of the functions together with their first derivatives on compacta), we see that (16.3) holds for all $\phi \in C_0^1$. Thus u is a solution of our problem, and we have observed earlier that u satisfies (16.4) and (16.5), and $|u| \leq \|u_0\|_{L_\infty}$.   □

§B. Uniqueness of the Entropy Solution

We shall show that solutions of (16.1), (16.2) which satisfy the entropy condition (16.4) are unique. It then follows from Theorem 16.1 that the solution which we constructed via the finite-difference scheme (16.6), is the unique solution of our problem. We call the solution which satisfies (16.4), the *entropy solution*.

The method of proof which we shall give is really a nonlinear version of the Holmgren method (see Chapter 5). Recall that for linear operators A on a Hilbert space, since $\eta_A \subset (\mathscr{R}_{A^*})^\perp$ (where η_A and $\mathscr{R}_{A^*}$ are the null space and range of A and A^*, respectively), in order to show that $\eta_A = 0$, it suffices to show that $\mathscr{R}_{A^*}$ is everywhere dense. Thus if u and v are solutions of (16.1) and (16.2), in order to show $u = v$ almost everywhere in $t > 0$, it suffices to show

$$\iint_{t \geq 0} (u - v)\phi = 0 \tag{16.38}$$

for every $\phi \in C_0^1$. Now if $\psi \in C_0^1$, we have both

$$\iint_{t \geq 0} u\psi_t + f(u)\psi_x + \int_{t=0} u_0 \psi = 0$$

and

$$\iint_{t \geq 0} v\psi_t + f(v)\psi_x + \int_{t=0} u_0 \psi = 0.$$

If we subtract these two equations we get

$$\iint_{t \geq 0} (u - v)\psi_t + [f(u) - f(v)]\psi_x = 0,$$

or

$$\iint_{t \geq 0} (u - v)\left[\psi_t + \frac{f(u) - f(v)}{u - v}\psi_x\right] = 0. \tag{16.39}$$

Next, if we define F by

$$F(x, t) = \frac{f(u) - f(v)}{u - v},$$

then (16.39) can be written as

$$\iint_{t \geq 0} (u - v)[\psi_t + F\psi_x] = 0. \tag{16.40}$$

Now if we could solve the linear (adjoint!) equation

$$\psi_t + F\psi_x = \phi, \tag{16.41}$$

for arbitrary $\phi \in C_0^1(t > 0)$ with $\psi \in C_0^1$, then (16.40) would imply (16.38) and we could conclude that $u = v$, almost everywhere. There is however an obstruction to this approach; namely, F is not smooth (or even continuous) in general, and so it is not clear that (16.41) has a solution $\psi \in C_0^1$. The way around this difficulty is to approximate u and v (and hence F), by smooth functions u_m, v_m and solve the corresponding linear equations

$$\psi_t^m + F_m\psi_x^m = \phi, \tag{16.42}$$

with smooth coefficients, for $\psi^m \in C_0^1$, where

$$F_m(x, t) = \frac{f(u_m) - f(v_m)}{u_m - v_m}.$$

Then

$$\iint_{t\geq 0} (u - v)\phi = \iint_{t\geq 0} (u - v)\psi_t^m + \iint_{t\geq 0} (u - v)F_m\psi_x^m$$

$$= -\iint_{t\geq 0} (f(u) - f(v))\psi_x^m + \iint_{t\geq 0} (u - v)F_m\psi_x^m,$$

so that

$$\iint_{t\geq 0} (u - v)\phi = \iint_{t\geq 0} (u - v)(F_m - F)\psi_x^m.$$

Then if $F_m \to F$, locally in L_1, and if ψ_x^m is bounded, independently of m, we could pass to the limit on the right-hand side of this last equation and conclude that (16.38) holds. This procedure will be carried out below, whereby the entropy condition (16.4) will be used to obtain control of ψ_x^m.

Theorem 16.10. Let $f \in C^2$, $f'' > 0$, and let u and v be two solutions satisfying the entropy condition (16.4). Then $u = v$ almost everywhere in $t > 0$.

Proof. For every positive integer m, let

$$u_m = u * \omega_m \quad \text{and} \quad v_m = v * \omega_m,$$

where ω_m is the usual averaging kernel of radius $1/m$, and $*$ denotes convolution product.[3] We define

$$F_m(x, t) = \int_0^1 f'(\theta u_m + (1 - \theta)v_m) \, d\theta$$

$$= \frac{1}{u_m - v_m} \int_{v_m}^{u_m} f'(s) \, ds = \frac{f(u_m) - f(v_m)}{u_m - v_m}, \tag{16.43}$$

and solve (16.42) for $\psi^m(x, t)$, subject to the boundary condition $\psi^m(x, T) = 0$, where $T > 0$ is chosen so large that $\phi = 0$ if $t \geq T$. It is not too hard to verify that the solution of this problem is given by

$$\psi^m(x, t) = \int_T^t \phi(x_m(s; x, t), s) \, ds, \tag{16.44}$$

where $x_m(s; x, t)$ is the unique solution of the characteristic ordinary differential equation

$$\frac{dx_m}{ds} = F_m(x_m, s), \quad \text{with } x_m(t) = x. \tag{16.45}$$

Now

$$|u_m(x, t)| \leq \int_R |u(x - y)| \omega_m(y) \, dy \leq M \int_R \omega_m(y) \, dy = M,$$

and similarly, $|v_m(x, t)| \leq M$. Thus since $f \in C^2$, we see that there is a constant $M_1 > 0$, independent of m, such that

$$|F_m(x, t)| < M_1. \tag{16.46}$$

This estimate enables us to show that $\psi^m \in C_0^1 (t \geq 0)$. First, it is clear that ψ^m is in C^1; to show that $\psi^m \in C_0^1 (t \geq 0)$, we proceed as follows. Let $S \subset \{t > 0\}$ denote the support of ϕ; S being compact. Now consider Figure 16.3. R denotes a region in $t \geq 0$ which is bounded by the two lines of slope $\pm 1/M_1$, and the lines $t = 0$, $t = T$, and R contains S in its interior. The claim is that spt $\psi^m \subset R$, for every m. To see this, note that if $t \geq T$, $\phi = 0$ so that $\psi^m = 0$, in view of (16.44). Furthermore, (16.46) shows that every solution of (16.45) which starts at a point P outside of R in the region $t < T$ must meet the line $t = T$ at a point not in R. It follows that all along this trajectory, $\phi = 0$. Thus ψ^m satisfies the homogeneous equation $\psi_t^m + F_m \psi_x^m = 0$, so

---

[3] See the appendix to Chapter 7.

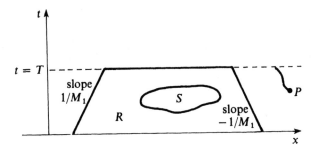

Figure 16.3

ψ^m is constant along trajectories of (16.45). Since the trajectory meets $t = T$, ψ^m is zero at $t = T$, and hence ψ^m is zero on the entire trajectory, and in particular at P. This shows that $\psi^m \in C_0^1$, and moreover, that the ψ^m have supports lying in the set R, independently of m.

It follows, as above, that

$$\iint_{t \geq 0} (u - v)\phi = \iint_{t \geq 0} (u - v)(F_m - F)\psi_x^m. \tag{16.47}$$

We shall next show that $F_m \to F$, locally in L_1. Let $c = \max\{f''(u) : |u| \leq M\}$; $c > 0$. Then

$$F(x, t) - F_m(x, t) = \int_0^1 [f'(\theta u + (1 - \theta)v) - f'(\theta u_m + (1 - \theta)v_m)]\, d\theta$$

$$= \int_0^1 f''(\xi)[\theta(u - u_m) + (1 - \theta)(v - v_m)]\, d\theta,$$

where ξ is between $\theta u + (1 - \theta)v$ and $\theta u_m + (1 - \theta)v_m$, so that $|\xi| \leq M$. Thus

$$|F(x, t) - F_m(x, t)| \leq c \int_0^1 \{\theta|u - u_m| + (1 - \theta)|v - v_m|\}\, d\theta$$

$$= \frac{c}{2}\{|u - u_m| + |v - v_m|\},$$

Accordingly, if K is any compact set in $t \geq 0$,

$$\iint_K |F(x, t) - F_m(x, t)| \leq c \iint_K |u_m - u| + |v_m - v|,$$

and this latter integral tends to zero as $m \to \infty$ (see Lemma B1 in Chapter 7).

It remains to show that we can control ψ_x^m. It is here where we shall use the entropy condition (16.4).

Thus, let $\alpha > 0$ be arbitrary. Then for each fixed $t \geq \alpha$, the function $u(x, t) - Ex/\alpha$ is nonincreasing in x. This follows from the entropy condition (16.4), since if $a > 0$,

$$u(x + a, t) - \frac{E(x + a)}{\alpha} - u(x, t) + \frac{Ex}{\alpha} = Ea\left(\frac{1}{t} - \frac{1}{\alpha}\right) \leq 0.$$

From this we find that

$$\omega_m * \left(u - \frac{Ex}{\alpha}\right) = u_m - \frac{E(\omega_m * x)}{\alpha}$$

is also nonincreasing in x, as one can easily check. Since this latter function is smooth and

$$\frac{\partial}{\partial x}\left[\omega_m * \left(u - \frac{Ex}{\alpha}\right)\right] = \frac{\partial u_m}{\partial x} - \frac{E}{\alpha},$$

we see that the following are true:

$$\frac{\partial u_m}{\partial x} \leq \frac{E}{\alpha}, \quad \text{and} \quad \frac{\partial v_m}{\partial x} \leq \frac{E}{\alpha}. \tag{16.48}$$

Next, from (16.43),

$$\frac{\partial F_m}{\partial x} = \int_0^1 f''(\theta u_m + (1 - \theta)v_m)\left[\theta \frac{\partial u_m}{\partial x} + (1 - \theta)\frac{\partial v_m}{\partial x}\right] d\theta,$$

and since $f'' > 0$, we get from (16.48),

$$\frac{\partial F_m}{\partial x} \leq \int_0^1 f''(\theta u_m + (1 - \theta)v_m)\frac{E}{\alpha}(\theta + 1 - \theta) \, d\theta$$

$$= \frac{E}{\alpha}\int_0^1 f''(\theta u_m + (1 - \theta)v_m) \, d\theta.$$

Therefore,

$$\frac{\partial F_m}{\partial x} \leq K_\alpha, \tag{16.49}$$

where

$$K_\alpha = \frac{E}{\alpha} \max_{|u| \leq M} f''(u),$$

so that K_α is independent of m.

Now let

$$a_m(s) = \frac{\partial x_m}{\partial \bar{x}} (s; \bar{x}, \bar{t}),$$

where $(\bar{x}, \bar{t})$ is a fixed point in the upper-half plane. Note that $x_m(\bar{t}; \bar{x}, \bar{t}) = \bar{x}$, so that

$$a_m(\bar{t}) = \frac{\partial x_m}{\partial \bar{x}} (\bar{t}; \bar{x}, \bar{t}) = 1. \tag{16.50}$$

Thus,

$$\frac{\partial a_m}{\partial s} = \frac{\partial}{\partial s} \frac{\partial x_m}{\partial \bar{x}} = \frac{\partial}{\partial \bar{x}} \frac{\partial x_m}{\partial s} = \frac{\partial}{\partial \bar{x}} F_m(x_m, s)$$

$$= \frac{\partial}{\partial \bar{x}} F_m(x_m(s; \bar{x}, \bar{t}), s) = \frac{\partial F_m}{\partial x} \frac{\partial x_m}{\partial \bar{x}} = \frac{\partial F_m}{\partial x} a_m.$$

Then from (16.50), we obtain the formula

$$a_m(s) = \exp\left(\int_{\bar{t}}^{s} \frac{\partial F_m}{\partial x} (x_m(\tau), \tau) \, d\tau \right).$$

Since $\alpha \leq \bar{t} \leq s \leq T$, we have from (16.49),

$$|a_m(s)| = a_m(s) \leq e^{K_\alpha(T - \alpha)}.$$

But from (16.44)

$$\frac{\partial \psi^m}{\partial x} = \int_T^{\bar{t}} \frac{\partial \phi}{\partial x_m} \frac{\partial x_m}{\partial x} \, ds = \int_T^{\bar{t}} \frac{\partial \phi}{\partial x} a_m \, ds.$$

Thus

$$\left| \frac{\partial \psi^m}{\partial x} \right| \leq \bar{K}_\alpha \quad \text{if } t \geq \alpha, \tag{16.51}$$

where $\bar{K}_\alpha$ is independent of m.

We now must investigate the behavior of ψ^m in the region $0 \le t \le \alpha$. To this end, we define

$$V_t(\psi^m) = \int_{-\infty}^{\infty} \left| \frac{\partial \psi^m}{\partial x} \right| dx,$$

the total variation of ψ^m as a function of x, for each fixed $t > 0$. From (16.51), and the fact that the ψ^m are in C_0^1 with their supports being contained in a region independent of m, we have

$$V_t(\psi^m) \le C_\alpha \quad \text{if } t \ge \alpha, \tag{16.52}$$

where C_α does not depend on m.

The last estimate we need is

$$\exists N \in \mathbf{Z}_+ \ni: n > N \Rightarrow V_t(\psi^m) \le C_{1/n}, \quad \forall t, \qquad 0 < t \le \frac{1}{n} < \frac{1}{N}. \tag{16.53}$$

To prove this, note that since ϕ has compact support in $t > 0$, there is an $N \in \mathbf{Z}_+$ such that $\phi(x, t) = 0$ if $t < 1/N$. Thus from (16.42),

$$\frac{\partial \psi^m}{\partial t} + F_m \frac{\partial \psi^m}{\partial x} = 0 \quad \text{if } t < \frac{1}{N}. \tag{16.54}$$

Now let $n > N$, and let $\sigma_t : \mathbf{R} \to \mathbf{R}$ be the bijection defined for $t < 1/n$ by the solution of the characteristic equation, (16.45); i.e., $\sigma_t(x) = x_m(1/n; x, t)$; see Figure 16.4.

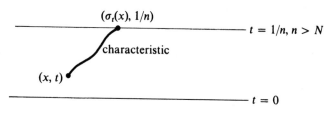

Figure 16.4

Now let t be any number, $0 < t \le 1/n < 1/N$. Then for any finite sequence $x_1 < x_2 < \ldots < x_p$, we have

$$\sum_{k=1}^{p-1} |\psi^m(x_{k+1}, t) - \psi^m(x_k, t)| = \sum_{k=1}^{p-1} \left| \psi^m\left(\sigma_t(x_{k+1}), \frac{1}{n} \right) - \psi^m\left(\sigma_t(x_k), \frac{1}{n} \right) \right|.$$

This holds since ψ^m is constant along characteristics, in view of (16.54). Thus using (16.52),

$$\sum_{k=1}^{p-1} |\psi^m(x_{k+1}, t) - \psi^m(x_k, t)| \le V_{1/n}(\psi^m) \le C_{1/n},$$

and this proves (16.53).

We can now complete the proof of the theorem. Let $\varepsilon > 0$ be arbitrary. With N defined as in (16.53), choose $\alpha > 0$ so small that

$$\alpha < \frac{1}{n} \le \frac{1}{N} \quad \text{and} \quad 4MM_1 C_{1/n} \alpha < \frac{\varepsilon}{2}. \tag{16.55}$$

For this α, choose $\tilde{M}$ so large that

$$\iint_{t \ge \alpha} |u - v||F_m - F||\psi_x^m| < \frac{\varepsilon}{2}, \quad \text{if } m \ge \tilde{M}. \tag{16.56}$$

This can be done since $|u - v| \le 2M$, $F_m \to F$ locally in L_1, $\psi^m \in C_0^1(t \ge 0)$ and (16.51) holds. Then from (16.47)

$$\left| \iint_{t \ge 0} (u - v)\phi \right| \le \iint_{t \ge \alpha} |u - v||F_m - F||\psi_x^m| + \iint_{t < \alpha} |u - v||F_m - F||\psi_x^m|. \tag{16.57}$$

Now since $\alpha < 1/n \le 1/N$,

$$\iint_{t < \alpha} |u - v||F_m - F||\psi_x^m| \le 2M \cdot 2M_1 \iint_{t < \alpha} |\psi_x^m| = 4MM_1 \int_0^\alpha \int_{-\infty}^\infty |\psi_x^m|$$

$$= 4MM_1 \int_0^\alpha V_t(\psi^m) \, dt$$

$$\le 4MM_1 C_{1/n} \cdot \alpha \quad \text{(by (16.53))},$$

so that

$$\iint_{t < \alpha} |u - v||F_m - F||\psi_x^m| < \frac{\varepsilon}{2}. \tag{16.58}$$

Thus using (16.56) and (16.58) in (16.57), we get

$$\left| \iint_{t \geq 0} (u - v)\phi \right| < \varepsilon,$$

and from the arbitrariness of ε,

$$\iint_{t \geq 0} (u - v)\phi = 0,$$

for all $\phi \in C_0^1 \, (t > 0)$. Thus $u = v$ almost everywhere in $t > 0$. This completes the uniqueness proof. □

As a corollary, we can easily show that the "entropy" solution, which we now know to be unique, is an "irreversible" solution (cf. §E in Chapter 15). Thus for $t > 0$, define the mapping

$$T_t : L_\infty(\mathbf{R}) \to L_1(|x| \leq X), \qquad X > 0,$$

given by

$$T_t(u_0) = u(\cdot, t),$$

where u is the unique solution of (16.1), with initial data u_0. Here is the "irreversibility" theorem; it implies the noncontinuity of T_t^{-1}.

Theorem 16.11. T_t *is a compact operator.*

Proof. Let $\{u_0^i\}$ be a sequence in $L_\infty(\mathbf{R})$, with

$$\|u_0^i\|_{L_\infty} \leq M, \qquad M \text{ independent of } i.$$

For each i, we use the difference scheme (16.6) to construct the unique solution u_i of (16.1) having initial data u_0^i. From Lemmas 16.2 and 16.4

$$\|u_i(\cdot, t)\|_{L_\infty(|x| \leq X)} \leq M,$$

and

$$\underset{|x| \leq X}{\text{Tot. Var}} (u_i(\cdot, t)) \leq M',$$

where M and M' are independent of i. By Helley's theorem, there is a subsequence of $\{u_i(\cdot, t)\}$ which converges in $L_1(|x| \leq X)$. Thus T_t is compact. □

§C. Asymptotic Behavior of the Entropy Solution

In this section we shall study the asymptotic form of the solution for large time. There are two distinctly different cases which we shall consider; namely, the case where the initial data u_0 is periodic and the case where u_0 has compact support. In the periodic case we shall show that u tends to the mean value of u_0 (over one period), at a rate t^{-1}, uniformly in x. If u_0 has compact support, then u tends uniformly to zero at a rate $t^{-1/2}$, and tends in the L_1-norm, to a particular function called an N-wave, again at a rate $t^{-1/2}$.

At the heart of these decay results is the fact that u satisfies the entropy condition (16.4). It is not difficult to get a feeling as to why this should be true. Thus, suppose that u is differentiable in a region R contained in $0 \leq t \leq T$, and let x_1 and x_2 be two characteristics in R. Since u is constant along the characteristics, we see that the variation of $u(\cdot, T)$ on the interval $[x_1(T), x_2(T)]$ equals the variation of $u(\cdot, 0)$ on the interval $[x_1(0), x_2(0)]$. Assume now that there is a shock y present in u between the two characteristics x_1 and x_2 (see Figure 16.5). Since u satisfies the entropy condition (16.4), we know

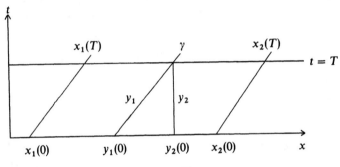

Figure 16.5

that there are characteristics y_1 and y_2 which impinge on the shock from both sides at time $t = T$ (see (15.18)). It follows that

$$\underset{[x_1(T),x_2(T)]}{\text{Tot. Var }u(\cdot, T)} = \underset{[x_1(0),y_1(0)]}{\text{Tot. Var }u(\cdot, 0)} + \underset{[y_2(0),x_2(0)]}{\text{Tot. Var }u(\cdot, 0)}$$

$$< \underset{[x_1(0),x_2(0)]}{\text{Tot. Var }u(\cdot, 0)}. \tag{16.59}$$

Thus, in the presence of shocks, the total variation of u between two characteristics decreases with time. Our goal is to give a quantitative statement of this decrease.

We begin by considering the case where u_0 has compact support; say $u_0(x) = 0$ if $x \notin [s_-, s_+]$. We assume again that f is convex; i.e., $f'' > 0$. For simplicity, we assume

$$f(0) = f'(0) = 0, \tag{16.60}$$

since in (16.1), we may replace $f(u)$ by $f(u) + a$, and u by $u + b$. Also, we let k and μ be defined by

$$k = f''(0) \quad \text{and} \quad \mu = \inf_{|u| \le M} f''(u) > 0,$$

where $\|u_0\|_{L_\infty} = M$. Finally, we define the two important quantities

$$q = \max_y \int_y^\infty u_0(x)\, dx \quad \text{and} \quad -p = \min_y \int_{-\infty}^y u_0(x)\, dx. \qquad (16.61)$$

Both q and p are finite since the function

$$Q(y) = \int_y^\infty u_0(x)\, dx$$

is bounded and continuous.

Now let $u(x, t)$ be the unique solution of (16.1), with initial data $u_0(x)$. We assume that u is piecewise smooth.[4] Since u_0 has compact support, the same is true for $u(\cdot, t)$, for each fixed $t > 0$; this follows easily from (16.5) with $v_0 \equiv v \equiv 0$. Now set

$$s^+(t) = \inf\{y : u(x, t) = 0, \forall x > y\}, \qquad s^+(0) = s_+.$$

We define

$$u^+(t) = u(s^+(t) - 0, t),$$

and note that $u(s^+(t) + 0, t) \equiv 0$. It follows from the entropy condition that $u^+(t) \ge 0$ for every $t > 0$ (see the discussion after (15.17)). Moreover, if $u^+(t) = 0$, then $(s^+(t), t)$ is a point of continuity for u, and conversely.

Now since $u^+(t) \ge 0$, the slope of the characteristic emanating from the point $(s^+(t) - 0, t)$ in backwards time is positive, since $f'(0) = 0$. This shows that

$$\frac{ds^+(t)}{dt} \ge 0. \qquad (16.62)$$

We need the following proposition.

Proposition 16.12. *If $q = 0$ (see (16.61)), then $u^+(t) \equiv 0$, $t > 0$.*

Proof. Suppose $u^+(\bar{t}) > 0$, for some $\bar{t} > 0$. Let R be the region depicted in Figure 16.6, defined by the two characteristics Γ_1, Γ_3

---

[4] This is always true if $f'' > 0$, as follows, for example from [Dp 3].

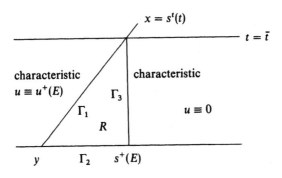

Figure 16.6

on both sides of the curve $x = s^+(t)$, and the region they cut off on the x-axis. Note that the entropy inequality (15.18) shows that these characteristics can be extended down to $t = 0$, as depicted.

Now since u is piecewise smooth, we have[5]

$$\int_{\partial R} u \, dx - f(u) \, dt = 0. \tag{16.63}$$

Also

$$\int_{\Gamma_1} u \, dx - f(u) \, dt = \int_0^{\bar{t}} \left(u \frac{dx}{dt} - f \right) dt = \int_0^{\bar{t}} (uf'(u) - f) \, dt,$$

and since $u \equiv u^+(\bar{t})$ along Γ_1, we have

$$\int_{\Gamma_1} u \, dx - f(u) \, dt = \bar{t}[\bar{u}f'(\bar{u}) - f(\bar{u})],$$

where we are using the notation $\bar{u} = u^+(\bar{t})$. Next,

$$\int_{\Gamma_2} u \, dx - f(u) \, dt = -\int_y^{s^+(\bar{t})} u_0(x) \, dx = -\int_y^{\infty} u_0(x) \, dx,$$

and

$$\int_{\Gamma_3} u \, dx - f(u) \, dt = 0,$$

---

[5] In the considered class of functions, this follows easily from the divergence theorem, using (16.3).

since $u = 0$ on Γ_3, and $f(0) = 0$. Using these in (16.63) gives

$$\bar{t}[\bar{u}f'(\bar{u}) - f(\bar{u})] = \int_y^\infty u_0(x)\, dx. \qquad (16.64)$$

But

$$0 = f(0) = f(\bar{u}) + f'(\bar{u})(0 - \bar{u}) + \tfrac{1}{2}f''(\xi)(0 - \bar{u})^2,$$

where ξ is intermediate to 0 and $\bar{u}$. Using this in (16.64) gives

$$0 < \tfrac{1}{2}f''(\xi)\bar{u}^2\bar{t} = \int_y^\infty u_0(x)\, dx \leq q. \qquad (16.65)$$

This shows that $q > 0$ and the proof of the proposition is complete. $\square$

From (16.65), it follows that if $u^+(t) > 0$, then $\mu u^+(t)^2 t/2 \leq q$, so that

$$u^+(t) \leq \left(\frac{2q}{\mu}\right)^{1/2} t^{-1/2}. \qquad (16.66)$$

Since $\dot{s}^+(t)(u^+(t) - 0) = f(u^+(t)) - f(0)$, we have, in view of (16.60),

$$\dot{s}^+(t) = \frac{f(u^+(t))}{u^+(t)} = \frac{f''(\theta)u^+(t)}{2}, \qquad (16.67)$$

where $0 < \theta < u^+(t)$. Thus from (16.66),

$$\dot{s}^+(t) \leq \frac{c}{2}\sqrt{\frac{2q}{\mu}}\, t^{-1/2}, \qquad (16.68)$$

where $c = \max_{|u| \leq M} f''(u)$. It follows that

$$s^+(t) \leq s_+ + c\sqrt{\frac{2q}{\mu}}\, t^{1/2}. \qquad (16.69)$$

If now we define

$$s^-(t) = \sup\{y: u(x, t) = 0, \forall x < y\}, \qquad s^-(0) = s_-,$$

then a similar argument gives

$$s^-(t) \geq s_- - c\sqrt{\frac{2p}{\mu}}\, t^{1/2}. \qquad (16.70)$$

Next, let $(x_\pm, t) \equiv (x \pm 0, t)$; then since u is constant along characteristics,

$$u(x_\pm, t) = u_0(y_\pm(x, t)),$$

where

$$x = y_\pm + t f'(u(x_\pm, t)).$$

However, since $s^-(t) \le x \le s^+(t)$, and $s_- \le y_\pm \le s_+$, we have for large t,

$$\left| f'(u(x_\pm, t)) \right| \le \left| \frac{x - y_\pm}{t} \right|$$

$$\le \frac{1}{t} \max(s^+(t) - s_-, s_+ - s^-(t))$$

$$\le \text{const.} \, t^{-1/2},$$

because of (16.69) and (16.70). Now (16.60) implies that $|f'(u)| \ge \mu |u|$, and thus

$$|u(x, t)| \le \text{const.} \, t^{-1/2}. \tag{16.71}$$

We have therefore proved the following decay theorem.

Theorem 16.13. *Assume that $f'' > 0$, $f(0) = f'(0) = 0$, and that $u_0(x)$ is a bounded, measurable function having compact support. Then if u is the unique (entropy) solution of (16.1), (16.2), u decays to zero as $t \to \infty$, uniformly in x, at a rate $t^{-1/2}$.*

We can now obtain the asymptotic shape of the solution as $t \to \infty$. To this end, we observe that from (16.67) and (16.65), we have

$$\dot{s}^+(t) = \frac{f''(\theta) u^+(t)}{2}$$

$$\le \frac{1}{2} \frac{f''(\theta)}{\sqrt{f''(\xi)}} \sqrt{\frac{2q}{t}}.$$

But as $t \to \infty$, it follows from (16.71), that both θ and ξ tend to zero. Thus, since $k = f''(0)$,

$$f''(\theta) = k + O(|\theta|) \quad \text{and} \quad f''(\xi) = k + O(|\xi|).$$

Now we have $\theta = O(|u|) = O(t^{-1/2})$, and $\xi = O(t^{-1/2})$, in view of (16.71). Thus

$$\frac{f''(\theta)}{\sqrt{f''(\xi)}} = \sqrt{k}\,\frac{1 + O(t^{-1/2})}{\sqrt{1 + O(t^{-1/2})}} = \sqrt{k}(1 + O(t^{-1/2})).$$

It follows that

$$\dot{s}^+(t) \le \sqrt{\frac{kq}{2}}\,t^{-1/2} + O(t^{-1}),$$

and thus

$$s^+(t) \le s_+ + \sqrt{2kq}\,t^{1/2} + O(\ln t) = s_+ + [\sqrt{2kq} + O(t^{-1/2}\ln t)]t^{1/2}.$$

$$(16.72)$$

Similarly

$$s^-(t) \ge s - [\sqrt{2kp} + O(t^{-1/2}\ln t)]t^{1/2}. \tag{16.73}$$

Now define the functions $w(x, t)$ and $\tilde{w}(x, t)$ by

$$w(x, t) = \begin{cases} x/kt, & \text{if } s_- - \sqrt{2kp}\,t^{1/2} < x < s_+ + \sqrt{2kq}\,t^{1/2}, \\ 0, & \text{otherwise,} \end{cases} \tag{16.74}$$

and

$$\tilde{w}(x, t) = \begin{cases} x/kt, & \text{if } s_- - [\sqrt{2kp} + O(t^{-1/2}\ln t)]t^{1/2} < x \\ & \quad\quad < s_+ + [\sqrt{2kq} + O(t^{-1/2}\ln t)]t^{1/2}, \\ 0, & \text{otherwise.} \end{cases}$$

Then since $u(x, t) = u_0(y(x, t))$, where $s_- < y(x, t) < s_+$, and $tf'(u(x, t)) = x - y(x, t)$, we get

$$f''(0)u(x, t) + O(|u|^2) = \frac{x - y(x, t)}{t}$$

so that since $k = f''(0)$,

$$u(x, t) = \frac{x}{kt} - \frac{y(x, t)}{kt} + O(|u|^2) = \frac{x}{kt} + O(t^{-1}), \tag{16.75}$$

again using (16.71). Hence for large t,

$$\int_{-\infty}^{\infty} |u(x, t) - \tilde{w}(x, t)| \, dx = \int_{s_- - \sqrt{2kp}\,t^{1/2} + O(\ln t)}^{s_+ + \sqrt{2kq}\,t^{1/2} + O(\ln t)} \left| u(x, t) - \frac{x}{kt} \right| dx$$

$$\leq \text{const.}\, t^{-1}[1 + t^{1/2} + \ln t] \quad \text{(by (16.75))}$$

$$\leq \text{const.}\, t^{-1/2}.$$

This shows that

$$\| u(\,\cdot\,, t) - \tilde{w}(\,\cdot\,, t) \|_{L_1(\mathbb{R})} = O(t^{-1/2}) \quad \text{as } t \to \infty.$$

Similarly,

$$\| w(\,\cdot\,, t) - \tilde{w}(\,\cdot\,, t) \|_{L_1(\mathbb{R})} = O(t^{-1/2}) \quad \text{as } t \to \infty.$$

We can thus conclude that

$$\| u(\,\cdot\,, t) - w(\,\cdot\,, t) \|_{L_1(\mathbb{R})} = O(t^{-1/2}) \quad \text{as } t \to \infty. \tag{16.76}$$

The function $w(x, t)$, defined by (16.74) is called an "N-wave," because of its profile at each fixed $t > 0$; see Figure 16.7. We have thus shown that the entropy solution decays to an N-wave in $L_1(\mathbb{R})$ as $t \to \infty$, provided that the hypotheses of Theorem 16.13 hold.

We turn our attention now to the case where u_0 is a periodic function, say of period p. We assume, for simplicity, that u_0 is also piecewise monotonic. We then have the following theorem.

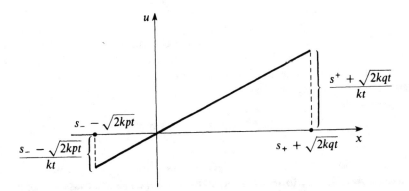

Figure 16.7

Theorem 16.14. *Let* $f'' > 0$, *and let* $u_0 \in L_\infty(\mathbf{R})$ *be a piecewise monotonic periodic function of period p. Let u be the solution of (16.1), (16.2) which satisfies the entropy condition (16.4). Then*

$$|u(x, t) - \bar{u}_0| \leq \frac{2p}{ht}, \tag{16.77}$$

where $\bar{u}_0 = (1/p) \int_0^p u_0(x)\, dx$, *and* $h = \min\{f''(u): |u| \leq \|u_0\|_{L_\infty}\}$.

Thus, u decays to the mean value of u_0 at a rate t^{-1}; this is actually at a faster rate than the case when u_0 has compact support; (see (16.71)).

Proof. Let $x_1(t)$ and $x_2(t)$ be any two characteristics of (16.1). Then the x_i satisfy the ordinary differential equation

$$\frac{dx}{dt} = f'(u(x, t)).$$

Define the "width" $D(t)$ of the strip $0 \leq t \leq T$ bounded by these two characteristics, as

$$D(t) = x_1(t) - x_2(t).$$

Then

$$D'(t) = x_1'(t) - x_2'(t) = f'(u_2) - f'(u_1),$$

where $u_i = u(x_i(t), t)$ are constants, $i = 1, 2$. Since $f'' \geq h > 0$, $u_2 > u_1$ implies $f'(u_2) - f'(u_1) > h(u_2 - u_1)$, so that

$$D'(t) \geq h(u_2 - u_1).$$

If we integrate this with respect to t, we find

$$D(T) > D(0) + h(u_2 - u_1)T, \qquad u_2 > u_1. \tag{16.78}$$

Let I be any interval on the x-axis. Since u_0 is piecewise monotone, we can divide I into subintervals by points $y_0 < y_2 < \ldots < y_n$ such that u_0 is alternately increasing and decreasing on the subintervals $[y_{i-1}, y_i]$. Let $y_i(t)$ be the characteristic through y_i; i.e.,

$$\frac{dy_i}{dt} = f'(u(y_i, t)), \qquad y_i(0) = y_i.$$

We make the convention that if $y_j(t)$ meets a shock, y_j is continued as that shock. Now it is easy to see that for each $t > 0$, $u(x, t)$ is alternately increasing and decreasing on the intervals $(y_{i-1}(t), y_i(t))$.

Let $I(t)$ be the interval $[y_0(t), y_n(t)]$, and let $L(t)$ be its length. We define

$$V_+(t) = \text{total increasing variation of } u(\cdot, t) \text{ on } I(t).$$

(Here by increasing variation on an interval, we mean the total variation of the function over the intervals on which the function is increasing.) We make the convention that $u(\cdot, t)$ is increasing on $(y_{i-1}(t), y_i(t))$, if and only if i is odd.
Now for i odd let (cf. Figure 16.8),

$$u_{i-1}(t) = \text{value of } u \text{ on the right edge of } y_{i-1}(t),$$

$$u_i(t) \quad = \text{value of } u \text{ on the left edge of } y_i(t).$$

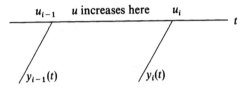

Figure 16.8

Since the entropy condition (16.4) implies that u decreases across shock, we have

$$V_+(t) = \sum_{i \text{ odd}} [u_i(t) - u_{i-1}(t)]. \tag{16.79}$$

If $y_i(t) = y_{i-1}(t)$, then the characteristics come together at the point $(y_i(t), t)$, so that this point lies on a shock, and the contribution of the ith term in (16.79) is zero. On the other hand, if $y_{i-1}(t) < y_i(t)$, then there exist characteristics χ_{i-1} and χ_i starting at $t = 0$ inside $[y_{i-1}, y_i]$ which, at time t, run into $y_{i-1}(t)$ and $y_i(t)$, respectively (of course, $\chi_j(t)$ will equal $y_j(t)$ if the $y_j(t)$ is not continued as a shock). If we let

$$D_i(t) = \chi_i(t) - \chi_{i-1}(t),$$

we have from (16.78),

$$D_i(t) \geq D_i(0) + h(u_i - u_{i-1})t$$

if i is odd, so that

$$\sum_{i \text{ odd}} D_i(t) \geq \sum_{i \text{ odd}} D_i(0) + ht \sum_{i \text{ odd}} (u_i - u_{i-1})$$

$$= \sum_{i \text{ odd}} D_i(0) + ht \, V_+(t).$$

Since the intervals $[\chi_{i-1}(t), \chi_i(t)]$ are disjoint and lie in $I(t)$, their total length is at most $L(t)$. Thus

$$L(t) \geq ht\, V_+(t),$$

and

$$V_+(t) \leq \frac{L(t)}{ht}. \tag{16.80}$$

At this point we need a lemma.

Lemma 16.15. *Let u_0 satisfy the hypotheses of Theorem 16.14, and let u be the unique (entropy) solution of (16.1), (16.2). Then for each $t > 0$, $u(\,\cdot\,, t)$ is periodic of period p.*

We postpone the proof of the lemma, and continue the above argument.

Since $u(\,\cdot\,, t)$ is periodic, of period p, we let $L(0) = p$; then $L(t) = p$, since $y_0(t) \equiv y_n(t)$. So from (16.80), we obtain

$$V_+(t) \leq \frac{p}{ht}.$$

Since the increasing variation of a periodic function per period equals half its total variation. we have

$$V(t) \leq \frac{2p}{ht}, \qquad t > 0, \tag{16.81}$$

where $V(t)$ denotes the variation of u over one period.

Now let Γ be the contour pictured in Figure 16.9. Then from (16.63), we have

$$0 = \int_0^p u_0(x)\, dx - \int_0^t f(u(p, \tau))\, d\tau + \int_p^0 u(x, t)\, dx - \int_t^0 f(u(0, \tau))\, d\tau.$$

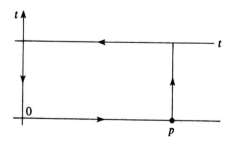

Figure 16.9

Since u is periodic of period p, the second and fourth terms cancel, and we obtain

$$\int_0^p u(x, t)\, dx = \int_0^p u_0(x)\, dx.$$

Thus the mean value of u is independent of t. It follows that

$$\begin{aligned}
|u(x, t) - \bar{u}_0| &= \left| u(x, t) - \frac{1}{p}\int_0^p u(z, t)\, dz \right| \\
&= \left| \frac{1}{p}\int_0^p [u(x, t) - u(z, t)]\, dz \right| \\
&\le \frac{1}{p}\int_0^p |u(x, t) - u(z, t)|\, dz \\
&\le \frac{1}{p}\int_0^p V(t)\, dz \\
&= V(t).
\end{aligned}$$

Hence from (16.81) we obtain the desired conclusion (16.77).

It remains to prove Lemma 16.15. To this end, define $v(x, t) = u(x + p, t)$. Then for $\phi \in C_0^1$ ($t > 0$), if $\tilde{\phi}(x, t) = \phi(x - p, t)$,

$$\begin{aligned}
\iint_{t>0} v\phi_t + f(v)\phi_x + \int_{t=0} u_0\phi &= \iint_{t>0} u(x + p, t)\phi_t + f(u(x + p, t))\phi_x \\
&= \iint_{t>0} u(x, t)\tilde{\phi}_t + f(u(x, t))\tilde{\phi}_x \\
&= 0.
\end{aligned}$$

Thus v is a solution of (16.1), (16.2). Since v obviously satisfies the entropy condition (16.4), we have $v(x, t) \equiv u(x, t)$ by the uniqueness theorem, (Theorem 16.10). This completes the proof of the theorem. $\square$

§D. The Riemann Problem for a Scalar Conservation Law

We recall from Chapter 15, §A, that the Riemann problem for (16.1) is the initial-value problem with initial data of the form

$$u_0 = \begin{cases} u_l, & x < 0, \\ u_r, & x > 0, \end{cases} \tag{16.82}$$

where u_l and u_r are constants. Our goal in this section is to explicitly solve this problem, when $f'' > 0$.

If $u(x, t)$ is a solution of (16.1), (16.82), then for every constant $\lambda > 0$, the function u_λ given by

$$u_\lambda(x, t) = u(\lambda x, \lambda t),$$

is also a solution. Since we are seeking the unique solution, it is natural to consider only solutions which depend on the ratio x/t.

The solutions $u = u(\xi)$, $\xi = x/t$ will be made up of three types of elementary waves (solutions) in the region $t > 0$. Namely they will be one of the following types of functions:

(a) *Constant states*; i.e., $u(\xi) = $ const. Obviously, these will be genuine (classical) solutions.
(b) *Shock waves*; i.e., solutions of the form

$$u(x, t) = \begin{cases} u_0, & x < st, \\ u_1, & x > st, \end{cases}$$

where, of course, $s(u_0 - u_1) = f(u_0) - f(u_1)$. In addition, we require that the entropy inequalities hold; i.e., $f'(u_0) > s > f'(u_1)$.
(c) *Rarefaction waves*; these are continuous solutions $u = u(\xi)$, $\xi = x/t$, of (16.1). Hence they must satisfy the ordinary differential equation

$$-\xi u_\xi + f(u)_\xi = 0,$$

or

$$(f'(u(\xi)) - \xi)u_\xi = 0.$$

Thus, if $u_\xi \neq 0$, $f'(u(\xi)) = \xi$. (On the other hand, if $f'(u(\xi)) = \xi$, then if we differentiate this with respect to ξ, we find $f''(u(\xi))u'(\xi) = 1$, so that $u'(\xi) \neq 0$.) We observe that the equation $f'(u(\xi)) = \xi$ defines a unique function $u(\xi)$, since $f'' > 0$.

We say that u_1 is connected to u_0 on the right by a *rarefaction wave*, if

$$f'(u_1) > f'(u_0) \quad \text{and} \quad \xi = f'(u(\xi)) \quad \text{if } f'(u_1) > \xi > f'(u_0).$$

With these notions under our command, we can solve the Riemann problem. There are only two cases to consider; namely, (i) $u_r > u_l$, and (ii) $u_l > u_r$.

Suppose $u_r > u_l$; then since $f'' > 0$, we have $f'(u_r) > f'(u_l)$, and the equation $\xi = f'(u(\xi))$ has a solution $u(\xi)$ where $f'(u_r) > \xi > f'(u_l)$. Assuming

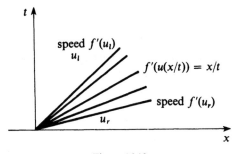

Figure 16.10

that $f'(u_l) > 0$, the solution can be depicted as in Figure 16.10. In the "fan-like" region $f'(u_r) > x/t > f'(u_l)$, the solution is given explicitly by solving the equation

$$f'\left(u\left(\frac{x}{t}\right)\right) = \frac{x}{t}$$

for u. This, of course, is possible since $f'' > 0$. The solution is smooth everywhere in $t > 0$, but is only Lipschitz-continuous on the two lines of speed $f'(u_l)$ and $f'(u_r)$. Moreover, u is constant on the rays $x/t = $ const., and $u(x, t)$ is an increasing function of x/t in the "fan-like" region, since $u'(\xi)f''(u(\xi)) = 1$ in this region. Note that u is discontinuous at the origin.

Now we consider the case $u_l > u_r$. We set

$$s = \frac{f(u_l) - f(u_r)}{u_l - u_r},$$

then $f'(u_l) > s > f'(u_r)$, and so the solution is a shock wave of speed s, connecting the two states u_l and u_r. It is depicted in Figure 16.11.

It is clear that the above two solutions are the unique solutions of our problem satisfying the entropy condition (16.4).

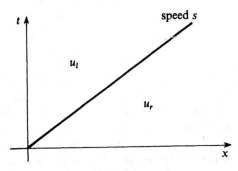

Figure 16.11

NOTES

The finite-difference equation (16.6) first appears in Lax' paper [Lx 1], and the existence Theorem 16.1, is due to Oleinik [O 1]; the proof given here is adapted from her's. By changing the mesh condition $h/l \geq \delta$ in Lemma 16.5 to the condition $l^2/h \to \varepsilon$, Oleinik shows in [O 1] that the difference scheme (16.6) can be used to construct smooth solutions u_ε of the equation $u_t + f(u)_x = \varepsilon u_{xx}$. She further obtains compactness of the set $\{u_\varepsilon\}$ so as to recover the (entropy) solution of (16.1) by passing to the limit as $\varepsilon \to 0$. A crucial ingredient in her proof is the use of the maximum principle for solutions of the parabolic equation; see Chapter 9.

The entropy condition (16.4) is due to Oleinik [O 1]; it is the heart of the existence and uniqueness proofs, as well as all results on the asymptotic behavior and decay of solutions. An analogue of this for systems is unknown. One consequence of (16.4) is however known for systems; namely (15.31). In all attempts to use the difference scheme (16.6) (or rather, its analogue) for systems of conservation laws, one is stopped very quickly since Lemma 16.2 fails to be true, and its replacement has yet to be found.

The elegant proof of uniqueness given in §B is due to Oleinik [O 1]; earlier proofs were somewhat less general; see Germain and Bader [GB, L], Douglis [Dol], and Wu [Wu]. The analogous proof for an important class of systems has also been given by Oleinik in [O 2]; (see Chapter 20, §E). In principle, the technique is applicable to general systems; in practice there is the obstruction of not knowing the replacement of (16.4) for such systems.

The study of the behavior of the solution as $t \to \infty$ goes back to the earliest papers on the subject; see [Hf 2] and [Lx 2]. The proof of Theorem 16.13, as well as the estimate (16.76), was shown to me by E. Conway in a private communication. The original theorem is due to Lax [Lx 2]. Theorem 16.14 seems to be first stated by Glimm and Lax [GL], as a preparatory example for the far more difficult theorem in the case of systems.

There are other approaches to the existence problem for (16.1), (16.2). In [CoH], Conway and Hopf show how to reduce the problem to the Hamilton–Jacobi equation, $v_t + f(v_x) = 0$. This latter equation is solved via a calculus of variations approach. They give an explicit representation of the solution which enables them to develop a fairly complete study of the solution, as concerns both qualitative and quantitative properties. Oleinik has given an extension of the entropy condition in the case of nonconvex, nonconcave functions f, [O 3]; this enables her to extend the uniqueness proofs to this case. Quinn in [Q], uses this extension to give a new uniqueness proof, as well as to show that the (entropy) solutions of (16.1) form an L_1-contraction semigroup. This fact has been used by Crandall [Cr], as well as Flashka [Fk], to obtain an existence proof via techniques from theory of nonlinear semigroups. Krushkov [Kv 2] has shown how to use the viscosity method directly to solve (16.1), (16.2); see also [L]. The viscosity method relies on techniques from the theory of second-order scalar parabolic equations (which satisfy a maximum principle). Next, in a paper of Douglis [Do 1], the existence problem has been solved via the method of characteristics, and in [Do 2] and [Ku] one finds proofs of existence via a "layering" method, related to finite-difference schemes. Similar work can be found in Dafermos [Df 4].

The existence proofs have been extended to the case of nonconvex f in [O 1] and [Ba 1]. In particular, in [Ba 2], Ballou gives an example of a Riemann problem with a nonconvex f, for which the entropy solution is discontinuous on a dense subset of $t > 0$; see also [Wu].

The results of this chapter have been extended to the case of several space variables. First, the analogue of (16.6) has been used by Conway and Smoller [CoS 1], to give an existence proof if $u_0 \in BVC$. Here BVC is the class of functions of bounded variation in the sense of Cesari. A function is in this class if its derivatives in $\mathscr{D}'$ (see Chapter 7), are in fact, finite measures. This class is rich enough to contain shock-wave discontinuities in $\mathbf{R}^n$, and it contains functions which are discontinuous on sets of positive measure [Gf]. This raises the interesting question as to whether a solution u of a system of conservation laws can evolve in time to a state in which it becomes discontinuous on a set of positive measure. An affirmative answer would have interesting interpretations. In [CoS 1] it is shown that for the single equation the solution remains in BVC both in space and time; see also Kotlow [Kw]. An existence proof in this class via the viscosity method has been given by Volpert in [V]. In this work is included a "calculus" on the class BVC which should find many applications. These existence results were extended by Kruskov [Kv 2] to the case in which the data is merely in L_∞. His method is via "vanishing viscosity." In this paper, he also gives a uniqueness criterion, which extends the entropy condition to the case of a single conservation law in n-space variables. Kruskov's result is used in [Cr] and [Fk] to apply semi-group techniques to the existence problem. The methods of Douglis [Do 2] and Kuznetsov [Ku] are valid for many space variables. The results on asymptotic behavior of the solution in the case of several space variables was given first by Conway [Coy], and then more generally by Hoff in [Hof 1]; see the references therein for earlier work.

A general approach to scalar difference schemes is given by Crandall and Majda in [CM].

Chapter 17

The Riemann Problem for Systems of Conservation Laws

We shall begin this chapter by studying the Riemann problem for an important class of equations which we designate as p-systems. This class includes as a special case the equations of isentropic as well as isothermal gas dynamics. For these equations one can give a complete solution of the Riemann problem for any two constant states. In §B we shall study the general properties of shocks and rarefaction waves, while in §C we shall solve the Riemann problem for general hyperbolic systems of conservation laws, but only when the initial states are sufficiently close.

§A. The p-System

A model for isentropic ($=$ constant entropy) or polytropic gas dynamics is given by the following pair of conservation laws (in Lagrangian coordinates):

$$v_t - u_x = 0, \qquad u_t + \left(\frac{k}{v^\gamma}\right)_x = 0, \qquad t > 0, \quad x \in \mathbf{R}.$$

Here $k > 0$ and $\gamma \geq 1$ are constants. These equations represent the conservation of mass and momentum [cf. (15.2)]. (Since the temperature is held constant, energy must be added to the system; whence there is no conservation of energy equation.) In these equations, v denotes the specific volume, i.e., $v = \rho^{-1}$, where ρ is the density, and u denotes the velocity. γ is the adiabatic gas constant, and $1 < \gamma < 3$ for most gases.

We shall consider a somewhat more general class of equations which includes the above class. These are systems of the form

$$v_t - u_x = 0, \qquad u_t + p(v)_x = 0, \qquad t > 0, \quad x \in \mathbf{R}, \tag{17.1}$$

where $p' < 0$, $p'' > 0$. We refer to (17.1) as the *p-system*. If we choose $p(v) = kv^{-\gamma}$, then we recover the isentropic gas dynamics equations.

We let

$$U = (v, u), \qquad F(U) = (-u, p(v)),$$

so that (17.1) can be written in the form

$$U_t + F(U)_x = 0. \tag{17.2}$$

We note that (17.2) is a hyperbolic system since the Jacobian matrix,

$$dF = \begin{pmatrix} 0 & -1 \\ p'(v) & 0 \end{pmatrix}$$

has real and distinct eigenvalues

$$\lambda_1 \equiv -\sqrt{-p'(v)} < 0 < \sqrt{-p'(v)} \equiv \lambda_2. \tag{17.3}$$

The Riemann problem for (17.2) is the initial-value problem with data of the form

$$U(x, 0) = U_0(x) \equiv \begin{cases} U_l = (v_l, u_l), & x < 0, \\ U_r = (v_r, u_r), & x > 0. \end{cases} \tag{17.4}$$

We shall solve this problem in the class of functions consisting of constant states, separated by either shock waves or rarefaction waves. Before doing this however, we must develop a deeper understanding of these two classes of waves.

We begin by studying shock waves. According to our discussion in Chapter 15, §D, there are two distinct types of shock waves for (17.2); namely 1-shocks and 2-shocks. Using (15.31), the 1-shocks satisfy the inequalities

$$s < \lambda_1(U_l), \qquad \lambda_1(U_r) < s < \lambda_2(U_r), \tag{17.5}$$

while the 2-shocks satisfy

$$\lambda_1(U_l) < s < \lambda_2(U_l), \qquad \lambda_2(U_r) < s. \tag{17.6}$$

Since $\lambda_1 < 0 < \lambda_2$, these inequalities show that $s < 0$ for 1-shocks and $s > 0$ for 2-shocks. We shall sometimes call these *back shocks* and *front shocks*, respectively.

In view of (17.3), we see that (17.5) and (17.6) can be written as

$$-\sqrt{-p'(v_r)} < s < -\sqrt{-p'(v_l)} \quad \text{(back-shocks)}, \tag{17.7}$$

and

$$\sqrt{-p'(v_r)} < s < \sqrt{-p'(v_l)} \quad \text{(front shocks)}. \tag{17.8}$$

We now consider the following question. Given a state $U_l = (v_l, u_l)$, what are the possible states $U = (v, u)$ which can be connected to U_l on the right by a back shock? More precisely, we ask what are the states U, for which the Riemann problem (17.2) with data (U_l, U) is solvable by a *single* back shock? The possible states U must satisfy the jump condition (15.11), which in this case reads

$$s(v - v_l) = -(u - u_l), \qquad s(u - u_l) = p(v) - p(v_l). \qquad (17.9)$$

Eliminating s from these equations we obtain

$$u - u_l = \pm\sqrt{(v - v_l)(p(v_l) - p(v))}. \qquad (17.10)$$

We now determine the sign in (17.10). To do this, first note that (17.7) must hold, so

$$-\sqrt{-p'(v)} < -\sqrt{-p'(v_l)}.$$

This gives $p'(v_l) > p'(v)$, and since $p'' > 0$, we have $v_l > v$. Since $s < 0$, the first equation in (17.9) implies that $u < u_l$. This shows that we must take the minus sign in (17.10). Thus the set S of states which can be connected to U_l by a 1-shock on the right must lie in the curve[1]

$$S_1 : u - u_l = -\sqrt{(v - v_l)(p(v_l) - p(v))} \equiv s_1(v; U_l), \qquad v_l > v. \qquad (17.11)$$

We call this curve the *back-shock curve*, or the *1-shock curve*.

Next, we calculate ds_1/dv:

$$\frac{ds_1}{dv} = \frac{v - v_l}{2\sqrt{(v - v_l)(p(v_l) - p(v))}} \left\{ p'(v) + \frac{p(v_l) - p(v)}{v_l - v} \right\} > 0.$$

A tedious, but straightforward calculation, shows that the curve $u - u_l = s_1(v; U_l)$ in the region $v_l > v$ is starlike with respect to the point U_l; i.e., any ray through U_l meets this curve in at most one point. Thus the back-shock curve can be depicted as in Figure 17.1(a). If U_r is any point on this curve, then the Riemann problem for (17.2) with initial data (U_l, U_r) can be solved by a back shock, as in Figure 17.1(b). The speed s of the shock can be obtained from the equation $s(v_r - v_l) = -(u_r - u_l)$, as follows from (17.9). Moreover, by construction, we know that (17.5) is valid for these things.

---

[1] The curve determined in (17.10) when the plus sign is chosen corresponds to the so-called "rarefaction" shock waves; they do not satisfy the correct entropy conditions (17.5). We shall return to these things in Chapter 20, §D.

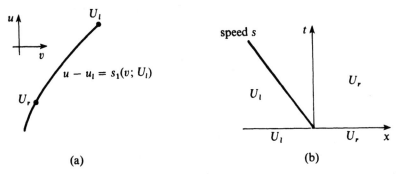

(a)   (b)

Figure 17.1

With a similar analysis, we can construct the curve S_2 consisting of all those states which can be connected to the state U_l by a 2-shock on the right. We find

$$S_2 : u - u_l = -\sqrt{(v - v_l)(p(v_l) - p(v))} \equiv s_2(v; U_l), \qquad v_l < v; \qquad (17.12)$$

we call this the *front-shock curve*, or the *2-shock curve*. A calculation as above shows that $ds_2/dv < 0$, and that the curve $u - u_l = s_2(v; U_l)$ is also starlike with respect to U_l in the region $v_l < v$. We can depict the front-shock curve in Figure 17.2(a). As before, if U_r is any point on this curve, then the Riemann problem for (17.2) with data (U_l, U_r) is solvable by a 2-shock, as depicted in Figure 17.2(b). The speed of the shock is obtained as above from the equation $s(v_r - v_l) = -(u_r - u_l)$, and (17.6) is valid for this solution.

We now turn to the study of rarefaction-wave solutions of (17.2); see Chapter 16, §D. We recall that a *rarefaction wave* is a continuous solution of (17.2) of the form $U = U(x/t)$. There are two families of rarefaction waves, corresponding to either characteristic family λ_1 or λ_2. Thus a k-rarefaction wave must satisfy the additional condition that the kth characteristic speed increases as x/t increases, $k = 1, 2$. In other words, we require that $\lambda_k(U(x/t))$ increases as x/t increases.

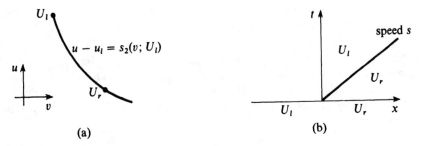

(a)   (b)

Figure 17.2

Now if we let $\xi = x/t$, then we see that $U = U(x/t)$ satisfies the ordinary differential equation

$$-\xi U_\xi + F(U)_\xi = 0,$$

or

$$(dF - \xi I)U_\xi = 0.$$

If $U_\xi \neq 0$, then U_ξ is an eigenvector of dF for the eigenvalue ξ.[2] Since dF has two real and distinct eigenvalues, $\lambda_1 < \lambda_2$, we see that there are two families of rarefaction waves, front- (or 2-)rarefaction waves, and back- (or 1-)rarefaction waves.

Let's first consider the case of 1-waves. The eigenvector $U_\xi = (v_\xi, u_\xi)^t$ satisfies

$$\begin{pmatrix} -\lambda_1 & -1 \\ p'(v) & -\lambda_1 \end{pmatrix} \begin{pmatrix} v_\xi \\ u_\xi \end{pmatrix} = \begin{pmatrix} 0 \\ 0 \end{pmatrix},$$

which gives $\lambda_1 v_\xi + u_\xi = 0$. Or, since $v_\xi \neq 0$, we must have $u_\xi/v_\xi = -\lambda_1$, so

$$\frac{du}{dv} = -\lambda_1(v, u) = \sqrt{-p'(v)}.$$

We can integrate this to obtain

$$R_1 : u - u_l = \int_{v_l}^{v} \sqrt{-p'(y)}\, dy \equiv r_1(v; U_l), \qquad v > v_l. \qquad (17.13)$$

The requirement $\lambda_1(v) > \lambda_1(v_l)$ gives $p'(v) > p'(v_l)$, so $v > v_l$, since $p'' > 0$. Finally, by direct calculation, we find

$$\frac{dr_1}{dv} = \sqrt{-p'(v)} > 0, \quad \text{and} \quad \frac{d^2 r_1}{dv^2} = \frac{-p''(v)}{2\sqrt{-p'(v)}} < 0.$$

We can thus depict the curve $u - u_l = r_1(v; U_l)$ in Figure 17.3(a). The corresponding solution of the Riemann problem for (17.2) with data (U_l, U_r) is depicted in Figure 17.3(b). The solution varies smoothly in the "fan," and every value of U between U_l and U_r on the curve $u - u_l = r_1(v; U_l)$ in Figure 17.3(a) moves with speed $\lambda_1(U)$ in the solution drawn in Figure 17.3(b). More precisely, if $\lambda_1(U_l) < x/t < \lambda_1(U_r)$, we obtain $U(x/t)$ as follows: from the equation $x/t = \lambda_1(U(x/t)) = -\sqrt{-p'(v(x/t))}$, we find $v(x/t)$, and then we use (17.13) to obtain $u(x/t)$.

---

[2] If $U_\xi \equiv 0$, then, of course, U is a constant.

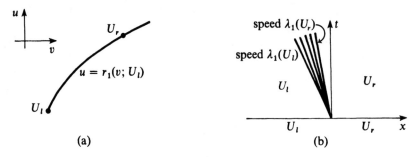

Figure 17.3

Finally, the 2-rarefaction wave curve is given by

$$R_2 : u - u_l = -\int_{v_l}^{v} \sqrt{-p'(y)}\, dy \equiv r_2(v; U_l), \qquad v_l > v, \qquad (17.14)$$

and $dr_2/dv < 0$, $d^2 r_2/dv^2 > 0$. This curve is depicted in Figure 17.4.

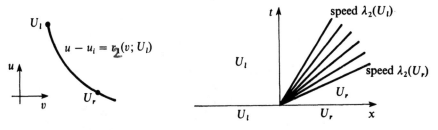

Figure 17.4

We can put all of these curves together in the $v - u$ plane to obtain a picture as in Figure 17.5. This shows that the $v - u$ plane is divided into four disjoint open regions I, II, III, and IV as depicted.

We remark that the curves $u - u_l = r_1(v; U_l)$ and $u - u_l = s_1(v; U_l)$ have second-order contact at U_l; i.e., their first two derivatives are equal at this point. The same is true for the two corresponding curves of the second-

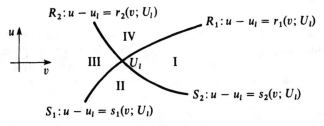

Figure 17.5

characteristic family. We shall not prove these facts here, since they are a consequence of a more general theorem, Theorem 17.15 below.

Now consider the general Riemann problem (17.2), (17.4). We consider U_l as fixed, and allow U_r to vary. If U_r lies on any of the above four curves; i.e., if $u_r - u_l = r_i(v_r; U_l)$, or if $u_r - u_l = s_i(v_r; U_l)$, $i = 1, 2$, then we have seen how to solve the problem. We thus assume that U_r lies in one of the four *open* regions I, II, III, or IV, drawn in Figure 17.5.

We define, for $\bar{U} \in \mathbf{R}^2$,

$$S_i(\bar{U}) = \{(v, u): u = s_i(v; \bar{U})\}, \qquad i = 1, 2,$$

$$R_i(\bar{U}) = \{(v, u): u = r_i(v; \bar{U})\}, \qquad i = 1, 2,$$

and

$$W_i(\bar{U}) = S_i(\bar{U}) \cup R_i(\bar{U}), \qquad i = 1, 2.$$

For fixed $U_l \in \mathbf{R}^2$. We consider the family of curves

$$\mathscr{F} = \{W_2(\bar{U}): \bar{U} \in W_1(U_l)\};$$

see Figure 17.6.

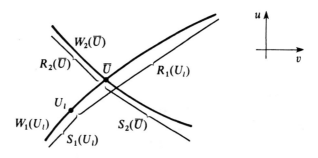

Figure 17.6

Let us assume for the moment, that the $v - u$ plane is covered univalently by the family of curves $\mathscr{F}$; i.e., through each point U_r, there passes exactly one curve $W_2(\bar{U})$ of $\mathscr{F}$. Then the solution to the Riemann problem (17.2), (17.4) is given as follows: we connect $\bar{U}$ to U_l on the right by a backward (shock, or rarefaction) wave, and then we connect U_r to $\bar{U}$ on the right by a forward (shock or rarefaction) wave. The particular type of occurring waves depends of course, on the position of U_r.

For example, if U_r is in region III (see Figure 17.5), then consider Figure 17.7(a). For each such U_r, there is a unique point $\bar{U}$, for which the curve $W_2(\bar{U})$ is in $\mathscr{F}$ and passes through U_r. Since $\bar{U} \in S_1(U_l)$, $\bar{U}$ is connected to U_l on the right by a back shock. Since $U_r \in R_2(\bar{U})$, U_r is connected to $\bar{U}$ on the right by a front rarefaction wave; see Figure 17.7(b).

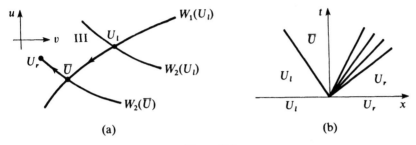

Figure 17.7

In Figure 17.8, we illustrate the four various possibilities.

It remains to determine whether the curves in $\mathscr{F}$ cover the $v - u$ plane univalently. We split the proof up into two cases; namely, in the first case, we assume that U_r lies in one of the regions I, II, or III; and in the second case, U_r lies in region IV (see Figure 17.5).

Suppose first that U_r lies in region I. Let the vertical line $v = v_r$ meet $W_1(U_l)$ at A, and let it meet $W_2(U_l)$ at B. Referring to Figure 17.9, we observe that the subfamily of curves in $\mathscr{F}$, consisting of the set $\{W_2(U) \equiv W_2(\bar{v}, \bar{u}) : v_l \le \bar{v} \le v_0\}$, induces a continuous mapping $p \to \phi(p)$ from the arc $U_l A$ to the line segment AB. This follows by transversality, since each of the S_2 curves have negative slope. Now in the region $U_l AB$, the slopes of these curves are bounded; thus we see that points P sufficiently close to A must map into points *above* U_r. Since U_l maps into B, which is *below* U_r, we see by continuity, that there must be a point U on the arc $U_l A$ which maps into U_r. This shows that the region I is covered by curves in $\mathscr{F}$. Since a similar argument works if U_r lies in regions II or III, we see that regions I, II, and III are covered by members of $\mathscr{F}$. This proves the *existence* of a solution of the Riemann problem for (17.2), (17.4), if U_r lies in regions I, II, or III, with respect to U_l.

We shall now show that the curves in $\mathscr{F}$ cover regions I, II, and III, univalently; i.e., that through each point U_r belonging to any of the regions I, II, or III, there passes *exactly one* element of $\mathscr{F}$.

Again let's suppose that U_r lies in region I. Referring to Figure 17.10, we see that it suffices to show that $\partial u / \partial \bar{v} > 0$. Now using the two equations

$$\bar{u} = u_l + \int_{v_l}^{\bar{v}} \sqrt{-p'(y)} \, dy \quad \text{and} \quad u = \bar{u} - \sqrt{(v_r - \bar{v})(p(\bar{v}) - p(v_r))},$$

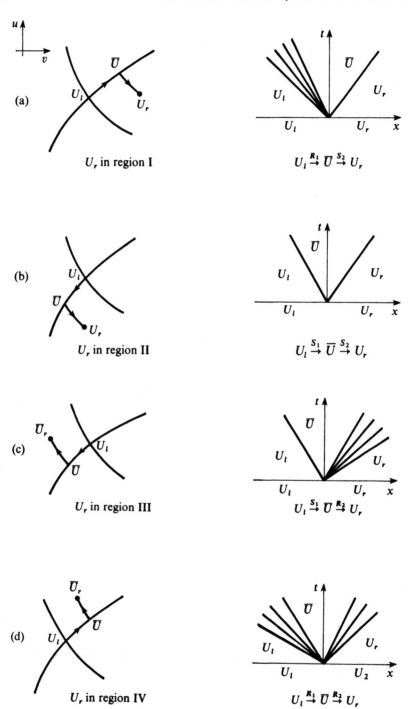

Figure 17.8

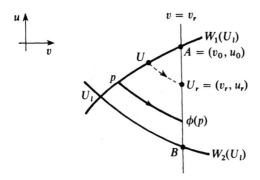

Figure 17.9

we compute

$$\frac{\partial u}{\partial \bar{v}} = \frac{\partial \bar{u}}{\partial \bar{v}} - \frac{1}{2\sqrt{(v_r - \bar{v})(p(\bar{v}) - p(v_r))}} \{(v_r - \bar{v})p'(\bar{v}) - (p(\bar{v}) - p(v_r))\}$$

$$= \sqrt{-p'(\bar{v})} - \frac{(v_r - \bar{v})}{2\sqrt{(v_r - \bar{v})(p(\bar{v}) - p(v_r))}} \left\{ p'(\bar{v}) + \frac{p(\bar{v}) - p(v_r)}{\bar{v} - v_r} \right\} > 0.$$

This implies the uniqueness result in region I; the proofs for regions II and III are similar.

We turn our attention now to the case where U_r lies in region IV. It is perhaps surprising that not every point in this region can always be covered by an element of $\mathscr{F}$. For example, if

$$u_0 \equiv \int_{v_l}^{\infty} \sqrt{-p'(y)} \, dy < \infty, \tag{17.15}$$

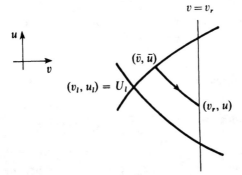

Figure 17.10

then it is easy to see that the curve

$$u = r_1(v; v_l, u_l) = u_l + \int_{v_l}^{v} \sqrt{-p'(y)}\, dy$$

has a horizontal asymptote; namely, the line $u = u_l + u_0$. Referring to Figure 17.11, we take $U_r = (v_l, u_r)$ to be in region IV, where u_r is chosen so

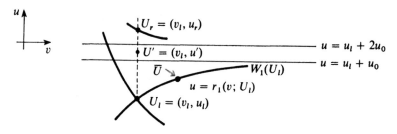

Figure 17.11

large that $u_r > u_l + 2u_0$. Let $\bar{U} = (\bar{v}, \bar{u})$ be any point on $W_1(U_l)$; then

$$\bar{u} = u_l + \int_{v_l}^{\bar{v}} \sqrt{-p'(y)}\, dy \leq u_l + u_0.$$

The point U' on $W_2(\bar{U})$ with abscissa v_l, has ordinate

$$u' = \bar{u} - \int_{\bar{v}}^{v_l} \sqrt{-p'(y)}\, dy = \bar{u} + \int_{v_l}^{\bar{v}} \sqrt{-p'(y)}\, dy$$

$$\leq \bar{u} + \int_{v_l}^{\infty} \sqrt{-p'(y)}\, dy \leq u_l + 2u_0 < u_r.$$

Thus U_r cannot lie on any curve in $\mathscr{F}$!

We have thus shown that if (17.15) holds, then the region IV is not covered by curves in $\mathscr{F}$. Generally speaking however, (17.15) is true in the interesting examples. Thus, for the case where

$$p(v) = \frac{k}{v^\gamma}, \qquad \gamma \geq 1 \quad (k = \text{const.} > 0),$$

an easy calculation shows that (17.15) holds if and only if $\gamma > 1$ (the physically relevant range).

The convergence of the integral (17.15) can be given a nice physical interpretation; namely, it corresponds to the appearance of a vacuum. For example, in the shock-tube problem, if the relative velocities on both sides of the membrane in the shock tube are sufficiently large, then a vacuum, or a region void of gas is formed. In this case, $\rho = 0$, or equivalently, $v = \rho^{-1}$ is infinite. To see this mathematically, note that the R_2 curve through U_r is given by

$$u - u_r = -\int_{v_l}^{v} \sqrt{-p'(y)}\, dy = r_2(v; U_r),$$

and, as above, u has a horizontal asymptote in the region $u < u_r$ (see Figure 17.11). Thus, along this curve $-\sqrt{-p'(v)} = \lambda_1$, tends to zero as $v \to \infty$, since $p'' > 0$. Similarly, $\lambda_2 = \sqrt{-p'(v)} \to 0$ as $v \to \infty$, if v lies on the R_1-curve through U_l (see Figure 17.11). A "solution" of the problem (17.2), (17.4) in this case is given in Figure 17.12, where we connect U_l on the right by a *complete* back-rarefaction wave, and we connect U_r on the left by a *complete* front-rarefaction wave. The solution is undefined on the line $x = 0$, since $v = +\infty$ (i.e., $\rho = 0$) there, and, of course, u is undefined.

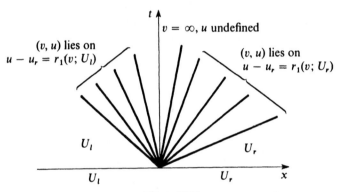

Figure 17.12

On the other hand, even if (17.15) holds, the vacuum does not necessarily appear, and it is possible to solve the Riemann problem if U_r is in region IV, provided that U_l and U_r are close; i.e., provided that $|U_l - U_r|$ is small. This solution is depicted in Figure 17.13 (see also Theorem 17.18 in §C of this chapter).

We have now shown how to solve the Riemann problem for the p-system (17.1), in the class of (at most three) constant states separated by shocks and rarefaction waves.

It is interesting to note that in the case where $p(v) = k/v^\gamma$, $\gamma = 1$; i.e., the isothermal gas case, the vacuum does not appear, since the integrals

$$\int_{v_l}^{\infty} \sqrt{-p'(y)}\, dy$$

all diverge.

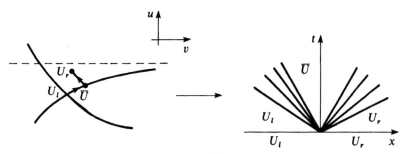

Figure 17.13

If the solution of the Riemann problem does not assume the vacuum state, then an examination of all the possible cases shows that the intermediate state in the solution lies between the rarefaction-wave curves determined by the initial states. That is, the solution satisfies the inequalities

$$r \geq \min(r_1, r_r) \equiv \min(r(v_l, u_l), r(v_r, u_r)),$$

$$s \leq \max(s_l, s_r) \equiv \max(s(v_l, u_l), s(v_r, u_r)),$$

where

$$r(v, u) = u - \int_{\bar{v}}^{v} \sqrt{-p'(s)}\, ds,$$

$$s(v, u) = u + \int_{\bar{v}}^{v} \sqrt{-p'(s)}\, ds.$$

Knowing how to solve the Riemann problem enables us to actually solve certain types of *interactions*. For example, suppose that we consider the equations (17.1) with the following initial data:

$$(v, u)(x, 0) = \begin{cases} (v_l, u_l), & x < x_1, \\ (v_m, u_m), & x_1 \leq x \leq x_2, \\ (v_r, u_r), & x > x_2, \end{cases}$$

and suppose too that the discontinuity (v_l, u_l), (v_m, u_m) is resolved by a front shock S_1 of speed s_1, and that the discontinuity (v_m, u_m), (v_r, u_r) is also resolved by a front shock S_2 of speed s_2; see Figure 17.14. From (17.6), we find $0 < s_2 < \lambda_2(v_m, u_m) < s_1$, from which it follows that S_1 overtakes S_2 at some time $T > 0$. Notice at $t = T$, we again have a Riemann problem with data (v_l, u_l), (v_r, u_r). In order to solve this problem, we must determine in what "quadrant" (v_r, u_r) lies in with respect to (v_l, u_l); see Figure 17.15.

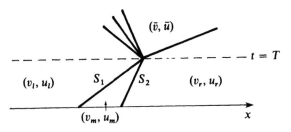

Figure 17.14

Our claim is that (v_r, u_r) lies in the first quadrant (not the second), so that the Riemann problem is resolved by a back-rarefaction wave and a front shock, as depicted in Figure 17.14. To see this, note that (v_m, u_m) lies on the front-shock curve S_2 starting at (v_l, u_l), and (v_r, u_r) lies on the front-shock curve $\tilde{S}_2$ starting at (v_m, u_m). If we can show that $\tilde{S}_2$ always lies above S_2 (as depicted in Figure 17.15), we will have proved our claim. To this end, we shall prove that: (i) the slope of $\tilde{S}_2$ at (v_m, u_m) is greater than the slope of S_2 at (v_m, u_m), so that $\tilde{S}_2$ "breaks into" the depicted region; and (ii), $\tilde{S}_2$ never meets S_2 for $v > v_m$.

To show (i), note that S_2 is given by the equation

$$u - u_l = -\sqrt{(v - v_l)p(v_l) - p(v)},$$

so that the slope of S_2 at (v_m, u_m) is

$$\frac{p'(v_m) + \dfrac{p(v_l) - p(v_m)}{v_l - v_m}}{2\sqrt{-\dfrac{p(v_l) - p(v_m)}{v_l - v_m}}} = \frac{p'(v_m) + p'(\xi)}{2\sqrt{-p'(\xi)}}, \qquad v_l < \xi < v_m.$$

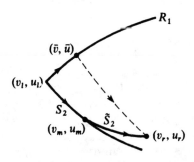

Figure 17.15

The slope of $\tilde{S}_2$ at (v_m, u_m) is $-\sqrt{-p'(v_m)}$ since the (normalized) right eigen-vector at (v_m, u_m) is $(1, \sqrt{-p'(v_m)})^t$. Thus, since

$$-\sqrt{-p'(v_m)} > \frac{p'(v_m) + p'(\xi)}{2\sqrt{-p'(\xi)}},$$

statement (i) follows. To prove (ii), we define a new function $\phi(x, y) = \sqrt{(x - y)(p(y) - p(x))}$. It is not hard to show that if $x > y > z$, then $\phi(x, z) > \phi(x, y) + \phi(y, z)$. Then, if there were a point $(v, u) \in S_2 \cap \tilde{S}_2$, with $v > v_m$, then $v > v_m > v_l$, and

$$u - u_m = -\phi(v, v_m), \qquad u - u_l = -\phi(v, v_l).$$

This gives

$$\phi(v_l, v_m) = u_l - u_m = \phi(v, v_l) - \phi(v, v_m) > \phi(v_m, v_l),$$

which is a contradiction, and the proof is complete.

§B. Shocks and Simple Waves

In this section we shall formulate a general theory of shock and rarefaction waves. It will turn out that to each characteristic family, λ_k, $k = 1, 2, , , , , n$, and to each point $u_l \in \mathbf{R}^n$, there corresponds a unique curve in u-space, this curve being the set of states that can be connected to u_l on the right by a k-shock, or a k-rarefaction wave. These curves are analogous to the corres-ponding curves $u - u_l = r_i(v; U)$ and $u - u_l = s_i(u; U_l)$, $i = 1,2$, which were constructed in §A. In the general case studied here however, these curves are only defined locally; i.e., near u_l. If then u_r is sufficiently close to u_l, we shall show that the Riemann problem is solvable. The proof is actually a nice application of the inverse function theorem, and is based on the hyperbolicity of the equations. Before giving the details, we must develop a general theory of shock and rarefaction waves.

We consider the system of n equations

$$u_t + f(u)_x = 0, \qquad x \in \mathbf{R}, \quad t > 0. \tag{17.16}$$

Here $u = (u_1, \ldots, u_n)$, $f(u) = (f_1(u), f_2(u), \ldots, f_n(u))$ is smooth in a neighbor-hood $N \subset \mathbf{R}^n$, and the Jacobian matrix, $df(u)$, has n real and distinct eigen-values $\lambda_1(u) < \cdots < \lambda_n(u)$ in N. Corresponding to each $\lambda_i(u)$, we have a right (column) eigenvector $r_i(u)$ and a left (row) eigenvector $l_i(u)$. These eigenvalues and eigenvectors are smooth functions of $u \in N$. We shall sometimes refer to the points in N as *states*.

We began our study by considering a general class of solutions called *centered rarefaction waves*. These are solutions which depend only on the ratio $(x - x_0)/(t - t_0)$, where (x_0, t_0) is called the *center* of the wave. We begin with a definition.

Definition 17.1. A *k-Riemann invariant* is a smooth function $w: N \rightarrow \mathbf{R}$ such that if $u \in N$,

$$\langle r_k(u), \nabla w(u) \rangle = 0.$$

(Here $\langle\ ,\ \rangle$ denotes the usual inner product in $\mathbf{R}^n$.)

Proposition 17.2. *There are* $(n - 1)$ *k-Riemann invariants whose gradients are linearly independent in* N.

Proof. Consider the vector field $R_k = r_k \cdot \nabla$, defined by

$$R_k(f) = \sum_i r_k^i \frac{\partial f}{\partial u_i}, \quad \text{where } r_k = (r_k^1, \ldots, r_k^n).$$

In some coordinate system $(Z_1, \ldots, Z_n)$, we can write $R_k = \partial/\partial Z_1$. Put $w_j = Z_{j+1}, j = 1, 2, \ldots, n - 1$. Then $R_k w_j = 0, 1 \le j \le n - 1$, and in this coordinate system, the $\nabla w_j, 1 \le j \le n - 1$, are linearly independent. $\square$

EXAMPLE 1. Consider the *p*-system (17.2)

$$v_t - u_x = 0, \qquad u_t + p(v)_x = 0,$$

where $p' < 0$. Here $f = (-u, p(v))$, and df has eigenvalues $\lambda_1 = -\sqrt{-p'(v)}$, $\lambda_2 = \sqrt{-p'(v)}$. The right eigenvector corresponding to, say, λ_2, is $R_2 = (-1, \sqrt{-p'(v)})^t$. Since $n = 2$, we should be able to find one Riemann invariant w^2. It must satisfy the equation $R_2 \cdot \nabla w^2 = 0$, or

$$-w_v^2 + \sqrt{-p'(v)}\, w_u^2 = 0,$$

and is given explicitly by

$$w^2(v, u) = u + \int^v \sqrt{-p'(y)}\, dy. \tag{17.17}$$

A similar calculation shows that the Riemann invariant corresponding to $R_1 = (1, \sqrt{-p'(v)})^t$, is given by

$$w^1(v, u) = u - \int^v \sqrt{-p'(y)}\, dy. \tag{17.18}$$

EXAMPLE 2. Consider the gas dynamics equations in Eulerian coordinates.

$$\rho_t + (\rho u)_x = 0,$$

$$u_t + uu_x + p_x/\rho = 0, \qquad \begin{pmatrix} \rho \\ u \\ s \end{pmatrix} = \begin{pmatrix} \text{density} \\ \text{velocity} \\ \text{entropy} \end{pmatrix},$$

$$s_t + us_x = 0,$$

where $p = p(\rho, s)$, $p_\rho > 0$. We denote the sound speed c by $c = \sqrt{p_\rho}$. The matrix

$$\begin{pmatrix} u & \rho & 0 \\ p_\rho/\rho & u & p_s/\rho \\ 0 & 0 & u \end{pmatrix}$$

has eigenvalues $u - c$, u, and $u + c$, with corresponding right eigenvectors $(\rho, -c, 0)^t$, $(p_s, 0, -p_\rho)^t$, and $(\rho, c, 0)^t$. The three pairs of Riemann invariants associated with these eigenvectors can be taken as

$$\{s, u + h\}, \qquad \{u, p\} \quad \text{and} \quad \{s, u - h\},$$

where $h = h(\rho, s)$ satisfies $h_\rho = c/\rho$; h is called the *enthalpy*. For example, if we consider the eigenvector $(\rho, c, 0)^t$, then the Riemann invariants w must satisfy the equation $\rho w_\rho + c w_u = 0$. Clearly $w^1 = s$ satisfies this equation, and also if $w^2 = u - h$, we find $\rho w_\rho^2 + c w_u^2 = \rho(-h_\rho) + c = 0$. Since $\nabla w^1 = (0, 0, 1)$ and $\nabla w^2 = (-h_\rho, 1, -h_s)$, we see that ∇w^1 and ∇w^2 are linearly independent.

Definition 17.3. Let u be a C^1 solution of (17.16) in a domain D, and suppose that all k-Riemann invariants are constant in D. Then u is called a k-*simple wave* (or sometimes, a k-*rarefaction wave*).

EXAMPLE 3. Let us consider again the p-system (17.1). We shall show that the rarefaction waves which we have discussed in §A satisfy the conditions of this definition. To this end, it suffices to show that the functions w^1 and w^2, as defined in (17.17) and (17.18), are constant along the curves (17.13) and (17.14), respectively. This is trivial; consider, e.g., w^1 defined on the curve given by (17.13). Letting this curve be parametrized by v, we have

$$\frac{dw^1}{dv} = \frac{\partial w^1}{\partial v} + \frac{\partial w^1}{\partial u} \frac{du}{dv} = -\sqrt{-p'(v)} + 1 \cdot \sqrt{-p'(v)} = 0.$$

A similar calculation is valid for w^2.

This last example illustrates a simple fact; namely, that for a k-simple wave u in a region R, the image, $u(R)$, is actually a curve in $\mathbf{R}^n$. To see this, note that in R, we have, by definition, $w_i(u) = c_i$, $i = 1, 2, \ldots, n - 1$, where the w_i are the $(n - 1)$-Riemann invariants, whose gradients are linearly independent. Thus the matrix (having n-rows and $(n - 1)$ columns)

$$M = [\nabla w_1^t \nabla w_2^t \cdots \nabla w_{n-1}^t]$$

has rank $(n - 1)$. This implies that not all the $(n - 1) \times (n - 1)$ minors in M are singular. Thus, the $(n - 1)$-equations $w_i(u) = c_i$, in n-variables, $u = (u_1, \ldots, u_n)$, are equivalent to a system of the form $u_j = \phi_j(u_i)$, $j = 1, 2, \ldots, n, j \neq i$, for some i.

We call attention to the fact that the curve determined by the intersection of the surfaces $w_i(u) = w_i(u_l)$, $i = 1, 2, \ldots, n - 1$, where u is a point in N, is precisely the integral curve of r_k passing through u_l. Thus, if $v(\theta)$ is such an integral curve, and w is a k-Riemann invariant, $dw(v(\theta))/d\theta = \nabla w \cdot v' = \nabla w \cdot r_k = 0$. On the other hand, if $\phi(\theta)$ is a (local) curve along which all the k-Riemann invariants are constant, then $dw/d\theta = \nabla w \cdot \phi' = 0$ for every k-Riemann invariant w. Thus ϕ' is orthogonal to the $(n - 1)$-dimensional space spanned by the ∇w_i, $i = 1, \ldots, n - 1$; i.e., ϕ' lies in the span of r_k.

To continue our development, it is useful to state explicitly the following lemma.

Lemma 17.4. *Let $d/dk = \partial/\partial t + \lambda_k \partial/\partial x$ denote differentiation in the λ_k direction. Then u is a classical solution of (17.16) if and only if $l_k \, du/dk = 0$, for $k = 1, 2, \ldots, n$, where l_k denotes the kth left eigenvector of $df(u)$ corresponding to the eigenvalue λ_k.*

Proof. u is a classical solution of (17.16) if and only if $u_t + df(u)u_x = 0$. If we multiply this on the left by l_k, we get

$$0 = l_k(u_t + df(u)u_x) = l_k(u_t + \lambda_k u_x) = l_k \frac{du}{dk},$$

as desired. $\square$

We next give a special property of simple waves. This result is analogous to the result for scalar equations which asserts that solutions are constant along characteristics, and that the characteristics are straight lines.

Theorem 17.5. *Let u be a k-simple wave in a domain D. Then the characteristics of the kth field (i.e., the curves $dx/dt = \lambda_k(u(x, t))$, are straight lines along which u is constant.*

Proof. From the last lemma, we have

$$l_k \frac{du}{dk} = 0. \tag{17.19}$$

Also, if $w_1, \ldots, w_{n-1}$ are $(n-1)$ k-Riemann invariants, then they are all constant in D. Thus

$$0 = \frac{dw_j}{dk} = \nabla w_j \cdot \frac{du}{dk}, \qquad j = 1, 2, \ldots, n-1.$$

These equations, together with (17.19) gives

$$\begin{bmatrix} l_k \\ \nabla w_1 \\ \nabla w_2 \\ \vdots \\ \nabla w_{n-1} \end{bmatrix} \frac{du}{dk} = 0.$$

Since the ∇w_j's are linearly independent, we see that the above $(n \times n)$ matrix is nonsingular, because $l_k r_k \neq 0$. Thus $du/dk = 0$, so u is constant in the kth characteristic direction, and it follows that the kth characteristic curves are straight lines. ◻

Note that the lines drawn to form the "fan" of a rarefaction wave are these straight line characteristic curves along which the solution is constant; cf. Figure 17.3(b).

Definition 17.6. A *centered simple wave*, centered at (x_0, t_0), is a simple wave depending only on $(x - x_0)/(t - t_0)$.

We now make an important definition; it corresponds to the same notion for systems as does convexity in the scalar case.

Definition 17.7. The kth characteristic family is said to be *genuinely nonlinear* in a region $D \subseteq \mathbf{R}^n$ provided that $\nabla \lambda_k \cdot r_k \neq 0$ in D. If this is the case, we shall normalize r_k by $\nabla \lambda_k \cdot r_k = 1$.

Notice that for a scalar equation, $u_t + f(u)_x = 0$, $\lambda = f'(u)$, $r = 1$ and $\nabla \lambda \cdot r = f''(u)$. Thus in this case, the notion of genuine nonlinearity reduces to convexity.

For the p-system (17.1), we have

$$\lambda_2 = \sqrt{-p'(v)}, \qquad r_2 = (-1, \sqrt{-p'(v)})^t \quad \text{and}$$

$$\nabla \lambda_2 \cdot r_2 = \left(\frac{-p''}{2\sqrt{-p'}}, 0 \right)^t \cdot (-1, \sqrt{-p'(v)})^t = \frac{p''}{2\sqrt{-p'}},$$

so that $p'' > 0$ implies that the second-characteristic family is genuinely non-linear. A similar calculation gives $\nabla \lambda_1 \cdot r_1 \neq 0$.

We now want to describe the states u which can be connected to a given state u_l on the right by a k-centered simple wave. Here is the main lemma.

Lemma 17.8. *Suppose that the kth characteristic field is genuinely nonlinear in N, and let $u_l \in N$. There exists a smooth one-parameter family of states $u(\gamma)$, defined for $|\gamma|$ sufficiently small, which can be connected to u_l on the right by a k-centered simple wave.*

Proof. Let $v(\gamma)$ be the solution of the problem

$$\frac{dv}{d\gamma} = r_k(v(\gamma)), \qquad v(\lambda_k(u_l)) = u_l, \qquad \gamma > \lambda_k(u_l).$$

The function $v(\gamma)$ exists on the interval $\lambda_k(u_l) \leq \gamma \leq \lambda_k(u_l) + a$, for sufficiently small a. Also

$$\frac{d}{d\gamma} \lambda_k(v(\gamma)) = \frac{dv}{d\gamma} \cdot \nabla \lambda_k = r_k \cdot \nabla \lambda_k = 1.$$

Thus $\lambda_k(v(\gamma)) = \gamma$, since $v(\lambda_k(u_l)) = u_l$. Define $u(x, t)$ by

$$u(x, t) = v(x/t), \qquad \lambda_k(u_l) \leq \frac{x}{t} \leq \lambda_k(u_l) + a.$$

Then if $\lambda_k(u_l) < x/t < \lambda_k(u_l) + a$, w is a k-Riemann invariant, and $\xi = x/t$, we have

$$\frac{dw}{d\xi} = \nabla w \cdot \frac{du}{d\xi} = \nabla w \cdot r_k = 0,$$

so w is constant in the region. Hence u defines a k-simple wave. Also, $d\lambda_k/d\xi = 1$, so that λ_k increases along this wave. Finally, since u is smooth, we have

$$
\begin{aligned}
u_t + df(u)u_x &= (-x/t^2)\frac{dv}{d\gamma} + \frac{df(v)(dv/d\gamma)}{t} \\
&= \frac{-x}{t^2}r_k(v) + \frac{df(v)r_k(v)}{t} \\
&= \frac{-x}{t^2}r_k(v) + \frac{\lambda_k(v)r_k(v)}{t} \\
&= \frac{-x}{t^2}(r_k(v)) + \frac{(x/t)r_k(v)}{t} = 0,
\end{aligned}
$$

so u is a solution of the conservation laws. This completes the proof. $\quad\square$

The states which are connected to u_l by a centered k-simple wave thus form a one-parameter family $u(\gamma)$. Since the k-Riemann invariants w_i are constant, we have $w_i(u) = w_i(u_l)$, $i = 1, 2, \ldots, n-1$. We introduce a parameter ε by $\lambda_k(u) = \lambda_k(u_l) + \varepsilon$ (this amounts to writing $\gamma = \varepsilon + \lambda_k(u_l)$ in the last theorem). We now have n-equations to solve for $u(\varepsilon)$; namely $F(u) = 0$, where F maps $\mathbf{R}^n$ into itself and is given by

$$
F(u) = (w_1(u) - w_1(u_l), \ldots, w_{n-1}(u) - w_{n-1}(u_l), \lambda_k(u) - \lambda_k(u_l) - \varepsilon), \quad u \in \mathbf{R}^n.
$$

Since the Jacobian matrix, $dF = (\nabla w_1, \ldots, \nabla w_{n-1}, \nabla \lambda_k)$ is nonsingular, the equation $F = 0$ defines a curve $u = u(\varepsilon; u_l)$ depending on u_l, for $|\varepsilon|$ sufficiently small. This is just a consequence of the implicit function theorem. This curve is (by uniqueness), the curve $v(\gamma)$ of Lemma 17.8, with a different parametrization. We note that since $d\lambda_k/d\varepsilon = 1$, the condition $\lambda_k(u_l) < \lambda_k(u(\varepsilon))$ is satisfied only for $\varepsilon > 0$. Moreover, for any k-Riemann invariant w,

$$
0 = \frac{dw}{d\varepsilon} = \nabla w \cdot \frac{du}{d\varepsilon}, \tag{17.20}
$$

so that $du/d\varepsilon$ is in the set $(\nabla w_1, \ldots, \nabla w_{n-1})^\perp$. It follows that $du/d\varepsilon = \alpha r_k$, for some α. But $\alpha = 1$ since

$$
1 = \frac{d\lambda_k}{d\varepsilon} = \nabla \lambda_k \cdot \frac{du}{d\varepsilon} = \alpha(\nabla \lambda_k \cdot r_k) = \alpha.
$$

Thus $du/d\varepsilon = r_k$. If we differentiate (17.20) with respect to ε, and denote differentiation with respect to ε by dot, we get

$$
0 = \nabla w \cdot \ddot{u} + \dot{\nabla w} \cdot \dot{u} = \nabla w \cdot \ddot{u} + \dot{\nabla w} \cdot r_k.
$$

We consider the relation $\nabla w \cdot r_k = 0$ along $u = u(\varepsilon)$, and obtain $\nabla w \cdot \dot{r}_k + \dot{\nabla} w \cdot r_k = 0$. These two equations give $\nabla w \cdot (\ddot{u} - \dot{r}_k) = 0$. It follows that $\ddot{u} - \dot{r}_k = \beta r_k$, for some constant β. But since $d\lambda_k/d\varepsilon = 1$, we have

$$
\begin{aligned}
0 = \frac{d^2 \lambda_k}{d\varepsilon^2} &= \frac{d}{d\varepsilon}\left(\nabla \lambda_k \cdot \frac{du}{d\varepsilon}\right) = \nabla \lambda_k \cdot \frac{d^2 u}{d\varepsilon^2} + \dot{\nabla} \lambda_k \left(\frac{du}{d\varepsilon}, \frac{du}{d\varepsilon}\right) \\
&= \nabla \lambda_k (\dot{r}_k + \beta r_k) + \dot{\nabla} \lambda_k (r_k, r_k) \\
&= \beta(\nabla \lambda_k \cdot r_k) + \nabla \lambda_k \cdot \dot{r}_k + \dot{\nabla} \lambda_k (r_k, r_k) \\
&= \beta + (\nabla \lambda_k \cdot r_k)^{\cdot} = \beta.
\end{aligned}
$$

Thus $\ddot{u} = \dot{r}_k$. We have therefore proved the following theorem, which in fact, completes our theoretical discussion of simple waves.

Theorem 17.9. *Let the kth characteristic field of the system* (17.16) *be genuinely nonlinear in N, and normalized so that $\nabla \lambda_k \cdot r_k = 1$. Let u_l be any point in N. There exists a one-parameter family of states $u = u(\varepsilon)$, $0 \le \varepsilon < a$, $u(0) = u_l$, which can be connected to u_l on the right by a k-centered simple wave. The parametrization can be chosen so that $\dot{u} = r_k$ and $\ddot{u} = \dot{r}_k$.*

at $\varepsilon = 0$

We turn our efforts now to developing a general theory of shock waves for systems (17.16). We recall from Chapter 15, §B, that if u suffers a discontinuity across a curve $x = x(t)$, and $x'(t) = s$, we must have the following jump conditions satisfied:[3]

$$
s(u_l - u_r) = f(u_l) - f(u_r). \tag{17.21}
$$

Furthermore, recall that a discontinuity satisfying (17.21) is called a k-shock if (see equation (15.31))

$$
\lambda_{k-1}(u_l) < s < \lambda_k(u_l),
$$

$$
\lambda_k(u_r) < s < \lambda_{k+1}(u_r). \tag{17.22}
$$

Now in analogy to what we have just done for centered simple waves, we pose the following problem. Given a state $u_l \in N$, describe the states u_r which can be connected to u_l on the right by a k-shock wave. Assuming that u_l is a known given state, we can view (17.21) as n-equations for the $(n + 1)$-unknowns u_r, s. This indicates that the answer to our problem should be a curve in u-space. This is indeed true, but[4] it requires a proof. We shall show

---

[3] If the system (17.16) is linear, $u_t + Au_x = 0$, then (17.21) becomes $s[u] = [Au] = A[u]$. Since $[u] \ne 0$, it follows that s is an eigenvalue of A, $s = \lambda_k$. Hence the curves of discontinuity for a linear system must lie along the characteristics. It is worth mentioning that this result is also true for linearly degenerate fields, as will be shown in Theorem 17.17 below.

[4] Note that we cannot get this result directly from the implicit function theorem since $u \equiv u_l$ is a solution of (17.21).

that under the assumptions of hyperbolicity and genuine nonlinearity the equations (17.21) do indeed define a one-parameter family of states near u_l which satisfy (17.22).

Before carrying out this program, we prove a simple lemma.

Lemma 17.10. *Let l_k and r_k denote the left and right eigenvectors of the hyperbolic system (17.16) corresponding to the eigenvector λ_k. Then*

$$l_k \cdot r_k (\nabla \lambda_k \cdot r_k) = l_k d^2 f(r_k, r_k), \qquad (17.23)$$

where $d^2 f$ is the second derivative (bilinear form) of the mapping f.

Recall that if $f = (f_1, f_2, \ldots, f_n)$, where each $f_i = f_i(u)$, and $H(f_i)$ denotes the Hessian matrix of f_i, then $d^2 f(r_i, r_i)$ is the column vector defined by

$$d^2 f(r_i, r_i) = \begin{bmatrix} r_i^t H(f_1) r_i \\ r_i^t H(f_2) r_i \\ \vdots \\ r_i^t H(f_n) r_i \end{bmatrix}$$

Proof of Lemma 17.10. Let $A = df$; then $Ar_k = \lambda_k r_k$. If we differentiate this equation in the direction r_k, we get

$$dA(r_k)r_k + A\, dr_k(r_k) = d\lambda_k(r_k)r_k + \lambda_k\, dr_k(r_k).$$

Multiplying on the left by l_k gives

$$l_k\, dA(r_k)r_k + l_k A\, dr_k(r_k) = d\lambda_k(r_k)l_k r_k + \lambda_k l_k\, dr_k(r_k),$$

or

$$l_k\, dA(r_k)r_k + \lambda_k l_k\, dr_k(r_k) = d\lambda_k(r_k)l_k r_k + \lambda_k l_k\, dr_k(r_k),$$

so that (17.23) holds. $\square$

Now since $l_k r_k \neq 0$, (17.23) shows that the kth characteristic field is genuinely nonlinear if and only if $l_k\, d^2 f(r_k, r_k) \neq 0$. Thus, the notion of genuine nonlinearity is now obviously seen to be a condition on the second derivative of f.

From now on, in this chapter, we assume that if the system (17.16) is genuinely nonlinear in the kth characteristic field, so that $\nabla \lambda_k \cdot r_k \neq 0$, then r_k is normalized by $\nabla \lambda_k \cdot r_k = 1$ and (then) l_k is normalized by $l_k r_k = 1$.

Theorem 17.11. *Let the system (17.16) be hyperbolic in N, and let $u_l \in N$. Then there are n smooth one-parameter family of states $u = u_k(\varepsilon)$, $k = 1, 2, \ldots, n$*

defined for $|\varepsilon| < a_k$, where $u_k(0) = u_l$, all of which satisfy the jump condition
(17.21).

Proof. Since

$$f(u) - f(u_l) = \int_0^1 \frac{d}{d\sigma} f(u_l + \sigma(u - u_l))\, d\sigma$$

$$= \int_0^1 df(u_l + \sigma(u - u_l))(u - u_l)\, d\sigma$$

$$\equiv G(u)(u - u_l),$$

we may write the jump condition (17.21) in the form

$$H(u, s) \equiv [G(u) - s](u - u_l) = 0. \qquad (17.24)$$

We change to polar coordinates (r, Ω); namely, set $(u - u_l) = r\Omega$, where $r = |u - u_l|$ and $\Omega \subset S^{n-1}$. Then (17.24) can be written as

$$H(r, \Omega, s) = rK(r, \Omega, s), \qquad (17.25)$$

where $K(r, \Omega, s) = [G(r, \Omega) - s]\Omega$. The solution set for (17.25) is given by $r = 0$ and $K = 0$, where $r = 0$ gives the trivial solution $u \equiv u_l$. Now $K = 0$ has the solution $(r, \Omega, s) = (0, r_i, \lambda_i)$, where $|r_i| = 1$, $i = 1, 2, \ldots, n$. At these points, K has full rank since $K_s(0, r_i, \lambda_i) = -r_i$ (has range r_i), and as $(1/r)(\partial/\partial\Omega) = \partial/\partial u$, $K_\Omega(0, r_i, \lambda_i) = f'(u_l) - \lambda_i I$ has range $\{r_j, 1 \le j \le n; j \ne i\}$. Thus the implicit function theorem implies that near the point $(0, r_i, \lambda_i)$, $K = 0$ can be written as $(\Omega, s) = (\Omega_i(r), s_i(r))$, so that

$$u_i(r) = r\Omega_i(r), \qquad i = 1, 2, \ldots, n. \quad \square \qquad (17.26)$$

We shall refer to the curves which we have just obtained, as the *k-shock curves*; $k = 1, 2, \ldots, n$.

Corollary 17.12. *The kth shock curve satisfies*

$$\dot{u}_k(0) = r_k, \quad \text{where } r_k = r_k(u_l). \qquad (17.27)$$

Proof. At $r = 0$, (17.26) gives $du_k/dr = \Omega_k(0) = r_k$. $\quad \square$

Corollary 17.13. *Along the k-shock curve, if the kth characteristic field is genuinely nonlinear, we can choose a parametrization so that $\dot{u}_k(0) = r_k$, and $\ddot{u}_k(0) = \dot{r}_k$, where $r_k = r_k(u_l)$. Moreover, with this parametrization $s(0) = \lambda_k(u_l)$, and $\dot{s}(0) = \frac{1}{2}$.*

Proof. We consider the equations (17.21), specialize to $u = u_k(\varepsilon)$, and differentiate twice with respect to ε. This gives

$$s\dot{u}_k + \dot{s}(u_k - u_l) = df\dot{u}_k \tag{17.28}$$

and

$$s\ddot{u}_k + 2\dot{s}\dot{u}_k + \ddot{s}(u_k - u_l) = df\ddot{u}_k + d\dot{f}\dot{u}_k. \tag{17.29}$$

If we consider (17.28) at $\varepsilon = 0$, we find

$$s\dot{u}_k = df\dot{u}_k$$

so that from the last corollary, $s(0) = \lambda_k(0) \equiv \lambda_k(u_l)$, since $\dot{u}_k(0) = r_k$.

Now consider the equation $df(u)r_k(u) = \lambda_k(u)r_k(u)$. If we take $u = u_k(\varepsilon)$ and differentiate with respect to ε, we find

$$\dot{\lambda}_k r_k + \lambda_k \dot{r}_k = d^2 f(r_k, r_k) + df\dot{r}_k.$$

Since $\dot{\lambda}_k = \nabla\lambda_k \cdot r_k = 1$ at $\varepsilon = 0$ by genuine nonlinearity, we have, at $\varepsilon = 0$,

$$r_k + \lambda_k \dot{r}_k = d^2 f(r_k, r_k) + df\dot{r}_k. \tag{17.30}$$

Using (17.29) at $\varepsilon = 0$, we find

$$\lambda_k \ddot{u}_k + 2\dot{s}r_k = df\ddot{u}_k + d^2 f(r_k, r_k). \tag{17.31}$$

Multiply both (17.30) and (17.31) by $l_k = l_k(u_l)$ on the left and subtract to get

$$1 = \dot{\lambda}_k(0) = 2\dot{s}(0),$$

so that $\dot{s}(0) = \frac{1}{2}$. Now subtract (17.30) from (17.31) to get (at $\varepsilon = 0$)

$$\lambda_k(\ddot{u}_k - \dot{r}_k) = df(\ddot{u}_k - \dot{r}_k).$$

Hence $\ddot{u}_k - \dot{r}_k = c'r_k$ at $\varepsilon = 0$. We can change our parametrization[5] so as to achieve $c' = 0$, and thus $\ddot{u}_k(0) = \dot{r}_k$. This completes the proof. $\square$

We next investigate the validity of the shock conditions (17.22) along the k-shock curve. It turns out that they are satisfied only along "half" of the

---

[5] Let $\delta - \frac{1}{2}\varepsilon\delta^2$ define the new parameter.

curve. The part of the curve in which they fail to hold consists of the so-called (inadmissible) *rarefaction shocks*, in analogy with gas dynamics. Specializing our results below to the gas dynamics equations, we will see later that the "bad" half consists of discontinuous solutions in which the entropy *decreases* across the discontinuity.

Theorem 17.14. *The shock inequalities* (17.22) *hold along the curve* $u = u_k(\varepsilon)$ *if and only if* $\varepsilon < 0$.

Proof. We write $\lambda_j(\varepsilon) = \lambda_j(u_k(\varepsilon))$, and $s(\varepsilon) = s(u_k(\varepsilon))$. The shock conditions (17.22) can be written as

(a) $\lambda_{k-1}(0) < s(\varepsilon) < \lambda_k(0)$,
(b) $\lambda_k(\varepsilon) < s(\varepsilon) < \lambda_{k+1}(\varepsilon)$.

Let $\phi(\varepsilon) = \lambda_k(\varepsilon) - s(\varepsilon)$. Then $\phi(0) = 0$, $\phi'(0) = \nabla\lambda_k \cdot r_k - \dot{s}(0) = 1 - \frac{1}{2} > 0$. Thus if (b) holds we see $\varepsilon < 0$. On the other hand, if $\varepsilon < 0$, then we see $\phi(\varepsilon) < 0$ so $\lambda_k(\varepsilon) < s(\varepsilon)$. Also $\dot{s}(0) = \frac{1}{2}$ and $\lambda_k(0) = s(0)$ imply $\lambda_k(0) > s(\varepsilon)$. Since $s(\varepsilon) \to \lambda_k(0) > \lambda_{k-1}(0)$ as $\varepsilon \to 0$, we have $s(\varepsilon) > \lambda_{k-1}(0)$ for small ε. Finally $\lambda_{k+1}(0) > \lambda_k(0) = s(0)$ gives $\lambda_{k+1}(\varepsilon) > s(\varepsilon)$ for small ε. This completes the proof. □

Using this theorem, together with Theorem 17.9, allows us to form the *composite* curves through $u_l \in N$, as follows: For each k, $1 \le k \le n$, define

$$U_k(\varepsilon) = \begin{cases} \bar{u}_k(\varepsilon), & \varepsilon \le 0, \\ \bar{\bar{u}}_k(\varepsilon), & \varepsilon \ge 0, \end{cases} \tag{17.32}$$

where $\bar{u}_k$ is the k-shock curve, and $\bar{\bar{u}}_k$ is the k-rarefaction-wave curve. Then Theorem 17.9 and Corollary 17.13 yield the following theorem.

Theorem 17.15. *The curves* $U_k(\varepsilon)$, $k = 1, 2, \ldots, n$, *have two continuous derivatives at* $\varepsilon = 0$.

This is also a classical result for the gas dynamics equations.

As a final result along these lines we have the following theorem, which again was classically known for the gas dynamics equations.

Theorem 17.16. (a) *The shock speed of a k-shock is the arithmetic average of the k-characteristic speeds on both sides of the shock, up to second-order terms in* ε.

(b) *The change in a k-Riemann invariant across a k-shock is of third order in* ε.

Since we may view ε as a measure of the strength *of a shock, these results say that for sufficiently weak shocks, the shock speed is well approximated by the average of the two characteristic speeds which impinge on the shock from both sides, and that the k-Riemann invariants are almost constant across a weak k-shock.*

Proof. Along the k-shock curve, we have from Corollary 17.13,

$$s(\varepsilon) = s(0) + \dot{s}(0)\varepsilon + O_2(\varepsilon) = \lambda_k(u_l) + \tfrac{1}{2}\varepsilon + O_2(\varepsilon), \qquad (17.33)$$

and

$$\lambda_k(u_r) \equiv \lambda_k(\varepsilon) = \lambda_k(0) + \dot{\lambda}_k(0)\varepsilon + O_2(\varepsilon) = \lambda_k(u_l) + \varepsilon + O_2(\varepsilon).$$

From the last equation we find

$$\varepsilon = \lambda_k(u_r) - \lambda_k(u_l) + O_2(\varepsilon),$$

and using this in (17.33) gives the conclusion in (a). Since w is constant along $\bar{u}_k$, we have $\dot{w}(0) = \ddot{w}(0) = 0$. □

We now consider the final type of elementary waves, namely, the so-called contact discontinuities. These arise when one characteristic family is not genuinely nonlinear; in fact, when it is as far from being genuinely nonlinear as possible. Thus we suppose that the kth characteristic field is *linearly degenerate* in N; that is, $\nabla\lambda_k \cdot r_k \equiv 0$ in N.

EXAMPLE 4. We consider the equations of gas dynamics which (as we shall show later), we may write in the form

$$v_t - u_x = 0,$$

$$u_t + p_x = 0,$$

$$S_t = 0,$$

where we take v, u, and S as variables, and $p = p(v, S)$. The associated matrix is

$$\begin{bmatrix} 0 & -1 & 0 \\ p_v & 0 & p_s \\ 0 & 0 & 0 \end{bmatrix}, \qquad p_v < 0, \quad p_{vv} > 0,$$

with eigenvalues $\lambda_1 = -\sqrt{-p_v}$, $\lambda_2 = 0$, $\lambda_3 = \sqrt{-p_v}$. The corresponding right eigenvectors are

$$r_1 = (1, -\lambda_1, 0)^t, \qquad r_2 = (p_s, 0, -p_v)^t, \qquad r_3 = (1, \lambda_1, 0)^t.$$

Then $\nabla \lambda_1 \cdot r_1 = -(\partial/\partial v)\sqrt{-p_v} \neq 0$ and similarly, $\nabla \lambda_3 \cdot r_3 \neq 0$. However, $\nabla \lambda_2 \cdot r_2 \equiv 0$, and thus the second characteristic family is linearly degenerate.

Now suppose that the kth characteristic field of the system (17.16) is linearly degenerate; i.e., $\nabla \lambda_k \cdot r_k \equiv 0$ in N. Then by definition, λ_k is a k-Riemann invariant. Thus if $u(\varepsilon)$, $|\varepsilon| < a$ is the solution of the problem

$$\frac{du}{d\varepsilon} = r_k(u(\varepsilon)), \qquad u(0) = u_l,$$

then $\nabla \lambda_k \cdot r_k = 0$, implies that λ_k is constant along this curve; i.e., $\lambda_k(u(\varepsilon)) = \lambda_k(u(0)) = \lambda_k(u_l)$, $|\varepsilon| < a$. Now if $|\varepsilon| < a$, define a function v by

$$v(x, t) = \begin{cases} u_l, & x < t\lambda_k(u_l), \\ u(\varepsilon), & x \geq t\lambda_k(u_l). \end{cases}$$

The claim is that v solves the equation (17.16) with the initial conditions

$$u(x, 0) = \begin{cases} u_l, & x < 0, \\ u(\varepsilon), & x > 0. \end{cases}$$

It is obvious that v takes on the correct data. To see that v is a solution, we need only check that the jump conditions (17.21) hold across the line of discontinuity. To this end, we set $s = \lambda_k(u_l)$; then

$$\frac{d}{d\varepsilon}\{f(u(\varepsilon)) - su(\varepsilon)\} = df\dot{u} - s\dot{u} = (df - \lambda_k)r_k = 0.$$

Thus $f(u(\varepsilon)) - su(\varepsilon) = f(u_l) - su_l$. This last equation is the jump condition, for solutions with a discontinuity along a curve $x = t\lambda_k(u_l)$. A solution of this type, where the shock speed equals the characteristic speed on one side, is called a *contact discontinuity*. We can now prove the following theorem.

Theorem 17.17. *If two nearby states u_l and u_r have the same k-Riemann invariants with respect to a linearly degenerate field, then they are connected to each other by a contact discontinuity of speed $s = \lambda_k(u_l) = \lambda_k(u_r)$. If the kth characteristic field is linearly degenerate in N_1 then if $u_l \in N$, there exists a* one-parameter family of states connected to u_l on the right by a contact discontinuity of speed $s = \lambda_k(u_l)$.

For, if $w_i(u_r) = w_i(u_l)$, $i = 1, 2, \ldots, n - 1$, where the w_i are k-Riemann invariants whose gradients are linearly independent, then we have seen in the remark just before Lemma 17.4, that u_l and u_r lie on the kth-characteristic curve through u_l, as long as u_r is close to u_l.

§C. Solution of the General Riemann Problem

In this section we apply the theory developed in §B to solve the Riemann problem for (17.16). We assume that (17.16) is hyperbolic, and that in N each characteristic field is either genuinely nonlinear or linearly degenerate, so that $\nabla\lambda_k \cdot r_k \equiv 1$ or $\nabla\lambda_k \cdot r_k \equiv 0$ in N, $1 \le k \le n$.

We consider the equation (17.16) with data

$$u(x, 0) = \begin{cases} u_l, & x < 0, \\ u_r, & x > 0. \end{cases} \tag{17.34}$$

We shall solve this problem uniquely, in the class of (at most) $(n + 1)$ constant states separated by shock waves, centered simple waves, and contact discontinuities, provided that $|u_l - u_r|$ is small.

In the remainder of this section, we shall assume that $|\varepsilon|$ is so small, that the curves $U_k(\varepsilon)$, defined by (17.32) all exist provided that the kth characteristic field is genuinely nonlinear, and that if the kth characteristic field is linearly degenerate, the curves satisfying $du/d\varepsilon = r_k(u(\varepsilon))$ all exist.

Theorem 17.18. *Let $u_l \in N$ and suppose that the system (17.16) is hyperbolic and that each characteristic field is either genuinely nonlinear or linearly degenerate in N. Then there is a neighborhood $\tilde{N} \subset N$ of u_l such that if $u_r \in \tilde{N}$, the Riemann problem (17.16), (17.34) has a solution. This solution consists of at most $(n + 1)$-constant states separated by shocks, centered simple waves or contact discontinuities. There is precisely one solution of this kind in $\tilde{N}$.*

Proof. From Theorems 17.15 and 17.17, we know that for each $k = 1, 2, \ldots, n$, there exists a one-parameter family of transformations

$$T_{\varepsilon_k}^k : N \to \mathbf{R}^n, \qquad |\varepsilon_k| < a,$$

which is C^2 in ε_k, with the property that any $u \in N$ can be joined to $T_{\varepsilon_k}^k u$ on the right by either a shock, centered simple wave, or contact discontinuity (depending on whether the kth characteristic field is genuinely nonlinear or linearly degenerate, and in the former case, depending on whether $\varepsilon_k < 0$ or $\varepsilon_k > 0$).

Now let u_l be any point in N, and define $U = \{(\varepsilon_1, \ldots, \varepsilon_n) \in \mathbf{R}^n : |\varepsilon_i| < a, i = 1, \ldots, n\}$. We consider the composite transformation $T : U \to \mathbf{R}^n$ given by $T(\varepsilon) = T_{\varepsilon_n}^n T_{\varepsilon_{n-1}}^{n-1} \ldots T_{\varepsilon_1}^1 u_l$, $\varepsilon = (\varepsilon_1, \ldots, \varepsilon_n)$. Our goal is to show that there is an $\bar{\varepsilon}$ in U such that $T(\bar{\varepsilon})u_l = u_r$, provided that $|u_r - u_l|$ is small. To this end, we define a mapping $F : U \to \mathbf{R}^n$ by

$$F(\varepsilon) = T(\varepsilon)u_l - u_l.$$

Then $F(0, \ldots, 0) = 0$, and since (from Theorem 17.15, for example)

$$T_{\varepsilon_k}^k u = u + \varepsilon_k r_k(u) + O_2(\varepsilon_k), \qquad k = 1, \ldots, n,$$

we have

$$F(\varepsilon_1, \ldots, \varepsilon_n) = \sum_{j=1}^{n} \varepsilon_j r_j(u_l) + O_2(\varepsilon).$$

This shows that $dF(0, \ldots, 0) = (r_1(u_l), \ldots, r_n(u_l))$. Since this latter matrix is nonsingular (by hyperbolicity), we can invoke the inverse function theorem to conclude that F is a homeomorphism of a neighborhood of $\varepsilon = 0$ onto a neighborhood of $u = 0$. Therefore, if $|u_r - u_l|$ is small, there is a unique $\bar{\varepsilon} = (\bar{\varepsilon}_1, \ldots, \bar{\varepsilon}_n)$ such that $F(\bar{\varepsilon}_1, \ldots, \bar{\varepsilon}_n) = u_r - u_l$. In other words,

$$T_{\bar{\varepsilon}_n}^n T_{\bar{\varepsilon}_{n-1}}^{n-1} \cdots T_{\bar{\varepsilon}_1}^1 u_l - u_l = u_r - u_l,$$

or

$$T_{\bar{\varepsilon}_n}^n \cdots T_{\bar{\varepsilon}_1}^1 u_l = u_r.$$

This completes the proof. $\square$

NOTES

The results in §A have been extended by Smoller [Smo], to systems of two equations having "big" data. The equations are of the form $u_t + F(u)_x = 0$, where $F = (f, g)$, $u = (u_1, u_2)$. It is assumed that $f_{u_2} g_{u_1} > 0$, a condition stronger than hyperbolicity, and also that $l_i d^2 F(r_j, r_j) > 0$, $i, j = 1, 2$, under the usual normalizations of l_i and r_i, the left and right eigenvectors of dF, respectively. If $f_{u_2} g_{u_1} < 0$, the shock curves can exhibit strange behavior, and the Riemann problem may fail to be solvable; see Borovikov, [Bv]. The results in §A have been extended to weakly hyperbolic systems by many authors [Is, KK, Df 2, 3, DL]. The appearance of the vacuum is studied by Liu and Smoller in [LS]. Shock interactions are considered from a general point of view in Smoller–Johnson [SJ].

The results in §B are adapted from Lax' fundamental paper [Lx 2]. It is there where one first encounters the basic ideas in the subject: the shock inequalities (17.22), the notion of genuine nonlinearity, the one-parameter families of shock- and rarefaction-wave curves, as well as the solution to the Riemann problem.

The proof of Theorem 17.11 is due to Conlon [Cn], with a further simplification due to K. Zumbrun (personal communcation); see also Foy [Fo], and Conley–Smoller [CS 2], for earlier different proofs.

Chapter 18

Applications to Gas Dynamics

In earlier sections, we have written the equations of gas dynamics in several different forms (see equations (18.3) and (18.3)' of this chapter, and Examples 1 and 2 in Chapter 15). In many standard texts, it is shown that they are all equivalent for classical solutions; i.e., they determine the same classical solutions. What we shall do here first is to prove that the computation of eigenvalues, Riemann invariants, and the genuine nonlinearity or linear degeneracy of the characteristic fields, are all independent of our choice of equations. That is, they are invariant under coordinate changes.

Thus let's consider two sets of coordinates u and w in $\mathbf{R}^n$, and suppose that $w = g(u)$ for some smooth invertible function g. Let the original system of equations be

$$w_t + \frac{\partial}{\partial x} f(w) = 0. \tag{18.1}$$

We shall derive the equation satisfied by u. Let $A = df(g(u))$, and $B = dg(u)$. Then if we consider "classical" solutions of (18.1), we see that (18.1) is equivalent to

$$Bu_t + ABu_x = 0,$$

or

$$u_t + B^{-1}ABu_x = 0. \tag{18.2}$$

We shall show that the Riemann invariants and eigenvalues of (18.2) are the same as those of (18.1). It is important to observe here, that these notions have meaning irrespective of whether or not the equations are in conservation form.

That the eigenvalues of (18.1) and (18.2) are the same is clear, since $B^{-1}AB$ and A always have the same eigenvalues.

Suppose now that r is a right eigenvector of $B^{-1}AB$ corresponding to the eigenvalue λ. Then, of course, Br is the corresponding eigenvector for A. If ϕ is a smooth function $\phi : \mathbf{R}^n \to \mathbf{R}$, let $\psi(w) = \phi(g^{-1}(w))$. Then

$$d\psi(w) = d\phi(g^{-1}(w)) \, dg^{-1}(w) = d\phi(g^{-1}(w))B^{-1} = d\phi(u)B^{-1},$$

so that

[handwritten marginalia: "too fast! Here dφr means nothing." "Ψbφp" does-]

$$d\phi r = d\phi B^{-1}(Br) = d\psi(Br).$$

Thus, ϕ is a Riemann invariant for (18.2) if and only if ψ is a Riemann invariant for (18.1).

Finally, Lemma 17.10 shows that the notions of genuine nonlinearity and linear degeneracy are both independent of the choice of coordinates.

It follows then, that the transformed system may be used for the purposes of deciding eigenvalues and Riemann invariants, as well as for computing whether a characteristic field is genuinely nonlinear or linearly degenerate.

§A. The Shock Inequalities

We write the equations of gas dynamics in Lagrangian coordinates:

$$v_t - u_x = 0 \quad \text{(conservation of mass)},$$
$$u_t + p_x = 0 \quad \text{(conservation of momentum)}, \qquad (18.3)$$
$$(e + \tfrac{1}{2}u^2)_t + (pu)_x = 0 \quad \text{(conservation of energy)}.$$

Here $v = \rho^{-1}$ is the specific volume, ρ = density, u = velocity, p = pressure, $p = p(E, v)$, $E = e + \tfrac{1}{2}u^2$ is the energy, and e = internal energy. The relation $p = p(E, v)$ is called the *equation of state*. It depends on the particular gas under consideration.

We assume, as is customary in thermodynamics, that given any two of the thermodynamics variables ρ, p, e, T, and S, we can obtain the remaining three variables.

The second law of thermodynamics asserts that

$$T\, dS = de + p\, dv, \qquad (18.4)$$

where T = temperature, and S = entropy. This implies that

$$e_S = T, \qquad e_v = -p,$$

where we are now assuming $e = e(S, v)$. Since

$$e_t = e_S S_t + e_v v_t = TS_t - pv_t = TS_t - pu_x,$$

we have

$$(e + \tfrac{1}{2}u^2)_t = e_t + uu_t = e_t - up_x = TS_t - pu_x - up_x = TS_t - (pu)_x.$$

Thus we can replace the third equation in (18.3) by $TS_t = 0$, and since[1] $T > 0$, we can write the system (18.3) as

$$v_t - u_x = 0,$$
$$u_t + p_x = 0, \qquad (18.3)'$$
$$S_t = 0,$$

where $p = p(v, S)$, $p_v < 0$. In these coordinates, the relevant Jacobian matrix is

$$\begin{pmatrix} 0 & -1 & 0 \\ p_v & 0 & p_S \\ 0 & 0 & 0 \end{pmatrix},$$

and the characteristic equation is $\lambda(\lambda^2 + p_v) = 0$. This gives for the eigenvalues

$$\lambda_1 = -\sqrt{-p_v}, \qquad \lambda_2 = 0, \qquad \lambda_3 = \sqrt{-p_v}. \qquad (18.5)$$

As we have seen in Chapter 17, §B, the first and third characteristic families are genuinely nonlinear, while the second is linearly degenerate. Thus there are only two families of shock curves, namely the first and the third. The system (18.3)′ is equivalent to (18.3) *only* for smooth solutions; the two cannot be equivalent for weak solutions since entropy is not conserved across shocks.

We shall use (18.3) to calculate the shock curves. In these coordinates, the jump conditions are

$$\sigma[v] = -[u],$$
$$\sigma[u] = [p], \qquad (18.6)$$
$$\sigma[e + \tfrac{1}{2}u^2] = [pu],$$

where the brackets denote the change across the shock, and σ is the shock speed. Let (v_0, u_0, e_0) denote the state on the left. Then

$$\sigma(e - e_0) + \frac{\sigma}{2}(u - u_0)(u + u_0) = pu - p_0 u_0.$$

But $\sigma(u - u_0) = p - p_0$, so

$$\sigma(e - e_0) + \tfrac{1}{2}(pu_0 - p_0 u) = \tfrac{1}{2}pu - \tfrac{1}{2}p_0 u_0.$$

---

[1] The equations are derived assuming $T > 0$.

Since $u = -\sigma(v - v_0) + u_0$, we get if $\sigma \neq 0$,

$$H(v, p) \equiv e - e_0 + \tfrac{1}{2}(p + p_0)(v - v_0) = 0. \tag{18.7}$$

Equation (18.7) is called the *Hugoniot equation*, and considering e as a function of v and p, $e = e(v, p)$, the points (v, p) with $H(v, p) = 0$ are referred to as the *Hugoniot curve*; or *shock adiabatic*. The equation $H = 0$ is the condition relating the thermodynamic variables across a shock transition, and it characterizes all those states p, v which can be connected to p_0, v_0 by a shock wave.

If we write $p = p(v, S)$, we make the assumptions

$$p_v < 0, \qquad p_{vv} > 0, \qquad p > 0. \tag{18.8}$$

We shall consider the shock curves in (v, u, S) space; as in Chapter 17, we can write them in the form $(v(\varepsilon), u(\varepsilon), S(\varepsilon))$, where, as we recall, $\varepsilon < 0$. Our object is to see how the thermodynamic variables change across a shock wave.

We have seen earlier that there are two families of shock curves, corresponding to the first and third characteristic families (see (18.5)). For either of these families we showed in the last chapter that the entropy S is a Riemann invariant. Thus S is constant along one branch of the composite C^2 curves defined in Equation (17.32). It follows from Theorem 17.16(b), that

$$S'(0) = S''(0) = 0, \tag{18.9}$$

where prime denotes differentiation with respect to ε.

Now from (18.4), $TS' = e' + pv'$, and differentiating (18.7) with respect to ε we get

$$e' = -\tfrac{1}{2}p'(v - v_0) - \tfrac{1}{2}(p + p_0)v'. \tag{18.10}$$

It follows that

$$TS' = \tfrac{1}{2}v'(p - p_0) - \tfrac{1}{2}(v - v_0)p'. \tag{18.11}$$

Now $p = p(v, S)$, so that $p' = p_v v' + p_s S'$, and using (18.9) we find

$$p'(0) = p_v v'(0). \tag{18.12}$$

We shall now consider the consequences of the shock inequalities (17.22). We know that they hold locally for genuinely nonlinear characteristic families (cf. Theorem 17.14). Let's consider first the case of a 3-shock. Here (17.22) gives

$$\sqrt{-p_v(\varepsilon)} < \sigma < \sqrt{-p_v(0)}, \tag{18.13}$$

where $p_v(\varepsilon) = p_v(v(\varepsilon), S(\varepsilon))$. From (18.13) it follows that $p_v(\varepsilon) > p_v(0)$, and since

$$p_v(\varepsilon) = p_v(0) + \varepsilon p_{vv} \dot{v}(0) + O_2(\varepsilon), \qquad \varepsilon < 0, \qquad (18.14)$$

we see from (18.8) that $\dot{v}(0) < 0$, and hence $v(\varepsilon) > v(0)$. It follows that

$$\rho(0) > \rho(\varepsilon) \quad \text{across a 3-shock.} \qquad (18.15)$$

Now (18.13) shows that $\sigma > 0$ across a 3-shock. Thus, if we consider Figure 18.1, we see that 3-shocks are *compressive*, in the sense that the density increases after the shock passes.

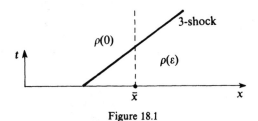

Figure 18.1

Next, from (18.10), we find $e'(0) > 0$, so

$$e(0) > e(\varepsilon), \qquad (18.16)$$

and thus the internal energy increases after the passage of a 3-shock. Finally, (18.12) and (18.8) give $p'(0) > 0$, so that

$$p(0) > p(\varepsilon), \qquad (18.17)$$

and thus the pressure also must increase across a 3-shock.

We now investigate the change in entropy across a 3-shock. In view of (18.9), we must compute $S'''(0)$. We differentiate (18.11) with respect to ε to get

$$TS'' + T'S' = \tfrac{1}{2}v''(p - p_0) - \tfrac{1}{2}(v - v_0)p''.$$

Then differentiating this expression and using (18.12) we obtain, at $\varepsilon = 0$,

$$2TS''' = v''p' - v'p''$$
$$= v''(p_v v') - v'p''.$$

But $p' = p_v v' + p_s S'$ so that we have $p'' = p_{vv}(v')^2 + p_v v''$, again from (18.9). Thus, at $\varepsilon = 0$,

$$2TS''' = -p_{vv}(v')^3. \qquad (18.18)$$

But we have seen earlier that $v'(0) < 0$, so that

$$S'''(0) > 0.$$

Writing

$$S(\varepsilon) = S(0) + \frac{S'''(0)}{6}\varepsilon^3 + O_4(\varepsilon), \tag{18.19}$$

we find

$$S(0) > S(\varepsilon), \tag{18.20}$$

and thus the entropy increases after the passage of a 3-shock.

We call attention to the fact that this conclusion is a consequence of the shock inequalities (18.13), which in turn follow from our general inequalities (17.22). We remark that similar calculations show that the density, pressure, and entropy also increase across a 1-shock.

Conversely, if we consider the 3-shock curve $(v(\varepsilon), u(\varepsilon), S(\varepsilon))$, and assume that (18.20) holds, then since (18.9) always holds, we have from (18.19) $S'''(0)\varepsilon^3 < 0$ so (18.18) gives $0 > -p_{vv}(\varepsilon v')^3$, whence sgn ε = sgn $v'(0)$. Then from (18.14) $p_v(\varepsilon) > p_v(0)$ or $\lambda_3(\varepsilon) < \lambda_3(0)$. Since $\lambda'_3(0) = \nabla\lambda_3 \cdot r_3 = 1$, we see $\varepsilon < 0$ and since $\sigma'(0) = \frac{1}{2}$, we have $\lambda_3(0) > \sigma(\varepsilon) > \lambda_3(\varepsilon)$. Since similar results hold for 1-shocks, we have the following theorem.

Theorem 18.1. *Consider the equations of gas dynamics* (18.3), *where the equation of state* $p = p(v, S)$ *satisfies* (18.8). *Then for sufficiently weak shocks, the inequalities* (17.22) *are equivalent to the increase of entropy across a shock.*

We shall show that the shock inequalities (17.22) actually hold globally along the shock curve, under an additional assumption on the equation of state; namely that

$$p_s > 0. \tag{18.21}$$

Again let's only consider 3-shocks. Writing the equations (18.3) in the form $U_t + F(U)_x = 0$, the jump conditions become $\sigma[U] = [F]$, where $[\phi]$ denotes the difference of ϕ on both sides of the shock, $[\phi] = \phi(\varepsilon) - \phi(0)$. We assume that the jump conditions define the shock curve $U = U(\varepsilon)$, (where U is nonsingular; i.e., $\dot{U} \neq 0$) with shock speed $\sigma = \sigma(\varepsilon)$, $\varepsilon \leq 0$. We know that for *small* $\varepsilon < 0$, $\lambda_3(\varepsilon) < \sigma(\varepsilon) < \lambda_3(0)$ (by Theorem 17.14). We shall show that these inequalities hold everywhere along the shock curve, provided that (18.21) holds.

Thus, suppose ε_1 is the first point where $\lambda_3(\varepsilon) = \sigma(\varepsilon)$, $\varepsilon_1 < 0$. Since

$$\sigma'[U] + \sigma U' = dFU', \tag{18.22}$$

if we consider this at $\varepsilon = \varepsilon_1$, and multiply by the left eigenvector $l_3(\varepsilon_1)$, we get $\sigma' l_3 \cdot [U] = 0$. *Suppose* now, that

$$l_3 \cdot [U] \neq 0; \qquad (18.23)$$

then $\sigma'(\varepsilon_1) = 0$; hence at ε_1, $\lambda_3 U' = dFU'$ so that $U' = r_3$. Then

$$\frac{d}{d\varepsilon}(\sigma - \lambda_3)\Big|_{\varepsilon=\varepsilon_1} = -\frac{d\lambda_3}{d\varepsilon}\Big|_{\varepsilon=\varepsilon_1} = -\nabla\lambda_3 \cdot r_3\big|_{\varepsilon=\varepsilon_1} = -1.$$

Thus $\lambda_3 = \sigma$ for some ε_2, $\varepsilon_1 < \varepsilon_2 < 0$; this contradicts the definition of ε_1. We conclude that if (18.23) holds, then $\sigma(\varepsilon) > \lambda_3(\varepsilon)$ for all ε. Furthermore, if $\sigma(\varepsilon_1) = \lambda_3(0)$ for some $\varepsilon_1 < 0$, then there is an ε_2 with $\varepsilon_1 < \varepsilon_2 < 0$ such that $\sigma'(\varepsilon_2) = 0$. Now $\sigma(\varepsilon_2) > \lambda_3(\varepsilon_2) > 0$ so (18.22) at ε_2 gives $\sigma U' = dFU'$. Thus necessarily $U'(\varepsilon_2) = r_3(\varepsilon_2)$, and $\sigma(\varepsilon_2) = \lambda_3(\varepsilon_2)$. Thus $\sigma'(\varepsilon)$ doesn't change sign, so sgn $\sigma'(\varepsilon) = $ sgn $\sigma'(0) = \frac{1}{2}$, from Corollary 17.13. Hence $\sigma' > 0$ so $\sigma(\varepsilon) < \lambda_3(0)$, if $\varepsilon < 0$.

To recapitulate, we have shown that if (18.23) holds, then the shock inequalities (18.13) hold. We shall now prove (18.23). Using (18.3)′ an easy calculation shows $l_3 = (p_v, \lambda_3, p_s)$. Thus

$$l_3 \cdot ([v], [u], [S]) = p_v[v] + \lambda_3[u] + [S]p_s. \qquad (18.24)$$

Now (18.6) implies $[v] \neq 0$; otherwise there is no discontinuity. Thus from (18.15), $[v] > 0$ so that

$$p_v[v] < 0. \qquad (18.25)$$

Next, from (18.6), we see that $[u] \neq 0$ implies $\sigma \neq 0$. Hence if $[u] \neq 0$, we find from (18.6) again,

$$\sigma^2 = -\frac{[p]}{[v]}.$$

Differentiating this along the shock curve gives

$$TS' = \sigma\sigma'[v]^2, \qquad (18.26)$$

where we have used (18.11). Now $[S] \neq 0$; otherwise there is a first point where $S' = 0$, so that from (18.26), $\sigma' = 0$. This gives the same contradiction as above. Thus, if $[u] \neq 0$, $\sigma > 0$, $[S] < 0$ and $[u] = [p]/\sigma$. Since $[p] \neq 0$, $[p] < 0$ by (18.17). Thus using $p_s > 0$, (18.25) shows $l_3 \cdot ([v], [u], [S]) < 0$. If $[u] = 0$, at a first point ε_1, so $\sigma(\varepsilon_1) = 0$, then if $[S]p_s + p_v[v] = 0$, at $\varepsilon = \varepsilon_1 < 0$, we have $[S] = 0$ for some ε_2, $\varepsilon_1 < \varepsilon_2 < 0$. Then $S' = 0$ at some ε_3, $\varepsilon_2 < \varepsilon_3 < 0$, so that from (18.26) either $\sigma'(\varepsilon_2) = 0$, in which case we get the same contradiction as above, or else $\sigma(\varepsilon_2) = 0$, and hence from (18.6) we find $[u] = 0$ at ε_2. Again this is a contradiction. Thus (18.23) holds. It follows that if $p_s > 0$, then the shock inequalities hold. We have therefore proved the following theorem.

Theorem 18.2. *For the gas dynamics equations* (18.3), *assume that the equation of state* $p = p(v, S)$ *satisfies*

$$p_v < 0, \qquad p_{vv} > 0, \qquad p_s > 0, \qquad p > 0.$$

Then the shock inequalities (17.22) *hold everywhere along the shock curves, provided that they are nonsingular.*

We end this section by showing that the assumptions on p given in this last theorem imply that S is monotone everywhere along the Hugoniot curve (18.7).

Theorem 18.3. *Consider the Hugoniot curve* $H(v, p) = 0$ *defined by* (18.7), *and assume that the equation of state* $p = p(v, S)$ *satisfies* $p_v < 0$, $p_{vv} > 0$, $p_s > 0$. *If* $dH \neq 0$ *along* $H = 0$, *then* S *is monotone all along this curve.*

Proof. To determine the behavior of S on $H = 0$, we consider S on those curves γ where the 1-form

$$\omega = dH - T \, dS$$

vanishes. Using (18.4)

$$
\begin{aligned}
dH &= de + \tfrac{1}{2}(p + p_0) \, dv + \tfrac{1}{2}(v - v_0) \, dp \\
&= -p \, dv + T dS + \tfrac{1}{2}(p + p_0) \, dv + \tfrac{1}{2}(v - v_0) \, dp \\
&= \tfrac{1}{2}(p_0 - p) \, dv + \tfrac{1}{2}(v - v_0) \, dp + T dS.
\end{aligned}
$$

Hence

$$\omega = \tfrac{1}{2}(p_0 - p) \, dv + \tfrac{1}{2}(v - v_0) \, dp,$$

and thus the curves γ are solutions of the differential equations

$$\dot{v} = v - v_0, \qquad \dot{p} = p - p_0. \tag{18.27}$$

This system has only one critical point, (v_0, p_0) which is easily seen to be a repellor.

Let γ be any integral curve of (18.27). Since $dH = \omega + T \, dS$, we see that the critical points of $H|_\gamma$ and $S|_\gamma$ coincide; i.e., H and S are critical together on γ.

At this point we need a lemma.

Lemma 18.4. *S has at most one critical point on* γ.

Proof. On γ we have

$$p - p_0 = \dot{p} = p_v \dot{v} + p_s \dot{S}.$$

We differentiate this along γ and evaluate it at a critical point of S along γ; i.e., at a point where $\dot{S} = 0$. This gives

$$p_v \dot{v} = p_{vv}(\dot{v})^2 + p_v \ddot{v} + p_s \ddot{S}.$$

Since $\ddot{v} = \dot{v} = v - v_0$, we get

$$p_{vv}(\dot{v})^2 + p_s \ddot{S} = 0,$$

which shows $\ddot{S} < 0$ when $\dot{S} = 0$. Thus S can have at most, and so precisely, one critical point on γ; this being a maximum. This proves the lemma. $\square$

We can now complete the proof of the theorem. Since $dH \neq 0$, we can write $H = 0$ as a curve $v = v(\sigma)$, $p = p(\sigma)$. Let

$$\frac{d}{d\mu} = \frac{dv}{d\sigma}\frac{\partial}{\partial v} + \frac{dp}{d\sigma}\frac{\partial}{\partial p}$$

denote differentiation along $H = 0$.

Suppose S were critical at (v_1, p_1), where $H(v_1, p_1) = 0$; then $dS/d\mu = 0$ at (v_1, p_1). Now (18.4) gives at (v_1, p_1),

$$\frac{de}{d\mu} = T\frac{dS}{d\mu} - p\frac{dv}{d\mu} = -p\frac{dv}{d\mu},$$

so from (18.7), we have, at (v_1, p_1),

$$0 = \frac{dH}{d\mu} = \frac{de}{d\mu} + \tfrac{1}{2}(p + p_0)\frac{dv}{d\mu} + \tfrac{1}{2}(v - v_0)\frac{dp}{d\mu}$$

$$= -p\frac{dv}{d\mu} + \tfrac{1}{2}(p + p_0)\frac{dv}{d\mu} + \tfrac{1}{2}(v - v_0)\frac{dp}{d\mu}$$

$$= \tfrac{1}{2}(p_0 - p)\frac{dv}{d\mu} + \tfrac{1}{2}(v - v_0)\frac{dp}{d\mu}.$$

Thus

$$\frac{d}{d\mu}\binom{v}{p} \perp \binom{p_0 - p_1}{v_1 - v_0},$$

so that at (v_1, p_1),

$$\frac{d}{d\mu}\binom{v}{p} = k\binom{v_1 - v_0}{p_1 - p_0} = k\binom{\dot{v}}{\dot{p}}, \qquad k = \text{const.} \qquad (18.28)$$

Note that $k \neq 0$ since $dH \neq 0$ at (v_1, p_1). Hence from (18.28) we see that the integral curve of (18.27) through (v_1, p_1) is tangent to $H = 0$ at (v_1, p_1). Thus at this point,

$$\dot{S} = \frac{\partial S}{\partial v}\dot{v} + \frac{\partial S}{\partial p}\dot{p} = k^{-1}\left[\frac{\partial S}{\partial v}\frac{dv}{d\mu} + \frac{\partial S}{\partial p}\frac{dp}{d\mu}\right] = k^{-1}\frac{dS}{d\mu} = 0.$$

If γ is the orbit of (18.27) through (v_1, p_1), we have shown that $H = 0$ at two points on γ; namely at (v_0, p_0) and at (v_1, p_1). Thus H on γ must be critical at some intermediate point on γ. So S too must be critical at this intermediate point. But since S is critical at (v_1, p_1), this implies that S is critical twice on γ, and thus violates Lemma 18.4. It follows that S is never critical on $H = 0$, and the proof is complete. $\square$

§B. The Riemann Problem in Gas Dynamics

In this section we shall solve the Riemann problem for the equations of gas dynamics for ideal polytropic gases without any restriction on the magnitude of the initial states. Of course, we will have to rule out the vacuum, as was necessary in Chapter 17, §A.

We assume that the gas is *ideal*, so that the equation of state is given by

$$p = R\rho T, \qquad (18.29)$$

and that it is *polytropic*, so that $e = c_v T$, and

$$p = ke^{S/c_v}\rho^\gamma. \qquad (18.30)$$

Here R, k, c_v, and γ are positive constants, and $1 < \gamma$. Thus $p_\rho > 0$, $p_{\rho\rho} > 0$ and $p_s > 0$. This implies from Theorem 18.2, that the shock inequalities (17.22) are everywhere valid along the shock curves.

We shall consider the gas dynamics equations in Eulerian coordinates:

$$\frac{\partial \rho}{\partial t} + \frac{\partial}{\partial x}(\rho u) = 0,$$

$$\frac{\partial}{\partial t}(\rho u) + \frac{\partial}{\partial x}(p + \rho u^2) = 0, \tag{18.31}$$

$$\frac{\partial}{\partial t}[\rho(\tfrac{1}{2}u^2 + e)] + \frac{\partial}{\partial x}[\rho u(\tfrac{1}{2}u^2 + e) + pu] = 0.$$

There are three characteristic families corresponding to the eigenvalues $\lambda_1 < \lambda_2 < \lambda_3$. The relevant facts are summarized in the following table (the calculations are quite straightforward, and we omit them).

	$i = 1$	$i = 2$	$i = 3$
λ_i	$u - c$	u	$u + c$
R_i	$(\rho, -c, 0)^t$	$(p_s, 0, -p_\rho)^t$	$(\rho, c, 0)^t$
$R_i \cdot \nabla \lambda_i$	$-c - \rho c_\rho$	0	$c + \rho c_\rho$
Riemann invariants	$\left\{ s, u + \dfrac{2}{\gamma - 1} \cdot c \right\}$	$\{u, p\}$	$\left\{ s, u - \dfrac{2}{\gamma - 1} c \right\}$

Here $c^2 = \gamma p / \rho \equiv \partial h / \partial \rho$, and c is called the *sound speed*.

We shall explicitly compute the one-parameter families of shocks, simple waves, and contact discontinuities. Since the formulas which we shall obtain are explicit, we shall not bother with the normalizations $R_i \cdot \nabla \lambda_i = 1$ which were needed in the general case (Chapter 17, §B), in order to make the analysis easier.

The jump conditions for this system are

$$\sigma[\rho] = [\rho u],$$

$$\sigma[\rho u] = [p + \rho u^2], \tag{18.32}$$

$$\sigma[\rho(\tfrac{1}{2}u^2 + e)] = [pu + \rho u(\tfrac{1}{2}u^2 + e)].$$

We shall rewrite these in a more convenient form by introducing the variables

$$v = u - \sigma, \qquad m = \rho v.$$

In (18.32) we eliminate u; this gives for the first equation in (18.32),

$$\sigma(\rho - \rho_0) = \rho u - \rho_0 u_0 = \rho(v + \sigma) - \rho_0(v_0 + \sigma) = m - m_0 + \sigma(\rho - \rho_0).$$

The first equation can thus be written as $[m] = 0$, and similarly the second equation becomes $\sigma(\rho_0 v_0 - \rho v) - (vm - v_0 m_0) = p - p_0$. Since $[m] = 0$, we have $[p + mv] = 0$. The third equation in (18.32) becomes

$$\rho(\tfrac{1}{2}u^2 + e)(\sigma - u) - \rho_0(\tfrac{1}{2}u_0^2 + e_0)(\sigma - u_0) = pu - p_0 u_0,$$

or

$$[pu + m(e + \tfrac{1}{2}u^2)] = 0.$$

But $pv = (\gamma p/\rho)(\rho v/\gamma) = mc^2/\gamma$, and $e = c^2/\gamma(\gamma - 1)$. This gives the following jump relations:

$$[m] = 0,$$

$$[p + mv] = 0,$$

$$m\left[\frac{2}{\gamma - 1}c^2 + v^2\right] = 0,$$

(18.33)

where the last equation comes from $[m([2/(\gamma - 1)]c^2 + v^2)] = 0$, since $m = \text{const}$.

We now see what further implications come from the shock inequalities, knowing that they hold all along the shock curve. For 1-shocks we have

$$\sigma < u_l - c_l, \qquad u_r - c_r < \sigma < u_r,$$

so that $c_l < v_l$ and $v_r - c_r < 0 < v_r$, or $0 < v_r < c_r$. Hence

$$c_l < v_l \quad \text{and} \quad 0 < v_r < c_r \quad \text{for 1-shocks.} \tag{18.34}$$

Since $c_l > 0$, we have $u_l > \sigma$ and $u_r > \sigma$. Thus the gas speed on both sides of the shock is greater than the shock speed, so for 1-shocks particles cross the shock from left to right.

For 3-shocks, the shock inequalities give

$$u_l < \sigma < u_l + c_l, \qquad u_r + c_r < \sigma,$$

so that $v_l < 0 < v_l + c_l$ or

$$-c_l < v_l < 0, \qquad v_r < -c_r < 0 \quad \text{for 3-shocks.} \tag{18.35}$$

Thus $\sigma > u_l, u_r$ implies that for 3-shocks the shock speed is greater than the gas speed on both sides of the shock; particles cross a 3-shock from right to left.

Note that for both shock families, $v_l \neq 0$, and $v_r \neq 0$, so that $m = \rho v \neq 0$.

Now let us denote by the subscript 1 the state of a particle just before it reaches the shock, and by the subscript 2 the state of a particle just after the shock. Hence

$$\text{for 1-shocks,} \quad l = 1, r = 2;$$

$$\text{for 3-shocks,} \quad l = 2, r = 1.$$

Then in this notation, for 1-shocks

$$v_1 > c_1 > 0, \qquad c_2 > v_2 > 0 \quad \text{so } v_1^2 > c_1^2, \quad c_2^2 > v_2^2,$$

while for 3-shocks,

$$v_1 < -c_1 < 0, \qquad -c_2 < v_2 < 0 \quad \text{so } v_1^2 > c_1^2 \quad \text{and} \quad c_2^2 > v_2^2.$$

Thus in both shock families we have

$$v_1^2 > c_1^2, \qquad c_2^2 > v_2^2.$$

Next $m \neq 0$, so that the third equation in (18.33) gives $[2c^2/(\gamma - 1) + v^2] = 0$. It follows that

$$\frac{2}{\gamma - 1} c_1^2 + c_1^2 < \frac{2}{\gamma - 1} c_1^2 + v_1^2 = \frac{2}{\gamma - 1} c_2^2 + v_2^2 < \frac{2}{\gamma - 1} c_2^2 + c_2^2,$$

so $c_1^2 < c_2^2$ or $c_1 < c_2$, and thus $v_1^2 > v_2^2$. This gives

$$c_2 > c_1 \quad \text{and} \quad |v_1| > |v_2|. \tag{18.36}$$

Since $m = \rho v$ is constant, $\rho_1 v_1 = \rho_2 v_2$ and (18.36) shows $\rho_2 > \rho_1$. From (18.33), $p_1 + mv_1 = p_2 + mv_2$ so that $p_2 > p_1$; thus shocks are compressive (we have observed this fact in §A. This is a consequence of the shock inequalities, just as before).

We now explicitly calculate the one-parameter family of shocks. We begin with 1-shocks and define the quantities

$$\pi = \frac{p_2}{p_1}, \qquad z = \frac{\rho_2}{\rho_1}, \qquad \beta = \frac{\gamma + 1}{\gamma - 1}, \qquad \tau = \frac{\gamma - 1}{2\gamma}. \tag{18.37}$$

Note that our above calculations show $\pi > 1$ and $z > 1$.

Now $c^2 = \gamma p / \rho$ so that $(c_2/c_1)^2 = (\gamma p_2/\rho_2)(\rho_1/\gamma p_1) = (p_2/p_1)(\rho_2/\rho_1)^{-1}$ or

$$\left(\frac{c_2}{c_1}\right)^2 = \frac{\pi}{z}. \tag{18.38}$$

Similarly $\rho_1 v_1 = \rho_2 v_2$ gives $v_2/v_1 = \rho_1/\rho_2$ so

$$\frac{\rho_1}{\rho_2} = \frac{v_2}{v_1} = \frac{1}{z}. \tag{18.39}$$

Using (18.38) and (18.39) in (18.33) gives

$$\frac{2}{\gamma - 1} c_1^2 + v_1^2 = \frac{2}{\gamma - 1} \frac{\pi}{z} c_1^2 + \frac{v_1^2}{z^2}, \quad \text{or}$$

$$\left(\frac{v_1}{c_1}\right)^2 = \frac{2z}{\gamma - 1} \frac{z - \pi}{(1 - z^2)}. \tag{18.40}$$

Also, from (18.33), $p_1 + mv_1 = p_2 + mv_2$, and since $m = \rho v$ and $p = c^2 \rho / \gamma$ we find $\rho_1(c_1^2/\gamma + v_1^2) = \rho_2(c_2^2/\gamma + v_2^2)$, and from (18.39),

$$\frac{c_1^2}{\gamma} + v_1^2 = z\left(\frac{c_2^2}{\gamma} + v_2^2\right).$$

Using (18.38) we can write

$$\frac{c_1^2}{\gamma} + v_1^2 = z\left(\frac{c_1^2 \pi}{\gamma z} + \frac{v_1^2}{z^2}\right) = \frac{c_1^2 \pi}{\gamma} + \frac{v_1^2}{z}, \quad \text{or}$$

$$\frac{zc_1^2}{\gamma}(1 - \pi) = v_1^2(1 - z).$$

Comparing this with (18.40) gives

$$z = \frac{1 + \pi \beta}{\pi + \beta}. \tag{18.41}$$

Note that this implies $z < \beta$, and since $1 < z$, we find $\rho_1 < \rho_2 < \beta \rho_1$. This gives a bound on the density ρ_2 in terms of ρ_1.

If we use (18.41) in (18.40) we get

$$\frac{v_1}{c_1} = \pm\left[\frac{(\beta - 1)z}{\beta - z}\right]^{1/2}, \tag{18.42}$$

where the $+$ sign is for 1-shocks, and the $-$ sign is for 3-shocks (since $v_1 > 0$ for 1-shocks, $v_1 < 0$ for 3-shocks, and $c_1 > 0$). Using $v = u - \sigma$ we obtain the following formula for the shock speed:

$$\sigma = u_1 \pm c_1\left[\frac{(\beta - 1)z}{\beta - z}\right]^{1/2}, \tag{18.43}$$

with the $+$ sign for 3-shocks, and the $-$ sign for 1-shocks.

Now (18.39) implies $(u_2 - \sigma) = (u_1 - \sigma)/z$ and so

$$u_2 - u_1 = \frac{z - 1}{z}(\sigma - u_1) = \pm c_1\frac{(\beta - 1)^{1/2}(z - 1)}{\sqrt{z(\beta - z)}}.$$

Finally from (18.41), we get the equation relating the change in velocity across a shock transition:

$$u_2 - u_1 = \pm c_1\sqrt{\frac{2}{\gamma(\gamma - 1)}}\frac{\pi - 1}{\sqrt{1 + \pi\beta}}, \tag{18.44}$$

with the $+$ sign for 3-shocks and the $-$ sign for 1-shocks.

We shall use (18.44) for 1-shocks. This will give an expression for $(u_r - u_l)/c_1$. If we would use (18.44) for 3-shocks we would obtain an expression for $(u_l - u_r)/c_r$. Now in order to again get an expression for $(u_r - u_l)/c_l$, we can proceed as follows. First define $\pi = p_1/p_2, z = \rho_1/\rho_2$, then find the expression for $(c_1/c_2)^2 = \pi/z$ and follow the procedure as above to obtain $v_2/c_2 = \pm[(\beta - 1)z/(\beta - z)]^{1/2}$, with the $-$ sign for 3-shocks. This gives

$$u_1 - u_2 = \pm c_2\sqrt{\frac{2}{\gamma(\gamma - 1)}}\frac{\pi - 1}{\sqrt{1 + \pi\beta}}. \tag{18.44'}$$

In this equation we only take the $+$ sign; this corresponds to 3-shocks.

The expressions $p_2/p_1 = \pi$, $\rho_2/\rho_1 = z$, together with (18.41) and (18.44) give the formulas for the shock curves. To make these somewhat more explicit, we introduce a new parameter x where

$$x = -\log \pi. \tag{18.45}$$

Notice that $e^{-x} = \pi = p_2/p_1 > 1$, so that $x \le 0$. In terms of this parametrization, we have the following formulas for the shock curves. (Recall that $\tau = (\gamma - 1)/2\gamma$, and $\beta = (\gamma + 1)/(\gamma - 1)$.)

$$
\text{1-shock curve} \atop (x \le 0)
\left\{
\begin{aligned}
&\frac{p_r}{p_l} = e^{-x}, \\[2mm]
&\frac{\rho_r}{\rho_l} = \frac{1 + \beta e^{-x}}{\beta + e^{-x}} = \frac{e^x + \beta}{1 + \beta e^x}, \\[2mm]
&\frac{u_r - u_l}{c_l} = \frac{2\sqrt{\tau}}{\gamma - 1} \frac{1 - e^{-x}}{\sqrt{1 + \beta e^{-x}}},
\end{aligned}
\right.
\tag{18.46}
$$

$$
\text{3-shock curve} \atop (x \le 0)
\left\{
\begin{aligned}
&\frac{p_r}{p_l} = e^{x}, \\[2mm]
&\frac{\rho_r}{\rho_l} = \frac{1 + \beta e^{x}}{e^x + \beta}, \\[2mm]
&\frac{u_r - u_l}{c_l} = \frac{2\sqrt{\tau}}{\gamma - 1} \frac{e^x - 1}{\sqrt{1 + \beta e^{x}}}.
\end{aligned}
\right.
\tag{18.47}
$$

We now calculate the simple-wave curves. Let us only consider the 1-simple waves; the details for the 3-simple waves are analogous.

Since the 1-Riemann invariants are constant in a 1-simple wave, we have

$$
S_r = S_l,
\tag{18.48}
$$

and

$$
u_r + \frac{2}{\gamma - 1} c_r = u_l + \frac{2}{\gamma - 1} c_l.
\tag{18.49}
$$

From (18.30) $S/c_v = \log(p/\rho^\gamma k)$, so (18.48) gives $p_r/p_l = (\rho_r/\rho_l)^\gamma$. Also from $c^2 = \gamma p/\rho$, we get $(c_r/c_l)^2 = (p_r/p_l)(\rho_r/\rho_l) = (p_r/p_l)^{(\gamma - 1)/\gamma}$ so that

$$
\frac{p_r}{p_l} = \left(\frac{c_r}{c_l}\right)^{2\gamma/(\gamma - 1)} = \left(\frac{\rho_r}{\rho_l}\right)^\gamma.
\tag{18.50}
$$

Also by (18.49),

$$
\frac{u_r - u_l}{c_l} = \frac{2}{\gamma - 1}\left(1 - \frac{c_r}{c_l}\right).
\tag{18.51}
$$

But $\lambda_1 = u - c$ must increase in a 1-rarefaction wave, so $\lambda_{1,r} \geq \lambda_{1,l}$ gives $u_r - u_l \geq c_r - c_l$. Thus from (18.51),

$$\frac{c_r - c_l}{c_l} \leq \frac{2}{\gamma - 1}\left(1 - \frac{c_r}{c_l}\right), \qquad 0 \leq \left(\frac{2}{\gamma - 1} + 1\right)\left(1 - \frac{c_r}{c_l}\right),$$

so that $0 < c_r/c_l \leq 1$. Using this in (18.50), we find

$$0 < \frac{p_r}{p_l} \leq 1. \tag{18.52}$$

We can thus introduce a parameter x by

$$x = -\log\left(\frac{p_r}{p_l}\right) \geq 0. \tag{18.53}$$

Then we can write the formulas for a 1-simple wave using (18.50) and (18.51) as

$$\text{1-simple wave curve} \atop (x \geq 0) \quad \left\{ \begin{aligned} &\frac{p_r}{p_l} = e^{-x}, \\[2mm] &\frac{\rho_r}{\rho_l} = e^{-x/\gamma}, \\[2mm] &\frac{u_r - u_l}{c_l} = \frac{2}{\gamma - 1}(1 - e^{-\tau x}). \end{aligned} \right. \tag{18.54}$$

Similarly, for the 3-family we have the formulas

$$\text{3-simple wave curve} \atop (x \geq 0) \quad \left\{ \begin{aligned} &\frac{p_r}{p_l} = e^{x}, \\[2mm] &\frac{\rho_r}{\rho_l} = e^{x/\gamma}, \\[2mm] &\frac{u_r - u_l}{c_l} = \frac{2}{\gamma - 1}(e^{\tau x} - 1). \end{aligned} \right. \tag{18.55}$$

We turn next to the contact discontinuity. This type of wave comes from the linear degeneracy of the second characteristic family. There are no shocks or rarefaction waves in this family, but instead a one-parameter family of contact discontinuities. The 2-Riemann invariants u and p are constant, and the density can have an arbitrary jump. Thus we have the following formulas for the one-parameter family of contact discontinuities:

$$\frac{p_r}{p_l} = 1,$$

$$\frac{\rho_r}{\rho_l} = e^x, \qquad -\infty < x < \infty, \qquad (18.56)$$

$$u_r - u_l = 0.$$

Now we can put together our formulas for the one-parameter families of curves, as follows:

1-family, for $x \in \mathbf{R}$,

$$\frac{p_r}{p_l} = e^{-x},$$

$$\frac{\rho_r}{\rho_l} = f_1(x) \equiv \begin{cases} e^{-x/\gamma}, & x \geq 0, \\ \dfrac{\beta + e^x}{1 + \beta e^x}, & x \leq 0, \end{cases}$$

$$\frac{u_r - u_l}{c_l} = h_1(x) \equiv \begin{cases} \dfrac{2}{\gamma - 1}(1 - e^{-\tau x}), & x \geq 0, \\ \dfrac{2\sqrt{\tau}}{\gamma - 1}\dfrac{1 - e^{-x}}{(1 + \beta e^{-x})^{1/2}}, & x \leq 0; \end{cases}$$

2-family, for $x \in \mathbf{R}$,

$$\frac{p_r}{p_l} = 1, \qquad \frac{\rho_r}{\rho_l} = e^x, \qquad u_r - u_l = 0;$$

3-family, for $x \in \mathbf{R}$,

$$\frac{p_r}{p_l} = e^x,$$

$$\frac{\rho_r}{\rho_l} = f_3(x) \equiv \frac{1}{f_1(x)},$$

$$\frac{u_r - u_l}{c_l} = h_3(x) \equiv \begin{cases} \dfrac{2}{\gamma - 1}(e^{\tau x} - 1), & x \geq 0, \\ \dfrac{2\sqrt{\tau}}{\gamma - 1}\dfrac{e^x - 1}{\sqrt{1 + \beta e^x}}, & x \leq 0. \end{cases}$$

Concerning the functions h_1, h_3, and f_1, we need the following simple lemma, whose proof is completely straightforward, and omitted.

Lemma 18.5. (a) $h_1' > 0$, and $h_1(\mathbf{R}) = (-\infty, 2/(\gamma - 1)]$.

(b) $h_3(x) = \sqrt{f_1(x)}\, e^{x/2} h_1(x)$.

It is useful to express the above one-parameter families in general notation. Thus let

$$v = (v_1, v_2, v_3) \equiv (\rho, p, u),$$

and define transformations $T_x^{(i)}$, $i = 1, 2, 3$, as follows:

$$T_x^{(1)}v = \left(f_1(x)v_1,\, e^{-x}v_2,\, v_3 + \left(\frac{\gamma v_2}{v_1} \right)^{1/2} h_1(x) \right),$$

$$T_x^{(2)}v = (e^x v_1,\, v_2,\, v_3),$$

$$T_x^{(3)}v = \left(f_3(x)v_1,\, e^x v_2,\, v_3 + \left(\frac{\gamma v_2}{v_1} \right)^{1/2} h_3(x) \right),$$

where the f_i's and h_i's are defined as above. Note that in this notation, we really mean that the state v can be connected to the state $T_x^{(i)}v$, by an i-shock or i-rarefaction wave if $i = 1$ or 3 and $x < 0$ or $x > 0$, respectively, and if $i = 2$, it can be connected to the state $T_x^{(2)}v$, by a contact discontinuity. We can now prove the main theorem of this chapter.

Theorem 18.6. *Consider the equations of gas dynamics (18.31) for an ideal polytopic gas whose equation of state is given by (18.29) and (18.30). Let v_l and v_r be any two states (not necessarily close). Then there is a unique[2] solution to the Riemann problem with these initial states, if and only if*

$$u_r - u_l < \frac{2}{\gamma - 1}(c_l + c_r). \tag{18.57}$$

If (18.57) is violated, then a vacuum is present in the solution.

---

[2] In the class of shocks, centered simple waves, and contact discontinuities separating constant states.

Proof. We let $v_l = (\rho_l, p_l, u_l)$ and $v_r = (\rho_r, p_r, u_r)$, where the ρ's and p's are positive. To solve the Riemann problem with this data means finding real numbers x_1, x_2, and x_3 for which

$$v_r = T^{(3)}_{x_3} T^{(2)}_{x_2} T^{(1)}_{x_1} v_l, \tag{18.58}$$

or more precisely,

$$
\begin{pmatrix} \rho_r \\ p_r \\ u_r \end{pmatrix} = \begin{pmatrix} f_1(x_1) e^{x_2} f_3(x_3) \rho_l \\ e^{x_3 - x_1} p_l \\ u_l + c_l \left[h_1(x_1) + \sqrt{\dfrac{e^{-(x_1 + x_2)}}{f_1(x_1)}}\, h_3(x_3) \right] \end{pmatrix}, \tag{18.59}
$$

where we have explicitly computed the composition. We define

$$A = \frac{\rho_r}{\rho_l}, \qquad B = \frac{p_r}{p_l}, \qquad C = \frac{u_r - u_l}{c_l}. \tag{18.60}$$

The second component in (18.59) shows that

$$x_3 - x_1 = \log B. \tag{18.61}$$

The first component gives

$$f_1(x_1) e^{x_2} f_3(x_3) = A, \tag{18.62}$$

and from the third component,

$$C = h_1(x_1) + \sqrt{\frac{e^{-(x_1 + x_2)}}{f_1(x_1)}}\, h_3(x_3).$$

Thus from (18.62)

$$C = h_1(x_1) + \sqrt{\frac{e^{-x_1}}{A f_1(x_3)}}\, h_3(x_3),$$

where we have also used the fact that $f_1 f_3 = 1$. Using Lemma 18.5(b) we find

$$h_1(x_1) + A^{-1/2} e^{(x_3 - x_1)/2} h_1(x_3) = C,$$

so from (18.61),

$$h_1(x_1) + \sqrt{\frac{B}{A}} h_1(x_1 + \log B) = C. \tag{18.63}$$

Now Lemma 18.5(a) shows that this last equation has a (unique) solution if and only if

$$\frac{2}{\gamma - 1}\left(1 + \sqrt{\frac{B}{A}}\right) > C.$$

This is (18.57). The procedure is to first find x_1 from (18.63), then (18.61) gives x_3, and finally (18.62) gives x_2. This completes the proof of the theorem. □

We remark again that this is a *global* theorem, in that the two states are not required to be close to each other. If (18.57) is violated, then the relative velocities on both sides of the membrane are so large that a vacuum is formed. Thus, strictly speaking, there is even existence in this case; we simply put a vacuum ($\rho = 0$) in between the two parts of the gas, and leave the other variables undefined. This is similar to what we found for the *p*-system in Chapter 17, §A.

We now show how these methods provide simple criteria for telling which of the two possibilities, shocks or simple waves, occurs in the 1-family and 3-family. That is, we show how to obtain the exact qualitative form of the solution.

Corollary 18.7. *Consider the solution of the Riemann problem obtained in Theorem 18.6. Then the following statements hold:*

(i) *The 1-component of the solution is a simple wave if and only if*

$$\sqrt{\frac{B}{A}} h_1(\log B) < C < \frac{2}{\gamma - 1}\left[1 + \sqrt{\frac{B}{A}}\right],$$

   and is a shock otherwise.

(ii) *The 3-component of the solution is a simple wave if and only if*

$$h_1(-\log B) < C < \frac{2}{\gamma - 1}\left[1 + \sqrt{\frac{B}{A}}\right]$$

and is a shock otherwise. (Recall that A, B, and C are defined in (18.60).)

Proof. We have seen in the theorem that the inequalities on the right must hold in both (i) and (ii).

We consider first the 1-family. If $\phi(x) \equiv h_1(x) + \sqrt{B/A}\, h_1(x + \log B)$, then $\phi(0) = \sqrt{B/A}\, h_1(\log B)$, and $\phi' > 0$. Thus (18.63) shows that the 1-wave is a simple wave if and only if $C > \phi(0)$, and this gives (i). For (ii) we use (18.61) in (18.63) to get

$$h_1(x_3 - \log B) + \sqrt{\frac{B}{A}}\, h_1(x_3) = C.$$

This shows that the 3-wave is a simple wave if and only if $C > h_1(-\log B)$. $\square$

§C. Interaction of Shock Waves

The results in §B can be used to solve the shock interaction problem for an ideal polytropic gas; i.e., for the equations (18.31), where the equation of state is given by (18.29) and (18.30).

We formulate the problem as follows. Given three constant states v_l, v_m, v_r, such that

$$v_m = T_t^{(3)} v_l, \qquad v_r = T_s^{(3)} v_m, \qquad s, t < 0,$$

we want to solve the Riemann problem for (18.31) with data v_l, v_r at the time of interaction; see Figure 18.2. That these two 3-shocks do indeed interact with each other is a consequence of the shock inequalities (18.13). We leave the details to the reader; cf. Chapter 17, §A, where the same problem is considered for the p-system (17.1). Our first theorem is the following.

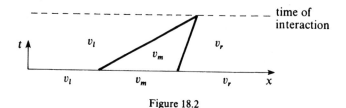

Figure 18.2

Theorem 18.8. *The interaction of two sufficiently weak 3-shocks produces a 1-rarefaction wave if $\gamma < \frac{5}{3}$, and a 1-shock if $\gamma > \frac{5}{3}$. (The same result holds for 1-shocks provided that we interchange 1 and 3.)*

Proof. We give the proof for 3-shocks and drop the subscript 3.

We have, using the formulas in §B,

$$T_t(v_l) = \left(f_3(t)\rho_l, e^t p_l, u_l + \left(\frac{\gamma p_l}{\rho_l}\right)^{1/2} h_3(t) \right).$$

and

$$T_s T_t(v_l) = \left(f_3(s) f_3(t)\rho_l, e^{s+t} p_l, u_l + \left(\frac{\gamma p_l}{\rho_l}\right)^{1/2} h_3(t) + \left(\frac{\gamma e^t p_l}{f_3(t)\rho_l}\right)^{1/2} h_3(s) \right).$$

This gives the equations

$$\rho_r = f_3(s) f_3(t)\rho_l,$$

$$p_r = e^{s+t} p_l,$$

$$u_r = u_l + \left(\frac{\gamma p_l}{\rho_l}\right)^{1/2} h_3(t) + \left(\frac{\gamma e^t p_l}{f_3(t)\rho_l}\right)^{1/2} h_3(s).$$

Now recalling definitions of A, B, and C from (18.60), we have

$$A = f_3(s) f_3(t), \qquad B = e^{s+t}, \qquad C = h_3(t) + h_3(s) \sqrt{\frac{e^s}{f_3(t)}}.$$

Using Corollary 18.7, the 1-wave is a rarefaction wave if and only if $C > \sqrt{(B/A)}\, h_1(\log B)$. From Lemma 18.5(b), this is equivalent to

$$h_1(t + s) < h_1(s) + \frac{h_1(s) h_1(t)}{h_3(s)}. \tag{18.64}$$

We define

$$\psi(x) = \frac{1 - e^{-x}}{\sqrt{1 + \beta e^{-x}}},$$

and

$$G(t, s) = \psi(t + s)\psi(-s) - \psi(s)\psi(-s) + \psi(t)\psi(s);$$

then (18.64) is the same as

$$G(t, s) < 0 \quad \text{for } t < 0, s < 0. \tag{18.65}$$

Defining $\phi(x) = \psi(x)/x$, we can write

$$\frac{G(t, s)}{s^2 t(s + t)} = 2\left[\frac{\phi(s) - \phi(-t)}{s + t}\right]\left[\frac{\phi(t) - \phi(-t)}{2t}\right]$$

$$- \phi(-s)\left[\frac{\dfrac{\phi(s + t) - \phi(s)}{t} - \dfrac{\phi(t) - \phi(-s)}{s + t}}{s}\right]$$

But

$$\phi(s + t) - \phi(s) = \int_s^{s+t} \phi'(x)\, dx = t\int_0^1 \phi'(ty + s)\, dy,$$

and

$$\phi(t) - \phi(-s) = \int_{-s}^t \phi'(x)\, dx = (s + t)\int_0^1 \phi'((s + t)y - s)\, dy,$$

so that

$$\frac{\phi(s + t) - \phi(s)}{t} - \frac{\phi(t) - \phi(-s)}{s + t} = \int_0^1 [\phi'(ty + s) - \phi'((s + t)y - s)]\, dy$$

$$= \int_0^1 \phi''(\xi)(2 - y)s\, dy,$$

where ξ is an intermediate point. It follows that

$$\lim_{s,t \to 0} \frac{\dfrac{\phi(s + t) - \phi(s)}{t} - \dfrac{\phi(t) - \phi(-s)}{s + t}}{s} = \tfrac{3}{2}\phi''(0).$$

Thus

$$\lim_{s,t \to 0} \frac{G(t, s)}{s^2 t(s + t)} = 2\phi'(0) - \tfrac{3}{2}\phi(0)\phi''(0)$$

$$= \frac{3}{64}\frac{\gamma^2 - 1}{\gamma^2}(\gamma - \tfrac{5}{3}).$$

This shows that for small s and t, $G(t, s) < 0$ if $\gamma < \tfrac{5}{3}$ and $G(t, s) > 0$ if $\gamma > \tfrac{5}{3}$. □

We remark that the condition $\gamma < \frac{5}{3}$ is true for most media; in particular $\gamma = 1.4$ for air.

We shall now develop a qualitative method for solving shock interaction problems, again for ideal, polytropic gases. The idea is to focus attention on the $u - p$ plane, and to consider the projections of the shock- and rarefaction-wave curves in this plane. Since u and p do not change across contact discontinuities (see (18.56)), we may effectively ignore the contact discontinuities which are produced in the resulting interactions; i.e., two states separated by a contact discontinuity represent the same point in $u - p$ space. This allows us to treat the gas dynamics equations in a manner completely analogous to the way in which we treated the isentropic gas dynamics equations (Chapter 17, §A); namely, we can study Riemann problems and shock interaction problems in a two-dimensional plane.

We begin by obtaining expressions for the shock- and rarefaction-wave curves in the $u - p$ plane. For ideal polytropic gases we can write

$$e = \frac{1}{\gamma - 1} p\tau, \qquad \tau = \rho^{-1}.$$

If we use this in the Hugoniot relation (18.7) and let

$$\mu^2 = \beta^{-1},$$

we find $(\tau_1 - \mu^2 \tau_0)p_1 = (\tau_0 - \mu^2 \tau_1)p_0$, so that

$$\frac{p_1}{p_0} = \frac{\tau_0 - \mu^2 \tau_1}{\tau_1 - \mu^2 \tau_0} = \frac{\rho_1 - \mu^2 \rho_0}{\rho_0 - \mu^2 \rho_1}. \tag{18.66}$$

Next, from (18.33), $p_1 + \rho_1 v_1^2 = p_0 + \rho_0 v_0^2$ so that if $v_1 = m\tau_1$, $v_0 = m\tau_0$, we get

$$-m^2 = \frac{p_1 - p_0}{\tau_1 - \tau_0}. \tag{18.67}$$

This, together with (18.66) gives

$$m^2 = \frac{p_1 + \mu^2 p_0}{(1 - \mu^2)\tau_0} = \frac{p_0 + \mu^2 p_1}{(1 - \mu^2)\tau_1}. \tag{18.68}$$

Again from (18.33), $m(v_1 - v_0) = p_0 - p_1$ so $m(u_1 - u_0) = p_0 - p_1$, and in view of (18.68) we obtain

$$(u_1 - u_0)^2 = (p_1 - p_0)^2 \frac{(1 - \mu^2)\tau_0}{p_1 + \mu^2 p_0} = (p_1 - p_0)^2 \frac{(1 - \mu^2)\tau_1}{p_0 + \mu^2 p_1}. \tag{18.69}$$

This equation and (18.68) imply

$$m = \frac{p_0 - p_1}{u_1 - u_0} = \pm\sqrt{\frac{p_1 + \mu^2 p_0}{(1 - \mu^2)\tau_0}},$$

and thus we get the desired formula,

$$u_1 - u_0 = \pm\phi_0(p_1), \tag{18.70}$$

where ϕ_0 is defined by

$$\phi_0(p) = (p - p_0)\sqrt{\frac{(1 - \mu^2)\tau_0}{p + \mu^2 p_0}}. \tag{18.71}$$

In this connection, we take the $+$ sign in (18.70) for 3-shocks, and $-$ sign for 1-shocks; this holds since for both shock families, $u_l > u_r$, while for 3-shocks $p_l > p_r$ and for 1-shocks, $p_r > p_l$, as we have observed earlier.

It is easy to check that the function $\phi_0(p)$ satisfies the following properties:

    (i) $\phi_0' > 0.$
    (ii) $\phi_0(p_1) = -\phi_1(p_0).$
    (iii) $\phi_0(p) \to \infty$ as $p \to \infty.$
    (iv) $\phi_0'(p) \to 0$ as $p \to \infty.$

We may depict the relations (18.69) in Figure 18.3(a) and 18.3(b) below. In Figure 18.3(a) we show the set of states connected to (u_l, p_l, τ_l) on the right by a shock wave S_i of the ith family, $i = 1$ or 3; i.e., here we are given the state on the left. In Figure 18.3(b), we are given the state on the right.

We can do a similar analysis to obtain the simple-wave curves. Namely, from (18.49) we know that the quantity $u + 2c/(\gamma - 1)$ is constant for a 1-simple waves. Similarly for a 3-simple wave $u - 2c/(\gamma - 1)$ is constant.

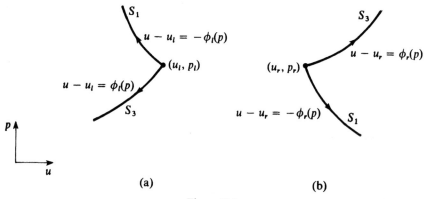

(a)   (b)

Figure 18.3

From (18.50) we have

$$\sqrt{\tau_1 p_1}\, p_1^{-(\gamma-1)/2\gamma} = \sqrt{\tau_0 p_0}\, p_0^{-(\gamma-1)/2\gamma},$$

so that

$$u_1 - u_0 = \pm \frac{2}{\gamma-1}(\sqrt{\gamma\tau_0 p_0} - \sqrt{\gamma\tau_1 p_1})$$

$$= \frac{\pm\sqrt{1-\mu^4}}{\mu^2}\sqrt{\tau_0}\, p_0^{1/2\gamma}(p_0^{(\gamma-1)/2\gamma} - p_1^{(\gamma-1)/2\gamma}).$$

This yields the following formula for the simple-wave curves:

$$u_1 - u_0 = \pm\psi_0(p_1), \tag{18.72}$$

where

$$\psi_0(p) = \frac{\sqrt{1-\mu^4}}{\mu^2}\sqrt{\tau_0}\, p_0^{1/2\gamma}(p^{(\gamma-1)/2\gamma} - p_0^{(\gamma-1)/2\gamma}). \tag{18.73}$$

In this connection, we take the $+$ sign in (18.72) for 3-simple waves, and the $-$ sign for 1-simple waves ($u_l < u_r$; and $p_r > p_l$ for 3-waves, $p_r < p_l$ for 1-waves).

It is easy to check that $\psi_0(p)$ has the following properties:

(i) $\psi_0' > 0$.
(ii) $\psi_0(p_1) = -\psi_1(p_0)$.
(iii) $\psi_0(p) \to \infty$ as $p \to \infty$.
(iv) $\psi_0'(p) \to 0$ as $p \to \infty$.

We can depict (18.72) in Figure 18.4, which is completely analogous to Figure 18.3.

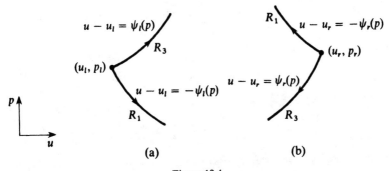

Figure 18.4

The formulas (18.70) and (18.72) are the desired analytical form of the projections of the shock- and rarefaction-wave curves in the $u - p$ plane. They give us Figures 18.3 and 18.4 which we will use to study shock-wave interactions in a qualitative manner.[3]

We consider the case of two forward-facing shocks (i.e., two 3-shocks) catching up with each other; see Figure 18.2. We can depict this situation in the $u - p$ plane in Figure 18.5, where we have $p_r < p_m < p_l$ and $u_l > u_m > u_r$, as noted above. We know that if $\gamma < \frac{5}{3}$, the 1-wave produced when the shocks interact is a simple wave, provided that the shocks are weak; this follows from Theorem 18.8. Using Figure 18.5, this means that the shock curve through (u_m, p_m) starts below the shock curve through (u_r, p_r), in the region $u > u_m$, as depicted. We shall now obtain the following *global* theorem.

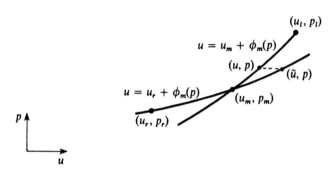

Figure 18.5

Theorem 18.9. *If $\gamma \leq \frac{5}{3}$, the interaction of two 3-shocks produces a 1-simple wave. (A similar statement is true for the interaction of 1-shocks; just interchange 1 and 3.)*

Proof. As we have observed above, it suffices to show that $\tilde{u} > u$ (see Figure 18.5).

We have from (18.70),

$$u_m = u_r + \phi_r(p_m),$$

$$\tilde{u} = u_r + \phi_r(p),$$

$$u = u_m + \phi_m(p),$$

so that

$$\tilde{u} - u = u_r - u_m + \phi_r(p) - \phi_m(p).$$

[3] One can also use these to solve the Riemann problems qualitatively; we leave the (easy) details to the reader.

But $u_m = u_r + \phi_r(p_m)$, and so we must show

$$\phi_r(p) > \phi_m(p) + \phi_r(p_m), \qquad p > p_m > p_r. \tag{18.74}$$

To this end, let $x = p/p_m$, $y = p_r/p_m$, and define

$$g(r, s) = (r - s)\sqrt{\frac{1}{\beta r + s} \cdot \frac{s + \beta}{\beta s + 1}}.$$

From (18.68), we can write

$$\frac{\rho_r}{\rho_m} = \frac{\tau_m}{\tau_r} = h(p_r/p_m),$$

where $h(t) = (\beta t + 1)/(t + \beta)$. This gives

$$\tau_r = \frac{\tau_m}{h(y)} = \tau_m \frac{y + \beta}{\beta y + 1},$$

which can be used in the expressions for $\phi_r(p)$ and $\phi_r(p_m)$. An easy calculation shows

$$\phi_r(p) = \sqrt{(\beta - 1)p_m\tau_m}\, g(x, y),$$

$$\phi_m(p) = \sqrt{(\beta - 1)p_m\tau_m}\, g(x, 1),$$

$$\phi_r(p_m) = \sqrt{(\beta - 1)p_m\tau_m}\, g(1, y).$$

Thus (18.74) is equivalent to showing that

$$0 > g(x, 1) - g(1, 1) - g(x, y) + g(1, y), \tag{18.75}$$

if $x > 1 > y > 0$. Now the right-hand side of (18.75) is equal to

$$\int_1^x \left(\int_y^1 g_{xy}(x, y)\, dy \right) dx,$$

so we will be done if we can show $g_{xy} < 0$ if $x > 1 > y > 0$. We find

$$g_{xy} =$$
$$\frac{\beta}{2} \frac{[(1 + 2\beta)\beta x - (2 + \beta)y](\beta y + 1)(\beta + y) - (\beta x + y)[\beta x + (2 + \beta)y](\beta^2 - 1)}{(\beta x + y)^{5/2}(\beta + y)^{1/2}(\beta y + 1)^{3/2}}$$

The numerator can be rewritten as

$$-\beta(\beta^2 + 1)(\beta - 4)(x - y)y - \beta^2(\beta^2 - 1)(x - 1)^2$$

$$-\beta^2(2\beta^2 - 2\beta - 3)(x - 1)(x - y) - \beta^2(\beta^2 - 2\beta - 2)(y - 1)^2$$

$$-\beta(\beta + 2)(y - 1)^2y - \beta^2(2\beta + 1)(x - 1)(y - 1)^2,$$

which is negative if $\beta \geq 4$; i.e., if $\gamma \leq \frac{5}{3}$. This completes the proof. $\square$

If $\gamma < \frac{5}{3}$, we may depict the interaction of two forward shocks as in Figure 18.6, where $V_l = (u_l, p_l, \rho_l)$, $V_m = (u_m, p_m, \rho_m)$, and $V_r = (u_r, p_r, \rho_r)$. The different values $\bar{\rho}_1$ and $\bar{\rho}_2$ of the density on both sides of the contact discontinuity are obtained from (18.69) and (18.68), once $\bar{u}$ and $\bar{p}$ are known.

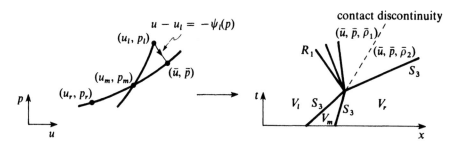

Figure 18.6

Finally, we note that using these curves in the $u - p$ plane, it is easy to see that the collision of a forward and backward shock produces a backward and forward shock separated by a contact discontinuity; see Figure 18.7. This qualitative statement is not easily proved in the general context in which Theorem 18.8 was obtained, even for weak shocks.

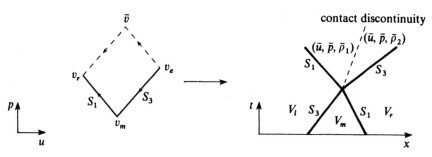

Figure 18.7

NOTES

The computations giving the Hugoniot curve are classical, and can be found in many standard textbooks; see [CF], [LL], and [ZR]. The conditions $p_{vv} > 0$, $p_s > 0$ are found in Weyl's paper [Wy], as well as in Bethe's paper [Be 1]. The inequalities (18.15)–(18.17) are classical, but are usually obtained differently. Theorem 18.2 is due to Wendroff [We 1]. The increase of S along the Hugoniot curve, Theorem 18.3, is due to Weyl [Wy] and Bethe [Be 1, 2]; see also [We 1, 2]. The proof given here follows Conley–Smoller [CS 7], where a more general theorem, applicable to magnetohydrodynamics, is given.

The global solution of the Riemann problem is part of the "folklore" in the subject, but is difficult to find in print. The book [CF], by Courant and Friedrichs does give an outline of the solution. The proof given in §B, for ideal polytopic gases, is due to Conway and Rosencrans in an unpublished manuscript [CR]. These results have been extended to the case of more general equations of state by Smith [St]. It is shown there that assuming (18.8) and (18.21), the Riemann problem is always solvable, but, somewhat surprisingly, the ("correct") solution need not be unique unless one imposes an additional condition on the equation of state. Thus Smith's result implies that either all gases satisfy this additional condition, or that there must be a way of choosing the "physically relevant" solution, which differs from the requirement that the entropy S increases across the shock. See also [We 1, 2] for the Riemann problem in gas dynamics.

The proof of Theorem 18.8 is taken from [CR]. The advantages of the $u - p$ plane to solve the Riemann problem and to study interactions of waves was first recognized by Courant and Friedrichs [CF]. The statement (without proof) of Theorem 18.9 is also found in [CF]; see also [Nn 1].

The book [CF] is an excellent source for the work on gas dynamics prior to 1948, and includes the research on shock waves done during the Second World War, when the development of supersonic aircraft, and the creation of the atomic bomb demanded a better theoretical understanding of shock phenomena. The standard work [LL], is a good reference for the physics of shock waves.

Chapter 19

The Glimm Difference Scheme

We consider a general system of conservation laws

$$u_t + f(u)_x = 0, \qquad x \in \mathbf{R}, \quad t > 0, \tag{19.1}$$

where $u = (u_1, \ldots, u_n)$, with initial data

$$u(x, 0) = u_0(x), \qquad x \in \mathbf{R}. \tag{19.2}$$

The system (19.1) is assumed to be hyperbolic and genuinely nonlinear in each characteristic field, in some open set $U \subset \mathbf{R}^n$ (see Definition 17.7). We let $\lambda_1(u) < \cdots < \lambda_n(u)$ denote the eigenvalues of $df(u)$. Concerning $u_0(x)$, we assume that

$$\text{T.V.}(u_0)$$

is sufficiently small, where by T.V.$(\cdot)$ we mean the total variation. With these assumptions, we shall show that the above problem has a solution which exists for all $t > 0$.

The proof will depend on four main ingredients:

(i)   construction of approximate solutions,
(ii) interaction estimates,
(iii) compactness of a subsequence of approximate solutions,
(iv) showing that the limit is indeed a solution.

The approximate solutions will be constructed by suitably "glueing together" solutions of Riemann problems. As we have seen in Chapter 17, §C, we know that these exist only if the states are sufficiently close; this is one reason for the smallness restriction on T.V. u_0.

To be a bit more explicit, the approximate solutions will be solutions of a difference scheme which involves a random choice. This can be roughly illustrated as follows. We divide the line $t = n \, \Delta t$ into subintervals by the points $k\Delta x$, $k = 0, \pm 1, \pm 2, \ldots$, where we suppose that for each k we are given a constant state u_n^k in the interval $k \, \Delta x \leq x \leq (k + 2) \, \Delta x$, where $(n + k) \equiv 0 \pmod 2$. We solve each of the Riemann problems for (19.1) with data

$$u(x, n\,\Delta t) = \begin{cases} u^n_{k-2}, & x < k\,\Delta x, \\ u^k_n, & x > k\,\Delta x, \end{cases}$$

in the region $n\,\Delta t < t \le (n+1)\,\Delta t$; call this solution $v_k(x, t)$. If $\Delta x/\Delta t > C$, where $C = \max|\lambda_j|$ (λ_j are the eigenvalues of df), then the solutions in each interval will not interact with each other. We can thus unambiguously define

$$u(x, (n+1)\,\Delta t) = v_k((k+1)\,\Delta x + \theta_{n+1}\,\Delta x, (n+1)\,\Delta t),$$

$$\text{if } k\,\Delta x \le x < (k+2)\,\Delta x,$$

where $k + n + 1 \equiv 0 \pmod 2$, and θ_{n+1} is some randomly chosen point in the interval $[-1, 1]$. This is the essence of the difference scheme. Of course, it is not a-priori obvious that the difference schemes is even well-defined, since we do not know that the states remain sufficiently close, in order that the Riemann problems are all solvable, or that C stays bounded. In other words, we must prove that the *approximate* solutions can be defined!

To do this, it is, of course, necessary to obtain bounds on the solutions of the Riemann problem, and, in fact to study the interaction problem. That is, we must obtain a quantitative estimate on the "strengths" of the emerging new waves, in terms of the "strengths" of the original interacting waves. Once this is accomplished, it is possible, with the help of certain functionals which decrease in time, to simultaneously show that the difference scheme is well defined, and that the difference approximations are uniformly bounded, and have uniformly bounded total variation. This implies that a subsequence of the approximate solutions, depending on the random choice, is compact in a strong enough topology so as to enable us to prove that the limit is indeed a solution, for "almost all" choices of the random points.

§A. The Interaction Estimate

We consider the Riemann problem for (19.1) with data

$$u(x, 0) = \begin{cases} u_l, & x < 0, \\ u_r, & x > 0, \end{cases}$$

which, for short, we shall refer to as "the Riemann problem (u_l, u_r)." This will cause no ambiguity, since the equation (19.1) will be fixed throughout our discussion. We will also denote by $(u_l = u_0, u_1, \ldots, u_n = u_r)$, the solution of this Riemann problem, where, we recall from Chapter 17, §B, u_k is connected to u_{k-1} by a k-shock, or a k-rarefaction wave, $k = 1, 2, \ldots, n$. That is, u_k lies on the composite k-shock/rarefaction-wave curve through u_{k-1}, so that in the notation of Chapter 17, §C, $u_k = T^{(k)}_{\varepsilon_k} u_{k-1}$, where $\varepsilon_k < 0$ if u_k is connected to u_{k-1} on the right by a k-shock, and $\varepsilon_k > 0$ if u_k is connected to u_{k-1} on the right by a k-rarefaction wave. We shall gather all this information in the concise notation:

$$(u_l, u_r) = [(u_0, u_1, \ldots, u_n)/(\varepsilon_1, \ldots, \varepsilon_n)]. \qquad (19.3)$$

We call $|\varepsilon_k|$ the *strength* of the k-wave connecting u_k to u_{k-1} in the above solution. We obviously can also unambiguously speak of the k-wave in u_{k-1}/u_k, and with a slight abuse of notation, we can also call this u_k.

Now let $u_l, u_m,$ and u_r be three states near a given state $\bar{u}$ and let

$$(u_l, u_m) = [(u_0', u_1', \ldots, u_n')/(\gamma_1, \ldots, \gamma_n)],$$

$$(u_m, u_r) = [(u_0'', u_1'', \ldots, u_n'')/(\delta_1, \ldots, \delta_n)], \qquad (19.4)$$

denote the solutions of the respective Riemann problems. We say that j-wave u_j' and the k-wave u_k'' are *approaching waves* if either (i) $j > k$, or (ii) if $j = k, \gamma_k < 0,$ or $\delta_k < 0$. Thus two waves are approaching, if (i) either the one on the left belongs to the larger characteristic family; or (ii) if both waves come from the same family, and at least one wave is a shock. In particular, rarefaction waves of the same family do not approach each other. Thus waves of different families approach if and only if the wave on the left travels faster than the wave on the right, while for waves of the same family, one of the two waves must be a shock wave. This is a perfectly reasonable definition if one is interested in studying interactions of waves, since it isolates those types of interactions that are possible; adjacent rarefaction waves of the same family cannot interact with each other since the "head" of one wave travels with the same speed as the "tail" of the other wave.

Proposition 19.1. *If $u_l, u_m,$ and u_r are three states near $\bar{u}$, and if (19.3) and (19.4) hold, we have*

$$\varepsilon_i = \gamma_i + \delta_i + O(|\gamma||\delta|). \qquad (19.5)$$

If there is a coordinate system $\{w_j\}$ near $\bar{u}$ with $R_i w_j \equiv r_i \cdot \nabla w_j = 0, i \neq j$, then

$$\varepsilon_i = \gamma_i + \delta_i + O([|\gamma| + |\delta|]^3). \qquad (19.6)$$

Remark. We shall show in Chapter 20, that (19.6) always holds in the case $n = 2$; the associated coordinates are called *Riemann invariants*.

In what follows, we use the notation $R_j = r_j \cdot \nabla$; where ∇ is the gradient with respect to u.

Could it hold if $n > 2$? (looks like it can't...)

Proof. Using the results of Chapter 17, Corollary 17.13, we can write

$$u_r - u_m = \sum_j \delta_j r_j + \sum_{j \leq i} \delta_j \delta_i R_j r_i (1 - \tfrac{1}{2}\delta_{ij}) + O(|\delta|^3)$$

and

$$u_l - u_m = \sum_j -\gamma_j r_j + \sum_{j \geq i} \gamma_j \gamma_i R_j r_i (1 - \tfrac{1}{2}\delta_{ij}) + O(|\gamma|^3), \tag{19.7}$$

where the coefficients are all evaluated at u_m, and δ_{ij} is the usual Kronecker delta. Thus

$$u_r - u_l = \sum_i (\delta_i + \gamma_i) r_i + O([|\gamma| + |\delta|]^2). \tag{19.8}$$

But also

$$u_r - u_l = \sum_i \varepsilon_i r_i + \sum_{j \leq i} \varepsilon_j \varepsilon_i R_j r_i (1 - \tfrac{1}{2}\delta_{ij}) + O(|\varepsilon|^3), \tag{19.9}$$

where the coefficients are evaluated at u_l. Now considering ε as a function of γ and δ, then since $\varepsilon = 0$ if $\delta = \gamma = 0$, we have $\varepsilon = O(|\gamma| + |\delta|)$. Also, since

$$r_i(u_l) = r_i(u_m) - \sum_j \gamma_j R_j r_i(u_m) + O(|\gamma|^2), \tag{19.10}$$

we get as in (19.7),

$$u_r - u_l = \sum_i \varepsilon_i r_i(u_m) + O([|\gamma| + |\delta|]^2).$$

Comparing this with (19.8) gives

$$\varepsilon_i = \gamma_i + \delta_i + O([|\gamma| + |\delta|]^2). \tag{19.11}$$

If we use (19.10) and (19.11) in (19.9), we find

$$u_r - u_l = \sum_i \varepsilon_i r_i(u_m) + \sum_{j \leq i} \varepsilon_j \varepsilon_i R_j r_i(u_m)(1 - \tfrac{1}{2}\delta_{ij})$$
$$- \sum_{i,j} \varepsilon_j \gamma_i R_i r_j(u_m) + O([|\gamma| + |\delta|]^3).$$

Finally, if we compare this with (19.7), we get

$$\sum_i (\varepsilon_i - \gamma_i - \delta_i) r_i = \sum_{j < i} \gamma_i \delta_j [R_i r_j - R_j r_i] + O([|\gamma| + |\delta|]^3),$$

or

$$\sum_i (\varepsilon_i - \gamma_i - \delta_i) R_i = \sum_{j < i} [R_i, R_j] \gamma_i \delta_j + O([|\gamma| + |\delta|]^3).$$

This proves (19.5). If a coordinate system $\{w_j\}$ exists with $R_i w_j = 0, i = j$, then in this coordinate system $R_j = \partial/\partial w_j$, and $[R_i, R_j] = 0$; this proves (19.6). $\quad\square$

The principal result of this section is the following improvement of (19.5) and (19.6).

Theorem 19.2. *Under the hypotheses of Proposition 19.1,*

$$\varepsilon_i = \gamma_i + \delta_i + D(\gamma, \delta)O(1) \quad \text{as } |\gamma| + |\delta| \to 0, \tag{19.12}$$

and if (19.6) holds,

$$\varepsilon_i = \gamma_i + \delta_i + D(\gamma, \delta)O(|\gamma| + |\delta|) \quad \text{as } |\gamma| + |\delta| \to 0. \tag{19.13}$$

Here $D(\gamma, \delta) = \sum |\gamma_i||\delta_j|$ *where the sum is over all pairs for which the i-wave from* u' *and the j-wave from* u'' *are approaching.*

Proof. We consider first the case where $D = 0$. If k is the largest index such that $\gamma_k \neq 0$ (so $\gamma_{k+1} = \cdots = \gamma_n = 0$), then $\delta_1 = \delta_2 = \cdots = \delta_{k-1} = 0$, and either (a) $\gamma_k < 0$ so $\delta_k = 0$; or (b) $\gamma_k > 0$ and $\delta_k > 0$. In case (a), $v_k' = v_m = v_k''$, so that the waves fit together to give the unique solution; see Figure 19.1.

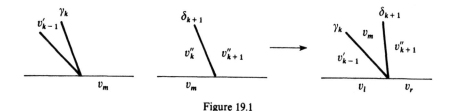

Figure 19.1

That is, $\varepsilon_i = \gamma_i$, $i \leq k$, $\varepsilon_i = \delta_i$, $i > k$. In case (b), the waves again fit together; see Figure 19.2. Thus $\varepsilon_i = \gamma_i$ if $i \leq k - 1$, $\varepsilon_k = \gamma_k + \delta_k$, and $\varepsilon_i = \gamma_i$ if $i > k$. In both cases (19.12) holds; this proves the theorem if $D = 0$.

We shall now prove the general version. The proof proceeds by induction on the number of nonzero waves in δ. Namely, we assume that the estimate is valid for all δ of the form $\delta = (\delta_1, \delta_2, \ldots, \delta_{p-1}, 0, \ldots, 0)$, and we shall prove it

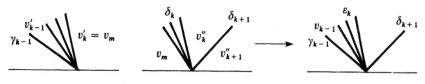

Figure 19.2

true for any δ of the form $(\delta_1, \ldots, \delta_{p-1}, \delta_p, 0, \ldots, 0)$. If $\delta = (0, \ldots, 0)$, then $D = 0$, and we have just observed that the theorem holds. We need only prove the induction step.

We define

$$\Delta = (\delta_1, \delta_2, \ldots, \delta_{p-1}, 0, \ldots, 0), \qquad \Delta_0 = (\underbrace{0, \ldots, 0}_{p-1}, \delta_p, 0, \ldots, 0), \quad (19.14)$$

and

$$\delta = (\delta_1, \delta_2, \ldots, \delta_{p-1}, \delta_p, 0, \ldots, 0) = \Delta + \Delta_0.$$

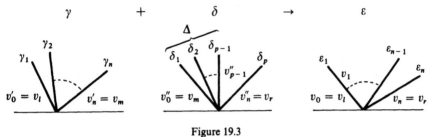

Figure 19.3

We indicate the dependence of ε on v_m by writing $\varepsilon = \varepsilon(\gamma, \delta; v_m)$.

We first let γ and Δ interact.[1] We define $\mu = (\mu_1, \ldots, \mu_n)$ and $\nu = (\nu_1, \ldots, \nu_n)$, by

$$\mu_i = \begin{cases} \varepsilon_i(\gamma, \Delta; v_m), & \text{if the } i\text{-wave } (\gamma_i) \text{ in } v', \text{ and the} \\ & p\text{-wave } (\delta_p) \text{ in } v'', \text{ do not approach,} \\ 0, & \text{otherwise,} \end{cases}$$

$$\nu_i = \begin{cases} 0, & \text{if } \mu_i = \varepsilon_i(\gamma, \Delta; v_m), \\ \varepsilon_i(\gamma, \Delta; v_m), & \text{if } \mu_i = 0. \end{cases}$$

Notice that

$$\mu_i = 0 \quad \text{if } i > p, \qquad \nu_i = 0 \quad \text{if } i < p,$$

$\mu + \nu = \varepsilon(\gamma, \Delta; v_m)$, and the interaction of γ with Δ produces the ε-waves; i.e., $\gamma + \Delta \to \mu + \nu$. We denote by $\tilde{v}_m$, the state joining μ and ν; i.e., if γ_p or δ_p is a shock, then $\tilde{v}_m = v_{p-1}$, while if both $\gamma_p \geq 0$ and $\delta_p \geq 0$, $\tilde{v}_m = v_p$; this follows from our earlier remark.

---

[1] We are being here a little loose with our terminology; when we say that we are letting waves α and β interact with each other to produce waves ζ (and write $\alpha + \beta \to \zeta$), we really mean that we have three Riemann problems (v_l, v_m), (v_m, v_r), and (v_l, v_r), which have solutions $(v_l, v_m) = [(v_l = v_0', \ldots, v_n' = v_m)/(\alpha_1, \ldots, \alpha_n)]$, $(v_m, v_r) = [(v_m = v_0'', \ldots, v_n'' = v_r)/(\beta_1, \ldots, \beta_n)]$ and $(v_l, v_r) = [(v_l = v_0, \ldots, v_n = v_r)/(\zeta_1, \ldots, \zeta_n)]$, respectively, as in (19.3).

We next allow v and Δ_0 to interact with each other; this requires v''_{p-1} to play the role of v_m, and we have $v + \Delta_0 \to \pi = \varepsilon(v, \Delta_0; v''_{p-1})$. Now we let π interact with $\mu, \mu + \pi \to \varepsilon(\mu, \pi; \tilde{v}_m)$. Since the last resolution joins v_l to v_r, and since we have uniqueness of this Riemann problem (provided that the states are close; see Theorem 17.18), we must have

$$\varepsilon(\gamma, \delta; v_m) = \varepsilon(\mu, \pi; \tilde{v}_m). \tag{19.15}$$

Now by our induction hypothesis,

$$\varepsilon_i(\gamma, \Delta; v_m) = \gamma_i + \Delta_i + D(\gamma, \Delta)O(1).$$

But $D(\gamma, \Delta) \le D(\gamma, \delta)$, and so we can write

$$\varepsilon_i(\gamma, \Delta; v_m) + \delta_{ip}\delta_p = \gamma_i + \delta_i + D(\gamma, \delta)O(1), \tag{19.16}$$

where δ_{ip} is the Kronecker symbol. If we now use (19.5), we can write (cf. Figure 19.4)

$$\pi_i = v_i + \delta_{ip}\delta_p + |v||\delta_p|O(1). \tag{19.17}$$

At this point, it is convenient to prove a lemma.

Lemma 19.3. $|v||\delta_p| = D(\gamma, \delta)O(1).$

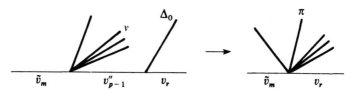

Figure 19.4

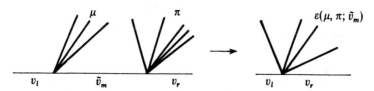

Figure 19.5

Proof. By definition, $v_i = 0$ if $i < p$; thus $|v||\delta_p| = \sum_{i \geq p}|v_i||\delta_p|$. Now if $i = p$, either $v_i = 0$, or since $v_p = \varepsilon_p(\gamma, \Delta; v_m)$, we can use (19.16) to obtain $v_p = \gamma_p + D(\gamma, \delta)O(1)$. Thus

$$
\begin{aligned}
|v_p||\delta_p| &= |\gamma_p||\delta_p| + D(\gamma, \delta)O(1)|\delta_p| \\
&\leq D(\gamma, \delta)[1 + O(1)|\delta_p|] \\
&= D(\gamma, \delta)O(1).
\end{aligned}
$$

Finally, if $i > p$, (19.16) gives $v_i = \gamma_i + D(\gamma, \delta)O(1)$, so that here too $|v_i||\delta_p| = D(\gamma, \delta)O(1)$. This completes the proof of the lemma. $\square$

We use the lemma to improve (19.17) to

$$
\pi_i = v_i + \delta_{ip}\delta_p + D(\gamma, \delta)O(1). \tag{19.18}
$$

Since $\varepsilon(\gamma, \Delta; v_m) = \mu + v$, (19.16) gives

$$
\gamma_i + \delta_i = \mu_i + v_i + \delta_{ip}\delta_p + D(\gamma, \delta)O(1). \tag{19.19}
$$

Now define $\tilde{\pi} = v + \Delta_0$; then $D(\mu, \tilde{\pi}) = 0$, for, if $i < p$, $\tilde{\pi}_i = 0$, if $i > p$, $\mu_i = 0$, and if $i = p$, either $\mu_p = 0$, or else $v_p = 0$ and μ_p does not approach δ_p. Thus, since the theorem holds if $D = 0$, we have

$$
\varepsilon(\mu, \tilde{\pi}; \tilde{v}_m) = \mu + \tilde{\pi},
$$

so that $\varepsilon_i(\mu, \tilde{\pi}; \tilde{v}_m) = \mu_i + v_i + \delta_{ip}\delta_p$. If we use this in (19.19), we get

$$
\varepsilon_i(\mu, \tilde{\pi}; \tilde{v}_m) = \gamma_i + \delta_i + D(\gamma, \delta)O(1). \tag{19.20}
$$

Since ε is a C^1 function, we have

$$
|\varepsilon(\mu, \pi; \tilde{v}_m) - \varepsilon(\mu, \tilde{\pi}; \tilde{v}_m)| = |\pi - \tilde{\pi}|O(1).
$$

But from (19.18), $|\pi - \tilde{\pi}| = D(\gamma, \delta)O(1)$. Thus

$$
\varepsilon(\mu, \pi; \tilde{v}_m) = \gamma_i + \delta_i + D(\gamma, \delta)O(1),
$$

and (19.15) implies (19.12).

In case (19.6) holds, then using (19.6) instead of (19.5) (above), we obtain

$$
\begin{aligned}
|\varepsilon(\mu, \pi; \tilde{v}_m) - \varepsilon(\mu, \tilde{\pi}, \tilde{v}_m)| &= |\pi - \tilde{\pi}|O(1) \\
&\leq D(\gamma, \delta)(|\gamma| + |\delta|).
\end{aligned}
$$

Thus we can replace (19.20) by

$$\varepsilon_i(\mu, \tilde{\pi}; \tilde{v}_m) = \gamma_i + \delta_i + D(\gamma, \delta)O(|\gamma||\delta|), \qquad (19.21)$$

and the rest of the proof proceeds in a similar fashion to give (19.13). The proof is complete. □

§B. The Difference Approximation

In all existence proofs, the main step is to obtain estimates on certain approximate solutions, in some norm. The question of which approximation to use is often only one of convenience. To obtain compactness of the approximating solutions u, we must obtain estimates on the approximations, and on their "derivatives." In order for a limit to be a solution we need strong convergence for both the approximations u, and all derivatives of u which occur nonlinearly, together with weak convergence for the higher derivatives of u. Thus, if k is the order of the highest nonlinear derivative which occurs in the equations, we need bounds on $D^{k+1}u$. In our case, $k = 0$, so that we must obtain bounds on both u and Du. Of course, the bounds on Du are more delicate than those on u. We remark that for linear equations, Du itself solves a linear equation, so that a method which yields bounds on u is likely to yield bounds on Du too. For conservation laws, it may happen that the conservation laws themselves give a bound on u, in that some conserved quantity, such as the energy, may be definite. But this fact doesn't help much as far as Du is concerned. However, in our case, we shall show that Du satisfies an "approximate" conservation law in the sense that the total variation of u is approximately conserved.

The norms which we choose on u and Du are

$$\|Du\| = \text{T.V.}(u),$$
$$\|u\| = \|u\|_{L^\infty} \le |u(\infty)| + |u(-\infty)| + \text{T.V.}(u),$$

where T.V. denotes the total variation.

We shall now turn to the construction of the approximate solutions. As mentioned earlier, these will be solutions of a difference scheme. We choose mesh lengths Δx and Δt such that $\Delta t \le$ const. Δx, and we approximate the initial data by piecewise constant data. We then solve the corresponding Riemann problems; this propagates the solution for one time step. Our choice of Δt will be made so that nearby waves do not interact with each other; namely,

$$\frac{\Delta x}{\Delta t} > \lambda,$$

where λ is an upper bound for λ_k, $1 \le k \le n$.

Observe now that on the line $t = \Delta t$, the solution is no longer piecewise constant. Thus in order to continue our procedure (of solving Riemann problems), we need a mechanism for choosing a constant value in each interval. We accomplish this by choosing a new constant state as the value of the solution at a randomly chosen point in each mesh interval; see Figure 19.6.

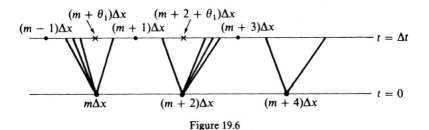

Figure 19.6

To do this we let $\theta = \{\theta_0, \theta_1, \ldots\}$, be a random sequence in the interval $[-1, 1]$. For each such θ, we get an approximate solution $u_{\theta, \Delta x}$. $u_{\theta, \Delta x}$ is clearly an exact solution except across the lines $t = n \Delta t$, and by construction,

$$u_{\theta, \Delta x}(x, n \Delta t + 0) = u_{\theta, \Delta x}((m + 1)\Delta x + \theta_n \Delta x, n \Delta t - 0),$$

for $m \Delta x \leq x < (m + 2) \Delta x$, with $m + n \equiv 0 \pmod 2$.

To understand this procedure better, we consider an example of a single shock wave solution (u_l, u_r) of speed s, propagating out of the origin. At $t = \Delta t$, there are two possibilities depending on the random point, namely $\theta_1 = \alpha$ or $\theta_1 = \beta$; see Figure 19.7. We can depict these two solutions in

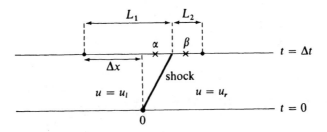

Figure 19.7

Figure 19.8.[2] If we set $L_1 = \Delta x + s \Delta t$, and $L_2 = \Delta x - s \Delta t$, then the probability that solution (α) occurs is

$$\frac{L_1}{L_1 + L_2} = \frac{\Delta x + s \Delta t}{2 \Delta x} = \tfrac{1}{2} + s \frac{\Delta t}{2 \Delta x},$$

[2] Note that for each random choice, the approximate solution is a shock wave, as it should be; what we have lost is only the *position* of the shock.

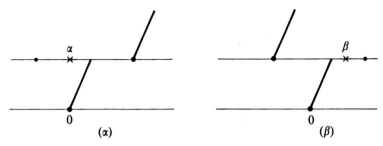

Figure 19.8

while the probability that solution (β) occurs is

$$\frac{L_2}{L_1 + L_2} = \frac{\Delta x - s\,\Delta t}{2\,\Delta x} = \tfrac{1}{2} - s\frac{\Delta t}{2\,\Delta x}.$$

Hence, the expected change in the shock position is

$$|(\Delta x)(\text{Prob. } \alpha) - (\Delta x)(\text{Prob. } \beta)| = \Delta x\left(\frac{1}{2} + s\frac{\Delta t}{2\,\Delta x}\right) - \Delta x\left(\frac{1}{2} - s\frac{\Delta t}{2\,\Delta x}\right)$$

$$= s\,\Delta t.$$

We thus conclude that the shock speed is correct, on the average, so that the conservation law is preserved, on the average. However, the total variation of u_θ equals the total variation of the true solution u; i.e., the total variation is preserved exactly.

Now let's consider many time steps, and introduce the notation

$$u_m^n = u_{\theta,\Delta x}((m + 1)\,\Delta x + \theta_n\,\Delta x, n\,\Delta t),$$

$$m\,\Delta x \le x < (m + 2)\,\Delta x, \quad \text{with } n + m \equiv 0 \ (\text{mod } 2).$$

Since the initial data is either u_l or u_r, we see that these are the only values assumed by the approximate solution. A typical approximate solution (depending on θ), can be depicted in Figure 19.9, (heavy lines).

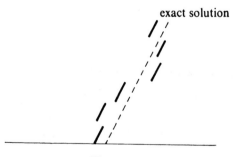

Figure 19.9

Now it is easy to see that

$$u_m^n = \begin{cases} u_l, & \text{if } m < J_n, \\ u_r, & m \geq J_n, \end{cases} \qquad [m + n \equiv 0 \,(\text{mod } 2)],$$

where $J_n =$ (number of $\theta_1, \ldots, \theta_n$ which are less than $s\,\Delta t/\Delta x$) − (number of $\theta_1, \ldots, \theta_n$ which are greater than $s\,\Delta t/\Delta x$). But according to the strong law of large numbers, [Rs, p. 258],

$$\frac{J_n}{n} \to \frac{s\,\Delta t}{\Delta x} \quad \text{as } n \to \infty.$$

(Indeed, if we let

$$H(x) = \begin{cases} 1, & \text{if } x > 0, \\ -1, & \text{if } x < 0, \end{cases}$$

then

$$\tilde{H}(x) \equiv H\left(x - \frac{s\,\Delta t}{\Delta x}\right) = \begin{cases} 1, & \text{if } \dfrac{s\,\Delta t}{\Delta x} < x, \\ -1, & \text{if } \dfrac{s\,\Delta t}{\Delta x} \geq x, \end{cases}$$

and $J_n = \sum_{i=1}^n H(\theta_i)$. Thus according to the strong law of large numbers,

$$\frac{J_n}{n} = \frac{\displaystyle\sum_{i=1}^n \tilde{H}(\theta_i)}{n} \to E(\tilde{H}(\theta)) \quad \text{as } n \to \infty,$$

where $E(\tilde{H}(\theta))$ is the expected value; i.e.,

$$E(\tilde{H}(\theta)) = P\left(\theta < \frac{s\,\Delta t}{\Delta x}\right) - P\left(\theta > \frac{s\,\Delta t}{\Delta x}\right),$$

where P denotes probability. Assuming $s > 0$, we have

$$P\left(\theta < \frac{s\,\Delta t}{\Delta x}\right) - P\left(\theta > \frac{s\,\Delta t}{\Delta x}\right) = \frac{1 + \dfrac{s\,\Delta t}{\Delta x}}{2} - \frac{1 - \dfrac{s\,\Delta t}{\Delta x}}{2} = \frac{s\,\Delta t}{\Delta x}.$$

The discontinuity in the approximate solution at time $t = n\,\Delta t$ is located at the point

$$x(t) = J_n\,\Delta x = \left[\frac{sn\,\Delta t}{\Delta x} + O(n)\right]\Delta x$$

$$= st[1 + (\Delta x)O(n)] \to st \quad \text{as } n \to \infty.^3$$

That is, $u_{\theta,\Delta x} \to u$ as $\Delta x \to 0$. In the case of more general data, we shall prove the convergence of the approximate solutions for "almost all" θ as $\Delta x \to 0$.

---

³ By taking $\theta_n \equiv n\sqrt{2}$, for instance, one can get optimal rate of convergence, whereby $O(1/\sqrt{n})$ is replaced by $O(1/n \log n)$; see [Ku N].

We now shall describe the difference scheme in precise detail. We may assume, without loss of generality, that $\bar{u} = 0$. Let U_1 be a compact neighborhood of 0, and let $U_3 \subset U_2 \subset U_1$ be neighborhoods of 0 such that if u_l and u_r are in U_3, then there is a unique solution of the Riemann problem (u_l, u_r), with intermediate states $u_1, u_2, \ldots, u_{n-1}$, in U_2. We choose Δx and Δt such that the stability condition,

$$\sup\{|\lambda_j(u)| : u \varepsilon U_2, 1 \leq j \leq n\} < \frac{\Delta x}{\Delta t} \qquad (19.22)$$

holds, and we let $\Delta x / \Delta t = c$ be held fixed so $\Delta t = c^{-1}\Delta x$. Let

$$Y = \{(m, n) \in \mathbf{Z}^2 : m + n \equiv 0 \,(\text{mod } 2), n \geq 0\},$$

and

$$\Phi = \prod_{(m,n) \in Y} \{[(m - 1)\Delta x, (m + 1)\Delta x] \times \{n \Delta t\}\}.$$

We put the same probability mass on each interval ($(2\Delta x)^{-1}$ times Lebesgue measure), and let Φ have the product measure $d\Phi$. We choose a point $a_{m,n}$ ($m \Delta x + \theta_n \Delta x, n \Delta t$) where θ_n is randomly chosen in $[-1, 1]$; this is our random choice, and the points $a_{m,n}$ will be the mesh points.

We can now define the difference scheme. This is done inductively by assuming that $u = u(x, t)$ has been defined at $a_{m-1,n-1}$ and $a_{m+1,n-1}$. To define u at $a_{m,n}$, we solve the Riemann problem

$$v_t + f(v)_x = 0, \quad (m - 1)\Delta x \leq x \leq (m + 1)\Delta x, \quad (n - 1)\Delta t \leq t < n\Delta t,$$

with data

$$v(x, (n - 1)\Delta t) = \begin{cases} u(a_{m-1,n-1}), & (m - 1)\Delta x \leq x < m \Delta x, \\ u(a_{m+1,n-1}), & m \Delta x < x \leq (m + 1)\Delta x. \end{cases}$$

We then define $u(a_{m,n}) = v(a_{m,n})$. Notice that this can be done if both $u(a_{m-1,n-1})$ and $u(a_{m+1,n-1})$ lie in U_3. The process can be repeated provided that the resulting intermediate states also lie in U_3; thus it is not a-priori clear that the difference scheme can be defined at all levels $n\Delta t$. This will be proved simultaneously with the proof of the estimates on $u_{\theta,\Delta x}$ and T.V.$(u_{\theta,\Delta x})$.

We find it convenient to set

$$u(x, t) = v(x, t), \quad (m - 1)\Delta x \leq x \leq (m + 1)\Delta x, \quad (n - 1)\Delta t \leq t < n\Delta t.$$

Then u is a solution in this rectangle. In view of our stability condition (19.22), we see that if x is near $(m - 1)\Delta x$ (resp. $(m + 1)\Delta x$), then $u(x, t) = u(a_{m-1,n-1})$ (resp. $u(a_{m+1,n-1})$). Thus, if u is defined in the strip $(n - 1)\Delta t \leq t < n\Delta t$, $-\infty < x < \infty$, then u is constant across the lines $x = (m - 1)\Delta x$, $m + n \equiv 0 \,(\text{mod } 2)$, and u is a solution in this strip, since waves don't interact with each other across the lines $x = (m - 1)\Delta x$.

In order to obtain the desired estimates, it is convenient to consider not horizontal lines, but rather, curves consisting of line segments joining $a_{m,n}$ to both $a_{m+1,n\pm1}$. Thus, we cover the upper-half plane $t \geq 0$ by "diamonds," the corners of which are the random points in the mesh intervals; see Figure 19.10.

Definition 19.4. A *mesh curve* is a (nonbounded) piecewise linear curve lying on diamond boundaries going from W to N or S, see Figure 19.11.

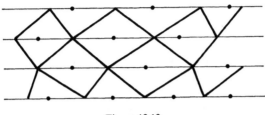

Figure 19.10

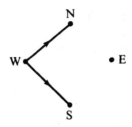

Figure 19.11

If I is any mesh curve, then I divides the half plane $t \geq 0$ into an I^+ and I^- part; the I^- part being the one containing $t = 0$. This allows us to partially order the mesh curves by saying $I_1 > I_2$ if every point of I_1, is either on I_2 or contained in I_2^+. We call I an *immediate successor* to J if $I > J$ and every mesh point of I except one is on J; see Figure 19.12. Finally we let O be the (unique) mesh curve which passes through the mesh points on $t = 0$ and $t = \Delta t$.

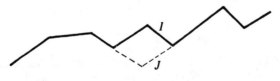

Figure 19.12

We shall obtain our estimates by considering certain functionals F defined on the restriction of u to a mesh curve J; i.e., $F(J) \equiv F(u|_J)$, where $u|_J$ consists of those shocks and rarefaction waves separating constant states, which intersect the mesh curve J.

Let α be a j-wave crossing J, and let β be a k-wave crossing J. We say that α and β *approach* if either

(a) $j \neq k$, and the wave belonging to the faster family (higher index) lies to the left of the other on J; or,

(b) if $j = k$, we have $\alpha \neq \beta$ and at least one is a shock wave.

(This definition is consistent with our earlier definition of approaching waves in §A.)

Given an approximate solution $u_{\theta,\Delta x}$, we define functionals Q and L on the mesh curve J by

$$Q(J) = \sum \{|\alpha|\,|\beta| : \alpha, \beta \text{ cross } J \text{ and approach}\},$$

and

$$L(J) = \sum \{|\alpha| : \alpha \text{ crosses } J\},$$

where α and β are waves in the function $u_{\theta,\Delta x}$.

We can now prove the main theorem in this section.

Theorem 19.5. *Let I and J be two mesh curves with $J > I$, and suppose that I is in the domain of definition of $u_{\theta,\Delta x}$. If $L(I)$ is sufficiently small, then J is in the domain of definition of $u_{\theta,\Delta x}$, $Q(I) \geq Q(J)$, and $L(I) + kQ(I) \geq L(J) + kQ(J)$ for some constant k independent of J. If T.V.(u_0) is small, $u_{\theta,\Delta x}$ can be defined in $t \geq 0$.*

Proof. We first assume that J is an immediate successor to I, so that J and I differ by a single diamond Δ; see Figure 19.13. Let $I = I_0 \cup I'$, and $J = I_0 \cup J'$. We have

$$L(I) = L(I_0) + L(I') = L(I_0) + \sum_{I'} |\gamma_i| + \sum_{I'} |\delta_i|,$$

and

$$L(J) = L(I_0) + L(J') = L(I_0) + \sum_{J'} |\varepsilon_i|,$$

and from Theorem 19.2,

$$L(J) \leq L(I) + k_0 Q(I'), \tag{19.23}$$

where k_0 is the $O(1)$ in Theorem 19.2. Next, both

$$Q(I) = Q(I_0) + Q(I') + Q(I_0, I'),$$

and

$$Q(J) = Q(I_0) + Q(I_0, J')$$

hold, where, e.g., by $Q(I_0, I')$, we denote the sum of the products of two approaching waves, one crossing I_0 and the other crossing I'. Now again from Theorem 19.2,

$$Q(I_0, J') = |\varepsilon_i|\,|\alpha| \leq \sum (|\gamma_i| + |\delta_i|)|\alpha| + k_0 Q(I')L(I_0)$$

(If ε_i and α have the same index, and α is a rarefaction wave, so that $\varepsilon_i < 0$, then if γ_i and/or δ_i is a rarefaction wave, it will not approach α. However, in this case, $|\varepsilon_i| = |\gamma_i + \delta_i + O(1)D(\gamma, \delta)| < |\delta_i + O(1)D(\gamma, \delta)|$, since $\varepsilon_i < 0$ and $\gamma_i > 0$. If $\delta_i > 0$ too, then $|\varepsilon_i| < |O(1)D(\gamma, \delta)|$.)

$$\leq Q(I_0, I') + k_0 Q(I')L(I_0)$$
$$\leq Q(I_0, I') + \tfrac{1}{2}Q(I'),$$

provided that $L(I_0)$ is sufficiently small; i.e., $k_0 L(I_0) < \tfrac{1}{2}$. Thus

$$Q(J) - Q(I) = Q(I_0, J') - Q(I_0, I') - Q(I')$$
$$\leq \tfrac{1}{2}Q(I') - Q(I') = -\tfrac{1}{2}Q(I') \leq 0.$$

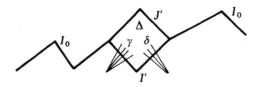

Figure 19.13

This shows that Q is monotone decreasing. Next, using (19.23), and the last inequality,

$$L(J) + kQ(J) \leq L(I) + k_0 Q(I') + kQ(I) - \tfrac{1}{2}kQ(I')$$
$$\leq L(I) + kQ(I),$$

if $k \geq 2k_0$. Thus the inequalities hold in case J is an immediate successor to I. Moreover, we have in this case

$$L(J) + kQ(J) \leq L(0) + kQ(0)$$
$$\leq L(0) + kL(0)^2$$
$$\leq 2L(0),$$

if $L(0) < 1/k$. This shows that if the total variation of the initial data is small, the variation of $u_{\theta,\Delta}$ on J is small. This means that $u_{\theta,\Delta}$ can be defined on an immediate successor to J. Thus if J is now any arbitrary mesh curve, with $J > I$, we can pass from I to J by immediate successors where at each stage Q and $L + kQ$ are monotone nonincreasing, and $u_{\theta,\Delta}$ can be defined on J. If T.V.(u_0) is small then $L(J) + kQ(J) < 2L(0)$, so that

$$\sup\{|\lambda_j(u_{\theta,\Delta x})|, 1 \leq j \leq n\} < \frac{\Delta x}{\Delta t},$$

and $u_{\theta,\Delta x}$ can be defined in $t \geq 0$. $\square$

Corollary 19.6. *If* T.V.(u_0) *is small, then on any mesh curve* I,

$$\text{osc } u_{\theta,\Delta x} \leq \text{T.V.}(u_{\theta,\Delta x}) \leq \text{const. } L(I) \leq \text{const. } L(0)$$
$$\leq \text{const. T.V.}(u_0),$$

where the constants are independent of θ *and* Δx.

Proof. osc $u_{\theta,\Delta x} \leq$ T.V. $u_{\theta,\Delta x}$ is always true, while T.V.$(u_{\theta,\Delta x}) \leq$ const. $L(I)$ holds since T.V. and L are equivalent metrics. Also $L(I) \leq L(I) + kQ(I)$ $\leq 2L(0) \leq$ const. T.V.(u_0) if T.V.(u_0) is small. $\square$

Corollary 19.7. *If* T.V.(u_0) *is small, then*

$$\text{T.V.}[u_{\theta,\Delta x}(x, n\,\Delta t)] + \sup_x u_{\theta,\Delta x}(x, nt) < C\,\text{T.V.}(u_0)$$

where the constant C is independent of n, θ, Δx and Δt.

Proof. Since T.V.$(u_0) < \infty$, we know that $\lim_{x \to \infty} u_0(x)$ exists; call it l. From the definition of $u_{\theta,\Delta x}$, $\lim_{x \to \infty} u_{\theta,\Delta x}(x, n\,\Delta t) = l$, for each n. Corollary 19.6 thus implies that $\sup_x u_{\theta,\Delta x}(x, n\,\Delta t)$ is bounded by C_1 T.V.(u_0). $\square$

We next show that these estimates imply that our approximating solutions are locally L_1 Lipschitz continuous in time.

Corollary 19.8. *If* T.V.(u_0) *is sufficiently small, then*

$$\int_{-\infty}^{\infty} |u_{\theta,\Delta x}(x, t) - u_{\theta,\Delta x}(x, t')|\,dx \leq C[|t - t'| + \Delta t],$$

where C and C_1 are independent of θ and Δx.

Proof. Let $t_2 > t_1$, and $t_0 = \sup\{t \leq t_1 : t = n\,\Delta t$ for some $n\}$. Let $S = [(t_2 - t_0)/\Delta t] + 1$; then $S\,\Delta t \leq t_2 - t_1 + 2\,\Delta t$. Now $u(x, t_i)$ is determined by the Cauchy data

$$\{u(y, t_0) : y \in I\}, \quad \text{where } I = [x - S\,\Delta x, S + \Delta x].$$

Thus, by what has already been shown,

$$|u(x, t_2) - u(x, t_1)| \leq \text{const. } \sup\{|u(y, t_0) - u(x, t_0)| : y \in I\}$$
$$\leq \text{const. T.V.}\{u(y, t_0) : y \in I\}.$$

Hence

$$\int_{m\Delta x}^{(m+2)\Delta x} |u(x, t_2) - u(x, t_1)|\,dx \leq \text{const. } \Delta x \text{ T.V.}\{u(y, t_0) : y \in I\}$$
$$\leq \text{const. } \Delta x \text{ T.V.}\{u(y, t_0) : y \in I_m\},$$

where $I_m = [(m - S)\Delta x, (m + 2 + S)\,\Delta x]$, so that

$$\int_{-\infty}^{\infty} |u(x, t_2) - u(x, t_1)|\,dx = \sum_m \int_{m\Delta x}^{(m+2)\Delta x} |u(x, t_2) - u(x, t_1)|\,dx$$
$$\leq \text{const. } \Delta x \sum_m \text{T.V.}\{u(y, t_0) : y \in I_m\},$$
$$\leq \text{const. } \Delta x \sum_m \text{T.V.}\{u(y, t_0) : y \in I'_m\},$$

where $I'_m = [m\,\Delta x, (m + 2 + S)\,\Delta x]$. It follows that

$$\int_{-\infty}^{\infty} |u(x, t_2) - u(x, t_1)|\, dx \leq \text{const. } \Delta x(2S + 2)\ \text{T.V. } u(\cdot, t_0)$$

$$\leq \text{const. } \Delta x(2S + 2)\ \text{T.V. } u(\cdot, 0)$$

$$= \text{const. } \frac{\Delta x}{\Delta t}(2S + 2)\,\Delta t$$

$$\leq \text{const.}[(t_2 - t_1) + \Delta t].$$

This completes the proof.   □

§C. Convergence

The compactness of the difference approximations is really a consequence of Corollaries 19.7 and 19.8, and is similar to the corresponding result for the scalar conservation law in Chapter 16.

Consider the following three hypotheses on a set of functions $\{u_\alpha(x, t)\}$:

$$\|u_\alpha(\cdot, \cdot)\|_{L_\infty} \leq M_1, \tag{H_1}$$

$$\text{T.V. } u_\alpha(\cdot, t) \leq M_2, \tag{H_2}$$

$$\|u_\alpha(\cdot, t_1) - u_\alpha(\cdot, t_2)\|_{L_1} \leq M_3|t_1 - t_2|, \tag{H_3}$$

where M_1 is independent of α, M_2 is independent of t and α, and M_3 is independent of α, t_1, and t_2.

Theorem 19.9. *Let $\{u_\alpha\}$ be a family of functions satisfying* (H_1), (H_2), *and* (H_3). *Then a subsequence* $\{u_n\}$ *converges in* L_1^{loc} *to a function u.*

The proof is exactly the same as the corresponding result in Chapter 16, Lemma 16.8. In view of Corollaries 19.7 and 19.8, we can apply this theorem to the family $\{u_{\theta, \Delta x} : \theta \in \Phi, \Delta x > 0\}$, to obtain the desired compactness. However, we must show that the limit functions are solutions.[4] As a first step we shall prove the following corollary.

Corollary 19.10. *Let $\{u_n\}$ be as in the theorem. Then there exists a subsequence* $\{u_{n_j}\} \subset \{u_n\}$ *such that* $f(u_{n_j}) \to f(u)$ *in* L_1^{loc} *for every continuous function f.*

Proof. Since $u_n \to u$ in L_1^{loc}, there is a subsequence $\{u_{n_j}\}$ of $\{u_n\}$ such that $u_{n_j} \to u$, a.e. Thus $f(u_{n_j}) \to f(u)$, a.e. But $\{u_{n_j}\}$ is uniformly bounded, so that $\{f(u_{n_j})\}$ too is uniformly bounded in L_1^{loc}. The result now follows by the Lebesgue bounded convergence theorem.   □

----

[4] This is highly nontrivial. It is here where we shall make essential use of the *random-choice* method.

Let's fix $\theta \in \Phi$, and temporarily write $u_{\Delta x}$ for $u_{\theta,\Delta x}$. If $\phi \in C_0^1$ $(t > 0)$ is a test function, we define a functional $\mathscr{L}_\phi$ by

$$\mathscr{L}_\phi(u, f(u)) = \int_0^\infty \int_{-\infty}^\infty (u\phi_t + f(u)\phi_x)\, dx\, dt + \int_{-\infty}^\infty \phi(x, 0)u_0(x, 0)\, dx. \quad (19.24)$$

Our goal is to produce a u for which $\mathscr{L}_\phi(u, f(u)) = 0$ for each test function ϕ.

Now we know that $u_{\Delta x}$ is a weak solution in each time strip $l\,\Delta t \le t \le (l + 1)\,\Delta t$, so that for each test function ϕ,

$$\int_{l\Delta t}^{(l+1)\Delta t} \int_{-\infty}^\infty \{\phi_t u_{\Delta x} + \phi_x f(u_{\Delta x})\}\, dx\, dt + \int_{-\infty}^\infty \phi(x, l\,\Delta t)u_{\Delta x}(x, l\,\Delta t + 0)\, dx$$

$$-\int_{-\infty}^\infty \phi(x, (l + 1)\,\Delta t)u_{\Delta x}(x, (l + 1)\,\Delta t - 0)\, dx = 0.$$

If we sum this over l we get

$$\mathscr{L}_\phi(u_{\Delta x}, f(u_{\Delta x})) = \int_0^\infty \int_{-\infty}^\infty \{\phi_t u_{\Delta x} + \phi_x f(u_{\Delta x})\}\, dx\, dt$$

$$+\int_{-\infty}^\infty \phi(x, 0)u_{\Delta x}(x, 0)\, dx$$

$$+\sum_{l=1}^\infty \int_{-\infty}^\infty \phi(x, l\,\Delta t)[u_{\Delta x}](x, l\,\Delta t)\, dx = 0. \quad (19.25)$$

Comparing (19.24) with (19.25), we see that the following corollary of Theorem 19.9 and Corollary 19.10 is valid.

Corollary 19.11. *Let $\theta \in \Phi$; then $u = \lim_{\Delta x_j \to 0} u_{\theta, \Delta x_j}$ is a solution of (19.1), (19.2) provided that the following two conditions hold (as $\Delta x_j \to 0$):*

$$u_{\theta, \Delta x_j}(\cdot, 0) \to u_0 \quad \text{weakly, and} \quad (19.26)$$

$$\sum_{l=1}^\infty [u_{\theta, \Delta x_j}(\cdot, l\,\Delta t)] \equiv \sum_{l=1}^\infty [u_{\theta, \Delta x_j}]_l \to 0 \quad \text{weakly.} \quad (19.27)$$

Thus our existence theorem will be proved if we can show both (19.26) and (19.27). Since (19.26) can be easily achieved, we confine our attention to (19.27). To this end, we fix the following: a test function ϕ, $\theta \in \Phi$, and $\Delta x > 0$. We define

$$J_l = \int_{\mathbf{R}} \phi(x, l\,\Delta t)[u_{\theta, \Delta x}]_l\, dx \equiv J_l(\theta, \Delta x, \phi),$$

and

$$J = \sum_{l=1}^{\infty} J_l \equiv J(\theta, \Delta x, \phi).$$

Lemma 19.12. *Let* $\theta \in \Phi$, $\phi \in C_0 \cap L_\infty$, *and assume that* (H_2) *holds with* $\alpha = (\theta, \Delta x)$. *Then there are constants M and M_1, independent of ϕ, θ, and Δx such that*

$$|J_l(\theta, \Delta x, \phi)| \leq M(\Delta x)\|\phi\|_\infty, \quad \text{and} \tag{19.28}$$

$$|J(\theta, \Delta x, \phi)| \leq M_1 (\text{diam. support } \phi)\|\phi\|_\infty. \tag{19.29}$$

Proof. (19.29) is a consequence of (19.28) since if the support of ϕ is in $[a, b] \times [0, T]$, then J_l is nonzero only if $l < T/\Delta t$; i.e., there are only $(\Delta t)^{-1}(\text{diam. support } \phi) = O((\Delta x)^{-1})(\text{diam. support } \phi)$, nonzero terms in the sum. It thus suffices to prove (19.28). We have

$$|J_l(\theta, \Delta_x, \phi)| \leq \sum_{m=-\infty}^{\infty} \int_{(m-1)\Delta x}^{(m+1)\Delta x} |\phi(x, l\,\Delta t)[u_{\theta,\Delta x}]_l|\, dx$$

$$\leq \|\phi\|_\infty \sum_m \int_{(m-1)\Delta x}^{(m+1)\Delta x} |u_{\theta,\Delta x}((m + \theta_l)\,\Delta x, l\,\Delta t - 0)$$
$$- u_{\theta,\Delta x}(x, l\,\Delta t - 0)|\, dx$$

$$\leq \|\phi\|_\infty \sum_m \text{osc}\{u_{\theta,\Delta x}(\,\cdot\,, l\,\Delta t - 0)$$
$$\text{on } [(m-1)\,\Delta x, (m+1)\,\Delta x]\} \int_{(m-1)\Delta x}^{(m+1)\Delta x} dx$$

$$\leq \|\phi\|_\infty \left(\sum_m \text{T.V.}\{u_{\theta,\Delta x}(\,\cdot\,, l\,\Delta t - 0) \right.$$
$$\left. \text{on } [(m-1)\,\Delta x, (m+1)\,\Delta x] \right) 2\,\Delta x$$

$$= M\,\Delta x \|\phi\|_\infty.$$

This completes the proof. $\square$

The measure space Φ depends on Δx; however there is an obvious isomorphism, $\Phi \approx \prod[0, 1]$, of Φ with a countable product of copies of the unit interval. By means of this isomorphism, we can consider the random points θ being defined in a fixed probability space Φ, independent of Δx. In the next theorem we shall find a null set N in Φ independent of Δx, such that if $\theta \in \Phi \backslash N$, there is a sequence $\Delta x_i \to 0$ for which (19.26) and (19.27) hold. But before doing this, we need one final lemma.

Lemma 19.13. *Let ϕ have compact support, and suppose that ϕ is piecewise constant on segments of the form $[(m-1)\,\Delta x, (m+1)\,\Delta x] \times l\,\Delta t$. Then if*

$l_1 \neq l_2$, $J_{l_1}(\cdot, \Delta x, \phi) \perp J_{l_2}(\cdot, \Delta x, \phi)$, *where the orthogonality is with respect to the L_2 inner product on Φ.*

Proof. The idea is that independent random variables with mean zero are orthogonal. Thus, if $l_1 < l_2$, then J_{l_1} is independent of θ_{l_2}, and we assert that

$$\int_\Phi J_{l_2} \, d\theta_{l_2} = 0. \tag{19.30}$$

Assuming (19.30), for the moment, we have

$$\langle J_{l_1}, J_{l_2} \rangle = \int \left(\int J_{l_1} J_{l_2} \, d\theta_{l_2} \right) \prod_{l \neq l_2} d\theta_l$$

$$= \int J_{l_1} \left(\int J_{l_2} \, d\theta_{l_2} \right) \prod_{l \neq l_2} d\theta_l$$

$$= 0.$$

We now prove (19.30). Let $t_l = l \Delta t - 0$; then

$$\int_\Phi J_{l_2}(\theta, \Delta x, \phi) \, d\theta_{l_2} = \int_\Phi \sum_{m=-\infty}^{\infty} \int_{(m-1)\Delta x}^{(m+1)\Delta x} [u_{\theta, \Delta x}((m + \theta_l)\Delta x, t_l)$$

$$- u_{\theta, \Delta x}(x, t_l)] \, dx \, d\theta_l.$$

But, by definition, we have

$$\sum_{m=-\infty}^{\infty} \left\{ \int_\Phi \int_{(m-1)\Delta x}^{(m+1)\Delta x} u(x, t_l) \, dx \, d\theta_l - \int_{(m-1)\Delta x}^{(m+1)\Delta x} \int_\Phi u((m + \theta_l)\Delta x, t_l) \, d\theta_l \, dx \right\} = 0.$$

The proof of the lemma is complete. $\square$

We now consider only Δx of the form 2^{-k}, and note that if ϕ satisfies the hypotheses of Lemma 19.13 for some Δx, it satisfies the hypotheses for all smaller Δx.

Theorem 19.14. *Suppose that* (H_2) *holds where* $\alpha = (\theta, \Delta x)$; *then there is a null set* $N \subset \Phi$ *and a sequence* $\Delta x_i \to 0$ *such that for any* $\theta \in \Phi \backslash N$ *and* $\phi \in C_0^1 (t > 0)$, $J(\theta, \Delta x_i, \phi) \to 0$ *as* $i \to \infty$.

Remark. $J(\theta, \Delta x_i, \phi) \to 0$ is equivalent to (19.27).

Proof. Let ϕ satisfy the hypotheses of the last lemma; then

$$\|J(\cdot, \Delta x_i, \phi)\|_2^2 = \sum_l \|J_l(\cdot, \Delta x_i, \phi)\|_2^2$$
$$\leq \sum_l \|J_l(\cdot, \Delta x_i, \phi)\|_\infty^2, \qquad (19.31)$$

since $\int_\Phi d\theta = 1$. But from (19.28),

$$\sum_l \|J_l(\cdot, \Delta x_i, \phi)\|_\infty^2 \leq \sum_{l \in \Lambda} M_l^2 (\Delta t_i)^2 \|\phi\|_\infty^2,$$

where $\Lambda = \{l: (\mathbf{R} \times l \,\Delta t_i) \cap (\text{support } \phi) \text{ is nonvoid}\}$, so that

$$\|J(\cdot, \Delta x_i, \phi)\|_2^2 \leq M_1^2 \,\Delta t_i (\text{diam. support } \phi) \|\phi\|_\infty^2. \qquad (19.32)$$

It follows that for each piecewise constant ϕ with compact support, there is a sequence $\Delta x_i \to 0$ such that $J(\cdot, \Delta x_i, \phi)$ tends to zero in L_2.

Next, for each $\phi \in L^\infty \cap C_0$, (19.29) gives

$$\|J(\cdot, \Delta x_i, \phi)\|_2 \leq \|J(\cdot, \Delta x_i, \phi)\|_\infty \leq \text{const.} \|\phi\|_\infty. \qquad (19.33)$$

Let $\{\phi_\nu\}$ be a sequence of piecewise constant functions with compact support which are L_∞-dense in the space of test functions.[5] For each ϕ_ν, there is a null set $N_\nu \subset \Phi$ and a sequence $\Delta x_{i_k} \to 0$ such that

$$J(\theta, \Delta x_{i_k}, \phi_\nu) \to 0 \quad \text{if } \theta \in \Phi \backslash N_\nu.$$

Let $N = \bigcup_\nu N_\nu$ and let $\theta \in \Phi \backslash N$; by a diagonal process, we can find a subsequence, call it Δx_i again, such that for each ν

$$J(\theta, \Delta x_i, \phi_\nu) \to 0 \quad \text{as } i \to \infty. \qquad (19.34)$$

Now let $\tilde\phi$ be any test function; then if $\theta \in \Phi \backslash N$,

$$|J(\theta, \Delta x_i, \tilde\phi)| \leq |J(\theta, \Delta x_i, \tilde\phi - \phi_\nu)| + |J(\theta, \Delta x_i, \phi_\nu)|$$
$$\leq \text{const.} \|\tilde\phi - \phi_\nu\|_\infty + o(1),$$

as $i \to \infty$, by (19.33) and (19.34). Choosing ν first to make $\|\tilde\phi - \phi_\nu\|_\infty$ small, we then choose i large to make the second term small. This completes the proof. $\square$

Our main theorem is really a corollary of Theorem 19.14.

---

[5] For example, we could choose the Haar functions; see [GP].

Theorem 19.15. *If T.V.(u_0) is sufficiently small, there exists a null set $N \subset \Phi$ and a sequence $\Delta x_i \to 0$ such that if $\theta \in \Phi \backslash N$, $u_\theta = \lim u_{\theta, \Delta x_i}$ is a solution of the problem (19.1), (19.2).*

Proof. First choose a subsequence of $\{\Delta x_i\}$ such that (19.26) holds. Then choose a further subsequence such that for $\theta \in \Phi \backslash N$ where N is as in Theorem 19.4 $J(\theta, \Delta x_i, \phi) \to 0$ as $i \to \infty$; this can be done by the last theorem. Then choose another subsequence so that $\{u_{\theta, \Delta x_j}\}$ converges; this can be done by Theorem 19.9. Then Corollary 19.11 finishes the proof. $\square$

NOTES

The entire contents of this chapter is taken, with some minor modifications, from Glimm's profound paper [G1]. This work constituted the major breakthrough in the subject, and it is fair to say that all of the subsequent work on systems of conservation laws is based upon it. In particular, we mention the paper of Glimm and Lax [GL] which deals with the decay of the solutions. In this work the estimates (19.10) and (19.11),

$$\varepsilon_i = \gamma_i + \delta_i + \begin{cases} D(\gamma, \delta)O(1), & \text{if } n > 2, \\ \\ D(\gamma, \delta)O(|\gamma| + |\delta|), & \text{if } n = 2, \end{cases} \qquad (*)$$

are also quite crucial. For the existence theorem the bound $|\varepsilon_i| \leq \{|\gamma_i| + |\delta_i| + (\text{error})\}$ suffices, but the decay comes from the cancellations in (*): if sgn $\gamma_i \neq$ sgn δ_i (i.e., if one is a rarefaction wave and the other a shock wave), then $|\varepsilon_i| = \{\|\gamma_i| - |\delta_i\| + \text{error}\}$.

The papers of DiPerna [Dp 2, 4], Liu [Lu 4, 7], and Greenberg [Gb 3] are concerned with rates of decay, as well as the asymptotic form of the solution as $t \to +\infty$. The problem of uniqueness is still not completely resolved, but see the papers of Oleinik [O 2], Liu [Lu 10], and Di Perna [Dp 5]. A deterministic version of the difference scheme presented here has been given by Liu; in it he proves that the difference scheme converges for any equidistributed sequence $\theta \in \Phi$; see [Lu 2].

For special equations, or general equations with certain restrictions on the data, see Nishida [N], Nishida and Smoller [NS 1], DiPerna [Dp 1], Smoller and Johnson [SJ], Greenberg [Gb 1, 2], and Moler and Smoller [MS]. In particular, Nishida obtains a global existence theorem for the p-system, with $p(v) = \text{const.} v^{-1}$, with arbitrary data of bounded variation. Nishida and Smoller [NS 2], consider the mixed (i.e., piston) problem; see also Liu [Lu 6]. The papers of Liu [Lu 5] and Temple [Te] deal with the full system of gas dynamics equations. DiPerna in [Dp 3], studies the regularity of solutions, and Liu and Smoller [LS] study the problem of the vacuum. An important unsolved problem is to extend Glimm's theorem to the case where the initial total variation is not small. Here one obstruction, say for the p-system, is the development of solutions whose density comes near to zero; the main estimates in §A no longer hold; see [LS].

Chapter 20

Riemann Invariants,
Entropy, and Uniqueness

There are better results known for pairs of conservation laws than for systems with more than two equations. We have already seen an example of this in the last chapter; namely, the interaction estimates are stronger when $n = 2$ than when $n > 2$. This was due to the existence of a distinguished coordinate system called Riemann invariants, which in general exists only for two equations. We shall study the implications one can draw using these coordinates. It turns out that the equations take a particularly nice form when written in terms of the Riemann invariants, and using this we can prove that for genuinely nonlinear systems, global classical solutions generally do not exist. (We only know this now for a single conservation law; see Chapter 15, §B.)

The second advantage of these coordinates is a bit more subtle. The point is that if we use them to measure the strengths of waves, then this allows for a great simplification in the form of the Glimm functionals $F(J)$ as described in the last chapter. We take advantage of this to give a global existence theorem for the isentropic gas dynamics equations where the data can have large total variation. Finally, we shall use Riemann invariants to obtain a precise instability theorem. Thus, we will show that shocks violating the entropy conditions (17.22) are unstable in a strong mathematical sense; namely, they are not closed under regularization; i.e., smoothing.

Another goal in this chapter is to define a general notion of entropy from which we can deduce some nice quantitative features about solutions. The most important one is that solutions which are limits of viscous equations, as well as solutions constructed via the Glimm difference scheme of the last chapter, must satisfy the shock conditions (17.22).

In the final section we shall prove a fairly general uniqueness theorem for the p-system. The result says that solutions with a finite number of shocks and centered rarefaction waves are uniquely determined by their initial data. The method of proof is the same as for the scalar case, and relies on solving the "adjoint" system. Only now there is no entropy inequality and the estimates are much more difficult to obtain.

§A. Riemann Invariants

We have studied certain particular Riemann invariants earlier in Chapters 17 and 18. Here we give a general approach, followed by several applications.

Thus consider the pair of conservation laws

$$u_t + f(u, v)_x = 0, \qquad v_t + g(u, v)_x = 0, \qquad (20.1)$$

where u and v are in $\mathbf{R}$. We let $U = (u, v)$, and $F(U) = (f, g)$, so that the equations (20.1) can be written as

$$U_t + dF(U)U_x = 0, \qquad (20.2)$$

where $dF(U)$ is the Jacobian matrix of F. Since the system is assumed to be hyperbolic, dF has real and distinct eigenvalues $\lambda < \mu$. Let l_λ, l_μ, and r_λ, r_μ denote the corresponding left and right eigenvectors of dF.

Consider the Riemann invariant $w = w(u, v)$ corresponding to λ. From Chapter 17, §B, we know that it satisfies the equation

$$\nabla w \cdot r_\lambda = 0. \qquad (20.3)$$

This says that w is constant along trajectories of the vector field r_λ. We solve (20.3) by taking a curve $\mathscr{C}$ nowhere tangent to r_λ, and we assign arbitrary values for w along $\mathscr{C}$. We then solve the pair of characteristic (ordinary differential) equations through $\mathscr{C}$, $\dot{U} = r_\lambda(U)$, and make w constant along each orbit. If w is taken to be strictly increasing along $\mathscr{C}$, then it has distinct values along distinct orbits.

If we now define $z = z(u, v)$ to be the Riemann invariant associated with μ, then

$$\nabla z \cdot r_\mu = 0.$$

Since r_λ and r_μ are linearly independent, any trajectory of $\dot{U} = r_\mu(U)$ could serve for $\mathscr{C}$ as above. Let's fix a particular one, Γ_μ. Now fix a particular trajectory Γ_λ passing through Γ_μ, and choose z to be strictly increasing along Γ_λ; see Figure 20.1. If we show that trajectories of distinct families (i.e., those corresponding to r_λ and r_μ), can meet in at most one point, then the mapping $(v, u) \to (w, z)$ is bijective, and we can take (w, z) as a coordinate system. To see that this is always possible, suppose for example, that an r_μ trajectory Γ'_μ meets an r_λ trajectory Γ'_λ, at two points P and Q; see Figure 20.2. The vector field r_μ must point towards opposite sides of Γ'_λ at both P and Q. This implies that r_μ must be tangent to r_λ somewhere on Γ'_λ between P and Q. But at this point r_λ and r_μ would be linearly dependent; this is a contradiction. Thus no r_μ trajectory meets an r_λ trajectory more than once. We have thus proved the following theorem.

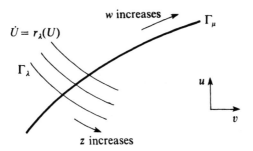

Figure 20.1

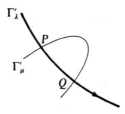

Figure 20.2

Theorem 20.1. *The mapping $(u, v) \to (w, z)$, as described above, is bijective in any simply connected region.*

Recall that left and right eigenvectors corresponding to distinct eigenvalues are orthogonal.[1] Thus since $\nabla w \cdot r_\lambda = 0$, we see that the equation $l_\mu r_\lambda = 0$ shows that ∇w is a multiple of l_μ. That is, ∇w is a left eigenvector of dF with eigenvalue μ:

$$\nabla w \, dF = \mu \nabla w. \tag{20.4}$$

Similarly we find

$$\nabla z \, dF = \lambda \nabla z. \tag{20.5}$$

Now multiply (20.2) on the left by ∇w, and use (20.4) to get

$$0 = \nabla w U_t + \nabla w \, dF(U)U_x = w_t + \mu \nabla w U_x,$$

or

$$w_t + \mu w_x = 0. \tag{20.6}$$

---

[1] If $dFr = \lambda r$, $l\,dF = \mu l$, then $\mu lr = l\,dFr = \lambda lr$ so $lr = 0$ if $\lambda \neq \mu$.

Similarly, we find that z satisfies the equation

$$z_t + \lambda z_x = 0. \tag{20.7}$$

Equations (20.6) and (20.7) can be considered equivalent to (20.1) only for smooth (i.e., classical) solutions. (In fact, since the former are not in conservation form, it is not apparent how one would interpret nonsmooth solutions; see [V].) Nevertheless these equations are quite useful, as we shall soon see.

Observe that (20.6) implies that w is constant along μ-characteristics, and (20.7) shows that z is constant along λ-characteristics. This is a nice analogy with the scalar conservation law $u_t + f'(u)u_x = 0$, where u is a constant along the $f'(u)$ characteristics. This fact was used to prove the nonexistence of a global smooth solution for the single conservation law; the corresponding statement will be used for proving the same result for the *system* (20.1).

We remark that *by definition* the Riemann invariants are constant across rarefaction waves of the corresponding characteristic family. Thus, if we consider z and w as coordinates, then in this frame, the rarefaction wave curves (see Chapter 17, §B), become straight lines. This fact too can often be useful; see §C.

We now assume that (20.1) is genuinely nonlinear in the μ-characteristic field. Recall from Chapter 17, §B, that this means

$$\nabla \mu \cdot r_\mu \neq 0. \tag{20.8}$$

We wish to write (20.8) in the (w, z) coordinates. For this, let T be the bijective mapping

$$T : (u, v) \to (w, z),$$

which we have discussed above. If $k = \det(dT)$, then $k \neq 0$ and

$$(dT)^{-1} = \frac{1}{k} \begin{bmatrix} z_v & -w_v \\ -z_u & w_u \end{bmatrix} = \begin{bmatrix} u_w & u_z \\ v_w & v_z \end{bmatrix}.$$

Since $cl_\lambda = \nabla z = (z_u, z_v)$, we see that we may choose $r_\mu = (z_v, -z_u)$. Thus

$$\nabla \mu \cdot r_\mu = \nabla \mu \cdot (z_v, -z_u) = k \nabla \mu \cdot (u_w, v_w)$$

$$= k \mu_w.$$

Similarly, $\nabla \lambda \cdot r_\lambda = k \lambda_z$. We record these facts in the following lemma.

Lemma 20.2. *The λ (resp. μ) characteristic field is genuinely nonlinear if and only if $\lambda_z \neq 0$ (resp. $\mu_w \neq 0$).*

In Chapter 15, §B, we showed that global classical solutions of a scalar conservation law are usually impossible to find. Our proof was geometric and consisted in showing that the solution was constant along characteristics, and that the characteristics were straight lines. In the case of two equations, equations (20.6) and (20.7) show that z and w are constant along characteristics. However, the characteristics are not straight lines, so that the scalar non-existence proof cannot be extended to our case. We present now a different nonexistence proof for the scalar case which is capable of generalization to pairs of conservation laws.

Let u satisfy the equation $u_t + f(u)_x = 0$, and let $a = f'$. Then $u_t + a(u)u_x = 0$, and differentiating with respect to x gives

$$u_{tx} + au_{xx} + a_u(u_x)^2 = 0.$$

If "prime" denotes differentiation in the direction $\partial/\partial t + a\partial/\partial x$, and $u_x \equiv q$, then we can write

$$q' + a_u q^2 = 0.$$

This equation can be integrated explicitly and we find

$$q(x, t) = \frac{q_0(x)}{1 + kt\, q_0(x)},$$

where $q_0(x) = q(x, 0)$, and $k = a_u(u)$ is constant along the characteristic. Thus if $f''(u(x, t))u_0'(x) = kq_0(x) < 0$ for some x, then $u_x(x, t) = q(x, t)$ blows up in finite time.

We shall extend this proof to the pair of conservation laws (20.1). For this purpose, we may consider the equivalent system (20.6), (20.7) for smooth solutions.

Theorem 20.3. *If (20.1) is genuinely nonlinear in the μth characteristic field; i.e., $\mu_w > 0$, and both $w(x, 0)$ and $z(x, 0)$ are bounded, then if $w_x(x, 0) < 0$, for some x, w_x becomes infinite in finite time. (A similar statement is true upon replacing μ by λ and w by z.)*

Proof. We assume that the system (20.1) has a smooth solution defined in $t > 0$. Then (20.6) and (20.7) are everywhere valid. Thus we can differentiate (20.6) with respect to x to get

$$w_{tx} + \mu w_{xx} + \mu_w w_x^2 + \mu_z w_x z_x = 0,$$

or if $w_x = r$,

$$r_t + \mu r_x + \mu_w r^2 + \mu_z r z_x = 0. \qquad (20.9)$$

Now let "prime" denote differentiation in the direction $\partial/\partial t + \mu \partial/\partial x$; then (20.9) becomes

$$r' + \mu_w r^2 + \mu_z r z_x = 0. \qquad (20.10)$$

If "dot" denotes differentiation in the direction $\partial/\partial t + \lambda \partial/\partial x$, then (20.7) shows $\dot{z} = 0$. Thus

$$z' = z_t + \mu z_x = (\dot{z} - \lambda z_x) + \mu z_x = (\mu - \lambda)z_x,$$

or $z_x = z'/(\mu - \lambda)$. If we use this in (20.10) we get

$$r' + \mu_w r^2 + \frac{\mu_z}{\mu - \lambda} z' r = 0. \qquad (20.11)$$

Let $a = a(w, z)$ be a function satisfying

$$a_z = \frac{\mu_z}{.\mu - \lambda}.$$

Since $w' = 0$, we find

$$a' = a_w w' + a_z z' = a_z z' = \frac{\mu_z z'}{\mu - \lambda}.$$

Thus (20.11) can be written as

$$r' + \mu_w r^2 + a'r = 0. \qquad (20.12)$$

We can actually prove the blow-up of r at this stage, but it is perhaps easier to get rid of the linear term. To this end, multiply (20.12) by e^a and let $p = re^a$. This gives

$$0 = e^a r' + e^a a' r + \mu_w e^a r^2 = p' + \mu_w e^a r^2,$$

or $p' + \mu_w e^{-a} p^2 = 0$. If we let $k = \mu_w e^{-a}$, then the last equation becomes

$$p' + kp^2 = 0.$$

The solution of this equation is given by

$$p(x, t) = \frac{p(x, 0)}{1 + p(x, 0)K(t)}, \qquad K(t) = \int_0^t k(s)\, ds, \qquad (20.13)$$

where the integration is along the μ-characteristic.

If w and z are bounded at $t = 0$, then they satisfy the same bounds for all $t > 0$, since they are constant along characteristics. Thus k is bounded from below, $k \geq k_0$; whence $K(t) \geq k_0 t$, $t \geq 0$. Since by hypothesis $p(\bar{x}, 0) < 0$ for some $\bar{x}$, (20.13) shows that $p(\bar{x}, t)$ becomes unbounded for some $t > 0$. This contradiction concludes the proof. $\square$

Corollary 20.4. *Suppose that* $\nabla \mu \cdot r_\mu \neq 0$, *and both* $u(x, 0)$ *and* $v(x, 0)$ *are bounded. If* $w_x(x, 0) \not\equiv 0$, *then there is a* $T > 0$ *such that if* $|t| > T$, *a classical solution of* (20.1) *cannot exist.*

The proof of this corollary is a consequence of (20.13) and the argument which follows this equation.

We shall give further applications involving Riemann invariants in subsequent sections.

§B. A Concept of Entropy

We have seen in Chapter 15 that physically nonacceptable discontinuous functions can arise as solutions of conservation laws. Such discontinuities were rejected on the basis of "entropy" conditions. We shall now introduce a concept of entropy which both generalizes and unifies all of the previous notions. As a byproduct, we shall also see how these ideas can be used to obtain some a-priori bounds on solutions of conservation laws.

Consider the system

$$u_t + f(u)_x = 0, \tag{20.14}$$

in n-dependent variables $u = (u_1, \ldots, u_n)$. We may assume that $f(0) = 0$; then if $|u|$ tends to zero sufficiently fast as $|x| \to \infty$, we can integrate (20.14) over $\mathbf{R}$ to obtain

$$\int_{\mathbf{R}} u_t \, dx = 0.$$

Thus the integral of u is conserved (for smooth solutions); i.e.,

$$\int_{\mathbf{R}} u(x, t) \, dx = \int_{\mathbf{R}} u(x, 0) \, dx, \qquad t > 0.$$

Viewing this phenomenon more generally, we ask the following question. When does the system (20.14) imply the existence of an additional conservation law?; that is, an equation of the type

$$U_t + F_x = 0, \tag{20.15}$$

where $U = U(u)$, and $F = F(u)$ are real-valued functions?

When (20.15) holds, we call U an *entropy* for (20.14), in analogy with the equations of gas dynamics where we know that the equation $S_t = 0$ can be (formally) derived. Here S is the "physical" entropy; see (18.3)'.

To analyze this, we carry out the differentiations in (20.15) and find

$$dU u_t + dF u_x = 0.$$

In order to get this from (20.14), we rewrite (20.14) as

$$u_t + df u_x = 0.$$

Multiplying this equation by dU on the left gives

$$dU u_t + dU\, df u_x = 0.$$

Thus (20.15) holds if and only if

$$dU\, df = dF. \tag{20.16}$$

This is a system of n partial differential equations for the unknown quantities U and F. Thus if $n > 2$, this system is usually overdetermined and has no solutions. However, there are some very important special cases in which a nontrivial solution exists.

EXAMPLE 1. Suppose that f is a gradient, that is, $f = \nabla_u \phi$. In this case df is symmetric, and (20.14) can be written as

$$u_t + (\nabla \phi)_x = 0.$$

We set

$$U(u) = \tfrac{1}{2}|u|^2, \qquad F(u) = \langle u, \nabla \phi \rangle - \phi,$$

and compute:

$$\begin{aligned}
F_x &= \langle u_x, \nabla \phi \rangle + \langle u, (\nabla \phi)_x \rangle - \langle \nabla \phi, u_x \rangle \\
&= \langle u, (\nabla \phi)_x \rangle = -\langle u, u_t \rangle = -U_t.
\end{aligned}$$

EXAMPLE 2. Here we take $n = 2$, and consider the "antigradient" system

$$u_t + \phi_v(u, v)_x = 0, \qquad v_t + \phi_u(u, v)_x = 0.$$

We put $U = \phi$, $F = \phi_u \phi_v$, and then

$$U_t = \phi_u u_t + \phi_v v_t = -[\phi_u(\phi_v)_x + \phi_v(\phi_u)_x] = -(\phi_u \phi_v)_x = -F_x.$$

Note that systems of the form

$$u_t + f(v)_x = 0, \qquad v_t + g(u)_x = 0,$$

are special cases of antigradient systems. This is easily seen by writing

$$\phi(u, v) = \int^v f(s) \, ds + \int^u g(s) \, ds.$$

In particular, if $f(v) = p(v)$, and $g(u) = -u$, we recover the p-system (cf. Chapter 17, §A). In this case,

$$\phi(u, v) = -\frac{u^2}{2} + \int^v p(s) \, ds.$$

The role of the entropy conditions is to differentiate "physically" realizable discontinuous solutions from the others. Another way in which the former can be characterized is to recover them from equations containing viscous terms, i.e., second-order derivatives. The equation we consider is the following modification of (20.14):

$$u_t + f(u)_x = \varepsilon A u_{xx}, \qquad \varepsilon > 0, \tag{20.17}$$

where A is an $n \times n$ positive semidefinite matrix with (for simplicity), constant coefficients. We multiply (20.17) on the left by dU, and assuming (20.16), we get

$$U_t + F_x = \varepsilon \, dU A u_{xx},$$

or

$$U_t + F_x = \varepsilon (dU A u_x)_x - \varepsilon u_x^t \, d^2 U A u_x, \tag{20.18}$$

where $d^2 U$ denotes the hessian matrix of U. We assume that U and A are *compatible* in the sense that

$$d^2 U A \geq 0. \tag{20.19}$$

Observe that if A is the identity matrix, (20.19) means that U is convex.
Now if we integrate (20.18) over the region $S_T = \{(x, t) \in \mathbf{R}^2 : 0 \leq t \leq T\}$, we find

$$\int_{-\infty}^{\infty} U(u(x, T)) \, dx - \int_{-\infty}^{\infty} U(u(x, 0)) \, dx = -\varepsilon \iint_{S_T} u_x^t \, d^2 U A u_x, \tag{20.20}$$

provided that $u(x, t)$ vanishes sufficiently fast at infinity.[2] Then from (20.19) we can obtain the estimate

$$\int_{-\infty}^{\infty} U(u(x, T)) \, dx \le \int_{-\infty}^{\infty} U(u(x, 0)) \, dx. \tag{20.21}$$

Notice that for the p-system, $U = -u^2/2 + \int^v p(s) \, ds$, and (20.21) gives the interesting estimate

$$\int_{-\infty}^{\infty} \left[-\frac{u^2}{2} + \int^v p(s) \, ds \, dx \right] \le \int_{-\infty}^{\infty} \left[\frac{-u(x, 0)^2}{2} + \int^{v(x,0)} p(s) \, ds \right] dx.$$

Furthermore, if instead of (20.19), we assume the stronger condition

$$d^2 U A \ge cI, \quad ?? \tag{20.22}$$

where c is a positive constant, (20.20) yields

$$\int_{-\infty}^{\infty} U(u(x, t)) \, dx + \varepsilon c \iint_{S_T} u_x^2 \le \int_{-\infty}^{\infty} U(u(x, 0)) \, dx.$$

If U is non-negative, then this last inequality gives the bound

$$\varepsilon \iint_{S_T} u_x^2 \le \text{const.,} \tag{20.23}$$

where the constant depends only on c and the initial entropy; i.e., $U(u(x, 0))$. Consequently, if u is bounded, we see that the term $\varepsilon(dU A u_x)_x$ tends to zero in the sense of distributions as $\varepsilon \to 0$. Thus from (20.18) we can conclude

$$U_t + F_x \le 0, \tag{20.24}$$

in the sense of the theory of distributions. We thus have

Theorem 20.5. *Suppose that the system* (20.14) *admits an additional conservation law* (20.15). *Let u be a (weak) solution of* (20.14) *which is the limit, boundedly (i.e., L_∞ on compacta), together with its x-derivatives, of solutions of the "viscosity" equation* (20.17). *If* (20.22) *holds, and $U \ge 0$, then u satisfies* (20.24) *in the sense of distributions.*

In the case $A = I$, the identity, we can strengthen the theorem as follows.

---

[2] In view of (20.16), there is no loss in generality to assume $F(0) = 0$.

Corollary 20.6. *If U is convex and (20.14) admits the additional conservation law (20.15), then any solution u of (20.14) which is the limit, boundedly, of solutions of* $u_t + f(u)_x = \varepsilon u_{xx}$, *satisfies (20.24) in the sense of distributions.*

For, in this case we can conclude from (20.18) that $U_t + F_x \leq \varepsilon(dUu_x)_x = \varepsilon U_{xx}$. If $\varepsilon \to 0$, then εU_{xx} tends weakly to zero, and the result follows. □

Corollary 20.7. *Under the hypotheses of Theorem 20.5 or Corollary 20.6, if u is a piecewise continuous solution, then across each discontinuity u satisfies*

$$\text{False!} \quad \to \quad s[U_l - U_r] - [F(U_l) - F(U_r)] \leq 0, \tag{20.25}$$

where s is the speed of the discontinuity, and U_l *and* U_r *are, respectively, the states on the left and right side of the discontinuity.*

The proof of this follows by integrating (20.24) around a small portion of the discontinuity, as we have done in Chapter 15, §B.

We shall refer to both (20.24) and (20.25) as *entropy conditions*. The following theorem justifies this.

Theorem 20.8. *Suppose that the system (20.14) is hyperbolic and genuinely nonlinear, and U is a strictly convex function satisfying the additional conservation law (20.15). Let u be a solution of (20.14) which contains a weak shock of speed s. Then the entropy inequalities (17.22) hold if and only if (20.25) holds.*

Proof. From the results in Chapter 17, §B, we know that the totality of states u_r which can be connected to a state u_l by a k-shock, forms a one-parameter family, $u_r = u(\varepsilon)$, $\bar{\varepsilon} < \varepsilon \leq 0$. Let

$$E(\varepsilon) = s(\varepsilon)[U(\varepsilon) - U(u_l)] - [F(\varepsilon) - F(u_l)], \tag{20.26}$$

where $(U(\varepsilon), F(\varepsilon)) \equiv (U(u(\varepsilon)), F(u(\varepsilon)))$. To prove the theorem it suffices to show that $E > 0$ for small ε, iff $\varepsilon < 0$, since we have seen in Theorem 17.14 that the inequalities (17.22) hold if and only if $\varepsilon < 0$.

To this end, we differentiate (20.26) with respect to ε, and get (with the obvious notation),

$$\dot{E} = \dot{s}(U - U_l) + sU'\dot{u} - F'\dot{u}.$$

But from (20.16), $F' = U'f'$, so that

$$\dot{E} = \dot{s}(U - U_l) + sU'\dot{u} - U'f'\dot{u}. \tag{20.27}$$

If we evaluate this at $\varepsilon = 0$, and use Corollary 17.13, we find $\dot{E}(0) = \lambda U'r - U'f'r = \lambda U'r - U'(\lambda r) = 0$, where r is the right eigenvector of df corresponding to the kth eigenvalue λ.

Now $s(u - u_l) = f(u) - f(u_l)$, so that $\dot{f} = s\dot{u} + \dot{s}(u - u_l)$, and

$$\dot{F} = F'\dot{u} = U'f'\dot{u} = U'\dot{f} = sU'\dot{u} + \dot{s}U'(u - u_l).$$

If we use this in (20.27), we get

$$\dot{E} = \dot{s}(U - U_l) - sU'(u - u_l).$$

We differentiate this with respect to ε and find

$$\ddot{E} = \ddot{s}(U - U_l) - \ddot{s}U'(u - u_l) - \dot{s}\dot{U}'(u - u_l),$$

so that $\ddot{E}(0) = 0$. If we differentiate again, we get

$$\dddot{E} = \dddot{s}(U - U_l) - \dddot{s}U'(u - u_l) - 2\ddot{s}\dot{U}'(u - u_l) - \dot{s}\ddot{U}'(u - u_l) - \dot{s}\dot{U}'\dot{u}.$$

Thus again using Corollary 17.13, we find

$$\dddot{E}(0) = -\dot{s}\dot{U}'\dot{u}\big|_{\varepsilon = 0} = -\tfrac{1}{2}\dot{r}^t U''\dot{r} < 0, \tag{20.28}$$

by the strict convexity of U. Since $\varepsilon < 0$, it follows that (20.28) implies $E > 0$ for small ε iff $\varepsilon < 0$. The proof is complete. $\square$

We remark that it is possible to give another proof of one part of the last theorem using the results in Chapter 24, §A. Namely, if the shock conditions (17.22) hold, then we shall show that the shock wave solution (u_l, u_r, s) of (20.14) is a limit of travelling wave solutions of the system $u_t + f(u)_x = \varepsilon u_{xx}$. Therefore Corollary 20.6 implies (20.25).

We conclude this section by showing that the solution constructed in Chapter 19 via the Glimm difference scheme satisfies the entropy conditions provided that the system admits an additional conservation law. Precisely, we have the following theorem.

Theorem 20.9. *Suppose that* (20.14) *is hyperbolic and admits an additional conservation law* (20.15) *where U is strictly convex. Then any solution u of* (20.14) *constructed by the Glimm difference scheme (Chapter 19), satisfies the entropy inequalities* (17.22) *across shock waves.*

Proof. It suffices to show that u satisfies (20.24) since this implies (20.25), and the result follows from Theorem 20.8.

Recall that u is a limit of approximate solutions $u_{\theta, \Delta}$ each one being piecewise continuous in each strip $k\,\Delta t < t < (k + 1)\,\Delta t$, and having only shocks as discontinuities. It follows from this together with Theorem 20.8, that (20.24) holds for each approximate solution in each strip. We multiply (20.24) by a positive test function ϕ and integrate over each strip. If we integrate by parts with respect to t over each strip, and then sum over all strips, we get

$$\sum_{k=1}^{\infty} \int \phi(x, k\,\Delta t)[U_{\theta,\Delta}(x, k\,\Delta t + 0) - U_{\theta,\Delta}(x, k\,\Delta t - 0)]\,dx$$

$$- \iint (\phi_t U_{\theta,\Delta} + \phi_x F_{\theta,\Delta})\,dx\,dt \le 0. \quad (20.29)$$

where $(F_{\theta,\Delta}, U_{\theta,\Delta}) \equiv (F(u_{\theta,\Delta}), U(u_{\theta,\Delta}))$. The sum in (20.29) tends to zero, at least for a suitable subsequence of mesh lengths, for almost all random points, by Theorem 19.14.

Since the approximations converge in a topology strong enough to pass the limit through U and F (by Corollary 19.10), we see that (20.29) implies

$$- \iint_{t>0} (\phi_t U + \phi_x F) \le 0,$$

for all positive test functions. This gives (20.24), and the proof is complete. $\square$

§C. Solutions With "Big" Data

In this section we shall consider the system of isentropic gas dynamics equations (cf. Chapter 17, §A) in Lagrangian coordinates,

$$v_t - u_x = 0, \qquad u_t + p(v)_x = 0, \qquad x \in \mathbf{R}, \quad t > 0, \qquad (20.30)$$

where $p(v) = k^2/v^\gamma$, and γ is a constant, $\gamma = 1 + 2\varepsilon$, and $\varepsilon \ge 0$. Together with (20.30) we consider the initial values

$$(v(x, 0), u(x, 0)) = (v_0(x), u_0(x)), \qquad x \in \mathbf{R}, \qquad (20.31)$$

which we assume to be bounded, and to have bounded total variation. In addition, we assume

$$0 < \underline{v} \le v_0(x) \le \bar{v} < \infty, \qquad x \in \mathbf{R}, \qquad (20.32)$$

for some constants $\underline{v}$, $\bar{v}$. With these assumptions, we shall prove that the problem (20.30), (20.31) has a global solution provided that

$$\varepsilon \, \text{T.V.}\{v_0, u_0\}$$

is sufficiently small. This allows us to consider data with large total variation, but only for small ε; i.e., only when γ is near 1. If $\gamma = 1$, then there is no restriction on the size of the initial total variation. The method which we shall use is a modification of the method described in Chapter 19, and involves a study of the global properties of the shock curves in $v - u$ space. The proof makes considerable use of the special form of the equations.

The system (20.30) is hyperbolic in $v > 0$ and the matrix $(d(-u, p(v))$ has eigenvalues

$$\lambda_\pm = \pm \frac{k\sqrt{\gamma}}{v^{1+\varepsilon}} = \pm k\sqrt{\gamma}\rho^{1+\varepsilon},$$

where $\rho = v^{-1}$, with corresponding Riemann invariants,

$$r = u - \frac{k\sqrt{\gamma}(\rho^\varepsilon - 1)}{\varepsilon}, \qquad s = u + \frac{k\sqrt{\gamma}(\rho^\varepsilon - 1)}{\varepsilon}.$$

It is easy to check that the mapping $(u, \rho) \to (r, s)$ is bijective when $\rho > 0$, and $s - r > -2k\sqrt{\gamma}/\varepsilon$; thus we may use (u, ρ) or (r, s) as variables, according to our convenience.

The Riemann problem for (20.30), with initial values

$$(\rho_0(x), u_0(x)) = \begin{cases} (\rho_-, u_-), & x < 0, \\ (\rho_+, u_+), & x > 0, \end{cases}$$

where $\rho_\pm > 0$, has been considered in Chapter 17, §A. The solution satisfies the estimates

$$r(x, t) \equiv r(u(x, t), \rho(x, t)) \geq \min(r_-, r_+),$$

$$s(x, t) \equiv s(u(x, t), \rho(x, t)) \leq \max(s_-, s_+), \tag{20.33}$$

where $(r_\pm, s_\pm) = (r(u_\pm, \rho_\pm), s(u_\pm, \rho_\pm))$, provided that $s_- - r_+ > -2k\sqrt{\gamma}/\varepsilon$ (i.e., provided that the solution does not assume the value $\rho = 0$). The estimates (20.33) merely state that the solution lies between the extreme Riemann invariants (or rarefaction-wave curves); see chapter 17, §A.

If we consider the shock curves for (20.30) we find $\sigma[v] = -[u]$, $\sigma[u] = [p]$, where σ is the shock speed, and $[\cdot]$ denotes the difference of the quantities across both sides of the shock, going from left to right (in the $x - t$ plane). Eliminating σ gives

$$u - u_0 = \pm k\sqrt{\frac{(\rho - \rho_0)(\rho^\gamma - \rho_0^\gamma)}{\rho\rho_0}}.$$

The shock curve S_1 corresponding to λ_- can be written in the form

$$
S_1 : \begin{cases}
r_0 - r = k\rho_0^\varepsilon \left\{ \sqrt{\dfrac{(\alpha - 1)(\alpha^\gamma - 1)}{\alpha}} + \sqrt{\gamma}\dfrac{\alpha^\varepsilon - 1}{\varepsilon} \right\}, \\[4mm]
s_0 - s = k\rho_0^\varepsilon \left\{ \sqrt{\dfrac{(\alpha - 1)(\alpha^\gamma - 1)}{\alpha}} - \sqrt{\gamma}\dfrac{\alpha^\varepsilon - 1}{\varepsilon} \right\}, \quad \alpha \geq 1,
\end{cases}
$$

where $(r_0, s_0) = (r(u_0, \rho_0), s(u_0, \rho_0))$ is the state on the left, and $\alpha = \rho/\rho_0 \geq 1$, while S_2, corresponding to λ_+, takes the form

$$
S_2 : \begin{cases}
s_0 - s = k\rho_0^\varepsilon \left\{ \sqrt{\dfrac{(1 - \alpha)(1 - \alpha^\gamma)}{\alpha}} + \sqrt{\gamma}\dfrac{1 - \alpha^\varepsilon}{\varepsilon} \right\}, \\[4mm]
r_0 - r = k\rho_0^\varepsilon \left\{ \sqrt{\dfrac{(1 - \alpha)(1 - \alpha^\gamma)}{\alpha}} - \sqrt{\gamma}\dfrac{1 - \alpha^\varepsilon}{\varepsilon} \right\}, \quad \alpha \leq 1,
\end{cases}
$$

where $0 < \alpha = \rho/\rho_0 \leq 1$. We can depict these curves in the $r - s$ plane as in Figure 20.3. We note that we may write $s_0 - s = g_1(r_0 - r, \rho_0)$ and $r_0 - r = g_2(s - s_0, \rho_0)$, for S_1 and S_2, respectively.

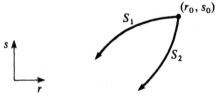

Figure 20.3

Of course, the lines $r = $ const. and $s = $ const. correspond to rarefaction-wave curves.

With these preliminaries out of the way, we can prove our first lemma.

Lemma 20.10. *The shock curve S_1 starting at (r_0, s_0) can be written as*

$$
s_0 - s = g_1(r_0 - r, \rho_0) \equiv \int_0^{r_0 - r} h_1(\alpha) \Big|_{\alpha = \alpha_1(\beta/k\rho_0^\varepsilon)} d\beta, \quad r < r_0, \quad (20.34)_1
$$

where $0 \leq \partial g_1(\beta, \rho_0)/\partial\beta < 1$ and $0 \leq \partial^2 g_1(\beta, \rho_0)/\partial\beta^2$. A similar representation is valid for the S_2-curves.

Proof. Let

$$h_1(\alpha) = \frac{\partial(s_0 - s)/\partial\alpha}{\partial(r_0 - r)/\partial\alpha},$$

and compute

$$h_1(\alpha) = \frac{(y-1)^2}{(y+1)^2}, \quad \text{where } y = \sqrt{\frac{\gamma\alpha^\gamma(\alpha-1)}{\alpha^\gamma - 1}}, \quad \alpha \geq 1.$$

If we define f_0 by

$$\frac{\beta}{k\rho_0^\varepsilon} = \sqrt{\frac{(\alpha-1)(\alpha^\gamma-1)}{\alpha}} + \sqrt{\gamma}\frac{\alpha^\varepsilon - 1}{\varepsilon} \equiv f_0(\alpha),$$

then $f_0'(\alpha) > 0$ for $\alpha > 1$ so the implicit function theorem gives $\alpha = \alpha_1(\beta/k\rho_0^\varepsilon)$. The calculations for g_1 are now straightforward, and we omit them. □

The next lemma is our main estimate; it shows how the Riemann invariants change across two shock waves of the same family.

Lemma 20.11. *Let $\varepsilon \geq 0$, $s_1 > s_0$, and consider the two S_1 curves originating at the points $(r_0, s_1) = (u_1, \rho_1)$ and $(r_0, s_0) = (u_0, \rho_0)$, $\rho_0, \rho_1 > 0$, which are continued to the points (r, s_2) and (r, s), respectively (cf. Figure 20.4). Then there is a constant C depending only on $\underline{\rho}$ and $\bar{\rho}$, such that*

$$0 \leq (s_0 - s) - (s_1 - s_2) \leq C\varepsilon(s_1 - s)(r_0 - r). \tag{20.35}$$

Proof. Set $z^1 = s_1 - s_2$, $z^0 = s_0 - s$, $w = r_0 - r$; see Figure 20.4.

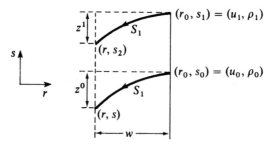

Figure 20.4

Using the last lemma and the mean-value theorem we find

$$z^0 - z^1 = \int_0^w \left[h_1\left(\alpha_1 \left(\frac{\beta}{k\rho_0^\varepsilon} \right) \right) - h_1\left(\alpha_1 \left(\frac{\beta}{k\rho_1^\varepsilon} \right) \right) \right] d\beta$$

$$= \int_0^w \left. \frac{dh_1(\alpha)}{d\alpha} \right|_{\alpha = \alpha(\theta)} \cdot \frac{d\alpha(\theta)}{d\theta} \cdot \left(\frac{\beta}{k\rho_0^\varepsilon} - \frac{\beta}{k\rho_1^\varepsilon} \right) d\beta,$$

where $\beta/k\rho_1^\varepsilon < \theta < \beta/k\rho_0^\varepsilon$. Since the integrand is nonnegative, we see $z^0 - z^1 \geq 0$. If y is as in the proof of the last lemma, and $\alpha = \alpha(\theta)$, we have

$$\frac{dh_1(\alpha)}{d\alpha} \cdot \frac{d\alpha}{d\theta} = \frac{4(y-1)\sqrt{\gamma}\alpha^{(\gamma+1)/2}}{(y+1)^3(\alpha^\gamma - 1)^2} \frac{[\alpha^{\gamma+1} - (\gamma+1)\alpha + \gamma]}{(y+1)^2}$$

$$\leq \frac{4(y-1)\sqrt{\gamma}\alpha^{(\gamma+1)/2}}{(y+1)^3(\alpha^\gamma - 1)^2} \frac{(y+1)(\alpha - 1)(\alpha^\gamma - 1)]}{y^2}$$

$$= \frac{4(y-1)\sqrt{\gamma}\alpha^{(\gamma+1)/2}}{(y+1)^3(\alpha^\gamma - 1)^2} \frac{(y+1)(\alpha^\gamma - 1)^2}{\gamma\alpha^\gamma}$$

$$= \frac{4(\gamma+1)\alpha^{(1-\gamma)/2}(y-1)}{\sqrt{\gamma}(y+1)^3}.$$

Thus

$$z^0 - z^1 \leq \frac{4(\gamma+1)}{k\sqrt{\gamma}\rho_0^\varepsilon \rho_1^\varepsilon} (\rho_1^\varepsilon - \rho_0^\varepsilon) \int_0^w \beta\alpha^{(1-\gamma)/2} \frac{y-1}{(y+1)^3} d\beta$$

$$\leq C(\gamma, \rho_1, \rho_0)(\rho_1^\varepsilon - \rho_0^\varepsilon) \int_0^w \beta\alpha^{-(1+\gamma)/2} d\beta,$$

where we have used the inequalities

$$0 \leq (y-1)(y+1)^{-1} \leq 1 \quad \text{and} \quad (y+1)^{-2} \leq y^{-2} \leq \alpha^{-1}.$$

Since $y \geq 1$, the first inequality is easily seen to be valid, while the second follows from

$$(y+1)^{-2} \leq y^{-2} = \frac{\alpha^\gamma - 1}{\gamma\alpha^\gamma(\alpha - 1)} \leq \frac{\gamma\alpha^{\gamma-1}(\alpha - 1)}{\gamma(\alpha - 1)\alpha^\gamma} = \alpha^{-1}.)$$

Since $\beta/k\rho_1^\varepsilon < 0$ and $d\alpha/d\theta > 0$, we have

$$\left. \alpha^{-(1+\gamma)/2} \right|_{\alpha = \alpha(\theta)} \leq \left. \alpha^{-(1+\gamma)/2} \right|_{\alpha = \alpha(\beta/k\rho_1^\varepsilon)}.$$

But

$$\frac{\beta}{k\rho_1^\varepsilon} = f_1(\alpha) \le 2\sqrt{\frac{(\alpha - 1)(\alpha^\gamma - 1)}{\alpha}} \le 2\alpha^{\gamma/2},$$

so that

$$\alpha^{-(1+\gamma)/2}\big|_{\alpha = \alpha(\theta)} \le \left(\frac{\beta}{2k\rho_1^\varepsilon}\right)^{-(\gamma+1)/\gamma}.$$

This gives

$$z^0 - z^1 \le C(\gamma, \rho_1, \rho_0)(\rho_1^\varepsilon - \rho_0^\varepsilon) \cdot \begin{cases} \displaystyle\int_0^w \beta \, d\beta, & \text{for } w \le 1, \\[2ex] \displaystyle\int_0^1 \beta \, d\beta + \int_1^w \beta^{1-(\gamma+1)/\gamma} \, d\beta, & \text{for } w > 1, \end{cases}$$

$$\le C(\rho_1^\varepsilon - \rho_0^\varepsilon) \cdot \begin{cases} \tfrac{1}{2}w^2, & \text{for } w \le 1, \\[2ex] \tfrac{1}{2} + \dfrac{1 + 2\varepsilon}{2\varepsilon}(w^{2\varepsilon/(1+2\varepsilon)} - 1), & \text{for } w > 1. \end{cases}$$

Since $\rho_1^\varepsilon - \rho_0^\varepsilon = \varepsilon(s_1 - s_0)/2k\sqrt{\gamma}$, and

$$\left.\begin{array}{ll} \text{for } 0 \le w \le 1, & \tfrac{1}{2}w^2 \\[2ex] \text{for } w > 1, & \tfrac{1}{2} + \dfrac{1 + 2\varepsilon}{\varepsilon}(w^{2\varepsilon/(1+2\varepsilon)} - 1) \end{array}\right\} \le \frac{w}{(1+\varepsilon)^{1/2\varepsilon}},$$

we conclude that

$$z^0 - z^1 \le C\varepsilon(s_1 - s_0)w,$$

and the proof is complete. $\square$

We define $\tilde{r}$ and $\tilde{s}$ by

$$[\tilde{r}, \tilde{s}] = \left[\inf_{x \in \mathbf{R}} r(u_0(x), \rho_0(x)), \sup_{x \in \mathbf{R}} s(u_0(x), \rho_0(x))\right],$$

and choose mesh lengths Δx and Δt to satisfy

$$\frac{\Delta t}{\Delta x} = \left(k\sqrt{\gamma}\left[1 + \frac{\varepsilon(\tilde{s} - \tilde{r})}{2k\sqrt{\gamma}}\right]^{(1+\varepsilon)/\varepsilon}\right)^{-1}. \tag{20.36}$$

We now construct our approximating solutions as in Chapter 19.

From our assumption (20.32), together with (20.33), we see that there is an $\varepsilon_0 > 0$, such that if $0 \le \varepsilon \le \varepsilon_0$, then the approximate solution $(u_{\theta,\Delta x}, \rho_{\theta,\Delta x})$ satisfies $\rho_{\theta,\Delta x} \ge \tilde{\rho} > 0$ in the strip $0 \le t < \Delta t$.

In order that the method of the last chapter works, we must find an analogue of Theorem 19.2, the interaction estimates. In our case, we must study the interaction of only two families of waves. Thus if R_1 and R_2 are rarefaction waves corresponding to the first and second characteristic families, respectively, then we must study the following six nontrivial interactions (here the "first" wave is considered to be on the left of the "second" wave):

(i) S_2 interacts with S_1;
(ii) S_2 interacts with R_1 (or R_2 interacts with S_1);
(iii) S_2 interacts with S_2 (or S_1 interacts with S_1);
(iv) S_2 interacts with R_2 (or R_1 interacts with S_1);
(v) R_2 interacts with S_2 (or S_1 interacts with R_1);
(vi) R_2 interacts with R_1;

see Figure 20.5. (The interactions obtained by interchanging the indices 1 and 2 can be treated similarly.)

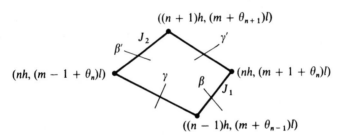

Figure 20.5

Let J_2 and J_1 be mesh curves, with J_2 an immediate successor to J_1, and let $\theta \in \Phi$, cf. Chapter 19, §B. Let β (resp. γ) be an S_1 (resp. S_2) shock on J_1, and let β' (resp. γ') denote an S_1 (resp. S_2) shock on J_2. Let the absolute value in terms of the Riemann invariant r (resp. s), denote the strengths of the S_1 (resp. S_2) shocks; see Figure 20.6. Observe that with this definition of strength of shocks, the variation always decreases across shocks. This is one advantage of using the $r - s$ coordinates; see the definition of the functionals L, Q, and F in (20.38), below.

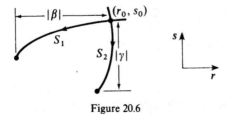

Figure 20.6

In the next lemma, we use the notation $\gamma + \beta \to \beta' + \gamma'$ to denote the interaction of an S_2 with an S_1 which produces an S_1 and an S_2; the other cases are written similarly. Rarefaction waves will be denoted by 0.

Lemma 20.12. *Let $\varepsilon \geq 0$, $0 < \rho < \bar{\rho} < \infty$, and suppose that all waves considered below lie in the strip $[\underline{\rho}, \bar{\rho}]$, $u \in \mathbf{R}$, in the $r - s$ plane. Then for the interactions given below, the corresponding estimates are valid, where C and C_0 are independent of ε, β, γ and $\rho_0 \in [\underline{\rho}, \bar{\rho}]$:*

(i) $\gamma + \beta \to \beta' + \gamma'$:
   (a) $|\beta'| \leq |\beta| + C\varepsilon|\beta||\gamma|$, $|\gamma'| \leq |\gamma| + C\varepsilon|\beta||\gamma|$, or there exist η, ζ such that
   (b) $0 \leq |\beta'| = |\beta| - \zeta, |\gamma'| \leq |\gamma| + C\varepsilon|\beta||\gamma| + \eta$ where $0 \leq \eta \leq g_1'(|\beta|, \rho_0)\zeta < \zeta$, or
   (c) $0 \leq |\gamma'| = |\gamma| - \zeta, |\beta'| \leq |\beta| + C\varepsilon|\beta||\gamma| + \eta$, where $0 \leq \eta \leq g_1'(|\gamma|, \rho_0)\zeta < \zeta$.
(ii) $\gamma + 0 \to 0 + \gamma' : |\gamma'| = |\gamma|$.
(iii) $\gamma_1 + \gamma_2 \to 0 + \gamma' : |\gamma'| = |\gamma_1| + |\gamma_2|$.
(iv) $\gamma + 0 \to \beta' + \gamma'$, or $\gamma + 0 \to \beta' + 0$:

   There exist β_0, γ_0 such that $\gamma_0 + \beta_0 \to \beta' + \gamma'$ is the same interaction as in (i), and

$$|\beta_0| + |\gamma_0| \leq |\gamma| - C_0|\beta_0|.$$

(v) $0 + \gamma \to \beta' + \gamma'$, or $0 + \gamma \to \beta' + 0$:

$$|\beta'| + |\gamma'| \leq |\gamma| - C_0|\beta'|.$$

(iv) $0 + 0 \to 0 + 0$.

(Note that the qualitative form of the interactions is obtained by solving Riemann problems for this system (see Chapter 17, §A), and considering all possible choices of "random points." Thus, for example, we cannot have $S_1 + S_2 \to R_1 + R_2$, etc.)

Proof. We shall only give the proof for a single typical case, namely case (iv); for the other cases, we refer the reader to [NS 1]. To this end, consider Figure 20.7.

From the figure we see that

$$|\beta_0| = r_1 - r = r_0 - r - (r_0 - r_1) = g_2(|\gamma|, \rho_0) - g_2(|\gamma_0|, \rho_0)$$
$$= (|\gamma| - |\gamma_0|)g_2'(\theta, \rho_0),$$

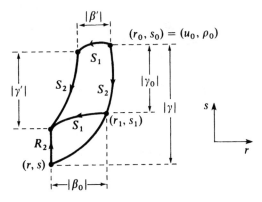

Figure 20.7

where θ is between $|\gamma|$ and $|\gamma_0|$. Thus

$$|\gamma_0| + |\beta_0| - |\gamma| = |\beta_0| - \frac{|\beta_0|}{g_2'} = -\frac{|\beta_0|(1 - g_2')}{g_2'}. \qquad (20.37)$$

But since $g_2'(|\gamma|, \rho_0) \to 1$ as $|\gamma| \to \infty$ (see $(20.34)_2$), we have

$$\lim_{|\gamma| \to \infty} \frac{1 - g_2'(|\gamma|, \rho_0)}{g_2'(|\gamma|, \rho_0)} = 0,$$

so that

$$\min_{\substack{\gamma \subset \{(r, s) : \underline{\rho} \leq \rho \leq \bar{\rho}\} \\ \underline{\rho} \leq \rho_0 \leq \bar{\rho}}} \frac{1 - g_2'(|\gamma|, \rho_0)}{g_2'(|\gamma|, \rho_0)} = C_0 > 0.$$

It follows from (20.37) that $|\gamma_0| + |\beta_0| - |\gamma| \leq -C_0|\beta_0|$. This proves
(iv). $\square$

We remark that we can treat the interactions of three waves in a similar
way to obtain similar estimates. (The interaction of three waves in a diamond
does occur.) As in Chapter 19, in order to investigate the convergence, we
introduce functionals on mesh curves. In our context, if J is a mesh curve, we
define

$$\begin{cases} L(J) = \sum \{|\alpha| : \alpha \text{ is a shock crossing } J\}, \\ Q(J) = \sum \{|\beta||\alpha| : \beta \in S_1, \gamma \in S_2, \beta, \gamma \text{ cross } J \text{ and approach}\}, \text{ and} \\ F(J) = L(J) + KQ(J), \end{cases} \qquad (20.38)$$

where $K > 0$ will be chosen below. It suffices to only consider shock waves in the above functionals since only shocks contribute to the decreasing variation of the solution across J, and the total variation is majorized by twice the decreasing variation, plus the difference in the value of the functions at $\pm \infty$.

Now if ε_0 is defined as above (after (20.36)), then if $\varepsilon < \varepsilon_0$, (20.33) shows that

$$L(O) \le \text{T.V.}[r_0(x), s_0(x)],$$

where O is the unique mesh curve passing through the mesh points on $t = 0$ and $t = \Delta t$.

Let $(r_\pm, s_\pm) = \lim_{x \to \pm \infty} (r_0(x), s_0(x))$, and define

$$\mathcal{U} = \{(r, s): \text{dist}[(r, s), (r_-, s_-)] + \text{dist}[(r, s), (r_+, s_+)]$$

$$\le 8\text{T.V.}[(r_0(x), s_0(x)] + \text{dist}[(r_-, s_-), (r_+, s_+)]\},$$

where $\text{dist}[(r_1, s_1), (r_2, s_2)] = |r_1 - r_2| + |s_1 - s_2|$.

Now choose ε_1 so that $0 < \varepsilon_1 \le \varepsilon_0$ and $\mathcal{U} \cap \{\rho = 0\} = \phi$ for $0 \le \varepsilon \le \varepsilon_1$. Let $K = 4C\varepsilon$, where C is the constant of Lemma 20.12 for this region $\mathcal{U}$. We take ε_1 so small that $4C\varepsilon_1 L(O) \le 1$; then if $0 \le \varepsilon \le \varepsilon_1$,

$$F(O) = L(O) + KQ(O) \le L(O) + KL(O)^2 \le 2L(O).$$

Lemma 20.13. *If $\varepsilon F(O)$ is sufficiently small, then $F(J_2) \le F(J_1)$ where J_1 and J_2 are mesh curves and J_2 is an immediate successor to J_1.*

Proof. In order to prove this lemma, we must consider all the possible types of interactions of waves in "diamonds"; cf. Lemma 20.12. We shall content ourselves here in considering just a typical two-wave interaction; namely, $S_2 + S_1 \to S_1 + S_2$; for other cases, the reader should consult [NS 1]. Thus suppose we have the situation depicted in Figure 20.8; namely,

$$S_2 + S_1 \to S_1 + S_2.$$

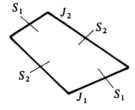

Figure 20.8

Then

$$L(J_2) = \sum |\alpha_k| + |\beta'| + |\gamma'|, \qquad L(J_1) = \sum |\alpha_k| + |\beta| + |\gamma|,$$

where α_k crosses J_i $(i = 1, 2)$ and lies outside of the diamond for every k, and

$$Q(J_2) = \sum |\beta_k||\gamma_l| + \sum |\beta_k||\gamma'| + \sum |\beta'||\gamma_l|,$$
$$Q(J_1) = \sum |\beta_k||\gamma_l| + \sum |\beta_k||\gamma| + \sum |\beta||\gamma_l|,$$

where β_k and γ_l cross J_i $(i = 1, 2)$, and lie outside of Δ for every k and l. Then

$$F(J_2) - F(J_1) = |\beta'| + |\gamma'| - (|\beta| + |\gamma|)$$
$$+ K[\sum |\beta_k|(|\gamma'| - |\gamma|) + \sum |\gamma_l|(|\beta'| - |\beta|) - |\beta||\gamma|].$$

Since we are considering S_2 interacting with S_1, we may apply Lemma 20.12(i), in view of our previous choice of ε_1.

Now suppose (ia) holds; then

$$F(J_2) - F(J_1) \le 2C\varepsilon|\beta||\gamma| + KC\varepsilon|\beta||\gamma| F(J_1) - K|\beta||\gamma|$$
$$= C\varepsilon|\beta||\gamma|[2 + KF(J_1) - 4]$$
$$\le C\varepsilon|\beta||\gamma|[KF(0) - 2] \le 0,$$

if we take $F(0) < 2/K$. If (i(b)) (or analogously, (i(c)) holds, we have

$$F(J_2) - F(J_1) \le -\zeta + C\varepsilon|\gamma||\beta| + \eta$$
$$+ K[\sum |\beta_k|(C\varepsilon|\beta||\gamma| + \eta) - \sum |\gamma_l|(\zeta) - |\beta||\gamma|]$$
$$\le -\zeta + C\varepsilon|\gamma||\beta| + \eta - \tfrac{1}{2}K|\beta||\gamma|$$
$$+ K[C\varepsilon|\gamma||\beta|\sum |\beta_k| + \eta \sum |\beta_k| - \tfrac{1}{2}|\beta||\gamma|]$$
$$\le -\zeta + \eta + \tfrac{1}{2}K(2C\varepsilon F(0) - 1)|\gamma||\beta| + K\eta \sum |\beta_k|$$
$$\le \zeta[-1 + g_1'(|\beta|, \rho_0) + Kg_1'(|\beta|, \rho_0) \sum |\beta_k|]$$
$$\le \zeta g_1'(|\beta|, \rho_0)\left[KF(J_1) - \frac{1 - g_1'(|\beta|, \rho_0)}{g_1'(|\beta|, \rho_0)}\right]$$
$$\le \zeta g_1'(|\beta|, \rho_0)[KF(0) - C_0] \le 0$$

if $KF(0) \le C_0$. $\quad\square$

We can now prove the main theorem of this section.

Theorem 20.14. *Let the initial data (u_0, ρ_0) have bounded total variation, and suppose $0 < \underline{\rho} \le \rho(x) \le \bar{\rho} < \infty$ for all $x \in \mathbf{R}$. There exists a constant C, depending only on $\underline{\rho}$ and $\bar{\rho}$ such that if*

$$(\gamma - 1)\, \text{T.V.}\{u_0, \rho_0\} < C, \tag{20.39}$$

then the problem (20.30), (20.31) has a global solution.

The proof of this follows exactly the procedure in Chapter 19 once Lemma 20.13 is known to be true.

Note that if $\gamma = 1$, (20.39) affords no restriction on the total variation of the data, save that it be bounded. We can say that when γ is near 1, the "isothermal" gas case, one can allow "big" data, and conversely, as γ increases, one must take correspondingly smaller data.

Finally, in view of Theorem 20.8, and Example 2 in §B, we see that our solution satisfies the entropy inequalities (17.22) across shock waves.

§D. Instability of Rarefaction Shocks

In this short section we shall show that for the p-system,

$$v_t - u_x = 0, \qquad u_t + p(v)_x = 0, \qquad p' < 0, \quad p'' > 0, \qquad (20.40)$$

shocks which do not satisfy the shock conditions (17.7) and (17.8) are unstable relative to smoothing of the data. That is, for such a shock, a so-called *rarefaction shock*, if we smooth the data, and solve the corresponding problems with the smooth data then the limiting solution is *not* the original shock.

Recall from Chapter 17, §A, that there are two families of rarefaction shocks corresponding to the two characteristics $\lambda_1 = -\sqrt{-p'(v)} < \sqrt{-p'(v)} = \lambda_2$; we call these S_1' and S_2' shocks, respectively. Moreover, these shocks satisfy the equations

$$S_1' : u - u_l = \sqrt{(v - v_l)(p(v_l) - p(v))}, \qquad v < v_l,$$

$$S_2' : u - u_l = \sqrt{(v - v_l)(p(v_l) - p(v))}, \qquad v > v_l.$$

We depict these two curves in Figure 20.9. Furthermore, from Chapter 17,

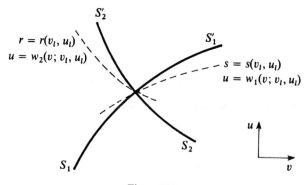

Figure 20.9

§A, we know that if

$$r = u + \int^v \sqrt{-p'(x)}\,dx \quad \text{and} \quad s = u - \int^v \sqrt{-p'(x)}\,dx$$

are a pair of Riemann invariants for (20.40), then the S_1' and S_2' curves lie between the curves $r = r(v_l, u_l)$ and $s = s(v_l, u_l)$, at least for points near (v_l, u_l). However, the curves $s = s(v_l, u_l)$ and $r = r(v_l, u_l)$ are really the rarefaction-wave curves

$$u = w_1(v; v_l, u_l) = u_l + \int_{v_l}^v \sqrt{-p'(x)}\,dx \quad \text{and}$$

$$u = w_2(v; v_l, u_l) = u_l - \int_{v_l}^v \sqrt{-p'(x)}\,dx,$$

respectively. But from the Schwarz inequality,

$$\int_{v_l}^v \sqrt{-p'(x)}\,dx \le \sqrt{(v - v_l)(p(v_l) - p(v))},$$

so that the S_1' and S_2' curves *always* lie between the curves $r = r(v_l, u_l)$ and $s = s(v_l, u_l)$.

Now suppose

$$(v, u)(x, t) = \begin{cases} (v_-, u_-), & x < \sigma t, \\ (v_+, u_+), & x > \sigma t, \end{cases}$$

is a rarefaction shock-wave solution of (20.40). Then we must have

$$r_- < r_+ \quad \text{and} \quad s_- < s_+, \tag{20.41}$$

where the notation is clear. Now let us smooth the initial data

$$r(x, 0) \equiv r_0(x) = \begin{cases} r_-, & x < 0, \\ r_+, & x > 0, \end{cases} \qquad s(x, 0) \equiv s_0(x) = \begin{cases} s_-, & x < 0, \\ s_+, & x > 0. \end{cases}$$

This gives a sequence of smooth data functions $r_n^0 = r_0 * \omega_n$, $s_n^0 = s_0 * \omega_n$, where ω_n is the usual averaging kernel of radius $1/n$, and $*$ denotes convolution product (cf. the appendix in Chapter 7). It is easy to check that both r_n^0 and s_n^0 are nondecreasing functions, for each n. Using (20.13), we see that the corresponding solutions (r_n, s_n) are globally defined smooth solutions for each n, and furthermore, for any $T > 0$, if $0 \le t \le T$, these functions are uniformly bounded, (by (20.6) and (20.7)), and their x-derivatives are also uniformly bounded (again by (20.13)). Thus the sequence is compact, and any limit (r, s) is continuous on each line $t = $ const. > 0. Thus, no limit can be equal to our original rarefaction shock. We have thus proved the following theorem.

Theorem 20.15. *Any rarefaction shock-wave solution of* (20.40) *is unstable relative to smoothing of the initial data.*

§E. Oleinik's Uniqueness Theorem

We have not said very much about the uniqueness of solutions of conservation laws except for the case of a single equation (Theorem 16.11). The reason for this is simple—there is just not too much known about the problem. We present here a uniqueness theorem for the p-system. The proof uses the same idea as in the scalar case; namely, it consists of showing that the "adjoint" system has many solutions (Holmgren's method). Of course, here the estimates are more difficult to obtain since we do not have an "entropy" condition, and we must work merely with the shock inequalities (17.7) and (17.8).

Theorem 20.16. *Consider the p-system* (20.40), *where $p \in C^2$, $p' < 0$, $p'' > 0$, $p(v) \to \infty$ as $v \to 0$, and $p'(v) \to 0$ as $v \to \infty$. Suppose that the initial data $(v(x, 0), u(x, 0)) = (u_0(x), v_0(x))$ is piecewise continuous. Then in the strip $0 \le t \le T$, there is a unique solution in the class of piecewise continuous functions, all of whose discontinuities are shock waves, having a finite number of centered rarefaction waves and shock waves in each compact subset of $t \ge 0$.*

We remind the reader that by "shock wave" we mean a discontinuity satisfying both the jump conditions (17.9) and the shock conditions (17.7), (17.8). The latter are equivalent to the two inequalities:

$$u(x + 0, t) < u(x - 0, t) \quad \text{and} \quad v(x, t + 0) < v(x, t - 0). \qquad (20.42)$$

Of course, the two inequalities in (20.42) are also equivalent to each other. This follows, for example, from considering Figures 17.1 and 17.2.

We shall first prove a preliminary uniqueness theorem. To this end, let D be a compact region in $t \ge 0$ which is bounded by a segment of the line $t = 0$. Let

$$(v(x, 0), u(x, 0)) = (v_0(x), u_0(x)), \qquad x \in \mathbf{R}, \qquad (20.43)$$

and suppose that $(v(x, t), u(x, t))$ is a solution of (20.40), (20.43) which satisfies for $x_1 > x_2$,

$$u(x_1, t) - u(x_2, t) < k(x_1 - x_2), \qquad (20.44)$$

where k is a constant. Note that (20.44) implies (20.42). Also observe that (20.44) cannot hold if (v, u) contains a centered rarefaction wave. For, if (x_0, t_0) is the center of the rarefaction wave, then

$$\frac{u\left(\dfrac{x_1 - x_0}{t_1 - t_0}\right) - u\left(\dfrac{x_2 - x_0}{t_2 - t_0}\right)}{x_1 - x_2} \to \infty$$

as $x_1 \to x_2$ since in the wave u is constant along the rays $(x - x_0)/(t - t_0) = \text{const.}$

Theorem 20.17. Let (v_1, u_1) and (v_2, u_2) be a pair of bounded and measurable solutions of (20.40) with bounded and measurable initial data (20.43). If both solutions do not contain any centered rarefaction waves; i.e., both satisfy (20.44), then the solutions are equal (a.e.) in the subregion $D_\tau \subset D$. Here D_τ is the region bounded by the lines $t = 0$, $t = \tau$, $x = At + a$, and $x = -At + b$, where

$$A = \max_D \sqrt{-P(x, t)}, \qquad P(x, t) = \frac{p(v_1(x, t)) - p(v_2(x, t))}{v_1(x, t) - v_2(x, t)} \qquad (20.45)$$

and a and b are constants.

Proof of Theorem 20.17. If $\omega_h = \omega_h(x, t)$ is the usual regularizing kernel depending only on the distance $\overline{PP_0}$ and h, we let (v_1^h, u_1^h) and (v_2^h, u_2^h) denote the regularizations of (v_1, u_1), and (v_2, u_2), respectively. These functions will be defined in D_τ with $t \geq \alpha$ if h is sufficiently small. Let $D_{\tau\alpha} = D_\tau \cap (t \geq \alpha)$.

Now we know that for any solution of our problem, if ϕ, ψ are in $C_0^1(D)$, both of the following equations hold:

$$\iint_D v\psi_t - u\psi_x + \int_a^b v_0\psi = 0,$$

$$\iint_D u\phi_t + p(v)\phi_x + \int_a^b u_0\phi = 0. \qquad (20.46)$$

If we let $\psi = \omega_h$ in the first equation, we get

$$\frac{\partial v^h}{\partial t} - \frac{\partial u^h}{\partial x} = 0. \qquad (20.47)$$

But (20.44) gives $\partial u_1^h/\partial x < K_1$ and $\partial u_2^h/\partial x < K_2$, where K_1 and K_2 are constants independent of h and α. Thus from (20.47) we conclude that

$$\frac{\partial v_i^h}{\partial t} < K_i, \qquad i = 1, 2. \qquad (20.48)$$

Now consider the equations

$$\frac{\partial \phi}{\partial t} - \frac{\partial \psi}{\partial x} = f_1, \qquad \frac{\partial \psi}{\partial t} + P_h \frac{\partial \phi}{\partial x} = P_h f_2, \tag{20.49}$$

where f_1 and f_2 are smooth functions which vanish in some neighborhood of ∂D_τ, and

$$P_h(t, x) = \frac{p(v_1^h) - p(v_2^h)}{v_1^h - v_2^h} = \int_0^1 p'(v_1^h + s(v_2^h - v_1^h)) \, ds. \tag{20.50}$$

The system (20.49) is hyperbolic, and its (distinct) characteristics satisfy the equations

$$\frac{dx}{dt} = \pm \sqrt{-P_h}.$$

If $\varepsilon > 0$ is a given small positive number, then for h sufficiently small,

$$\max_{D_\tau} \sqrt{-P_h} \leq \max_D \sqrt{-P} + \varepsilon.$$

This implies that in $D_{\tau\alpha}$ the system (20.49) has a solution ϕ, ψ (depending on h), which vanishes in a neighborhood of that part of $\partial(D_{\tau\alpha})$ consisting of segments of the lines $t = \tau$, $x = a + At$, $x = b - At$ (see [Ga, p. 121 ff]; the proof is analogous to what we have seen in the scalar case, Theorem 16.11).

Now differentiating equations (20.49) with respect to t gives

$$\frac{\partial}{\partial t} \phi_t - \frac{\partial}{\partial x} \psi_t = f_{1t} \quad \text{and} \quad \frac{1}{P_h} \frac{\partial}{\partial t} \psi_t + \frac{\partial}{\partial x} \phi_t + \left(\frac{1}{P_h} \right)_t \psi_t = f_{2t}.$$

We multiply the first equation by $e^{\theta t} \phi_t$, the second by $-e^{\theta t} \psi_t$, where $\theta > 0$ will be chosen later, and integrate these equations over $D_{\tau\alpha}$. We find

$$\iint_{D_{\tau\alpha}} \left\{ \frac{1}{2} \frac{\partial}{\partial t} (\phi_t^2 e^{\theta t}) - \frac{\theta}{2} e^{\theta t} \phi_t^2 - \frac{\partial}{\partial x} (e^{\theta t} \phi_t \psi_t) - \frac{1}{2} \frac{\partial}{\partial t} [e^{\theta t} (P_h)^{-1} \psi_t^2] \right.$$

$$\left. + \frac{\theta}{2} e^{\theta t} (P_h)^{-1} \psi_t^2 - \frac{1}{2} e^{\theta t} \frac{\partial}{\partial t} [(P_h)^{-1}] \psi_t^2 \right\} dx \, dt$$

$$= \iint_{D_{\tau\alpha}} [\phi_t f_{1t} - \psi_t f_{2t}] e^{\theta t} \, dx \, dt.$$

Applying the divergence theorem and using the fact that ϕ_t and ψ_t vanish on $\partial(D_{\tau\alpha})$ gives

$$\int_{\Gamma_\alpha} \left[\frac{1}{2} e^{\theta t}(\phi_t^2 - (P_h)^{-1}\psi_t^2) \right] dx + \iint_{D_{\tau\alpha}} \frac{\theta}{2} e^{\theta t}[\phi_t^2 - (P_h)^{-1}\psi_t^2] \, dx \, dt$$

$$= - \iint_{D_{\tau\alpha}} \frac{1}{2} e^{\theta t} \frac{\partial}{\partial t}[(P_h)^{-1}]\psi_t^2 \, dx \, dt + \iint_{D_{\tau\alpha}} [\psi_t f_{2t} - \phi_t f_{1t}]e^{\theta t} \, dx \, dt,$$

$$\tag{20.51}$$

where $\Gamma_\alpha = D_{\tau\alpha} \cap \{t = \alpha\}$. Since $p'' > 0$, (20.48) and (20.50) show that

$$\frac{1}{2}\frac{\partial}{\partial t}(P_h)^{-1} > c, \tag{20.52}$$

where c is independent of h. Using the inequality $2\alpha\beta \le \alpha^2 + \beta^2$, we see that the right side of (20.51) does not exceed

$$\iint_{D_{\tau\alpha}} e^{\theta t}\left[-c\psi_t^2 + \frac{1}{2}\psi_t^2 + \frac{1}{2}\phi_t^2 \right] dx \, dt + \iint_{D_{\tau\alpha}} \frac{1}{2} e^{\theta t}[f_{1t}^2 + f_{2t}^2] \, dx \, dt,$$

so that

$$\int_{\Gamma_\alpha} \frac{1}{2} e^{\theta t}(\phi_t^2 - P_h^{-1}\psi_t^2) + \iint_{D_{\tau\alpha}} \phi_t^2\left[\frac{\theta}{2} - \frac{1}{2} \right]e^{\theta t} + \psi_t^2\left[-\frac{\theta}{2}P_h^{-1} + c - \frac{1}{2} \right]e^{\theta t}$$

$$\le \iint \frac{1}{2} e^{\theta t}[f_{1t}^2 + f_{2t}^2]. \tag{20.53}$$

Now let $\theta > 3$ and $\theta \max|P_h|^{-1} > \max(3 - 2c, 2)$; then

$$\tfrac{1}{2}e^{\theta t}(-P_h)^{-1} \ge \frac{\theta}{2}(-P_h)^{-1} > 1, \qquad \tfrac{1}{2}(\theta - 1)e^{\theta t} > \tfrac{1}{2}(\theta - 1) > 1,$$

and

$$\tfrac{1}{2}(-\theta P_h^{-1} + 2c - 1) > 1.$$

Thus from (20.5) we get

$$\int_{\Gamma_\alpha} (\phi_t^2 + \psi_t^2) \, dx + \iint_{D_{\tau\alpha}} (\phi_t^2 + \psi_t^2) \, dx \, dt \le M_1, \tag{20.54}$$

where M_1 is independent of h and α. Next, (20.54) together with (20.49) implies that

$$\int_{\Gamma_\alpha} (\phi_x^2 + \psi_x^2)\, dx + \iint_{D_{\tau\alpha}} (\phi_x^2 + \psi_x^2)\, dx\, dt \leq M_2, \qquad (20.55)$$

where M_2 is independent of h and α.

We extend the functions ϕ and ψ into the region $0 \leq t \leq \alpha$, by setting $(\phi(x, t), \psi(x, t)) = (\phi(x, \alpha), \psi(x, \alpha))$. Now take α so small that both f_1 and f_2 vanish for $t < \alpha$. From (20.49), we get

$$\iint_{D_\tau} [f_1(u_1 - u_2) + P_h f_2(v_1 - v_2)]\, dx\, dt$$

$$= \iint_{D_{\tau\alpha}} [(\phi_t - \psi_x)(u_1 - u_2) + (\psi_t + P_h \phi_x)(v_1 - v_2)]\, dx\, dt,$$

$$(20.56)$$

and (20.46) implies that

$$\iint_{D} (P_h - P)\phi_x(v_1 - v_2) - \iint_{D \backslash D_{\tau\alpha}} P_h \phi_x(v_1 - v_2) - \psi_x(u_1 - u_2)$$

$$= \iint_{D_{\tau\alpha}} [(\phi_t - \psi_x)(u_1 - u_2) + (\psi_t + P_h \phi_x)(v_1 - v_2),$$

since $\phi_t = \psi_t = 0$ if $t \leq \alpha$. Thus from (20.56),

$$\iint_{D_\tau} [f_1(u_1 - u_2) + P_h f_2(v_1 - v_2)] = \iint_{D} (P_h - P)\phi_x(v_1 - v_2)$$

$$- \iint_{D \backslash D_{\tau\alpha}} P_h \phi_x(v_1 - v_2) - \psi_x(u_1 - u_2). \qquad (20.57)$$

Now v_1, v_2, u_1, u_2 and P are uniformly bounded. Thus if we use the Schwarz inequality, together with (20.55) and the convergence in the mean of P_h to P as $h \to 0$, we see that the right-hand side of (20.57) tends to zero as $h, \alpha \to 0$. Thus

$$\iint_{D_\tau} f_1(u_1 - u_2) + P f_2(v_1 - v_2) = 0,$$

from which it follows that $u_1 = u_2$ and $v_1 = v_2$ a.e. in D_τ. This completes the proof. $\square$

We turn now to the proof of Theorem 20.16. We assume that for the class of solutions (v, u) as stated in the hypothesis, we have (as in (20.42)),

$$v(x, t_1) - v(x, t_2) < K(t_1, t_2, x)(t_1 - t_2) \qquad (20.58)$$

for all (x, t_1), (x, t_2) in the strip $0 \le t \le T$, where $T > 0$ is any fixed positive number. Here K is continuous and everywhere bounded in the strip, except at a finite number of points, $P_1, P_2, \ldots, P_n$, the centers of rarefaction waves.

In a neighborhood of any $P_k = P_k(x_0, t_0)$, if $t \ge t_0$, there are (at most) two sectors of the form $\lambda_1 \le \lambda \le \lambda_2$, $(\lambda = (x - x_0)/(t - t_0)$, $\text{sgn } \lambda_1 = \text{sgn } \lambda_2)$, in which (v, u) can be represented as follows:

$$
\begin{aligned}
v(x, t) &= \alpha(\lambda^2) + \delta(x, t), \\
u(x, t) &= \sigma(\lambda^2) + \delta_1(x, t) + C_1, && \text{if } \lambda_1 > 0, \\
u(x, t) &= -\sigma(\lambda^2) + \delta_2(x, t) + C_2, && \text{if } \lambda_1 < 0.
\end{aligned}
\qquad (20.59)
$$

Here $\alpha^{-1}(s) = -p'(s)$, $\alpha' < 0$, $\sigma' > 0$, $\delta, \delta_1, \delta_2$ are continuous and tend to zero as $(x, t) \to (x_0, t_0)$, and C_1 and C_2 are constants.[3] In other words, the solution (20.59) is a rarefaction wave in the sector $\lambda_1 \le \lambda \le \lambda_2$. If (x, t) is outside of these sectors, the function K is continuous and bounded near each P_k. Finally, for $\lambda > \lambda_1$, if $\lambda_1 > 0$ (and for $\lambda < \lambda_2$ if $\lambda_1 < 0$), if (x, t) is near P_k and $t \ge t_0$, we shall require that the following inequality holds:

$$v(x, t_1) - v(x, t_2) > H(t_1 - t_2), \qquad (20.60)$$

where H is some constant. This inequality is again a consequence of our results in Chapter 17, §A; see, e.g., Figure 17.4.

With these preliminary things out of the way, we can proceed with the proof.

First let us recall that in a small neighborhood of any point $P = (x_0, t_0)$, either, (i) the solution is continuous at P, or (ii) the solution has at most one rarefaction wave centered at P in each of the regions $x \le x_0$ and $x \ge x_0$, or (iii) the solution has at most one shock wave passing through P in each of these regions. These facts follow easily from the shock conditions (17.7), and the definition of rarefaction waves; namely, no two rarefaction waves of the same family can have the same center.

---

[3] This is consistent with our results in Chapter 17, §A; namely, we showed there that along a rarefaction wave, $\lambda^2 = -p(v(\lambda))$, where $\lambda = (x - x_0)/(t - t_0)$, and

$$u(\lambda) - u_0 = \pm \int_{v_0}^{v(\lambda)} \sqrt{-p'(s)} \, ds \equiv \sigma(\lambda^2),$$

since $v(\lambda)$ is a function of λ^2. The presence of the δ's is due to the fact that we consider a small neighborhood of the centers.

Now suppose that (v_1, u_1) and (v_2, u_2) are two solutions of (20.40) satisfying the conditions of Theorem 20.16. Let the centers of the rarefaction waves of these solutions be on the lines $t = t_j, j = 1, \ldots, n, 0 \le t_j < T$. If there are no t_j's on $t = 0$, then Theorem 20.17 shows that the two solutions are equal (a.e.) in $0 \le t \le t_1$. If $t_1 = 0$, then in view of Theorem 20.17, it suffices to prove that the solutions are the same in a neighborhood of a center of a rarefaction wave.

Thus, let $P = (x_0, 0)$ be the center of a rarefaction wave. We shall assume that at P, for both solutions (since they both satisfy the same initial conditions), there emerges a back shock and a front rarefaction wave; the argument is similar for the other cases. Let λ_1 and λ_2, respectively, denote the speed of the "tail" and "head" of the rarefaction wave.

Let Ω be a small region bounded by $t = 0$ and containing P in the interior of that segment, and let the rest of the boundary be described by the equation $t = \mu(x)$, where $|\mu'(x)|$ is sufficiently small. Denote by $\tilde{v}_1^h$ and $\tilde{v}_2^h$ functions defined in Ω which are regularizations of v_1 and v_2 with the kernel $\omega(h, r)$, outside of a strip of width $2h$ around the line $(x - x_0)/t = \lambda_1$, and equal to a constant V inside this strip. We define $v_1 \equiv v_0 \equiv v_2$ and $u_1 \equiv u_0 \equiv u_2$ in $t \le 0$. We choose the constant V such that $V > \max_\Omega(v_1, v_2)$ and

$$-p'(V) < \lambda_1^2. \tag{20.61}$$

From our assumptions on the boundedness and continuity of $K(t_1, t_2, x)$ in (20.58), we see that for $(x - x_0)/t = \lambda < \lambda_1$ and $(x, t) \in \Omega$, we have

$$\frac{\partial \tilde{v}_1^h}{\partial t} < K_1 \quad \text{and} \quad \frac{\partial \tilde{v}_2^h}{\partial t} < K_2, \tag{20.62}$$

where K_1 and K_2 do not depend on h. For $\lambda > \lambda_1$, (20.60) gives

$$\frac{\partial \tilde{v}_1^h}{\partial t} > H_1 \quad \text{and} \quad \frac{\partial \tilde{v}_2^h}{\partial t} > H_2, \tag{20.63}$$

where H_1 and H_2 and also independent of h.

Now consider the system

$$\phi_t - \psi_x = f_1, \quad \psi_t + \tilde{P}_h \phi_x = \tilde{P}_h f_2, \tag{20.64}$$

where f_1 and f_2 are arbitrary smooth functions equal to zero in a neighborhood of $\partial\Omega$ and

$$\tilde{P}_h = \frac{p(\tilde{v}_1^h) - p(\tilde{v}_2^h)}{\tilde{v}_1^h - \tilde{v}_2^h}.$$

In analogy to what we have seen in the proof of the last theorem, we find that (20.64) has solutions (ϕ, ψ) in Ω, which vanish on the curve $t = \mu(x)$. Since $K(t_1, t_2, x)$ in (20.58) is continuous in Ω for $t \geq \sigma > 0$ (σ small), and consequently bounded, it follows that for any ϕ and ψ in $\Omega \cap (t \geq \sigma)$, the estimates (20.54) and (20.55) hold, for $\lambda \geq \lambda_1$, where M_1 and M_2 depend on σ, but not on h.

We shall now estimate the derivatives of ϕ and ψ in the region $\lambda \leq \lambda_1$. For this we shall proceed as in the proof of the last theorem. Thus, let $\Omega_\sigma = \Omega \cap (t \geq \sigma) \cap (\lambda \leq \lambda_1)$; we have

$$\iint_{\Omega_\sigma} \frac{1}{2} \theta e^{\theta t} [\phi_t^2 - (\tilde{P}_h)^{-1} \psi_t^2] + \int_{t=\sigma} \frac{1}{2} e^{\theta t} [\phi_t^2 - (\tilde{P}_h)^{-1} \psi_t^2] \, dx$$

$$+ \int_{(x-x_0)/t = \lambda_1} e^{\theta t} \phi_t \psi_t \, dt + \frac{1}{2} e^{\theta t} [\phi_t^2 - (\tilde{P}_h)^{-1} \psi_t^2] \, dx$$

$$= \iint_{\Omega_\sigma} e^{\theta t} [\phi_t f_{1t} - \psi_t f_{2t}] - \frac{1}{2} \iint_{\Omega_\sigma} e^{\theta t} \frac{\partial}{\partial t} [(\tilde{P}_h)^{-1}] \psi_t^2. \tag{20.65}$$

Let us consider the last integral on the left. Since $\tilde{v}_1^* = \tilde{v}_2^* = V$ and $\tilde{P}_h = p'(V)$ on $(x - x_0)/t = \lambda_1$, we obtain by (20.61)

$$\int_{(x-x_0)/t = \lambda_1} e^{\theta t} \phi_t \psi_t \, dt + \frac{1}{2} e^{\theta t} [\phi_t^2 - (P_h)^{-1} \psi_t^2] \, dx$$

$$\geq \int_{(x-x_0)/t = \lambda_1} e^{\theta t} \phi_t \psi_t \, dt + \frac{1}{2} e^{\theta t} \left[\phi_t^2 + \frac{1}{\lambda_1^2} \psi_t^2 \right] \lambda_1 \, dt$$

$$= \int_{(x-x_0)/t = \lambda_1} e^{\theta t} \left(\sqrt{\lambda_1} \phi_t + \frac{1}{\sqrt{\lambda_1}} \psi_t \right)^2 dt \geq 0.$$

Therefore in Ω_σ we can again derive estimates of the form (20.54) and (20.55), where M_1 and M_2' are independent of h and σ. Also, from the estimate of the type (20.55), it follows that ϕ and ψ are uniformly bounded in h in the region $\Omega \cap (\lambda < \lambda_1)$; this follows from a simple application of the Schwarz inequality and writing

$$\phi(x, t) = \int_{p_t}^{x} \phi_x(s, t) \, ds, \qquad \phi(p_t, t) = 0.$$

We now estimate the functions ϕ, ψ in Ω for $(x - x_0)/t \geq \lambda_1$. Let $\Omega^\sigma = \Omega \cap (t \geq \sigma) \cap (\lambda \geq \lambda_1)$. If we multiply the first equation (20.64) by $e^{\theta t}\phi$, and the second by $-e^{\theta t}\psi$, integrate over Ω^σ, and add, we get

$$\iint_{\Omega^\sigma} \left\{ \frac{1}{2}\frac{\partial}{\partial t}(e^{\theta t}\phi^2) - \frac{\partial}{\partial x}(e^{\theta t}\psi\phi) - \frac{1}{2}\frac{\partial}{\partial t}(e^{\theta t}(\tilde{P}_h)^{-1}\psi^2) - \frac{\theta}{2}e^{\theta t}\phi^2 \right.$$

$$\left. + \frac{\theta}{2}e^{\theta t}(\tilde{P}_h)^{-1}\psi^2 + \frac{1}{2}e^{\theta t}\frac{\partial}{\partial t}(\tilde{P}_h)^{-1}\psi^2 \right\}$$

$$= \iint_{\Omega^\sigma} [f_1 e^{\theta t}\phi - f_2 e^{\theta t}\psi]. \tag{20.66}$$

If we choose θ sufficiently large and observe that (20.63) implies $\partial[\tilde{P}_h^{-1}]/\partial t \leq \tilde{C}$, then since the functions ϕ and ψ are uniformly bounded in h on $\lambda = \lambda_1$, we may apply the divergence theorem to (20.66) to conclude that ψ and ϕ are bounded uniformly in σ and h, in the space $L_2[(t = \sigma) \cap \Omega^\sigma]$. From (20.46) and (20.64), we find

$$\iint_{\Omega\backslash(t\leq\sigma)} [f_1(u_1 - u_2) + \tilde{P}_h f_2(v_1 - v_2)]\, dx\, dt$$

$$= \iint_{\Omega\backslash(t\leq\sigma)} (\tilde{P}_h - P)\phi_x(v_1 - v_2)\, dx\, dt$$

$$- \int_{t=\sigma} [u_1(x, \sigma) - u_2(x, \sigma)]\phi(x, \sigma)\, dx$$

$$- \int_{t=\sigma} [v_1(x, \sigma) - v_2(x, \sigma)]\psi(x, \sigma)\, dx. \tag{20.67}$$

As $\sigma \to 0$, the last two integrals in (20.67) tend to zero since each v_i and u_i is piecewise continuous, $(v_i, u_i)(x, 0) = (v_0(x), u_0(x))$ for $i = 1, 2$, and ϕ and ψ are uniformly bounded in L_2, independent of h and σ, on $(t = \sigma) \cap \Omega^\sigma$. If h is small, then since $\tilde{P}_h$ tends to P in L_2 as $h \to 0$, and we have estimates of the type (20.55) for fixed σ, the first integral on the right in (20.67) tends to zero as $h \to 0$. Consequently, first letting $h > 0$, and then $\sigma \to 0$, gives

$$\iint [f_1(u_1 - u_2) + Pf_2(v_1 - v_2)]\, dx\, dt = 0,$$

so that in view of the arbitrariness of f_1 and f_2 we conclude that

$$(v_1, u_1) = (v_2, u_2) \quad \text{a.e.} \tag{20.68}$$

in Ω. Thus (20.68) is valid in $t_1 \leq t \leq t_2$. In the same way, we can prove that (20.68) holds in each strip $t_j \leq t \leq t_{j+1}, j = 2, \ldots, n - 1$. This completes the proof of the theorem. $\square$

We note that the previous analysis does not give a complete proof of Theorem 20.16. This is because claims (20.62) and (20.63) are not consequences of (20.58) and (20.60) unless additional structures are assumed on the solutions. The problem is that in, general, K_i and H_i ($i = 1, 2$) may depend on h, if the wave fans in (u_1, v_1) and (u_2, v_2) overlap and the line $(x - x_0)/t = \lambda_1$ lies in the overlap. A different proof, using a new entropy condition can be found in: P. LeFloch, and Z.-P. Xin, Uniqueness Via the Adjoint Problem for Systems of Conservation Laws (*Comm. Pure Appl. Math.*, **46**, 1993, 1499–1533).

NOTES

The results on Riemann invariants for systems of two equations is classical. Their use for "nonexistence" proofs was first recognized by Lax [Lx 3]. Our proof of Theorem 20.4 is taken from this paper; see also John [J2], and Liu [Lu 8]. The notion of generalized entropy is due to Lax [Lx 4]; see Friedrichs and Lax [FL], and also Godunov [Go], Dafermos [Df 1], and Hsiao [Hs], for related results. Section B is taken from Lax [Lx 4]; see also DiPerna [Dp 6]. Using results in Conley–Smoller [CS 2], one can extend Theorem 20.8 for two equations, to the case where the shocks are not necessarily weak; this was noticed by Lax in [Lx 4]. The theorem in §C is taken from Nishida–Smoller [NS 1], following the work of Nishida [N] concerning the case $\gamma = 1$. These results have been extended to the "piston" problem (i.e., system (20.30) in regions $x, t > 0$ with boundary data on $x = 0$), in [NS 2], and to the full gas dynamics equations by Liu [Lu 6] and Temple [Te]. Theorem 20.15 is the only known stability (or rather, instability) result for the full initial-value problem. The proof given here is a specialization of the one given by Conway–Smoller [CoS 2], for a more general class of systems. It would be interesting to extend this result to $n > 2$ equations. Theorems 20.17 and 20.18 are due to Oleinik [O 2]. Some extensions have been given by DiPerna [Dp 5], and by Liu [Lu 11].

Chapter 21

Quasi-Linear Parabolic Systems

We have studied second-order quasi-linear parabolic systems in Chapter 14, where it was assumed that the equations admitted a bounded invariant region. For the gas dynamics equations with all of the dissipative mechanisms taken into account (viscosity and thermal conductivity), and for various models of these, there may exist invariant regions, but they are usually unbounded. Thus we cannot conclude that the solution is a-priori bounded, and global existence theorems become more difficult to prove. One way to overcome this problem is to obtain "energy" inequalities in the unknown function and its derivatives, in a manner somewhat analogous to what we have done for linear hyperbolic equations in Chapter 4. In order to obtain these estimates for nonlinear equations, certain additional restrictions must be imposed: small data, special forms of the equations, restrictions on the data at infinity, and so on.

We shall prove three typical theorems using the energy method. The last theorem will include as a special case the equations of isentropic gas dynamics with viscosity. Since the viscosity matrix is only semidefinite for these systems, a-priori sup-norm bounds are not sufficient for a global existence theorem. It is necessary to obtain additional estimates on certain Hölder norms of the function and its derivatives.

§A. Gradient Systems

We begin by illustrating the energy method for a fairly general class of systems in "gradient" form, where the nonlinear function doesn't grow too fast at infinity. Thus let ϕ be smooth, $\phi : \mathbf{R}^n \to \mathbf{R}$, and assume that ϕ satisfies

$$\lim_{M \to \infty} \frac{M^2}{1 + \max_{|u| \le M} \| \phi''(u) \|} = \infty. \tag{21.1}$$

Here $\| \phi''(u) \|$ denotes any convenient matrix norm of the hessian ϕ'', of ϕ. We note that (21.1) implies that $|\phi|$ grows slower than $|u|^4$ at infinity. Now consider the system

$$u_t + (\phi_u)_x = B u_{xx}, \qquad x_1 < x < x_2, \quad t > 0, \tag{21.2}$$

where B is a positive definite constant matrix, and ϕ_u denotes the gradient of ϕ. We study the initial-boundary value problem for (21.2) with data

$$u(x, 0) = u_0(x), \qquad u(x_1, t) = u(x_2, t) = 0, \qquad x_1 \leq x \leq x_2, \quad t \geq 0. \quad (21.3)$$

We assume that u_0 is smooth and satisfies the compatibility conditions $u_0(x_1) = u_0(x_2) = 0$. There is no loss of generality in assuming $\phi(0) = 0$.

Theorem 21.1. *If ϕ, u_0, and B satisfy the above conditions, then the problem (21.2), (21.3) has a global solution. This solution tends to zero uniformly in the interval $x_1 \leq x \leq x_2$, as $t \to \infty$.*

Proof. We take the inner product of (21.2) with u, and use the identities

$$\langle u, (\phi_u)_x \rangle = \frac{\partial}{\partial x} \{ \langle u, \phi_u \rangle - \phi(u) \},$$

and

$$\langle u, u_t \rangle = \frac{1}{2} \frac{\partial}{\partial t} |u|^2,$$

to get

$$\frac{1}{2} \frac{\partial}{\partial t} |u|^2 + \frac{\partial}{\partial x} \{ \langle u, \phi_u \rangle - \phi(u) \} = \langle u, B u_{xx} \rangle.$$

We take $T > 0$ to be arbitrary, and let R_T be the rectangle $\{x_1 \leq x \leq x_2, 0 \leq t \leq T\}$. Integrating the last equation over R_T gives successively

$$\frac{1}{2} \int_{x_1}^{x_2} |u(x, T)|^2 \, dx - \frac{1}{2} \int_{x_1}^{x_2} |u_0(x)|^2$$

$$+ \int_0^t \{ \langle u, \phi_u \rangle - \phi(u) \}_{x_1}^{x_2} \, dt = \iint_{R_T} \langle u, B u_{xx} \rangle,$$

$$\frac{1}{2} \int_{x_1}^{x_2} |u(x, T)|^2 \, dx - \frac{1}{2} \int_{x_1}^{x_2} |u_0(x)|^2 \, dx = - \iint_{R_T} \langle u_x, B u_x \rangle,$$

in view of the boundary conditions in (21.3). Since $\langle u_x, B u_x \rangle \geq b |u_x|^2$ for some $b > 0$, we have

$$\frac{1}{2} \int_{x_1}^{x_2} |u(x, T)|^2 + b \iint_{R_T} |u_x|^2 \leq \frac{1}{2} \int_{x_1}^{x_2} |u_0(x)|^2 \, dx. \qquad (21.4)$$

Next, we differentiate (21.2) with respect to x, and let $v = u_x$; this gives the equation

$$v_t + (\phi_u)_{xx} = Bv_{xx}.$$

We take the inner product with v and integrate over R_T to get

$$\frac{1}{2}\int_{x_1}^{x_2} |v(x, T)|^2 - \frac{1}{2}\int_{x_1}^{x_2} |v(x, 0)|^2 + \iint_{R_T} \langle v, (\phi_u)_{xx} \rangle = \iint_{R_T} \langle v, Bv_{xx} \rangle,$$

and after integrating by parts,

$$\frac{1}{2}\int_{x_1}^{x_2} |v(x, T)|^2 - \frac{1}{2}\int_{x_1}^{x_2} |v(x, 0)|^2 + \iint_{R_T} \langle v_x, Bv_x \rangle$$

$$= \iint_{R_T} \langle u_{xx}, (\phi_u)_x \rangle + \int_0^T \langle Bv_x - (\phi_u)_x, v \rangle \big|_{x_1}^{x_2}.$$

The boundary term on the right can be rewritten as

$$\int_0^T \langle Bu_{xx} - (\phi_u)_x, u_x \rangle \big|_{x_1}^{x_2} = \int_0^T \langle u_t, u_x \rangle \big|_{x_1}^{x_2} = 0,$$

again in view of the boundary conditions in (21.3). Thus

$$\frac{1}{2}\int_{x_1}^{x_2} |u_x(x, T)|^2 + b \iint_{R_T} |u_{xx}|^2 \leq \frac{1}{2}\int_{x_1}^{x_2} |u_{0,x}(x)|^2 + \iint_{R_T} \langle u_{xx}, (\phi_u)_x \rangle$$

$$\leq \frac{1}{2}\int_{x_1}^{x_2} |u_{0,x}|^2 + \frac{b}{2}\iint_{R_T} |u_{xx}|^2 + \frac{1}{2b}\iint_{R_T} |(\phi_u)_x|^2$$

$$\leq \frac{1}{2}\int_{x_1}^{x_2} |u_{0,x}|^2 + \frac{b}{2}\iint_{R_T} |u_{xx}|^2$$

$$+ \frac{1}{2b}\iint_{R_T} \|\phi''(u)\|^2 |u_x|^2,$$

since

$$\langle (\phi_u)_x, u_{xx} \rangle \leq \tfrac{1}{2}b|u_{xx}|^2 + \frac{1}{2b}|(\phi_u)_x|^2. \tag{21.5}$$

Therefore

$$\frac{1}{2}\int_{x_1}^{x_2} |u_x(x, T)|^2 + \frac{b}{2}\iint_{R_T} |u_{xx}|^2 \le \frac{1}{2}\int_{x_1}^{x_2} |u_{0,x}|^2 + \frac{1}{2b}\iint_{R_T} \|\phi''(u)\|^2 |u_x|^2.$$

Let $M = \max\{|u(x, t)| : 0 \le t \le T\}$, and $|\phi''(M)| \equiv \max_{|u| \le M} \|\phi''(u)\|$. Then from the last inequality and (21.4) we get

$$\int_{x_1}^{x_2} |u_x(x, t)|^2 + b\int_0^t \int_{x_1}^{x_2} |u_{xx}|^2 \le \int_{x_1}^{x_2} |u_{x,0}|^2 + \frac{1}{b}|\phi''(M)|^2 \iint_{R_T} |u_x|^2$$

$$\le \int_{x_1}^{x_2} |u_{x,0}|^2 + \frac{1}{2b^2}|\phi''(M)|^2 \int_{x_1}^{x_2} |u_0|^2.$$

(21.6)

Now since $u(t, x_1) = 0$, we have

$$|u|^2 = \langle u, u \rangle = \int_{x_1}^x \frac{\partial}{\partial x}\langle u, u \rangle \, dx = 2\int_{x_1}^x \langle u_x, u \rangle \, dx$$

$$\le 2\left(\int_{x_1}^{x_2} |u_x|^2\right)^{1/2}\left(\int_{x_1}^{x_2} u^2\right)^{1/2},$$

(21.7)

again by the Schwarz inequality. Then using (21.4) and (21.6), we see that we have an estimate of the form

$$|u(x, t)|^2 \le \frac{1}{b}C|\phi''(M)| + C_1,$$

(21.8)

where C and C_1 depend only on the L_2-norms of u_0 and $u_{0,x}$, and not on t. From (21.8), it follows that

$$M^2 \le \frac{1}{b}C|\phi''(M)| + C_1,$$

(21.9)

which, in view of (21.1) implies that $M \le M_0$, with M_0 independent of t. Thus for any (x, t) in the strip $\{x_1 \le x \le x_2, t \ge 0\}$, we have

$$|u(x, t)| \le M_0.$$

The existence of a global solution follows from this inequality; see Theorem 14.4, for a similar statement.

We now consider the asymptotic behavior of our solution. We shall show that $\|u(\cdot, t)\|_\infty \to 0$ as $t \to \infty$. For this, it suffices to show that

$$z(t) = \int_{x_1}^{x_2} |u_x(x, t)|^2 \, dx \to 0 \quad \text{as } t \to \infty. \tag{21.10}$$

This follows at once from (21.7), since (21.4) implies that $u(\cdot, t)$ is uniformly bounded in $L_2(x_1 \leq x \leq x_2)$.

From (21.4), we see that

$$\int_0^\infty z(t) \, dt = \int_0^\infty \int_{x_1}^{x_2} |u_x|^2 \leq \frac{1}{2b} \int_{x_1}^{x_2} |u_0(x)|^2 < \infty, \tag{21.11}$$

since the estimate (21.4) was valid for any $T > 0$. Also, since $u_t = 0$ on the boundary, we have after integrating by parts and using (21.2) together with (21.5),

$$|z'(t)| = 2 \left| \int_{x_1}^{x_2} \langle u_x, u_{xt} \rangle \, dx \right| = 2 \left| \langle u_x, u_t \rangle \big|_{x_1}^{x_2} - 2 \int_{x_1}^{x_2} \langle u_{xx}, u_t \rangle \, dx \right|$$

$$= \left| -2 \int_{x_1}^{x_2} \langle u_{xx}, Bu_{xx} \rangle \, dx + 2 \int_{x_1}^{x_2} \langle u_{xx}, (\phi_u)_x \rangle \right|$$

$$\leq +2b \int_{x_1}^{x_2} |u_{xx}|^2 + 2 \int_{x_1}^{x_2} |\langle u_{xx}, (\phi_u)_x \rangle| \, dx, \quad (b = |B|),$$

$$\leq +2b \int_{x_1}^{x_2} |u_{xx}|^2 + b \int_{x_1}^{x_2} |u_{xx}|^2 + \frac{1}{b} \int_{x_1}^{x_2} \|\phi''(u)\|^2 |u_x|^2.$$

Therefore, from (21.6)

$$\int_0^\infty |z'(t)| \, dt \leq \frac{1}{b} |\phi''(M_0)|^2 \int_0^\infty \int_{x_1}^{x_2} |u_x|^2 + C_0$$

$$\leq C < \infty. \tag{21.12}$$

Thus z has finite total variation so it has a finite limit at $t = \infty$. In view of (21.11), this limit is zero, and thus (21.10) is proved. This completes the proof of the theorem. $\square$

We remark that we could also assume that u satisfies the somewhat more general boundary conditions $u(x_1, t) = u(x_2, t) = c$, since we can replace u by $u' = u - c$, and observe that this transformation doesn't change the form of (21.2). It is also interesting to notice that if $B = \varepsilon B_1$ where $\varepsilon > 0$ and B_1 is positive definite, then M_0 is of the form M_1/ε, where M_1 is independent of t and ε. Thus one *cannot* use this technique in a "vanishing viscosity" approach to prove an existence theorem for the system $u_t + (\phi_u)_x = 0$.

§B. Artificial Viscosity

We next consider the p-system with "artificial" viscosity terms; namely

$$v_t - u_x = \varepsilon v_{xx}, \qquad u_t + p(v)_x = \varepsilon u_{xx}, \qquad \varepsilon > 0. \qquad (21.13)$$

Here, $x_1 < x < x_2$, $t > 0$, and as usual we assume $p' < 0$ and $p'' > 0$ in $v > 0$.

We consider (21.13) together with initial- and boundary-value conditions

$$(v(x, 0), u(x, 0)) = (v_0(x), u_0(x)), \ x_1 < x < x_2, \ (v, u)(x_i, t) = (h, 0),$$

$$t \geq 0, \quad i = 1, 2, \qquad (21.14)$$

where $v_0(x) \geq \delta' > 0$, and h is a constant, $h \geq \delta'$. We assume that v_0 and u_0 are bounded and satisfy

$$\int_{x_1}^{x_2} \{u_0^2 + v_0^2 + (u_0')^2 + (v_0')^2\} \, dx < \infty. \qquad (21.15)$$

Theorem 21.2. *Under the above assumptions, the initial-boundary-value problem for (21.13) has a global solution defined in $t > 0$.*

We shall obtain a-priori bounds for u and v from which as above, the global existence theorem will follow ([CCS]). To this end, we recall, from Example 5 in Chapter 14, §B, that we have the important a-priori estimate

$$v(x, t) \geq \delta > 0, \qquad (x, t) \in \mathbf{R} \times \mathbf{R}_+,$$

for some $\delta \leq \delta'$. (Furthermore, in the case where $p(v) = kv^{-\gamma}$, $\gamma > 1$, we recall from the same example that we also have a bound $|u(x, t)| \leq M$ in $\mathbf{R} \times \mathbf{R}_+$, so that in this case it is only necessary to bound v from above.)

Let $T > 0$ be arbitrary, and let $0 \leq t \leq T$. If we multiply the first equation in (21.13) by u, the second by $[p(h) - p(v)]$, and add we get

$$\frac{\partial}{\partial t}\left[\frac{u^2}{2} + \int_h^{v(x,t)} (p(h) - p(s)) \, ds\right] + \frac{\partial}{\partial x}(up(v)) = \varepsilon(uu_{xx} - p(v)v_{xx}) + p(h)v_t.$$

If we integrate over $R_T \equiv \{0 \leq t \leq T, x_1 \leq x \leq x_2\}$, we find

$$\int_{x_1}^{x_2}\left[\frac{u^2}{2} + \int_h^v (p(h) - p(s)) \, ds\right]\bigg|_{t=T} dx + \int_0^T up(v)\bigg|_{x_1}^{x_2} dt$$

$$= \int_{x_1}^{x_2}\left[\frac{u_0^2}{2} + \int_h^{v_0} (p(h) - p(s)) \, ds\right] dx + \varepsilon \iint_{R_T} uu_{xx} + (p(h) - p(v))v_{xx}.$$

From the boundary conditions $u = 0$ on $x = x_1$ and $x = x_2$, the second term on the left vanishes. Also, we can integrate the last term by parts to get

$$\int_{x_1}^{x_2} \left[\frac{u^2}{2} + \int_h^{v(x,t)} (p(h) - p(s)) \, ds \right]\Bigg|_{t=T} dx$$

$$- \varepsilon \iint_{R_T} p'(v)v_x^2 + \varepsilon \iint_{R_T} u_x^2 \equiv \bar{k}, \quad (21.16)$$

where $\bar{k}$ depends only on the initial data.

Next we differentiate (21.13) with respect to x, then multiply the first equation by v_x, the second by u_x, add the results and integrate over R_T. We obtain

$$\frac{1}{2} \int_{x_1}^{x_2} (u_x^2 + v_x^2)|_{t=T} + \iint_{R_T} u_x p_{xx} - v_x u_{xx}$$

$$= k_1 + \varepsilon \iint_{R_T} u_x u_{xxx} + v_x v_{xxx}, \quad (21.17)$$

where k_1 depends only on the data. Then integrating by parts gives

$$\iint_{R_T} u_x p_{xx} - \varepsilon u_x u_{xxx} = \int_0^T (u_x p_x - u_x \varepsilon u_{xx})|_{x_1}^{x_2} - \iint_{R_T} u_{xx} p_x - \varepsilon u_{xx}^2$$

$$= \int_0^T u_x(-u_t)|_{x_1}^{x_2} - \iint_{R_T} u_{xx} p_x - \varepsilon u_{xx}^2$$

$$= \iint_{R_T} u_{xx} p_x - \varepsilon u_{xx}^2,$$

and similarly,

$$\iint_{R_T} - \varepsilon v_x v_{xxx} - v_x u_{xx} = - \iint_{R_T} v_{xx}^2 - v_{xx} u_x.$$

Thus from (21.17),

$$\frac{1}{2} \int_{x_1}^{x_2} (u_x^2 + v_x^2)|_{t=T} \, dx + \varepsilon \iint_{R_T} u_{xx}^2 + v_{xx}^2 = k_1 + \iint_{R_T} u_{xx} p_x - v_{xx} u_x$$

$$\leq k_1 + \iint_{R_T} \frac{\varepsilon}{2} u_{xx}^2 + \frac{p_x^2}{2\varepsilon} + \frac{\varepsilon}{2} v_{xx}^2 + \frac{1}{2\varepsilon} u_x^2,$$

so that

$$\int_{x_1}^{x_2} (u_x^2 + v_x^2)|_{t=T}\, dx + \varepsilon \iint_{R_T} u_{xx}^2 + v_{xx}^2 \le 2k_1 + \frac{1}{\varepsilon} \iint_{R_T} p_x^2 + u_x^2. \qquad (21.18)$$

Now since $p' < 0$, $p'' > 0$, $p_x^2 = p'(v)^2 v_x^2 \le p'(\delta)^2 v_x^2$, and thus we have

$$\frac{1}{\varepsilon} \iint_{R_T} p_x^2 + u_x^2 \le -\frac{p'(\delta)}{\varepsilon} \iint_{R_T} -p'(v)v_x^2 + \frac{1}{\varepsilon} \iint_{R_T} u_x^2$$

$$\le -\frac{p'(\delta)\,\tilde{k}}{\varepsilon}\,\frac{}{\varepsilon} + \frac{\tilde{k}}{\varepsilon}, \qquad (21.19)$$

by (21.16). Hence (21.18) implies

$$\int_{x_1}^{x_2} (u_x^2 + v_x^2)|_{t=T}\, dx + \varepsilon \iint_{R_T} u_{xx}^2 + v_{xx}^2 \le 2k_1 + \frac{\tilde{k}}{\varepsilon^2}(1 - p'(\delta)), \qquad (21.20)$$

where k_1 and $\tilde{k}$ depend only on the data.

We define an auxiliary function f by

$$f(v) = \int_h^v \left[\int_h^y (p(h) - p(s))\, ds\right]^{1/2} dy.$$

Note that $f(h) = 0$; thus

$$|f(v)| = \left|\int_{x_1}^x \frac{\partial f}{\partial x}\, dx\right| = \left|\int_{x_1}^x f'(v)v_x\, dx\right|$$

$$\le \left(\int_{x_1}^{x_2} f'(v)^2\, dx\right)^{1/2} \left(\int_{x_1}^{x_2} v_x^2\right)^{1/2}$$

$$= \left\{\int_{x_1}^{x_2} \left(\int_h^v [p(h) - p(v)]\, ds\right) dx\right\}^{1/2} \left(\int_{x_1}^{x_2} v_x^2\right)^{1/2},$$

so that in view of (21.16) and (21.20), we have

$$|f(v)| \le \tilde{k}^{1/2}\left(2k_1 + (1 - p'(\delta))\frac{\tilde{k}}{\varepsilon^2}\right)^{1/2}, \qquad (21.21)$$

where k_3 depends only on the data.

Let $A = \{(x, t) \in R_T : v > 3h\}$; then on A

$$|f(v)| = \int_h^v \left[\int_h^y (p(h) - p(s)) \, ds \right]^{1/2} dy$$

$$> \int_{2h}^v \left[\int_{2h}^y (p(h) - p(s)) \, ds \right]^{1/2} dy$$

$$> \int_{2h}^v \left[\int_{2h}^y (p(h) - p(2h)) \, ds \right]^{1/2} dy$$

$$= \sqrt{p(h) - p(2h)} \int_{2h}^v (y - 2h)^{1/2} \, dy$$

$$= \tfrac{2}{3}\sqrt{p(h) - p(2h)}(v - 2h)^{3/2}$$

$$\geq c_1 v^{3/2},$$

where c_1 depends only on h. Thus from (21.21), if $v > 3h$,

$$c_1 v^{3/2} \leq \tilde{k}^{1/2}\left(2k_1 + (1 - p'(\delta))\frac{\tilde{k}}{\varepsilon^2} \right)^{1/2}.$$

This implies an estimate of the form $v \leq M_0$, where M_0 is independent of T. Thus since T was arbitrary, we have the bounds

$$\delta \leq v(x, t) \leq M_0, \qquad x \in \mathbf{R}, \quad t > 0. \tag{21.22}$$

We now estimate u. Using (21.16) and (21.20), we have

$$u^2 = \int_{x_1}^x \frac{\partial}{\partial x} u^2 \, dx = 2 \int_{x_1}^x u u_x$$

$$\leq 2\left(\int_{x_1}^{x_2} u^2 \, dx \right)^{1/2} \left(\int_{x_1}^{x_2} u_x^2 \right)^{1/2}$$

$$\leq 2(2\tilde{k})^{1/2}\left(2k_1 + \frac{\tilde{k}}{\varepsilon^2}(1 - p'(\delta)) \right)^{1/2}.$$

This yields the required a-priori bound on u, which, in conjunction with (21.22), completes the proof. □

§C. Isentropic Gas Dynamics

The equations we study here have the form

$$v_t - u_x = 0, \qquad u_t + p(v)_x = (k(v)u_x)_x, \qquad x \in \mathbf{R}, \quad t > 0, \quad (21.23)$$

where $p' < 0$, $p'' > 0$, and $k(v) > 0$, in $v > 0$. This system describes isentropic gas motion with viscosity, when we take $p(v) = c_1 v^{-\gamma}$, and $k(v) = c_2 v^{-1}$, where $c_1, c_2 > 0$ are constants, and $\gamma > 1$. We assume that (21.23) satisfies the initial conditions

$$(v(x, 0), u(x, 0)) = (v_0(x), u_0(x)), \qquad x \in \mathbf{R}. \qquad (21.24)$$

Here v_0 and u_0 are smooth, and satisfy both

$$v_0(x) \geq \delta > 0 \quad \text{for } x \in \mathbf{R}, \quad \text{and}$$

$$\lim_{|x| \to \infty} (v_0(x), u_0(x)) = (h, 0), \qquad (21.25)$$

where $\delta > 0$ and $h > 0$ are constants. (We can consider the boundary condition $u_0 \to c$ as $|x| \to \infty$, but since the transformation $u \to u - c$, $v \to v$ leaves (21.23) invariant, we may as well assume $c = 0$.) We also assume that the following integral converges:

$$\int_{\mathbf{R}} [u_0^2 + (u_0')^2 + (v_0')^2 + P(v_0, \delta)] \, dx < \infty, \qquad (21.26)$$

where $P(v, \delta)$ is defined by

$$P(v, \delta) = \int_{\delta}^{v} (p(\delta) - p(s)) \, ds \geq 0.$$

In the sequel, we shall let C_0 and α_0 denote constants depending only on the data, while $C_0(T)$ and $\alpha_0(T)$ denote constants depending on T and the data.

The existence of a solution to problem (21.23), (21.24) can be proved by the following method. Let $T > 0$ be arbitrary and let $R_T = \{(x, t) : x \in \mathbf{R}, 0 \leq t \leq T\}$. We define Hölder norms in R_T by

$$|f|_0^T = \sup_{R_T} |f(x, t)|,$$

$$|f|_\alpha^T = |f|_0^T + \sup_{(x,t),\,(x',t') \in T_T} \frac{|f(x, t) - f(x', t')|}{|x' - x|^\alpha + |t' - t|^{\alpha/2}}.$$

$$|f|_{1+\alpha}^T = |f|_0^T + |f_x|_\alpha^T,$$

$$|f|_{2+\alpha}^T = |f|_0^T + |f_{xx}|_\alpha^T + |f_t|_\alpha^T,$$

where $0 < \alpha < 1$. Suppose that in some R_τ, $0 < \tau < T$, there exists a solution such that $v > v_1 = \text{const.} > 0$, and $|u|^\tau_{1+\alpha} + |u|^\tau_{2+\alpha} + |v|^\tau_{1+\alpha} < \infty$. Such a solution exists for sufficiently small τ (see [Na]). In order to prove that there is a solution in the large, it suffices to obtain a-priori estimates of the form

$$|u|^\tau_{1+\alpha} + |u|^\tau_{2+\alpha} + |v|^\tau_{1+\alpha} < C_0(T), \qquad 0 < \alpha < \alpha_0(T) < 1, \qquad (21.27)$$

and

$$v \geq h_1 > 0, \qquad h_1 = \text{const.} \qquad (21.28)$$

Indeed, if we take $(v(x, \tau), u(x, \tau))$ as data on $t = \tau$, we can extend the solution to the strip $R_{\tau+\sigma}$, where σ depends only on $C_0(T)$ and h_1, see [Na]. In this way we can extend the solution to R_T, and in view of the arbitrariness of T, the solution exists in the entire upper-half plane $t > 0$. The required estimates (21.27), (21.28) will be obtained via the "energy" method. All arguments will be carried out in the strip R_τ.

We multiply the second equation in (21.23) by u, the first by $p(\delta) - p(v)$, add the results, and integrate over the rectangle $R = \{|x| \leq X, 0 \leq t \leq t_1\}$, $t_1 \leq \tau$, where X is an arbitrary positive number. This gives

$$\iint_R \frac{\partial}{\partial t}(\tfrac{1}{2}u^2 + P(v, \delta)) + ku_x^2 = \iint_R \frac{\partial}{\partial x}[u(ku_x + P(v, \delta))],$$

so that

$$\int_{|x|<X} (\tfrac{1}{2}u^2 + P(v, \delta))|_{t=t_1}\, dx + \iint ku_x^2 = \int_{|x|\leq X} (\tfrac{1}{2}u_0^2 + P(v_0, \delta))|_{t=0}\, dx$$

$$+ \int_0^{t_1} u(ku_x + p(\delta) - p(v)|^X_{-X}\, dt. \qquad (21.29)$$

In this equation, the right-hand side is bounded independently of X, and the integrands on the left-hand side are nonnegative. Thus

$$\int_R (\tfrac{1}{2}u^2 + P(v, \delta))\, dx + \int_0^t \int_{-\infty}^\infty k(v)u_x^2 < \infty,$$

for every t, $0 \leq t \leq \tau$. Since $u(\cdot, t) \in L_2(\mathbf{R})$, and $(u^2)_x$ is bounded, it follows that $u(x, t) \to 0$ as $|x| \to \infty$ for every such t. (To see this, suppose that for some such t and $\varepsilon > 0$, $u^2(x_i, t) \geq 2\varepsilon > 0$, for a sequence $x_i \to \infty$. Then for each i, there is an interval I_i about x_i where $u^2(I_i, t) \geq \varepsilon > 0$. Since $u^2 \in L_1(\mathbf{R})$, diam. $I_i \to 0$ as $i \to \infty$, and this violates the boundedness of $(u^2)_x$.)

Now since u_t is bounded in R_t, $u(x, t) \to 0$ uniformly in t as $|x| \to \infty$. (Take $\{t_n\}$ countable and dense in $[0, \tau]$, and if $|u_t| \le M$, choose t_n such that $M|t - t_n| < \varepsilon/2$; then choose K such that $|u(x, t_n)| < \varepsilon/2$ if $|x| > K$. Then if $0 \le t \le \tau$, $|u(x, t)| \le |u(x, t) - u(x, t_n)| + |u(x, t_n)| < \varepsilon$ if $|x| > K$.) It follows that we can pass to the limit as $X \to \infty$ in (21.29) to get

$$\int_R (\tfrac{1}{2} u^2 + P(v, \delta))|_{t=t_1} \, dx + \int_0^{t_1} \int_{-\infty}^{\infty} k u_x^2 = \int_{-\infty}^{\infty} (\tfrac{1}{2} u_0^2 + P(v_0, \delta)) \, dx. \quad (21.30)$$

We shall now obtain an energy estimate for v_x. First we define

$$K(v) = \int_\delta^v k(s) \, ds,$$

and then rewrite the second equation in (21.23) as

$$u_t - K_{xt} + p'(v)v_x = 0.$$

Multiply this by K_x and integrate over the rectangle $|x| \le X, 0 \le t \le t_1 \le \tau$. After integrating by parts and using the first equation in (21.23), we get

$$\frac{1}{2} \int_{|x| \le X} k^2 v_x^2 |_{t=t_1} - \int_0^{t_1} \int_{|x| \le X} p'(v) k v_x^2 = \int_{|x| \le X} k v_x u |_{t=t_1} + \int_0^{t_1} (Ku_t)|_{-X}^{X} \, dt$$

$$+ \int_0^{t_1} \int_{|x| \le X} k u_x^2 \, dx \, dt - (uK)|_{-X}^{X} + \int_{|x| \le X} u_0' K(v_0) + \frac{1}{2} \int_{|x| \le X} k^2(v_0) v_{0,x}^2. \quad (21.31)$$

Now from (21.29) and the inequality

$$\int_{|x| \le X} ukv_x \le \frac{1}{4} \int_{|x| \le X} k^2 v_x^2 + \int_{|x| \le X} u^2,$$

we obtain from (21.31),

$$\frac{1}{4} \int_{|x| \le X} k^2 v_x^2 |_{t=t_1} - \int_0^{t_1} \int_{|x| \le X} p' k v_x^2$$

$$\le -u(x, t_1) K(v(x, t_1))|_{-X}^{X} + \int_0^{t_1} (Ku_t)|_{-X}^{X} + \int_{|x| \le X} u_0' K(v_0)$$

$$+ \frac{1}{2} \int_{|x| \le X} k^2(v_0) v_0' + \int_{|x| \le X} [u_0^2 + 2P(v_0, \delta)]$$

$$+ \int_0^{t_1} [kuu_x + u(p(\delta) - p(v))]|_{-X}^{X}.$$

In this inequality we let $X \to \infty$, and bearing in mind that $u(x, t) \to 0$ as $|x| \to \infty$, $v \ge$ const. > 0, and

$$\int_0^{t_1} Ku_t = uK|_0^{t_1} - \int_0^{t_1} ukv_t \to 0 \quad \text{as } |x| \to \infty,$$

we find that for $0 \le t_1 \le \tau < T$,

$$\frac{1}{4} \int_{\mathbf{R}} k^2 v_x^2 \big|_{t=t_1} - \int_0^{t_1} \int_{\mathbf{R}} p'(v)kv_x^2 \le C_0. \tag{21.32}$$

Now define an auxiliary function f by

$$f(v) = \int_\delta^v k(s)(P(s, \delta))^{1/2} \, ds. \tag{21.33}$$

Using (21.30), we see that $P(v(\cdot, t), \delta)$ is integrable on $\mathbf{R}$, and since $\partial P/\partial x = (p(\delta) - p(v))v_x$ is bounded, we may conclude as above, that $P \to 0$ as $|x| \to \infty$; i.e., $v \to \delta$ as $|x| \to \infty$. Thus $f(v) \to 0$ as $|x| \to \infty$, so we may write

$$f(v) = \int_{-\infty}^x \frac{\partial f}{\partial x} \, dx = \int_{-\infty}^x v_x k(v)(P(v, \delta))^{1/2} \, dx.$$

Therefore, using (21.30) and (21.32) together with the Schwarz inequality, we see that

$$|f(v)| \le \left(\int_{\mathbf{R}} k^2 v_x^2 \, dx \right)^{1/2} \left(\int_{\mathbf{R}} P(v, \delta) \, dx \right)^{1/2} < C_0. \tag{21.34}$$

In what follows, we shall make the following two assumptions; namely,

$$\int_0^\delta k(v)(P(v, \delta))^{1/2} \, dv = \infty, \tag{21.35}$$

and in addition

$$\int_\delta^\infty k(v)(P(v, \delta))^{1/2} \, dv = \infty. \tag{21.36}$$

We shall not pause here to discuss these conditions; this will be done below.

If we compare (21.35) and (21.36) with (21.33), we see that from (21.34), it follows that there are constants $\underline{v}$ and $\bar{v}$, depending only on the data, such that

$$0 < \underline{v} \le v(x, t) \le \bar{v} < \infty, \qquad (x, t) \in R_\tau. \tag{21.37}$$

These inequalities imply that there are constants k_i, $i = 1, 2, 3$, depending only on the data, such that

$$0 < k_1 < k(v) < k_2, \qquad 0 < k_3 < -p'(v)k(v). \tag{21.38}$$

We shall now estimate u_x. In order to do this, we shall first obtain an "energy inequality" for the difference quotient $\Delta u / \Delta x$, where we define $\Delta u = u(x + \Delta x, t) - u(x, t)$.

From (21.23), we get

$$\frac{\partial}{\partial t}\left(\frac{\Delta u}{\Delta x}\right) + \frac{\Delta p_x}{\Delta x} = \frac{\partial}{\partial x}\left(\frac{\Delta(ku_x)}{\Delta x}\right).$$

We multiply this equation by $\Delta u / \Delta x$, and integrate over the rectangle $R = \{|x| \le X, 0 \le t \le t_1 \le \tau\}$. In so doing, we make use of the identity

$$\frac{\Delta(ku_x)}{\Delta x} - \frac{\Delta u_x}{\Delta x} = \frac{\Delta u_x}{\Delta x}\left[k(v(x + \Delta x, t))\frac{\Delta u_x}{\Delta x} + \frac{\Delta k}{\Delta x}u_x(x, t) \right],$$

and inequalities of the form $ab \le \varepsilon a^2/2 + b^2/2\varepsilon$. This gives

$$\frac{1}{2}\int_{|x| \le X}\left(\frac{\Delta u}{\Delta x}\right)^2 + (k_1 - \varepsilon)\iint_R\left(\frac{\Delta u_x}{\Delta x}\right)^2 \le \frac{1}{2}\int_{|x| \le X}\left(\frac{\Delta u_0}{\Delta x}\right)^2$$

$$+ \frac{1}{2\varepsilon}\max_{\underline{v} \le v \le \overline{v}}|p'(v)|^2 \iint_R\left(\frac{\Delta v}{\Delta x}\right)^2$$

$$+ \max_{\underline{v} \le v \le \overline{v}}|k'(v)|^2\frac{1}{2\varepsilon}\iint_R\left(\frac{\Delta v}{\Delta x}\right)^2 u_x^2$$

$$+ \int_0^{z_1}\left[\frac{\Delta u}{\Delta x}\left(\frac{\Delta(ku_x)}{\Delta x} - \frac{\Delta p}{\Delta x}\right)\Big|_{-X}^X\right].$$

If we let $\Delta x \to 0$, we obtain

$$\frac{1}{2}\int_{|x| \le X}u_x^2 + (k_1 - \varepsilon)\iint_R u_{xx}^2 \le \frac{1}{2}\int_{|x| \le X}u_{0,x}^2 + \frac{1}{2\varepsilon}\max_{\underline{v} \le v \le \overline{v}}|p'(v)|\iint_R v_x^2$$

$$+ \max_{\underline{v} \le v \le \overline{v}}|k'(v)|^2\frac{1}{2\varepsilon}\iint_R v_x^2 u_x^2$$

$$+ \int_0^{z_1}(u_x(ku_x)_x - p_x)\big|_{-X}^X x. \tag{21.39}$$

Since u_x, u_{xx}, and v_x are bounded in R_τ, we see from (21.30), (21.32), (21.37), and (21.38), that the right-hand side of (21.39) is bounded independently of X. Thus the two integrals

$$\int_R u_x^2(x, t_1)\, dx, \qquad \int_0^{t_1} \int_R u_{xx}^2(x, t_1)\, dx\, dt$$

both converge. The convergence of the first of these integrals together with the fact that $(u_x^2)_x$ is bounded, implies that $u_x \to 0$ as $|x| \to \infty$. Thus, passing to the limit in (21.39) as $X \to \infty$, we find for $0 < \varepsilon < k_1$,

$$\frac{1}{2}\int_R u_x^2(x, t_1) + (k_1 - \varepsilon)\int_0^{t_1}\int_R u_{xx}^2 \le \frac{1}{2}\int_R (u_{0,x})^2 + \frac{1}{2\varepsilon}\max_{\underline{v} \le v \le \bar{v}} |p'(v)|^2 \int_0^{t_1}\int_R v_x^2$$

$$+ \frac{1}{2\varepsilon}\max_{\underline{v} \le v \le \bar{v}} |k'(v)|^2 \int_0^{t_1}\int_R u_x^2 v_x^2. \quad (21.40)$$

We see from (21.38) and (21.32), that the second term on the right in (21.40) is bounded by a constant depending only on the data. We consider now the third term.

Since $u_x \to 0$ as $x \to -\infty$, we can write

$$u_x^2 = \int_{-\infty}^x \frac{\partial}{\partial x}(u_x^2)\, dx = 2\int_{-\infty}^x u_x u_{xx} \le 2\left(\int_R u_x^2\right)^{1/2}\left(\int_R u_{xx}^2\right)^{1/2}$$

$$\le \frac{1}{\varepsilon^2}\int_R u_x^2 + \varepsilon^2 \int_R u_{xx}^2, \qquad \varepsilon > 0.$$

Therefore

$$\frac{1}{\varepsilon}\int_0^{t_1}\int_R (u_x^2)v_x^2 \le \frac{1}{\varepsilon}\int_0^{t_1}\left(\int_R u_x^2\right)\left(\frac{1}{\varepsilon^2}\int_R v_x^2\right) dt + \varepsilon^2 \int_0^{t_1}\left(\int_R u_{xx}^2\right)\left(\int_R v_x^2\right) dt$$

$$\le \frac{1}{\varepsilon^3} C_0 + \varepsilon C_0 \int_0^{t_1}\int_R u_{xx}^2, \quad (21.41)$$

by (21.29), (21.32), and (21.38). Thus from (21.40),

$$\frac{1}{2}\int_R u_x^2(x, t_1) + (k_1 - \varepsilon)\int_0^{t_1}\int_R u_{xx}^2 \le \frac{1}{\varepsilon^3} C_0 + \varepsilon C_0 \int_0^{t_1}\int_R u_{xx}^2.$$

Accordingly, if ε is sufficiently small,

$$\int_{\mathbf{R}} u_x^2(x, t_1)\, dx + \int_0^{t_1} \int_{\mathbf{R}} u_{xx}^2\, dx\, dt < C_0, \qquad 0 \le t_1 \le \tau. \qquad (21.42)$$

Now $u = 0$ at $|x| = \pm\infty$, so that

$$u^2 = 2\int_{-\infty}^{x} u u_x \le 2\left(\int_{\mathbf{R}} u^2\right)^{1/2}\left(\int_{\mathbf{R}} u_x^2\right)^{1/2},$$

and thus from (21.29) and (21.42),

$$|u(x, t)| \le C_0. \qquad (21.43)$$

We now estimate $|v_x|$. For this we multiply the first equation in (21.23) by $k(v)$ to get $K_t = k u_x$, where K is as above. Then

$$K(v) = K(v_0) + \int_0^{t_1} k u_x\, dt,$$

$$K_x = k v_x = k(v_0)v_0' + \int_0^{t} (k u_x)_x\, dt,$$

so from the second equation in (21.23),

$$k v_x = k(v_0)v_0' + u - u_0 + \int_0^{t_1} p' v_x. \qquad (21.44)$$

Then using (21.37), (21.38), and (21.43), we see

$$|v_x| \le C_0\left(1 + \int_0^{t_1} |v_x|\, dt\right),$$

and therefore by Gronwall's inequality,

$$|v_x| \le C_0(T). \qquad (21.45)$$

Now we can consider the second equation in (21.23) as linear equation for u. In this regard, the coefficients are all bounded (because of (21.37), (21.38), and (21.45)), so that we can use the Schauder-type estimates (cf. [Kv 1]), to obtain an estimate of the form

$$|u|_{1+\alpha}^{\tau} < C_0(T), \qquad 0 < \alpha < \alpha_0(T). \qquad (21.46)$$

It still remains to estimate $|v_{xx}|$ and $|u|^{\tau}_{2+\alpha}$. We begin with v_{xx}. We take difference quotients of both sides of (21.44) to get

$$k(v(x, t))\frac{\Delta v_x}{\Delta x} = -v_x(x, t)\frac{\Delta k}{\Delta x} + \frac{\Delta(v'_0 k(v_0))}{\Delta x} - \frac{\Delta u_0}{\Delta x} + \frac{\Delta u}{\Delta x}$$

$$+ \int_0^t p'(v(x, t))\frac{\Delta v_x}{\Delta x} dt + \int_0^t v_x(x, t)\frac{\Delta p'}{\Delta x} dt. \qquad (21.47)$$

Then using (21.37), (21.38), (21.45), and (21.46), we have

$$\frac{\Delta v_x}{\Delta x} \le C_0(T)\left(1 + \int_0^t \frac{\Delta v_x}{\Delta x} dt\right),$$

so that again using Gronwall's inequality, we get

$$\frac{\Delta v_x}{\Delta x} \le C_0(T).$$

It follows that v_{xx} exists a.e. in R^{τ}, and

$$|v_{xx}| < C_0(T). \qquad (21.48)$$

We now estimate $|u|^{\tau}_{2+\alpha}$. For this, we will work with the averaged function v^r, where r denotes the radius of the averaging kernel. Let us first define v in $t \le 0$, by $v(x, t) = v_0(x) + u_{0,x}(x)t$. Let u^r satisfy

$$\frac{\partial u^r}{\partial t} + \frac{\partial}{\partial x} p(v^r) = \frac{\partial}{\partial x}\left(k(v^r)\frac{\partial u^r}{\partial x}\right), \qquad u^r(x, 0) = u_0(x). \qquad (21.49)$$

From inequalities (21.37), (21.38), (21.45), and (21.48), we have estimates of the form

$$\tfrac{1}{2}\bar{v} < v^r < 2\bar{v}, \qquad \tfrac{1}{2}k_1 < k(v^r) < 2k_2 \qquad (21.50)$$

$$|v^r_x| < C_0(T), \qquad |v^r_{xx}| < C_0(T), \qquad (21.51)$$

for sufficiently small r. Now let's regard (21.49) as a linear equation for u^r. It follows from the maximum principle for this equation, [IKO], together with (21.50) and (21.51) that

$$|u^r| \le C_0(T). \qquad (21.52)$$

Using (21.50)–(21.52) together with the Schauder-type estimates in [Kv 1], we obtain as above

$$|u^r|_{1+\alpha}^\tau < C_0(T), \qquad 0 < \alpha < \alpha_0(T). \tag{21.53}$$

Now differentiate (21.49) with respect to x; this gives

$$\frac{\partial u_x^r}{\partial x} - k(v^r)\frac{\partial^2 u_x^r}{\partial x^2} - 2k'(v^r)v_x^r\frac{\partial u_x^r}{\partial x} = [k''(v^r)(v_x^r)^2 + k'(v^r)v_{xx}^r]u_x^r - \frac{\partial^2 p(v^r)}{\partial x^2}.$$

We regard this as a linear equation for u_x^r. Applying again the Schauder-type estimates of [Kv 1], and using (21.50)–(21.52), we obtain

$$|u^r|_{2+\alpha}^\tau < C_0(T), \qquad 0 < \alpha < \alpha_0(T). \tag{21.54}$$

Now set $\Delta u = u^r - u$, and subtract the second equation in (21.33) from (21.49) to get

$$\frac{\partial \Delta u}{\partial t} - k(v)\frac{\partial^2 \Delta u}{\partial x^2} - k'(v)v_x\frac{\partial \Delta u}{\partial x} = f_r(x, t), \tag{21.55}$$

where

$$f_r(x, t) = (k(v^r) - k(v))\frac{\partial^2 u^r}{\partial x^2} + (k'(v^r)v_x^r - k'(v)v_x)u_x^r + p'(v)v_x - p'(v_r)v_x^r.$$

We shall show that $f_r(x, t) \to 0$ as $r \to 0$, uniformly in the strip R_τ. We have already seen that $v \to \delta$ as $|x| \to \infty$. Since $v_t = u_x$ is bounded, this convergence is uniform with respect to $t, 0 \le t \le \tau$. Hence $v^r \to v$ as $r \to 0$, uniformly in R_τ.

From (21.32) and (21.38), $v_x(\cdot, t) \in L_2(\mathbf{R})$ for $0 \le t \le \tau$. From $(v_x^2)_x$ being bounded, it follows that $v_x \to 0$ as $|x| \to \infty$, $0 \le t \le \tau$, and since $v_{xt} = u_{xx}$ is also bounded, this convergence is uniform in $t, 0 \le t \le \tau$. Therefore $v_x^r \to v_x$ as $r \to 0$, uniformly in R_τ. Also, (21.53) and (21.54) show that u_x^r and u_{xx}^r are uniformly bounded with respect to r. Thus since (v^r, v_x^r) converges uniformly to (v, v_x) in R_τ, it follows that $f_r \to 0$ as $r \to 0$, uniformly in R_τ.

Now consider (21.55) as a linear equation for Δu. Since $\|f_r\|_{L_\infty(R_\tau)} \to 0$ as $r \to 0$, and $\Delta u(x, 0) = 0$, the maximum principle (see [IKO]), implies that $\Delta u \to 0$ as $u^r \to u$, uniformly in R_τ. This, together with (21.54) gives the desired estimate; namely,

$$|u|_{2+\alpha}^\tau < C_0(T), \qquad 0 < \alpha < \alpha_0(T).$$

In particular, u_{xx} is bounded, so since $v_{xt} = u_{xx}$, it follows from (21.37), (21.45), and (21.48), that we can obtain our final estimate:

$$|v|_{1+\alpha}^\tau \le C_0(T).$$

We have therefore proved the following theorem.

Theorem 21.3. *Consider the system* (21.23) *together with the data* (21.24). *We assume that* (21.25) *and* (21.26) *hold, and in addition we require that* (21.35) *and* (21.36) *hold. Then a global solution exists in t > 0.*

We remark that if $p(v) = c_1 v^{-\alpha}$, and $k(v) = c_2 v^{-\delta}$, then (21.35) holds if $2\delta + \alpha \geq 3$, and (21.36) holds if $\delta \leq \frac{3}{2}$. In particular, if $\delta = 1$, then all we require is $\alpha \geq 1$. Thus, the hypotheses (21.35), and (21.36) are valid for the most important example; namely, isentropic gas dynamics.

NOTES

The theorems in this chapter are all due to Kanel' [K 1, 2]; our proofs are taken from these papers. Theorem 21.3 has been extended to the full gas dynamics equations by Itaya [It], Kazhikhov and Shelukhin [KS], Kawashima and Nishida [KN] and by Matsumura and Nishida [MN 1, 2] for very general equations in 3-space variables. These authors require certain "smallness" hypotheses on the data or very specific equations of state. In the last four papers, the method is the same; namely to obtain a-priori estimates via energy methods. Itaya uses a different method but for a particular equation of state; he does not require the data to be in L_2. It would be interesting to obtain a general existence theorem where the data is nonzero at infinity. A step in this direction was taken by Nishida and Smoller in [NS 3].

Part IV

The Conley Index

Chapter 22

The Conley Index

In any theoretical investigation of a real physical system, one is always forced to make simplifying assumptions concerning the true nature of the system. Since such idealizations are inevitable, it is reasonable to inquire as to how far one can go in this direction and still obtain satisfactory results. In certain cases, for example the motion of the planets, the equations of celestial mechanics provide a quite accurate model of the real physical system. In other situations, such as ecological or chemical interactions, or the study of large scale atmospheric phenomena, one either writes down certain reasonable relations between the quantities involved and their rates of change, or one tremendously reduces the number of actual equations involved. If such "leaps of faith" are to be of any use, it is necessary to study "rough" equations in "rough" terms. This in a nutshell, is our aim in this chapter and the next one. In other words, we want to fit these vague notions into a precise mathematical framework.

We have come across such ideas earlier in Chapter 12, where we defined the notion of the degree of a mapping. This integer is a topological invariant which depends solely on the boundary values of the mapping. Furthermore, the boundary values themselves need not be given precisely. Thus, if the mapping is deformed to another one in such a way that throughout the deformation the image of the boundary is nonzero, then the degrees of the two mappings are the same. It follows that the degree is perfectly well defined, even though the mapping is only "roughly" known. Our approach in this chapter will be of a similar nature. That is, we shall obtain other algebraic-topological invariants which have similar "stability" properties with respect to changes in the equations.

A salient feature of these techniques is that they do not require difficult computations and they are, therefore, of an intrinsically qualitative nature. On the other hand, in order to show that these techniques are applicable to a given situation, it is often necessary to bring to bear upon the problem something of a quantitative nature. For example, one might have to prove an a-priori estimate, or compute the homotopy type of some space, and so on.

We shall develop here a powerful topological tool, the Conley index, which is a generalization of the Morse index in somewhat the same spirit that the Leray–Schauder degree is a generalization of the Brouwer degree. The Conley index, however, is different from other "indices" in that it has two uses. Thus it can be used as in the case of most "index-theories," to prove the *existence* of certain distinguished solutions, but secondly, since it is a "Morse-type" index, it is an invariant which carries *stability* information. Furthermore, it is a true extension of the Morse index in that they both agree when the latter is defined; i.e., for nondegenerate rest points. But note that even in applications to gradient flows, it is desirable to have an index for sets other than nondegenerate critical points. This is easy to see from the point of view of bifurcation theory, since at the "time" of bifurcation, the critical point becomes degenerate, and thereafter may continue to a set which does not consist of only critical points; the Morse index doesn't apply but the Conely index of all of these sets is perfectly well defined.

We shall begin with a purely descriptive section, in which we describe the basic notions, by giving a fairly careful analysis of an easily understood example. We use it to illustrate the important theorems, and then we present two examples which show how the index is used. In §B, and §C we give a complete development of the theory for flows defined by differential equations, except that we omit a proof of the continuation theorem; this will be done in greater generality in the next chapter.

§A. An Impressionistic Overview

The purpose of this section is to convey to the reader the general ideas of the Conley index theory, and to indicate some ways in which it can be applied. As the title indicates, we shall be rather descriptive; the rigorous justifications will be given in the succeeding sections.

We consider the ordinary differential equation

$$x' = f(x), \qquad x \in \mathbf{R}^n, \qquad ' = \frac{d}{dt}. \tag{22.1}$$

Here f is assumed to be a locally Lipschitz-continuous function. We shall use the notation $x \cdot t$ to denote that point in $\mathbf{R}^n$ which lies on the orbit (solution curve) starting at the point x, which runs for t units of "time"; thus $x \cdot 0 = x$ and $x \cdot (s + t) = (x \cdot s) \cdot t$ for all $s, t \in \mathbf{R}$. A (complete) *solution curve* is a set of the form $x \cdot \mathbf{R} = \{x \cdot t : t \in \mathbf{R}\}$. A set $I \subset \mathbf{R}^n$ is called an *invariant set* if I is the union of solution curves; evidently I is invariant if and only if $I \cdot \mathbf{R} = I$. The objects of study in this chapter are the special class of invariant sets which we term isolated invariant sets according to the following definition.

Definition 22.1. A invariant set S is an *isolated invariant set* if S is the maximal invariant set in some bounded open neighborhood of itself. A compact such neighborhood N is called an *isolating neighborhood*.

Thus $S \subset \text{int } N = N$, and S is the maximal invariant set in cl(N).

For example, consider the origin in Figure 22.1(a) and (b). The hyperbolic point in Figure 22.1(a) is an isolated invariant set, and an isolating neighborhood is indicated. On the other hand, the origin in Figure 22.1(b) is an isolated rest point, but it is not an isolated invariant set since every neighborhood of the origin contains an invariant set different from 0; namely, a periodic orbit.

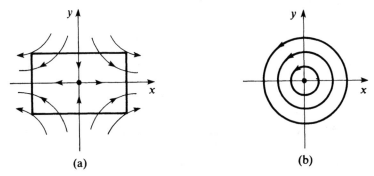

(a)   (b)

Figure 22.1

Note that disjoint unions of isolated invariant sets are themselves isolated invariant sets since the union of the isolating neighborhoods will serve as an isolating neighborhood for the union of the sets. This is not true in general, however. In Figure 22.2, the flow on the disk has three rest points and all other solutions run downward as indicated. Both orbits connecting the middle rest point to another rest point are isolated invariant sets, but their union obviously is not.

The interest in isolated invariant sets comes from the fact that they are *stable* objects in the sense that they can be continued to nearby equations. Here the notion of continuation is defined in terms of isolating neighborhoods. Thus if N is an open set having compact closure, N is an isolating neighborhood for the isolated invariant set S if and only if $S \subset \text{int}(N)$, and S is the maximal invariant set in cl(N). This latter property clearly holds provided

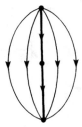

Figure 22.2

that no point on ∂N is on a solution curve contained completely in cl(N); otherwise S would not be the maximal invariant set in cl(N). Now this condition is stable in the sense that if f is perturbed slightly to $\tilde{f}$, then orbits of $\tilde{f}$ which pass through ∂N also cannot be contained in N; hence N is an isolating neighborhood of $\tilde{f}$. We have thus shown:

> If N is an isolating neighborhood of some equation, it is one
> for all "nearby" equations.

The corresponding isolated invariant sets so determined are called the *continuations* of S. Specifically, if N is an isolating neighborhood for a "connected" set of equations, the corresponding isolated invariant sets are said to be *related by continuation*. We can extend this relation to non-nearby equations by making the relation transitive.

Let us pause here to illustrate these ideas by means of a simple but quite instructive example. Thus consider the scalar equation on $\mathbf{R}$ given by

$$x' = x(1 - x^2) - \lambda \equiv f(x, \lambda), \qquad x \in \mathbf{R},$$

where λ is to be thought of as a parameter. In Figure 22.3, we have sketched the curve $f(x, \lambda) = 0$, and in addition we have marked off some points together with certain intervals containing them. Observe that the curve $f = 0$ meets each horizontal line in the set of critical points of the equation with the corresponding value of λ; for each fixed $\lambda = \lambda_0$, the horizontal line $\lambda = \lambda_0$ is the "phase space" of the equation $x' = f(x, \lambda_0)$.

At each of the three λ-levels, the intervals which are marked off are easily seen to be isolating neighborhoods; namely, neither boundary point lies on a solution curve which stays in the interval. The rest points are all examples of isolated invariant sets. But more generally, any interval each of whose endpoints is a rest point, is also an isolated invariant set: a slightly larger interval will serve as an isolating neighborhood. So for example, the closed

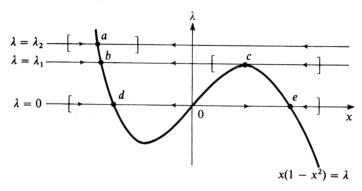

Figure 22.3

interval $d0$ is an isolated invariant set for the $\lambda = 0$ equations. Since disjoint unions of isolated invariant sets are again isolated sets, we can easily find all of them. Thus for the $\lambda = \lambda_2$ equations, there is only one (nonvoid[1]) isolated invariant set; namely, the point a. Similarly, for the $\lambda = \lambda_1$ equations there are four isolated invariant sets: b, c, the set $\{b, c\}$, and the interval $[b, c]$. Finally, the $\lambda = 0$ equation has twelve isolated invariant sets: d, 0, e, $[d, 0]$, $[0, e]$, $[d, e]$, $\{d, e\}$, $\{0, e\}$, $\{d, 0\}$, $\{d, 0, e\}$, $\{[d, 0], e\}$, and $\{d, [0, e]\}$.

If we choose N to be the interval marked off on $\lambda = \lambda_2$, we see that the left-hand rest points in each phase portrait are related by continuation; namely, "translate N down." If we choose N to be the interval depicted on $\lambda = 0$, we see that a is related by continuation to the full set of bounded orbits in the other two phase portraits; it follows that d is related by continuation to the set of bounded orbits for the $\lambda = 0$ equations. The same statement is true for the rest point e. But not all isolated invariant sets are in this class; for example, 0 is not, since no choice of N will continue 0 to a or e. We shall soon see another, perhaps better, reason for this.

As we have mentioned several times before, the Conley index is a generalization of the classical Morse index (Chapter 12, §C). It takes the form of the homotopy type of a pointed topological space. For our purposes here, it suffices to think of a pointed space simply as a pair (X, x), where X is a topological space, and $x \in X$; x is often called the "distinguished" point. The pointed spaces (X, x) and (Y, y) are said to be homotopically equivalent, written $(X, x) \sim (Y, y)$, if there is a homotopy (roughly speaking, a continuous deformation), from X into Y which takes x into y; homotopy type is defined similarly; see the appendix to Chapter 12.

The Conley index is computed from special isolating neighborhoods which are called *isolating blocks*, and they have the property that the solution through each boundary point (immediately) leaves the neighborhood in one or the other time direction. (Notice that for flows in $\mathbf{R}$, all isolating neighborhoods are isolating blocks.) If B is an isolating block, the subset of ∂B consisting of points which leave B in positive time (the *exit set*), is denoted by b^+. Denoting by S the maximal invariant set in B, the *Conley index* of S, $h(S)$, is defined to be the homotopy equivalence class of the quotient space B/b^+; i.e., $h(S) = [B/b^+]$. Of course, we may view B/b^+ as the pointed space (B, b^+).

Let us return to the last example and compute the indices of some of the isolated invariant sets. Consider first the rest point 0. Taking for B any proper subinterval $[\alpha, \beta]$ of the interval $[d, e]$ (see Figure 22.4(a)), we find that b^+ is the complete boundary, $\{\alpha, \beta\}$. If we identify these two endpoints, we see that B/b^+ has the homotopy type of a (pointed) circle which we denote as Σ^1; i.e., $h(0) = \Sigma^1$. Now consider the rest point a. Taking for B the interval indicated in Figure 22.4(b), we see that b^+ is void. Now when the empty set of a space is collapsed to a point, the resulting space is homeomorphic to the disjoint union of the space and a point (Figure 22.5(a)). We may deform

---

[1] The empty set $\varnothing$ is always an isolated invariant set; $\varnothing$ is its own isolating neighborhood.

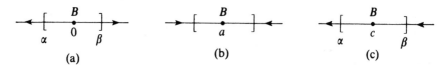

Figure 22.4

Figure 22.5

the interval to a point without changing the homotopy type; this gives (cf. Figure 22.5(b)) a (pointed) two-point space. Therefore $h(a) = \Sigma^0$, the pointed zero-sphere. We shall show in Chapter 23, §C that the Conley index is invariant under continuation in the sense that if S and $\tilde{S}$ are isolated invariant sets which are related by continuation, then $h(S) = h(\tilde{S})$. This will be called the continuation theorem. Since the zero sphere is not homotopically equivalent to the one-sphere (they have different cohomology groups; see the appendix to Chapter 12), we see that a cannot be continued to 0.

Next observe that all of the rest points in the $\lambda = 0$ phase space (Figure 22.3), are nondegenerate critical points of the gradient system

$$x' = \nabla F, \qquad F = \frac{x^2}{2} - \frac{x^4}{4},$$

and so they all have classical Morse indices. These are (cf. Chapter 12, §C) the number of positive eigenvalues of the linearization. The linearized equations are

$$\dot{y} = (1 - 3x^2)y$$

and when $x = 0$, $1 - 3x^2 > 0$, while at $x = d$ or $x = c$, $1 - 3x^2 < 0$. Thus the Morse indices of 0, d, and e are 1, 0, and 0, respectively. This agrees with the dimensions of the pointed spheres which we obtained for the Conley indices. We shall prove later that this is a general fact; namely, the Conley index agrees with the Morse index whenever the latter is defined; see Chapter 23, §D, 4.

Finally, let's consider the rest point c (Figure 22.3) for the $\lambda = \lambda_1$ equations. It is clear that this point is a degenerate rest point of these equations ($1 - 3x^2 = 0$ at c; see footnote 7); thus the Morse index is not even defined for this critical point. But c is an isolated invariant set so we may compute its Conley index. Referring to Figure 22.4(c), we see that $b^+ = \alpha$. Thus B/b^+ has the homotopy type of a pointed interval. Since a simpler representative can be obtained by collapsing the interval to a point, we see that $h(c) = \bar{0}$, a (pointed) one-point space. Now consider the interval about c as depicted in Figure 22.3. If we raise this interval (i.e., increase λ), we see that c continues to the empty set. As we have noted earlier, the empty set is an isolated invariant set and it can be taken to be its own isolating block. The exit set is thus empty and the space obtained by collapsing the exit set to a point is therefore the one-point (pointed) space, $\bar{0}$. This agrees with our above calculation since we have continued the degenerate rest point c to the empty set.

Let us dwell a little longer on the $\lambda = \lambda_1$ flow, the middle portrait in Figure 22.3. If we increase λ above λ_1, then the isolated invariant set c continues to the empty set $\varnothing$. Now suppose that we *start* with λ slightly larger than λ_1, and decrease λ to λ_1. Here we see that $\varnothing$ continues to the isolated invariant set c; i.e., we may alternately say that c *bifurcates out of the empty set*. For λ slightly *below* λ_1, we see that c itself bifurcates into two nondegenerate rest points which continue to 0 and e. These rest points therefore have indices Σ^1 and Σ^0. But we have not taken into consideration still another isolated invariant set; namely, as λ decreases from λ_1, the critical point c actually continues to the isolated invariant set consisting of the two nondegenerate critical points, *together with the entire interval between them*. This isolated invariant set, on the other hand, continues to the interval $[0, e]$. It follows then that $h([0, e]) = \bar{0}$. Similarly, $h([d, 0]) = \bar{0}$. Note that these latter two isolated invariant sets both consist of two rest points together with the orbit connecting them. We shall see later that these simple remarks are actually illustrations of a very general theorem (Theorem 22.33).

We have noted above that the disjoint union of isolated invariant sets again forms isolated invariant sets. It is therefore reasonable to expect that there should exist relations between these various indices. This is indeed the case, as we shall now demonstrate. Consider first the case of the disjoint union of two isolated invariant sets, S_1 and S_2. Using the fact that one can always construct an isolating block for a given isolated invariant set (Theorem 22.18), we may assume that each S_i is contained in an isolating block B_i, with $B_1 \cap B_2 = \varnothing$. Then $B_1 \cup B_2$ will be a block for $S_1 \cup S_2$. To calculate $h(S_1 \cup S_2)$, we collapse the exit set of each block individually, and then collapse the resulting two distinguished points to a single point. The first collapse gives the disjoint union of B_1/b_1^+ with B_2/b_2^+. The second, upon identifying b_1^+ with b_2^+ gives a space which we denote by $B_1/b_1^+ \vee B_2/b_2^+$. This is called the *wedge* or *sum* of the two pointed spaces, and is the pointed space which results from "glueing" the two pointed spaces together at their distinguished points; see Figure 22.6. Thus

$$h(S_1 \cup S_2) = h(S_1) \vee h(S_2). \tag{22.2}$$

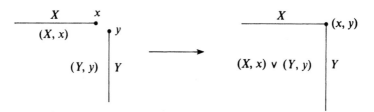

Figure 22.6

This operation is always well defined on pointed spaces (see [Sp]); the topology on $(X, x) \vee (Y, y)$ is the obvious one: a neighborhood of (x, y) is the union of a neighborhood of x in X with a neighborhood of y in Y. To illustrate this by a specific example, consider the set in Figure 22.7, consisting of two rest points. We can take as a block, the disjoint union of the depicted blocks for each of the rest points. Since $B_1/b_1^+ = \Sigma^1$, and $B_2/b_2^+ = \Sigma^0$, the index of $\{p_1, p_2\}$ is $h(\{p_1, p_2\}) = \Sigma^1 \vee \Sigma^0$, the sum of a pointed one-sphere and a (pointed) zero sphere.

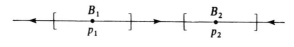

Figure 22.7

Note that the "zero" element is just $\bar{0}$, i.e., $(X, x) \vee \bar{0} = (X, x)$, as follows easily from the definition; this corresponds to the relation $S \cup \varnothing = S$, where S is any isolated invariant set. However, there are no inverses under the wedge operation. In fact the index is nonnegative in the sense that if the sum (i.e., wedge) of two indices is $\bar{0}$, then each factor is equal to $\bar{0}$ (Lemma 22.27).

Indices (and more generally, pointed spaces) can also be "multiplied"; the *(smash) product* $(X, x) \wedge (Y, y)$ of two pointed spaces is defined by

$$(X, x) \wedge (Y, y) = X \times Y/(X \times y) \cup (x \times X),$$

i.e., the space obtained from the topological product upon collapsing to a point the set of pairs, either of whose entries is a distinguished point. The product of isolated invariant sets is isolated, and the index is the product of the indices of its factors (see [Sp]). To see how one might use this, consider the equations $dx/dt = \nabla F(x)$, $dy/dt = \nabla G(y)$, $x \in \mathbf{R}^m$, $y \in \mathbf{R}^n$, and assume that the origin is a nondegenerate critical point for both F and G. Thus each equation admits the origin as an isolated invariant set, with isolating blocks $\{|x| \le \varepsilon\}$, and $\{|y| \le \varepsilon\}$, respectively, for some $\varepsilon > 0$. For the "x-equations," the origin has index Σ^k, the pointed k-sphere, and for the "y-equations" it has index Σ^l, the pointed l-sphere, where k and l are the Morse indices of 0 for the x and y equations, respectively. The product of these two isolated invariant sets is the origin in $\mathbf{R}^{m+n}$; the product of the two blocks is the product block in $\mathbf{R}^{m+n}$, and the index $h(0, 0) = \Sigma^k \wedge \Sigma^l = \Sigma^{k+l}$ (see [Sp]). If, for example, we modify the equations by considering system

$$\frac{dx}{dt} = \nabla F(x) + f(x, y), \qquad \frac{dy}{dt} = \nabla G(y) + g(x, y),$$

where $f(0, 0) = g(0, 0) = 0$, and both f and g are (say) C^2-small near $(0, 0)$, then the same calculations apply, and $h(0, 0) = \Sigma^{k+l}$. Note that the pointed zero sphere is the multiplicative identity, not just on pointed spheres, but in general; it is sometimes denoted by $\bar{1}$.

We can use these ideas to demonstrate the correspondence between the classical Morse index and the Conley index for nondegenerate rest points. This in fact can be done somewhat more generally, as we shall now show. Thus let $dx/dt = f(x)$ be an equation on $\mathbf{R}^n$ which admits x_0 as a rest point, and let the linearized equations about x_0 be given by $d\xi/dt = A\xi$. Our assumption is that the matrix A has no eigenvalues with zero real part; i.e., x_0 is a *hyperbolic* rest point. Note that this is always true for gradient systems if 0 is not an eigenvalue of the linearization.

Assertion. *The rest point x_0 is an isolated invariant set of the equation $dx/dt = f(x)$, and $h(x_0) = \Sigma^k$, the pointed k-sphere, where k is the number of eigenvalues of A having positive real parts.*

To see this, first note that the origin $\xi = 0$ is the only bounded solution of the linearized equations and is thus an isolated invariant set (it is the maximal invariant set in any ball of positive radius centered at the origin). Since the linearized equations provide a good approximation to the nonlinear ones in a sufficiently small ball about x_0, the rest point $x = x_0$ is also an isolated invariant set, having the same index as $\xi = 0$. In order to compute the index, we observe that A can be deformed to a diagonal matrix having diagonal entries ± 1, in such a way that throughout the deformation, no eigenvalue has zero real part. Thus the number of positive entries in the diagonal matrix is the number of eigenvalues of A having positive real part. The rest point of the diagonal matrix is a continuation of that of A since the unit ball can be taken as the isolating neighborhood throughout the deformation. The final equation is completely de-coupled; namely, it is a product of equations $dy/dt = \pm y$ on $\mathbf{R}$. The origin, being an isolated invariant set which is a product of those of the one-dimensional systems, has index which is a product of the one-dimensional indices. If $dy/dt = y$, the index is Σ^1, while if $dy/dt = -y$, it is Σ^0. The product is therefore a pointed sphere Σ^k, where k is the number of eigenvalues of A having positive real parts. This result will be carried out in detail in Chapter 23, §D4.

We shall close this section by giving two nice consequences of the continuation theorem, both of which are not easy to prove without these techniques. The first is what might be termed a result in bifurcation theory and takes the following form. Suppose we are given a one-parameter family of differential equations in $\mathbf{R}^n$

$$\frac{dx}{dt} = f(x, \lambda), \qquad |\lambda| \leq 1,$$

and that the origin is a rest point for all λ; i.e., $f(0, \lambda) = 0$, $|\lambda| \leq 1$. Suppose too that this rest point is an attractor if $\lambda < 0$, and a repeller if $\lambda > 0$. Thus if $\lambda \neq 0$, the origin is an isolated invariant set, and we have

$$h(0) = \begin{cases} \Sigma^n, & \lambda > 0, \\ \Sigma^0, & \lambda < 0. \end{cases}$$

This change of index reflects a change in structure of the solution set as λ crosses zero. In fact, if N is a compact neighborhood at zero for the $\lambda = 0$ equation, N cannot be an isolating neighborhood of zero for the $\lambda = |\varepsilon|$ equations for $|\varepsilon|$ small enough. This would violate the continuation theorem. Thus if N is any compact neighborhood of the origin in $\mathbf{R}^n$, the $\lambda = \varepsilon$ or $\lambda = -\varepsilon$ equations must contain a complete solution curve different from $x = 0$ which stays in N. In the case $n = 2$ and the eigenvalues are complex, it is not too hard to show that N must in fact, contain a periodic orbit; i.e., a "*Hopf bifurcation*" occurs at $\lambda = 0$. The corresponding (weaker) statement for general n can also be considered as a "bifurcation" theorem.

We can also view these results somewhat differently. Namely, suppose that N is any compact neighborhood of the origin in $\mathbf{R}^n$ which is an isolating neighborhood for the origin when $\lambda = \pm 1$ and 0 is an attractor (resp. repellor) for the $\lambda = +1$ (resp. $\lambda = -1$) equations. Then there must be some λ, $|\lambda| < 1$ such that there is a solution of the equations with this λ which stays in N for all $t \in \mathbf{R}$ and which passes through a boundary point of N; in particular, it is not the rest point. (Otherwise N would define a continuation of the origin to itself for $\lambda = \pm 1$, and this would force the indices to be the same.) Observe that the vast freedom of choice in N implies the existence of many solutions which lie near the origin, but not necessarily for the same value of λ.

Our next application is a bit more special but it again serves to illustrate the power of the continuation theorem, together with the "addition" (sum) theorem. Consider the system of equations in $\mathbf{R}^n$,

$$\frac{dy_i}{dt} = y_{i+1}, \quad i = 1, 2, \ldots, n - 1; \qquad \frac{dy_n}{dt} = y_1^2 - 1. \tag{22.3}$$

The problem is to show that there are nonconstant bounded solutions. Observe that the system obviously has the two rest points $(\pm 1, 0, \ldots, 0)$.

In order to analyze these equations, we make the change of variables

$$z_i = \varepsilon^{n-1+i} y_i, \qquad \tau = \frac{t}{\varepsilon};$$

then if $1 \leq i < n$

$$\frac{dz_i}{d\tau} = \frac{dz_i}{dy_i} \frac{dy_i}{dt} \frac{dt}{d\tau} = \varepsilon^{n-1+i} y_{i+1} \varepsilon = \varepsilon^{n+i} y_{i+1} = z_{i+1},$$

and

$$\frac{dz_n}{d\tau} = \varepsilon^{2n-1}(y_1^2 - 1)\varepsilon = \varepsilon^{2n}(y_1^2 - 1) = (\varepsilon^n y_1)^2 - \varepsilon^{2n} = z_1^2 - \varepsilon^{2n}.$$

Thus we arrive at the system

$$\frac{dz_i}{d\tau} = z_{i+1}, \quad i = 1, \ldots, n-1; \qquad \frac{dz_n}{d\tau} = z_1^2 - \varepsilon^{2n}. \qquad (22.4)$$

Consider now the related system

$$\frac{dz_i}{d\tau} = z_{i+1}, \quad i = 1, \ldots, n-1, \qquad \frac{dz_n}{d\tau} = z_1^2. \qquad (22.5)$$

We claim that $(0, \ldots, 0)$ is the only bounded solution of (22.5). To see this, note that the function $z_n(\tau)$ is nondecreasing on solutions so that on bounded solutions it must have limits at $\tau = \pm\infty$. If the limit were nonzero in either case, the solution would be unbounded as we see from the equation $dz_{n-1}/d\tau = z_n$. Thus $z_n(\tau)$ must vanish identically on bounded solutions, so that the equation $dz_{n-1}/d\tau = z_n(\tau)$ implies that $z_{n-1}(\tau)$ is constant on bounded solutions. Were this constant nonzero, the equation $dz_{n-2}/d\tau = z_{n-1}(\tau)$ would imply that z_{n-2} is unbounded. Working upward in this way, we see that all the z_i's must be zero on any bounded solution. This proves our claim.

It follows from this that the origin is an isolated invariant set for (22.5). Taking the unit ball N as an isolating neighborhood, we see that for small ε this set continues to an isolating neighborhood for the system

$$\frac{dz_j}{d\tau} = z_{j+1}, \quad j = 1, \ldots, n-1; \qquad \frac{dz_n}{d\tau} = z_1^2 + \varepsilon^{2n}. \qquad (22.6)$$

This system has *no* bounded solutions since $dz_n/d\tau > 0$; hence the maximal invariant set of (22.6) in N has index $\bar{0}$. The same is thus true for the isolated invariant set of (22.5); i.e., $h(0) = \bar{0}$. This set on the other hand, continues for small ε to an isolated invariant set of (22.4) in N, having index $\bar{0}$.

For small ε, N contains the two rest points $(\pm \varepsilon^n, 0, \ldots, 0)$ of (22.4). The linearized equations are

$$\dot{w}_i = w_{i+1}, \quad i = 1, \ldots, n-1; \qquad \dot{w}_n = 2w_1,$$

and at the rest points the linearized matrix is

$$\begin{bmatrix} 0 & 1 & 0 & & 0 \\ 0 & 0 & 1 & & 0 \\ 0 & 0 & 0 & \cdots & 0 \\ \vdots & \vdots & \vdots & & \vdots \\ \pm 2\varepsilon & 0 & 0 & & 0 \end{bmatrix}.$$

This matrix has eigenvalues satisfying $\lambda^n = \pm 2\varepsilon$. This shows that not both rest points have zero index. The sum formula (22.2) implies that there must be a complete solution in the unit ball N which is different from the rest points.

In this section we have tried to give an overview of the main features of the Conley index, together with an indication of how it can be applied. In the next sections we shall develop these properties in detail.

§B. Isolated Invariant Sets and Isolating Blocks

Our purpose here is to describe some aspects of the index in greater detail, and to give proofs of the important theorems relating the notions described in the title.

Throughout this section, unless otherwise noted, X will be a compact metric space; the reader can think of a bounded closed subset in $\mathbf{R}^n$, or even in $\mathbf{R}^2$.

Let $f: X \times \mathbf{R} \to X$ be continuous; denote $f(x, t)$ by $x \cdot t$. We give a general definition of when f is a flow.

Definition 22.2. If (a) $x \cdot 0 = x$, and (b) $(x \cdot t) \cdot s = x \cdot (t + s)$, both hold for all $x \in X$, and all t, s in $\mathbf{R}$, then f is called a *flow* on X.

It follows easily from the definition that if f is a flow on X, then the function $f^t: X \to X$ defined by $x \to x \cdot t$ is a homeomorphism of X onto itself for all $t \in \mathbf{R}$.

We put the *compact-open (C-O) topology* on the set of functions $X \times \mathbf{R} \to X$; i.e., $f_n \to f$ if and only if f_n converges to f uniformly on compact subsets of $X \times \mathbf{R}$. We let F be the (closed) topological subspace consisting of the flows on X endowed with the (relativized) (C-O) topology.

It is easy to see that if $\varepsilon > 0$, and f is a flow, then f is completely determined by the restriction

$$f_\varepsilon = f \,|\, X \times [-\varepsilon, \varepsilon].$$

On the other hand, the topology on F corresponds precisely to the topology of uniform convergence of the f_ε's; thus F is a complete metric space.

In this section (and in fact, in this entire chapter), we will be mainly interested in those properties of flows which are shared by all "nearby" flows, where in this case, "nearby" means with respect to the (C-O) topology.

We shall use the following notation; namely, if $Y \subseteq X$, and $J \subseteq \mathbf{R}$, the set $f(Y, J)$ will be denoted by $Y \cdot J$. In these terms a subset $Y \subseteq X$ is called *invariant* (*under f*) if $Y \cdot \mathbf{R} = Y$, or equivalently, if $Y \cdot t \subseteq Y$ for all $t \in \mathbf{R}$. If[2] Y is invariant under f then so is $Cl(Y)$ (closure) and $Comp(Y)$ (complement); the same statements are true for the union and intersection over any collection of invariant sets. If $x \in X$, the *orbit through x* means the set $x \cdot \mathbf{R}$; the orbit is obviously an invariant set. It is easy to check that a set is invariant if and only if it is a union of orbits.

We pause now to illustrate these notions. First, in these terms, if $x \cdot \mathbf{R} = x$, then x is called a *rest point* of f. If there is some t such that $x \cdot t = x$, and if x is not a rest point, then x is called a *periodic point*, and t is called a *period* of x. The set of periods of x is a closed subset of $\mathbf{R}$, and if it contains t, it also contains nt for all integers n. It follows that the set of periods of x must be a discrete set, and hence contains a minimum positive period called the *fundamental period*. If x is periodic, then so are all points of $x \cdot \mathbf{R}$, and moreover they have the same fundamental period. Periodicity is therefore a property of orbits. Note that if x is a periodic point or a rest point, then $x \cdot \mathbf{R}$ is compact, since it is a closed subset of a compact space.

If $x \in X$, we define the following important sets:

$$\alpha(x) = \bigcap_{t < 0} cl[x \cdot (-\infty, t)],$$

$$\omega(x) = \bigcap_{t > 0} cl[x \cdot (t, \infty)].$$

$\alpha(x)$ and $\omega(x)$ are called, respectively, the α- and ω-*limit sets of x*. One thinks of these as the "asymptotic limits" of the orbit through x. If x and y lie on the same orbit, then they obviously have the same limit sets. Both $\alpha(x)$ and $\omega(x)$ are closed invariant sets, and since we have assumed X to be compact, these limit sets are compact, connected, and nonvoid (see, e.g., [NeS]).

We are now ready to state the main definitions in this section; they concern the objects which will be of primary interest to us.

---

[2] The reader having difficulty proving these statements and similar ones given below, should consult [NeS], for example.

Definition 22.3. (a) The closure of a bounded open subset $N \subset X$ is called an *isolating neighborhood* for f if for each $x \in \partial N$, there is a $t \in \mathbf{R}$ such that $x \cdot t \notin N$.

(b)   A (closed) invariant set I is called an *isolated invariant set* if it is the maximal invariant set in some isolating neighborhood.

It is important to notice that $I \cap \partial N = \emptyset$, and thus I is the maximal invariant set in some neighborhood of itself. Also, it is clear that the intersection of isolating neighborhoods is again an isolating neighborhood.

If $N \subset X$ is the closure of an open set we let

$$\mathscr{U}(N) = \{f \in F : N \text{ is an isolating neighborhood for } f\}.$$

Since ∂N is compact and $\mathrm{Comp}(N)$ is open, it follows that $\mathscr{U}(N)$ is open in F. This allows for a natural "continuation" of isolated invariant sets of f to nearby flows.

Thus let $C(X)$ denote the collection of closed subsets of X (without topology for the moment); here is the key space.

Definition 22.4. The set $\mathscr{S} \subset F \times C(X)$ is defined by

$$\mathscr{S} = \{(f, I) : I \text{ is an isolated invariant set of } f\}.$$

We want to put a topology on $\mathscr{S}$ which is suitable for the continuation results alluded to above. Thus, if N is the closure of an open set in X, we define the map $\sigma_N : \mathscr{U}(N) \to \mathscr{S}$ by

$$\sigma_N(f) = (f, I),$$

where I is the maximal f-invariant set in N. The maps σ_N are "sections" of $\mathscr{S}$ in the sense that $\pi \circ \sigma_N : \mathscr{U}(N) \to \mathscr{U}(N)$ is the identity; here π is the projection onto the first factor. We put the topology on $\mathscr{S}$ which is generated by the sets $\sigma_N(U)$, where $U \subset \mathscr{U}(N)$ is an open subset of F.[3]

Sets in the same component of $\mathscr{S}$ are to be thought of as "continuations" of each other, and are said to be *related by continuation*. We shall see later that they share some of the same properties; notably that their "index" (as yet undefined), is the same.

Theorem 22.5. *The projection* $\pi : \mathscr{S} \to F$ *is a local homeomorphism* (*i.e.,* $(\mathscr{S}, F, \pi)$ *is a "sheaf"*[4])

---

[3] Thus the sets $\sigma_N(U)$ form a subbase for the topology in the sense that the family of all finite intersections of these sets gives a basis for the topology, i.e., every open set is a union of these basis elements.

[4] See, e.g., [Sp].

Note that this theorem tells us how to recognize when two points $\sigma_N(f)$ $= (f, I)$ and $\sigma_N(f') = (f', I')$ in $\mathcal{S}$ are related by continuation. For example, if f and f' lie in the same component C of $\mathcal{U}(N)$, then since $\pi^{-1}(C)$ is a component in $\mathcal{S}$, (f, I), and (f', I') are related by continuation.

We shall need a few preliminary results before giving the proof.

Lemma 22.6. *Suppose N_1 and N_2 are isolating neighborhoods for f which isolate the same invariant set I, and $N_1 \supset N_2$. Then there is an open set V with $f \in V \subset \mathcal{U}(N_1) \cap \mathcal{U}(N_2) \subset F$ such that $\sigma_{N_1}|V = \sigma_{N_2}|V$.*

Proof. Since $I \subset \text{int } N_2$ is the maximal invariant set in N_1, it follows that for each $x \in \text{Cl}(N_1 \backslash N_2)$, there is a t such that $x \cdot t \notin N_1$. Such a statement obviously holds for all f' in some open set V in the (C-O) topology on flows, since $\text{cl}(N_1 \backslash N_2)$ is compact and $\text{Comp}(N_1)$ is open. This V meets the desired requirements. $\square$

Corollary 22.7. *Let N_i be closures of open subsets of X and let $U_i \subset \mathcal{U}(N_i)$ be open sets, $i = 1, \ldots, n$. Then if $(f, I) \in \bigcap_1^n \sigma_{N_i}(U_i)$, there exists a closed set N and an open set $V \subset \mathcal{U}(N)$ such that $(f, I) \in \sigma_N(V) \subset \bigcap_1^n \sigma_{N_i}(U_i)$.*

Proof. Let $U = \bigcap_1^n U_i$, and let $N = \bigcap_1^n N_i$. Then $U \subset \mathcal{U}(N)$ since if $f \in U$, $f \in U_i$ for all i so $f \in \mathcal{U}(N_i)$ for all i. Thus if $x \in \partial N$, $x \in \partial N_i$ for some i so $x \cdot t \notin N_i$ for some t so $x \cdot t \notin N$. Furthermore $(f, I) \in \sigma_N(U)$, since (f, I) being in $\sigma_{N_i}(U_i)$, implies $f \in U_i$ and I is maximal in N_i for each i; thus $f \in U$ and I is maximal in N. In particular, the N_i, as well as N are all isolating neighborhoods of I for the flow f. Using the last lemma, we can find $V \subset \mathcal{U}(N)$ such that $\sigma_N|V = \sigma_{N_i}|V$, $i = 1, 2, \ldots, n$, and the result follows. $\square$

Note that this corollary gives us some insight concerning the topology on $\mathcal{S}$. Thus, if (f, I) and (g, J) lie in $\bigcap_1^n \sigma_{N_i}(U_i)$, (a typical basic open set), then the corollary implies that (f, I) and (g, J) both lie in a set of the form $\sigma_N(V)$. Thus f and g belong to $\mathcal{U}(N)$ so that N is an isolating neighborhood for both f and g. Thus, two points that are "close" in this topology means that their flows are "close" in F, and their maximal invariant sets are "close," in the sense that they both lie in N.

Corollary 22.8. *$\pi : \mathcal{S} \to F$ is an open map.*

Proof. Let O be open in $\mathcal{S}$, and let $f \in \pi(O)$. Then by definition of the topology on $\mathcal{S}$, O is the union of finite intersections of sets of the form $\sigma_N(U)$, where U is open and $U \subset \mathcal{U}(N)$. Thus there exists an isolated invariant set I of f such that (f, I) is in such a finite intersection. By Corollary 22.7, O contains a set of the form $\sigma_N(V)$ which contains (f, I). Thus $\pi(O)$ must contain the open set V containing f, and $\pi(O)$ is open. $\square$

Proof of Theorem 22.5. We must show that given $s \in \mathscr{S}$, there is a neighborhood O of s such that $\pi : O \to \pi(O)$ is a homeomorphism. Since π is open, we need only find O such that $\pi : O \to \pi(O)$ is one-to-one and continuous. Thus let $s = (f, I)$ and let N be an isolating neighborhood of I (for f). Choose V to be an open set containing f with $V \subset \mathscr{U}(N)$. Let $O = \sigma_N(V)$; then if $s_i = (f_i, I_i)$ is in $\sigma_N(V)$ for $i = 1, 2$, and $f_1 = \pi(s_1) = \pi(s_2) = f_2$, it follows that $I_1 = I_2$ since they are both maximal invariant sets in N. This shows π is one-to-one on O. On the other hand, if U is open in F and $U \subset \pi(O)$, then $\pi^{-1}(U) \cap O = \{(f, I) \in \sigma_N(V) : f \in V\} \cap O$ is open in O by definition of the topology. Thus π is continuous on O and the theorem is proved. $\square$

We shall now turn to the notion of an isolating block. In preparation for this, we need the following definition.

Definition 22.9. Let $S \subset X$. Given $\delta > 0$, we define $h_\delta : S \times (-\delta, \delta) \to X$ by $(x, t) \to x \cdot t$. If for some δ, h_δ is a homeomorphism with open range in X, then S is called a *local section*.

This notion is illustrated in Figure 22.8(a). S is not a local section in Figure 22.8(b) near p, since no mapping h_δ is one-to-one near the point p. Note that if S is a local section, then if $x \in S \cdot (-\delta, \delta)$, there is a unique $t \in \mathbf{R}$, $|t| < \delta$, such that $x \cdot t \in S$.

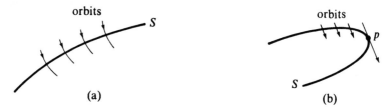

(a)   (b)

Figure 22.8

Given $S \subset X$, let $p_s : X \to [0, \infty]$ be defined by

$$p_s(x) = \sup\{t \geq 0 : x \cdot [0, t] \cap S = \varnothing\}. \tag{22.7}$$

Thus $p_s(x)$ measures "how long" it takes to get to S along the orbit through x. Obviously $p_s(x) = 0$ if $x \in S$, and conversely.

Lemma 22.10. *Let S be a local section (with respect to δ), and let $U \subset X$ be such that for some $\delta' > 0$, $U \subset X \backslash (S \cdot (0, \delta'))$. Then:*

(a) p_s *is upper semicontinuous on U.*[5]

---

[5] Recall what this means: if $\xi \in U$, and $\varepsilon > 0$ is given, there is an $\alpha > 0$ such that if $x \in \mathscr{U}$ and $|x - \xi| < \alpha$, then $p_s(x) \geq p_s(\xi) - \varepsilon$.

(b) If $x \in U$ is a point of discontinuity of p_s, then there is a $t, 0 < t < p_s(x)$ such that $x \cdot t \in \partial S \equiv \text{cl}(S)\backslash S$.

(c) If $x \in U$ and $p_s(x) < \infty$ then $x \cdot p_s(x) \in S$.

Proof. In order to prove the semicontinuity of p_s, first note that $p_s|S \cdot (-\delta, 0]$ is continuous. Namely, if π is the projection of $S \times (-\delta, 0]$ onto $(-\delta, 0]$, then $p_s|S \cdot (-\delta, 0] = \pi \circ h_\delta^{-1}$.

Since $S \cdot (-\delta, 0]$ is relatively open in U, we have only to show semicontinuity at points x where $p_s(x) \geq \delta$. Furthermore, if $p_s(x) = +\infty$, then x is automatically a point of semicontinuity. We may thus assume that $\delta \leq p_s(x) < \infty$.

Now given $\varepsilon > 0$ there is a $t, 0 < t - p_s(x) < \varepsilon$, with $x \cdot [0, t] \cap S \neq \varnothing$, since $t > p_s(x)$. If x' is close to x, and $x' \in U$, then using the continuity of the flow together with the fact that S is a section (so $S \cdot (-\varepsilon, \varepsilon)$ contains a neighborhood of S), we have that $x' \cdot [0, t] \cap S \cdot (-\varepsilon, \varepsilon) \neq \varnothing$. It follows that there exist $\tau, 0 < \tau < t$ and $t_1, |t_1| < \varepsilon$ such that $x' \cdot \tau = s \cdot t_1$ for some $s \in S$. Thus $x' \cdot (\tau - t_1) \in S$, and if $t' = \tau - t_1, x' \cdot t' \in S$.

Now choose $\varepsilon < \delta/2$; then since $x' \notin S \cdot (0, \delta)$ (since $x' \in U$), $x' \cdot (-\delta, 0) \notin S$ so $t' \notin (-\delta, 0)$. Furthermore, $t' > 0$ (if $t' < -\delta$, then $\tau - t_1 < -\delta$ so $\tau < t_1 - \delta < \varepsilon - \delta < 0$), and thus $t' \geq p_s(x')$. Then

$$t' - p_s(x) = (\tau - p_s(x)) - t_1 \leq t - p_s(x) + \varepsilon < 2\varepsilon,$$

and $p_s(x') \leq t' < p_s(x) + 2\varepsilon$; this shows that $p_s|U$ is upper semicontinuous.

To prove (b), suppose that $x \in U$ is a discontinuity point of p_s. Then there exist $x_n \in U, x_n \to x$ with $p_s(x_n) \to t' < p_s(x)$ (by the upper semicontinuity). Therefore $x_n \cdot p_s(x_n) \to x \cdot t' \in \text{cl}(S)$. But $x \cdot t' \notin S$ since $t' < p_s(x)$.

Finally, that $x \cdot p_s(x) \in S$ if $p_s(x) < \infty$ is obvious; this gives (c) and the proof is complete. □

We can now define an isolating block. Let B be the closure of an open subset of X, and let S^+ and S^- be disjoint local sections related to B as follows (see Figure 22.9):

(i) $[\text{cl}(S^\pm)\backslash S^\pm] \cap B = \varnothing$.

(ii) $S^- \cdot (-\delta, \delta) \cap B = (S^- \cap B) \cdot [0, \delta)$.

(iii) $S^+ \cdot (-\delta, \delta) \cap B = (S^+ \cap B) \cdot (-\delta, 0]$.

(iv) If $x \in \partial B\backslash(S^+ \cup S^-)$, then there exist $\varepsilon_1 < 0$ and $\varepsilon_2 > 0$ such that $x \cdot [\varepsilon_1, \varepsilon_2] \subset \partial B$ and $x \cdot \varepsilon_1 \in S^-, x \cdot \varepsilon_2 \in S^+$.

Definition 22.11. If B satisfies the above requirements then B is called an *isolating block* for the flow f.

Observe that both S^+ and S^- meet B only in ∂B, and that neither intersection is void unless $B = X$ and $S^+ = S^- = \varnothing$.

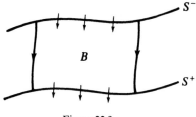

Figure 22.9

Note that according to our definitions the isolating neighborhood $\tilde{B}$ as depicted in Figure 22.10(a) is not an isolating block. However, it is "almost" one; namely, it can be modified near P and Q to give the block as depicted in Figure 22.10(b). On the other hand, given any block B, we can construct a set $\tilde{B}$ as in Figure 22.10(a) by using the flow to identify the orbit segments $\gamma \cap B$ and $\delta \cap B$ to points, without changing the maximal invariant set in B. That is, both B and $\tilde{B}$ have the same isolated invariant sets and in fact as we shall show below, they have the same Conley index. Thus we could have developed the theory using sets of the form $\tilde{B}$; we have chosen our way because it is technically a little easier.

Proposition 22.12. *Isolating blocks are isolating neighborhoods.*

Proof. If $p \in \partial B$, then the orbit through p meets S^+ or S^- (both if $p \notin S^+ \cup S^-$), and thus B is an isolating neighborhood. $\square$

If B is an isolating block, we define the following sets (cf. Figure 22.11):

$$b = \partial B,$$

$$b^+ = B \cap S^+ \subset b,$$

$$b^- = B \cap S^- \subset b,$$

$$\tau = \mathrm{cl}[b \backslash (b^+ \cup b^-)],$$

$$A^+ = \{x \in B : x \cdot \mathbf{R}_- \subset B\}, \qquad (\mathbf{R}_- = -\mathbf{R}_+),$$

$$A^- = \{x \in B : x \cdot \mathbf{R}_+ \subset B\},$$

$$I = A^+ \cap A^-,$$

$$A = A^+ \cup A^-,$$

$$a^+ = A^+ \cap b \subset b^+,$$

$$a^- = A^- \cap b \subset b^-.$$

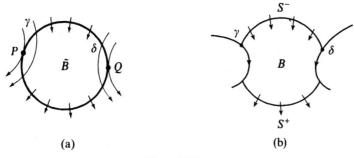

Figure 22.10

Concerning these sets, we have the following result.

Lemma 22.13. *The sets $b^{\pm}$, $A^{\pm}$, $a^{\pm}$, and I are closed, $a^{\pm} \subset \text{int } b^{\pm}$ (relative to the topology on $S^{\pm}$), and $I \subset \text{int } B$. I is an isolated invariant set, maximal in the isolating neighborhood B.*

Proof. Since B is closed, the sets $b^{\pm}$ are relatively closed in $S^{\pm}$; since $B \cap [\text{cl}(S^{\pm})\backslash S^{\pm}] = \varnothing$, b^+ and b^- are closed in X. If $x \notin A^+$, then there is a t such that $x \cdot t \notin B$; thus the same holds for x' near x. Hence $B\backslash A^+$ is relatively open in B, so A^+ is relatively closed, and so A^+ is closed in X. Similarly, A^- is closed and it follows that $a^{\pm} = A^{\pm} \cap b$ are closed, as is $I = A^+ \cap A^-$.

Now I is the union of those orbits of the flow contained in B; thus I is the maximal invariant set contained in B. Since no boundary point of B stays in B for all time, we see $I \subset \text{int}(B)$.

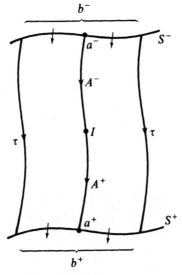

Figure 22.11

Finally, using (iii) in Definition 22.11 above, if $x \in \partial b^+$ (relative to S^+), then $x \in \tau$. Hence $x \notin a^+$ and $a^+ \subset \text{int}(b^+)$. Since a similar result is true for a^-, the proof is complete. $\quad\square$

Before stating the next lemma, we recall the notion of a strong deformation retract. Thus, we say $A \subset X$ is a *strong deformation retract* (sdr) of X if there is a continuous function $r: X \times [0, 1] \rightarrow X$ such that $r(x, t) = x$ for all $x \in A$, $(r(x, 1) \in A$ for all x, and $r(x, 0) = x$ for all x. One thinks of X being "squeezed" to A whereby points of A stay fixed at each stage of the deformation.

We can now state the following important result.

Lemma 22.14. *The sets b^+ and b^- are strong deformation retracts of the sets $B \backslash A^-$ and $B \backslash A^+$, respectively, and $b^+ \backslash a^+$ is homeomorphic to $b^- \backslash a^-$. In particular, if $p(x) \equiv p_{s^+}(x)$ (cf. (22.7)), then the function $r: (B \backslash A) \times [0, 1] \rightarrow B \backslash A^-$, defined by $r(x, t) = x \cdot (t \cdot p(x))$ provides the deformation retraction from $B \backslash A^-$ to b^+ and $r|[(b^- \backslash a^-) \times \{1\}]$ is the homeomorphism from $b^- \backslash a^-$ onto $b^+ \backslash a^+$.*

Note that since $b^+ \backslash a^+$ is homeomorphic to $b^- \backslash a^-$ under the flow, we see, in particular, that the mapping from $b^- \backslash a^-$ to $b^+ \backslash a^+$ defined by the flow is continuous. If we identify $b \backslash (S^+ \cup S^-)$ to a point (by using the flow; cf. Figure 22.10), then there cannot be any internal tangencies in the resulting set. Thus, for example, if there was an internal tangency as depicted in Figure 22.12, then the mapping from $b^- \backslash a^-$ to $b^+ \backslash a^+$ would obviously *not* be continuous at P.

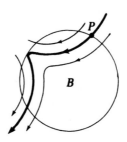

Figure 22.12

Proof of Lemma 22.14. Since $B \subset X \backslash [S^+ \cdot (0, \delta]]$, Lemma 22.10 shows that $p|B$ is upper semicontinuous, and for any point x of discontinuity there is a t', $0 < t' < p(x)$, such that $x \cdot t' \in \text{cl}(S^+) \backslash S^+$. On the other hand, since $[\text{cl}(S^+) \backslash S^+] \cap B = \emptyset$ (by (i) in Definition 22.11), the orbit segment from x to $x \cdot t$ would have to meet ∂B. From (iii) and (iv) of Definition 22.11, the intersection would have to include a point of S^+ which would mean $p(x) < t'$. This contradiction shows that $p|B$ is continuous. Also, $p|B = \infty$ exactly on the set A^- (from the definition of A^-). Thus $x \in B \backslash A^-$ implies $x \cdot p(x) \in S^+$, and r is continuous.

If $x \in B \backslash A^-$, $r(x, 1) = x \cdot p(x) \in S^+$, and since $x \in B$, $x \cdot p(x) \in b$ so $r(x, 1) \in S^+ \cap b = b^+$. If $x \in b^+ \subset S^+$, then $p(x) = 0$ so $r(x, t) = x$. Thus b^+ is an sdr of $B \backslash A^-$. That b^- is an sdr of $B \backslash A^+$ follows by symmetry and a time reversal. Finally, if $x \in b^- \backslash a^-$, then function $x \to r(x, 1)$ is continuous and maps onto $b^+ \backslash a^+$. This holds since if $x \in b^- \backslash a^-$, $r(x, 1) = x \cdot p(x) \in b^+$. If $x \cdot p(x) \in a^+ = A^+ \cap b$, then $x \cdot p(x) \in A^+$ so $x \cdot p(x) \cdot R_- \subset B$ so $x \cdot (-\delta, 0) \subset B$. But $x \in b^- = B \cap S^-$; this is a contradiction. Thus $r(x, 1) \in b^+ \backslash a^+$.

To see that $b^- \backslash a^-$ gets mapped onto $b^+ \backslash a^+$, let $y \in b^+ \backslash a^+$. We are to find $x \in b^- \backslash a^-$ such that $r(x, 1) = y$; i.e., $x \cdot p(x) = y$. Since $y \notin a^+$, there is a $t < 0$ such that $y \cdot t \notin B$. But $y \cdot [-\delta, 0) \subset B$ since $y \in b^+$. Thus there is an $x \in b^-$ such that $x \cdot t_1 = y$, for some $t_1 > 0$. By definition, $t_1 \le p(x)$, but $t_1 < p(x)$ implies $x \cdot t_1 \notin S^+$. Thus $t_1 = p(x)$ and the map is onto. Finally, if $r(x, 1) = r(y, 1)$, then $x \cdot p(x) = y \cdot p(y)$ so $x = y$. Hence $b^- \backslash a^-$ gets mapped onto $b^+ \backslash a^+$, injectively. Finally, the fact that the inverse map is continuous is obtained again by symmetry and time reversal. This completes the proof. $\square$

The following corollary is immediate.

Corollary 22.15. *If $r: (B \backslash A^+) \times [0, 1] \to B \backslash A^+$ is an sdr of $B \backslash A^+$ to b^-, then the map $r: B \backslash A^+ \to b^-$ defined by $r(x) = r(x, 1)$ is continuous.*

We shall pause here to give an application. Consider a flow which is defined in a neighborhood of a disk crossed with $[0, 1]$ as depicted in Figure 22.13 below. That is, orbits go out on the sides, and come in on the top and bottom. Let $\gamma \subset b^+$ be as depicted. Here b^- is not connected and since each end of γ lies in a different component of b^-, the corollary implies that not all of γ gets carried back to b^- in negative time. Thus there is a point on γ which stays in B for all $t \le 0$; whence $A^+ \ne \emptyset$. It follows that there is a point in B whose α-limit set lies in B, so $I \ne \emptyset$.

Local sections are usually obtained as level surfaces of some function; the relevant definition and lemma follow.

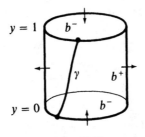

Figure 22.13

Definition 22.16. Let $U \subset X$ be open, and let $g: U \to \mathbf{R}$. The flow f on X is called *gradient-like* with respect to g in U if $x \cdot [0, t] \subset U$ and $x \neq x \cdot t$ imply $g(x \cdot t) > g(x)$.

Thus g is strictly increasing on nonconstant orbit segments in U; in particular, f can have no periodic orbits in U.

For example in $\mathbf{R}^n$ consider the equation $\dot{x} = B \nabla F(x)$, where B is a positive definite matrix, and F is a smooth real-valued function. Then $\dot{F} = \langle \nabla F, \dot{x} \rangle = \langle \nabla F, B \nabla F \rangle > 0$ if $\nabla F \neq 0$. Thus the flow is gradient-like with respect to F.

For another example, consider the equation in $\mathbf{R}^1$, $\ddot{x} = k\dot{x} + f(x)$, $k = \text{const.} > 0$. If $\dot{x} = y$, $\dot{y} = ky + f(x)$, and $F' = f$, then the equations are gradient-like with respect to $H(x, y) = y^2/2 - F(x)$, the "total energy." This holds since $\dot{H} = ky^2$.

Lemma 22.17. *Let f be gradient-like with respect to g in an open set $U \subset X$, and assume that for some constant c, the set $S \subset \{x : g(x) = c\}$ is relatively open and $\mathrm{cl}(S) \subset U$. If S contains no rest points of f, then S is a local section for f.*

Proof. Since $\mathrm{cl}(S) \subset U$, there exists $\delta > 0$ such that $\mathrm{cl}(S) \cdot [-\delta, \delta] \subset U$. Also since f is gradient-like with respect to g in U, there is no orbit segment in U which meets S in more than one point. Combining this with the fact that S contains no rest points we see that the function $h_\delta : \mathrm{cl}(S) \times [-\delta, \delta] \to \mathrm{cl}(S) \times [-\delta, \delta]$ defined by $x \to x \cdot t$ is one-to-one. Since it is also continuous, it is a homeomorphism (both spaces are compact metric). Thus $h_\delta | S \times (-\delta, \delta) \to S \cdot (-\delta, \delta)$ is a homeomorphism.

It remains to show that $S \cdot (-\delta, \delta)$ is open in X. Let $x \in S \cdot (-\delta, \delta)$, and let $x \cdot t = y \in S$, where $|t| < \delta$. Let J be an open interval containing both 0 and t, such that $\mathrm{cl}(J) \subset (-\delta, \delta)$. Note that $g | x \cdot J$ increases from below c to above c.

Now let $x_n \to x$; we shall show that for large n, $x_n \in S \cdot (-\delta, \delta)$, and this will imply $S \cdot (-\delta, \delta)$ is open. We have $x_n \cdot J \to x \cdot J$ and, in particular, $g(x_n \cdot J) \to g(x \cdot J)$. It follows that for large n, $x_n \cdot J$ meets the set $\{g(x) = c\}$ in a point, say y_n. Now the y_n's must converge to y (since $x_n \cdot t_n = y_n$, $g(y_n) = c$ so $x_{n_k} \cdot t_{n_k} \to \bar{y} = x \cdot \bar{t}$, and $g(x \cdot t) = c = g(\bar{y}) = g(x \cdot \bar{t})$; thus $t = \bar{t}$ so $\bar{y} = x \cdot \bar{t} = x \cdot t = y$). But since S is relatively open in $\{g(x) = c\}$, the points y_n must eventually be in S. Hence the points $x_n \in y_n \cdot (-\delta, \delta)$ must eventually be in $S \cdot (-\delta, \delta)$; thus $S \cdot (-\delta, \delta)$ is open in X, and the proof is complete. $\square$

We can now state the first fundamental theorem about isolated invariant sets.

Theorem 22.18. *Every neighborhood of an isolated invariant set contains one in the form of an isolating block.*

Proof. We may assume that the given neighborhood is an isolating neighborhood N for the isolated invariant set I. We define

$$A^{\pm} = \{x \in N : x \cdot \mathbf{R}_{\mp} \subset N\}, \qquad I = A^+ \cap A^-.$$

For any set Y, let $O(x, Y)$ denote the component of $x \cdot \mathbf{R} \cap Y$ which contains x. Let $O^+(x, Y)$ and $O^-(x, Y)$ denote respectively, the positive and negative half-orbits (from x) of $O(x, Y)$; both halves are to contain x.

Observe that if $x_n \to x$ from within Y, then if Y is open, $\underline{\lim}\, O^{\pm}(x_n, Y) \supset O^{\pm}(x, Y)$, and if Y is closed, $\overline{\lim}\, O^{\pm}(x_n, Y) \subset O^{\pm}(x, Y).^6$

Now let $\rho : N \to \mathbf{R}$ be continuous and nonnegative, with $\rho(x) = 0$ if and only if $x \in I$. Define $l^{\pm} : N \to \mathbf{R}$ by

$$l^+(x) = \inf\{\rho(x') : x' \in O^+(x, N)\},$$

$$l^-(x) = \inf\{\rho(x') : x' \in O^-(x, N)\}.$$

Observe that l^+ (resp. l^-) is nondecreasing (resp. nonincreasing) as t increases on orbit segments.

We define $V_\varepsilon = \{x \in N : \rho(x) < \varepsilon\}$; then we have

Lemma 22.19. l^+ *is lower semicontinuous. Furthermore, if* $x \in V_\varepsilon$ *is a point of discontinuity of* l^+, *then there is an* $x' \in \partial N$ *such that* $O^{\pm}(x', N) \cap V_\varepsilon \neq \varnothing$.

Proof. To see that l^+ is lower semicontinuous, note that the values of $l^+(x_n)$ are determined by those of ρ in a neighborhood of $O^+(x_n, N)$, while $l^+(x)$ is determined by ρ on $O^+(x, N)$. Since $O^+(x, N) \supset \overline{\lim}[O^+(x_n, N)]$, it follows that $l^+(x) \leq \underline{\lim}\, l^+(x_n)$, and thus l^+ is lower semicontinuous.

Now suppose that l^+ is discontinuous at x. Then there are $x_n \to x$ with $\lim l^+(x_n) = l^+(x) + \delta$, for some $\delta > 0$. Let $x' \in O^+(x, N)$ be such that $\rho(x') \leq \min(\varepsilon, l^+(x) + \delta/2)$; such an x' exists since $\rho(x) < \varepsilon$. Then $x' \notin \overline{\lim}\, O^+(x_n, N) \supset \underline{\lim}\, O^+(x_n, \text{int } N) \supset O^+(x, \text{int } N)$, and $x' \in V_\varepsilon$. It follows that the segment xx' meets ∂N and the proof is complete. $\square$

Define

$$T_\varepsilon = \inf\{|t| : \exists x \in \partial N, \text{ and } \exists x' \in V_\varepsilon, \text{ such that } x \cdot t = x'\};$$

then $T_\varepsilon \to \infty$ as $\varepsilon \to 0$, since otherwise some orbit would run from ∂N to I in finite time.

---

[6] Recall that $\underline{\lim}\, S_n = \{p : \text{any neighborhood of } p \text{ meets all but finitely many } S_n\}$, and $\overline{\lim}\, S_n = \{p : \text{any neighborhood of } p \text{ meets infinitely many } S_n\}$.

Lemma 22.20. *There is an open set V containing I in which $l^{\pm}$ are continuous.*

Proof. If not, there would be a discontinuity point x of l^{+} (say), in each neighborhood $V_{\varepsilon} \supset I$. Let $x \in \partial N$ be such that $O^{+}(x_{\varepsilon}, N)$ and $O^{-}(x_{\varepsilon}, N)$ both meet V_{ε}. Then $x_{\varepsilon} \cdot [-T_{\varepsilon}, T_{\varepsilon}] \subset N$. Let x be a limit point of $\{x_{\varepsilon}\}$ as $\varepsilon \to 0$. Then $x \cdot \mathbf{R} \subset N$ since $T_{\varepsilon} \to \infty$. But this contradicts the fact that N is an isolating neighborhood. □

Now define $\tilde{l}(x) = \max(l^{-}(x), l^{+}(x))$ for $x \in V$. Then $\tilde{l}$ is nonnegative and continuous on V and $\tilde{l}(x) = 0$ if and only if $x \in I$. Also $\tilde{l} \mid O(x, V)$ is:

    (a) $\equiv 0$ if $x \in I$,
    (b) $= l^{+}$ (hence nondecreasing) if $x \in A^{+}$,
    (c) $= l^{-}$ (hence nonincreasing) if $x \in A^{-}$,
    (d) $= l^{-}$ on an initial segment, and $= l^{+}$ on an (overlapping) final segment, if $x \notin A = A^{-} \cup A^{+}$.

Choose $c > 0$ such that $\tilde{l} > c$ in some neighborhood of ∂V, and let $W = \{x : \tilde{l}(x) < c\}$. Define $l : \text{cl}(V) \to \mathbf{R}$ by $l \mid W = \tilde{l} \mid W$, and $l = c$ otherwise.

In view of (a)–(d) above, $O^{\pm}(x, \text{cl}(W)) = \text{cl } O^{\pm}(x, W)$, and this enables us to write the right-hand equality in the following definitions of the functions k^{+} and $k^{-} : \text{cl}(W) \to [0, \infty]$; namely

$$k^{+}(x) \equiv \int_{O^{-}(x, \text{cl}(W))} (c - l(x \cdot s)) \, ds = \int_{O^{-}(x, W)} (c - l(x \cdot s)) \, ds,$$

$$k^{-}(x) \equiv \int_{O^{+}(x, \text{cl}(W))} (c - l(x \cdot s)) \, ds = \int_{O^{+}(x, W)} (c - l(x \cdot s)) \, ds.$$

Let $k(x) = \min(k^{+}(x), k^{-}(x))$.

Lemma 22.21. k and $k^{\pm}$ are continuous on $\text{cl}(W)$ and nonzero on W. k^{+} (resp. k^{-}) is strictly increasing (resp. decreasing) as a function on the orbit segments in W. $k^{+}(x) = \infty$ if and only if $x \in A^{+}$, while $k^{-}(x) = \infty$ if and only if $x \in A^{-}$. $k(x) = \infty$ if and only if $x \in I$, while $k(x) = 0$ if and only if $x \in \partial W$.

Proof. Let $x_{n} \to x$; using the facts that

$$O^{+}(x, W) \subset \underline{\lim} \, O^{+}(x_{n}, W) \subset \overline{\lim} \, O^{+}(x_{n}, \text{cl}(W)) \subset O^{+}(x, \text{cl}(W)),$$

$c - l > 0$ in W, and l is continuous on W, we find

$$\int_{O^{+}(x, W)} (c - l(x \cdot s) \, ds \leq \underline{\lim} \int_{O^{+}(x_{n}, W)} (c - l(x \cdot s)) \, ds$$

$$\leq \overline{\lim} \int_{O^{+}(x_{n}, \text{cl } W)} (c - l(x \cdot s)) \, ds \leq \int_{O^{+}(x, \text{cl } W)} (c - l(x \cdot s)) \, ds.$$

Since the two ends are equal, k^- must be continuous; similarly, k^+ and thus k are continuous. Also, k^+ and k^- can be differentiated along orbit segments; the first has derivative $c - l > 0$ while the second has derivative $-(c - l) < 0$. $k^+|_{A^+}$ is infinite, since if $x \in A^+$, $O^-(x, W) = x \cdot \mathbf{R}^-$. Similarly, $k^-|_{A^-}$ is infinite. Of course, $x \in I$ if and only if $x \in A^+ \cap A^-$ so $k = \min(k^+, k^-)$ is infinite on I. If $x \in \partial W$, then either $O^+(x, \mathrm{cl}(W))$ or $O^-(x, \mathrm{cl}(W))$ consists of just one point, depending on whether $l^+ > l^-$ or $l^- > l^+$ at x. In the first case, since l^+ is nondecreasing on V, x must leave $\mathrm{cl}(W)$ before re-entering W in the forward direction. A similar argument is valid for the backward direction if $l^- > l^+$. In any case, either $k^-(x) = 0$ or $k^+(x) = 0$, so $k(x) = 0$. This completes the proof of the lemma.   □

Finally, we can describe an isolating block. To this end, let $\varepsilon > 0$ be given, and define

$$\tilde{B} = \{x : k(x) \geq \varepsilon \text{ and } \sup k | O(x, W) \geq 2\varepsilon\}.$$

Lemma 22.22. *Let* $B = \mathrm{cl}(\mathrm{int}\,\tilde{B})$; *then* B *is an isolating block.*

Proof. Define

$$S^- = \{x : k(x) = k^+(x) = \varepsilon \text{ and } k^+(x) < k^-(x)\},$$

$$S^+ = \{x : k(x) = k^-(x) = \varepsilon \text{ and } k^-(x) < k^+(x)\}.$$

By the last lemma, the flow is gradient-like with respect to k^+ and k^- in W. Also, $\mathrm{cl}(S^+)$ and $\mathrm{cl}(S^-)$ are contained in W ($k = \varepsilon$ on the closure). Furthermore, the sets S^- and S^+ are relatively open in the sets $\{k^+ = \varepsilon\}$ and $\{k^- = \varepsilon\}$, respectively, since the sets $\{k^+ < k^-\}$ and $\{k^- < k^+\}$ are open. Thus S^+ and S^- are local sections. Since $k^+ = \varepsilon$ on S^- but not on S^+, we see that S^+ and S^- are disjoint.

Now if $x \in \mathrm{cl}(S^-) \backslash S^-$, then $k^+(x) = k^-(x) = \varepsilon$, so $k \leq \varepsilon$ on $O(x, W)$ and $x \notin \tilde{B}$ so $x \notin B$ since $\tilde{B} \neg B$.

Choose $\delta > 0$ so that $x \in \mathrm{cl}(S^-) \cdot [-\delta, \delta]$ implies $|k^+(x) - \varepsilon| < \varepsilon/2$. If $x \in S^- \cdot (-\delta, 0)$, then $k^+(x) < \varepsilon$ and $x \notin \tilde{B}$, so $x \notin B$. If $x \in S^- \cdot [0, \delta)$, then $x \in \tilde{B}$ if and only if there is an $x' \in O(x, W)$ such that $k(x') = 2\varepsilon$. But this is equally true for the point in S^- from which x' comes. Thus $\tilde{B} \cap \{S^- \cdot [0, \delta)\} = (\tilde{B} \cap S^-) \cdot [0, \delta)$. This implies the same result for B (cf. the statement that S is relatively open in $\{g(x) = c\}$ implies that $S \cdot (-\delta, \delta)$ is open, in the proof of Lemma 22.17). The proof of the last statement (iv), in Definition 22.11 is left as an easy exercise. The proof of the lemma, and hence of Theorem 22.18 is considered complete.   □

§C. The Homotopy Index

By an "index," we will mean a function which is constant on components of $\mathcal{S}$, or in other words, a property of invariant sets which is invariant under continuation.

The index is defined in terms of isolating blocks. To show that it really is a function on $\mathcal{S}$, we shall show that it is independent of the block. In a later section, we will show that the index is invariant under continuation.

For the state of completeness, and for the convenience of the reader, we shall review some definitions and results from topology.

A *topological pair* is an ordered pair (X, A) of spaces such that A is a closed subspace of X. If $A = \varnothing$, (X, A) will sometimes be written X.

A map $f : (X, A) \to (Y, B)$ is a continuous function from X into Y which takes A into B. 1_X will denote the identity map from (X, A) to (X, A) for any A.

The pair $(X, A) \times (Y, B)$ is $[X \times Y, (X \times B) \cup (A \times Y)]$. From now on, J will denote the unit interval.

Let $f_0, f_1 : (X, A) \to (Y, B)$; $f_0 \sim f_1$ means that there exists a continuous function $F : (X, A) \times J \to (Y, B)$ such that $F(x, 0) = f_0(x)$, and $F(x, 1) = f_1(x)$. We think of "continuously deforming" f_0 to f_1 through the maps $f_t : x \to F(x, t)$. It is easy to check that $\sim$ is an equivalence relation.

Let (X, A) and (Y, B) be pairs; then $(X, A) \sim (Y, B)$ means that there are maps $f : (X, A) \to (Y, B)$ and $g : (Y, B) \to (X, A)$ such that $f \circ g \sim 1_Y$ and $g \circ f \sim 1_X$. (It is instructive to compare this with the statement that if X and Y are groups, and there exist homomorphisms $f : X \to Y$, and $g : Y \to X$ such that $g \circ f$ and $f \circ g$ are isomorphisms, then f and g are also isomorphisms.)

It is straightforward to check that this last relation also is an equivalence relation. We call the equivalence classes *homotopy equivalence classes*, and we say that all elements in a given such class are *homotopically equivalent*.

Lemma 22.23. *If A is a strong deformation retract of X, then $A \sim X$. The inclusion $A \subset X$ gives a homotopy equivalence.*

Proof. Let $r : X \times J \to X$ be the deformation retraction and let i denote the inclusion $A \subset X$. Define $f : X \to A$ by $f(x) = r(x, 1)$. Then $f \circ i = 1_A$ by definition of r. Also $(i \circ f)(x) = r(x, 1)$, and r is the homotopy between $g \circ f$ and 1_X. The second statement follows from this. $\quad\square$

We recall the notion of a pointed space; cf., Chapter 12, §D. Thus a pointed space (X, x_0) is a pair where x_0 is a point in X. Suppose (X, A) is a pair; let $\sim$ be the equivalence relation on X defined by $x \sim y$ if and only if $x = y$ or x and y both are in A. Let $X/\sim$ denote the set of equivalence classes $[x]$ and let $\pi : X \to X/\sim$ be the projection $x \to [x]$. A topology on $X/\sim$ is obtained by defining $U \subset X/\sim$ to be open if there is an open set V in X such that $V \cap A \neq \varnothing$ or $V \supset A$, and $\pi(V) = U$.

The pointed space X/A is defined to be the pair $(X/\sim, [A])$, provided $A \neq \varnothing$. If $A = \varnothing$, X/A means the pair $(X \amalg x_0, x_0)$, where x_0 is a point, and $X \amalg x_0$ denotes disjoint union; i.e., $x_0 \notin X$.

Lemma 22.24. *Let $f: (X, A) \to (Y, B)$, and define $[f]: X/A \to Y/B$ by $[f]([x]) = [f(x)]$. Then $[f]$ is well defined and continuous.*

Proof. Since $f(A) \subset B$, $[f]$ is well defined. Also if V is open in Y/B, then there is an open set $\tilde{V} \subset Y$ such that $\pi(\tilde{V}) = V$ and either $\tilde{V} \supset B$ or $\tilde{V} \cap B = \varnothing$. Let $\tilde{U} = f^{-1}(\tilde{V})$; then $\tilde{U}$ is open in X and either $\tilde{U} \supset A$ or $\tilde{U} \cap A = \varnothing$. Thus $\pi(\tilde{U}) \equiv U$ is open in X/A, and $U = [f]^{-1}(V)$. $\square$

Now suppose that f_0 and f_1 are homotopic, i.e., $f_0 \sim f_1$ and both map (X, A) into (Y, B). Let $F: (X, A) \times J \to (Y, B)$ be the homotopy. An argument similar to the last one shows that the mapping $[F]: (X/A) \times J \to Y/B$, defined by $([x], t) \to [F(x, t)]$ is well defined and continuous. Thus $[f_0] \sim [f_1]$, and this proves the next lemma.

Lemma 22.25. *If $(X, A) \sim (Y, B)$, then $X/A \sim Y/B$.*

From now on, X/A denotes the homotopy class of X/A.

For two pointed spaces (X, x_0) and (Y, y_0), their "*sum*" or *wedge* is the space $(X, x_0) \vee (Y, y_0) \equiv X \amalg Y/\{x_0, y_0\} \equiv ((X, x_0) \cup (Y, y_0), (x_0, y_0))$, obtained by "glueing" together the pointed spaces at their distinguished points, see Figure 22.6. The next lemma shows that this operation is well defined on homotopy equivalence classes.

Lemma 22.26. *If $(X, x_0) \sim (X', x_0')$, and $(Y, y_0) \sim (Y', y_0')$, then $(X, x_0) \vee (Y, y_0) \sim (X', x_0') \vee (Y', y_0')$.*

Proof. We have

$$(X \amalg Y, (x_0, y_0)) = ((x_0 \times Y) \cup (X \times y_0), (x_0, y_0))$$
$$\sim ((x_0' \times Y') \cup (X' \times y_0'), (x_0', y_0')) = (X' \amalg Y', (x_0', y_0')),$$

and the result follows from Lemma 22.25. $\square$

We define the *one-point pointed space* $\bar{0}$ by $\bar{0} = x_0/x_0$. Concerning this, we have the next lemma.

Lemma 22.27. *If $X/A \vee Y/B = \bar{0}$, then $X/A = \bar{0}$, and $Y/B = \bar{0}$.*

Proof. $X/A \sim \bar{0}$ if and only if there are maps $f_1: \bar{0} \to X/A$, $f_2: X/A \to \bar{0}$ with both $f_1 \circ f_2 \sim \mathrm{id}$, and $f_2 \circ f_1 \sim \mathrm{id}$.

Now if $X/A \vee Y/B = \bar{0}$, then there is a map $F : (X/A \vee Y/B) \times [0,1] \to X/A \vee Y/B$ with $F(z,0) = z$, $F(z,1) = [A] \times [B]$ for all $z \in X/A \vee Y/B$. If j is the inclusion map $X/A \times [0,1] \subset (X/A \vee Y/B) \times [0,1]$, and k is the projection $(X/A \vee Y/B) \to X/A$, then

$$k \circ F \circ j : X/A \times [0,1] \to X/A$$

is easily seen to be a homotopy from the identity to the constant map, whence $X/A \sim \bar{0}$; similarly $Y/B \sim \bar{0}$. $\quad\square$

We are now ready to show that the Conley index is independent of the particular isolating block containing it. In order to do this, we shall introduce the notions of "shaves" and "squeezes."

Let B be an isolating block for the invariant set I and let $Y \subset B$. We define

$$O(Y, B) = \bigcup_{y \in Y} O(y, B),$$

where we recall from the last section that $O(y, B)$ is the component of $(y \cdot \mathbf{R}) \cap B$ containing y. Suppose that $U \subset b^-$ is an open neighborhood of a^-, and let $Y = b^- \backslash \mathrm{cl}(U)$. Then the set $B_1 = B \backslash O(Y, B)$, obtained by "shaving off" Y is a block for I with $b_1^- \subset \mathrm{cl}(U)$; see Figure 22.14(a). Note that

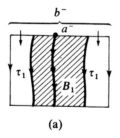
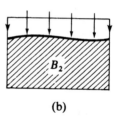

(a)   (b)

Figure 22.14

$B_1 \cup b^-$ is a deformation retract of B (take a neighborhood U of B_1, outside of U, let the flow carry points up to b^-, and on $U \backslash B_1$, make the map continuous—one must be careful near τ_1). Hence the inclusion $(B_1 \cup b^-, b^-) \subset (B, b^-)$ is a homotopy equivalence. Since the inclusion $B_1/b_1^- \subset (B_1 \cup b^-)/b^-$ is a homeomorphism, it follows that the inclusion $B_1/b_1^- \subset B/b^-$ is a homotopy equivalence; similarly, the inclusion $B_1/b_1^+ \subset B/b^+$ is a homotopy equivalence. Thus we have shown that the block B_1 obtained by "shaving" B has the same "index" as the block B_1; see Definition 22.30.

Now let $x \in B \backslash A$; then there are numbers $t_1^x > 0$, $t_2^x < 0$ such that $x \cdot t_i^x$ $\in \partial B$; let $t_x = t_1^x - t_2^x$. If $x \in A$ let $t_x = +\infty$. We define $T = \inf\{t_x : x \in B\}$ $\equiv |O(x, B)|$. Then if $T' < T$, the set $B_2 = B \backslash b^- \cdot [0, T')$, obtained by "*squeezing*" B is a block, and b^-, as well as $b_2^- = b^- \cdot T'$ are strong deformation retracts of $B^- \equiv b^- \cdot [0, T']$; see Figure 22.14(b). It follows that the inclusions $(B_2, b_2^-) \subset (B, B^-) \supset (B, b^-)$ are homotopy equivalences, and that (B_2, b_2^-) $\sim (B, b^-)$ and $B_2/b_2^- \sim B/b^-$; similarly, $B_2/b_2^+ \sim B/b^+$. Thus the block B_2 obtained by "squeezing" B also has the same "index" as B. We call T as defined above the *amount of the squeeze*.

Lemma 22.28. *Let B be an isolating block, and let there be given neighborhoods V of A and W of I. Then there exist blocks $B_1 \subset V$ and $B_2 \subset W$ such that B_1 is obtained by shaving B, and B_2 is obtained by squeezing B_1.*

Proof. First note that there is a $T > 0$ such that $A^+ \cdot (-T)$ and $A^- \cdot T$ are contained in $W' \equiv W \cap V$, since $\bigcap \{A^- \cdot t : t > 0\} = I = \bigcap \{A^+ \cdot t : t < 0\}$. Choose a neighborhood $U^- \subset b^-$ of a^- such that $V^- \equiv U^- \cdot [0, T) \subset V$, and a neighborhood $U^+ \subset b^+$ of a^+ with $V^+ \equiv U^+ \cdot (-T, 0] \subset V$. Let $V' = V^+ \cup W' \cup V^-$, and note that $V' \subset V$ is a neighborhood of A.

Now observe that if $x_n \in B$, and $x_n \to a \in A$, then $\overline{\lim} \, O(x_n, B) \subset A$; namely, as $x_n \to a, |O(x_n, B)| \to \infty$, and this implies that for $x \in \overline{\lim} \, O(x_n, B), |O(x, B)| = \infty$ so $x \in A$. Thus we can choose $U \subset b^-$ to be an open set about a^- so small that if $x \in \text{cl}(U)$, $O(x, B) \subset V'$ and $|O(x, B)| > 2T$. Let $Y = b^- \backslash \text{cl} \, U$; then $B_1 \equiv B \backslash O(Y, B) \subset V'$ and $B_2 \equiv B_1 \backslash b_1^- \cdot [0, T) \cup b^+ \cdot (-T, 0] \subset U$, the last inclusion being true since $x \in U^-$ (or U^+) forces $x \cdot [-T, 0] \not\subset B$ (or $x \cdot [0, T] \not\subset B$). This completes the proof. $\square$

Here is the main theorem in this section.

Theorem 22.29. *Let B_1 and B_2 be isolating blocks for I. Then $B_1/b_1^- \sim B_2/b_2^-$, and $B_1/b_1^+ \sim B_2/b_2^+$.*

Proof. We shall give the details only for the first statement; the proof for the second is similar. By a shave and a squeeze, we can assume that $B_2 \supset B_1$; thus $A_2 \supset A_1$. Let V be a neighborhood of A_2 such that different components of $A_2 \cap B_1$ are contained in different components of $V \cap B_1$. Let W be the union of those components of $V \cap B_1$ which meet A_1, and let $B_3 \subset V$ be obtained by shaving B_2, and having the property that it can be squeezed to B_4 in W. Let T be the amount of the squeeze.

Now let $x \in b_4^-$, so $x \cdot (-T) \in b_2^-$, and x is the unique point on $O(x, B_2)$ $\supset O(x, B_1)$ with this property. It follows that the orbit segments $O^-(x, B_1)$ meet B_4 only in x. These segments are precisely the paths traversed by x under the deformation retract of $B_1 \backslash A_1^+$ to b_1^-; in particular, $B^- \equiv \bigcup \{O^-(x, B_1) : x \in b_4^-\}$ is homeomorphic to $b_4^- \times J$; similarly $B^+ \equiv \bigcup \{O^+(x, B_1) : x \in b_4^+\}$ is a product.

It follows then that $B_5 \equiv B_4 \cup B^+ \cup B^-$ is a block and it is in fact obtained from B_1 by shaving off the set $Y = b_1^- \backslash (B_5 \cap b_1^-)$; namely, if $x \in B_5$, $O(x, B_5) = O(x, B_1)$ so B_5 is obtained from B_1 by removing a collection of orbit segments $O(x', B_1)$.

Since $(B_4, b_4^-) \subset (B_5, B^-)$, and $(B_5, b_5^-) \subset (B_5, B^-)$ are homotopy equivalences, we have $(B_4, b_4^-) \sim (B_5, b_5^-)$ and thus

$$B_2/b_2^- \sim B_4/b_4^- \sim B_5/b_5^- \sim B_1/b_1^-,$$

as desired. □

Definition 22.30. Let I be an isolated invariant set and let B be any isolating block for I; we define the *Conley Index* of I by $h(I) = [B/b^+]$; i.e., the homotopy equivalence class of the pointed space B/b^+.

The last theorem shows that $h(I)$ is well defined. Our next result is the important "addition" property of the Conley index.

Theorem 22.31. *If I_1 and I_2 are disjoint isolated invariant sets, then $I = I_1 \amalg I_2$ is an isolated invariant set, and*

$$h(I_1 \amalg I_2) = h(I_1) \vee h(I_2) \tag{22.8}$$

Proof. A block B for $I_1 \amalg I_2$ is obtained by taking the union of two disjoint blocks B_1 and B_2 for I_1 and I_2, respectively. Then $B/b^+ = B_1/b_1^+ \vee B_2/b_2^+$, and this is well defined in view of Lemma 22.26. □

As an application of this theorem, we shall give a proof of a result stated in Chapter 12, §C. Namely, suppose that (12.18) holds; then f cannot have exactly two critical points both of which are relative minima. Namely, (12.18) implies that the set S of bounded orbits of the gradient flow $\dot{x} = \nabla F(x)$ is compact and thus S is an isolated invariant set having index Σ^n, the index of a repellor. If the only critical points in S were the two relative minima, then these would have to comprise the entire set S (any bounded orbit of a gradient flow must tend to rest points both time directions). Then (22.8) would give $\Sigma^n = \Sigma^0 \vee \Sigma^0$, the desired contradiction.

Next, note that $h(\varnothing) = \bar{0}$; thus we have the following useful result.

Theorem 22.32. *If N is an isolating neighborhood for the isolated invariant set S, and if $h(S) \neq \bar{0}$, then $S \neq \varnothing$; i.e., N contains a complete orbit.*

Finally we end this section with a useful application of (22.8); the so-called "connecting orbit" theorem. First, we recall that a flow is called *gradient-like* with respect to a real-valued function F, if F increases on nonconstant orbits of the flow (Definition 22.16).

Theorem 22.33. *Let $\dot{x} = f(x)$ be gradient-like in an isolating neighborhood N, and let N contain precisely two rest points x_1, x_2, of f, not both of which are degenerate.*[7] *Let S(N) be the maximal invariant set in N. If $h(S(N)) = \bar{0}$, then there is an orbit of f connecting the two rest points.*

Proof. It suffices to show that $S(N)$ contains an orbit γ different from x_1 and x_2; γ would then tend to different rest points in both time directions, due to the gradient-like nature of the flow. If $S(N) = \{x_1, x_2\}$, then (22.8) gives $\bar{0} = h(x_1) \vee h(x_2)$. But $h(x_1) \vee h(x_2) \neq \bar{0}$, by nondegeneracy. This is a contradiction, and the proof is complete. $\square$

NOTES

The subject matter in §A is adapted from Conley's basic monograph [Cy 2], with some modifications and expansions. The material in §B and §C is taken from unpublished class notes of Conley; here too I have made minor changes. The papers of Conley and Easton [CE], Churchill [Chr], and Montgomery [Mt], contain the original proofs. Theorem 22.33 is found in Conley and Smooler [CS 3].

---

[7] Recall that this means that zero is not an eigenvalue of the linearized matrix df at the critical point.

Chapter 23

Index Pairs and the Continuation Theorem

In this chapter we shall consider the Conley index from a more general point of view, one which allows us to apply the theory to a wide variety of equations including in particular, systems of reaction–diffusion equations. For such equations, it is not at all clear that the equations even define a flow. To get around such problems, we introduce the concept of a local flow and develop the theory in this setting. Roughly speaking, a local flow is a subset of the underlying space which is locally invariant for positive time; one thinks of a subspace of a function space, say L_2, which is invariant under the equations for small $t > 0$.

The new idea in this chapter is the concept of an "index pair" (N_1, N_0), in an isolating neighborhood N, which generalizes the notion of the isolating block given in the last chapter. Thus, in particular, if B is an isolating block, and b^+ is the exit set in ∂B, then (B, b^+) is an example of an index pair. If N is a given isolating neighborhood and S is the maximal invariant set in N we shall show three things:

(i) index pairs in N exist;
(ii) if (N_1, N_0) and $(\tilde{N}_1, \tilde{N}_0)$ are index pairs in N then $N_1/N_0 \sim \tilde{N}_1/\tilde{N}_0$ so that the Conley index $h(S) = [N_1/N_0]$ is well-defined as the homotopy equivalence class of N_1/N_0; and
(iii) $h(S)$ is "invariant under continuation."

The first two properties are analogous to what we have proved in Chapter 22 for flows; namely, that isolating blocks exist, and that the Conley index of S doesn't depend on the particular block which isolates it. The third property was alluded to in the last chapter, but never proved; the proof will be given in §C.

The development given here has been done with an eye towards the applications in the next chapter. Thus we introduce the concept of a Morse decomposition of an isolated invariant set, and construct our index pairs at this level of generality. This gives us the results discussed above, but in addition, it makes applicable powerful algebraic techniques. Thus, the Morse decompositions allow us to construct in a fairly standard and straightforward way, exact sequences of cohomology groups involving the index of S with the indices of the elements of the Morse decompositions. One use of such structures is to calculate indices, but there are more subtle applications as we shall show in the next chapter.

The final section is devoted to a few miscellaneous remarks. Here we have tried to both clarify some aspects of the theory, as well as to "set things up" for Chapter 24.

§A. Morse Decompositions and Index Pairs

Let Γ be a metric space, and let F be a flow on Γ. As before, we denote $F(x, t)$ by $x \cdot t$. If N is a subset of Γ, we denote by $I(N)$ the maximal invariant set in N:

$$I(N) = \{\gamma \in N : \gamma \cdot \mathbf{R} \subset N\}. \tag{23.1}$$

Definition 23.1. Assume that I is a compact invariant set in Γ. A *Morse decomposition* of I is a finite collection $\{M_i : 1 \leq i \leq n\}$, of disjoint compact invariant subsets of I which can be ordered $(M_1, M_2, \ldots, M_n)$ in such a way that if $\gamma \in I \setminus \cup \{M_i : 1 \leq i \leq n\}$, then there are indices $i < j$ such that $\omega(\gamma) \subset M_i$ and $\alpha(\gamma) \subset M_j$. Such an ordering will be called *admissible*. The elements M_i of the Morse decomposition of I will be called *Morse sets* of I.

We pause to give some examples.

EXAMPLE 1. Consider the equations in $\mathbf{R}^2$:

$$\dot{x} = y, \qquad \dot{y} = \theta y - x(x - \tfrac{1}{3})(1 - x).$$

For various values of θ, the complete set of bounded orbits is depicted in Figure 23.1 below. In all four cases, this set is an isolated invariant set.

(a) $\theta = 0$.

(b) $1 \gg \theta > 0$.

(c) $\theta = \theta^*$.

(d) $\theta \gg 1$.

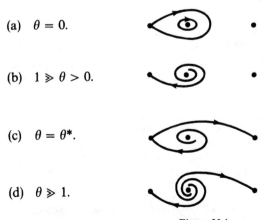

Figure 23.1

In (a), the Morse decompositions are:

$$\{\,\langle\!\!\!\bigcirc\!\!\!\rangle,\cdot\,\} \quad \text{and} \quad \{\,\cdot,\langle\!\!\!\bigcirc\!\!\!\rangle\,\}.$$

In (b), they are:

$$\{\,\bigcirc,\cdot\,\}, \quad \{\,\cdot,\bigcirc\,\}, \quad \{0,\tfrac{1}{3},1\}, \quad \{0,1,\tfrac{1}{3}\}, \quad \text{and} \quad \{1,0,\tfrac{1}{3}\}.$$

In (c), they are:

$$\{\,\cdot,\bigcirc\,\}, \quad \{1,0,\tfrac{1}{3}\}, \quad \text{and} \quad \{\,\longrightarrow\,,\tfrac{1}{3}\}.$$

In (d), they are:

$$\{0,1,\tfrac{1}{3}\} \quad \text{and} \quad \{1,0,\tfrac{1}{3}\}.$$

For an admissible ordering $(M_1,\ldots,M_n)$ of a Morse decomposition I we define subsets $M_{ij} \subset I, j \geq i$, by

$$M_{ij} = \{\gamma \in I: \omega(\gamma) \text{ and } \alpha(\gamma) \text{ are in } M_i \cup M_{i+1} \cup \cdots \cup M_j\}; \quad (23.2)$$

in particular, $M_{jj} = M_j$. For example, in (b) above, for the Morse decomposition $\{0,\tfrac{1}{3},1\}$: $M_{12} = I\backslash\{1\}$, $M_{23} = \{\tfrac{1}{3},1\}$, and $M_{13} = I$.

The following statement follows at once from the definition.

Lemma 23.2. *Let* $(M_1,\ldots,M_n)$ *be an admissible ordering of a Morse decomposition of* I; *then* $(M_1,\ldots,M_{i-1},M_{ij},M_{j+1},\ldots,M_n)$ *is an admissible ordering of a Morse decomposition of* I. *Also,* $(M_i,M_{i+1},\ldots,M_{j-1},M_j)$ *is an admissible ordering of a Morse decomposition of* M_{ij}.

Our aim is to introduce algebraic invariants of the Morse sets of a Morse decomposition of I, and to relate them to algebraic invariants of I. These invariants will depend on the flow near I. In order that we can apply these methods to partial differential equations, we introduce the notion of a local flow.

Definition 23.3. Let X be a locally compact subset of Γ. X is called a *local flow* if for every $\gamma \in X$, there is an $\varepsilon > 0$, and a neighborhood $U \subset \Gamma$ of γ such that $(X \cap U) \cdot [0,\varepsilon) \subset X$. If U and ε can be chosen so that $(X \cap U)(-\varepsilon,\varepsilon) \subset X$, the local flow is called *two-sided*.

Thus for a local flow, the orbit, as well as a neighborhood of itself, is required to stay in X, locally; i.e., for small time.

EXAMPLE 2. Consider the Hamiltonian system on $\mathbf{R}^2$: $\dot{x} = H_y$, $\dot{y} = -H_x$. Then $\dot{H} = 0$, so the "energy surfaces," $H_c = \{(x, y) : H(x, y) = c\}$ are all two-sided local flows.

We wish to mention here that we can construct local flows for reaction–diffusion equations which are not (apparently!) defined on locally compact spaces. This will be demonstrated in §D where we shall construct a local flow in some function spaces.

Definition 23.4. Let $N \subset X$ be a compact subset of a local flow X. If $I(N) \subset$ int N (relative to X), then N is called an *isolating neighborhood* in X and $I(N)$ is called an *isolated invariant set*.

Proposition 23.5. *Let $S = I(N)$ be an isolated invariant set in the local flow X, and let $\{M_i\}$ be a Morse decomposition of S. Then each M_i is an isolated invariant set in X.*

Proof. By hypothesis, there is a compact set $N \supset S$ with $I(N) = S \subset$ int N. Let N_i be a compact X-neighborhood of M_i such that $N_i \cap M_k = \varnothing$ if $i \neq k$ and $N_k \subset N$; then N_i is an isolating neighborhood of M_i. If $\gamma \in I(N_i)$, then $\gamma \cdot \mathbf{R} \subset N_i \subset N$ so $\gamma \in S$. Since $\omega(\gamma)$ and $\alpha(\gamma)$ are in N_i, the only Morse set containing them is M_i. Thus by definition of Morse decompositions, $\gamma \in M_i$ so $I(N_i) = M_i \subset$ int N_i. $\quad\square$

We now turn to the important concept of index pairs for Morse decompositions; as we have remarked earlier, these will be generalizations of the pair (B, b^+) where B is an isolating block and b^+ is the exit set on ∂B. Before giving the definition we need some preliminary notions.

If $Z \subset Y \subset \Gamma$, we call Z *positively invariant relative to* Y, if $\gamma \in Z$, and $\gamma \cdot [0, t] \subset Y$, together imply that $\gamma \cdot [0, t] \subset Z$. By a *compact pair* (Z_1, Z_2), we mean an ordered pair of compact spaces with $Z_1 \supset Z_2$.

Definition 23.6. Let $S = I(N)$ be an isolated invariant set in the local flow X. A compact pair (N_1, N_0) in X is called an *index pair* for S if the following hold:

(i) $\mathrm{cl}(N_1 \backslash N_0)$ is an isolating neighborhood for S.
(ii) N_0 is positively invariant relative to N_1.
(iii) if $\gamma \in N_1$ and $\gamma \cdot \mathbf{R}_+ \not\subset N$, then there is a $t \geq 0$ such that $\gamma \cdot [0, t] \subset N_1$, and $\gamma \cdot t \in N_0$.

Note that orbits can leave N_1 only through N_0; thus N_0 does indeed play the role of b_+; in fact, if B is an isolating block for the isolated invariant set S, then it is easy to check that the pair (B, b^+) is an index pair for S.

We shall first show that index pairs do indeed exist; this is analogous to the construction of isolating blocks in the last chapter. After that, we shall show that the homotopy type of the space N_1/N_0 is independent of the index pair (N_1, N_0) and depends only on S; this too is a generalization of a result in Chapter 22. The Conley index of S will be defined as this homotopy equivalence class. Finally, we shall show that if S_1 and S_2 are isolated invariant sets which are related by continuation, then they have the same Conley index.

Concerning the existence of index pairs, we have the following more general theorem which actually gives the existence of index pairs for each element in a Morse decomposition of S; i.e., we in fact obtain a "(*Morse*) *filtration*" of an isolated invariant set.

Theorem 23.7. *Let S be an isolated invariant set and let $(M_1, M_2, \ldots, M_n)$ be an admissible ordering of a Morse decomposition of S. Then there exists an increasing sequence of compact sets (a (Morse) filtration of S),*

$$N_0 \subset N_1 \subset \cdots \subset N_n, \tag{23.3}$$

such that for any $i \leq j$, the pair (N_j, N_{i-1}) is an index pair for M_{ij}. In particular, (N_n, N_0) is an index pair for S, and (N_j, N_{j-1}) is an index pair for M_j. Furthermore, given any isolating neighborhood N of S, and any neighborhood U of S, the sets N_j can be chosen so that $\mathrm{cl}(N_n \backslash N_0) \subset U$ and each N_j is positively invariant relative to N.

The rest of this section is devoted to the proof of this theorem.

Let N be an isolating neighborhood of S, so $I(N) = S$, and for $j = 1, 2, \ldots, n$, define

$$I_j^+ = \{ \gamma \in N : \gamma \cdot \mathbf{R}_+ \subset N \text{ and } \omega(\gamma) \subset M_j \cup M_{j+1} \cup \cdots \cup M_n \},$$

$$I_j^- = \{ \gamma \in N : \gamma \cdot \mathbf{R}_- \subset N \text{ and } \alpha(\gamma) \subset M_1 \cup M_2 \cup \cdots \cup M_j \}. \tag{23.4}$$

Then I_j^+ (resp. I_j^-) stays in N in forward (resp. backward) time. Furthermore, if $i < j$, then $I_i^+ \cap I_j^- = M_{ij}$; in fact, if $\gamma \in I_i^+ \cap I_j^-$, then $\gamma \cdot \mathbf{R} \subset N$ so $\gamma \in S$. Also $\omega(\gamma) \subset M_i \cup \cdots \cup M_n$, so $\alpha(\gamma) \subset M_i \cup \cdots \cup M_n$. But $\alpha(\gamma) \subset M_1 \cup \cdots \cup M_j$; whence $\alpha(\gamma) \subset M_i \cup \cdots \cup M_j$. Similarly, $\omega(\gamma) \subset M_i \cup \cdots \cup M_j$, and thus $\gamma \in M_{ij}$. The inclusion $M_{ij} \subset I_i^+ \cap I_j^-$ is immediate.

Lemma 23.8. *The sets $I_j^\pm$ are compact.*

Proof. We break the proof up into three parts.

(a) The sets I_1^+ and I_n^- are compact: Since $I_1^+ = \{ \gamma \in N : \gamma \cdot \mathbf{R}_+ \subset N \}$, it follows that for $\gamma \notin I_1^+$, $\gamma \cdot t \notin N$ for some $t > 0$. Using the compactness of N and the continuity of the flow, there is an open neighborhood $U \subset \Gamma$ of γ such that $(U \cdot t) \cap N = \emptyset$. Thus if $\gamma \in U \cap N$, then $\gamma \notin I_1^+$, so $N \backslash I_1^+$ is open relative to N. Thus I_1^+ is compact; similarly, I_n^- is compact.

(b) Suppose $n = 2$: Let (M_1, M_2) be an admissible ordering of a Morse decomposition of S. By definition $I_2^+ \subset I_1^+$, and by (a) I_1^+ is compact. We must show that I_2^+ is closed. Thus let $\gamma_n \to \gamma$, $\gamma_n \in I_2^+$; then $\gamma \in I_1^+$ so $\omega(\gamma) \subset M_1 \cup M_2$, and we must show $\omega(\gamma) \subset M_2$. Suppose $\omega(\gamma) \subset M_1$. Since $M_1 \cap M_2 = \varnothing$ and both are compact, we can find open neighborhoods U_1 and U_2 of M_1 and M_2, respectively, with $\mathrm{cl}(U_1) \cap \mathrm{cl}(U_2) = \varnothing$. Since $\omega(\gamma_n) \subset M_2$ and $\omega(\gamma) \subset M_1$, there exist t_n' and t_n'' with $\gamma_n \cdot [t_n', \infty) \subset U_2$, and $\gamma_n \cdot t_n'' \in U_1$, $n = 1, 2, \ldots$. Thus we can find $\{t_n\}$ such that $\gamma_n \cdot [t_n, \infty) \subset N \backslash U_1$ and $\gamma_n \cdot t_n \in N \backslash (U_1 \cup U_2)$. Choose a subsequence such that $\tilde{\gamma} = \lim \gamma_n \cdot t_n$ exists. Then $\tilde{\gamma} \notin M_1 \cup M_2$ and $\tilde{\gamma} \cdot [0, \infty) \subset N \backslash U_1$ so $\omega(\tilde{\gamma}) \subset M_2$. If $\{t_n\}$ is bounded, then $\tilde{\gamma} \in \gamma \cdot \mathbf{R}$ so $\omega(\gamma) = \omega(\tilde{\gamma}) \subset M_2$, contradicting $\omega(\gamma) \subset M_1$. If $\{t_n\}$ is unbounded, then for any $t > 0$ $\tilde{\gamma} \cdot [-t, 0]$ is a limit of orbit segments $\gamma_n \cdot t_n \cdot [-t, 0] = \gamma_n \cdot [t_n - t, t_n]$. For large n, these segments lie in $\gamma_n \cdot \mathbf{R}_+ \subset N$, and thus $\tilde{\gamma} \cdot [-t, 0] \subset N$. Since this holds for each $t > 0$, $\tilde{\gamma} \cdot \mathbf{R}_- \subset N$, so $\tilde{\gamma} \cdot \mathbf{R}$ is in N so $\tilde{\gamma} \in S$. Since (M_1, M_2) is an admissible ordering of a Morse decomposition of S, $\omega(\tilde{\gamma}) \subset M_2$ implies $\tilde{\gamma} \in M_2$, and this contradicts $\tilde{\gamma} \notin M_1 \cup M_2$.

(c) The general case: Note that if $j > 1$, then I_j^+ is the set $\bar{I}_2^+$ where $\bar{I}_2^+$ corresponds to the Morse set $\bar{M}_2 = M_{jn}$ of the 2-decomposition $(\bar{M}_1 = M_{1(j-1)}, \bar{M}_2 = M_{jn})$ of S. Since a similar remark applies to I_{j-1}^- for $j \leq n$, we see that the result follows from (b). $\quad\square$

For a subset $Z \subset N$, we define the set $P(Z)$, where $Z \subset P(Z) \subset N$, by

$$P(Z) = \{\gamma \in N : \text{there exist } \gamma' \in Z, t' \geq 0 \text{ with } \gamma' \cdot [0, t'] \subset N \text{ and } \gamma' \cdot t' = \gamma\}. \tag{23.5}$$

Thus $P(Z)$ consists of those points in N which can be reached by orbit segments contained in N which "begin" in Z; $P(Z)$ is the set "swept-out" by Z. Obviously $P(Z)$ is positively invariant relative to N.

Lemma 23.9. *Let V be any Γ-neighborhood of I_j^-. Then there is a compact X-neighborhood Z of I_j^- such that $P(Z)$ is compact and $P(Z) \subset V$.*

Proof. I_j^- and I_{j+1}^+ being compact and disjoint implies that we can find open X-neighborhoods V^+ of I_{j+1}^+ and V^- of I_j^- such that $V^- \subset V^+$, and $\mathrm{cl}(V^+) \cap \mathrm{cl}(V^-) = \varnothing$.

Claim. *There is a $\tilde{t} > 0$ such that if $\gamma \in N \backslash V^-$, the arc $\gamma \cdot [-\tilde{t}, 0]$ contains a point in V^+ or a point in $\Gamma \backslash N$; i.e., $\gamma \in N \backslash V^-$ implies $\gamma \cdot [-\tilde{t}, 0] \not\subset N \backslash V^+$.*

To see this, note that if $\gamma \in N \backslash V^-$ and $\gamma \cdot \mathbf{R}_- \not\subset N$ then there is a t such that $\gamma \cdot (-t) \notin N$. If $\gamma \cdot \mathbf{R}_- \subset N$, then since $\gamma \notin I_j^-$, $\alpha(\gamma) \subset M_{j+1} \cup \cdots \cup M_n \subset I_{j+1}^+ \subset V^+$ and there is a t' with $\gamma \cdot (-t') \in V^+$. In either case there is a neighborhood W of γ such that $W \cdot t$, or $W \cdot t'$ lies in the complement of $N \backslash V^+$, and the claim follows since $N \backslash V^-$ is compact.

In order to define Z, let $\gamma \in I_j^-$; then $\gamma \cdot \mathbf{R}_- \subset I_j^- \subset V^-$, and we can find a compact neighborhood C_γ of γ such that $C_\gamma \cdot [-\bar{t}, 0] \subset V^-$. Since I_j^- is compact, a finite collection of such C_γ's cover I_j^-, and we let Z be their union. Z is a compact neighborhood of I_j^-, and the claim now is that $P(Z) \subset V^-$. Suppose not; then there is a $\gamma \in P(Z)$ with $\gamma \notin V^-$. By definition $\gamma = \gamma' \cdot t'$ for some $\gamma' \in Z$ and some $t' \geq 0$ with $\gamma' \cdot [0, t'] \subset N$. Pick τ such that $\gamma' \cdot [0, \tau) \subset V^-$ and $\tilde{\gamma} \equiv \gamma' \cdot \tau \in N \backslash V^-$. Then $\tilde{\gamma} \cdot [-\tau, 0) \subset V^-$ and $\tilde{\gamma} \cdot (-\tau) = \gamma' \in Z$. By definition of Z, $\gamma' \cdot [-\bar{t}, 0] \subset V^-$ so that $\tilde{\gamma} \cdot [-(\bar{t} + \tau), 0] \subset \mathrm{cl}(V^-) \subset N \backslash V^+$, and this contradicts our above claim. Thus $P(Z) \subset V^-$.

It remains to show $P(Z)$ is compact; this will be done by showing that the complement of $P(Z)$ in N is open. This let $\gamma \in N \backslash P(Z)$; then $\alpha(\gamma) \not\subset M_1 \cup \cdots \cup M_j$ so there is a t such that $\gamma \cdot (-t) \notin N \backslash V^+$. Let $t_1 = \sup\{t \geq 0 : \gamma \cdot [-t, 0] \subset N \backslash V^+\}$; then $\gamma \cdot [-t_1, 0] \subset N \backslash V^+$ since $N \backslash V^+$ is closed. Also $\gamma \cdot [-t_1, 0] \cap Z = \varnothing$ since $\gamma \notin P(Z)$. Using the compactness of Z, we can find $t_2 > t_1$ with $\gamma \cdot [-t_2, 0] \cap Z = \varnothing$ and $\gamma \cdot (-t_2) \notin N \backslash V^+$. Using the continuity of the flow and the compactness of $N \backslash V^+$, we can find a neighborhood W of γ such that $W \cdot [-t_2, 0] \cap Z = \varnothing$, and $W \cdot (-t_2) \cap (N \backslash V^+) = \varnothing$. Since $P(Z) \subset V^-$, we conclude that if $\gamma' \in W$, then there is no orbit segment from Z to γ', which lies in $N \backslash V^+$. Thus $\gamma' \notin P(Z)$ and the complement of $P(Z)$ in N is open so $P(Z)$ is compact. $\quad\square$

The constructions in this lemma are schematically illustrated in Figure 23.2.

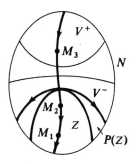

Figure 23.2

We shall now construct the index pair (N_n, N_0) for $S = I(N)$. From their definitions $I_1^+ \cap I_n^- = S \subset \mathrm{int}(N)$. Thus since I_1^+ and I_n^- are compact, if U is a neighborhood of S, we can choose open neighborhoods U^+ of I_1^+ in N and U^- of I_n^- in N such that $\mathrm{cl}(U^+ \cap U^-) \subset U \cap (\mathrm{int}\, N)$. We define

$$N_0 = P(N \backslash U^+); \tag{23.6}$$

then N_0 is positively invariant relative to N. We shall show that N_0 is compact.

Since $N \setminus U^+$ is compact and disjoint from $I_1^+ = \{\gamma \in N : \omega(\gamma) \subset M_1 \cup \cdots \cup M_n\}$, there is a $\tilde{t} > 0$ such that $\gamma \in N \setminus U^+$ implies $\gamma \cdot [0, \tilde{t}] \not\subset N$. Let $\tilde{\gamma}$ be a limit point of N_0; then there exist $\gamma_n \in N_0$ with $\gamma_n \to \tilde{\gamma}$. By definition $\gamma_n = \gamma_n' \cdot t_n'$ with $\gamma_n' \in N \setminus U^+$ and $\gamma_n' \cdot [0, t_n] \subset N$. Thus $0 \le t_n \le \tilde{t}$, and since $N \setminus U^+$ is compact, we can find $\gamma \in N \setminus U^+$, and $t \ge 0$ such that $\tilde{\gamma} = \gamma \cdot t$ with $\gamma \in N \setminus U^+$ and $\gamma \cdot [0, t] \subset N$. Thus $\tilde{\gamma} \in N_0$ and N_0 is compact.

In order to define N_n, we use Lemma 23.9 and find a compact neighborhood $N_n' \subset U^-$ of the set I_n^-, which is positively invariant relative to N, and define

$$N_n = N_n' \cup N_0. \tag{23.7}$$

By construction, N_n is positively invariant, and (N_n, N_0) is a compact pair.

Lemma 23.10. (N_n, N_0) *is an index pair for* S *and* $\mathrm{cl}(N_n \setminus N_0) \subset U$.

Proof. We shall verify the three conditions in Definition 23.6.

(i) S and N_0 being compact and disjoint implies that $N \setminus N_0$ is a neighborhood of S. N_n' is also a neighborhood of S, so that N_n, and thus $N_n \setminus N_0$ is a neighborhood of S. Also since $(N \setminus U^+) \subset N_0$ and $N_n' \subset U^-$, we conclude that $(N_n \setminus N_0) \subset U^- \cap U^+$, so $\mathrm{cl}(N_n \setminus N_0) \subset \mathrm{cl}(U^+ \cap U^-) \subset (U \cap \mathrm{int}\, N)$. Thus $\mathrm{cl}(N_n \setminus N_0)$ is an isolating neighborhood of S.

(ii) If $\gamma \in N_0$ and $\gamma \cdot [0, t] \subset N_n$, then $\gamma \cdot [0, t] \subset N$, so that $\gamma \cdot [0, t] \subset N_0$, since N_0 is positively invariant relative to N. Thus N_0 is positively invariant relative to N_n.

(iii) Suppose $\gamma \in N_n$ and $\gamma \cdot \mathbf{R}_+ \not\subset N_n$; if $\gamma \in N_0$ we are done. If $\gamma \notin N_0$, set $\tilde{t} = \sup\{t \ge 0 : \gamma \cdot [0, t] \subset N_n \setminus N_0\}$; then $\gamma \cdot \tilde{t} \in \mathrm{cl}(N_n \setminus N_0) \subset \mathrm{int}\, N$ (relative to X). We now use the fact that X is a local flow; namely, since $\gamma \cdot \tilde{t} \in X$, there is a Γ-neighborhood W of $\gamma \cdot \tilde{t}$ and $\varepsilon > 0$ such that $W \cap X \cdot [0, \varepsilon) \subset X$. Since $\gamma \cdot \tilde{t} \in \mathrm{int}\, N$ (relative to X), there is an $\varepsilon > 0$ such that $\gamma \cdot [\tilde{t}, \tilde{t} + \varepsilon] \subset N$. But N_n being positively invariant relative to N implies that this orbit segment is in N_n. From the definition of $\tilde{t}$ we see that there is a $t', \tilde{t} < t' < \tilde{t} + \varepsilon$ with $\gamma \cdot t' \in N_0$. Since $\gamma \cdot [0, t'] \subset N_n$, the third condition holds. □

Finally, we shall construct the desired filtration (23.3). Applying Lemma 23.9 to $N_n \subset N$ in place of N, we can find for every j, $1 \le j \le n - 1$, a compact neighborhood N_j' of I_j^- such that

(a) $I_j^- \subset N_j' \subset N_n$,
(b) $N_j' \cap I_{j+1}^+ = \varnothing$,
(c) N_j' is positively invariant relative to N_n.

(Recall that $I_j^- \cap I_{j+1}^+ = \varnothing$.) Now define successively

$$N_j = N_j' \cup N_{j-1}, \qquad 1 \le j \le n - 1; \tag{23.8}$$

see Figure 23.3.

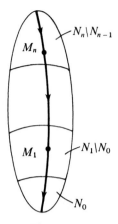

Figure 23.3

The next lemma completes the proof of Theorem 23.7.

Lemma 23.11. (N_j, N_{i-1}) *is an index pair for* M_{ij}.

Proof. Again we shall verify the conditions in Definition 23.6.

(i) Suppose $\gamma \cdot \mathbf{R} \subset \text{cl}(N_j\backslash N_{i-1})$; then $\gamma \in S$ and since $\gamma \notin I_{i-1}^-$, $\alpha(\gamma) \subset M_i$ $\cup \cdots \cup M_n$. Since $\gamma \notin I_{j+1}^+$, $\omega(\gamma) \subset M_1 \cup \cdots \cup M_j$, so $\gamma \in M_{ij}$, and $I(\text{cl}(N_j\backslash N_{i-1})) \subset M_{ij}$. Since $M_{ij} \subset I_j^- \subset N_j$, $M_{ij} \subset I_i^+$, and $I_i^+ \cap N_{i-1} = \varnothing$, we see $M_{ij} \subset N_j\backslash N_{i-1}$, and thus $\text{cl}(N_j\backslash N_{i-1})$ is an isolating neighborhood for M_{ij}.

(ii) By construction, N_{i-1} is positively invariant relative to N_n, and since $N_j \subset N_n$, N_{i-1} is positively invariant relative to N_j.

(iii) Suppose $\gamma \in N_j\backslash N_{i-1}$ and $\gamma \cdot \mathbf{R}_+ \not\subset N_j$; then $\gamma \cdot \mathbf{R}_+ \not\subset N_n$ since N_j is positively invariant relative to N_n. Thus by Lemma 23.10, there is a $t \geq 0$, such that $\gamma \cdot [0, t] \subset N_n$ and $\gamma \cdot t \in N_0$. Since $\gamma \in N_j$ we have $\gamma \cdot [0, t] \subset N_j$. Also by construction, $N_0 \subset N_{i-1}$. Thus we have shown that if $\gamma \cdot \mathbf{R}_+ \not\subset N_j$, there is a $t \geq 0$ with $\gamma \cdot [0, t] \subset N_j$ and $\gamma \cdot t \in N_{i-1}$; this proves the third condition.

Thus the proof of the lemma, and therefore of Theorem 23.7, is complete. $\square$

§B. The Conley Index of an Isolated Invariant Set

We recall from Chapter 22, §C the notions of a pointed space and quotient topology. The theorem that we shall prove here is that if (N_1, N_0) is an index pair for the isolated invariant set S, then the homotopy type of N_1/N_0 depends only on S. More precisely, we have the following theorem.

Theorem 23.12. *Let* (N_1, N_0) *and* $(\bar{N}_1, \bar{N}_0)$ *be two index pairs for the isolated invariant set* S. *Then the spaces* N_1/N_0 *and* $\bar{N}_1/\bar{N}_0$ *are homotopically equivalent.*

Thus, if $[\cdot]$ denotes the equivalence class of a pointed space, we can associate to S the unique homotopy equivalence class

$$h(S) = [N_1/N_0], \tag{23.9}$$

where (N_1, N_0) is any index pair for the isolated invariant set S. We call $h(S)$ the *Conley Index* of S. Note that if B is an isolating block for the isolated invariant set S, then as we have observed earlier, (B, b_+) is an index pair. Thus, in view of the theorem, the definition (23.9) agrees with our old definition (Definition 22.30) of the Conley index of S.

Before giving the proof of the theorem we define some useful sets.

If (N_1, N_0) is an index pair for the isolated invariant set S, we define for $t \geq 0$ the following subsets of N_1:

$$N_1^t = \{\gamma \in N_1 : \gamma \cdot [-t, 0] \subset N_1\},$$

$$\tag{23.10}$$

$$N_0^{-t} = \{\gamma \in N_1 : \text{there is a } \gamma' \in N_0$$
$$\text{and } t' \in [0, t] \text{ with } \gamma' \cdot [-t', 0] \subset N_1 \text{ and } \gamma' \cdot (-t') = \gamma\}.$$

It is easy to see that $N_0 \subset N_0^{-t}$, and that

$$N_1^t = \{\gamma \in N_1 : \text{there is a } \gamma' \in N_1 \text{ with } \gamma' \cdot [0, t] \subset N_1 \text{ and } \gamma' \cdot t = \gamma\}.$$

Roughly speaking, N_1^t is N_1 "pushed forward" for time t, and N_0^{-t} is N_0 "pulled backward" for time t; the definition of N_0^{-t} takes into account that the local flow X is not necessarily two-sided.

It is easy to check that N_1^t and N_0^{-t} are both compact and positively invariant relative to N_1. Furthermore, one can easily show that (N_1, N_0^{-t}) is an index pair for S, and that (N_1^t, N_0) is one also if X is a two-sided local flow.

The proof of Theorem 23.12 will follow from a sequence of lemmas. Let

$$i : (N_1^t, N_1^t \cap N_0) \to (N_1, N_0)$$

be the inclusion map, and let $\hat{\imath}$ be the induced map on the pointed spaces

$$\hat{\imath} : N_1^t/(N_1^t \cap N_0) \to N_1/N_0,$$

defined by $\hat{\imath}[x] = [i(x)]$. Of course, $\hat{\imath}$ is also continuous.

Lemma 23.13. *Let $t \geq 0$; then $\hat{\imath}$ is a homotopy equivalence.*

Proof. We define

$$F : (N_1/N_0) \times J \to N_1/N_0, \qquad J = [0, 1], \tag{23.11}$$

by

$$F([\gamma], \sigma) = \begin{cases} [\gamma \cdot \sigma t], & \text{if } \gamma \cdot [0, \sigma t] \subset N_1 \backslash N_0, \\ [N_0], & \text{otherwise.} \end{cases} \tag{23.12}$$

Note that if $\gamma \in N_0$, then if $\gamma \cdot [0, \sigma t] \subset N_1$, it follows that $\gamma \cdot [0, \sigma t] \subset N_0$, so $F([\gamma], \sigma) = [N_0]$; thus the map is well defined. We show that F is continuous. First assume $F([\gamma], \sigma) \neq [N_0]$; then from (ii) in Definition 23.6, $\gamma \cdot [0, \sigma t] \subset N_1 \backslash N_0$. Let U be any neighborhood of $\gamma \cdot \sigma t$ disjoint from N_0, and let V be any neighborhood of $\gamma \cdot [0, \sigma t]$ disjoint from N_0. From the continuity of the flow, there are neighborhoods W of γ and W' of σ such that if $(\gamma', \sigma') \in W \times W'$, then $\gamma' \cdot \sigma' t \in U$, and $\gamma' \cdot [0, \sigma' t] \subset V$. It then follows from Definition 23.6(iii), and the fact that $V \cap N_0 = \varnothing$, that $\gamma' \cdot [0, \sigma' t] \subset N_1 \backslash N_0$. Thus $F([\gamma], \sigma') = [\gamma' \cdot \sigma' t] \subset [U]$ and F is continuous at $([\gamma], \sigma)$. Suppose now that $F([\gamma], \sigma) = [N_0]$; then there are two possibilities; $\gamma \cdot [0, \sigma t] \not\subset N_1$ or $\gamma \cdot [0, \sigma t] \subset N_1$. If $\gamma \cdot [0, \sigma t] \not\subset N_1$, then for γ' near γ and σ' near σ, we have $\gamma' \cdot [0, \sigma' t] \not\subset N_1$ so $F([\gamma'], \sigma) = [N_0]$, and again F is continuous at $([\gamma], \sigma)$. Finally, if $F([\gamma], \sigma) = [N_0]$, and $\gamma \cdot [0, \sigma t] \subset N_1$, then $\gamma \cdot \sigma t \in N_0$. Let $\tilde{U}$ be a neighborhood of $[N_0]$ in N_1/N_0; then there is a neighborhood U of N_0 such that $[N \cap U] = \tilde{U}$. If (γ', σ') is near (γ, σ), then $\gamma' \cdot \sigma' t \in U$, by the continuity of the flow. If $\gamma' \cdot [0, \sigma' t] \subset N_1$, then $F(\gamma', \sigma') = [\gamma' \cdot \sigma' t] \in \tilde{U}$. If $\gamma' \cdot [0, \sigma' t] \not\subset N_1$, then $F([\gamma'], \sigma') = [N_0] \in \tilde{U}$. Thus in both cases, if (γ', σ') is close to (γ, σ), $F([\gamma'], \sigma') \in \tilde{U}$ and so here too F is continuous at $([\gamma], \sigma)$.

We shall use F to prove the homotopy equivalence. If $\sigma = 1$, the map $F(\cdot, 1)$ on N_1/N_0 has its range in $(N_1^t \cup N_0)/N_0 = N_1^t/(N_0 \cap N_1^t)$. Let f be the map $F(\cdot, 1)$, now considered as a map from N_1/N_0 into $N_1^t/(N_0 \cap N_1^t)$; then

$$\hat{\imath} \circ f = F(\cdot, 1) \sim \text{id.} \quad \text{on } N_1/N_0, \tag{23.13}$$

by definition of F ($\sim$ denotes "homotopically equivalent to"). On the other hand, since N_1^t is positively invariant relative to N_1, the restriction of F to $[(N_1^t \cup N_0)/N_0] \times J$ has range in $(N_1^t \cup N_0)/N_0$. Let F_r denote this restricted map as a map into $(N_1^t \cup N_0)/N_0$; then

$$f \circ \hat{\imath} = F_r(\cdot, 1) \sim \text{id.} \quad \text{on } N_1^t/(N_0 \cap N_1^t). \tag{23.14}$$

The lemma now follows from (23.13) and (23.14). $\square$

Next, for $t \geq 0$, define the map

$$g: N_1/N_0^{-t} \to N_1/(N_0 \cap N_1^t),$$

by

$$g([\gamma]) = \begin{cases} [\gamma \cdot t], & \text{if } \gamma \cdot [0, t] \subset N_1 \backslash N_0, \\ [N_0 \cap N_1^t], & \text{otherwise.} \end{cases} \tag{23.15}$$

To see that g is well defined, note that if $\gamma \in N_0^{-t}$, then $\gamma \cdot [0, t] \not\subset N_1 \backslash N_0$ so $g([\gamma]) = [N_0 \cap N_1^t]$; also, if $\gamma \cdot [0, t] \subset N_1 \backslash N_0$, then $\gamma \cdot t \in N_1^t$, so $g([\gamma]) = [\gamma \cdot t] \in N_1^t / (N_0 \cap N_1^t)$.

Lemma 23.14. *The map g defined by (23.15) is a homeomorphism.*

Proof. If $g([\gamma]) = [N_0 \cap N_1^t]$, then $\gamma \cdot [0, t] \not\subset N_1 \backslash N_0$, so by definition, $\gamma \cdot [0, t] \cap N_0 \neq \varnothing$, so $\gamma \in N_0^{-t}$, and thus $g^{-1}([N_0 \cap N_1^t]) = [N_0^{-t}]$. Also, if both $[\gamma_1]$ and $[\gamma_2]$ are different from $[N_0^{-t}]$ (in N_1 / N_0^{-t}), then $\gamma_i \cdot [0, t] \subset N_1 \backslash N_0, i = 1, 2,$ so $\gamma_1 \cdot t \neq \gamma_2 \cdot t,$ and thus $[\gamma_1 \cdot t] \neq [\gamma_2 \cdot t]$ in $N_1^t / (N_0 \cap N_1^t)$. Hence g is injective. To see that g is onto, let $\gamma \in N_1^t \backslash N_0$. Then there is a γ' in N_1 with $\gamma' \cdot [0, t] \subset N_1 \backslash N_0$, and $\gamma' \cdot t = \gamma$; thus $g([\gamma']) = [\gamma]$. Next, the continuity of g can be proved in a manner similar to the proof of the continuity of F in the last lemma. It follows then that g is a homeomorphism since it is a mapping between compact Hausdorff spaces. $\square$

Let $j: (N_1, N_0) \to (N_1, N_0^{-t})$ be the inclusion map, and let $\hat{j}$ be the induced map on the quotient spaces; i.e., $\hat{j}: N_1 / N_0 \to N_1 / N_0^{-t}$, and $j([\gamma]) = [j(\gamma)]$.

Lemma 23.15. *$\hat{j}$ is a homotopy equivalence.*

Proof. Consider the sequence of maps

$$N_1 / N_0 \underset{\hat{j}}{\rightrightarrows} N_1 / N_0^{-t} \underset{\hat{g}}{\rightrightarrows} N_1^t / (N_0 \cap N_1^t) \underset{\hat{i}}{\rightrightarrows} N_1 / N_0.$$

From the definitions, $g \circ \hat{j} = f$, where f is defined as above; thus $(\hat{i} \circ g) \circ \hat{j} = \hat{i} \circ f \sim \text{id.}$ on N_1 / N_0, in view of (23.13). On the other hand, using (23.14), $(g \circ \hat{j}) \circ \hat{i} = f \circ \hat{i} \sim \text{id.}$ on $N_1^t / (N_0 \cap N_1^t)$. Since g is a homeomorphism, we have

$$\hat{j} \circ (\hat{i} \circ g) \sim g^{-1} \circ [(g \circ \hat{j}) \circ \hat{i}] \circ g \sim \text{id.} \quad \text{on } N_1 / N_0^{-t},$$

and this proves the lemma. $\square$

In §A we showed that for any isolating neighborhood N of $S = I(N)$, there exists an index pair (N_1, N_0) for S such that $N_0 \subset N_1 \subset N$, with the property that N_1 and N_0 are positively invariant relative to N. We call such an index pair, *an index pair contained in N*. Note that for an index pair (N_1, N_0) contained in N, if

$$I^{\pm}(N) = \{\gamma \in N : \gamma \cdot \mathbf{R}_{\pm} \subset N\},$$

then

$$I^-(N) \subset N_1 \quad \text{and} \quad I^+(N) \cap N_0 = \emptyset.$$

Lemma 23.16. *Let (N_1, N_0) and $(\bar{N}_1, \bar{N}_0)$ be two index pairs in N for $S = I(N)$. Then there is a $t > 0$ such that*

$$(N_1^t, N_0 \cap N_1^t) \subset (\bar{N}_1, \bar{N}_0^{-t}),$$

and

$$(\bar{N}_1^t, \bar{N}_0 \cap \bar{N}_1^t) \subset (N_1, N_0^{-t}).$$

Proof. Since $I^-(N) \subset N_1$, $\gamma \in \text{cl}(N \backslash N_1)$ implies $\gamma \cdot \mathbf{R}_- \not\subset N$. Using the compactness of $\text{cl}(N \backslash N_1)$, we can find $t_1 \geq 0$ such that $\gamma \in \text{cl}(N \backslash N_1)$ implies $\gamma \cdot [-t_1, 0] \not\subset N$. Similarly, if $\gamma \in N_0$, then $\gamma \cdot \mathbf{R}_+ \not\subset N$ so there is a $t_0 > 0$ such that if $\gamma \in N_0$, then $\gamma \cdot [0, t_0] \not\subset N$. Let $\bar{t}_1$ and $\bar{t}_0$ be the corresponding numbers for the pair $(\bar{N}_1, \bar{N}_0)$, and set $t = \max(t_1, t_0, \bar{t}_1, \bar{t}_0)$. Let $\gamma \in N_1^t$; then $\gamma \cdot [-t, 0] \subset N_1 \subset N$ so $\gamma \notin \text{cl}(N \backslash \bar{N}_1)$ and thus $\gamma \in \bar{N}_1$, and $N_1^t \subset \bar{N}_1$. If $\gamma \in (N_0 \cap N_1^t) \subset \bar{N}_1$, then $\gamma \cdot [0, t] \not\subset N$ so by (iii) in Definition 23.6, there is a $\bar{t} \leq t$ such that $\gamma \cdot [0, \bar{t}] \subset \bar{N}_1$ and $\gamma \cdot \bar{t} \in \bar{N}_0$. Thus $\gamma \in \bar{N}_0^{-t}$, and $\gamma \in \bar{N}_0^{-t}$. The other inclusion is similar. $\square$

The next lemma shows that index pairs *contained in N* have homotopically equivalent quotient spaces.

Lemma 23.17. *Let (N_1, N_0), and $(\bar{N}_1, \bar{N}_0)$ be two index pairs for S contained in the isolating neighborhood N of $S = I(N)$. Then $[N_1/N_0] = [\bar{N}_1/\bar{N}_0]$.*

Proof. Let $t > 0$ be as in the last lemma, let i_1 and i_2 be the inclusion maps:

$$i_1 : (N_1^t, N_0 \cap N_1^t) \to (\bar{N}_1, \bar{N}_0^{-t}),$$

$$i_2 : (\bar{N}_1^t, \bar{N}_0 \cap \bar{N}_1^t) \to (N_1, N_0^{-t}),$$

and let $\bar{i}_1$ and $\bar{i}_2$ be the corresponding induced maps between the quotient spaces. Consider the sequence of maps

$$N_1^t/(N_0 \cap N_1^t) \underset{\bar{i}_1}{\to} \bar{N}_1/\bar{N}_0^{-t} \underset{\bar{g}}{\to} \bar{N}_1^t/(\bar{N}_0 \cap \bar{N}_1^t) \underset{\bar{i}_2}{\to} N_1/N_0^{-t} \underset{g}{\to} N_1^t/(N_0 \cap N_1^t)$$
$$\underset{\bar{i}_1}{\to} \bar{N}_1/\bar{N}_0^{-t},$$

where the map g is defined by (23.15), and $\bar{g}$ is the corresponding map $\bar{N}_1/\bar{N}_0^{-t} \to \bar{N}_1^t/(\bar{N}_0 \cap \bar{N}_1^t)$. Observe now that by definition

$$\bar{i}_2 \circ \bar{g} \circ \bar{i}_1 = \hat{\jmath} \circ \hat{\imath} \circ g \circ \hat{\jmath} \circ \hat{\imath},$$

since both maps take $[\gamma]$ onto $[\gamma \cdot t]$ or $[N_0^{-t}]$. Thus by Lemmas 23.13–23.15, $\hat{\imath}_2 \circ \bar{g} \circ \hat{\imath}_1$ is a homotopy equivalence. Similarly, $\hat{\imath}_1 \circ g \circ \hat{\imath}_2$ is a homotopy equivalence. Using Proposition 12.25, we see that $g \circ \hat{\imath}_1, \hat{\imath}_2$ and $\hat{\imath}_1 \circ g$ are homotopy equivalences. Thus by Lemma 23.14, $\hat{\imath}_1$ is one also. It follows from the sequence of maps

$$N_1/N_0 \underset{\hat{\imath}}{\to} N_1/N_0^{-t} \underset{g}{\to} N_1^t/(N_0 \cap N_1^t) \underset{\hat{\imath}_1}{\to} \bar{N}_1/\bar{N}_0^{-t} \underset{\bar{g}}{\to} \bar{N}_1^t/(\bar{N}_0 \cap \bar{N}_1^t) \underset{\hat{\imath}}{\to} \bar{N}_1/\bar{N}_0,$$

that $N_1/N_0 \sim \bar{N}_1/\bar{N}_0$. $\quad\square$

In order to prove Theorem 23.12, we must show that $N_1/N_0 \sim \bar{N}_1/\bar{N}_0$, for two *arbitrary* index pairs (N_1, N_0) and $(\bar{N}_1, \bar{N}_0)$ for S. In order to do this we shall show that they are equivalent to two index pairs both contained in a common isolating neighborhood, and then the last lemma will finish the proof.

Let (N_1, N_0) be an index pair for S, and let N' be an isolating neighborhood of S whose interior contains $\mathrm{cl}(N_1 \backslash N_0)$. Define the pair $(\tilde{N}_1, \tilde{N}_0) = (N' \cap N_1, N' \cap N_0)$; then the claim is that $(\tilde{N}_1, \tilde{N}_0)$ is an index pair for S contained in the isolating neighborhood $N = N' \cap N_1$ of S. To see this, note first that $\tilde{N}_1 \backslash \tilde{N}_0 = N_1 \backslash N_0$, so that $\mathrm{cl}(\tilde{N}_1 \backslash \tilde{N}_0)$ is an isolating neighborhood for S. Furthermore, to see that $\tilde{N}_0$ is positively invariant relative to $\tilde{N}_1$, suppose that $\gamma \in \tilde{N}_1 \backslash \tilde{N}_0$, and $\gamma \cdot [0, t] \not\subset \tilde{N}_1$. Let $\bar{t} = \sup\{s : \gamma \cdot [0, s] \subset \tilde{N}_1 \backslash \tilde{N}_0\}$; then $\gamma \cdot \bar{t} \in \mathrm{cl}(\tilde{N}_1 \backslash \tilde{N}_0) = \mathrm{cl}(N_1 \backslash N_0) \subset N'$. But $\gamma \cdot \bar{t}$ is not in the interior of $\mathrm{cl}(N_1 \backslash N_0)$, so $\gamma \cdot \bar{t} \in N_0$ and $\gamma \cdot \bar{t} \in N_0 \cap N' = \tilde{N}_0$.

Lemma 23.18. N_1/N_0 *is homeomorphic to* $\tilde{N}_1/\tilde{N}_0$.

Proof. Since $N_1 \backslash N_0 = \tilde{N}_1 \backslash \tilde{N}_0$ and $\tilde{N}_1 \subset N_1, \tilde{N}_0 \subset N_0$, the inclusion map $i : (\tilde{N}_1, \tilde{N}_0) \to (N_1, N_0)$ induces the desired homeomorphism. $\quad\square$

We can now complete the proof of Theorem 23.12. Let $\hat{N}$ be any isolating neighborhood of S contained in $N_1 \backslash N_0 = \tilde{N}_1 \backslash \tilde{N}_0$. Using Theorem 23.7, we can find an index pair $(\hat{N}_1, \hat{N}_0)$ in N_1 such that $\hat{N}_1$ and $\hat{N}_0$ are positively invariant relative to N and $\mathrm{cl}(\hat{N}_1 \backslash \hat{N}_0) \subset \mathrm{int}\,\hat{N}$. As we have seen above, $(\hat{N}_1 \cap \hat{N}, \hat{N}_0 \cap \hat{N})$ is in index pair for S in $\hat{N}$ and $(\hat{N}_1 \cap \hat{N})/(\hat{N}_0 \cap \hat{N})$ is homeomorphic to $\hat{N}_1/\hat{N}_0$ (by the last lemma). But $(\tilde{N}_1, \tilde{N}_0)$ is an index pair in N, so from Lemma 23.17, $N_1/N_0 \sim (\hat{N}_1 \cap \hat{N})/(\hat{N}_0 \cap \hat{N})$.

We have thus shown that N_1/N_0 has the homotopy type of an index pair in $\hat{N}$; namely, that of $(\hat{N}_1 \cap \hat{N})/(\hat{N}_0 \cap \hat{N})$. It follows then that if (N_1, N_0) is any index pair and $\hat{N} \subset \mathrm{int}(N_1 \backslash N_0)$ is any compact neighborhood of S, then N_1/N_0 has the same homotopy type as the index pairs of S in $\hat{N}$ (Lemma 23.17). If $(\bar{N}_1, \bar{N}_0)$ is another index pair for S, we simply choose $\hat{N}$ interior to $\mathrm{cl}(\bar{N}_1 \backslash \bar{N}_0)$, and using Lemma 23.17 again we find $\bar{N}_1/\bar{N}_0 \sim N_1/N_0$. The proof of Theorem 23.12 is thus complete. $\quad\square$

We can use our theorem to obtain algebraic invariants associated with an isolated invariant set S. In order to do this, note that if (X, A) is any pair, then, e.g., for the Čech-cohomology, $H^*(X, A) = H^*(X/A)$ (see [Sp]). If we use the fact that the cohomologies of two homotopically equivalent pairs are isomorphic (appendix to Chapter 12, §4C), we can obtain the following consequence of Theorem 23.12.

Corollary 23.19. *Let* (N_1, N_0) *and* $(\bar{N}_1, \bar{N}_0)$ *be two index pairs for the isolated invariant set* S. *Then* $H^*(N_1, N_0)$ *is isomorphic to* $H^*(\bar{N}_1, \bar{N}_0)$.

Thus the cohomology groups $H^*(N_1, N_0)$ are algebraic invariants associated with the isolated invariant set S; they are independent of the particular index pair we choose for S.

We shall conclude this section by proving "Morse-like" inequalities for a filtration; see §A. In analogy to the proof of Theorem 12.24, these will be a consequence of the axioms of cohomology theory. Thus, if $A \supset B \supset C$ are compact spaces, then there is a long exact sequence

$$0 \to H^0(A, B) \to H^0(A, C) \to H^0(B, C) \to$$
$$\xrightarrow[\delta^0]{} H^1(A, B) \to H^1(A, C) \to H^1(B, C) \to$$
$$\xrightarrow[\delta^1]{} H^2(A, B) \to \cdots. \tag{23.16}$$

We assume that the above groups are all finitely generated, and we denote by $r^k(X, Y)$ the rank of $H^k(X, Y)$, and by $d^k(A, B, C)$, the rank of the co-kernel of $\delta^k (= \text{rank}[H^{k+1}(A, B)/\text{Image}(\delta^k)])$. If (X, Y) is a compact pair, we can define the following formal power series with nonnegative integer coefficients:

$$p(t, X, Y) = \sum_{n \geq 0} r^n(X, Y)t^n$$

$$q(t, A, B, C) = \sum_{n \geq 0} d^n(A, B, C)t^n. \tag{23.17}$$

Now from the exactness of (23.16), we find for every $n \geq 0$,

$$r^0(A, B) - r^0(A, C) + r^0(B, C)$$

$$- r^1(A, B) + r^1(A, C) - r^1(B, C)$$

$$\vdots$$

$$+ (-1)^n r^n(A, B) - (-1)^n r^n(A, C) + (-1)^n r^n(B, C)$$

$$- (-1)^n d^n(A, B, C) = 0.$$

This gives

$$(-1)^r d^n(A, B, C) = (-1)^{n-1} d^{n-1}(A, B, C) + (-1)^r r^n(A, B)$$
$$- (-1)^r r^n(A, C) + (-1)^r r^n(B, C).$$

Multiplying by $(-1)^n t^n$ and adding gives

$$q(t, A, B, C) = -tq(t, A, B, C) + p(t, A, B) - p(t, A, C) + p(t, B, C),$$

or

$$p(t, A, B) + p(t, B, C) = p(t, A, C) + (1 + t)q(t, A, B, C). \tag{23.18}$$

We can use this to prove the following result.

Proposition 23.20. *Assume that $N_0 \subset N_1 \subset \cdots \subset N_n$ is a filtration of the Morse decomposition $(M_1, M_2, \ldots, M_n)$ of the isolated invariant set S. Then*

$$\sum_{j=1}^{n} p(t, N_j, N_{j-1}) = p(t, N_n, N_0) + (1 + t)Q(t),$$

where

$$Q(t) = \sum_{j=2}^{n} q(t, N_j, N_{j-1}, N_0).$$

Proof. We apply (23.18) to the triples $N_j \supset N_{j-1} \supset N_0, j \geq 2$ to get

$$p(t, N_j, N_{j-1}) + p(t, N_{j-1}, N_0) = p(t, N_j, N_0) + (1 + t)q(t, N_j, N_{j-1}, N_0).$$

Summing over $j \geq 2$ gives the result. $\square$

Now using Corollary 23.19, we may define

$$p(t, h(S)) = p(t, N_n, N_0), \tag{23.19}$$

where $h(S)$ is the Conley index of the isolated invariant set S, and (N_n, N_0) is any index pair for S. We then have the following consequence of Proposition 23.20, which generalizes the classical Morse inequalities.

Theorem 23.21. *Let S be an isolated invariant set in the local flow X, and let $(M_1, \ldots, M_n)$ be an admissible ordering of a Morse decomposition of S. Then*

$$\sum_{j=1}^{n} p(t, h(M_j)) = p(t, h(S)) + (1 + t)Q(t),$$

where $Q(t)$ is defined in Proposition 23.20. In particular, the coefficients of Q are nonnegative integers.

Proof. By Theorem 23.7, there is a filtration $N_0 \subset N_1 \subset \cdots \subset N_n$ for the Morse decomposition such that (N_n, N_0) is an index pair for S, and (N_j, N_{j-1}) is an index pair for M_j, $1 \leq j \leq n$. The result now follows from Proposition 23.20 and (23.19). $\square$

That this generalizes the usual Morse inequalities follows from the fact that the coefficients of $(1 + t)Q$ are nonnegative, so the coefficients of $\sum p(t, h(M_j)) - p(t, h(s))$ are nonnegative too.

§C. Continuation

In order to define the continuation of isolated invariant sets to nearby equations, it is first necessary to define what we mean by "nearby" equations. We have actually already done that in §A, but not for local flows. Moreover, we want our equations to depend on parameters which will define a family of local flows in which we can speak of "nearby local flows."

In order to motivate the next definition, suppose $J = [0, 1]$, X is an open subset of $\mathbf{R}^n$, and $f: X \times J \to \mathbf{R}^n$ is continuous. Assume too that for each $\lambda \in J, f(\cdot, \lambda)$ is Lipschitz continuous; then there is a unique curve $\gamma_\lambda(t; x)$ $\in \Gamma(X)$ ($=$ space of curves on X with the compact-open topology) such that $d\gamma_\lambda(t)/dt = f(\gamma_\lambda(t), \lambda), \lambda_\lambda(0) = x$. If $\phi: X \times J \to \Gamma(X)$, where $\phi(x, \lambda) = \gamma_\lambda(\cdot, x)$, then the continuity of the solution with respect to both initial conditions and parameters shows that ϕ is continuous. In fact the range of ϕ, $R(\phi)$, is a local flow in $\Gamma(X)$ (where we consider the translation flow on $\Gamma(X)$: namely $\bar{\phi}: \Gamma(X) \times \mathbf{R} \to \Gamma(X)$, defined by $(\gamma(\cdot, x), t) \to (\gamma \cdot t)(\cdot, x)$, where $(\gamma \cdot t)(s, x)$ $= \gamma(s + t, x)$). Now we can obviously write $R(\phi) = \cup \{\Phi_\lambda : \lambda \in J\}$, where $\Phi_\lambda = \{\gamma_\lambda(\cdot, x): x \in X\}$ is also a local flow. Moreover if $\lambda, \mu \in J$, then Φ_λ and Φ_μ are homeomorphic. Thus if N_λ is an isolating neighborhood in Φ_λ, and μ is close to λ, then if things are set up right, the corresponding N_μ (under the homeomorphism), should be an isolating neighborhood in Φ_μ, and the corresponding isolated invariant sets should have the same Conley index. This is the idea.

Now J can be replaced by any connected Hausdorff topological space Λ, $\Gamma(X)$ can be replaced by $\Gamma(X \times \Lambda)$, and $\gamma(\cdot)$, can be replaced by $(\gamma(\cdot), \lambda)$. The map $(x, \lambda) \to (\gamma(\cdot), \lambda)$, where $\gamma(0) = x$, from $X \times \Lambda$ into $\Gamma(X \times \Lambda)$ is again a homeomorphism. This leads to the following definition.

Definition 23.22. A *product parametrization* of a local flow $\Phi \subset \Gamma = \Gamma(X \times \Lambda)$ is a homeomorphism

$$\phi: X \times \Lambda \to \Phi$$

such that for each λ, $\phi(X \times \lambda) \equiv \Phi_\lambda$, is a local flow. We write $\phi_\lambda = \phi|_{X \times \{\lambda\}}$.

We think of ϕ_λ as a (local) flow depending on the parameter λ, and ϕ as a flow on the product space $X \times \Lambda$, or as a flow on $\Phi \subset \Gamma$. Thinking of ϕ in this latter way makes possible the application of these ideas to parameterized families of differential equations on a space U; namely, if $u_0 \in U$ is the initial condition, the λ-differential equation defines a curve $u(t, u_0, \lambda)$, and the map Φ takes (u_0, λ) into $u(\cdot, u_0, \lambda)$.

From now on X is assumed to be a locally compact subset of a metric space, and Λ is a connected Hausdorff space.

Lemma 23.23. *Let $\phi: X \times \Lambda \to \Phi$ be a product parametrization of Φ, let K be a compact subset of X, and let U be open in Γ. Then the set*

$$\Lambda(K, U) = \{\lambda: \phi_\lambda(K) \subset U\}$$

is open in Λ.

Proof. If $\lambda_0 \in \Lambda(K, U)$, then $\phi(K, \lambda_0) \subset U$; i.e., $\phi(k, \lambda_0) \in U$ if $k \in K$. Since ϕ is continuous and U is open, then for any $k \in K$ there is a neighborhood U_k of k and an $\varepsilon_k > 0$ such that if $\bar{k} \in U_k$, and $|\lambda - \lambda_0| < \varepsilon_k$, then $\phi(\bar{k}, \lambda) \in U$. Finitely many of the U_k cover K; say $U_{k_1}, \ldots, U_{k_n}$. Let $\varepsilon = \min \varepsilon_{k_i}$; then if $|\lambda - \lambda_0| < \varepsilon$, and $k \in K$, $k \in U_{k_i}$ for some i so since $|\lambda - \lambda_0| < \varepsilon_i$, $\phi(k, \lambda) \in U$. $\square$

Lemma 23.24. *Let K_1 and K_2 be compact subsets of X and let Q be a compact subset of $\mathbf{R}$. Then the set*

$$T = \{\lambda \in \Lambda: \phi_\lambda(K_1) \cdot Q \cap \phi_\lambda(K_2) = \varnothing\}$$

is open.

Proof. Let $\lambda \in T$. Now $\phi_\lambda(K_2)$ and $\phi_\lambda(K_1) \cdot Q$ are both compact, since ϕ is continuous. Thus there are disjoint open sets U_1 and U_2 with $\phi_\lambda(K_1) \subset U_1$, and $\phi_\lambda(K_2) \cdot Q \subset U_2$. By the previous lemma, we can find ε_1 and $\varepsilon_2 > 0$ such that if $|\lambda - \lambda_0| < \varepsilon_1$, then $\phi_{\lambda_0}(K_1) \subset U_1$, and if $|\lambda - \lambda_0| < \varepsilon_2$, then $\phi_{\lambda_0}(K_2) \cdot Q \subset U_2$. If $\varepsilon = \min(\varepsilon_1, \varepsilon_2)$, then $|\lambda - \lambda_0| < \varepsilon$ implies that $\lambda_0 \in T$. $\square$

We now define the space of isolated invariant sets of a family of (local) flows; cf. Definition 22.4. Thus let $\phi: X \times \Lambda \to \Phi$ be a product parametrization of the local flow Φ. We define

$$\mathscr{S} = \mathscr{S}(\phi) = \{(S_\lambda, \Phi_\lambda): S_\lambda \text{ is an isolated invariant set in } \Phi_\lambda\}.$$

For more compact notation, we will write (S, λ) for the isolated invariant set in Φ_λ.

If N is a compact subset of X, let

$$\Lambda(N) = \{\lambda \in \Lambda: \phi_\lambda(N) \text{ is an isolating neighborhood in } \Phi_\lambda\}.$$

(The reader should think here of $\phi_\lambda(N)$ as the isolating neighborhood N for the "λ-flow," if $\lambda \in \Lambda(N)$). Define

$$\sigma_N : \Lambda(N) \to \mathscr{S}, \quad \text{by}$$

$$\sigma_N(\lambda) = (S, \lambda) \equiv (S_\lambda, \Phi_\lambda),$$

where S is the maximal invariant set in $\phi_\lambda(N)$. We put the topology on $\mathscr{S}$ which is generated by the sets $\sigma_N(U)$, where U is open in $\Lambda(N)$. Observe that σ_N is a continuous function with this topology.

Here is the main definition in this section.

Definition 23.25. Let $p_1 = (S_1, \lambda_1)$, and $p_2 = (S_2, \lambda_2)$, be two points in $\mathscr{S}$; then p_1 is said to be *related by continuation* to p_2 if they both lie in the same quasi-component of $\mathscr{S}$; i.e., $\mathscr{S}$ is not the disjoint union of two open sets, each of which contains one of the points.

We sometimes say that p_1 and p_2 are *continuations of each other*, if they are related by continuation.

We need a few technical lemmas. Here is the first one.

Lemma 23.26. *Let $(S, \lambda) \in \mathscr{S}$, and let N be a compact subset of X such that $\phi_\lambda(N)$ is an isolating neighborhood for S. Then there exist compact subsets $N_1, N_2, \bar{N}_1, \bar{N}_2$ of N with $N_i \subset \text{int } \bar{N}_i$ (relative to N), $i = 1, 2$, $\text{cl}(\bar{N}_1 \backslash N_2) \subset \text{int } N$, and such that $\langle \phi_\lambda(N_1), \phi_\lambda(N_2) \rangle$ and $\langle \phi_\lambda(\bar{N}_1), \phi_\lambda(\bar{N}_2) \rangle$ are index pairs for (S, λ).*

Proof. Since λ is fixed here, we suppress the ϕ_λ notation. Using Theorem 23.7 (see also Lemma 23.10), we can find an index pair $\langle N_1', N_0' \rangle$ in N with $\text{cl}(N_1' \backslash N_0') \subset \text{int } N$.

Now choose an index pair $\langle \bar{N}_1, \bar{N}_0 \rangle$ in $\text{cl}(N_1' \backslash N_0')$ (which is an isolating neighborhood for S), with $\bar{N}_1 \supset N_0$, and, without loss of generality, $N_0 \subset \text{int } N_0'$ (relative to N). (This last condition can be achieved by replacing N_0 by N_0^t for sufficiently large t.) Choose $\langle \tilde{N}_1, \tilde{N}_0 \rangle$ an index pair in $\text{cl}(\bar{N}_1 \backslash N_0) \subset \text{int } \bar{N}_1$ (relative to N), and let $\langle N_1, N_0 \rangle = \langle \tilde{N}_1, N_0 \rangle$, and $\langle \bar{N}_1, \bar{N}_0 \rangle = \langle \bar{N}_1, N_0' \rangle$. We have $\bar{N}_1 \backslash N_0 \subset \text{int } N$, $N_1 \subset \text{int } \bar{N}_1$ (relative to N), and $N_0 \subset \text{int } \bar{N}_0$ (relative to N).

To see that $\langle N_1, N_0 \rangle$ is an index pair, note that N_1 and N_0 are positively invariant relative to N, $S \subset \text{int } \tilde{N}_1 \backslash N_0 = \text{int } N_1 \backslash N_0$, and if $\gamma \in N_1$, $\gamma \cdot t \notin N$, then $\gamma \in \tilde{N}_1$ so there is a t' such that $\gamma \cdot [0, t'] \subset \tilde{N}_1 = N_1$ and $\gamma \cdot t' \in \tilde{N}_0 \subset N_0$.

To see that $\langle \bar{N}_1, \bar{N}_0 \rangle$ is an index pair, again $\bar{N}_1$ and $\bar{N}_0$ are positively invariant relative to N, $S \subset \text{int } \bar{N}_1 \backslash N_0' = \text{int } \bar{N}_1 \backslash N_0$, and if $\gamma \in \bar{N}_1$, $\gamma \cdot t \notin N$, then there is a t' such that $\gamma \cdot [0, t'] \subset \bar{N}_1$, $\gamma \cdot t' \in N_0' = N_0$. This completes the proof. $\square$

Before proceeding, we need an important definition. For $Z \subset Y \subset \Gamma$, let

$$P(Z, Y) = \cap \{K \supset Z : K \text{ is compact and positively invariant relative to } Y\}.$$

Concerning this set we have the following lemma whose proof is immediate.

Lemma 23.27. *If Y is compact and $Z \subset Y$, then $P(Z, Y)$ contains Z, and it is the smallest such compact subset of Y which is positively invariant relative to Y.*

We turn now to another technical lemma.

Lemma 23.28. *Let $(S, \mu) \in \mathscr{S}$ and let N, $\phi_\mu(N)$, N_1, N_0, $\bar{N}_1$, $\bar{N}_0$ be as in Lemma 23.26 (with λ replaced by μ).*
  (A)   *There is a neighborhood W of μ in Λ such that $W \subset \Lambda(N)$.*
  (B)   *If*

$$P_i(\lambda) \equiv \phi_\lambda^{-1} P(\phi_\lambda(N_i), \phi_\lambda(N)), \qquad i = 0, 1,$$

  then if $\lambda \in W$, $P_i(\lambda) \subset \bar{N}_i$, and $\langle \phi_\lambda(P_1), \phi_\lambda(P_0) \rangle$ is an index pair for $S_\lambda = \sigma_N(\lambda)$ in $\phi_\lambda(N)$.

Note that the lemma implies in particular that $\Lambda(N)$ is open in Λ.

Proof. If $x \in \mathrm{cl}(N \backslash \bar{N}_1)$, then if $\alpha_\mu(x) \subset N$, it follows that $\alpha_\mu(x) \subset S \subset \bar{N}_1$, and from the positive invariance of $\bar{N}_1$, $x \in \bar{N}_1$. Thus $\alpha_\mu(x) \not\subset N$, so there is a $t_x > 0$ with $\phi_\mu(x) \cdot (-t_x) \notin \phi_\mu(N)$. From the positive invariance of $\bar{N}_i$, $\phi_\mu(x) \cdot [-t_x, 0] \cap \phi_\mu(\bar{N}_1) = \varnothing$, and using the continuity of the flow together with the compactness of $\bar{N}_1$, we can find a compact neighborhood K_x of x in N such that both of the following hold:

$$\phi_\mu(K_x) \cdot [-t_x, 0] \cap \phi_\mu(\bar{N}_1) = \varnothing, \quad \text{and}$$

$$\phi_\mu(K_x) \cdot (-t_x) \cap \phi_\mu(N) = \varnothing. \tag{23.20}$$

From Lemma 23.26, for each such x there is a neighborhood W_x of μ in Λ such that $\lambda \in W_x$ implies that (23.20) holds, with μ replaced by λ. Since $\mathrm{cl}(N \backslash \bar{N}_1)$ is compact, a finite number of the K_x cover $\mathrm{cl}(N \backslash \bar{N}_1)$; say $K_{x_1}, \ldots, K_{x_n}$. Then $W_1 \equiv \cap \{W_{x_i} : 1 \le i \le n\}$, is a neighborhood of μ in Λ such that (23.20) holds for all $\lambda \in W_1$. We claim that if $\lambda \in W_1$, $P(\phi_\lambda(N_1), \phi_\lambda(N)) \subset \phi_\lambda(\bar{N}_1)$. To see this, let $x \in P(\phi_\lambda(N_1), \phi_\lambda(N))$; then x lies in every compact set $K \supset N_1$ which is positively invariant (for λ), relative to N. If we show that $\bar{N}_1$ is positively invariant (for λ), relative to N, then we will have proved the claim since $\bar{N}_1$ is compact and $\bar{N}_1 \supset N_1$. Thus suppose $x \in \bar{N}_1$, and $\phi_\lambda(x) \cdot [0, t] \subset N$. If $\phi_\lambda(x) \cdot s \notin \bar{N}_1$, $0 < s \le t$, then $\phi_\lambda(x) \cdot s \in \mathrm{cl}(N \backslash \bar{N}_1)$, so from (23.20), there is a $t_x > 0$ with $\phi_\lambda(x) \cdot [s - t_x, s] \cap \bar{N}_1 = \varnothing$, and $\phi_\lambda(x) \cdot (s - t_x) \cap N = \varnothing$. But

$s - t_x < 0$, since otherwise, $\phi_\lambda(x) \cdot (s - t_x) \in N$ and thus $\phi_\lambda(x) \cdot [s - t_x, s]$ $\cap \overline{N}_1 = \varnothing$ implies $\phi_\lambda(x) \in \overline{N}_1$, a contradiction. Hence $\overline{N}_1$ is positively invariant (for λ) relative to N; this proves the claim. Thus W_1 is a neighborhood of μ in Λ such that if $\lambda \in W_1$, $P(\phi_\lambda(N_1), \phi_\lambda(N)) \subset \phi_\lambda(\overline{N}_1)$, so $P_1(\lambda) \equiv \phi_\lambda^{-1} P(\phi_\lambda(N_1), \phi_\lambda(N)) \subset \overline{N}_1$.

If now $x \in \overline{N}_0$, $\phi_\mu(x) \cdot \mathbf{R}_+ \not\subset \overline{N}_0$ (since $S \cap \overline{N}_0 = \varnothing$), so that by the positive invariance of $\overline{N}_0$, there is a $t'_x > 0$ with $\phi_\mu(x) \cdot t'_x \notin N$ and $\phi_\mu(x) \cdot [0, t'_x] \cap \mathrm{cl}(\phi_\mu(N) \backslash \phi_\mu(\overline{N}_0)) = \phi$. Thus as above there is a compact neighborhood K_x of x in N with

$$\phi_\mu(K_x) \cdot [0, t'_x] \cap \mathrm{cl}(\phi_\mu(N) \backslash \phi_\mu(\overline{N}_0)) = \varnothing \quad \text{and} \quad \phi_\mu(x) \cdot t'_x \notin \phi_\mu(N).$$

Thus again as before, there is a neighborhood W_0 of μ in Λ such that if $\lambda \in W_0$ and $x \in \overline{N}_0$,

$$\phi_\lambda(K_x) \cdot [0, t'_x] \cap \mathrm{cl}(\phi_\lambda(N) \backslash \phi_\lambda(\overline{N}_0)) = \varnothing, \qquad \phi_\lambda(x) \cdot t'_x \notin \phi_\lambda(N), \quad (23.21)$$

and $P_0(\lambda) \subset \overline{N}_0$. Let $\tilde{W} = W_1 \cap W_0$; then $\mu \in \tilde{W}$ so $\tilde{W}$ is a neighborhood of μ in Λ. If $\lambda \in \tilde{W}$, then $P_i(\lambda) \subset \overline{N}_i$, $i = 0, 1$, and (23.20), (23.21) hold. Hence, (A) holds. (We will cut $\tilde{W}$ down to W, below.)

Now $P_i \subset \overline{N}_i$, and by definition, $P_i \supset N_i$, both for $i = 0, 1$. Hence $P_1 \backslash P_0 \subset \overline{N}_1 \backslash N_0$ so $\mathrm{cl}(P_1 \backslash P_0) \subset \mathrm{cl}(\overline{N}_1 \backslash N_0) \subset \mathrm{int}\, N$.

Next we shall find a neighborhood W of μ in Λ such that (B) holds. The W we obtain will lie in $W_1 \cap W_2$; hence all of the above results will be valid for this W, and $N_i \subset P_i \subset \overline{N}_i$ will also hold.

If $x \notin (\mathrm{int}\, N_1) \backslash N_0$, then $x \in N_0$ so $x \notin S$. Thus there is a t_x such that either $\phi_\mu(x) \cdot t_x \notin \phi_\mu(N)$, or $\phi_\mu(x) \cdot (-t_x) \notin \phi_\mu(N)$. As before, there is a compact neighborhood K_x of x in N such that either $\phi_\mu(K_x) \cdot t_x \cap \phi_\mu(N) = \varnothing$, or $\phi_\mu(K_x) \cdot (-t_x) \cap \phi_\mu(N) = \varnothing$. Arguing again as before, there is a neighborhood W_3 of μ in Λ such that if $\lambda \in W_3$, and $x \notin \mathrm{int}(\phi_\lambda(N_1) \backslash \phi_\lambda(N_0))$, then either

$$\phi_\lambda(x) \cdot t_x \notin \phi_\lambda(N) \quad \text{or} \quad \phi_\lambda(x) \cdot (-t_x) \notin \phi_\lambda(N). \quad (23.22)$$

Let $W = \tilde{W} \cap W_3$; we show that $\lambda \in W$ implies that $\langle \phi_\lambda(P_1), \phi_\lambda(P_0) \rangle$ is an index pair for $S_\lambda = \sigma_N(\lambda)$ in $\phi_\lambda(N)$; see Definition 23.6. Again we shall verify the conditions in Definition 23.6.

1. To see that $\lambda \in W$ implies that $\sigma_N(\lambda) = S_\lambda$ is in $\mathrm{int}(\phi_\lambda(P_1) \backslash \phi_\lambda(P_0))$, we proceed as follows. Let $\lambda \in W$, $x \in S_\lambda$. If $x \in \phi_\lambda(P_0)$, then $P_0 \equiv P_0(\lambda) \subset \overline{N}_0$ implies that $x \in \phi_\lambda(\overline{N}_0)$. But from (23.21), there is a $t'_x > 0$ such that $\phi_\lambda(x) \cdot t'_x \notin N$, and this contradicts $x \in S_\lambda$. Thus $x \in S_\lambda$ implies $x \notin \phi_\lambda(P_0)$. To see that such an $x \in \mathrm{int}\, \phi_\lambda(P_1)$, we suppose that $x \notin \mathrm{int}\, \phi_\lambda(N_1)$; then $x \notin \mathrm{int}(\phi_\lambda(N_1)) \backslash \phi_\lambda(N_0)$ so by (23.22), $\phi_\lambda(x) \cdot \mathbf{R} \not\subset N$. This is a contradiction so $S_\lambda \subset \mathrm{int}\, \phi_\lambda(N_1) \subset \mathrm{int}\, P(\phi_\lambda(N_1), \phi_\lambda(N)) = \mathrm{int}\, \phi_\lambda(P_1)$, where we have used Lemma 23.27. Thus $S_\lambda \subset \mathrm{int}(\phi_\lambda(P_1) \backslash \phi_\lambda(P_0))$.

2. Since $\phi_\lambda(P_i) = P(\phi_\lambda(N_i), \phi_\lambda(N))$, Lemma 23.27 shows that $\phi_\lambda(P_i)$ is positively invariant relative to $\phi_\lambda(N)$, $i = 0, 1$.

3. Suppose $\phi_\lambda(x) \in \phi_\lambda(P_1)$, $t > 0$ and $\phi_\lambda(x) \cdot t \notin \phi_\lambda(N)$. We shall show that there is a $t' > 0$ such that $\phi_\lambda(x) \cdot [0, t'] \subset \phi_\lambda(P_1)$ and $\phi_\lambda(x) \cdot t' \in \phi_\lambda(P_0)$. Let $t_1 = \sup\{t > 0 : \phi_\lambda(x) \cdot [0, t] \subset \phi_\lambda(N)\}$; then by compactness $\phi_\lambda(x) \cdot t_1 \in \phi_\lambda(N)$, and in fact, $\phi_\lambda(x) \cdot t_1 \in \partial N$, by definition of t_1. The claim is that $\phi_\lambda(x) \cdot t_1 \in \phi_\lambda(\bar{N}_1)$. This holds since $\phi_\lambda(x) \in P(\phi_\lambda(N_1), \phi_\lambda(N))$ and $\phi_\lambda(x) \cdot [0, t_1] \subset \phi_\lambda(N)$; hence since $P(\phi_\lambda(N_1), \phi_\lambda(N))$ is positively invariant relative to $\phi_\lambda(N)$, it follows that $\phi_\lambda(x) \cdot [0, t_1] \subset P(\phi_\lambda(N_1), \phi_\lambda(N)) = \phi_\lambda(P_1) \subset \phi_\lambda(\bar{N}_1)$. Thus $\phi_\lambda(x) \cdot t_1 \in \phi_\lambda(\bar{N}_1)$.

Now if $\phi_\lambda(x) \cdot t_1 \notin \phi_\lambda(N_0)$, then $\phi_\lambda(x) \cdot t_1 \in \phi_\lambda(\bar{N}_1) \backslash \phi_\lambda(N_0) \subset \text{int } N$, an impossibility. Thus $\phi_\lambda(x) \cdot t_1 \in \phi_\lambda(N_0) \subset \phi_\lambda(P_0)$. Also, $\phi_\lambda(x) \cdot [0, t_1] \subset \phi_\lambda(N)$, $\phi_\lambda(x) \in \phi_\lambda(P_1)$ so since $\phi_\lambda(P_1)$ is positively invariant relative to $\phi_\lambda(N)$, we have $\phi_\lambda(x) \cdot [0, t_1] \subset \phi_\lambda(P_1)$. Thus the third condition for an index pair holds, and the proof is complete. $\square$

We define the projection $\pi : \mathscr{S} \to \Lambda$ by $\pi(S, \lambda) = \lambda$.

Theorem 23.29. The projection $\pi : \mathscr{S}(\phi) \to \Lambda$ is a local homeomorphism; in fact, if N is a compact subset of X, $\pi|_{\sigma_N(\Lambda(N))}$ is a homeomorphism with inverse σ_N.

As we have remarked after the statement of Theorem 22.5, this theorem tells when two points $(S, \lambda) = (S_\lambda, \Phi_\lambda)$, and $(S, \mu) = (S_\mu, \Phi_\mu)$ are related by continuation.

Proof. We are to show that for $(S, \mu) \equiv (S_\mu, \Phi_\mu) \in \mathscr{S}$, there exists an open set U containing (S, μ) such that $\pi|U$ is a homeomorphism onto its range.

Let (S, μ) be any point in $\mathscr{S}$. Take N to be a compact subset of X such that $\phi_\mu(N)$ is an isolating neighborhood for S in Φ_μ. Now $\mu \in \Lambda(N)$, so from the last lemma we see that $\sigma_N(\Lambda(N))$ is an open subset of $\mathscr{S}$ containing (S, μ). We set $U = \sigma_N(\Lambda(N))$, and we shall show that U has the required properties.

If $\pi(\sigma_N(\lambda_1)) = \pi(\sigma_N(\lambda_2))$, $\lambda_1, \lambda_2 \in \Lambda(N)$, then $\pi(S_{\lambda_1}, \Phi_{\lambda_1}) = \pi(S_{\lambda_2}, \Phi_{\lambda_2})$ implies $\lambda_1 = \lambda_2$. Thus $\pi|U$ is one-to-one; it is also continuous by definition of the topology on $\mathscr{S}$; namely, if V is open in U, then $\pi^{-1}(V) = \sigma_N(V)$ is open in $\mathscr{S}$.

The proof will be complete if we show $\pi|_U$ is open. To this end, note that it suffices to show that if $V_i = \sigma_i(U_i) \equiv \sigma_{N_i}(U_i)$, are two sub-basic open sets in $\mathscr{S}$, $i = 1, 2$, then $\pi(V_1 \cap V_2)$ is open in Λ. For if this were the case, then $\pi|$ (basic open) is open so if V is any open set in $\mathscr{S}$, then $V = \cup W_\alpha$, where W_α are basic open sets in $\mathscr{S}$, and $\pi(V)$ is open. Hence if $\pi(x) \in \pi(V)$, then $x \in W_\alpha$ for some α so $\pi(x) \subseteq \pi(W_\alpha) \subset \pi(V)$, and since $\pi(W_\alpha)$ is open, $\pi(V)$ must be open.

Thus let $V_i = \sigma_i(U_i)$, $i = 1, 2$, where U_i is open in $\Lambda(N_i)$, and N_i is compact; i.e., $\sigma_i = \sigma_{N_i}$. Now if $V_1 \cap V_2 = \varnothing$, then $\pi(V_1 \cap V_2)$ is certainly open in Λ; thus we may assume $(S_\lambda, \Phi_\lambda) \in V_1 \cap V_2$. Now $\phi_\lambda(N_i)$, $i = 1, 2$, both isolate S_λ in Φ_λ. Let $N_0 = N_1 \cap N_2$; then $\phi_\lambda(N_0)$ is an isolating neighborhood for

S_λ in Φ_λ. Also by the last lemma, there is a neighborhood W of λ in Λ such that if $v \in W$, then the maximal invariant set in $\phi_v(N_i)$ is in int $\phi_v(N_0)$. At this point we need a lemma.

Lemma 23.30. *Let C be closed in Λ and suppose that $\phi_\lambda(N)$ is an isolating neighborhood for each λ in C. Let $A = \{(\lambda, x): \lambda \in C, x \in S_\lambda\}$; then A is closed.*

Proof. If the lemma were false, then there would exist $(\lambda_n, x_n) \in A$ such $(\lambda_n, x_n) \to (\lambda, x) \notin A$. Thus $x \in S_\lambda$ so $\phi_\lambda(x) \cdot \mathbf{R} \not\subset N$. It follows that for large n, $\phi_{\lambda_n}(x) \cdot \mathbf{R} \not\subset N$; an impossibility. $\square$

Returning now to the proof of the theorem, we know $S_\lambda = \sigma_\lambda(N_1) = \sigma_\lambda(N_0)$, and $N_0 \subsetneqq N_1$. We claim that this implies that there is a neighborhood W of λ such that if $v \in W$, then $\sigma_v(N_1) = \sigma_v(N_0)$. To see this we argue by contradiction; thus assume that there exist $\lambda_n \in \Lambda$ with $\lambda_n \to \lambda$ and $S_n' \equiv \sigma_{\lambda_n}(N_0) \neq \sigma_{\lambda_n}(N_1) \equiv S_N$. Choose $x_n \in S_n \backslash S_n'$. By compactness of N_1, we may assume that $x_n \to \bar{x}$, and by our lemma, $\bar{x} \in S$ so by hypothesis, $\bar{x} \in S'$. Thus for n large, $x_n \in \phi_{\lambda_n}(N_0)$. Also $x_n \in S_n \backslash S_n'$ implies that there exist $t_n > 0$ such that $x_n \cdot t_n \in N_1 \backslash \text{int } N_0$, and by compactness, we may assume $\{x_n \cdot t_n\}$ converges; say $x_n \cdot t_n \to x \in N_1 \backslash \text{int } N_0$. Again by our lemma, $x \in S$ so $x \in S' \subset N_0$, a contradiction. This proves the claim.

It is now easy to see that the maps σ_{N_i}, all agree on W, $i = 0, 1, 2$. Namely, if $v \in W$, $\sigma_N(v) = (S_v, \Phi)$, $\sigma_{N_0}(v) = (S_v', \Phi_v)$, and by our claim $S_v = S_v'$, so $\sigma_{N_1}(v) = \sigma_{N_0}(v)$; similarly $\sigma_{N_2}(v) = \sigma_{N_0}(v) = \sigma_{N_1}(v)$. From this it follows that $\sigma_{N_0}(W) \subset V_1 \cap V_2$ (since if $v \in W$, then $\sigma_{N_i}(v) = \sigma_{N_0}(v)$, $i = 1, 2$). But $v \in W_i$ implies $\sigma_{N_i}(v) \in V_i$, and $\sigma_{N_1}(v) = \sigma_{N_2}(v)$; hence $\sigma_{N_i}(v) \in V_1 \cap V_2$ so $\sigma_{N_0}(v) \in V_1 \cap V_2$ and thus $\sigma_{N_0}(W) \subset V_1 \cap V_2$). Therefore, $W \subset \pi(V_1 \cap V_2)$ so that $\pi|_U$ is open and this map is a homeomorphism. This proves that π is a local homeomorphism and completes the proof of the theorem. $\square$

We now come to the main theorem in this section, the invariance of the Conley index under continuation.

Theorem 23.31. *If S_λ and S_μ are related by continuation then they have the same Conley index.*

Proof. We claim that it is only necessary to show that the set of points in $\mathscr{S}$ with a given index is open. To see this, suppose that $(S_\mu, \Phi_\mu) \in \mathscr{S}$ and $U = \{S_v, \Phi_v) \in \mathscr{S}: h(S_v) = h(S_\mu)\}$ is open. If $h(S_\alpha) \neq h(S_\lambda)$, then (S_α, Φ_α) clearly cannot lie in cl(U), so $(S_\alpha, \Phi_\alpha) \in \mathscr{S} \backslash \text{cl } U$, $(S_\lambda, \Phi_\lambda) \in U$, and since $\mathscr{S} \backslash \text{cl } U$ and U are disjoint open sets, (S_α, Φ_α) and (S_μ, Φ_μ) are not related by continuation. This proves the claim.

Now let $(S_\mu, \Phi_\mu) \in \mathscr{S}$ and let N be a compact subset of X such that $\phi_\mu(N)$ is an isolating neighborhood for S_μ in Φ_μ. Choose sets $N_i, \bar{N}_i, i = 0, 1$, as in Lemma 23.28; then there is a neighborhood W_1 of μ, $W_1 \subset \Lambda(N)$ such that if $\lambda \in W_1, (N_1, N_0) \subset (P_1^\lambda, P_0^\lambda) \subset (\bar{N}_1, \bar{N}_0)$, where $\langle \phi_\mu(N_1), \phi_\mu(N_0) \rangle$ and $\langle \phi_\mu(\bar{N}_1),$

$\phi_\mu(\bar{N}_0)\rangle$ are index pairs for S_μ in Φ_μ, and $\langle \phi_\lambda(P_1^\lambda), \phi_\lambda(P_0^\lambda)\rangle$ is an index pair for S_λ in Φ_λ.

Now $S_\mu \subset \text{int } N_1 \backslash N_0 \subset \text{int } N_1$, so N_1 is an isolating neighborhood for S_μ in $\phi_\lambda(N_1)$. We take N_i', $\bar{N}_i'$, $i = 0, 1$, as in Lemma 23.28, where $\langle \phi_\mu(N_1'),$ $\phi_\mu(N_0')\rangle$ and $\langle \phi_\mu(\bar{N}_1'), \phi_\mu(\bar{N}_0')\rangle$ are index pairs for S_μ in N_1. Again using Lemma 23.28, we can find a neighborhood W_2 of μ in $\Lambda(N_1)$ such that if $\lambda \in W_2$, then $(N_1', N_0') \subset (\tilde{P}_1^\lambda, \tilde{P}_0^\lambda) \subset (\bar{N}_1', \bar{N}_0')$, where $\tilde{P}_i^\lambda$ are the analogous "P's" associated with these latter two index pairs.

Let $W = W_1 \cap W_2$; then $\mu \in W$, W is open in Λ, and $W \subset \Lambda(N) \cap \Lambda(N_1)$ $\subset \Lambda(N)$. To complete the proof, we shall show that if $\lambda \in W$, then $h(S_\lambda) = h(S_\mu)$. Thus for $\lambda \in W$, consider the sequence of maps

$$N_1'/N_0' \underset{i_1}{\rightrightarrows} \tilde{P}_1^\lambda/\tilde{P}_0^\lambda \underset{i_2}{\rightrightarrows} \bar{N}_1'/\bar{N}_0' \underset{\psi}{\rightrightarrows} N_1/N_0 \underset{i_3}{\rightrightarrows} P_1^\lambda/P_0^\lambda,$$

where i_1, i_2, and i_3 are inclusion-induced maps, and ψ is the homotopy equivalence map between the index pairs $(\bar{N}_1', \bar{N}_0')$ and (N_1, N_0) (for the isolated invariant set S_μ), whose existence was proved in Theorem 23.12. We re-write this sequence as

$$N_1'/N_0' \underset{i_1}{\overset{i_3 \circ \psi \circ i_2}{\rightrightarrows}} \tilde{P}_1^\lambda/\tilde{P}_0^\lambda \underset{\psi \circ i_2}{\longrightarrow} N_1/N_0 \underset{i_3}{\rightrightarrows} \tilde{P}_1^\lambda/P_0^\lambda. \qquad (23.23)$$

$$\underset{\psi \circ i_2 \circ i_1}{}$$

For simplicity in notation, let's rewrite this last sequence in the form

$$A \underset{i}{\overset{f}{\rightrightarrows}} B \underset{j}{\overset{f_1}{\rightrightarrows}} C \overset{g_1}{\rightrightarrows} D,$$

$$\underset{g}{}$$

where i and j are inclusion induced maps, f and g are homotopy equivalences and f_1 is chosen such that $f = f_1 \circ i$, while g_1 is chosen so that $g = g_1 \circ j$. (To see how to construct f_1, for example, define $f_1(x) = f(x)$ if $x \in A$, and extend f_1 to B by the Tietze extension theorem; see [Sp].) From Proposition 12.25, it follows that i is a homotopy equivalence. Referring back to (23.23), this implies that

$$N_1'/N_0' \sim \tilde{P}_1^\lambda/\tilde{P}_0^\lambda,$$

so that $h(S_\mu) = h(S_\lambda)$, and the proof is complete. $\square$

§D. Some Further Remarks

In this final section we shall derive some consequences of our previous results which will be useful in the applications.

1. Recall from §B, that if A, B, and C are compact spaces, and $C \subset B \subset A$, then there is a (long) exact sequence of cohomology groups

$$\cdots \to H^{n-1}(B, C) \to H^n(A, B) \to H^n(A, C) \to H^n(B, C) \to \cdots. \qquad (23.24)$$

Now suppose that $N_0 \subset N_1 \subset \cdots \subset N_n$ is a Morse filtration of the isolated invariant set S (see Theorem 23.7). Then if $0 \leq p - 1 < p < p + 1 \leq n$, we have $N_{p-1} \subset N_p \subset N_{p+1}$, and from (23.24) we obtain the (long) exact sequence

$$\cdots \to H^{n-1}(N_p, N_{p-1}) \to H^n(N_{p+1}, N_p) \to H^n(N_{p+1}, N_{p-1})$$
$$\to H^n(N_p, N_{p-1}) \to \cdots.$$

Now as we have seen in §B, $H(X, A) = H(X/A)$ if (X, A) is a topological pair. If we use this in the above sequence we obtain the important (long) exact sequence

$$\cdots \to H^{n-1}(N_p/N_{p-1}) \to H^n(N_{p+1}/N_p)$$
$$\to H^n(N_{p+1}/N_{p-1}) \to H^n(N_p/N_{p-1}) \to \cdots.$$

relating the cohomologies of the indices of the Morse decomposition of S.

Now suppose $n = 2$, (M_1, M_2) is a Morse decomposition of S and $N_0 \subset N_1 \subset N_2$ is a filtration for (M_1, M_2). Then from Proposition 23.5 the M_k are isolated invariant sets, and from Theorem 23.7, (N_1, N_0) is an index pair for M_1, (N_2, N_1) is an index pair for M_2 and (N_2, N_0) is an index pair for S. We thus obtain the exact sequence

$$\cdots \to H^{n-1}(h(M_1)) \to H^n(h(M_2)) \to H^n(h(S)) \to H^n(h(M_1)) \to \cdots. \qquad (23.25)$$

In particular, (23.25) applies to the case where M_1 and M_2 are critical points connected by an orbit; here S is the isolated invariant set consisting of the two rest points together with the connecting orbit. The orbit is assumed to run from M_2 to M_1.

2. We point out that if we have a gradient-like equation in a local flow having a finite rest point set $C = \{x_1, x_2, \ldots, x_n\}$ contained in an isolating neighborhood N, then the rest points form a Morse decomposition of $I(N)$. To see this, let F be the gradient-like function, and let $x \in I(N) \backslash C$.

Then if $t > 0$, $F(x \cdot t) > F(x)$, so F is constant on both the α- and ω-limit set of x. Now we can consider the flow on the *compact* set $I(N)$. Since this flow is gradient-like, $\omega(x)$ and $\alpha(x)$ must be rest points. Then the result follows once we order C as $(x_{i_1}, x_{i_2}, \ldots, x_{i_n})$, in such a way that if $j > k$, then $F(x_{i_k}) \geq F(x_{i_j})$.

3. It is often important to know when an invariant set is isolated. This is a difficult problem in general, but here is one useful criterion.

Theorem 23.32. *Suppose that x_0 is an isolated rest point in a local flow X of a gradient-like equation. Then x_0 is an isolated invariant set.*

Note that the gradient-like hypothesis is crucial; a center for an equation in $\mathbf{R}^2$ is an obvious counter-example, since every neighborhood of such a point contains a nonconstant periodic solution.

Proof. Let N be a compact X-neighborhood of x_0 for which x_0 is the only rest point. Let U be any X-neighborhood of x_0, $U \subsetneqq N$. Then $N \backslash U$ is compact, and if $x \in N \backslash U$, $F(x \cdot 1) > F(x)$, so $F(x \cdot 1) - F(x) \geq \delta > 0$ for some $\delta > 0$ and all $x \in N \backslash U$. If $y \in N$ and $y \cdot \mathbf{R} \subset N$, then $\omega(y)$ is a rest point so $\omega(y) = x_0 \in I(N)$. Similarly $\alpha(y) \in I(N)$ and $\alpha(y) = x_0$ since $\alpha(y)$ is a rest point. Thus $y \cdot \mathbf{R} \subset U$ in view of the above inequality. Since this holds for all U, we see $y \cdot \mathbf{R} = \{x_0\}$. $\square$

4. We shall explicitly show how the Conley index generalizes the classical Morse index. To this end, consider the equation $\dot{x} = \nabla F$ defined on a closed compact manifold M of dimension k. This equation defines a flow on M and we take the local flow to be all of M.

Let the critical point set of F be $C = \{x_1, \ldots, x_n\}$. As we have seen in §2 above, the elements in C are all Morse sets, $M_j = \{x_j\}$. If x_j is a nondegenerate rest point (Chapter 12, §C), then the equation near x_j can be written as

$$\dot{x}^- = A_- x^- + O_2(x),$$

$$\dot{x}^+ = A_+ x^+ + O_2(x),$$

where $x = (x^-, x^+) \in V_- \times V_+ = \mathbf{R}^k$, and where $\langle A_- x^-, x^- \rangle < 0$ if $x^- \neq 0$, and $\langle A_+ x^+, x^+ \rangle > 0$ if $x^+ \neq 0$. Of course, dim V_+ is the Morse index of x_j.

For $\varepsilon > 0$, let

$$B_\varepsilon = \{x \in \mathbf{R}^k : |x^-| \leq \varepsilon \text{ and } |x^+| \leq \varepsilon\};$$

then

$$\partial B_\varepsilon = \{x \in B_\varepsilon : |x^-| = \varepsilon \text{ or } |x^+| = \varepsilon\}.$$

If $x \in \partial B_\varepsilon$ and $|x^-| = \varepsilon$, then

$$(\tfrac{1}{2}|x^-|^2)' = \langle x^-, A_- x^- \rangle + O_3(\varepsilon)$$

$$= \varepsilon^2 \left\langle \frac{x^-}{|x^-|}, \frac{A_- x^-}{|x^-|} \right\rangle + O_3(\varepsilon)$$

$$= -\varepsilon^2 |x^-|^{-2} |\langle x^-, A_- x^- \rangle| + O_3(\varepsilon).$$

It follows that for small ε, $(|x^-|^2/2)' < 0$ for all $x^- \in \partial B_\varepsilon$. Similarly, $(|x^+|^2/2)' > 0$ for $x^+ \in \partial B_\varepsilon$, if ε is small. Thus if $N_1 = B$ and $N_0 = \partial B_\varepsilon$ $\cap \{|x^+| = \varepsilon\}$, then (N_1, N_0) is an index pair for M_j. We shall prove that

$$h(M_j) = \Sigma^p, \quad \text{where } p = \dim V_+. \tag{23.26}$$

This shows that the Conley index carries the same information as the Morse index whenever the latter is defined.

To prove (23.26), write

$$N_1 = \{|x^-| \le \varepsilon\} \times \{|x^+| \le \varepsilon\},$$

$$N_0 = \{|x^-| \le \varepsilon\} \times \{|x^+| = \varepsilon\},$$

and observe that

$$N_1/N_0 \sim \{x \in B : |x^+| \le \varepsilon\}/\{x \in B : |x^+| = \varepsilon\},$$

a ball of dimension $(\dim V_+)$ modulo its boundary, or as is easily seen, a pointed sphere of dimension $(\dim V_+)$. This gives (23.26).

5. In this final section we shall consider a system of reaction–diffusion equations on a bounded domain Ω and show how to fit this situation into the general development of the index theory which we have given in this chapter.

Thus consider the initial-boundary-value problem for the system of reaction–diffusion equations:

$$u_t = D\Delta u + f(u), \quad (x, t) \in \Omega \times \mathbf{R}_+,$$

$$bu = 0, \qquad\qquad (x, t) \in \partial\Omega \times \mathbf{R}_+,$$

$$u(x, 0) = u_0(x), \qquad x \in \Omega.$$

Here $u = (u_1, u_2, \ldots, u_n)$, each $u_i = u_i(x, t)$, and Ω is a bounded domain in $\mathbf{R}^n$ having smooth boundary, $\partial\Omega$. $D = D(x)$ is a smooth poisitive diagonal $n \times n$ matrix-valued function, f is a smooth function, and b is a diagonal Dirichlet or Neumann boundary operator; i.e., for all i, $u_i = 0$ on $\partial\Omega \times \mathbf{R}_+$, or

$du_i/dn = 0$ on $\partial\Omega \times \mathbf{R}_+$, $i = 1, 2, \ldots, n$, where n is the outward pointing normal vector on $\partial\Omega$. The function u_0 is assumed to be of class C^1. Finally, we assume that the system admits arbitrarily large bounded invariant regions (see Chapter 14, §B).

In order to be able to apply the index theory to this system of partial differential equations, we can first write it as an evolution equation

$$u' = Au + f(u), \qquad u(0) = u_0, \tag{23.27}$$

where u belongs to the space F; namely, the intersection of the space C^1 and the domain of A with the space of functions satisfying the boundary conditions.

If Γ is the Banach space of continuous curves, $\gamma: \mathbf{R} \to L_2(\Omega)$, endowed with the compact-open topology, then we may define a continuous flow on Γ by setting, for every $\sigma \in \mathbf{R}$,

$$(\gamma \cdot \sigma)(t) = \gamma(\sigma + t) \quad \text{for all } t \in \mathbf{R}, \quad \gamma \in \Gamma.$$

This is called the translation flow on Γ.

We assume Σ is an invariant region for (23.27) such that $u_0(x) \in \Sigma$ for all $x \in \Omega$. Let $\phi(u)$ be a smooth compactly supported function satisfying $\phi(u) \equiv 1$ in a neighborhood of Σ, and set $F(u) = \phi(u)f(u)$. It is known that $W_2^2(\Omega)$ is continuously and densely embedded in $L_2(\Omega)$, and in fact, this embedding is compact; see [Am 2]. Set

$$M = \{u \in L^2(\Omega): u(x) \in \Sigma \text{ a.e. in } \Omega\};$$

then (23.27) is equivalent to

$$u' = Au + F(u), \qquad u(0) = u_0 \in M \cap W_2^2. \tag{23.28}$$

Also, if u is a solution, then $u(t) \in M \cap W_2^2(\Omega)$ for all $t \geq 0$, and u satisfies the integral equation

$$u(t) = e^{At}u_0 + \int_0^t e^{A(t-s)}F(u(s)) \, ds. \tag{23.29}$$

Since $u(t) \in M \cap W_2^2(\Omega)$ for all $t \geq 0$, it follows from (23.29), in view of the well-known properties of analytic semigroups ([Am 2], cf. Chap. 14, §D or [Mx]), that the following estimates hold for all $t, s \geq 0$:

$$\|u(t)\|_{W_2^2(\Omega)} \leq C(1 + \|u_0\|_{W_2^2(\Omega)})$$
$$\|u(t) - u(s)\|_{L_2(\Omega)} \leq C(1 + \|u_0\|_{W_2^2(\Omega)})|t - s|^\nu, \tag{23.30}$$

for some ν, $0 < \nu < 1$, where c and ν are independent of u_0, t, and s. Conversely, if $u(t)$ is a solution of (23.29) which satisfies (23.30), then it is not hard to show that $u(t)$ solves (23.28), cf. [HS 1, 2].

Now define $X \subset \Gamma$ as follows: Let $0 < v < 1$, and set

$$X = \{\gamma \in \Gamma : \gamma(t) \in W_2^2(\Omega), \gamma(0) \in M \cap W_2^2(\Omega), \|\gamma(t) - \gamma(s)\|_{L_2(\Phi)} \le C_\gamma' |t - s|^v,$$
$$\text{and } \|\gamma(t)\|_{W_2^2(\Omega)} \le C_\gamma, \text{ for all } s, t \in \mathbf{R}, \text{ and for } t \ge 0; \text{ and}$$
$$\gamma \text{ solves (23.28) with initial condition } \gamma(0)\}.$$

Here C_γ and C_γ' depend on γ but not on t.

Observe that the elements in X are solutions of (23.28) which have their "past histories" attached. The following lemma is needed to prove that X defines a local flow.

Lemma 23.33. *The embedding $X \subset \Gamma$ is compact; thus X is locally compact subset of Γ.*

Proof. Let $C_1 > 0$; we will show that the bounded subset $B = X \cap \{\gamma : \|\gamma(t)\|_{W_2^2(\Omega)} \le C_1, \text{ for all } t \in \mathbf{R}\}$ is compact in Γ. Since $W_2^2(\Omega)$ is compactly embedded in $L_2(\Omega)$, the Arzela–Ascoli theorem implies that $\bar{B}$ (closure in Γ) is a compact subset of Γ. To complete the proof, we show that $B = \bar{B}$. For this, observe first that the closed unit ball, $\{u \in W_2^2(\Omega) : \|u\|_{W_2^2(\Omega)} \le 1\}$, is closed in $L_2(\Omega)$. Now let $\gamma \in \bar{B}$; then $\exists \gamma_n \in B$ with $\gamma_n \to \gamma$ in Γ, and by our observation, we have

$$\|\gamma(t)\|_{W_2^2(\Omega)} \le C_1, \quad \text{and} \quad \|\gamma(t) - \gamma(s)\|_{L_2(\Omega)} \le C_\gamma' |t - s|^v,$$

for $t, s \in \mathbf{R}$. Moreover, M being closed in $L_2(\Omega)$ implies $\gamma(0) \in M$. Since γ satisfies (23.29), γ is a solution of (23.28), and this completes the proof. □

From the local existence and uniqueness of solutions of (23.28), we conclude that $(X \cap U) \cdot [0, \varepsilon) \subset X$ for every open set $U \subset \Gamma$ and every $\varepsilon > 0$. Since our lemma implies that X is locally compact, we conclude that X is a local flow in Γ.

We remark that the replacement of f by F, as described above, can be avoided if we choose to work in a scale of Banach spaces rather than $W_2^2(\Omega)$ and $L_2(\Omega)$; see [CZ] for details.

NOTES

The results in this chapter are due to Conley and are all contained in his important monograph [Cy 2]. In §A and §B, we have followed the somewhat different approach given by Conley and Zhender [CZ]. The development in this paper focuses upon the Morse decomposition. In §C, we have also followed Conley in [Cy 2], but we have expanded the version given there. The elegant development of the Morse inequalities as given in §B is also taken from [CZ]. The material in §D is taken both from [Cy 2] and [CZ]; in particular we have followed [CZ] in §5 with certain minor modifications.

Chapter 24

Travelling Waves

The Conley index is a double-edged sword: if it is ever shown to be nontrivial, then this implies the existence of an orbit which stays in the isolating neighborhood for all time; in this sense it gives an existence theorem. On the other hand, being a Morse-type index, it also carries stability information concerning the isolated invariant set. In this chapter we shall illustrate both of these properties for a special class of solutions of partial differential equations called travelling waves.

Travelling waves make up an important class of solutions of both reaction–diffusion equations and nonlinear hyperbolic equations with "viscosity." They are solutions of the form $u = u(x - ct)$, where c is a constant, the *speed* of the wave. Many phenomena arising in various physical or biological contexts can be modelled by travelling waves; for example, shock waves, nerve impulses, and various oscillatory chemical reactions. The nice mathematical feature associated with such solutions is that the problem often reduces to one in ordinary differential equations. Of course this is not meant to imply that the problem becomes trivial. Indeed, the phase space is usually of dimension $n \geq 3$, and typical problems are to find orbits with certain distinguished properties: ones that connect rest points, orbits homoclinic to a point, periodic solutions, etc.

In §A, we shall consider the so-called "shock-structure" problem for hyperbolic systems of conservation laws. The problem concerns the existence of an orbit of a flow in $\mathbf{R}^n$ which connects two rest points. We have already considered such a problem in a simple (scalar) case in Chapter 15, §C. Here we shall study the higher-dimensional case with the aid of the Conley index. It will become apparent that this is precisely the right tool for solving the problem. In §B we shall give the complete solution of the shock structure problem for an interesting physical case; namely, magneto-hydrodynamic shock waves of arbitrary strength, for gases with general equations of state. In this case the problem becomes a global one in $\mathbf{R}^6$, and the novelty is in the construction of the isolated invariant set. The next section deals with the existence of a periodic travelling wave for the Nagumo equations. Here too it is not at all obvious how to apply the Conley index since the construction of an isolating neighborhood is difficult—one simply doesn't know where to look for it. The technique is to construct a "singular" periodic orbit which is obtained by setting some parameters equal to zero. If the singular orbit is contained in an isolating block, then the block persists under perturbation. This shows one where (and how!) to build an isolating neighborhood. The periodic orbit is obtained by using the continuation property. In §D, we consider steady-state solutions of reaction–diffusion equations (travelling waves of *zero* speed), and we study the stability of these solutions, now considered as "rest points" of the full time dependent partial differential equation. Here we shall use the Conley index as a Morse-type index in order to obtain precise stability statements. This is done by using the continuation theorem together with certain exact sequences of cohomology groups determined by the Morse decompositions of the associated isolated invariant sets. We again make use of the gradient-like nature of the equations. The final section illustrates how neatly linearization techniques can be used to obtain stability (or rather, instability) results for solutions of reaction-diffusion equations.

§A. The Structure of Weak Shock Waves

We consider the hyperbolic genuinely nonlinear system of conservation laws (cf. Chapter 15) in n dependent variables

$$u_t + f(u)_x = 0, \qquad (x, t) \in \mathbf{R} \times \mathbf{R}_+, \tag{24.1}$$

where df_u, the Jacobian matrix of f at u, has real and distinct eigenvalues $\lambda_1(u) < \cdots < \lambda_n(u)$ with corresponding left and right eigenvectors l_i, r_i, $i = 1, 2, \ldots, n$. We let $(u_l, u_r; s)$ denote a k-shock wave solution of (23.1); i.e., a solution of the form

$$u(x, t) = \begin{cases} u_l, & \text{if } x - st < 0, \\ u_r, & \text{if } x - st > 0, \end{cases} \tag{24.2}$$

where the jump condition

$$s(u_l - u_r) = f(u_l) - f(u_r), \tag{24.3}$$

as well as the entropy conditions

$$\lambda_1(u_l) < \cdots < \lambda_{k-1}(u_l) < s < \lambda_k(u_l) < \cdots < \lambda_n(u_l),$$

$$\lambda_1(u_r) < \cdots < \lambda_k(u_r) < s < \lambda_{k+1}(u_r) < \cdots < \lambda_n(u_r), \tag{24.4}$$

are assumed to hold. The problem is to determine whether the shock-wave "admits structure." Namely, whether or not the equation (with "viscosity" added)

$$u_t + f(u)_x = \varepsilon u_{xx}, \qquad \varepsilon > 0, \quad (x, t) \in \mathbf{R} \times \mathbf{R}_+, \tag{24.5}$$

admits a travelling wave solution of the form

$$u = u\left(\frac{x - st}{\varepsilon}\right), \tag{24.6}$$

which tends to the given shock wave solution $(u_l, u_r; s)$ as $\varepsilon \to 0$. If u as defined by (24.6) is to be solution of (24.5), then u must satisfy the system of ordinary differential equations

$$-su_\xi + f(u)_\xi = u_{\xi\xi}, \qquad \xi = \frac{x - st}{\varepsilon}.$$

This equation can be integrated once to give

$$-su + f(u) + C = u_\xi, \tag{24.7}$$

where C is a constant. Now if $u(\xi)$ is to converge to the given shock wave as $\varepsilon \to 0$, then we see that we must have

$$\lim_{\xi \to -\infty} u(\xi) = u_l, \qquad \lim_{\xi \to +\infty} u(\xi) = u_r. \tag{24.8}$$

It follows that the left-hand side of (24.7) vanishes at both u_l and u_r. Hence using u_l we find $C = su_l - f(u_l)$, and (24.7) becomes

$$u_\xi = -s(u - u_l) + f(u) - f(u_l). \tag{24.9}$$

Observe that (24.3) shows that u_r is also a rest point of the system. If we use the notation

$$V(u) = -s(u - u_l) + f(u) - f(u_l),$$

then our problem is to find a solution of the ordinary differential equation

$$u_\xi = V(u), \tag{24.10}$$

which satisfies the "boundary" conditions (24.8). That is, we are to find an orbit of (24.10) which "connects" the rest points u_l and u_r; such a solution is called a *heteroclinic orbit*.

There is an alternate topological way of expressing the fact that two rest points are connected by an orbit; namely, we can say that the stable manifold[1] of one intersects the unstable manifold of the other. If this situation is

---

[1] Recall that the stable manifold of a point is the manifold composed of the totality of orbits which tend to the point in positive time; the unstable manifold is defined similarly.

to be "structurally stable" (i.e., remain true under small perturbations), then the sum of the dimensions of the stable and unstable manifolds must exceed that of the space. It is interesting that the entropy inequalities (24.4), allow us to explicitly compute these dimensions.

The linearized equations at u_l and u_r are easily calculated from (24.10) and turn out to be

$$y' = (-sI + df_u)y,$$

where either $u = u_l$, or $u = u_r$. The number of positive eigenvalues of the matrix $-sI + df_u$, gives the dimension of the unstable manifold at u, while the number of negative eigenvalues is the dimension of the stable manifold at u. Now using (24.4) we find that the unstable manifold at u_l has dimension $n - k + 1$, and the stable manifold at u_r has dimension k; their sum is $n + 1$. It seems reasonable to suspect that in this case the problem (24.8), (24.10) is solvable, and we shall indeed show that this contention is correct. Before doing this however we need the following two results. The first one is similar to Theorem 22.33.

Theorem 24.1. *Let V be a vector field in $\mathbf{R}^n$ which admits an isolating neighborhood N containing precisely two rest points x_1 and x_2. If $h(S(N)) \neq h(x_1) \vee h(x_2)$, then there is an orbit γ of V in N which is different from x_1 and x_2. If V is also gradient-like in N, then γ connects x_1 and x_2.*

Proof. If x_1 and x_2 were the only complete orbits in N, then $S(N)$, the maximal invariant set in N would be x_1 and x_2; whence by the addition formula (Theorem 22.31) $h(S(N)) = h(x_1) \vee h(x_2)$. If V is gradient-like in N, then there is a function P such that $V \cdot \nabla P > 0$ in $N \backslash \{x_1, x_2\}$. Since γ is compact, P achieves both its maximum and minimum on γ. Since ∇P vanishes at these points, they must be x_1 and x_2 (in some order); thus γ connects them. $\square$

In certain contexts we cannot show that our equations are gradient-like in the entire isolating neighborhood so this theorem is not directly applicable. Thus we must rely on a somewhat different strategy (albeit in the same spirit as above). Namely, we have the following theorem. (See Chapter 22, §B for the definitions.)

Theorem 24.2. *Let V be a vector field in $\mathbf{R}^n$ which admits an isolating block B. Assume that there is a hypersurface Γ separating B into two isolating blocks B_1 and B_2 and that $\Gamma = b_1^+ \cap b_2^-$. If $h(S(B)) \neq h(S(B_1)) \vee h(S(B_2))$ then there is an orbit γ of V whose ω-limit set is in B_2 and whose α-limit limit set is in B_1. If in addition each B_i contains precisely one rest point x_i, and V is gradient-like in each B_i, then γ runs from x_1 to x_2.*

Proof. The hypothesis on the indices implies that there is an orbit γ of V in B which is not contained in either B_i. Since $B_1 \cap B_2 = \Gamma$, γ crosses from B_1 into B_2 precisely once and has its α- and ω-limit sets in the required B_i's. The gradient-like nature of V forces the orbit to connect the two rest points in the stated direction. $\square$

We return now to the shock-structure problem. It is convenient to consider $u_l = u_0$ as fixed and allow $u_r = u(\rho)$ to vary along the shock curve; thus we may assume $\rho \leq 0$, ρ is sufficiently small, and $u_0 = u(0)$, (see Chapter 17, §B). We know that in addition to (24.4) the following conditions hold:

(i) $u(\rho) = \rho[r_k + \tilde{u}(\rho)]$, where $r_k = r_k(u_0)$,
(ii) $\tilde{u}(\rho) = O(\rho)$,
(iii) $s(\rho) = \lambda_k + \rho/2 + O(\rho^2)$, where $\lambda_k = \lambda_k(u_0)$,
(iv) $s(\rho)(u(\rho) - u_0) = f(u(\rho)) - f(u_0)$.

Furthermore there is no loss in generality if we assume that $u_0 = f(u_0) = 0$; then (24.9) becomes

$$u' = -s(\rho)u + f(u), \qquad ' = \frac{d}{d\xi}. \qquad (24.11)$$

Using (iv) we see $u = 0$, and $u = u(\rho)$ are rest points, while the simplicity of λ_k together with (iii) shows that there is a ball around $u = 0$ such that for ρ sufficiently small these are the only zeros of $f(u) - s(\rho)u$ in this ball. In what follows we shall restrict our attention to this ball and to these values of ρ.

Let $v = (v_1, \ldots, v_n)$ be coordinates relative to the basis $(r_1, \ldots, r_n)$ and define $v_- = (v_1, \ldots, v_{k-1})$, and $v_+ = (v_{k+1}, \ldots, v_n)$. Then (24.11) takes the form

$$v'_- = A_-(\rho)v^- + O_2(v),$$

$$v'_k = (\lambda_k - s(\rho))v_k + \tfrac{1}{2}l_k d^2 f_0(v, v) + O_3(v), \qquad (24.12)$$

$$v'_+ = A_+(\rho)v^+ + O_2(v).$$

From the entropy inequalities (24.4) (with $u_l = u_0$ and $u_r = u(\rho)$), we see that $df_0 - s(\rho)I$, restricted to the span of $r_1, \ldots, r_{k-1}$ has negative eigenvalues. We may thus assume that $A_-(\rho)$ is (uniformly) negative definite for all small ρ. Similarly, $A_+(\rho)$ can be assumed to be positive definite.

We change variables, $w = \rho^{-1}v$, and extend our decompositions by replacing v_+ and v_k by w_+ and w_k, respectively. Using the relations

$$l_k d^2 f(v, v) = (v_k)^2 l_k d^2 f(r_k, r_k) + (\|v_+\| + \|v_-\|)\|v\|O(1)$$

$$= \rho^2[(w_k)^2 + (\|w_-\| + \|w_+\|)(\|w\|)O(1)],$$

and

$$s(\rho) = \lambda_k + \frac{\rho}{2} + \rho^2 O(1),$$

(24.12) becomes

$$w'_- = A_- w_- + \rho O_2(w),$$

$$w'_k = \tfrac{1}{2}\rho[(w_k)^2 - w^k] + \rho[(\|w_+\| + \|w_-\|)\|w\|O(1)] + \rho^2 O_3(w),$$

$$w'_+ = A_+ w_+ + \rho O_2(w). \tag{24.13}$$

We shall now construct isolating blocks for these equations. For a given $\varepsilon > 0$ we define three closed sets in w-space by:

$$B = \{w : |w_-| \le \varepsilon, |w_+| \le \varepsilon, -\tfrac{1}{2} \le w_k \le \tfrac{3}{2}\},$$

$$B_1 = \{w \in B : w_k \le \tfrac{1}{2}\},$$

$$B_2 = \{w \in B : w_k \ge \tfrac{1}{2}\}.$$

Lemma 24.3. *For sufficiently small ρ and sufficiently small ε (ε independent of ρ), the sets B_1, B_2, and B are isolating blocks.*

Proof. Note that if $w \in \partial B$, then either $|w_-| = \varepsilon$ or $|w_+| = \varepsilon$ or else $w_k = -\tfrac{1}{2}$ or $w_k = \tfrac{3}{2}$.
On the set $|w_-| = \varepsilon$,

$$(|w_-|^2)' = 2 < w_-, A_- w_- > + \rho O(1) \le -c\varepsilon^2 + \rho O(1)$$

for some constant $c > 0$. Clearly this derivative is negative for sufficiently small ρ. Thus $\{|w_-| = \varepsilon\} \subset b^-$; similarly, $\{|w_+| = \varepsilon\} \subset b^+$. Next we have

$$w'_k = \rho[w_k^2 - w_k + (\|w_+\| + \|w_-\|)\|w\|O(1) + \rho O_3(w)]/2$$
$$= \rho[w_k^2 - w_k + \rho\varepsilon O(1)],$$

if $-\tfrac{1}{2} \le w_k \le \tfrac{3}{2}$ and $|w_+|^2 \le 2\varepsilon, |w_-|^2 \le 2\varepsilon$. Since $\rho < 0$, it is easy to check that for ε sufficiently small (independent of ρ), and ρ sufficiently small, that w'_k will be negative for $w_k = \tfrac{1}{2}$, positive for $w_k = -\tfrac{1}{2}$, and negative for $w_k = \tfrac{3}{2}$. This completes the proof. $\square$

Now let S_1, S_2, and S denote the maximal invariant sets in B_1, B_2, and B, respectively. We shall compute the indices of these sets. This is the main content of the next lemma.

Lemma 24.4. (a) $h(S_1) = \Sigma^{n-k-1}$, $h(S_2) = \Sigma^{n-k}$, and $h(S) = \bar{0}$.
  (b) $b_1^+ \cap b_2^- = B \cap \{w_k = \frac{1}{2}\}$.

Proof. From the last lemma we have

$$b_1^+ = B_1 \cap \{|w_+| = \varepsilon\} \cup B_1 \cap \{|w_{k-1}| = \frac{1}{2}\},$$

$$b_2^+ = B_2 \cap \{|w_+| = \varepsilon\},$$

$$b_1^- = B_1 \cap \{|w_-| = \varepsilon\},$$

$$b_2^- = B_2 \cap \{|w_-| = \varepsilon\} \cup B_2 \cap \{|w_k - 1| = \frac{1}{2}\},$$

$$b^+ = B \cap \{|w_+| = \varepsilon\} \cup B \cap \{w_k = -\frac{1}{2}\},$$

$$b^- = B \cap \{|w_-| = \varepsilon\} \cup B \cap \{w_k = \frac{3}{2}\}.$$

It is clear from this that (b) holds. Next, since the k-ball D^k is contractible (i.e., has the homotopy type of a point),

$$h(S_2) = [B_2/b_2^+] \sim (D^k \times D^{n-k})/(D^k \times S^{n-k-1}) \sim (D^{n-k}/S^{n-k-1}) \times D^k$$
$$\sim D^{n-k}/S^{n-k-1} = \Sigma^{n-k}.$$

Similarly,

$$h(S_1) = [B_1/b_1^+] \sim (D^{k+1} \times D^{n-k-1})/(D^{k+1} \times S^{n-k-2}) \sim \Sigma^{n-k-1}.$$

Finally to see that B/b^+ has trivial homotopy type, we note that b^+ is contractible in ∂B so that $h(S) = \bar{0}$. That is, the pair (B, b^+) is homeomorphic to $D^{k-1} \times (D^{n-k+1}, J)$, where J is the boundary of the $(n - k + 1)$-cube minus one of its faces. Thus D^{n-k+1} can be deformed into J along the lines of the standard projection ϕ from an exterior point; see Figure 24.1. $\square$

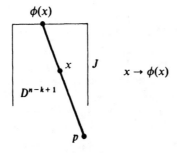

Figure 24.1

As a consequence of Lemma 24.4 and Theorem 24.2 we have the following corollary.

Corollary 24.5. *There is an orbit of (24.9) which has its ω-limit set in B_2 and its α-limit set in B_1.*

Observe now that $0 \in B_1$ and by moving the hyperplane $w_k = \frac{1}{2}$ closer to 0, if necessary, we may assume that $u(\rho)$ is in B_2. As we have remarked earlier, these are the only rest points in B. We can now state our theorem.

Theorem 24.6. *Let u_l and u_r be two states connected by a shock-wave solution of the hyperbolic genuinely nonlinear system (24.1). If $|u_l - u_r|$ is sufficiently small,[2] then this shock has structure; i.e., the problem (24.8), (24.9) has a solution.*

Proof. In view of the last corollary and Lemma 24.4 it suffices to show that V is gradient-like in each B_i, $i = 1, 2$. We shall only give the details for B_1; the assertion for B_2 will follow by a symmetric argument.

The function we choose is

$$P(w) = \langle w_-, A_- w_- \rangle + \langle w_+, A_+ w_+ \rangle + w_k^2.$$

If we differentiate P along orbits and use the fact that $A_\pm$ are nonsingular and that $A_\pm^2$ are positive definite, we find

$$P(w)' \geq c(|w_-|^2 + |w_+|^2 + \rho w_k^2) + \rho(|w_-| + |w_+|)|w|^2 O(1),$$

where the constant c depends on $A_\pm$, $A_\pm^2$, is well as on $w_k^2 - w_k$, in the middle equation in (24.13). The error term includes the error term in all three equations. Replacing ε by a smaller one if necessary, we see that

$$P(w)' \geq c(|w_-|^2 + |w_+|^2 + \rho w_k^2) + \rho\varepsilon|w|^2 O(1) > 0$$

for sufficiently small ε and ρ. This completes the proof. $\square$

§B. The Structure of Magnetohydrodynamic Shock Waves

As another application of the Conley index, in particular of Theorem 24.1, we will consider the structure problem for magnetohydrodynamic shock waves of arbitrary strength. An essential difference between this problem and the one considered in §A, is that in the latter we were able to explicitly construct an isolating block. This was due to the fact that a (weak)k-shock

---

[2] In this generality, we know that only "weak" shocks exist; see Chapter 17, Theorem 17.11.

singles out a specific direction. Namely, for $|u_r - u_l|$ sufficiently small, we know that u_r is approximately in the direction of the r_k eigenvector at u_l. In the problem which we consider here, the rest points are not close, the problem is a global one, and we must construct isolating neighborhoods in a more general way. Furthermore, there is the added complication of having four rest points rather than two.

We shall write down the equations and refer the reader to [KL] for details concerning their derivation. The problem is one in $\mathbf{R}^6$; namely let

$$u = (x_1, x_2, y_1, y_2, V, T) \equiv (x, y, V, T) \in \mathbf{R}^6,$$

where x_1 and y_1 are mechanical variables (velocity components), x_2 and y_2 are electromagnetic variables (components of the magnetic field), and V and T denote thermodynamic variables, the specific volume and temperature, respectively. Of course these latter two variables are to be positive.

Let A denote the 2×2 matrix

$$A = \begin{bmatrix} 1 & -\delta \\ \delta & V \end{bmatrix},$$

where δ is a positive constant. Q and P are real-valued functions defined by

$$Q = \tfrac{1}{2}\langle Ax, x \rangle + \tfrac{1}{2}\langle Ay, y \rangle + \varepsilon x_2 + \frac{V^2}{2} - JV + E - f(V, T),$$

$$P = \frac{Q}{T}, \tag{24.14}$$

where ε_2, J, and E are constants and $f(V, T)$ is the *Helmholtz free energy function* satisfying

$$f_V = p \quad \text{and} \quad f_T = s,$$

where p and s denote the pressure and entropy, respectively. In addition the *internal energy* e is obtained from the formula

$$e = f + Ts.$$

The dissipative mechanisms are components of the "viscosity" vector

$$\lambda = (\mu, \nu, \mu_1, \kappa),$$

where μ, ν, μ_1, and κ denote viscosity, second viscosity, magnetic viscosity and thermal conductivity, respectively; they are all taken to be positive constants. We define the "viscosity matrix" $B = B(T, \lambda)$ by

$$B = T^{-1} \operatorname{diag}(\mu, \nu, \mu, \nu, \mu_1, \kappa T^{-1}).$$

In these terms our differential equations take the form

$$Bu' = \nabla P(u). \tag{24.15}$$

In [Gr] it is shown that the system admits (at most!) four rest points u_i, $0 \le i \le 3$ all of which are nondegenerate, with respective indices $h(u_i) = \Sigma^{3-i}$, $0 \le i \le 3$. Furthermore, $P(u_i) < P(u_{i+1})$, for $i = 0, 1$, and 2. The problem is to show that there is an orbit of (24.15) running from u_0 to u_1, and also one running from u_2 to u_3. The transition $u_0 \to u_1$ is called the "slow" mhd shock, and $u_2 \to u_3$ is the "fast" mhd shock. Thus, we are to show that both *fast and slow mhd shocks, of arbitrary strength, have structure.*

Note that if we differentiate P along orbits of (24.15) we get

$$P' = (\nabla P, u') = (\nabla P, B^{-1}\nabla P) > 0$$

in $V, T > 0$, except at the critical points of P^3. We thus have

Fact 1. The system (24.15) is gradient-like in the region $V, T > 0$.

Before proceeding further we must place the following (rather reasonable) assumptions on the thermodynamic functions:

(H$_a$) p, e, and s are positive in $V, T > 0$.
(H$_b$)   For fixed $T > 0$, $p(V, T) \to \infty$ as $V \to 0$.
(H$_c$)   Given $V_0, \kappa > 0$, there is a $T_0 > 0$ such that if $0 < V \le V_0$, and $T \ge T_0$, then $e(V, T) > \kappa$.
(H$_d$)   For $0 < V \le V_0$, $s(V, T) \to 0$, uniformly in V as $T \to 0$.
(H$_e$)   If $p = p(V, s)$, then $p_V < 0$, $p_{VV} > 0$, and $p_s > 0$.

A few comments are in order concerning these hypotheses. In (H$_a$), p is naturally positive, and e and s are only defined up to an additive constant; thus there is no loss in generality to assume that they are positive. (H$_b$) is equivalent to the fact that for T fixed, $p \to \infty$ when the density tends to infinity. (H$_c$) says that for bounded volume, the internal energy can be made arbitrarily large for sufficiently high temperature. (H$_d$) is called Nernst's third law of thermodynamics. It can be replaced by

(H$'_d$)   If $s = s(V, T)$, then $s_V > 0$ and $s_T > 0$, and for fixed $V > 0$, $s(V, T)$ tends to a limit which is independent of V as $T \to 0$.

Finally, we have had occasion to use (H$_e$) in Chapter 18, §A; $p_V < 0$ is forced upon us by thermodynamics, while the latter two inequalities are usually referred to as Weyl's hypotheses.

---

[3] P can be considered as a "generalized entropy."

We can rewrite equations (24.15) as

$$B_0 x' = Ax + \bar{\varepsilon}, \qquad B_0 y' = Ay,$$

$$\mu_1 V' = \tfrac{1}{2}(x_2^2 + y_2^2) + V - J + p(V, T),$$

$$\kappa T' = -Q + Ts = -(Q + f) + e, \qquad (24.16)$$

where $B_0 = \operatorname{diag}(\mu, \nu)$, and $\bar{\varepsilon} = (0, \varepsilon)$. Using these we find if $\varepsilon \neq 0$, then at the rest points the following equations must hold:

$$V' = F_1(V, T) \equiv \tfrac{1}{2}\varepsilon^2(V - \delta^2)^{-2} + V - J + p(V, T) = 0,$$

$$T' = F_2(V, T) \equiv \tfrac{1}{2}\varepsilon^2(V - \delta^2)^{-1} - \tfrac{1}{2}V^2 + JV - E + e(V, T) = 0.$$

Since $\partial F_1/\partial T = p_T > 0$, and $\partial F_2/\partial T = e_T > 0$, it follows that $F_1 = 0$ and $F_2 = 0$ are both graphs of functions $T = T_i(V), i = 1, 2$. Using our hypotheses, they can be depicted as in Figure 24.2, where $\bar{u}_i$ is the projection of u_i onto $V - T$ space, $i = 1, 2, 3, 4$.

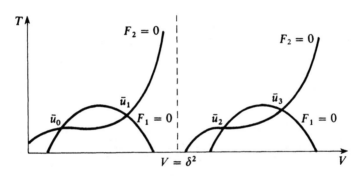

Figure 24.2

Observe that the constant E appears in the expression for F_2, but not in F_1. As $E \to +\infty$, both components of the curve $F_2 = 0$ pull off of the respective components of $F_1 = 0$, so that we have

Fact 2. For sufficiently large E the equations (24.15) admit no rest points.

Now fix both the viscosity vector λ and the constant E and define $S' = S'$ (λ, E) to be the set of points on complete bounded orbits of (24.15) contained in $V, T > 0$. We then have the following proposition.

Proposition 24.7. $S'(\lambda, E)$ is a bounded set, for each fixed λ and E.

Proof. Observe that if on some bounded orbit V exceeded J, then $V' > 0$ at that point so $V \to \infty$ on the orbit and thus the orbit is not in S'. Hence $V \leq J$ on S'. We shall show that $|x| + |y|$ is bounded on S' in the next lemma. Assuming this, it follows from (21.14) that $(Q + f)$ is bounded on S'. Then the equation $T' = -(Q + f) + e$, together with hypothesis (H_c) implies that T is bounded on S'. $\square$

Lemma 24.8. $|x| + |y|$ *is bounded on* $S' = S'(\lambda, E)$.

Proof. Since $V \leq J$, we see that the matrix A is bounded. It follows easily from this that x, say, can decrease at most exponentially on orbits in S'. Thus by taking $x(0)$ sufficiently large, $x(t)$ cannot get arbitrarily small in a pre-assigned time t. Hence given any $\kappa > 0$, and $t_0 > 0$, there is an $M > 0$ such that on any orbit segment starting at a point x_0, with $|x_0| > M$, which has time length t_0, either V sometimes exceeds J or $|x(t)| \geq \kappa$, $0 \leq t \leq t_0$.

Now let $\theta = \arctan(x_2/x_1)$, $\theta \equiv 0 \,(\mathrm{mod}\, 2\pi)$ on the positive x_1-axis. A straightforward calculation gives

$$\theta' = \left[-\frac{\delta}{v} \cos^2 \theta + \left(\frac{V}{v} - \frac{1}{\mu} \right) \sin \theta \cos \theta + \frac{\delta}{\mu} \sin^2 \theta \right] + \frac{\varepsilon}{v|x|} \cos \theta.$$

Now for $|x|$ sufficiently large, say $|x| \geq \kappa$, we can find constants $c_1 < 0$, and $\alpha > 0$, such that $\theta' \leq c_1 < 0$, if $|\theta| < \alpha$; see Figure 24.3.

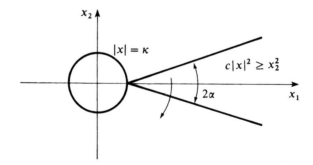

Figure 24.3

Thus if $V \leq J$ and $|x| \geq \kappa$, $\theta - \theta_0 \leq c_1 t$, so in the cone $x_2^2 \leq c|x|^2$ (i.e., $|\theta| < \alpha$), $t < 2\alpha/(-c_1)$. That is, any orbit segment, on which $V \leq J$, and $|x| \geq \kappa$, either stays out of the cone or else spends a time in the cone which is bounded independently of the length of the orbit segment. Having crossed out of the cone, it cannot re-enter it.

We choose κ so large that $c\kappa^2 \geq 2(J + \mu_1)$, and let $t_0 = J + 2\alpha/(-c_1)$. Then by what we have said above, we can find an $M > 0$ such that if $|x_0| > M$, then along the orbit segment starting at x_0 and running for a time length t_0, either V sometimes exceeds J or $|x(t)| \geq \kappa$, $0 \leq t \leq t_0$. But in S', we know that the former possibility cannot occur; hence $|x(t)|^2 \geq \kappa^2 \geq 2(J + \mu_1)/c$, $0 \leq t \leq t_0$, if $|x(0)| = |x_0| > M$.

We shall show $|x| \leq M$ on S'. Suppose not; then there is an orbit in S' which starts at $x(0) = x_0$ with $|x_0| > M$, and hence by what we have just shown, $|x(t)|^2 \geq \kappa^2 \geq 2(J + \mu_1)/c$, $0 \leq t \leq t_0$. If the orbit segment is ever in the cone, it gets out in time $2\alpha/(-c_1)$ and thus since $t_0 = J + 2\alpha/(-c_1)$, $x_2^2(t)/c \geq |x(t)|^2 \geq 2(J + \mu_1)/c$ for a time interval at least J. If the orbit segment is always outside the cone, then $x_2(t)^2 \geq c|x(t)|^2 \geq 2(J + \mu_1)$ for a time interval at least J; i.e., $\frac{1}{2}x_2(t)^2 \geq J + \mu_1$ for a time interval at least J. Thus in every case, if $|x(0)| > M$, then $\frac{1}{2}x_2(t)^2 \geq J + \mu_1$ for an interval of time equal to J. Therefore,

$$\mu_1 V' = \tfrac{1}{2}x_2^2 + \tfrac{1}{2}y_2^2 + V + p - J \geq \tfrac{1}{2}x_2^2 - J \geq \mu_1$$

for a J-time interval; i.e., $V' \geq 1$ for a J-time interval. Thus, for some a,

$$\int_a^{a+J} V' \geq J \quad \text{so } V(a + J) > J.$$

Such an orbit cannot lie in S'. Hence $|x| \leq M$ (and similarly, $|y| \leq M$), on S' This completes the proof. $\square$

We now shall construct the isolated invariant set. For this we let R denote the set of rest points of (24.15). If $u \in R$ then $T'(u) = 0$ so $-Q + Ts = 0$; whence at u, $s = Q/T = P$. Thus $\min s|_R = \min P|_R$ is bounded away from zero since R is a finite set. Using (H_d), we can find a neighborhood U of the interval $[0, J]$ in $V - T$ space, and a real number κ such that

$$\min s|_R > \kappa > \max s|_U. \tag{24.17}$$

For fixed (λ, E) let $S = S(\lambda, E)$ be the set of all points on complete orbits of (24.15) contained in the set $\{u: V > 0, T > 0, P > \kappa\}$.

Proposition 24.9. *For each fixed (λ, E), $S(\lambda, E)$ is an isolated invariant set and the sets $\{S(\lambda, E)\}$ are related by continuation.*

Proof. Fix the pair (λ, E) and observe that $S(\lambda, E) \subset S'(\lambda, E)$. Thus $S(\lambda, E)$ is a bounded invariant set. If we can show that $S(\lambda, E)$ is also closed, then it will be compact and thus isolated. Now in order to show that $S(\lambda, E)$ is closed it suffices to show that it has no limit points on the sets $\{V = 0\}$, $\{T = 0\}$, or $\{P = \kappa\}$.

Suppose first that $S = S(\lambda, E)$ has a limit point in the set $\{V = 0\}$. Then there is a sequence $\{u'_n\} \subset S$ such that the V-coordinates converge to zero.

Since these points lie on orbits in S, we can find a subsequence $\{u_n\} \subset \{u'_n\}$ whose V-coordinates tend to zero and moreover $V'(u_n) = 0$. From (24.16) we have at u_n,

$$0 = \tfrac{1}{2}(x_2^2 + y_2^2) - J + V_n + p(V_n, T_n). \tag{24.18}$$

Since $u_n \in S$, $T_n \leq K$ for some K independent of n. If $\{T_n\}$ did not converge to zero, then some subsequence of the T_n's would converge to some $\bar{T} > 0$. Then (H_b) would imply that $p \to +\infty$ along this sequence, thus violating (24.18). It follows that there is a sequence of points whose T coordinates converge to zero and at which $T' = 0$. Since $T' = -Q + Ts$, we have $s = QT^{-1} = P$, at these points. But these points have their (V, T) coordinates eventually in U, so that s, and hence P, is less than κ at these points—they are therefore not in S. We see then that S has no limit points in either of the sets $V = 0$ or $T = 0$.

Now suppose that S has a limit point in the set $\{P = \kappa\}$. At such a point, the derivative of P along the orbit is well defined (since neither V nor T is zero) and positive, since (24.17) shows that this point is obviously not a critical point of P. Hence this point, as well as a neighborhood of it, leaves the set $P \geq \kappa$ as t decreases. Thus no points of S can lie in this neighborhood; this contradiction completes the proof that S is closed.

We have therefore proved that $S = S(\lambda, E)$ is compact for each (λ, E). Let $N = N(\lambda, E)$ be a bounded neighborhood of $S(\lambda, E)$; $\bar{N}$ is compact and $S \subset \text{int } N$. Thus N is an isolating neighborhood and S is the maximal invariant set in N. It follows that S is an isolated invariant set. Since it is obvious that N is an isolating neighborhood for all nearby flows, the sets $S(\lambda, E)$ are all related by continuation. This completes the proof of the proposition. $\square$

For large E, $S(\lambda, E)$ has no rest points of (24.15) (Fact 2), and thus using the gradient-like nature of the flow (Fact 1), $S(\lambda, E)$ is empty for sufficiently large E. It follows that $h(S(\lambda, E)) = \bar{0}$, for all E and all λ.

Now choose a constant c such that $P(u_1) < c < P(u_2)$, and define

$$S_{01} = S \cap \{P < c\}, \qquad S_{23} = S \cap \{P > c\}.$$

Then the sets S_{01} and S_{23} are isolated invariant sets, and have index $\bar{0}$. Moreover, $\{u_0, u_1\} \subset S_{01}$ and $\{u_2, u_3\} \subset S_{23}$. Since

$$h(u_0) \vee h(u_1) = \Sigma^0 \vee \Sigma^1 \neq \bar{0},$$

and similarly, $h(u_2) \vee h(u_3) \neq \bar{0}$, it follows from Theorem 24.1 that there exists an orbit of (24.15) running from u_0 to u_1, as well as one running from u_2 to u_3. We have therefore proved the following theorem.

Theorem 24.10. *Assume that hypotheses (H_a)–(H_e) hold, and that λ has positive entries. Then fast and slow mhd shocks of arbitrary strength have structure.*

§C. Periodic Travelling Waves

The example which we shall discuss here concerns the Nagumo equations:

$$u_t = \varepsilon v, \qquad v_t = v_{xx} + f(v) - u, \qquad \varepsilon > 0, \qquad (24.19)$$

and the problem is to find a periodic travelling wave solution. That is, we seek a periodic solution of (24.19) which depends only on the variable $\xi = x + \theta t$. The significant difference now from the previous examples is that here θ is not given; we must deal with a one-parameter family of flows.

Our approach to the problem is based on the fact that we can assume ε is small so that v changes more rapidly than u. The idea is to first consider a limiting case where $\varepsilon = 0$, so that the slow variable u doesn't change at all. The effect of this is that we are able to consider a lower-dimensional system, which however, depends on more parameters. We shall construct "singular" orbits for the limiting system; these will be contained in isolating blocks which persist under perturbation to small $\varepsilon > 0$. The isolating blocks will be shown to have nontrivial index so that they contain complete orbits in their interior. A continuation argument is needed to show that periodic orbits indeed exist.

Let $\xi = x + \theta t$; then in this variable if prime denotes $d/d\xi$, (24.19) becomes

$$\theta u' = \varepsilon v, \qquad \theta v' = v'' + f(v) - u.$$

If we let $\sigma = \varepsilon/\theta$, and $w = v'$, the equations can be written as the following first-order system:

$$u' = \sigma v, \qquad v' = w, \qquad w' = \theta w + u - f(v). \qquad (24.20)$$

We assume that f is a cubic polynomial; f and its integral F are depicted in Figure 24.4. We note that (24.20) admits the origin as the only rest point.

Now let's consider the limiting case $\varepsilon = 0$. The equations become

$$u' = 0, \qquad v' = w, \qquad w' = \theta w + u - f(v), \qquad (24.21)$$

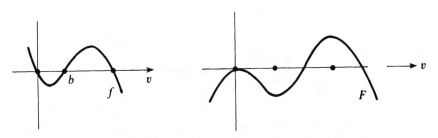

Figure 24.4

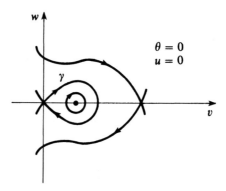

Figure 24.5

which has the rest point set $\{(u, v, w): w = 0, u = f(v)\}$. We now discuss the *pair* of equations obtained from (24.21),

$$v' = w, \qquad w' = \theta w + u - f(v), \tag{24.22}$$

where both θ and u will be thought of as parameters.

If $\theta = 0$, then these equations go over into the Hamiltonian system

$$v' = w, \qquad w' = u - f(v) \tag{24.23}$$

with Hamiltonian

$$H(v, w) = \frac{w^2}{2} + F(v) - uv. \tag{24.24}$$

For $u = 0$, they have the phase portrait depicted in Figure 24.5. Note that there is a bounded orbit γ which "begins" and "ends" at the hyperbolic rest point $(0, 0)$; such an orbit is called a *homoclinic* orbit. Now let $u = \tilde{u} > 0$ be chosen in such a way that the two maxima of $F - \tilde{u}v$ are the same (see Figure 24.6); this is equivalent to saying that the positive and negative

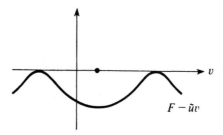

Figure 24.6

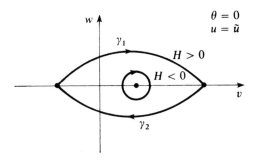

Figure 24.7

"humps" of $f - \tilde{u}$ have the same area. The phase portrait for equations (24.23), $(u = \tilde{u})$, is given in Figure 24.7. Note that here too there are bounded orbits, γ_1 and γ_2, which begin and end in hyperbolic rest points; i.e., the so-called *heteroclinic orbits*.

We next consider the equations (24.22), where $\theta > 0$. The Hamiltonian (24.24) now increases along orbits; i.e., $H' = \theta w^2$, and the phase portrait can be drawn using the level curves of H. For example, if θ is taken to be small and positive, then the phase portrait of

$$v' = w, \qquad w' = \theta w + \tilde{u} - f(v)$$

is drawn in Figure 24.8, where γ_1 and γ_2 are the dotted lines.

Observe that if $u > 0$, the graphs of $f(v) - u$ can be portrayed as in Figure 24.9. That is, the "middle" graph is $f(v) - \tilde{u}$, the higher one is $f(v) - \underline{u}$, $\underline{u} < \tilde{u}$ and the lower one is $f(v) - \bar{u}$, $\bar{u} > \tilde{u}$. It follows easily that since $H_{\tilde{u}} < 0$ in $v > 0$, that for small positive θ, there are unique values $\bar{u}(\theta)$ and $\underline{u}(\theta)$ with $\bar{u}(\theta) > \tilde{u} > \underline{u}(\theta)$ such that for these values of u there are homoclinic orbits

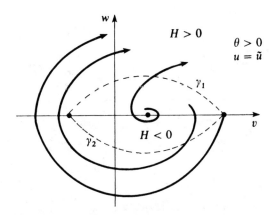

Figure 24.8

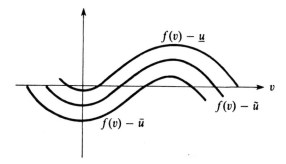

Figure 24.9

γ_1 and γ_2 which connect the two hyperbolic rest points and run in opposite directions; these are depicted in Figure 24.10(a) and (b). Here the level curves of H are the dotted lines, and the homoclinic orbits are as depicted.

Note that as $\theta \to 0$, both $\bar{u}(\theta)$ and $\underline{u}(\theta)$ tend to $\tilde{u}$, and these phase portraits tend to that of Figure 24.7. On the other hand as θ increases, $\bar{u}(\theta)$ increases and $\underline{u}(\theta)$ decreases until there is a value $\theta = \theta^*$ for which $\underline{u}(\theta^*) = 0$. *From now on, we assume that $0 < \theta < \theta^*$.*

For these values of θ we now consider the full set of equations (24.20). With reference to Figure 24.11, we have portrayed the rest point set when $\varepsilon = 0$ by the cubic curve and the orbits γ_1 and γ_2 are the ones from Figure 24.10(a) and (b) lying in their respectively u-levels. The depicted dotted simple closed curve is the "singular" periodic solution. Note that the portions in the cubic have the "correct" direction; this follows from the equation $\theta u' = \varepsilon v$. The contention is that for small $\varepsilon > 0$, there is a periodic orbit near the singular one, which follows it around in the same direction. To show this, we will build an isolating block around this singular orbit.

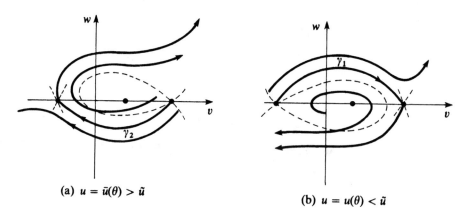

(a) $u = \bar{u}(\theta) > \tilde{u}$

(b) $u = \underline{u}(\theta) < \tilde{u}$

Figure 24.10

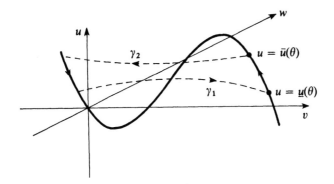

Figure 24.11

Consider first the right-hand vertical arc in Figure 24.11. Since the critical points for $\varepsilon = 0$ are hyperbolic, it follows that we can build an isolating block around this arc. Thus consider the tube with square cross-section around this arc, as shown in Figure 24.12(a); in Figure 24.12(b) is given a typical cross-section, the projection on $u = 0$. The fact that the arrows are drawn correctly in Figure 24.12(b), follows from a straightforward linearization argument. Thus, if $\varepsilon > 0$, the orbits cross vertically upward through the bottom and top of the tube (since $v > 0$ so $u' > 0$), and on the "sides" of the tube, orbits either strictly enter or strictly leave. A similar tube can be built around the left-hand vertical arc.

We now consider the horizontal orbits. In order to see how to construct tubes around these, it is necessary to determine how the orbits near the unstable manifolds of the hyperbolic rest points behave for u values close to $\bar{u}(\theta)$ and $\underline{u}(\theta)$. For example, if u is close to $\bar{u}(\theta)$, then from Figure 24.10(a), we see that the behavior can be depicted as in Figure 24.13. There is of course an analogous picture if u is near $\underline{u}(\theta)$.

We call attention to the fact that the "singular" flows are stable under perturbation, and in fact are "isolated." Thus it is quite reasonable to expect that for small $\varepsilon > 0$, there should be a periodic orbit near the singular one.

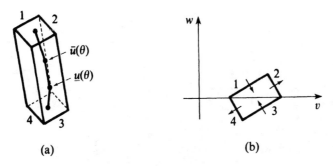

(a)   (b)

Figure 24.12

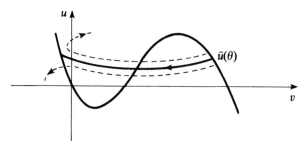

Figure 24.13

To build a tube around the "horizontal" arcs, we can consider Figure 24.14(a) (which is obtained from the last diagram). In this picture we show face 4 of the right-hand tube, and in it the vertical cubic arc consisting of points on the unstable manifold of the corresponding hyperbolic points at each u level (Figure 24.12). The point at level $\bar{u}(\theta)$ is carried into an entering face of the left-hand tube, and the rest of the arc at that level is carried by the flow in such a way that it stretches across this tube. In Figure 24.14(b) is shown a cross section of the right-hand (horizontal) tube (abcd). The "dots" are all at level $\bar{u}(\theta)$. We can make this construction in such a way that orbits (strictly) enter the sides c, c', a, and a', and (strictly) exit the sides b, b', d, and d'. Of course, we must check those points which lie on the boundary of both a vertical and horizontal tube; these can be a strict exit set for one tube and a strict entrance set for another tube. Such points could give internal tangencies for the union; cf. Figure 24.15. However, using Figures 24.12 and 24.13, we can check that such situations do not occur.

It follows that the union of the tubes is an isolating block B which has the topological type of a solid torus T. Furthermore, by examining the diagrams it is easy to see that the set of exit points, b^+, consists of two disjoint annuli on the torus, see Figure 24.16(a). Let S denote the maximal invariant set in B. In order to show that S is nonempty, it suffices to show that $h(S) \neq \bar{0}$ (Theorem 22.32). This is perhaps most easily done by showing that some cohomology group of B/b^+ is nonzero (recall that homotopic spaces have isomorphic

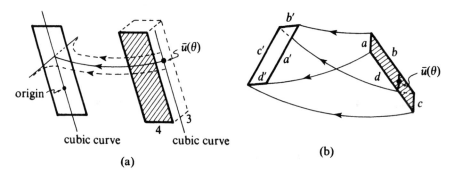

Figure 24.14

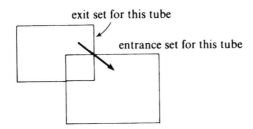

Figure 24.15

cohomology groups; see §D in Chapter 12). We shall in fact show that $H^2(B/b^+) \neq 0$. To this end, recall that if A is a closed subset of a topological space X, we have an exact sequence of cohomology groups (see Chapter 23, §B, equation (23.16))

$$\cdots \to H^n(X) \to H^n(A) \to H^{n+1}(X, A) \to \cdots.$$

Now if $n > 1$, $H^2(X/A) \approx H^2(X, A)$. In our case X is the solid torus, B, and A is the disjoint union of two circles, $A = S^1 \amalg S^1$. Now if $n = 1$, we have the exact sequence

$$H^1(X) \to H^1(A) \to H^2(X/A).$$

Since B is homotopic to a circle, it follows that $H^1(X) = \mathbf{Z}$. Also (Chapter 12, §D, 4C(vi)) $H^1(S^1 \amalg S^1) = \mathbf{Z} \oplus \mathbf{Z}$. Since these latter two groups are *not* isomorphic, we conclude that $H^2(B/b^+) = H^2(X/A) \neq \bar{0}$. This implies that $h(S) \neq \bar{0}$.

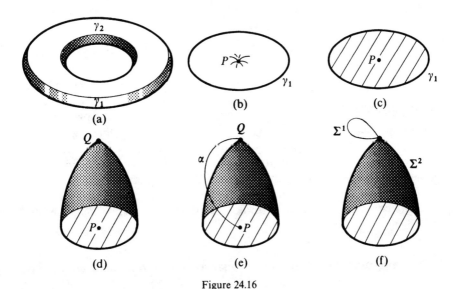

Figure 24.16

It is instructive to go a little further and actually compute the homotopy type of B/b^+. We can depict the two circles γ_1 and γ_2 on the boundary of T in Figure 24.16(a). (γ_1 is on the "outer" boundary, and γ_2 is on the "inner" one). We shall do the identification of b^+ to a point in two steps. First we collapse the inner circle to a point, P, obtaining a "pinched ball": (b). Then up to homotopy type, we obtain a pinched disk: (c). In order to collapse γ_1, the boundary of this disk, we adjoin a cone over γ_1: (d), where the point Q now is identified with γ_1. In order to identify P and Q, we attach an arc α between them: (e). Notice that we have Σ^2 (the shaded part), together with the arc α. Now if we move P up to the point Q, we finally obtain the desired homotopy type: (f); namely, $\Sigma^2 \vee \Sigma^1$; i.e., $h(S) = \Sigma^2 \vee \Sigma^1$.

Now let $\gamma \in S$, and note that orbits cannot turn around in B; i.e., they have positive angular velocity. Let D be a disk in B which is a surface of section for the flow, i.e., the Poincaré map T (the first return map; see Chapter 12, §A) is well defined on $S \cap D$. Let U be the set of points on D such that $u \in U$ implies $Tu \in U$. Then U is open relative to D, and U gets mapped into $D \backslash (D \cap \partial B)$. Thus the degree of T is well defined on U (see Chapter 12, §A), since B is a block.

Observe now that again since B is an isolating block, if we continue the equations always maintaining a block throughout the deformation, then the degree of the "continued" map T continues to be defined; this follows from the homotopy invariance of the degree. Note too that this degree stays constant throughout the deformation. Thus in order to prove the existence of a periodic travelling wave, it suffices to show that the degree of T is non-zero, and to do this, we shall continue our equations to a system in which we can make the computations; namely, to the van der Pol equations "crossed with a repelling critical point."

We consider the deformation

$$u' = \sigma(v - a), \qquad v' = w, \qquad w' = k(\theta w + u - f(v)),$$

where $0 \le a \le b$, and $k \ge 1$. We recall that $v = b$ is the smallest positive root of f; see Figure 24.4. If $a = b$ and k is very large then we see that our system continues to one in which w moves very quickly to the plane $w = 0$, and in this plane, the equations go over into the van der Pol equations. The corresponding set B continues to a neighborhood of the periodic orbit, crossed with an interval, and the degree of the corresponding Poincaré map is nonzero (it is well known that the van der Pol equations admit a unique attracting periodic orbit). The same is therefore true of $T|_U$, and thus our original equations admit a periodic travelling wave. This completes the proof. $\square$

We remark that one can use a similar technique to find a "homoclinic" travelling wave for the Nagumo equations (24.19); that is a solution of

(24.20) which tends to the rest point $(0, 0, 0)$ as $|\xi| \to \pm\infty$; see [Ca 1, Ha 2]. We shall outline this approach.

As before, we first consider the limiting case $\varepsilon = 0$; namely (24.22). For $\theta > 0$, the equations are gradient-like, and have 1, 2, or 3 rest points, depending on u. Consider those u-values such that there are precisely three rest points. Two of these are hyperbolic, and for some values of u and θ, these rest points are connected by an orbit of the corresponding equation. A "limit" homoclinic orbit is found which consists of two arcs connecting the hyperbolic points, and two arcs contained in the critical point set $\{w = 0, u = f(v)\}$, as before.

To find this "limit" homoclinic orbit, let $u = 0$, and find $\theta = \theta^*$ such that the flow looks like Figure 24.17(a). Now fix $\theta = \theta^*$, and find $u = u^*$ such

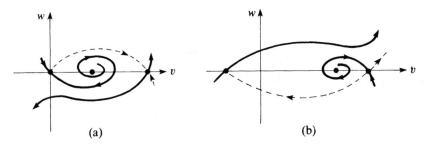

(a) (b)

Figure 24.17

that the corresponding flow looks like Figure 24.17(b). At this point we can see the "limiting" orbit in u, v, w space, with $\theta = \theta^*$; see Figure 24.18.

To show that this is a limit of homoclinic orbits as $\varepsilon \to 0$, one has to construct isolating blocks about the "limiting" orbit as see what happens as θ varies about θ^* and u varies about u^*. An argument similar to the previous one gives the desired result. We omit the details.

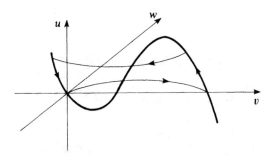

Figure 24.18

We remark that this construction also works for the Fitz-Hugh–Nagumo equations; see [Cy 1], for small ε. We have seen in Chapter 14, §E, that "small" initial data for these equations leads to solutions which tend to zero as $t \to \infty$; here we see that "big" solutions exist. This gives an example of the existence of a "threshold" phenomenon.

§D. Stability of Steady-State Solutions

We propose to now use the Conley index as a Morse-type index in order to obtain stability results. Thus we shall consider a scalar reaction–diffusion equation

$$u_t = u_{xx} + f(u), \qquad |x| < L, \quad t > 0, \tag{24.25}$$

in a single space variable. It will be especially convenient to consider this equation as an ordinary differential equation, $u' = A(u)$, in an appropriate infinite-dimensional space. The rest points of this latter equation are just the steady-state solutions of (24.25); i.e., solutions of

$$u'' + f(u) = 0, \qquad |x| < L, \tag{24.26}$$

where prime denotes differentiation with respect to x.

For certain interesting nonlinear functions f, we shall show that solutions of (24.26) are isolated rest points of (24.25). Under rather general homogeneous boundary conditions, the system (24.25) is gradient-like, so that these rest points are actually isolated invariant sets (Theorem 23.32). The index of the rest points will be obtained either by continuation or by some other topological technique such as computing it directly from an exact sequence of cohomology groups. This will allow us to find the *precise* dimension of the unstable manifolds of each of the steady-state solutions. Our techniques will be sufficiently general so as to allow us to obtain, qualitatively at least, the global picture of *all* solutions of (24.25), in certain cases.

We begin by considering (24.25) together with the homogeneous boundary conditions in $t > 0$,

$$\alpha u(\pm L, t) - \beta u_x(\pm L, t) = 0. \tag{24.27}$$

Here α and β are constants, $\alpha^2 + \beta^2 = 1$. If we let Φ be the functional

$$\Phi(u)(t) = \int_{-L}^{L} \left[\tfrac{1}{2} u u_{xx} + F(u) \right] dx, \tag{24.28}$$

where $F' = f$, then differentiating Φ with respect to t gives

$$\Phi_t = \int_{-L}^{L} (\tfrac{1}{2}uu_{xxt} + \tfrac{1}{2}u_t u_{xx} + f(u)u_t)\, dx$$

$$= \int_{-L}^{L} (\tfrac{1}{2}u_{xx}u_t + \tfrac{1}{2}u_t u_{xx} + f(u)u_t)\, dx + \tfrac{1}{2}uu_{tx}|_{-L}^{L} - \tfrac{1}{2}u_x u_t|_{-L}^{L}$$

$$= \int_{-L}^{L} [uu_{xx} + f(u)]^2\, dx.$$

The boundary terms vanish since $\alpha u - \beta u_x = 0$, so $\alpha u_t - \beta u_{xt} = 0$ at $x = \pm L$. Thus $\Phi_t > 0$ except on the solutions of (24.26); i.e., except at the "rest points" of (24.25). This proves the following proposition:

Proposition 24.11. *The equation* (24.25) *together with the boundary conditions* (24.27) *is gradient-like with respect to the function* Φ.

We now turn to the Dirichlet problem; i.e., we consider (24.25) together with the boundary conditions

$$u(\pm L, t) = 0, \qquad t > 0. \tag{24.29}$$

We take for f the cubic polynomial

$$f(u) = -u(u - a)(u - 1), \qquad 0 < a < \tfrac{1}{2}. \tag{24.30}$$

The steady-state solutions satisfy (24.26) together with the boundary conditions

$$u(\pm L) = 0. \tag{24.31}$$

We note that $u \equiv 0$ is always (i.e., for all L), a steady-state solution.

In Chapter 13, §D, we showed that all of the nonconstant solutions of (24.26), (24.31) can be depicted as in Figure 24.19. That is, if $L > L_0$, there are precisely two such solutions, u_1 and u_2; if $L < L_0$ there are no nonconstant solutions; while if $L = L_0$, there is precisely one, $\bar{u}$, which is obviously degenerate. If $L < L_0$, then $u \equiv 0$ is easily seen to be stable, as can be checked by linearization (see Chapter 11). Since the flow is gradient-like, and the interval $0 \leq u \leq 1$ is an attracting region for solutions of (24.25), (24.29) (by Corollary 14.8), we see that this zero solution is, in fact, a *global attractor* for all solutions of the partial differential equation if $L < L_0$. In *what follows, we shall assume that $L > L_0$.*

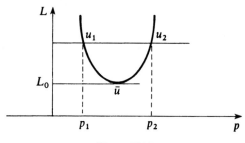

Figure 24.19

In this case, we let u_1 and u_2 be two typical nonconstant solutions of (24.26), (24.31), and we shall determine the dimensions of their unstable manifolds. To this end we first compute their Conley indices. In order to show that the theory of the last chapter is applicable, we must show that the local flow determined by (24.25), (24.29) can be embedded in a flow. However, this is precisely the content of part 5 of Chapter 23, §D.

With this technical point out of the way, we may proceed with our discussion. It will be convenient to reproduce the phase plane for (24.26); see Figure 24.20.

Since $L > L_0$, there are exactly three steady-state solutions; namely u_1 and u_2 as depicted in Figure 24.20, and $u_0 \equiv 0$. Let us now compute the Conley indices of these solutions. We begin with u_2, the "outer" nonconstant solution. The idea is to consider the nonlinear function f as a "parameter," and to use the invariance of the index under continuation. Thus referring to Figure 24.21, we first deform f to the function f_1 by "pulling up the valley," as in (b), then "straighten out" f_1, to obtain f_2 and then in (c), we slide f_2 to the left to obtain f_3, where $f_3(u) = -ku$, with k a positive constant. These deformations of f (from (a) to (b), for example), can be carried out in such a way as to ensure that the solution u_2 continues to a corresponding solution $\tilde{u}_2$, where $\tilde{u}_2$ is the "outer" solution corresponding to the phase plane where

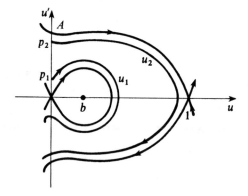

Figure 24.20

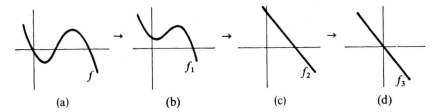

Figure 24.21

f is deformed to f_1. A sufficient condition for this to hold is that throughout the deformations, the solution u_2 stays an isolated rest point, i.e., u_2 does not bifurcate. If p_2 is close to A (Figure 23.20), it is not difficult to show this for the corresponding solution u_2. Namely, using (13.28), we can show that $T'(p_2)$ stays positive throughout the deformation. But using a theorem which we prove in the appendix to this chapter, we can show that when $T'(p_2) \neq 0$, the corresponding solution u_2 is nondegenerate in the sense that 0 is not in the spectrum of the linearized operator about u_2. This means that u_2 cannot bifurcate (Chapter 13, §A).

We see now that we have continued u_2 to the zero solution of the equation

$$u_t = u_{xx} - ku, \qquad |x| < L, \quad t > 0, \qquad (24.32)$$

with the same boundary data (24.29). From Theorem 14.19 we see that the zero solution is a *global* attractor for all solutions of (24.32); thus its index is Σ^0. It follows then by the continuation theorem, Theorem 23.31, that $h(u_2) = \Sigma^0$. On the other hand, since $g(x) \equiv f'(u_2(x))$ is bounded, the linearized operator $Q = d^2/dx^2 + g$, together with the boundary conditions (24.31) has at most a finite number, say k, of positive eigenvalues (Theorem 11.3).

In the appendix, we show that since $T'(p_2) \neq 0$ (see Figure 24.19), 0 is not in the spectrum of Q; i.e., u_2 is nondegenerate. It follows that u_2 has a k-dimensional unstable manifold for some integer $k \geq 0$. Thus $h(u_2) = \Sigma^k = \Sigma^0$; whence $k = 0$ and thus u_2 is an attractor. In a completely analogous manner, we can show that u_0 is also a stable solution; this follows from the deformation of f as given in Figure 24.22, whereby u_0 continues to the zero solution of (24.25). (Alternatively, we can simply compute the spectrum of u_0 by linearization.)

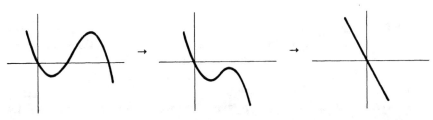

Figure 24.22

We shall now show that $h(u_1) = \Sigma^1$. In order to do this, we need the following lemma, which itself is of independent interest.

Lemma 24.12. *There exist solutions v_0 and v_2 of (24.25), (24.29), which connect u_1 to u_0 and u_1 to u_2, respectively; i.e.,*

$$\lim_{t \to -\infty} v_0(x, t) = u_1(x), \qquad \lim_{t \to \infty} v_0(x, t) = u_0(x), \quad and$$

$$\lim_{t \to -\infty} v_2(x, t) = u_1(x), \qquad \lim_{t \to \infty} v_2(x, t) = u_2(x),$$

uniformly for $|x| \le L$.

Proof. We first note again that the interval $\{u : 0 \le u \le 1\}$ attracts all solutions in the sense that all solutions tend to this set as $t \to \infty$; this follows from Corollary 14.8.

We now let

$$A = \{u \in C^2(|x| < L) : u(\pm L) = 0 \text{ and } \|u\|_\infty \le 1\}.$$

Then A is an attractor for the flow and obviously the rest points u_0, u_1, and u_2 are in A. The set A being an attractor, implies that it is an isolated invariant set; let U be an isolating neighborhood of A. Let $V \subset U$ be an isolating neighborhood for the attractor u_0, such that all points on ∂V are entrance points. Then $B = \text{cl}(U \backslash V)$ is an isolating neighborhood having only u_1 and

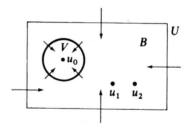

Figure 24.23

u_2 as rest points. Also (B, b^+) is an index pair in B, where $b^+ = \partial V$ and B/b^+ has the homotopy type of a point, since V is contractible to a point. Thus if S is the maximal invariant set in B, $h(S) = \bar{0}$. On the other hand, $h(u_1) \vee h(u_2) \ne \bar{0}$ since $h(u_2) = \Sigma^0$ (see Lemma 22.27). It follows from Theorem 22.33, that there is a solution v_2 connecting u_1 and u_2, as asserted. Similarly, we can find a solution v_1 which connects u_1 and u_0. This completes the proof. $\square$

Now it is easy to show that $h(u_1) = \Sigma^1$. The argument goes as follows. First, from equation (23.25), we have the (long) exact sequence of cohomology groups

$$\cdots \to H^{k-1}(h(S)) \to H^{k-1}(h(u_2)) \to H^k(h(u_1)) \to H^k(h(S)) \to \cdots$$

Since $h(S) = \bar{0}$ and $h(u_2) = \Sigma^0$, it follows that

$$H^k(h(u_1)) \approx H^{k-1}(h(u_2)) = \begin{cases} \mathbf{Z}, & k = 1, \\ 0, & k > 1. \end{cases}$$

Now as before $h(u_1) = \Sigma^m$ for some $m > 0$. The above calculation shows that $m = 1$, and it follows that u_1 has a one-dimensional unstable manifold. It is now possible to give the complete global picture of all solutions of (24.25), (24.29). This is done schematically in Figure 24.24. Thus the unstable manifold of u_1 has for its ω-limit set the two solutions u_0 and u_2, and one imagines that all solutions not in this picture (i.e., the ones in the infinite-dimensional space out of this "plane"), tend, as $t \to \infty$ to the rest points in this "plane." There are no other solutions.

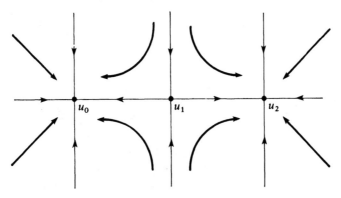

Figure 24.24

We can actually go a little further and obtain some more explicit information concerning the *stable* manifolds of the two attractors u_0 and u_1. Thus suppose that $u(x, 0) = \bar{u}(x)$, and $\bar{u}(\pm L) = 0$. Let $\bar{u}(x, t)$ be the corresponding solution of (24.25), (24.29) with this data. We claim that if $u_2(x) > \bar{u}(x) > u_1(x)$, for all x with $|x| < L$, then (the datum) $\bar{u}$ lies in the stable manifold of u_2, while if $0 < \bar{u}(x) < u_1(x)$ for all x, $|x| < L$, then $\bar{u}$ lies in the stable manifold of $u_0 \equiv 0$. To see this, suppose, e.g., $\bar{u}(x) > u_1(x)$, $|x| < L$. Then, by continuity, $\bar{u}(x, t_0) > u_1(x)$, $|x| < L$, for some $t_0 > 0$, and it follows that $\partial\bar{u}(-L, t_0)/\partial x > 0$ and $\partial\bar{u}(L, t_0)/\partial x < 0$. Thus if v_2 is the solution defined in Lemma 24.12, then

$$v_2(x, t_0 - \bar{t}) < \bar{u}(x, t_0), \qquad |x| < L,$$

for some $\bar{t} > 0$. If $w(x, t) \equiv v_2(x, t - \bar{t})$, then $w(x, t_0) < \bar{u}(x, t_0)$. Therefore, by the basic comparison theorem, Theorem 10.1, we have

$$w(x, t) < \bar{u}(x, t), \qquad t \geq t_0, \quad |x| < L.$$

But $w(x, t) \rightarrow u_2(x)$ as $t \rightarrow \infty$, so that the same is true for $\bar{u}(x, t)$ since $\bar{u}(x, t) \leq u_2(x)$, again by Theorem 10.1. Thus $\bar{u}(x)$ lies in the stable manifold of u_2. Similarly, if $0 < \bar{u}(x) < u_2(x)$, $|x| < L$, then the corresponding solution tends to zero as $t \rightarrow +\infty$. We summarize these results in the following theorem.

Theorem 24.13. *Let f be defined by (24.30), and let $L > L_0$. Then there are exactly three steady-state solutions of (24.25), (24.29): 0, u_1, and u_2 (cf. Figure 24.19). 0 and u_2 are stable, and u_1 has a one-dimensional unstable manifold which consists of orbits connecting u_1 to each of the other rest points. All solutions of the problem are depicted (qualitatively) in Figure 24.24. Initial data $u(x, 0)$ which satisfies $u_1(x) < u(x, 0) < u_2(x)$ (resp. $0 < u(x, 0) < u_1(x)$) on $|x| < L$ is in the stable manifold of u_2 (resp. 0).*

We turn now to a discussion of the Neumann problem for equation (24.25). Here the continuation theorem is not directly applicable, and we can use Morse decompositions and exact sequences of the corresponding cohomology groups in order to compute indices.

Thus consider again equation (24.25), now with homogeneous Neumann boundary conditions

$$u_x(\pm L, t) = 0, \qquad t > 0. \tag{24.33}$$

We also consider the corresponding steady-state equation (24.26) with boundary conditions

$$u'(\pm L) = 0. \tag{24.34}$$

To make the diagrams a bit simpler here, we take for f the cubic polynomial

$$f(u) = -u(u + b)(u - 1), \qquad 0 > b > -\tfrac{1}{2}. \tag{24.35}$$

Notice that the three roots of f are always solutions of the steady-state equations (24.26), (24.34). The equations (24.26) can be written as the first-order system $u' = v$, $v' = -f(u)$, and the phase plane is depicted in the figure below. Nonconstant solutions which satisfy the boundary conditions (24.34) correspond to orbit segments which "begin" and "end" on the u-axis ($v = 0$), and take parameter length $2L$ to make the trip. Thus they correspond to orbits which wind around the center $(0, 0)$ in Figure 24.25. It is easy to guess that as L increases, the number of solutions should also increase; these solutions correspond to orbits which wrap around the center many times. This is indeed the case, but we shall not prove this here; see [SW]. In fact, the global bifurcation picture of all solutions is depicted in Figure 24.26. The curves correspond to the "time" $p \rightarrow \beta(p)$ (Figure 24.25), in analogy

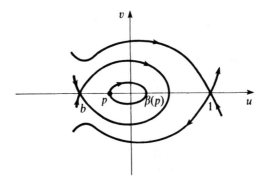

Figure 24.25

to what have just seen for the Dirichlet problem. Thus, if $L < L_1$, there are no nonconstant solutions. The curve C_1 corresponds to solutions whose total angular rotation is π; there are two such and they are symmetric about the u-axis; namely the positive solution $p \to \beta(p)$, and the negative one $\beta(p) \to p$; see Figure 24.25. Similarly the branch C_2 denotes solutions whose total angular rotation is $2\pi: p \to \beta(p) \to p$ and $\beta(p) \to p \to \beta(p)$. Continuing, we find that C_n consists of solutions whose total angular rotation is $n\pi$. It is easy to check that solutions on C_n have precisely n nodal points (see Chapter 13, §C, Example).

Next, it is shown in [SW], that for each n, both branches of C_n are smooth curves with finite nonzero derivatives all along the curves, except at the point L_n. Thus from the theorem in the appendix, each nonconstant solution is *nondegenerate* in the sense that 0 is *not* in the spectrum of the corresponding linearized operator. Finally, the numbers L_n correspond to those L for which the spectrum of the operator $d^2/dx^2 + f'(0)$ contains 0; these are the bifurcation points, as we have seen in Chapter 13, §A, Example 1. They can be computed explicitly, and one finds $L_n^2 = \pi^2(2n + 1)^2/4 f'(0)$.

We shall calculate the dimensions of the unstable manifold of each steady-state solution. First we consider the easy ones; namely, the three constant solutions. By the method which we used for the Dirichlet problem, it is easy to show that the solutions $u \equiv -b$ and $u \equiv 1$ are always stable, and $h(-b) = h(1) = \Sigma^0$. Furthermore, by what we have said above, one can check that

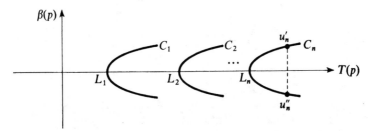

Figure 24.26

if $L_n < L < L_{n+1}$, then the solution $u \equiv 0$ has an $(n + 1)$-dimensional unstable manifold, and thus $h(0) = \Sigma^{n+1}$, $L_n < L < L_{n+1}$.

As shown in Chapter 13, §C, the nonconstant solutions bifurcate out of
0 when L crosses L_n. Thus let S_n denote the maximal isolated invariant set
containing 0, and the two nonconstant solutions u_n' and u_n'', as its only rest
points (Figure 24.26). These solutions are isolated rest points of (24.25),
(24.33) and so from Theorem 23.32, they are isolated invariant sets. Since u_n'
and u_n'' are bounded, the operators $A' = d^2/dx^2 + f'(u_n'(x))$, and $A'' = d^2/dx^2$
$+ f'(u_n''(x))$ (together with the boundary conditions (24.34)) have, at most,
a finite number of positive eigenvalues (Theorem 11.3). Since $u_n'(x) = u_n''(-x)$,
we see that A' and A'' have the same dimensional unstable manifold. Thus
$h(u_n') = h(u_n'')$.

Theorem 24.14. *Let* $L_n < L < L_{n+1}$; *then* $h(u_n') = h(u_n'') = \Sigma^n$, *and the rest
points* u_n', u_n'' *have n-dimensional unstable manifolds. Moreover each of these
rest points is connected to the rest point* $u \equiv 0$ *by a solution of* (24.25), (24.33).

Proof. Since the system is gradient-like (Proposition 24.11), Morse decompositions of S_n always exist (Chapter 23, §D,2). Let M_n', M_n'', M_n denote the
Morse sets which contain, respectively, the rest points u_n', u_n'', $0 \equiv u_n$, and let
h_n', h_n'', h_n denote their respective indices. As before, we know that $h_n' = h_n''$
$= \Sigma^k$, for some k; we shall show that $k = n$. Note that $h(S_n) = \Sigma^n$ and
$h(u_n) = \Sigma^{n+1}$.

We claim first that there must be orbits in S_n which connect critical points.
This follows from the addition formula for indices (Theorem 22.31); namely,
if such orbits didn't exist we would have $\Sigma^n = h(S_n) = h(u_n) \vee h(u_n') \vee h(u_n'')$
$= \Sigma^{n+1} \vee \Sigma^k \vee \Sigma^k$, which is impossible. Next, we claim that there cannot
be an orbit connecting u_n' and u_n''. This follows from the gradient-like nature
of the equations; namely,

$$\Phi(u_n') = \int_{-L}^{L} \{\tfrac{1}{2}u_n'(x)u_{n,xx}'(x) + F(u_n'(x))\}\, dx$$

$$= \int_{-L}^{L} \{-\tfrac{1}{2}u_n'(x)f(u_n'(x)) + F(u_n'(x))\}\, dx$$

$$= \int_{-L}^{L} \{-\tfrac{1}{2}(u_n'(-x))f(u_n'(-x)) + F(u_n'(-x))\}\, dx$$

$$= \int_{-L}^{L} \{-\tfrac{1}{2}u_n''(x)f(u_n''(x)) + F(u_n''(x))\}\, dx$$

$$= \int_{-L}^{L} \{\tfrac{1}{2}u_n''(x)u_{n,xx}''(x) + F(u_n''(x))\}\, dx$$

$$= \Phi(u_n'').$$

Thus there must be an orbit connecting u_n to one of u'_n, u''_n. Suppose that u_n and u'_n are connected by an orbit. Then there is a solution $u(x, t)$ such that, say,

$$\lim_{t \to -\infty} u(x, t) = 0, \qquad \lim_{t \to +\infty} u(x, t) = u'_n(x).$$

Hence if we define v by $v(x, t) = u(-x, t)$, then v is a solution and

$$\lim_{t \to -\infty} v(x, t) = 0, \qquad \lim_{t \to +\infty} v(x, t) = u'_n(-x) = u''_n(x).$$

Thus u_n and u''_n are also connected by an orbit (the proof is similar if the orbit u runs from u'_n to u_n).

We shall show that $h(u'_n) = \Sigma^n$, assuming that the connecting orbits run from u_n to u'_n and u''_n; then we shall show that this is the only case possible. Thus, in the assumed case, we have the Morse decomposition of S_n given by $(u'_n \perp\!\!\!\perp u''_n, u_n)$. From (23.25) we have the long exact sequence

$$\cdots \to H^k(h(s_n)) \to H^k(h(u'_n)) + H^k(h(u''_n)) \to H^{k+1}(h(u_n)) \to \cdots,$$

or, since $h(S_n) = \Sigma^n$ and $h(u_n) = \Sigma^{n+1}$,

$$\cdots \to H^k(\Sigma^n) \to H^k(h(u'_n)) + H^k(h(u''_n)) \to H^{k+1}(\Sigma^{n+1}) \to \cdots$$

If $k \ne n$, we have

$$0 \to H^k(h(u'_n)) + H^k(h(u''_n)) \to 0,$$

so that $H^k(h(u'_n)) = H^k(h(u''_n)) = 0$ for all $k \ne n$. Since $h(u'_n) = \Sigma^p$ for some p, we see $p = n$.

If now the connecting orbits ran from u'_n and u''_n to u_n, then $(u_n, u'_n \perp\!\!\!\perp u''_n)$ is a Morse decomposition of S_n. This gives the long exact sequence

$$\cdots \to H^{k-1}(\Sigma^n) \to H^{k-1}(\Sigma^{n+1}) \to H^k(h(u'_n)) + H^k(h(u''_n)) \to H^k(\Sigma^n) \to \cdots.$$

If $k = n + 2$, we have

$$0 \to \mathbf{Z} \to H^{n+2}(h(u'_n)) + H^{n+2}(h(u''_n)) \to 0, \qquad (24.36)$$

which gives the contradiction $\mathbf{Z} \approx H^{n+2}(h(u'_n)) + H^{n+2}(h(u''_n))$. This completes the proof. $\square$

Concerning the Dirichlet problem again, we have the following theorem.

Theorem 24.15. *Let f be a locally Lipschitzian function having a finite number of zeros*: $0 = a_1 < a_2 < \cdots < a_n = b$, *where*

(i) $\int_0^b f(u)\, du > 0$, *and*

(ii) $f'(0) < 0$, $f'(b) < 0$.

Then if the number of steady-state solutions of (24.25), (24.29) is finite in each connected component of the (graph of the) time map, there must be an even number of nondegenerate, nonconstant ones, $u_1 u_2, \ldots, u_{2k}$, where $u'_j(-L) < u'_{j+1}(-L)$, $j = 1, \ldots, 2k - 1$, and $h(u_{odd}) = \Sigma^1$, $h(u_{even}) = \Sigma^0$. The unstable solutions u_{2j+1} are connected to u_{2j} and u_{2j+2} by connecting orbits.

Note that the finiteness hypothesis is fulfilled if all but a finite number of steady-state solutions are nondegenerate, or if there is a "generic-type" hypothesis on the time map. For example, we could assume that the time map is a Morse function; i.e., it has a finite number of critical points, all nondegenerate. Thus each component of the time map has the form of Figure 24.27 (here we consider solutions like those corresponding to points $\bar{p}$, which are not relative extrema but satisfy $T'(\bar{p}) = 0 = T''(\bar{p})$ as being non-degenerate since they don't bifurcate, even though 0 may be in the spectrum of their linearization).

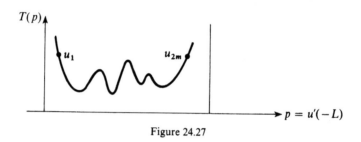

Figure 24.27

Proof. First observe that the hypotheses imply that the time map has the qualitative form of a finite number of components of the form like that of Figure 24.27. Thus the number of nondegenerate nonconstant steady-state solutions must be even. We may assume that the time map looks like Figure 24.27; namely, we can work with each component separately.

We denote by u_1 and u_{2m} the solutions with the smallest and largest derivative at $-L$; see Figure 24.27. As in the proof of Theorem 24.13, we can show that u_{2m} is an attractor and $h(u_{2m}) = \Sigma^0$. If L is very large, the isolated invariant set containing u_1, u_{2m}, and $u \equiv 0$ has index Σ^0, and there are orbits connecting 0 to both of the other critical points. As before, this implies $h(u_1) = \Sigma^1$ and that u_1 has a one-dimensional unstable manifold. The proof is completed by the following lemma (together with Theorem 22.33, for the last sentence in the statement of the theorem).

Lemma 24.16:

(A) *If $h(u_k) = \Sigma^q$ and $h(u_{k+1}) = \Sigma^p$, then $q = p \pm 1$.*

(B) *If $\bar{u}$ is a nonnegative solution of (24.26), (24.31), then the dimension of the unstable manifold of $\bar{u}$ is at most one.*

Proof. (A) Suppose that u_k and u_{k+1} bifurcate out of the degenerate rest point $\bar{u}$. Then they are the only rest points near $\bar{u}$ which are contained in an isolating neighborhood N, and $H(I(N)) = \bar{0}$. As in Lemma 24.12, u_k and u_{k+1} are connected by an orbit, which we may assume, runs from u_k to u_{k+1}. Using (23.25), we have to long exact sequence

$$\rightarrow H^{r-1}(I(N)) \rightarrow H^{r-1}(h(u_{k+1})) \rightarrow H^r(h(u_k)) \rightarrow H^r(I(N)) \rightarrow \cdots,$$

or

$$0 \rightarrow H^{r-1}(\Sigma^p) \rightarrow H^r(\Sigma^q) \rightarrow 0.$$

It follows that $H^{r-1}(\Sigma^p) \approx H^r(\Sigma^q)$. But $H^r(\Sigma^q) = \mathbf{Z}$ if $r = q$, while $H^{r-1}(\Sigma^p) = \mathbf{Z}$ if $r = p + 1$. Hence $q = p + 1$, and this proves (A).

(B) Consider the eigenvalue problem

$$v'' + f'(\bar{u}(x))v = \lambda v, \qquad |x| < L, \quad v(\pm L) = 0; \qquad (24.37)$$

we are to show that this problem has at most one positive eigenvalue.

If $w = \bar{u}'$, then

$$w'' + f'(\bar{u})w = 0.$$

Also, if $\lambda > 0$, $f'(\bar{u}(x)) - \lambda < f'(\bar{u}(x))$, so by the elementary Sturm comparison theorem,[4] we know that between any two zeros of v, there must be a zero of w. Since u is a nonnegative solution, it is easy to see (from elementary phase plane analysis), that w has precisely one zero in $|x| \leq L$.

Now if v had a zero in $|x| < L$ then w would have to be zero twice in this open interval. Thus $v > 0$ in $|x| < L$. It follows that λ must be the principal eigenvalue of (24.37) (see the discussion in Chapter 11, §A). This shows that (24.37) has at most one positive eigenvalue, and completes the proof. $\square$

Notice that this theorem is *not* obtainable from purely topological methods. For example, consider the time map in Figure 24.28. Here we see that $h(u_4) = \Sigma^0$, so $h(u_3) = \Sigma^1$, and $h(u_1) = \Sigma^1$. Thus it is topologically consistent that $h(u_2)$ be either Σ^0 or Σ^2. The latter possibility is ruled out by the last lemma.

We close this section by pointing out that our results here can be viewed as a generalization to reaction–diffusion equations of a familiar technique in ordinary differential equations. Thus in order to find the global "phase portrait," it is necessary to first find all of the rest points, then to show that they are nondegenerate, and then to compute the local flow near each rest

---

[4] If $u'' + pu = 0$, $v'' + qv = 0$, and $p \geq q$, then u has at least one zero between any two zeros of v.

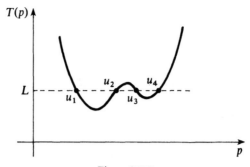

Figure 24.28

point by linearization. In our case, the difficult computation of the spectrum of the linearized equations is replaced by a topological technique, namely, by computing the Conley index of the rest point. For the case of cubic f; e.g., f defined by (24.30), we can actually describe the precise *global* flow for the Dirichlet problem by this technique; see Figure 24.24.

§E. Instability of Equilibrium Solutions of the Neumann Problem

It is usually easier to prove the instability of a solution than to prove its stability. Namely, in the former case one must only prove that one element of the spectrum of the linearized operator lies in the right half-plane $\operatorname{Re} z > 0$, while in the latter case, the *entire* spectrum must be shown to lie in $\operatorname{Re} z < 0$. In this section we shall prove that steady-state solutions of the Neumann problem for the Fitz-Hugh–Nagumo equations are always unstable. We shall then obtain the analogous result for a scalar equation in several space variables which satisfies homogeneous Neumann boundary conditions. Here it is required that the spatial domain be convex (or at least "nearly" convex).

We begin with the Fitz-Hugh–Nagumo Equations (see Chapter 14, §B)

$$v_t = v_{xx} + f(v) - u, \qquad u_t = \delta v - \gamma u, \qquad |x| < L, \quad t > 0, \quad (24.38)$$

with homogeneous Neumann boundary conditions

$$v_x(\pm L, t) = 0, \qquad t > 0. \tag{24.39}$$

The steady-state solutions satisfy the equations

$$v'' + f(v) - u = 0, \qquad \delta v - \gamma u = 0, \qquad |x| < L, \tag{24.40}$$

together with the boundary conditions

$$v'(\pm L) = 0. \tag{24.41}$$

Here f is the cubic polynomial defined in (24.30). We can eliminate u from (24.40), thereby obtaining the scalar problem

$$v'' + f(v) - \frac{\delta v}{\gamma} = 0, \qquad v'(\pm L) = 0. \tag{24.42}$$

We assume that δ/γ is so small that the cubic polynomial $f(v) - \delta v/\gamma$ has three real roots, 0, r, and s, where $0 < r < s/2$. Let $(v(x), \delta v(x)/\gamma)$ be a non-constant solution of (24.40), (24.41); thus v satisfies (24.42), and $u(x) = \delta v(x)/\gamma$. We shall prove that this steady-state solution is an unstable solution of (24.38), (24.39). This will be done by showing that the corresponding linearized operator contains an element in its spectrum which has positive real part.

Theorem 24.17. *Every nonconstant steady-state solution of (24.38), (24.39) is unstable.*

Proof. Consider the linearized equations for real $\lambda > 0$,

$$\lambda w = w'' + f'(v)w - z, \qquad \lambda z = \delta w - \gamma z, \qquad |x| < L, \tag{24.43}$$

where w satisfies the boundary conditions (24.41). We assume that $\lambda + \gamma \neq 0$; this will be verified later. Thus we can eliminate z from (24.43) thereby obtaining

$$\lambda w = w'' + f'(v)w - \frac{\delta w}{\lambda + \gamma}, \qquad w'(\pm L) = 0. \tag{24.44}$$

We denote by $A(\lambda)$ the operator defined by

$$A(\lambda)w = w'' + \left(f'(v) - \frac{\delta}{\lambda + \gamma}\right)w, \qquad w'(+L) = 0;$$

then (24.44) becomes, for $\lambda \geq 0$,

$$A(\lambda)w = \lambda w, \qquad w'(\pm L) = 0. \tag{24.45}$$

We shall show that there is a solution $w \not\equiv 0$ of this equation with $\lambda > 0$.
To this end, first note that (from invariant regions), the potential, $f'(v)$

$-\delta/(\lambda + \gamma)$, is uniformly bounded, independently of $\lambda \geq 0$. Thus let $\mu(\lambda)$ and $v(\lambda)$ denote the sup of the spectrum of the operator $A(\lambda)$ with homogeneous Neumann and Dirichlet boundary conditions, respectively (cf. Chap. 11, §A). Then from the elementary Sturm comparison theorem, we have

$$\mu(\lambda) > v(\lambda). \tag{24.46}$$

(In fact, if $\mu \leq v$, then letting q be the potential $q = f'(v) - \delta/(\lambda + \gamma)$, we would have $q - v \leq q - \mu$ so that the principal eigenfunction ϕ corresponding to μ has a zero in $|x| < L$. This is impossible since ϕ is of one sign by Theorem 11.10.) Moreover, since $v(x)$ is a nonconstant solution of (24.42), we know that $v'(x) \not\equiv 0$. Since

$$(v')'' + f'(v)v' - \left(\frac{\delta}{\gamma}\right)v' = 0, \qquad v'(\pm L) = 0,$$

we see that v' is an eigenvector of $A(0)$ with homogeneous Dirichlet boundary conditions corresponding to the eigenvalue $\lambda = 0$. It follows that $v(0) \geq 0$, so that (24.46) gives

$$\mu(0) > 0. \tag{24.47}$$

Next, from the results in Chapter 11, §A, we know that $\mu(\lambda)$ has a variational characterization; namely

$$\mu(\lambda) = \sup_{\substack{w \in W_2 \\ \|w\|_{L_2} = 1}} \langle A(\lambda)w, w \rangle.$$

Moreover, $\mu(\lambda)$ is a continuous function of λ (Theorem 11.7), and $\mu(\lambda)$ is always an eigenvalue of $A(\lambda)$ (Theorem 11.4).

Now if $\|w\|_{L_2} = 1$, and $w(\pm L) = 0$, then

$$\langle A(\lambda)w, w \rangle = \int_{-L}^{L} w'' w + \int_{-L}^{L} f'(v)w^2 - \int_{-L}^{L} \frac{\delta w^2}{\lambda + \gamma}$$

$$\leq - \int (w')^2 + c_1 \int_{-L}^{L} w^2 + c_2 \int_{-L}^{L} w^2$$

$$\leq c_1 + c_2 \equiv c,$$

where c is independent of λ. It follows that $\mu(\lambda)$ is bounded from above, uniformly in λ. Thus there exists a $\bar{\lambda} > 0$ such that $\mu(\bar{\lambda}) = \bar{\lambda}$ (see the figure below), and so there is a $\bar{w} \neq 0$ satisfying

$$A(\bar{\lambda})\bar{w} = \mu(\bar{\lambda})\bar{w} = \bar{\lambda}\bar{w}, \qquad \bar{w}(\pm L) = 0.$$

It follows that the linearized operator contains a positive element in its spectrum; namely $\bar{\lambda}$, so that the solution $(v(x), \delta v(x)/\gamma)$ is unstable. This concludes the proof, since $\bar{\lambda} + \gamma > 0$. □

We turn now to the case of a scalar equation in several space variables. Here too we shall prove that nonconstant equilibrium solutions of the Neumann problem are unstable.

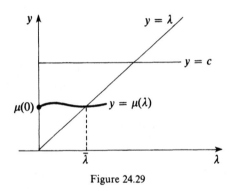

Figure 24.29

Consider the equation

$$u_t = \Delta u + f(u), \qquad (x, t) \in \Omega \times \mathbf{R}_+, \tag{24.48}$$

together with the homogeneous Neumann boundary conditions

$$\frac{du}{dn} = 0 \quad \text{on } \partial\Omega \times \mathbf{R}_+, \tag{24.49}$$

where n is the outward pointing normal on $\partial\Omega$. Here Ω is a smooth bounded domain in $\mathbf{R}^n$, and Δ denotes the n-dimensional Laplacian. The function f is assumed to be of class C^1. The corresponding equilibrium (steady-state) solutions satisfy

$$\Delta u + f(u) = 0, \qquad x \in \Omega, \qquad \frac{du}{dn} = 0 \quad \text{on } \partial\Omega. \tag{24.50}$$

If u is an equilibrium solution, we consider the corresponding eigenvalue problem

$$\Delta v + f'(u)v = \lambda v \quad \text{on } \Omega; \qquad \frac{dv}{dn} = 0 \quad \text{on } \partial\Omega. \tag{24.51}$$

Using the results in Chapter 11, §A, we know that the principal eigenvalue, λ_1 of (24.50) is given by (Theorem 11.4)

$$\lambda_1 = \sup_{\substack{w \in W_2(\Omega) \\ \|w\|_{L_2} = 1}} \int_\Omega \{-|\nabla w|^2 + f'(u)w^2\}\, dx.$$

Let

$$I(w) = \int_\Omega \{-|\nabla w|^2 + f'(u)w^2\}\, dx;$$

if we can find an "admissible" w for which $I(w) > 0$, then, of course, $\lambda_1 > 0$ and u is unstable. Thus for example, if u is an equilibrium solution and

$$\int_\Omega f'(u(x))\, dx > 0,$$

then $I(1) > 0$ and u is unstable. A much deeper result is given by the following theorem.

Theorem 24.18. Let Ω be a convex subset of $\mathbf{R}^n$, $n \geq 1$. Then any nonconstant equilibrium solution of (24.48), (24.49) is unstable.

Proof. Suppose first that $n \geq 2$. Let u be a nonconstant equilibrium solution. Then $\partial u/\partial x_i \equiv u_i$ satisfies the equation

$$\Delta u_i + f'(u)u_i = 0.$$

If we multiply this equation by u_i, integrate over Ω, and apply the divergence theorem, we find

$$I(u_i) = -\int_{\partial\Omega} u_i \frac{du_i}{dn},$$

so that

$$\sum_{i=1}^n I(u_i) = -\int_{\partial\Omega} \nabla u \cdot \frac{d(\nabla u)}{dn}. \tag{24.52}$$

If we show that for some i,

$$I(u_i) \geq 0, \qquad u_i \neq 0, \qquad (24.53)$$

then $\lambda_1 \geq 0$. Now, if $\lambda_1 = 0$, then $I(u_{x_i}) = 0$ and u_i is the principal eigenfunction of the operator $\Delta + f'(u)$ satisfying (24.49); see Theorem 11.4, Part 1. But $u_i = du/dn$ at some point on $\partial\Omega$, and so $u_i = 0$ at this point. But this is not possible, according to Lemma 24.19 below.

We establish (24.53) by proving that

$$\nabla u(x) \cdot \frac{d(\nabla u(x))}{dn} \leq 0 \qquad (24.54)$$

at each $x \in \partial\Omega$. If we replace x by $a \cdot x + b$, $a, b \in \mathbf{R}^n$, this does not affect the sign of (24.54), so we may assume that $0 \in \partial\Omega$, and we shall prove (24.54) at $x = 0$. Suppose that $x_n = g(x_1, \ldots, x_{n-1})$ is a convex function which describes $\partial\Omega$ near $x = 0$ and that at 0 the $-x_n$ axis is in the outward normal direction to Ω; thus $x = 0$ is a minimum point for g, and $g_{x_i}(0) = 0$, $1 \leq i \leq n - 1$. Then since $u_n(0) = \partial u(0)/\partial n = 0$, we have

$$\nabla u(0) \cdot \frac{\partial}{\partial n} \nabla u(0) = - \sum_{i=1}^{n-1} u_i(0) \frac{\partial u_i}{\partial x_n}(0)$$

$$= - \sum_{i=1}^{n-1} u_{x_i}(0) u_{x_i x_n}(0). \qquad (24.55)$$

Near $x = 0$, the condition $du/dn = 0$, can be written in these coordinates as

$$\sum_{i=1}^{n-1} u_{x_i}(\bar{x}, g(\bar{x}))g_{x_i}(\bar{x}) - u_{x_n}(\bar{x}, g(\bar{x})) = 0,$$

where $\bar{x} = (x_1, \ldots, x_{n-1})$. Differentiating this with respect to x_j, $1 \leq j \leq n - 1$, gives at $x = 0$

$$\sum_{i=1}^{n-1} u_{x_i}(0)g_{x_i x_j}(0) - u_{x_j x_n}(0) = 0,$$

since $g_{x_i}(0) = 0$, $1 \leq i \leq n - 1$. Substituting in (24.55) we get

$$\nabla u(0) \cdot \frac{\partial}{\partial n} \nabla u(0) = - \sum_{i,j=1}^{n-1} g_{x_i x_j}(0)u_{x_i}(0)u_{x_j}(0) \leq 0,$$

by the convexity of g. Thus proves the theorem if $n \geq 2$. If $n = 1$, we have $I(du/dx) = 0$ and $du/dx = 0$ on $\partial\Omega$, so a similar argument gives $\lambda_1 > 0$. $\square$

We prove the following lemma which was used in the proof of the last theorem.

Lemma 24.19. *Consider the operator*

$$L = \Delta + c(x) \quad \text{in } \Omega,$$

where Ω is a bounded domain in $\mathbf{R}^n$ with $\partial\Omega$ smooth, and c is a continuous function on $\bar{\Omega}$. If ϕ is the principal eigenfunction of L with homogeneous Neumann boundary conditions, then $\phi \neq 0$ in $\bar{\Omega}$.

Proof. Let λ be the principal eigenvalue associated to ϕ. We may assume $\phi < 0$ in Ω. Choose a constant $k < 0$ such that $k + c(x) - \lambda < 0$ in $\bar{\Omega}$. Then

$$L\phi + (k - \lambda)\phi = k\phi < 0 \quad \text{in } \Omega,$$

and the coefficient of ϕ on the left side is $c(x) + k - \lambda < 0$. Thus the strong maximum principle, Theorem 8.6, shows that if ϕ has a minimum on $\partial\Omega$, then $d\phi/dn < 0$ at that point. This is impossible so $\phi < 0$ on $\partial\Omega$ and the proof is complete. $\square$

The next theorem places a restriction on f but no restriction on the domain Ω.

Theorem 24.20. *Let u be a nonconstant equilibrium solution of (24.48), (24.49). If f'' is of one sign on the range of u, then u is unstable.*

Proof. Let's suppose $f''(u(x)) > 0$ for all $x \in \Omega$, and let $m = \min\{u(x) : x \in \bar{\Omega}\}$. We shall show that $I(u - m) > 0$.

Since $u \geq m$ on $\bar{\Omega}$, then if $u(\bar{x}) = m$, for $\bar{x} \in \Omega$, we have $\Delta u(\bar{x}) \geq 0$ so $f(m) = f(u(\bar{x})) \leq 0$. If $u = m$ occurs only on $\partial\Omega$, then the proof that $f(m) \leq 0$ is a bit more difficult; we postpone it till later.

Now

$$I(u - c) = \int_\Omega \{-|\nabla u|^2 + f'(u)(u - m)^2\} \, dx$$

$$= \int_\Omega \{uf(u) + f'(u)(u - m)^2\} \, dx$$

$$= \int_\Omega \{(u - m)f(u) + f'(u)(u - m)^2\} \, dx$$

$$= \int_\Omega (u - m)\{f(u) + f'(u)(u - m)\} \, dx.$$

But since $f'' > 0$, we have

$$0 \geq f(m) > f(u) + f'(u)(u - m)$$

and hence $I(u - c) > 0$.

The proof is completed by proving that $f(m) \leq 0$ even if m is achieved by u only at $\bar{x} \in \partial\Omega$. To do this we shall show that $\Delta u \geq 0$ at $\bar{x}$. We let s denote a local coordinate system on $\partial\Omega$ near $\bar{x}$, where $s = 0$ corresponds to the points on $\partial\Omega$. We also let t denote the normal direction to $\partial\Omega$ near $\bar{x}$; thus (s, t) is a local coordinate system near $\bar{x}$. We choose the parametrizations in such a way that $\bar{x}$ corresponds to $s = 0 = t$. u will be considered as a function of s and t; $u = u(s, t)$.

Now $\bar{x}$ is a critical point of u since $u_t = 0$ when $t = 0$, and m is a minimum for u on $s = 0$. Thus the hessian matrix is well defined at $\bar{x}$ in this coordinate system. From (24.49) it follows that $u_{st} = 0$ at $s = t = 0$. Furthermore,

$$u(0, t) = u(0, 0) + \frac{u_{tt}(0, 0)t^2}{2} + O(t^3),$$

and (with a slight abuse of notation),

$$u(s, 0) = u(0, 0) + u_{ss}(0, 0)(s, s) + O(|s|^3).$$

This shows that $u_{tt}(0, 0) \geq 0$ and the matrix $u_{ss}(0, 0)$ is positive semi-definitive at $(0, 0)$. It follows that $\Delta u \geq 0$ at $\bar{x}$ and then $f(m) = f(u(\bar{x})) = -\Delta u(\bar{x}) \leq 0$, as desired. $\square$

§F. Appendix: A Criterion for Nondegeneracy

We consider here the nondegeneracy problem for solutions of the equation

$$u'' + f(u) = 0, \qquad |x| < L, \tag{A1}$$

subject to homogeneous boundary conditions of the form

$$\alpha u(-L) + \beta u'(-L) = 0 = \alpha u(L) - \beta u'(L). \tag{A2}$$

As usual, we write (A1) as the first-order system

$$u' = v, \qquad v' = -f(u). \tag{A3}$$

If L_1 denotes the line $\alpha u + \beta v = 0$, and L_2 the line $\alpha u - \beta v = 0$, we assume that there is a solution $u(x, p)$ of (A2), (A3) with $u(o, p) = p$. Let $T(p)$ be the "time" (x-parameter length) of this orbit going from L_1 to L_2. It is obvious that $u(\cdot, p)$ bifurcates if and only if T is not monotone in a neighborhood of p. The purpose of this section is to prove the following strengthening of this remark.

Theorem A1. *With the above notation, suppose that $T'(\bar{p}) \neq 0$, and that the orbit cuts each L_i transversally; i.e., that both*

$$\alpha v(-T(\bar{p}), \bar{p}) + \beta v'(-T(\bar{p}), \bar{p}) \neq 0, \tag{A4}$$

and

$$\alpha v(T(\bar{p}), \bar{p}) - \beta v'(T(\bar{p}), \bar{p}) \neq 0.$$

Then the solution $u(\cdot, \bar{p})$ of (A1) and (A2) is nondegenerate in the strong sense that 0 is not in the spectrum of the linearized operator about $u(\cdot, \bar{p})$.

Proof. Differentiating (A1) with respect to x and p gives the two equations

$$u_p'' + f'(u)u_p = 0 \quad \text{and} \quad v'' + f'(u)v = 0.$$

Since $u_p(0, p) = 1$, and $v(0, p) = 0$, it follows easily that v and u_p are linearly independent. Thus any solution w satisfying

$$w'' + f'(u)w = 0, \qquad |x| < T(\bar{p}),$$

and the boundary conditions (A2) with $L = T(\bar{p})$, can be written in the form

$$w = C_1 u_p + C_2 v.$$

Our goal is to show $w \equiv 0$. Now using (A2), we get the equations

$$\alpha u_p(-\bar{T}, \bar{p}) + \beta u_p'(-\bar{T}, \bar{p}) - [\alpha v(-\bar{T}, \bar{p}) + \beta v(-\bar{T}, \bar{p})]T'(\bar{p}) = 0,$$

and

$$\alpha u_p(\bar{T}, \bar{p}) - \beta u_p'(\bar{T}, \bar{p}) + [\alpha v(\bar{T}, \bar{p}) - \beta v(\bar{T}, \bar{p})]T'(\bar{p}) = 0,$$

where $\bar{T} = T(\bar{p})$. Now

$$\begin{aligned}
0 = \alpha w(-\bar{T}) + \beta w'(-\bar{T}) &= \alpha[C_1 u_p(-\bar{T}) + C_2 v(-\bar{T})] \\
&\quad + \beta[C_1 u_p'(-\bar{T}) + C_2 v'(-\bar{T})] \\
&= C_1[\alpha u_p(-\bar{T}) + \beta u_p'(-\bar{T})] \\
&\quad + C_2[\alpha v(-\bar{T}) + \beta v'(-\bar{T})] \\
&= C_1[\alpha_1 v(-\bar{T}) + \beta_1 v'(-\bar{T})]T'(\bar{p}) \\
&\quad + C_2[\alpha v(-\bar{T}) + \beta v'(-\bar{T})] \\
&= (C_1 T'(\bar{p}) + C_2)[\alpha v(-\bar{T}) + \beta v'(-\bar{T})].
\end{aligned}$$

Thus from (A4),

$$C_1 T'(\bar{p}) + C_2 = 0. \tag{A5}$$

Similarly, using the equation $\alpha w(\bar{T}) - \beta w'(\bar{T}) = 0$, gives

$$C_1 T'(\bar{p}) - C_2 = 0. \tag{A6}$$

Hence since $T'(\bar{p}) \neq 0$, (A5) and (A6) yield $C_1 = C_2 = 0$, so that $w \equiv 0$, as asserted. □

NOTES

The material in §A is taken from Conley–Smoller [CS 2]. Other proofs have been given by Conlon [Cn], Foy [Fo], and Mock [Mk]. Related papers are [CS 1–5], where more general "viscosity" matrices are considered. In particular, there can exist positive definite self-adjoint matrices B for which no connecting orbit exists for $B\dot{u} = V(u)$, where $u \in \mathbf{R}^2$. Such matrices B are also "bad" for the viscosity method; see Smoller and Taylor, [ST]. The structure problem for mhd shocks is taken from Conley–Smoller [CS 6]. The important fundamental work upon which this development is based was done by Germain [Gr], who demonstrated the gradient-like nature of the equations, and proved that there are at most four rest points. He was also able to show that fast mhd shocks have structure. See the books [An], [Du], and [KL], where this structure problem is discussed. The results in §B have been extended by Conley–Smoller [CS 8], and more generally by Hesaraaki [Hki], to the case where some viscosity parameters are zero. The existence of a periodic travelling wave for the Nagumo equations is taken from Conley's paper [Cy 1]; see also Hastings [Ha 1, 2], for a different approach. In Gardner–Smoller [GS], these ideas have been extended to predator–prey equations. For related work, see Carpenter [Ca 1, 2], and Chueh [Ch]. The material in §D is taken from Conley–Smoller [CS 12, 13]. Theorem 24.15 appears here for the first time. In [CS 13], it is shown how to "continue" one boundary value problem to another. Here one sees some rather interesting Morse theoretic-type "cancellation" phenomena in an infinite-dimensional context. An extension of the results in §D to the Fitz-Hugh–Nagumo equations can be found in [CS 14]. Theorem 24.17 is due to Bardos–Smoller; see [BS], where a more general result is proved; see also Bardos–Matano–Smoller [BMS] for an extension to systems in several space variables. For related work, see the papers of Chafee and Infante, [Cf] and [CI].

For other applications of the Conley index see Conley and Gardner, [CG] and Conley and Zhender [CZ]. Theorems 24.18 and 24.20 are due to Casten–Holland [CaH]. If Ω is not convex, there may exist nonconstant stable equilibria, see Matano [Mo], and also Hale and Vegas [HV] for the beginnings of a theory. The nondegeneracy theorem in the appendix is adapted from Smoller–Tromba–Wasserman [STW].

Chapter 25

Recent Results

In this chapter we shall present a summary of what could be considered as some of the most significant results which have appeared since the appearance of the original edition of this book. Of course, limitations of space (and time!) forces us to give only statements of major results, and very brief outlines of some of their proofs. This chapter will be divided into four sections, the first two corresponding to Parts II and III of the text, and the last two related to Part IV of the text. The *numbered* references correspond to the new reference list given at the end of this chapter.

SECTION I. Reaction–Diffusion Equations

§A. Stability of the Fitz-Hugh–Nagumo Travelling Pulse

The Fitz-High–Nagumo equations were considered in Chapter 14, §B. They can be written in the form

$$u_t = u_{xx} + f(u) - w, \qquad w_t = \varepsilon(u - \gamma w), \tag{25.1}$$

where $f(u) = u(u - a)(1 - u)$, $0 < a < \frac{1}{2}$, and $1 \gg \varepsilon > 0$. A travelling wave solution for (25.1) is a function of the single variable $\xi = x - ct$; i.e., $U_\varepsilon(\xi) = (u(\xi), w(\xi))$ satisfies

$$-cu' = u'' + f(u) - w, \qquad -cw' = \varepsilon(u - \gamma w) \qquad (' = d/d\xi). \tag{25.2}$$

A "travelling pulse" ("homoclinic" orbit), is a travelling wave which satisfies $(u(\xi), w(\xi)) \to (0, 0)$ as $\xi \to \pm\infty$. The existence of such a solution for some value of c, for $\varepsilon \ll 1$, has been proved in [Ca 1, Cy 1, [28, 37]]. The point of taking ε small is that we can consider the singularly perturbed equations (25.2), with $\varepsilon = 0$, and the pulse is constructed by piecing together certain solutions of the reduced system; see Chapter 24, §C. It is an open question as to whether a travelling pulse exists for large ε.

We discuss the stability of U_ε in the space of bounded uniformly continuous functions BC(**R**) with the sup-norm topology. Thus, U_ε is called *stable* if there is a $\delta > 0$ such that if $U(\xi, t) = (u(\xi, t), w(\xi, t))$ is any solution of (25.1), which satisfies

$$\|U(\xi + k_1, 0) - U_\varepsilon(\xi)\|_\infty < \delta \qquad \text{for some } k_1,$$

there is a k_2 for which

$$\lim_{t \to \infty} \|U(\xi + k_2, t) - U_\varepsilon(\xi)\|_\infty = 0.$$

Thus, if a solution to (25.1) starts near some translate of the travelling wave, it tends to some other translate as $t \to \infty$. Here is the stability theorem of Jones [33, 34]:

Theorem I.1. *If ε is sufficiently small, $U_\varepsilon(\xi)$ is stable.*

A standard method for determining stability is to use linearization techniques (cf. Chapter 11, §B). If we linearize (25.1) about $U_\varepsilon(\xi)$, we obtain the operator

$$L\begin{pmatrix} p \\ r \end{pmatrix} = \begin{pmatrix} p'' + cp' + f'(u_\varepsilon)p - r \\ cr' + \varepsilon(p - \gamma r) \end{pmatrix}, \qquad p, r \in BC(\mathbf{R}). \tag{25.3}$$

Note that 0 lies in $\sigma(L)$ (the spectrum of L), because a translate of a travelling wave is again a travelling wave. The linearized criterion for stability is that both of the following hold:

(i) $\sigma(L)\backslash\{0\} \subset (\lambda: \text{Re } \lambda < a)$ for some $a < 0$.
(ii) 0 is a simple eigenvalue.

Evans [13] has laid the foundation for studying the (nonlinear) stability of travelling waves in this set-up. In [13, 14] he showed that if (i) and (ii) hold, then U_ε is (nonlinearly) stable; thus the theorem holds if (i) and (ii) are valid, and we now discuss some ideas in the proofs of (i) and (ii).

$\sigma(L)$ consists of two parts: the eigenvalues of finite multiplicity, and the essential spectrum, $\sigma_e(L)$. It is shown that $\sigma_e(L)$ lies in a half-plane $N = (\lambda: \text{Re } \lambda < a)$ for some $a < 0$; this essentially follows from proving that the system (25.1) is stable at $(0, 0)$. In [13, IV] Evans defined an analytic function $D(\lambda)$ on $\mathbf{C}\backslash N$ whose zeros are the eigenvalues of L, and the order of whose zero is the algebraic multiplicity of the eigenvalue. $D(\lambda)$ is used to approximately locate the eigenvalues of L; these lie close to the eigenvalues for a certain reduced system associated with some parts of the singular travelling wave ($\varepsilon = 0$). Thus the only danger to stability comes for the eigenvalues near zero. However, there can be at most two such eigenvalues; this follows from a computation of the degree of D in a small circle about 0 (cf. Chapter 12, §A).

Since D is analytic, the degree measures the number of zeros inside the circle, and it is proved that this degree is two. Since zero is an eigenvalue, the other eigenvalue must be real, and the proof is completed by showing that it is, in fact, negative. This is done by showing: (i) $D(\lambda) < 0$ for λ large and real; and (ii) $D'(0) < 0$. The latter inequality is proved by studying how the unstable manifold of the travelling wave equation "sweeps across" the stable manifold as c varies through the wave speed $\bar{c}$, the speed at which the pulse solution exists. Thus, important stability information is obtained by studying how the underlying wave varies as a function of the wave speed.

We turn now to a brief discussion of the function D.

We first write $(L - \lambda I)\begin{pmatrix} p \\ r \end{pmatrix} = 0$ as the first-order system

$$
\begin{aligned}
p' &= q, \\
q' &= -cq + (\lambda - f'(u_\varepsilon))p + r, \\
r' &= -(\varepsilon/c)p + [(\lambda + \varepsilon\gamma)/c]r,
\end{aligned}
\tag{25.4}
$$

which we abbreviate as $z' = Az$, $z = (p, q, r) \in \mathbb{C}^3$. It is not difficult to show that (25.4) has a one-dimensional set of bounded solutions; this follows from studying the asymptotic system; i.e., the system (25.4) with u_ε replaced by 0. Let this solution space be spanned by $\zeta(\lambda, \xi)$. $\zeta(\lambda, \xi)$ is the only candidate (up to scalar multiples) for an eigenfunction; $\lambda \in \mathbb{C}\backslash N$ is an eigenvalue if and only if $\zeta(\lambda, \xi)$ is bounded as $\xi \to \infty$.

To test the boundedness of ζ as $\xi \to \infty$, we consider the adjoint equation to (25.4)

$$
z^{*\prime} = Bz^*, \qquad B = -A^*. \tag{25.5}
$$

It is trivial to check that if z^* satisfies (25.5), and z satisfies (25.4), then $z^* \cdot z$ (scalar product in $\mathbb{C}^3$) is constant. Let $\eta(\lambda, \xi)$ be the unique (up to scalar multiple) solution of (25.5) which is bounded as $\xi \to \infty$. For $\xi \to \infty$, η is normal to the stable manifold for the asymptotic system related to (25.4). Thus ζ is bounded as $\xi \to \infty$ if and only if $\zeta(\lambda, \xi) \cdot \eta(\lambda, \xi) \to 0$ as $\xi \to \infty$. But as $D(\lambda) = \zeta \cdot \eta$ is independent of ξ, it follows that λ is an eigenvalue if and only if $D(\lambda) = 0$. Since both ζ and η can be chosen to be analytic in λ for each ξ, it follows that $D(\lambda)$ is analytic and its zeros in $\mathbb{C}\backslash N$ are eigenvalues. The proof consists of showing that in $\mathbb{C}\backslash N$, D has only one zero (at 0) and this zero is simple.

§B. Symmetry-Breaking

In this section we discuss the bifurcation of symmetric (radial) solutions of semilinear elliptic equations into asymmetric ones. This problem differs from the usually encountered bifurcation results (cf. Chapter 13) in several ways.

Namely, we are not bifurcating from a "constant" solution; moreover, we are not bifurcating from a "simple eigenvalue" (cf. Chapter 13, §A), and finally, we have the problem of showing that we actually *bifurcate* to an asymmetric solution. This last problem consists of two parts: first we must show that the kernel of the linearized operator contains an asymmetric element, and then we must show that bifurcation to an asymmetric solution actually occurs. The tool for handling this last problem is the Conley index.

The equations considered here are of the form

$$\Delta u(x) + f(u(x)) = 0, \qquad x \in D_R^n, \tag{25.6}$$

together with the boundary conditions

$$\alpha u(x) - \beta \, du(x)/dn = 0, \qquad x \in \partial D_R^n. \tag{25.7}$$

Here D_R^n is an n-ball of radius R, α and β are constants with $\alpha^2 + \beta^2 = 1$, and f is a real-valued C^1-function. Note that (25.6) is invariant under the orthogonal group $O(n)$; i.e., for any $g \in O(n)$ and any $x \in D_R^n$,

$$(\Delta u)(gx) + f(u(gx)) = \Delta(u \circ g)(x) + f(u(gx)).$$

This motivates us to consider radial solutions of (25.6) (i.e., solutions of the form $u = u(r)$, $r = |x|$), as the "symmetric" solutions, since they possess maximal symmetry; i.e., they are the only ones invariant under the full group $O(n)$. The nonradial solutions will be termed "asymmetric solutions."

Radial solutions of (25.6) satisfy the ordinary differential equation

$$u'' + \frac{n-1}{r} u' + f(u) = 0, \qquad 0 < r < R \qquad (' = d/dr), \tag{25.8}$$

together with the boundary conditions

$$u'(0) = 0 = \alpha u(R) - \beta u'(R). \tag{25.9}$$

The condition $u'(0) = 0$ is necessary in order that u be differentiable at $r = 0$. Rewriting (25.8) as a first-order system gives

$$u' = v, \qquad v' = -\frac{n-1}{r} v - f(u), \tag{25.10}$$

and we shall consider orbits of (25.10) satisfying the initial conditions

$$u(0) = p > 0, \qquad v(0) = 0. \tag{25.11}$$

The corresponding solution will be denoted by $(u(r, p), v(r, p))$, $r \geq 0$, and p will be taken as the bifurcation parameter.

Solutions of (25.9), (25.10) will be allowed to undergo many sign changes. Thus if

$$\theta_0 = \text{Tan}^{-1}(\beta/\alpha), \qquad -\pi \le \theta_0 < 0,$$

and k is a fixed nonnegative integer, we define the function $p \mapsto T(p)$ by the condition

$$\theta(T(p), p) = \theta_0 - k\pi, \tag{25.12}$$

where $\theta(r, p) = \text{Tan}^{-1}(v(r, p)/u(r, p))$. If (25.12) holds we say that $u(\cdot, p)$ *lies in the kth nodal class*. Note that since p is allowed to vary, this allows the radii $T(p)$ of the balls to also vary.

If we linearize (6) about a given radial solution $u(\cdot, p)$, we obtain

$$\Delta w + f'(u(r, p))w = 0, \qquad r < T(p), \tag{25.13}$$

$$\alpha w(r) - \beta \, dw(r)/dn = 0, \qquad r = T(p). \tag{25.14}$$

An easy application of the implicit function theorem shows that a necessary condition for $u(\cdot, p)$ to be a "bifurcation point" is that the equations (25.13), (25.14) admit a solution $w \ne 0$. Using the fact that any function on an n-ball admits a spherical harmonic decomposition [29], we may write

$$w(r, \theta) = \sum_{N=0}^{\infty} a_N(r)\Phi_N(\theta), \qquad r = |x|, \quad \theta \in S^{n-1}, \tag{25.15}$$

where $\Phi_N \in E_N$, the Nth-eigenspace of the Laplacian on S^{n-1}, corresponding to the Nth-eigenvalue λ_N. As is well known [29],

$$\lambda_N = -N(N + n - 2), \qquad \dim E_N = \binom{N + n - 2}{N}\left(\frac{n + 2N - 2}{n + N - 2}\right) \equiv l_N. \tag{25.16}$$

Furthermore, each summand in (25.15) lies in the kernel of the linearized operator, and using (25.15) in (25.13), we find that each a_N satisfies the equation

$$a_N'' + \frac{n-1}{r}a_N' + \left(f'(u(r, p)) + \frac{\lambda_N}{r^2}\right)a_N = 0, \qquad 0 < r < T(p), \tag{25.17}$$

together with the boundary conditions

$$\alpha a_N(T(p)) - \beta a_N'(T(p)) = 0, \qquad N \ge 0,$$

$$a_N'(T(p)) = 0, \qquad N > 0, \tag{25.18}$$

$$a_0'(T(p)) = 0.$$

Now if the symmetry breaks on $u(\cdot, p)$, the necessarily $a_N \neq 0$ for some $N > 0$ (see [8]). We thus say that:

(a) *the symmetry breaks infinitesimally on* $u(\cdot, p)$, provided that $a_N \neq 0$ for some $N > 0$; and

(b) *the symmetry breaks on* $u(\cdot, p)$ if $u(\cdot, p)$ bifurcates into an asymmetric solution.

In order to obtain symmetry-breaking results, we shall reduce the problem to one in finite dimensions via a Lyapunov–Schmidt reduction (cf. pp. 175–176). This will allow us to use a finite-dimensional version of the Conley Index Theory. In particular, we enrich this invariant by using the group structure, present from symmetry considerations, to obtain an equivariant Conley index. This equivariant Conley index is used to prove a general bifurcation theorem which can be applied to the symmetry-breaking problem for semilinear elliptic equations.

The framework can be described as follows. Let B_0 and B_1 be Banach spaces, and let H be a Hilbert space, where $B_1 \subset B_0 \subset H$ and the inclusions are all assumed to be continuous. Let $I = [\lambda_1, \lambda_2]$ be an interval in $\mathbf{R}$ and suppose that $M: B_1 \times I \to B_0$ is a smooth, gradient operator and that there exists a smooth curve $\{u_\lambda: \lambda \in I\} \subset B_1$ satisfying

$$M(u_\lambda, \lambda) = 0, \qquad \lambda \in I. \tag{25.19}$$

For $i = 1, 2$, denote by P_i the (closed) vector space spanned by

$$\{v \in B_1: d_u M(u_{\lambda_i}, \lambda_i)v = \mu v, \mu \geq 0\};$$

P_i is called the *peigenspace* of $(u_{\lambda_i}, \lambda_i)$. We first assume that the following conditions hold:

$$dM_i \equiv d_u M(u_{\lambda_i}, \lambda_i) \quad \text{is nonsingular for} \quad i = 1, 2, \tag{25.20}$$

and

$$\dim P_1 \neq \dim P_2. \tag{25.21}$$

The question we ask is: Does there exist λ, $\lambda_1 < \lambda < \lambda_2$, such that (u_λ, λ) is a bifurcation point for (25.19)? Notice that in the finite-dimensional case (i.e., B_1 and B_0 have finite dimensions), the Conley indices satisfy $h(u_{\lambda_1}) \neq h(u_{\lambda_2})$ so Conley's Continuation Theorem (Theorem 23.31) implies that our question has an affirmative answer.

More generally, suppose that G is a compact Lie group acting on H with B_0 and B_1 invariant under G, and that M is equivariant with respect to G in the sense that

$$M(gu, \lambda) = gM(u, \lambda), \qquad \forall g \in G, \quad u \in B_1, \quad \lambda \in I. \tag{25.22}$$

Then the set of solutions of $M = 0$ also admits a G-action. If $\{(u_\lambda, \lambda): \lambda \in I\}$

is a smooth curve of invariant solutions (i.e., symmetric solutions) under G (in the sense that $gu_\lambda = u_\lambda$, $\forall g \in G$, $\lambda \in I$), we ask the question as to whether the symmetry breaks along this curve; i.e., under what conditions does there exist bifurcation from this curve of invariant solutions to noninvariant (asymmetric) solutions?

In order to describe an approach to this problem, we first note that (25.22) implies that for all $g \in G$, $v \in B_1$, and each λ,

$$d_u M(gu, \lambda)gv = gd_u M(u, \lambda)v.$$

If now V denotes the eigenspace of $d_u M(u_\lambda, \lambda)$ corresponding to a given eigenvalue μ, then if $w \in V$, we have that $gw \in V$ for all $g \in G$. Indeed,

$$d_u M(u_\lambda, \lambda)w = \mu w \quad \text{implies} \quad gd_u M(u_\lambda, \lambda)w = \mu gw,$$

so that

$$d_u M(u_\lambda, \lambda)gw = d_u M(gu_\lambda, \lambda)gw = \mu gw,$$

since $gu_\lambda = u_\lambda$, so $gw \in V$. Thus if T_g is the operator on V defined by $T_g v = gv$, then the mapping $g \to T_g$ defines a *representation* of G. In the language of representation theory, we say that V *is a representation of* G. It follows that for each fixed λ, the peigenspace of $d_u M(u_\lambda, \lambda)$ defines a representation of G. This motivates us to replace condition (25.21) by the statement

$$P_1 \text{ is not isomorphic to } P_2 \text{ as a representation of } G. \qquad (25.21)_G$$

Now we ask this question: Suppose that (25.19), (25.20), and (25.21)$_G$ hold; does this imply that bifurcation occurs? Observe that condition (25.21)$_G$ is considerably weaker than (25.21). Indeed, it is possible that (25.21)$_G$ holds even though P_1 and P_2 are isomorphic vector spaces.

We will consider such questions from the Conley index point of view. This requires that we extend these ideas to the equivariant case. Moreover, the hypotheses (25.19), (25.20), and (25.21), or (25.21)$_G$ does not directly imply, in the infinite-dimensional case, that there is a change in Conley index at λ_1 and λ_2. Thus we shall pass to a (global) equivariant Lyapunov–Schmidt reduction, which will enable us to prove a general bifurcation theorem which answers affirmatively the above questions. This theorem then will be applied to the symmetry-breaking problem for semilinear elliptic equations defined on n-balls. Finally, we shall mention several related results, and also show how to clarify some subtle points in Chapter 24.

§C. A Bifurcation Theorem

We shall now describe a general bifurcation theorem. Our technique will be to reduce the problem, via a Lyapunov–Schmidt reduction, to one in finite dimensions, where we can apply Conley index techniques.

In order not to complicate the descriptions, we shall first describe a Lyapunov–Schmidt reduction in the nonequivariant case; the modifications necessary to extend our techniques to the equivariant case will be discussed in the next section.

Let $B_1 \subset B_0 \subset H$, where B_0, and B_1 are Banach spaces, H is a Hilbert space, and the embeddings are assumed to be continuous. (In applications, we often take $B_1 = C_0^2(D^n)$, $B_0 = C(D^n)$, and $H = L^2(D^n)$, where D^n is an n-ball centered at 0.) Let M be a smooth mapping

$$M: B_1 \times I \to B_0, \qquad I = [\lambda_1, \lambda_2] \subset R,$$

and assume that there is a family $\{u_\lambda : \lambda \in I\} \subset B_1$, depending smoothly on λ, which satisfies (25.19). For ease in notation, let

$$dM_\lambda = d_u M(u_\lambda, \lambda).$$

Definition. The *peigenspace* P_λ of dM_λ is the closed subvector space generated by the set

$$\{v \in B_1 : dM_\lambda v = \mu v \text{ for some } \mu \geq 0\}.$$

P_λ is thus seen to be the space generated by those eigenvectors of dM_λ corresponding to nonnegative eigenvalues.

If we replace u by $u + u_\lambda$, then with a slight abuse of notation, we write (25.19) as

$$M(0, \lambda) = 0, \qquad \lambda \in I. \tag{25.23}$$

For any $\varepsilon > 0$, let $P_{\lambda, \varepsilon}$ be the peigenspace of $dM_\lambda + \varepsilon I$. We assume that there is a fixed $\varepsilon > 0$ such that

$$\dim P_{\lambda, \varepsilon} < \infty, \qquad \forall \lambda \in I. \tag{25.24}$$

We now state the main result in this section.

Theorem I.2. *Let M be a gradient operator for each $\lambda \in I$, and assume that M satisfies (25.23) and (25.24). If*

$$dM_{\lambda_1} \text{ and } dM_{\lambda_2} \text{ are nonsingular}, \tag{25.25}$$

and

$$P_{\lambda_1} \text{ is not isomorphic to } P_{\lambda_2}, \tag{25.26}$$

then there is a λ_0, $\lambda_1 < \lambda_0 < \lambda_2$, for which $(0, \lambda_0)$ is a bifurcation point for $M(0, \lambda) = 0$.

Notice that since $P_\lambda \subset P_{\lambda,\varepsilon}$, condition (25.24) implies that dim $P_{\lambda_i} < \infty$, $i = 1, 2$; thus (25.26) means that dim $P_{\lambda_1} \neq$ dim P_{λ_2}. We have stated (25.26) in this slightly awkward way in anticipation of the results in the next section where this theorem will be extended to the equivariant case.

Proof. Since M is a gradient for each $\lambda \in I$, it follows that dM_λ is symmetric and so has real eigenvalues. Then (25.24) implies that the eigenvectors of dM_λ corresponding to eigenvalues greater than $-\varepsilon$ lie in a finite-dimensional space. Thus there is a finite-dimensional space $E_\lambda \subset B_1$ such that

$$(dM_\lambda e, e) \leq -\varepsilon |e|^2, \qquad \forall e \in E_\lambda.$$

(Here the norm and inner product are those in H.) By continuity, there is an open interval J_λ about λ such that

$$(dM_\mu e, e) \leq -\frac{\varepsilon}{2}|e|^2, \qquad \forall \mu \in J_\lambda, \quad \forall e \in E_\lambda.$$

A finite number of the intervals $J_{\bar\lambda_1}, \ldots, J_{\bar\lambda_s}$ cover I. Thus let

$$E = \left(\bigcap_{i=1}^{s} E_{\lambda_i}\right) \cap P_{\lambda_1}^\perp \cap P_{\lambda_2}^\perp;$$

then $E \subset B_1$ has finite codimension. Note that $E_\lambda = P_{\lambda,\varepsilon}^\perp \cap B_1$; thus

$$E = \bigcap_{i=1}^{s} P_{\lambda_{i,\varepsilon}}^\perp \cap P_{\lambda_1}^\perp \cap P_{\lambda_2}^\perp \cap B_1 = \left(\bigcup_{i=1}^{s} P_{\lambda_{i,\varepsilon}} \cup P_{\lambda_1} \cup P_{\lambda_2}\right)^\perp \cap B_1 \equiv F^\perp \cap B_1,$$

so that E is a closed subspace of B_1 of finite codimension and $P_{\lambda_1} \cup P_{\lambda_2} \subset F$. Now if $\lambda \in I$, then $\lambda \in J_{\bar\lambda_k}$ for some k, $1 \leq k \leq s$, and if $e \in F^\perp \cap B_1 = E$, $e \in E_{\bar\lambda_k}$ so $(dM_\lambda e, e) \leq -(\varepsilon/2)|e|^2$. Thus dM_λ restricted to $F^\perp \cap B_1$ is uniformly negative definite.

Next, write

$$H = F^\perp \oplus F \quad \text{and since} \quad F \subset B_1,$$

$$B_0 = (B_0 \cap F^\perp) \oplus F,$$

$$B_1 = (B_1 \cap F^\perp) \oplus F.$$

We can now describe our Lyapunov–Schmidt reduction. Thus for $h \in B_1$, write $h = (x, y)$, $x \in F^\perp$, $y \in F$. Then

$$M(h, \lambda) = M(x, y, \lambda) = (u(x, y, \lambda), v(x, y, \lambda)),$$

where $u \in F^\perp$ and $v \in F$. Since $M(0, 0, \lambda) = 0$, we have $u(0, 0, \lambda) = 0$. We claim that $u_x \equiv u_x(0, 0, \lambda)$, defined on $F^\perp \cap B_1$, is an isomorphism. First, if $u_x \xi = 0$, for some $\xi \neq 0$, then as u_x is strictly negative definite, the inequality $0 > (u_x \xi, \xi)$ yields a contradiction. Next, an easy calculation shows that u_x is symmetric. Finally, if $\xi \in H$, we have

$$(u_x(\pi_2 \xi), \pi_2 \xi) \leq -\frac{\varepsilon}{2} |\pi_2 \xi|^2,$$

where $\pi_2 =$ projection onto F. Thus the spectrum of u_x is uniformly bounded away from zero. It follows from these three observations that $u_x(0, 0, \lambda)$ is indeed an isomorphism. Therefore the implicit function theorem implies that the equation $u(x, y, \lambda) = 0$ can be solved for $x = x(y, \lambda)$ in a neighborhood of $(0, 0, \lambda)$ of the form $U_\lambda \times \mathcal{O}_\lambda$, where U_λ is a neighborhood of $(0, 0)$ and $\mathcal{O}_\lambda$ is a neighborhood of λ in I. By compactness, a finite number of the $\mathcal{O}_\lambda$ cover I, say $\mathcal{O}_{\lambda_1}, \ldots, \mathcal{O}_{\lambda_k}$. Therefore we have a unique solution $x = x(y, \lambda)$ on a neighborhood $U \times \mathcal{O} \equiv \bigcap_1^k U_{\lambda_i} \times \bigcup_{i=1}^k \mathcal{O}_{\lambda_i}$ of $(0, 0, \lambda)$. Now define

$$\phi(y, \lambda) = v(x(y, \lambda), y, \lambda);$$

this is our global Lyapunov–Schmidt reduction.   □

Lemma I.3. *Fix $\lambda \in I$ and consider the ordinary differential equation*

$$y_s = \phi(y(s), \lambda). \tag{25.27}$$

Then this equation is gradient-like.

Proof. If $\psi(x, y)$ is Lyapunov function for M, then

$$\overline{\psi}(y, \lambda) = \psi(x(y, \lambda), \lambda)$$

is easily seen to be a Lyapunov function for (25.27).   □

Lemma I.4. $P(d\phi_{(0, \lambda_i)})$ *is isomorphic to* P_{λ_i}, $i = 1, 2$.

We omit the somewhat technical proof of this result; see [53] for the details.

We can now finish the proof of the theorem. To this end, note first that solutions of $M = 0$ correspond in a one–one way to solutions of $\phi = 0$. It is easy to check that (25.24) implies that $d\phi_{(0, \lambda_i)}$ is nonsingular for $i = 1, 2$. It follows that the Conley indices for the rest points $(0, \lambda_i)$ of the equations (25.27) satisfy $h(0, \lambda_1) \neq h(0, \lambda_2)$. Thus if $\mathcal{V}$ is any neighborhood of 0 (in the finite-dimensional space), then 0 cannot be the maximal invariant set in $\mathcal{V}$ for each $\lambda \in I$. Thus there is a point $\lambda(\mathcal{V}) \in I$ for which 0 is not the maximal invariant set in $\mathcal{V}$ for the equations $y_s = \phi(y(s), \lambda(\mathcal{V}))$. The maximal invariant set in $\mathcal{V}$ for this equation must then contain a point $y \neq 0$. The gradient-like nature of the equation forces the α- and ω-limit sets of y to differ. Hence the equation admits another rest point in $\mathscr{V}$, different from 0. Now let the neighborhoods $\mathscr{V}$ shrink to 0. The $\lambda(\mathscr{V})$'s have a convergent subsequence $\lambda_i \to \lambda_0$, and $(0, \lambda_0)$ is easily seen to be a bifurcation point for $\phi(y, \lambda) = 0$ (so $\lambda_1 < \lambda_0 < \lambda_2$), and hence for $M = 0$.

§D. Equivariant Conley Index

We now show how to extend the Conley index ideas to the case where there is a group acting on the space. This will give us a finer index invariant which will prove quite useful in applications to symmetry-breaking problems. Again let B_0 and B_1 be Banach spaces, let H be a Hilbert space, and assume $B_1 \subset B_0 \subset H$, where the embeddings are continuous. Let $I = [\lambda_1, \lambda_2]$ be an interval in R, and let M be a smooth mapping $M: B_1 \times I \to B_0$. We again consider solutions of the equation $M(u, \lambda) = 0$.

We now assume the existence of a compact Lie group G acting on H with B_0 and B_1 invariant (so $gu \in B_1$ and $gv \in B_0$ for all $g \in G$, $u \in B_1$, $v \in B_0$). We further assume that M is *equivariant* with respect to G, in the sense that for all $u \in B_1$, $\lambda \in I$, $g \in G$, we have

$$M(gu, \lambda) = gM(u, \lambda).$$

Let (u_λ, λ) be a smooth curve of invariant (symmetric) solutions of $M = 0$; thus

$$M(u_\lambda, \lambda) = 0, \qquad \forall \lambda \in I, \tag{25.28}$$

and $gu_\lambda = u_\lambda$ for all $g \in G$ and all $\lambda \in I$. Now, as discussed above, each peigenspace P_λ is a representation of G. Since G is compact, it follows by the Peter–Weyl theorem [29], that each P_λ is a sum of finite-dimensional irreducible representations. (A representation V of G is called *irreducible* if V has no proper, closed invariant (under all operators T_g, $g \in G$) subspaces.)

Again write $dM_\lambda \equiv d_u M(u_\lambda, \lambda)$, and as before, we assume that there is an $\varepsilon > 0$ such that peigenspaces $P_{\lambda, \varepsilon}$ of $dM_\lambda + \varepsilon I$ satisfy (25.24).

Theorem I.5. *Let M be a gradient operator for each $\lambda \in I$, and assume that M satisfies (25.24) and (25.28). If*

$$dM_{\lambda_1} \text{ and } dM_{\lambda_2} \text{ are nonsingular}, \tag{25.29}$$

and for $i = 1, 2$

$$P_{\lambda_i} \text{ contains } k_i \text{ copies of a fixed irreducible} \\ \text{representation of } G, \text{ where } k_1 \neq k_2, \tag{25.30}$$

then there is a λ, $\lambda_1 < \lambda < \lambda_2$, such that (u_λ, λ) is a bifurcation point for $M = 0$.

The theorem is proved by extending the Conley index to the equivariant case, obtaining a G-Conley index $h_G(\cdot)$. This puts more structure on the Conley index and thus makes it more useful. For example, it is possible to have $h(I_1) = h(I_2)$ but $h_G(I_1) \neq h_G(I_2)$. Thus we can prove bifurcation theorems using the G-Conley index h_G, while no such statement is possible using the (usual) Conley index, h. Also, in the finite-dimensional case, (as in the nonequivariant case), at a nondegenerate critical point I, the Conley index of I reduces to the Morse index of I, i.e., we have an equivariant homotopy equivalence

$$h(I) \approx D(V)/S(V).$$

Here V is the peigenspace of I and $D(V)$ (resp. $S(V)$), denotes the unit ball (resp. unit sphere) in V. Now at distinct parameter values $\lambda_1 \neq \lambda_2$, we can have $\dim V_1 = \dim V_2$ and thus $h(I_1, \lambda_1) = h(I_2, \lambda_2)$. But if $G = O(n)$ (for example), and $V_1 \neq V_2$ as representations of $O(n)$, then $h_{O(n)}(I_1, \lambda_1) \neq h_{O(n)}(I_2, \lambda_2)$. It is thus of some interest to extend the Conley index to the equivariant case, and such an extension will now be outlined.

Let X be a metric space, and G a compact Lie group. Suppose that ψ is a G-flow (or local flow, in which case R is to be replaced by R_+) on X, i.e.,

$$\psi : (R \times G) \times X \to X,$$

ψ is continuous, and satisfies $\forall x \in X$, $\forall g_1, g_2 \in G$, $\forall t_1, t_2 \in R$,

$$\psi(0, e, x) = x \qquad (e = \text{id. in } G),$$

and

$$\psi(t_1, g_1, \psi(t_2, g_2, x)) = \psi(t_1 + t_2, g_1 g_2, x).$$

If $g = e$, $\psi_t(x) \equiv \psi(t, e, x)$ defines a flow on X, and if $t = 0$, ψ induces an action of G on X by

$$gx \equiv \psi(0, g, x).$$

Let X/G denote the *orbit space* of X with respect to G, i.e., X/G is the set of orbits in X under G. Thus, we define an equivalence relation on X by $x \sim y$ if $y = gx$ for some $g \in G$; then X/G is the set of equivalence classes. Let $\pi : X \to X/G$ be the canonical projection. We put the quotient topology on X/G, i.e., $A \subset X/G$ is open if $\pi^{-1}(A)$ is open in X. Then if $(\overline{N}_1, \overline{N}_2)$ is an index pair in X/G, $(\pi^{-1}(\overline{N}_1), \pi^{-1}(N_2))$ is an index pair in X. Moreover, if (N_1, N_2) is a G-invariant index pair in X, then $(\pi(N_1), \pi(N_2))$ is an index pair in X/G. If I is a G-invariant set in X, and I is also an isolated invariant set for the flow ψ_t, then $\pi(I)$ is an isolated invariant set for the induced flow on X/G, and we define $h_G(I)$, the G-Conley index of I, to be the equivalent homotopy type of the pointed space

$$h_G(I) = \pi^{-1}(\overline{N}_1)/\pi^{-1}(\overline{N}_2),$$

where $(\overline{N}_1, \overline{N}_2)$ is an index pair for $\pi(I)$. [Two spaces X and Y are said to be of the *same equivariant homotopy type with respect to* G if they are homotopy equivalent, and the group action commutes with all of the relevant maps. In other words there exist maps $f: X \to Y$ and $h: Y \to X$, such that $f \circ h$ and $h \circ f$ are homotopic to the identity through equivariant maps, i.e., maps which commute with the group action. It is also not too hard to show that there is an induced flow on X/G given by $[x] \cdot t = [x \cdot t]$ and if I/G is an isolated invariant set for the flow on X/G, and $(\overline{N}_1, \overline{N}_0)$ is an index pair for I/G, then $(\pi^{-1}(\overline{N}_1), \pi^{-1}(\overline{N}_0))$ is a G-invariant index pair for I.]

Let (N_1, N_0) and $(\overline{N}_1, \overline{N}_0)$ be G-invariant index pairs for $I \subset X$. Then there exist equivariant maps f and h (i.e., maps which commute with the G-action), satisfying

$$f: (N_1/N_0, N_0) \to (\overline{N}_1/\overline{N}_0, \overline{N}_0),$$

$$h: (\overline{N}_1/\overline{N}_0, \overline{N}_0) \to (N_1/N_0, N_0),$$

such that both $h \circ f$ and $f \circ h$ are equivariantly homotopic to the identity map; i.e., the pointed spaces $(N_1/N_0, N_0)$ and $(\overline{N}_1/\overline{N}_0, \overline{N}_0)$ are of the same equivariant homotopy type. We can thus unambiguously define the G-invariant Conley index of I be to be this equivariant homotopy type. Note that this definition reduces to the usual one when G is the trivial group. The definition given here is richer, since as the relevant spaces admit a group action, we can distinguish indices which have the same homotopy type as pointed spaces, but are not of the same homotopy type as pointed G-spaces.

Definition. If dM_λ is nonsingular for some $\lambda \in I$, *the reduced G-index of* u_λ, $h_G^R(u_\lambda, \lambda)$, is the equivariant homotopy type of the pointed G-space:

$$h_G^R(u_\lambda, \lambda) = D(P_\lambda)/S(P_\lambda). \tag{25.31}$$

Note that in the finite-dimensional case, this definition agrees with the equivariant formulation of the Conley index as given above. We now have

Theorem I.6. *With M as defined above (satisfying (24.24) and (25.28)), assume that M is a gradient for each $\lambda \in I$. Suppose that*

$$dM_{\lambda_i} \text{ is nonsingular for } i = 1, 2, \tag{25.32}$$

and

$$h_G^R(u_{\lambda_1}, \lambda_1) \not\simeq h_G^R(u_{\lambda_2}, \lambda_2) \tag{25.33}$$

(i.e., are not of the same homotopy type). Then there exists a λ, $\lambda_1 < \lambda < \lambda_2$, such that (u_λ, λ) is a bifurcation point.

The theorem is proved in exactly the same way as in the nonequivariant case (Theorem I.2). Next, we have the following theorem which is applicable to a wide class of groups G; we only give the statement for $G = O(n)$; (see [53]).

Theorem I.7. *If $G = O(n)$ and V and W are representations of $O(n)$, then the pointed $O(n)$-spaces $D(V)/S(V)$, $D(W)/S(W)$ are equivariantly homotopy equivalent if and only if V is isomorphic to W as $O(n)$ representations.*

Since $h_{O(n)}^R(u_{\lambda_i}, \lambda_i) = D(P_{\lambda_i})/S(P_{\lambda_i})$, we see that Theorems I.5 and I.6 are equivalent. The importance of Theorem 1.7 lies in the fact that one can verify (25.33) via a study of the linearized equations; this will be demonstrated in the next section.

§E. Application to Semilinear Elliptic Equations

In this section we shall apply the general bifurcation theorem of the last section, to study the symmetry-breaking problem for solutions of the semi-linear elliptic equation

$$\Delta u(x) + f(u(x)) = 0, \qquad x \in D_R^n, \tag{25.34}$$

with boundary conditions

$$\alpha u(x) - \beta \, du(x)/du = 0, \qquad x \in \partial D_R^n. \tag{25.35}$$

Here D_R^n denotes an n-ball of radius R, f is a smooth function, $\alpha^2 + \beta^2 = 1$ and d/dn denotes differentiation in the radial direction.

The function f is assumed to satisfy the following assumptions:
There exist points $b < 0 < \gamma$ such that

$$(H) \begin{cases} (H_1) & f(\gamma) = 0, f'(\gamma) < 0, \\ (H_2) & F(\gamma) > F(u) \text{ if } b < u < \gamma \text{ (here } F' = f \text{ and } F(0) = 0), \\ (H_3) & F(b) = F(\gamma), \\ (H_4) & \text{if } f(b) = 0, \text{ then } f'(b) < 0, \\ (H_5) & uf(u) + 2(F(\gamma) - F(u)) > 0 \text{ if } b < u < \gamma. \end{cases}$$

We refer the reader to [52] for a discussion of these conditions, and merely point out that (H_2) and (H_5) both hold if $uf(u) > 0$ for $u \in (b, \gamma)/\{0\}$; for example, if $f(u) = u(1 - u)$, then (H) holds.

For equation (25.33), the relevant symmetry group is $G = O(n)$, i.e., $O(n)$ is the largest group which leaves (25.34) invariant. Solutions of (25.34) which possess this maximal symmetry are the radial solutions—any nonradial (i.e., asymmetric) solution is called a *symmetry-breaking* solution.

In order to illustrate how representation theory is used in this problem, we first discuss some results in [52]. Thus it was shown that there is a $\sigma > 0$ such that the interval $(\gamma - \sigma, \gamma)$ lies in the domain of T; cf. §B. For $p \in (\gamma - \sigma, \gamma)$, consider the linearized operator

$$L^p(w) = \Delta w + f'(u(\cdot, p))w, \qquad |x| < T(p),$$

where

$$w \in \Phi_p = \{\phi \in C^2(|x| < T(p)): \alpha\phi(x) - \beta \, d\phi(x)/dn = 0, |x| = T(p)\}.$$

Since any function on the ball has a spherical harmonic decomposition, we may write

$$w(r, \theta) = \sum_{n \geq 0} \alpha_N(r)\Phi_N(\theta), \qquad 0 \leq r \leq T(p), \quad \theta \in S^{n-1},$$

where S^{n-1} denotes the unit sphere in R^n and $\Phi_N \in E_N$. This leads to the following decomposition of L^p as a direct sum:

$$L^p = \oplus \sum_{n \geq 0} L^p_N,$$

where each L^p_N is defined for $\phi \in \Phi_p$ by

$$L^p_N \phi = \phi'' + \frac{n-1}{r}\phi' + \left(f'(u(\cdot, p)) + \frac{\lambda_N}{r^2}\right)\phi, \qquad 0 < r < T(p).$$

Now it was proved in [52] that there exists a sequence $q_N \uparrow \gamma$ such that for large N, say $N \geq N_0$, $L^{q_N}_N$ has positive spectrum; i.e., there exist positive numbers $\mu_1^N, \mu_2^N, \ldots, \mu_{k_N}^N$, and nontrivial functions $\alpha_j^N \in \Phi_{q_N}, j = 1, 2, \ldots, k_N$, such that

$$L^{q_N}_N \alpha_j^N = \mu_j^N \alpha_j^N, \qquad j = 1, 2, \ldots, k_N.$$

Under the additional hypothesis that f is analytic (this assumption is needed only if $k > 1$, where k denotes the nodal class of radial solutions), we may assume that each radial solution $u(\cdot, q_N)$ is nondegenerate in the sense that $0 \notin \mathrm{sp}(L^{q_N})$. Furthermore, one may assume that $L^{q_N}_M$ has positive spectra if $M > N$. Using the fact that the λ_N's decrease monotically, it is not difficult to show that whenever L^p_N has a positive eigenvalue, the same is true for each operator L^p_M, $M < N$, see [52]. It follows that the peigenspace of each $u(\cdot, q_N)$ is a representation of $O(n)$ of the form

$$k_N E_N \oplus \cdots \oplus k_1 E_1 \oplus k_0 E_0,$$

where each k_j is a nonnegative integer, $j = 0, \ldots, N$. Therefore the dimension of the peigenspace of $u(\cdot, q_N)$ is

$$\dim P(u(\cdot, q_N)) = k_N l_N + \cdots + k_1 l_1 + k_0.$$

Thus it is a-priori possible that dim $P(u(\cdot, q_N)) = \dim P(u(\cdot, q_{N+1}))$; i.e.,

$$k_N l_N + \cdots + k_1 l_1 + k_0 = k'_{N+1} l_{N+1} + k'_N l_N + \cdots + k'_1 l_1 + k'_0,$$

so that the corresponding peigenspaces, P_N, P_{N+1} of $u(\cdot, q_N)$ and $u(\cdot, q_{N+1})$ are isomorphic. That is, the homotopy type of the corresponding pointed spaces satisfy

$$D(P_N)/S(P_N) \approx D(P_{N+1})/S(P_{N+1}).$$

Thus if we do not take account of the group structure we cannot apply Theorem I.2. Thus, in this case, we cannot use the ordinary (i.e., non-equivariant) Conley index to prove that bifurcation occurs—there is no index change at $u(\cdot, q_N)$ and $u(\cdot, q_{N+1})$. On the other hand, P_N and P_{N+1} differ as representations of $O(n)$, because P_N contains no copies of E_{N+1}. Thus Theorem I.5 is applicable and shows that for each $N \geq N_0$, there is a point p_N, $q_N < p_N < q_{N+1}$, for which $u(\cdot, p_N)$ is a bifurcation point. Now as was shown in [51], under the given hypotheses on f, for p near γ, no radial bifurcation is possible on $u(\cdot, p)$ if $T'(p) \neq 0$, and since that $T'(p) > 0$ if p is near γ [51], it follows that for all sufficiently large N the symmetry breaks on $u(\cdot, p_N)$. We thus have the following theorem:

Theorem I.8. *Let f satisfy hypotheses* (H) *and let k be a given integer representing a fixed nodal class of radial solutions of* (25.33), (25.34). *Assume that f is analytic if $k > 1$. Then there exists a sequence of points $p_N \uparrow \gamma$ such that the symmetry breaks on each radial solution $u(\cdot, p_N)$.*

Concluding Remarks

The question of structure of the set of bifurcating assymetric solutions is not fully resolved. We merely mention that at each symmetry-breaking bifurcation point there bifurcates out a family of $O(n - 1)$-invariant solutions; i.e., axisymmetric solutions. Moreover, one can prove that there are (degenerate) radial solutions for which there also bifurcates out (possibly among others), distinct solutions having symmetry groups (at least) $O(p) \times \tilde{O}(n - p)$, where $O(p)$ and $\tilde{O}(n - p)$ are $n \times n$ orthogonal matrices of the forms

$$O(p) = \begin{pmatrix} * & 0 \\ 0 & I_{n-p} \end{pmatrix}, \qquad \tilde{O}(n - p) = \begin{pmatrix} I_p & 0 \\ 0 & * \end{pmatrix},$$

respectively; for a proof see [52].

Next we point out an interesting technical difficulty arising in studying the symmetry-breaking problem for solutions of (25.33), (25.34). Namely, since p is taken as a parameter, the radius R of the ball on which the radial solution

$u(\cdot, p)$ is defined, changes with p; i.e., $R = T(p)$. The space

$$\left\{(\phi, p) \in C^2(0, T(p)) \times R_+ : \alpha\phi(x) - \beta\frac{d\phi(x)}{dn} = 0, |x| = T(p)\right\},$$

on which our bifurcation is presumed to occur, is *not* in the form of a product space $B \times \Lambda$ where B is a Banach space and Λ is the parameter space. If we change variables, writing $y = x/R$, then (25.33) and (25.34) become

$$\Delta u(y) + R^2 f(u(y)) = 0, \qquad |y| \leq 1,$$
$$\alpha u(y) - \beta R\, du(y)/dn = 0, \qquad |y| = 1.$$

Thus, if $\alpha\beta = 0$ (i.e., Dirichlet or Neumann boundary conditions), then the parameter R does not appear in the boundary conditions, and we have the desired product structure. However, if $\alpha\beta \neq 0$, then we lose the product structure. This problem is overcome by showing that the space

$$\left\{(u, \lambda) \in C^2(0, 1) \times R_+ : \alpha u(y) - \beta\lambda\frac{du(y)}{dn} = 0, |y| = 1\right\}$$

forms a vector bundle over R_+, and is thus *locally* a product space. However, this local product structure is sufficient for doing bifurcation theory; for details see [53].

We remark that the Lyapunov–Schmidt reduction to the peigenspace, as described Section II, is a good framework on which the application of the Conley Index Theory to reaction–diffusion equations should be performed. In particular, the development in Chapter 24, §D, should be redone in this setting.

Finally, we wish to mention the celebrated Gidas–Ni–Nirenberg theorem [26], which states that positive solutions of the Dirichlet problem for (25.34) on n-balls, which vanish on the boundary (or positive solutions of the Dirichlet problem for (25.34) in all of $\mathbf{R}^n$ which vanish at infinity), must be radially symmetric; i.e., they must be functions only of $r = |x|$. This remarkable theorem is proved using a device due to Alexandroff, of moving parallel planes to a critical position, and then showing that the solution is symmetric about the critical plane; see [26].

SECTION II. Theory of Shock Waves

§A. Compensated Compactness

As we have discussed in Chapter 19, §B, a major obstacle in solving nonlinear partial differential equations is to obtain a-priori estimates on approximating sequences, in order to obtain convergence of a subsequence to a solution. Such methods rely on compactness results, and for nonlinear equations the desired esimtates can be quite difficult to obtain. This is in sharp contrast to the case of linear equations, where "weak convergence" often suffices to prove existence theorems. This "lack of compactness" for nonlinear problems has restricted the use of weak convergence to linear equations. However, a deeper analysis can render such techniques useful for certain nonlinear problems, and this is the essence of the method of compensated compactness.

For example, a typical result is the following. Suppose:

(a) (u_n^i, v_n^i) tends weakly to (u^i, v^i) in $L^2(\Omega)$, $i = 1, \ldots, N$;
(b) $\{\text{div } u_n \equiv \sum_{i=1}^{N} \partial u_n^i / \partial x_i\}$ is bounded in $L^2(\Omega)$; and
(c) curl v_n is bounded in $L^2(\Omega)^{N^2}$, ($[\text{curl } v_n]_{ij} \equiv \partial v_n^i / \partial x_j - \partial v_n^j / \partial x_i$); then

$$\sum_{i=1}^{N} u_n^i v_n^i \to \sum_{i=1}^{N} u^i v^i \quad \text{in } \mathscr{D}'.$$

(Note that $u_n^i v_n^i$ does not converge strongly to $u^i v^i$, in general; hence the term "compensated" compactness.)

We shall show here how the method can be used to obtain global existence theorems for certain systems of pairs of conservation laws. For these equations the technique yields convergence via the method of vanishing viscosity, as well as convergence of certain difference approximations. The analysis is based on viewpoint which takes into consideration averaging techniques together with the weak topology.

We begin the discussion with a brief review of some standard functional–analytic results on weak convergence, cf. [Ru 3]. First, if B is a Banach space, and B' its dual, we define the weak topology on B as that generated by the family of seminorms $p_{x'}$, $x' \in B'$, where $p_{x'}(x) = |(x', x)|$, $x \in B$. If B' is separable, then on bounded sets of B this topology is metrizable, and $x_n \to x$ weakly in B means $(x_n, x') \to (x, x')$, $\forall x' \in B'$. The weak * topology on B' is that defined by the seminorms q_x, $x \in B$, given by $q_x(x') = |(x, x')|$, $x' \in B'$. As before, if B is separable, x_n' converges weak * to x' means that $(y, x_n') \to (y, x')$ for all $y \in B$. The topology determined by the norm on the unit ball in B' is compact in the weak * topology. Moreover, closed convex sets are weakly closed, and convex functions which are strongly lower semicontinuous, are weakly lower semicontinuous, as follows from the Hahn–Banach theorem.

We now consider specific spaces:

(a) $L^p(\Omega)$, $1 \leq p \leq \infty$; and
(b) $M(\Omega)$, the space of Radon measures on an open set Ω in $\mathbf{R}^N$.

Recall that $L^p(\Omega)' = L^{p'}(\Omega)$ where $1/p + 1/p' = 1$ $(1 < p < \infty)$, and that if f_n is a bounded sequence in $L^p(\Omega)$, and $1 < p < \infty$, then there is a subsequence f_m converging weakly to f; i.e., $\int_\Omega f_m \phi \to \int_\Omega f \phi$ for all $\phi \in L^{p'}(\Omega)$; for $p = \infty$ a subsequence converges weak * to f. $L^1(\Omega)$ is isometrically embedded in

$M_b(\Omega) = \{$measures of finite total mass$\} = C_b(\Omega)'$, where $C_b(\Omega)$ is the space of bounded continuous functions with the sup-norm topology. If f_n is bounded in $L^1(\Omega)$, it has a subsequence f_m which converges weak $*$ to a measure μ, in the sense that $\int_\Omega f_m \phi \to \langle \mu, \phi \rangle$ for all $\phi \in C_b(\Omega)$. The space of measures $M(\Omega)$ is not a Banach space, but for each compact set $K \subset \Omega$, if $C_K^0(\Omega) \equiv C_K$ denotes the space of continuous functions on Ω with support in K (with the sup-norm topology), then $M(\Omega)$ is a subspace of C_K' and so comes equipped with the weak $*$ topology. If we extract subseqences which converge weak $*$, we mean that we have a bounded sequence in each C_K' and that the subsequence converges weak $*$ in C_K' for a countable family of compacta α whose union is Ω. An example of a weakly converging sequence is given by oscillating periodic functions. Thus, let $y_1, \ldots, y_k$ be independent vectors in $\mathbf{R}^k$, and $Y = \{\sum_i \theta_i y_i : 0 \le \theta_1 \le 1, \forall i\}$. If $F(x, y)$ is a bounded continuous function on $\Omega \times \mathbf{R}^k$, which is Y-periodic ($F(x, y - y_j) = F(x, y), j = 1, \ldots, k$), then as $n \to \infty$, the sequence $f_n(x) = F(x, nx)$ converges weak $*$ in $L^p(\Omega)$ to f_∞ where $f_\infty(x) = [1/(\text{meas } Y)] \int_Y F(x, y) \, dy$. This shows why the weak $*$ limit of $\phi(f_n)$ is not $\phi(f_\infty)$ for nonaffine ϕ.

The following result is useful to compare limits for different functions:

Proposition II.1. *Let K be a bounded set in $\mathbf{R}^N$, and let u_n be a sequence of functions from Ω to $\mathbf{R}^N$ satisfying:*

(i) $u_n \to u$ *in* $L^\infty(\Omega)^N$ *weak* $*$; *and*

(ii) $u_n(x) \in K$ *a.e.*

Then

(iii) $u(x)$ *lies in* $\overline{\text{conv } K}$, *the closed convex hull of K, a.e.*

Conversely, if u satisfies (iii), *there exists a sequence u_n satisfying* (i) *and* (ii).

If (i) and (ii) hold, then since $\overline{\text{conv } K}$ is the intersection of half-spaces containing K, and $L(u_n) \ge 0$ implies $L(u) \ge 0$ a.e., for affine L, we easily show (iii). Conversely, to construct u_n, one first considers the case where u is a step function and uses a construction as in the above example in each set where u is constant.

A more precise analysis of the oscillations of a sequence u_n can be given if one knows all weak $*$ limits. This leads us to the "generalized functions" of Young [62]. Thus, to each function u_n in $L^\infty(\Omega)^k$ we associate a Radon measure μ_n on $\Omega \times \mathbf{R}^k$ defined by

$$\langle \mu_n, F \rangle = \int_\Omega F(x, u_n(x)) \, dx, \qquad (25.36)$$

where $F(x, \lambda)$ is continuous with compact support in x. Working with the μ_n instead of u_n leads to the following basic result:

Theorem II.2. *Let K be a bounded set in $\mathbf{R}^k$ and suppose $\{u_n\}$ is a sequence in $L^\infty(\Omega)^k$ with $u_n(x) \in K$ a.e., for each n. Let μ_n be defined by (25.36). Then there is a subsequence $\{\mu_m\}$ converging weak * to a measure μ, in the sense that for each continuous F, having compact support in x*

$$\langle \mu, F \rangle = \lim \int_\Omega F(x, u_m(x)) \, dx. \tag{25.37}$$

Moreover,

$$\mu \geq 0, \quad \text{spt}(\mu) \subset \Omega \times \overline{K}, \tag{25.38}$$

and if $F(x, \lambda) = G(x)$, G continuous with compact support, then

$$\langle \mu, F \rangle = \int_\Omega G(x) \, dx. \tag{25.39}$$

Conversely, if μ is a measure satisfying (25.38) and (25.39), there exists a sequence u_m in $L^\infty(\Omega)^k$ satisfying (25.37), with $u_m(x) \in K$ a.e.

Such a measure μ will be called a *"generalized function."*

Corollary II.3. *Under the hypotheses of the theorem, there exists a measurable family v_x of probability measures on $\mathbf{R}^k$ satisfying*

$$\text{spt}(v_x) \subset \overline{K} \quad a.e., \tag{25.40}$$

$$\langle \mu, F \rangle = \int_\Omega \langle v_x, F(x, \cdot) \rangle \, dx \text{ for all continuous functions F}$$
having compact support in x, \hfill (25.41)

and

*for each $\phi \in C(\mathbf{R}^k)$, $\phi(u_m)$ converges in $L^\infty(\Omega)$ weak * to a function ψ satisfying $\psi(x) = \langle v_x, \phi \rangle$ a.e.* \hfill (25.42)

Proof (Outline). The existence of μ follows from our quoted results concerning $M(\Omega \times \mathbf{R}^k)$, and the corollary follows from measure-theoretic results. The v_x can actually be constructed as follows. First extract a subsequence u_m satisfying $P(u_m) \to p$ in L^∞ weak *, for every polynomial P with rational coefficients. Since $\|P(u_m)\|_\infty \leq \max\{|P(x): x \in \overline{K}\}$, we have $|p(x)| \leq \max\{|P(x): x \in \overline{K}\}$ a.e., and these inequalities are all true outside a set of measure zero. For such x, the map $P \to p(x)$ is linear and defines a unique probability measure v_x supported in $\overline{K}$. Approximating continuous functions by polynomials completes the proof of the corollary. $\square$

It is important to know when no oscillations occur on some subset of Ω; this is related to the fact that v_x is a Dirac measure, as the following result shows:

Theorem II.4. *Let* $A \subset \Omega$; *then* $u_m \to u$ *in* $L^p_{loc}(A)^k$ *for all* $p < \infty$ *if and only if each* v_x *is a Dirac measure for* $x \in \Omega$ *(then* $v_x = \delta_{u(x)}$ *a.e.).*

Proof. $u^j_m \to u^j$ *strongly in* $L^p_{loc}(A)$ *iff* $u^j_m \to u^j$ *weakly and* $(u^j_m)^2 \to (u^j)^2$ *in* $L^\infty(A)$ *weak* *. This latter is equivalent to* $\mathrm{spt}(v_x) \subset u^j(x)$ *a.e. in* A. $\square$

We turn now to the case where we have some information on the derivatives of u_n. We assume the following:

$$u_n \to u \quad \text{weakly in } L^2(\Omega)^p, \tag{25.43}$$

$$\sum_{j,k} a_{ijk} \, \partial u^j_n / \partial x_k \quad \text{lies in a compact subset of} \quad H^{-1}_{loc}(\Omega), \quad i = 1, \ldots, q. \tag{25.44}$$

Here the a_{ijk} are real constants, each u_n is real-valued, and $H^{-1}_{loc}(\Omega)$ denotes the space of distributions of the form $f_0 + \sum_j \partial f_j / \partial x_j$, where each $f_j \in L^2_{loc}(\Omega)$. (Recall that bounded sets in $L^2_{loc}(\Omega)$ are relatively compact in $H^{-1}_{loc}(\Omega)$, and that if ϕ_n converges strongly in $L^2_{loc}(\Omega)$, then $\partial \phi_n / \partial x_j$ converges strongly in $H^{-1}_{loc}(\Omega)$.) We define the following characteristic sets:

$$\Lambda = \left\{ \lambda \in \mathbf{R}^p : \exists \xi \in \mathbf{R}^N \setminus \{0\} : \sum_{j,k} a_{ijk} \lambda_j \xi_k = 0, i = 1, \ldots, q \right\},$$

$$V = \left\{ (\lambda, \xi) \in \mathbf{R}^p \times (\mathbf{R}^N \setminus \{0\}) : \sum_{j,k} a_{ijk} \lambda_j \xi_k = 0, i = 1, \ldots, q \right\}.$$

Here is the basic result of compensated compactness; the proof can be found in [55].

Theorem II.5. *Let* Q *be a quadratic form on* $\mathbf{R}^p$ *satisfying*

$$Q(\lambda) \geq 0, \qquad \forall \lambda \in \Lambda, \tag{25.45}$$

and assume that (25.43), and (25.44) hold. If

$$Q(u_n) \to \mu \quad \text{weak * in } M(\Omega), \tag{25.46}$$

then

$$\mu \geq Q(u) \quad \text{in } M(\Omega). \tag{25.47}$$

Corollary II.6. *Suppose* Q *is a quadratic form on* $\mathbf{R}^p$ *satisfying*

$$Q(\lambda) = 0, \qquad \forall \lambda \in \Lambda, \tag{25.48}$$

and suppose u_n *satisfies (25.43) and (25.44). Then*

$$Q(u_n) \to Q(u) \quad \text{weak * in } M(\Omega). \tag{25.49}$$

Proof. Equation (25.43) implies $Q(u_n)$ is a bounded sequence in $L^1(\Omega)$. Thus we can find a subsequence u_m satisfying $Q(u_m) \to \phi$ in $M(\Omega)$. Applying the theorem to $\pm Q$, we see $\phi = Q(u)$. $\square$

EXAMPLE. Suppose Ω is an open subset of the (x, t)-plane. Assume v_n^j converges to v^j weak $*$ in $L^\infty(\Omega)$, $1 \le j \le 4$, and that both

$$\frac{\partial v_n^1}{\partial t} + \frac{\partial v_n^2}{\partial x} \quad \text{and} \quad \frac{\partial v_n^3}{\partial t} + \frac{\partial v_n^4}{\partial x}$$

lie in compact subsets of $H_{\text{loc}}^{-1}(\Omega)$. Then $v_n^1 v_n^4 - v_n^2 v_n^3$ converges to $v^1 v^4 - v^2 v^3$ weak $*$ in $L^\infty(\Omega)$, since $\Lambda = \{(\lambda^1, \lambda^2, \lambda^3, \lambda^4): \lambda^1\lambda^4 - \lambda^2\lambda^3 = 0\}$.

Now let us apply these ideas to systems of conservation laws

$$u_t + f(u)_x = 0, \tag{25.50}$$

where u is an m-vector. Assume that the system is hyperbolic. As in Part III of the text, we only consider solutions which satisfy the entropy conditions. We will find solutions of (25.50) via the viscosity method (cf. Chapter 25, §C). Thus consider the perturbed system

$$\frac{\partial u_\varepsilon}{\partial t} + \frac{\partial f(u_\varepsilon)}{\partial x} = \varepsilon \frac{\partial^2 u_\varepsilon}{\partial x^2}, \qquad \varepsilon > 0, \tag{25.51}$$

and assume that we have a sequence $u_n = u_{\varepsilon_n}$ of solutions satisfying

$$u_n \text{ is bounded in } L^\infty(\Omega)^p \quad \text{and} \quad \sqrt{\varepsilon_n}\frac{\partial u_n}{\partial x} \text{ is bounded in } L^2(\Omega)^p, \tag{25.52}$$

for some open set $\Omega \subset \mathbf{R} \times \mathbf{R}_+$. As discussed in Chapter 20, §B, the entropy conditions are

$$U_t + F_x \le 0,$$

where U is convex, and

$$dU \, df = dF.$$

From (20.18), we have

$$\frac{\partial}{\partial t} U(u_n) + \frac{\partial}{\partial x} F(u_n) = \varepsilon_n \frac{\partial^2}{\partial x^2} U(u_n) - \varepsilon_n U''(u_n, u_n). \tag{25.53}$$

Our goal is to use (25.53) in the context of (25.44) (indeed, to use an infinite number of them). Now by (25.53), $U(u_n)$ and $F(u_n)$ are bounded in $L^\infty(\Omega)$, and since $\varepsilon_n \, \partial^2 U(u_n)/\partial x^2 = \sqrt{\varepsilon_n}\partial[U'(u_n)\sqrt{\varepsilon_n} \, \partial u_n/\partial x]/\partial x$, this term converges strongly to zero in $H_{\text{loc}}^{-1}(\Omega)$. On the other hand, the term $\varepsilon_n U''(u_n, u_n)$ is only bounded in $L^1(\Omega)$, and we need the following lemma [43] to obtain the desired compactness in $H_{\text{loc}}^{-1}(\Omega)$.

Lemma II.7. *Let Ω be open in $\mathbf{R}^N$, and suppose $f_n \in \mathscr{D}'$ satisfies f_n is bounded in $W_p^{-1}(\Omega)$[1] for some $p > 2$, and $f_n = g_n + h_n$ where g_n lies in a compact subset of $H_{\text{loc}}^{-1}(\Omega)$, and h_n is in a bounded subset of $M(\Omega)$. Then f_n lies in a compact subset of $H_{\text{loc}}^{-1}(\Omega)$.*

Having this lemma, we can find a subsequence u_n' which by Theorem II.2 gives a generalized function defined by a family $v = v_{x,t}$ of probability measures on $\mathbf{R}^m$. Next, we apply the above example with $v_1 = U_1(u_n')$, $v_2 = F_1(u_n')$, $v_3 = U_2(u_n')$, $v_4 = F_2(u_n')$, where (U_i, F_i), $i = 1, 2$, are both entropy-pairs. Using Corollary II.3, together with the last example, we see that for almost all $(x, t) \in \Omega$

$$\langle v, U_1 F_2 - U_2 F_1 \rangle = \langle v, U_1 \rangle \langle v, F_2 \rangle - \langle v, U_2 \rangle \langle v, F_1 \rangle, \qquad (25.54)$$

where $v = v_{x,t}$. We now characterize such probability measures v and prove that this equation is satisfied only by Dirac measures. If $m = 1$, this can be done since any U is an entropy. One shows that $\text{spt}(v)$ lies in an interval in which f is affine, so this interval is a point if, say, f is convex; see Tartar [55].

In the case $m = 2$, the entropies are not easily constructed, as they satisfy linear second-order hyperbolic equations. We now turn to a study of these equations. Using the basis of eigenvectors of $f'(p)$ (p a point in Ω), we have

$$F'(a)r_j(a) = \lambda_j(a)U'(a)r_j(a), \qquad j = 1, \ldots, m.$$

If $m = 2$, we can go over to Riemann-invariant coordinates; i.e., w_j is a j-Riemann invariant if $w_j'(a)r_j(a) = 0$, for all a (see Chapter 20, §A). We assume that these functions define a global nonsingular coordinate transformation. Then (25.41) becomes

$$\frac{\partial F}{\partial w_1} - \lambda_2 \frac{\partial U}{\partial w_2} = 0, \qquad \frac{\partial F}{\partial w_2} - \lambda_1 \frac{\partial U}{\partial w_1} = 0,$$

and one finds solutions of the form (see [Lx 4])

$$\left.\begin{aligned} U_k(w_1, w_2) &= e^{kw_1}\left[A_0 + \frac{A_1}{k} + o\left(\frac{1}{k^2}\right)\right], \\ F_k(w_1, w_2) &= e^{kw_1}\left[B_0 + \frac{B_1}{k} + o\left(\frac{1}{k^2}\right)\right], \end{aligned}\right\}, \qquad |k| \to \infty.$$

---

[1] $W_p^{-1}(\Omega) = \{w : (1 + |x|^2)^{-1}w(x) \in L^2(\Omega)\}.$

(A similar family exists upon permuting the roles of w_1 and w_2 as well as λ_1 and λ_2.)

Now suppose v satisfies (25.54), and let R be the smallest rectangle $w_j^- \leq w_j \leq w_j^+$ containing spt(v). The main technical result is that if $w_1^- < w_1^+$ (or $w_2^- < w_2^+$), then each vertical (resp. horizontal) side of R contains a point where $\partial\lambda_2/\partial w_1 = 0$ (resp. $\partial\lambda_1/\partial w_2 = 0$). This yields the following result (cf. Lemma 20.2):

Theorem II.8. *If both characteristic fields are everywhere genuinely nonlinear* $(\partial\lambda_1/\partial w_2 \neq 0, \partial\lambda_2/\partial w_1 = 0)$, *then v is a Dirac measure.*

A somewhat finer analysis yields

Theorem II.9. *If both characteristic fields are genuinely nonlinear off a curve* $w_2 = \phi(w_1)$, *where ϕ is strictly monotone, then v is a Diract measure.*

As an example (see DiPerna [11]), consider the p-system (Chapter 17, §A)

$$v_t - u_x = 0, \qquad u_t + p(v)_x = 0, \qquad p' < 0. \qquad (25.55)$$

The characteristic fields are genuinely nonlinear if $p'' > 0$. The last result applies if p'' has at most one zero, as u being constant is equivalent to $w_2 - w_1$ being constant. Of course, this development is based on (25.52). Using invariant regions (cf. Chapter 14, §B), one proves the L^∞ estimate in (25.52) (the other part of (25.52) is not difficult to prove, in general), in the case of nonlinear elasticity; namely, when $p'(v) < 0$ if $v < v_0$, and $p' > 0$ if $v > v_0$. This proves the existence of a global bounded, entropy solution by the viscosity method, provided that the data is bounded, and no vacuum is present.

These methods have been extended by DiPerna [12], to the case of isentropic gas dynamics: the p-system with $p(v) = k^2/v^\gamma$, where $\gamma = 1 + 2/n$, $n \in \mathbf{Z}_+, n \geq 3$, for the case of smooth, bounded initial data.

DiPerna [12] has also shown how the method of compensated compactness can be used to demonstrate the convergence of certain finite difference approximations to systems of conservation laws. These methods have been extended by Ding Xiaxi, Chen Guigiang, and Luo Peizhu in a series of papers [9].

§B. Stability of Shock Waves

In this section we shall discuss the recent results concerning the nonlinear stability of travelling waves (i.e., viscous shock waves, see Chapter 24, §§A, B), for systems of viscous conservation laws of the form

$$u_t + f(u)_x = (B(u)u_x)_x, \qquad x \in \mathbf{R}^1, \quad t > 0, \qquad (25.56)$$

where $u = u(x, t)$ is an n-vector, f is a smooth vector-valued function, and $B(u)$ is a smooth $n \times n$ matrix. To (25.56) there is associated the system of conservation laws

$$u_t + f(u)_x = 0. \tag{25.57}$$

We assume that this latter system is hyperbolic, and that each characteristic field is either genuinely nonlinear, or linearly degenerate, see Chapter 17, §B. As in Chapter 24, if (u_l, u_r, s) denotes a shock-wave solution of (25.57), then if $|u_r - u_l|$ is small, the system (25.56) admits a travelling-wave solution $u(x, t) = \phi(\xi)$, $\xi = x - st$, with $\phi(-\infty) = u_l$, $\phi(\infty) = u_r$. The goal is to show that these solutions (properly translated; see below), are attractors for (25.56) in the sense that if the initial-data for (25.56) is "close" to (a certain translate of) ϕ, then the corresponding solution of (25.56) tends to ϕ as $t \to \infty$. Thus the study of viscous shock waves for equations of the form (25.56), (such as the compressible Navier–Stokes equations), can often be replaced by the study of shock waves for inviscid equations (25.57), (such as the compressible Euler equations).

We now turn to a more precise statement of the result, together with an indication of the proof; see Liu's memoir [38] for complete details. We consider k-shock-wave solutions $\phi_k(x - s_k t)$ of (25.56), $k = p_1, \ldots, p_l$, $1 \le p_1 < \cdots < p_l \le n$, and we assume that these can be superimposed, i.e., for some constant states $\tilde{u}_0, \tilde{u}_1, \ldots, \tilde{u}_l$,

$$\phi_{p_i}(\infty) = \phi_{p_{i+1}}(-\infty) \equiv u_i, \quad i = 1, \ldots, l-1; \quad \text{and} \quad \phi_{p_1}(-\infty) \equiv \tilde{u}_0. \tag{25.58}$$

We denote the linear superposition if these waves by

$$\sum_{i \in P} \phi_i(x - s_i t) \equiv \tilde{u}_0 + \sum_{j=1}^{l} [\phi_{p_j}(x - s_{p_j} t) - \tilde{u}_{j-1}], \tag{25.59}$$

$P = \{p_{i_1}, \ldots, p_{i_l}\}$. Of course if $l > 1$, this is not an exact solution of (25.56), but it may be regarded as an asymptotic solution, as we shall see. Now consider initial data $u(x, 0)$ for (25.56) which has the same limiting states at $\pm\infty$; namely, $u(\infty, 0) = \tilde{u}_l$, $u(-\infty, 0) = \tilde{u}_0$, or equivalently

$$u(x, 0) = \bar{u}(x) + \sum_{i \in P} \phi_i(x), \quad \bar{u}(\pm\infty) = 0. \tag{25.60}$$

The main result is that the solution of (25.56) with data (25.60), approaches the sum of the ϕ_i, $i \in P$, properly translated by an amount x_i; i.e.,

$$\lim_{t \to \infty} \sup_{x \in \mathbf{R}} \left| u(x, t) - \sum_{i \in P} \phi_i(x + x_i - s_i t) \right| = 0. \tag{25.61}$$

This is the strong stability statement for the viscous shock waves. The translations x_i, can be a-priori determined from the perturbation $\bar{u}$, using the "conservation of mass" principle. This can be seen as follows. First, (25.56) implies that

$$\int_{-\infty}^{\infty} \left[u(x, t) - \sum_{i \in P} \phi_i(x - s_i t) \right] dx = \int_{-\infty}^{\infty} \bar{u}(x)\, dx, \qquad t \geq 0.$$

Now any translate of ϕ_1 yields a net flow of mass parallel to $\phi_1(\infty) - \phi_i(-\infty) = \tilde{u}_i - \tilde{u}_{i-1}$, in the sense that for $i \in P$

$$\int_{-\infty}^{\infty} [\phi_i(x + x_i - s_i t) - \phi_i(x - s_i t)]\, dx = x_i(\tilde{u}_i - \tilde{u}_{i-1}).$$

It follows from these last two equations that

$$\int_{-\infty}^{\infty} \left[u(x, t) - \sum_{i \in P} \phi_i(x + x_i - s_i t) \right] dx = \int_{-\infty}^{\infty} \bar{u}(x)\, dx - \sum_{i \in P} x_i(\tilde{u}_i - \tilde{u}_{i-1}).$$

$$(25.62)$$

Using (25.62), we see that the convergence in (25.61) does not imply convergence in the integral sense if $P \neq \{1, 2, \ldots, n\}$, since for the general perturbation $\bar{u}$, the first term on the right-hand side of (25.62) is a general n-vector, while the second term lies in the l-dimensional subspace of $\mathbf{R}^n$ spanned by the $(\tilde{u}_i - \tilde{u}_{i-1})$, $i = 1, \ldots, l$. If $P = \{1, \ldots, n\}$, convergence in the integral sense can be proved since x_i could be determined by setting the right-hand side of (25.62) equal to zero, because the vectors $(\tilde{u}_i - \tilde{u}_{i-1})$, $i = 1, \ldots, n$, are linearly independent (for weak shocks), and hence the left-hand side is zero for all $t > 0$. If $P \neq \{1, \ldots, n\}$, as would be the case if one is studying the stability of a single shock wave, the right-hand side of (25.62) is nonzero, in general, for any choice of x_i's. Thus from (25.61), and (25.62), it can be shown that the perturbation $\bar{u}(x)$, in addition to translating the ϕ_i must also give rise to waves which decay pointwise, but carry a finite amount of mass. Such waves are called *diffusion waves* of complementary families, $i \notin P$. An i-diffusion wave with base state u_0^i (i.e., this diffusion wave is constructed near the constant state u_0^i and takes values along the ith-rarefaction-wave curve through u_0^i), can be shown to carry a net flow of mass parallel to $r_i(u_0^i)$ as $t \to \infty$ (the r_i are the right-eigenvectors of f'). In this case, $i \notin P$, and the base states are

$$u_0^i = \tilde{u}_j \quad \text{if } p_j < i < p_{j+1}, \quad j = 0, 1, \ldots, l; \quad p_{l+1} = n.$$

Thus the right-hand side of (25.62) will be a linear combination of the $r_i(u_0^i)$, $i \notin P$, so there exist constants α_i, $i \notin P$, such that

$$\int_{-\infty}^{\infty} \bar{u}(x)\, dx = \sum_{i \in P} x_i(\tilde{u}_i - \tilde{u}_{i-1}) + \sum_{i \notin P} \alpha_i r_i(u_0^i). \qquad (25.63)$$

It is this equation which determines the translations x_i, $i \in P$, as well as the strengths α_i, $i \notin P$, of the diffusion waves since the vectors $(\tilde{u}_i - \tilde{u}_{i-1})$, $i \in P$, and the $r_i(u_0^i)$, $i \notin P$, form a basis in $\mathbf{R}^n$ (if the shocks are weak).

For systems of conservation laws (25.57), the behavior of shock waves, linearly degenerate waves, and nonlinear diffusion waves (N-waves) is well understood [Lu 4]. Aside from the fact that one is discontinuous and the other is continuous, shock waves for (25.57) and viscous shock waves for (25.56) have essentially the same shape; the results given here show that both are nonlinearly stable. The analysis is based on the study of diffusion waves for (25.56), and the technique combines both the characteristic and energy methods and requires a deep understanding of the diffusion waves. The linear diffusion waves for (25.56) are governed, qualitatively, by solutions of the linear "heat" equation

$$u_t + cu_x = \varepsilon u_{xx}, \qquad u \in \mathbf{R}, \tag{25.64}$$

which behaves quite differently from the corresponding linear waves for (25.57), which are governed qualitatively by solutions of the linear wave equation

$$u_t + cu_x = 0, \qquad u \in \mathbf{R}. \tag{25.65}$$

Similarly, nonlinear diffusion waves for (25.56), governed qualitatively by Burgers equation,

$$u_t + uu_x = \varepsilon u_{xx}, \qquad u \in \mathbf{R}, \tag{25.66}$$

differ greatly from the corresponding N-waves for (25.57) which are governed qualitatively by the inviscid Burgers equation

$$u_t + uu_x = 0. \tag{25.67}$$

One usually studies system (25.56) with the view that it is an accurate approximation of (25.57) when the viscosity matrix $B(u)$ is small. Since the aforementioned differences between the wave behaviors for (25.56) and (25.57) hold independent of the strength of viscosity, the large time behavior of waves for (25.56), which is of great physical interest, can be accurately replaced by solutions of (25.57) only on the level of shock waves, and not on the level of diffusion waves, even if the viscosity is small.

As is well known (cf. Chapter 16, §C), the combination of nonlinear effects, together with the entropy condition, forces waves of each genuinely nonlinear characteristic family to combine and cancel. Thus, even though (25.57) is a hyperbolic system, it is also dissipative and admits diffusion waves; i.e., N-waves. For each genuinely nonlinear field, the system (25.56) carries nonlinear diffusion waves, the construction of which are based on the diffusion waves for (25.66). For this one assume that for all u under consideration, we have

$$r_i(u)^t B(u) r_i(u) = \alpha_i(u) > 0; \tag{25.68}$$

i.e., the diffusion matrix $B(u)$ has positive diagonal elements with respect to the basis of right eigenvectors of $f'(u)$. Many physical systems, such as the compressible Navier–Stokes equations, satisfy (25.68).

The main theorem concerning the nonlinear stability of viscous shock waves is the following:

Theorem II.10. *Assume that* (25.57) *is hyperbolic, that each characteristic field is either genuinely nonlinear or linearly degenerate, that $f'(u)$ admits eigenvectors $r_i(u)$, $i = 1, \ldots, n$, and that the viscosity matrix $B(u)$ of* (25.56) *satisfies*

$$\bar{B}(u) = (r_1(u)^t, \ldots, r_n(u)^t)B(u)(r_1(u), \ldots, r_n(u)) > 0, \qquad (25.69)$$

(i.e., $\bar{B}$ is positive definite), for each u under consideration. Then weak viscous shock waves for (25.56) *are nonlinearly stable in the following sense: Suppose we are given viscous shock waves of* (25.56),

$$\phi_i(x - s_i t), \qquad i \in P = \{p_1, \ldots, p_l\}, \quad 1 \le p_1 < p_2 < \cdots < p_l \le n,$$

corresponding to the p_ith characteristic field, where $\phi_{p_i}(-\infty) = \tilde{u}_{i-1}$, $\phi_{p_i}(\infty) = \tilde{u}_i$, $i = 1, \ldots, l$. Then the solution of (25.56) *with initial data* (25.60) *exists and tends to the translated shock waves in the sense of* (25.61), *provided that the viscous shock waves ϕ_{p_i}, $i = 1, \ldots, l$, are sufficiently weak, that the perturbation $\bar{u}$ is sufficiently small in L^∞, and that $\bar{u}(x) \to 0$ sufficiently rapidly as $|x| \to \infty$. The constants x_i, $i = 1, \ldots, l$, are uniquely determined by* (25.63).

The proof of this theorem can be divided into three steps. First, one studies some intrinsic geometric properties of viscous shock waves and diffusion waves. In particular, it can be shown that viscous shock waves are compressive (this is the main reason that they are stable), while diffusion waves are either compressive or weakly expansive. Second, by considering linear hyperbolic waves, the solution u can be decomposed as a sum of shock waves ϕ_i, $i \in P$, diffusion waves with strengths α_i, $i \notin P$, a linear hyperbolic wave, and a remainder term u^* which carries a net flow of zero mass; i.e., $\int_{-\infty}^{\infty} u^*(x, t)\, dx = 0$. This last equation is crucial as it allows one to work with the antiderivative v of u^*. Finally, by introducing a new characteristic-energy method, one can obtain the desired stability estimates for v.

The assumption that $\bar{B}(u)$ is positive-definite is not needed for the discussion of diffusion waves; for these (25.68) suffices. The compressible Navier–Stokes equations satisfy (25.68) but not (25.69). However, using some special features of the compressible Navier–Stokes equations, the ideas in the proof can be extended to this system as well (cf. [39]).

We remark that Liu's theory is based upon his important discovery that a general perturbation of a given viscous shock profile not only produces a phase shift in the profile, but also introduces diffusion waves (linear or nonlinear) in the transversal wave directions. We remark that Liu's approach applies only to a special class of initial perturbations which satisfy the condition that the initial difference between the perturbations and the viscous shock profile has the same asymptotic behavior as that of the linear hyperbolic wave as $|x| \to \infty$ (constructed by Liu in [38]), which has an algebraic decay rate as $|x| \to \infty$. In particular, initial compact perturbations of the viscous shock profile are excluded by Liu's theory. Recently, Szepessy and Xin (Nonlinear Stability of Viscous Shock Waves, *Arch. Rat. Mech. Anal.* **122**, (1993), 55–103), observed the following fact; namely, a general perturbation of a given shock wave produces not only a phase shift in the profile and diffusion waves in the transversal wave directions, but also produces resonant diffusion waves in the shock wave due to wave interactions. The effects on shock waves due to self-interactions of transversal diffusion waves is realized through a coupled diffusion wave. Based on this observation, Szepessy and Xin proved that a viscous shock profile is asymptotically nonlinear stable for general perturbations; for details, see the above paper.

The study of the stability of the viscous shock wave for a scalar equation was first considered in [Hf 2] and, more generally, in [35]. For systems, the papers [27] and [35] study this problem under the restrictive condition $\int_{-\infty}^{\infty} \bar{u}(x)\,dx = 0$; this assumption precludes the existence of diffusion waves.

Finally, the stability of rarefaction waves for systems of conservation laws has been considered by Liu and Xin; see [40].

§C. Miscellaneous Results

In this section we shall mention several recent results, without going into as much detail as in the last sections; the interested reader can consult the quoted references in order to find a more complete discussion.

A major source of research has been in studying systems of conservation laws of the form (25.57) which are either weakly hyperbolic (i.e., the eigenvalues of $f'(u)$ are real, but not necessarily distinct), or which have the property that some characteristic fields are neither genuinely nonlinear nor linearly degenerate. These systems are of interest because they arise in certain applied problems, such as oil reservoir simulation, viscoelasticity, and multiphase flow. These systems are also of mathematical interest as they raise interesting questions of well-posedness, structure of solutions, and admissibility criteria for weak solutions. A good general reference to these problems can be found in the collection of articles [36]. In [56], Temple studies a class of equations, coming from enhanced oil recovery problems, which exhibits a new phenomena; namely, the eigenvalues of f' degenerate with a $\left(\begin{smallmatrix} 1 & 1 \\ 0 & 1 \end{smallmatrix}\right)$ normal form at certain eigenvalues of the unknowns. Glimm's analysis (cf. Chapter 19) does not apply since the degeneracy leads to oscillations and hence to unbounded variation for any numerical method based on solutions of the Riemann problem. New techniques are developed in [56] to prove existence of a solution. In [57, 58] Temple presents an existence theory for systems of conservation laws which have coinciding shock- and rarefaction-wave curves; such systems arise in the study of gas dynamics, oil reservoir simulation, nonlinear motion of elastic strings, and multicomponent chromatography. In [59] some conservative finite difference schemes are proved to converge for certain such systems.

Next, in [61] Temple proves that for any 2×2 system of conservation laws, outside of the class of systems with coinciding shock and rarefaction curves (in particular, including the p-system, cf. Chapter 17, §A), no L^1-contractive metric exists. It follows that for such systems the usual semigroup methods are doomed to failure.

The problem of decay of solutions of systems of conservation laws with a rate independent of the support of the initial data is central to the issue of uniqueness and continuous dependence of the solutions on the data. In [60] Temple proves that for systems of two conservation laws one has an estimate of the form

$$\|u(\cdot, t)\|_{\infty} \leq F\left(\frac{t}{\|u(\cdot, 0)\|_{L^1}}\right),$$

where $F(\xi) \to 0$ as $\xi > \infty$; specifically, $F(\xi) = (\log \xi)^{-1/2}$. This is the first decay result with a rate independent of the support of the data, and it immediately implies the stability of the constant state in L^1_{loc}. Previous decay results, e.g., [DP 2, 4, GL, Lu 3, 4], give decay with a rate only in the case of compactly supported data. Moreover, these decay estimates are in the total-variation norm, and easy examples show that the total variation does not decay at a rate depending only on the L^1-norm of the data.

We mention a result in [30] in which it is shown that for the compressible Navier–Stokes equations, smoothing of initial discontinuities must occur for the velocity and energy but cannot occur for the density; cf. footnote 2 of Chapter 15.

Finally, there has been much interest lately in the study of detonation waves for combustible gases; i.e., the equations of a reactive gas flow. In Lagrangian coordinates, these take the form

$$v_t - u_x = 0,$$

$$u_t + p_x = (\mu v^{-1} u_x)_x,$$

$$(e + u^2/2)_t + (pu)_x = (\mu v^{-1} u u_x)_x + (\lambda v^{-1} T_x)_x,$$

$$z_t = D z_{xx} - \phi(T) z,$$

where v, p, u, T, and e denote the usual "gas dynamic" variables (cf. Chapter 18), z is the mass fraction of unburned gas, ϕ denotes the reaction rate, and λ, μ, D denote, respectively, heat conduction, viscosity, and species-diffusion coefficients. The last equation governs the mass fraction of the unburned gas. In [22] Gardner proves an existence theorem for strong and weak detonation waves for explicit ranges of λ, μ, and D. The problem reduces to finding an orbit of an associated system of four ordinary differential equations which connects two distinct critical points. The proof uses topological methods, including the Conley index.

SECTION III. Conley Index Theory

§A. The Connection Index

Suppose that a flow is generated by a parametrized system of differential equations

$$x' = f(x, \theta), \qquad x \in \mathbf{R}^n, \quad \theta_0 \le \theta \le \theta_1, \tag{25.70}$$

where f depends continuously on θ. If we append the equation $\theta' = 0$ to this system, the equations generate a flow on $X = \mathbf{R}^n \times [\theta_0, \theta_1]$, for which each slice $\theta = \text{const.}$ is invariant. Given any subset S of X, let S_θ denote those points in S whose last coordinate equals θ.

A homotopy invariant for the augmented equations can be defined as follows. Suppose that S, S', and S'' are invariant sets in X such that S_θ, S'_θ, and S''_θ are isolated invariant sets with respect to (25.70), for each $\theta \in [\theta_1, \theta_2]$, and the following hold:

(i) $S' \cup S'' \subset S$;
(ii) $S' \cap S'' = \phi$; and
(iii) $S_\theta = S'_\theta \cup S''_\theta$ if $\theta = \theta_1$ or $\theta = \theta_2$.

Then (S, S', S'') is called a *connection triple*. A homotopy invariant, called the *connection index*, denoted by $\bar{h}(S, S', S'')$ can be defined for connection triples; a complete definition, together with the relevant theorem, and examples is given in [CG, CS 11]. We content ourselves here with giving a somewhat imprecise, though instructive, description of this invariant. Let $N \subset X$ be a compact neighborhood of S such that $S(N_\theta) = S_\theta$ ($S(N_\theta)$ denotes the maximal invariant set in N_θ), $\theta_1 \le \theta \le \theta_2$. Let N^2 be a subset of N such that (N_θ, N_θ^2) is an index pair for S_θ (cf. Chapter 23, §A), and let $\bar{N}^2$ be N^2 together with the closure of all orbit segments in N_{θ_1} and N_{θ_2} which tend to S'_{θ_1} and S'_{θ_2}, respectively, in negative time. The *connection index*, $\bar{h} = \bar{h}(S, S', S'')$, is defined to be $[N/\bar{N}^2]$, the homotopy type of the quotient space $N/\bar{N}^2$. (This is not strictly correct since $S'_{\theta_i} \subset N/\bar{N}^2$, $i = 1, 2$, so $(N, \bar{N}^2)$ is not an index pair, as defined in Chapter 23, §A. The problem is remedied by suitably modifying the equation $\theta' = 0$ in neighborhoods of S'_{θ_i} and S''_{θ_i}.) It turns out that $[N/N^2]$ for the modified equations is the same as $[N/\bar{N}^2]$; thus $[N/\bar{N}^2]$ gives the "correct" homotopy type, but for a different set of equations.

This index depends only on the connection triple, and is invariant under continuation. Roughly speaking, $\bar{h}$ measures a change in the way solutions in the "unstable manifold" of S_θ' leave N_θ, when $\theta = \theta_1$ and θ_2. Thus, if $(\tilde{N}/\tilde{N}^2)$

is an index pair for some θ, $(\tilde{N}_\theta, \tilde{N}_\theta^2)$ is an index pair for S_θ. At $\theta = \theta_i$, $i = 1, 2$, $\tilde{N}_\theta^2$ also consists of points in the unstable manifold of S'_{θ_i}, $i = 1, 2$, which lie in the exit set. Thus $\bar{h}$ measures a change in the way the unstable manifold of S' leaves $\tilde{N}$ at $\theta = \theta_1$ and $\theta = \theta_2$. The modifications mentioned earlier in the θ' equation are such as to make this picture correct. If the index is "nontrivial" in a sense defined below, then the unstable manifold sweeps across S'' as θ varies from θ_1 to θ_2, and thus at some $\theta \in (\theta_0, \theta_1)$, there is a connection from S'_θ to S''_θ. In particular, the following theorem holds; see [CS 11]:

Theorem III.0. *Suppose $S = S' \cup S''$. Then $\bar{h}(S, S', S'') = (\Sigma^1 \wedge h') \vee h''$ where $h' = h(S')$, and $h'' = h(S'')$.*

If the index can be computed, and shown to differ from $(\Sigma^1 \wedge h') \vee h''$ (i.e., to be "nontrivial"), then it follows that $S' \cup S'' \subsetneqq S$; hence there exists a $\theta \in (\theta_1, \theta_2)$ for which $S'_\theta \cup S''_\theta \subsetneqq S_\theta$. If, for example, the flow is gradient-like in S_θ, it follows that S_θ contains an orbit connecting S' to S''.

Finally, consider the product system

$$x' = f(x, \theta), \qquad q'_1 = q_2,$$
$$\theta' = 0, \qquad q'_2 = q_1.$$

The q-equations are linear, with a saddle at the origin. It follows (see [CS 11]) that if (S, S', S'') is a connection-triple for the (x, θ) equations, then $(\tilde{S}, \tilde{S}', \tilde{S}'')$ is a connection triple for the product system, where $\tilde{S}, \tilde{S}', \tilde{S}''$ are, respectively, S, S', S'' augmented with two zeros for the q-components. Moreover, $\tilde{h}(\tilde{S}, \tilde{S}', \tilde{S}'') = \Sigma^1 \wedge \bar{h}(S, S', S'')$.

We shall illustrate these ideas with a simple two-dimensional example. Consider the system

$$w' = z, \qquad z' = -\theta z + c(w), \qquad (25.71)$$

where $c(w)$ is qualitatively a cubic with roots $0 < r_1 < r_2$; suppose too that $c'(0) < 0$ and $c'(r_2) < 0$. The critical points of (25.71) are $\bar{y}_0 = (0, 0)$, $\bar{y}_1 = (r_1, 0)$, $\bar{y}_2 = (r_2, 0)$. Depending on the value of θ, the rest point $\bar{y}_1$ can be a stable or unstable node, or a spiral or a center, while $\bar{y}_0$ and $\bar{y}_2$ are always saddles; thus $h(\bar{y}_0) = \Sigma^1 = h(\bar{y}_2)$.

If $\theta = 0$ (25.71) is a Hamiltonian system with "energy" $H = z^2/2 + C(w)$, $(C' = c)$, and $H' = 0$. If $\theta \neq 0$ (25.71) is gradient-like as H is now a Lyapunov function, since $H' = -\theta z^2$. The phase portrait of (25.71) consists of the level curves of H when $\theta = 0$. For $\theta \neq 0$, a rough description of the phase plane can be obtained from the $\theta = 0$ case and the fact that $H' \leq 0$; a typical case is indicated in the figure below. In particular, if $|\theta|$ is sufficiently large, there are no solutions which connect $\bar{y}_0$ to $\bar{y}_2$, say for $|\theta| = \theta_1$.

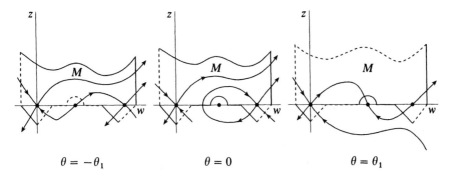

$$\theta = -\theta_1 \qquad\qquad \theta = 0 \qquad\qquad \theta = \theta_1$$

Note that if a solution is to connect $\bar{y}_0$ at $-\infty$ to $\bar{y}_2$ at $+\infty$, it must be that $w' = z > 0$ along this orbit. We thus consider the region M bounded by the dotted lines in the above diagrams. The curved edges of M are solution curves of (25.71) when $\theta = 0$. It is easy to check that M is an isolating neighborhood for (25.71) for every θ; namely, if $\theta \neq 0$, every orbit through ∂M is transverse to ∂M; if $\theta = 0$ solutions in the curved boundaries eventually leave M. Note, too, that for every θ, there are three components along ∂M where solution curves (eventually) leave M in positive time. In the cases $|\theta| = \theta_1$, these are the heavy lines in ∂M. For each θ, let M_θ^2 denote these subsets of ∂M.

From these remarks, it follows that if we define

$$S' = \{\bar{y}_0\} \times [-\theta_1, \theta_1], \qquad S'' = \{\bar{y}_2\} \times [-\theta_1, \theta_1],$$
$$S = S(M \times [-\theta_1, \theta_1]),$$

then (S, S', S'') is a connection triple for the equations (25.71), appended with $\theta' = 0$. Also, if $N = M \times [-\theta_1, \theta_1]$, then the set $\bar{N}^2$ as defined above consists of each of the sets (M_θ^2, θ) together with the closure of the unstable manifolds of $\bar{y}_0$ when $\theta = \pm\theta_1$. Since these manifolds connect different components of M_θ^2, $\theta = \pm\theta_1$, it follows that $\bar{N}^2$ is a connected set and is, in fact, contractible to a point; thus $\bar{h} = [N/\bar{N}^2] = \bar{0}$.

Suppose now that $S' \cup S'' = S$ and, in particular, that there is no orbit running from $\bar{y}_0$ to $\bar{y}_2$ for any θ. It would then follow from the above theorem that $\bar{h} = (\Sigma^1 \wedge \Sigma^1) \vee \Sigma^1 = \Sigma^2 \vee \Sigma^1$. Since this is not true, we have $S'_\theta \cup S''_\theta \subsetneqq S_\theta$ for some θ. Also, since $w' = z > 0$ along nonconstant solutions in S_θ, it follows that the nonconstant solution in S_θ must be a connecting orbit.

Finally, consider the product system

$$z' = w, \qquad\qquad q_1' = q_2,$$
$$w' = -\theta w - c(u), \qquad q_2' = q_1.$$

Let

$$\tilde{M} = M \times \{(q_1, q_2): |q_i| \le \varepsilon, i = 1, 2\}, \qquad \tilde{Y}_i = (\bar{y}_i, 0, 0), \qquad i = 0, 1, 2,$$

$$\tilde{S}' = \{\tilde{Y}_0\} \times [-\theta_1, \theta_1], \quad \tilde{S}'' = \{\tilde{Y}_2\} \times [-\theta_1, \theta_1], \quad \tilde{S} = S(\tilde{M} \times [-\theta_1, \theta_1]);$$

then $(\tilde{S}, \tilde{S}', \tilde{S}'')$ is a connection triple for the product flow with $\bar{h}(\tilde{S}, \tilde{S}', \tilde{S}'') = \bar{h}(S, S', S'') \wedge \Sigma^1 = \bar{0} \wedge \Sigma^1 = \Sigma^1$. Since this is a different homotopy type than $(\Sigma^1 \wedge h(S')) \vee h(S'') = (\Sigma^2 \wedge \Sigma^1) \vee \Sigma^2 = \Sigma^3 \vee \Sigma^2$, it follows that the connection index of $(\tilde{S}, \tilde{S}', \tilde{S}'')$ is also "nontrivial."

The existence of travelling waves for both competition and predator–prey type reaction diffusion systems is proved by this method; see [CG] and [19].

§B. Conley's Connection Matrix

An important question for dynamical systems with rest points is whether one can prove the existence of heteroclinic orbits. On a more general level, one can investigate whether, given a collection of invariant sets, does there exist connecting orbits between them? Conley introduced an algebraic object called the connection matrix which is designed to study such questions. The ideas involved are generalizations of earlier techniques based on the Conley index which allowed one to prove the existence of connections between attractor–repellor pairs [Cy 2]. The simplest description of the connection matrix is that it organizes the information provided by the homology (or cohomology) groups of the Conley indices associated to the isolated invariant sets. It consists of a collection of homomorphisms between the homology groups of the Conley indices of the sets in a Morse decomposition of a flow (cf. Chapter 23, §A). The maps are determined by the Morse-theoretic relations in the flow, and they contain information about the connecting orbits between the Morse sets. In its present form, the connection matrix can be applied to many interesting problems even by someone having little or no training in algebraic topology. In this section we briefly describe the connection matrix on a somewhat elementary level.

We assume the general set-up of Chapter 23; in particular, X is a locally compact Hausdorff space, $(x, t) \to x \cdot t$ is a local flow on X, and, as usual, $h(S)$ denotes the Conley index of the isolated invariant set S. Now, dealing with the homotopy equivalence classes of topological spaces is often quite difficult. Thus to simplify matters, we shall only consider the homology of the pointed topological spaces $h(S)$. In particular, we shall restrict our attention to the singular homology groups with $\mathbf{Z}_2$ coefficients. Thus given S, we shall study the algebraic object $H_*(h(S); \mathbf{Z}_2) = H_*(N_1/N_0; \mathbf{Z}_2) \approx H_*(N_1, N_0; \mathbf{Z}_2)$, where (N_1, N_0) is an index pair for S (cf. Chapter 23). The important fact is that $H_*(N_1, N_0; \mathbf{Z}_2) = \{H_n(N_1, N_0; \mathbf{Z}_2)\}: n = 0, 1, 2, \ldots\}$, where each $H_n(N_1, N_0; \mathbf{Z}_2)$ is a vector space over $\mathbf{Z}_2$. Thus $H_*(N_1, N_0; \mathbf{Z}_2)$ is an infinite collection of vector spaces over $\mathbf{Z}_2$, indexed by the nonnegative integers. In most applications only a finite number of these are nontrivial. The following proposition describes $H_*(h(S); \mathbf{Z}_2)$ in the case that S is a hyperbolic critical point or a periodic orbit; see [Cy 2]:

Proposition III.1.

(A)   *Let S be a hyperbolic critical point with exactly k eigenvalues having positive real parts; then*

$$H_n(h(S); \mathbf{Z}_2) \approx \begin{cases} \mathbf{Z}_2, & \text{if } n = k, \\ 0, & \text{otherwise.} \end{cases}$$

(B)   *Let S be a hyperbolic periodic orbit (i.e., no Floquet exponent has zero real part), with Poincaré map ϕ. Assume $d\phi$ has real positive eigenvalues, exactly k of which are greater than one. Then*

$$H_n(h(S); \mathbf{Z}_2) \approx \begin{cases} \mathbf{Z}_2, & \text{if } n = k \text{ or } k + 1, \\ 0, & \text{otherwise.} \end{cases}$$

In applications, one is often given a complicated isolated invariant set S, whose structure one wishes to determine. To do so, one needs to decompose S into smaller sets, and then prove the existence or nonexistence of connecting orbits. Given two isolated invariant sets, S_1 and S_2, we define the set of connections from S_1 to S_2 as (cf. Chapter 22, §B)

$$C(S_1, S_2) = \{x : \alpha(x) \subset S_1, \omega(x) \subset S_2\}.$$

Let $(P, >)$ be finite set with a partial order relation $>$, satisfying:

(i) $i > i$ never holds if $i \in P$; and
(ii) if $i > j$ and $j > k$, then $i > k$, for i, j, k in P.

Definition III.2 (cf. Definition 23.1). A *Morse decomposition* of S is a finite collection $M(S) = \{M(i) : i \in (P, >)\}$ of compact invariant sets in S indexed by P such that if $x \in S$, then either $x \in M(i)$ or $x \in C(M(i), M(j))$ where $i > j$.

The sets $M(i)$ are called *Morse sets*, and are isolated invariant sets; cf. Chapter 23, §A. To simplify the notation, we write $H(i) = H_*(h(M(i)); \mathbf{Z}_2)$ and $H_n(i) = H_n(h(M(i)); \mathbf{Z}_2)$. We can now define the connection matrix. Let S be an isolated invariant set and let $M(S) = \{M(i) : i \in (P, >)\}$ be a Morse decomposition of S. Let $\Delta = [\Delta_{ij}]$, $i, j \in P$, be a matrix whose entries Δ_{ij} are matrices over $\mathbf{Z}_2$.

Definition III.3. $\Delta : \bigoplus_{i \in P} H(i) \rightarrow \bigoplus_{i \in P} H(i)$ is a *connection matrix* if the following hold:

(a) $\Delta_{ij} = 0$ if $i < j$ (0 is the zero matrix).
(b) Δ is a boundary map; i.e., $\Delta^2 = 0$ and $\Delta_{ij} = 0$ except possibly as a map from $H_n(j)$ to $H_{n-1}(i)$.
(c) Let $\Delta_n = \Delta | \bigoplus_{i \in P} H_n(i)$ and let $H_n\Delta = \ker \Delta_n / \operatorname{Im} \Delta_{n+1}$; then $H_n\Delta \approx$ $H_n(h(S); \mathbf{Z}_2)$.

Theorem III.4 (cf. [19, 20, 41]). *Given S and $M(S)$, there always exists a connection matrix.*

Of course, connection matrices need not be unique. Moreover, the definition of connection matrix as given here is weaker than that of Franzosa [19], the difference being that we do not specify the isomorphism between $H_n\Delta$ and $H_n(h(S); \mathbf{Z}_2)$, whereas Franzosa does. This implies that the set of matrices satisfying our conditions may be larger than his; however, for most simple examples this is not the case.

The most important property of the connection matrix is the following result:

Proposition III.5. *Let $i < j$ and assume that there is no k such that $i < k < j$; then if $\Delta_{ij} \neq 0$ it follows that $C(M(j), M(i)) \neq \phi$.*

This means that in certain circumstances, a nonzero entry in Δ implies that a connection exists.

We consider next an example. Let S be an isolated invariant set with Morse decomposition $M(S) = \{M(i): i = 0, 1, \ldots, N\}$. Assume too that we have the following information for the homologies of the Morse sets:

$$H_n(i) = \begin{cases} \mathbf{Z}_2, & \text{if } n = 2i, 2i + 1, \\ 0, & \text{otherwise,} \end{cases} \qquad 0 \leq i < N,$$

$$H_n(N) = \begin{cases} \mathbf{Z}_2, & \text{if } n = 2N, \\ 0, & \text{otherwise,} \end{cases}$$

$$H_n(N + 1) = 0 \qquad \text{for all } n,$$

$$H_n(h(S); \mathbf{Z}_2) = \begin{cases} \mathbf{Z}_2, & \text{if } n = 0, \\ 0, & \text{otherwise.} \end{cases}$$

Then one can show $C(M(i), M(i - 1)) \neq \phi$ for $i = 1, \ldots, N$. Rather than give a proof for the general case, which might obscure the simplicity of this application of the connection matrix, we shall only consider the case $N = 2$.

Thus we are looking at $\Delta: \bigoplus_0^3 H(i) \to \bigoplus_0^3 H(i)$, and since $H_n(3) = 0$ for all n, we need only consider $\Delta: \bigoplus_0^2 H_n(i) \to \bigoplus_0^2 H_n(i)$. Similarly, ignoring $H_n(i)$ when $H_n(i) = 0$, we have that Δ maps

$$H_0(0) \oplus H_1(0) \oplus H_2(1) \oplus H_3(1) \oplus H_4(2)$$

into itself. From (a) in Definition III.3, if $i \leq j$, then $\Delta_{ij} = 0$; for example, $\Delta_{00}: H_0(0) \oplus H_1(0) \rightarrow H_0(0) \oplus H_1(0)$ can be written as

$$\Delta_{00} = \begin{matrix} & H_0(0) & H_1(0) \\ H_0(0) & \\ H_1(0) \end{matrix} \begin{bmatrix} 0 & 0 \\ 0 & 0 \end{bmatrix}.$$

Now consider $\Delta_{02}: H_4(2) \rightarrow H_0(0) \oplus H_1(0)$; by (b) in Definition III.3, $\Delta_{02} = 0$; similarly applying III.3(b) to the maps

$$\Delta_{01}: H_2(1) \oplus H_3(1) \rightarrow H_0(0) \oplus H_1(0),$$

and

$$\Delta_{12}: H_4(2) \rightarrow H_2(1) \oplus H_3(1),$$

gives

$$\Delta_{01} = \begin{bmatrix} 0 & 0 \\ * & 0 \end{bmatrix} \quad \text{and} \quad \Delta_{12} = \begin{bmatrix} 0 \\ * \end{bmatrix},$$

where $*$ denotes an unknown entry. Thus

	$H_0(0)$	$H_1(0)$	$H_2(1)$	$H_3(1)$	$H_4(2)$
$H_0(0)$	0	0	0	0	0
$H_1(0)$	0	0	*	0	0
$H_2(1)$	0	0	0	0	0
$H_3(1)$	0	0	0	0	*
$H_4(2)$	0	0	0	0	0

Finally, we use (c) in Definition III.3. Since $H\Delta = H_*(h(S); \mathbf{Z}_2)$, it must be that dim $H\Delta = 1$; i.e., dim ker Δ = rank $\Delta + 1$. But, clearly, dim ker $\Delta \geq 3$; so rank $\Delta = 2$ and thus both $*$ entries equal 1. In particular, $C(M(1), M(0))$ and $C(M(2), M(1))$ are nonempty by Proposition III.5.

We end this section by describing some other applications of the connection matrix. In [45] Reineck studies the qualitative behavior of solutions to 2-species ecological models, and he determines the connection matrices related to the various possible phase portraits. He also obtains results on a 3-species system modeling two predators and one prey. Mischaikow [42] classifies travelling wave solutions for systems of reaction–diffusion equations of the form $u_t = Du_{xx} + \nabla F(u)$ where D is a positive diagonal matrix; these results can be extended to obtain results on homoclinic orbits for the Hamiltonian system, $u' = v$, $v' = -\nabla F(u)$, as well as results on heteroclinic orbits for the related system, $u' = v$, $v' = \theta v - \nabla F(u)$. Finally, Franzosa [19, 20] gives a complete rigorous development of the Connection Matrix Theory.

§C. Miscellaneous Results

We describe some extensions of the Conley index which have appeared in the recent literature—no attempt is made to go into any detail.

In [46, 47] Rybakowski develops the Conley index theory for semi-flows on metric spaces. The isolated invariant set S is still assumed to be compact, but index pairs consist of closed, admissible sets. A set N is called admissible if:

(a) for every sequence $\{x_n\} \subset N$, and every sequence $\{t_n\} \subset \mathbf{R}^+$, where $t_n \to \infty$ and $x \cdot [0, t_n] \not\subset N$, the sequence $\{x_n \cdot t_n\}$ is precompact; and
(b) given $x \in N$ and $x \cdot t$ defined for all $t \in [0, w_x)$; if $w_x < \infty$, then $x \cdot [0, w_x) \subset N$.

This theory is applied to nonlinear Dirichlet boundary-value problems to obtain orbits which connect equilibria; see [48]. In [49] Rybakowski and Zehnder obtain generalized Morse "inequalities" in this setting (cf. Theorem 23.21). The Conley index has been applied by Conley and Zehnder [5] to study maps which arise in celestial mechanics, and to prove a certain conjecture of Arnold; see also [16, 17, 18], for related results.

Next, in [50], Salamon gives a slightly different approach to the Conley index theory; also in [3], Benci finds an alternate, in some ways even simpler approach, to the Conley index theory. Finally, in [17], Floer proves an extension of the important Continuation Theorem (Theorem 23.31) and gives some applications.

SECTION IV. Stability of Travelling Waves— A Topological Approach

§A. Introduction

As we have seen in Chapter 24, nonlinear parabolic systems of the form

$$u_t = Du_{xx} + f(u, u_x), \tag{25.72}$$

with $u \in \mathbf{R}^n$ and $x \in \mathbf{R}^1$ admit a physically important class of self-similar wave solutions, $u(x, t) = V(\xi)$, where $\xi = x - \theta t$, which satisfy an associated system of ordinary differential equations

$$-\theta V' = DV'' + f(u, u_\xi). \tag{25.73}$$

A typical problem is to locate solutions of (25.73) which tend to limits $V_\pm$ as $\xi \to \infty$. A central issue is to determine when such solutions are asymptotically stable. We shall consider them as steady states of

$$u_t = Du_{\xi\xi} + \theta_{u_\xi} + f(u, u_\xi), \tag{25.74}$$

the equation obtained from (25.72) under the change of variables $(x, t) \to$ (ξ, t). In this section, some recently developed geometric tools for analyzing the stability of such solutions will be described.

To begin with, we say that a travelling wave $V(\xi)$ is a *stable* solution of (25.74) in a prescribed norm $\|\cdot\|$, if there exists $\delta > 0$ so that, whenever $\|u(\xi, 0) - V(\xi)\| < \delta$, there is a $k \in \mathbf{R}$ for which $\|u(\xi, t) - V(\xi + k)\| \to 0$ as $t \to +\infty$. The choice of norm can be a crucial issue. For reaction–diffusion systems the sup norm usually suffices while for other types of problems, such as viscous shock profiles and periodic travelling waves, weighted norms are more appropriate. The choice of norm turns out to be closely associated with the properties of the spectrum $\sigma(L)$ of the linear operator L defined by linearizing (25.74) about $V(\xi)$,

$$Lp = Dp'' + (\theta + \alpha(\xi))p' + b(\xi)p, \tag{25.75}$$

(where $\alpha(\xi) = d_2 f(V(\xi), V'(\xi))$ and $b(\xi) = d_1 f(V(\xi), V'(\xi))$), and its associated eigenvalue problem

$$Lp = \lambda p. \tag{25.76}$$

Since the spatial domain is unbounded, L can have both point spectrum, $\sigma_p(L)$, and essential spectrum, $\sigma_e(L)$, either of which can cause instability. Typically for many reaction–diffusion systems (e.g., scalar reaction–diffusion equations (24.25), where f is as in (24.30)), there is a $\beta < 0$ such that

$$\text{Re } \lambda < \beta \qquad \text{for all} \quad \lambda \in \sigma_e(L); \tag{25.77}$$

for viscous shock profiles and periodic waves, the essential spectrum is always tangent to the imaginary axis. Each case has its own subtleties. In order to better appreciate this, we mention two well-known theorems.

Theorem 1 (Weyl's Lemma). *Let $L_\pm$ be the constant coefficient operators obtained from L by setting $a = a_\pm, b = b_\pm$ where $a_\pm, b_\pm$ are the limiting values of $a(\xi), b(\xi)$ at $\xi = \pm\infty$; then $\sigma(L_-) \cup \sigma(L_+)$ divides $\mathbf{C}$ into components, and the component containing $\text{Re } \lambda > 0$ is free from essential spectrum.*

Since the operators $L_\pm$ have constant coefficients, their spectra, and, hence, $\sigma_e(L)$ are easily computed (see §B).

Theorem 2 (Linearized Stability). *Suppose that L is a sectorial operator and that $\sigma_e(L)$ satisfies (25.77) for some $\beta < 0$. If, in addition, $\sigma_p(L)\backslash\{0\} \subset \text{Re } \lambda < 0\}$ and $\lambda = 0$ is a simple eigenvalue of L, then $V(\xi)$ is (nonlinearly) stable; more precisely, for some $c, M > 0$ and $k \in \mathbf{R}$, we have that*

$$\|V(\xi + k) - u(\xi, t)\|_\infty \leq M\|V(\xi) - u(\xi, 0)\|_\infty e^{-ct}.$$

(For a proof, see, e.g., Bates and Jones [2].) The latter result draws upon the theory of analytic semigroups and is therefore widely applicable. Regarding its application, the difficult part is to verify the conditions concerning the point spectrum of L; this usually depends sensitively upon the internal structure of the wave. In the result of Jones [15] for the Fitz-Hugh–Nagumo system (which we discussed above in Section I, §A), an approach to this question was developed drawing upon some general geometric methods in dynamical systems. Some of the constructions, however, are tied to the specific form and dimensionality of the Fitz-Hugh–Nagumo equations. In the next section, we shall describe some geometric machinery for finding the location and the multiplicies of the eigenvalues of L. The techniques are motivated by Jones' original insights, and are quite general. There are two important constructions in the theory:

- An analytic function $D(\lambda)$ with domain $\mathbf{C} \backslash \sigma_e(L)$ whose roots (counting order) coincide with the eigenvalues of L (counting algebraic multiplicity).
- A complex vector bundle $\mathscr{E}(K)$, where K is a simple closed curve in $\mathbf{C}$ with $K \cap \sigma(L) = \varnothing$.

Together they provide a vehicle for counting eigenvalues of L interior to K. In regard to checking the hypotheses of Theorem IV.2, it is frequently possible to bound the modulus $|\lambda|$ of any unstable eigenvalue a-priori; hence K is usually chosen to be a curve enclosing $\lambda = 0$, which has a sufficiently large portion of $\operatorname{Re} \lambda \geq 0$ in its interior. A key ingredient in both the construction and the application of $\mathscr{E}(K)$ and $D(\lambda)$ is the flow induced by the linearized equations on the space $\mathbf{CP}^{N-1}$, of complex lines in $\mathbf{C}^N$, or, more generally, the Grassmannian $G_k(\mathbf{C}^N)$ of k-planes in $\mathbf{C}^N$. We call the associated flow the *projectivized flow*; it will be described in some detail in the next section. We shall show how these very general tools are used to determine the stability of travelling waves for a wide class of parabolic systems.

§B. The Search for $\sigma_p(L)$

B.1. Preliminaries

We first write the linearized equations (25.75) as a first-order system:

$$p' = q,$$
$$q' = D^{-1}[-(\theta + \alpha(\xi))q + (\lambda I - b(\xi))p],$$

or, in abbreviated form, as

$$Y' = A(\xi, \lambda)Y, \tag{25.78}$$

where $Y = (p, q) \in \mathbf{C}^N$, $N = 2n$, and $A(\xi, \lambda)$ is the appropriate $N \times N$ coefficient matrix. Since the underlying wave $V(\xi)$ tends to limits at $\pm\infty$, the matrices $A(\xi, \lambda)$ have limits $A_\pm(\lambda)$ as $\xi \to \pm\infty$.

Definition IV.1. $\Omega \subset \mathbf{C}$ has *consistent splitting* if there exists k such that $A_{\pm}(\lambda)$ both have k eigenvalues with positive real part and $N - k$ eigenvalues with negative real part for all $\lambda \in \Omega$.

We remark that for bistable diffusive waves (solutions of the parabolic systems which tend at $\pm\infty$ to stable rest points of the system), there exists Ω with consistent splitting containing $\operatorname{Re} \lambda > 0$; in these cases it is easily seen that $k = N/2 = n$. We also point out that the curves in $\mathbf{C}$ where $A_{\pm}(\lambda)$ have pure imaginary eigenvalues define the boundary of the essential spectrum of L; in fact, $\sigma_e(L)$ is contained in the region to the left of the union of these curves.

B.2. The Evans Function

If Ω has consistent splitting, then it is not difficult to see that, for each $\lambda \in \Omega$, there exist k independent solutions $Y_i(\xi, \lambda)$, $1 \leq i \leq k$, which decay to zero as $\xi \to -\infty$, and $N - k$ independent solutions $Y_i(\xi, \lambda)$, $k + 1 \leq i \leq N$, which decay to zero as $\xi \to +\infty$. (A rigorous construction is given in the next section.) In view of the hyperbolicity of $A_{\pm}(\lambda)$, λ is an eigenvalue of L if and only if (25.76) possesses a nontrivial solution which decays to zero at both $-\infty$ and $+\infty$. Hence the span of either set satisfies half of the conditions at $\pm\infty$ required of eigenfunctions. Clearly, both conditions are satisfied if and only if these two subspaces intersect nontrivially. This suggests introducing the N-form

$$\eta(\xi, \lambda) = Y_1(\xi, \lambda) \wedge Y_2(\xi, \lambda) \wedge \cdots \wedge Y_N(\xi, \lambda).$$

Clearly, η vanishes for all ξ if and only if λ is an eigenvalue; furthermore, we have by Abel's formula that

$$\eta' = \text{trace } A(\xi, \lambda)\eta.$$

Definition IV.2. The Evans function $D(\lambda)$ is

$$D(\lambda) = e^{-\int_0^\xi \text{trace } A(s, \lambda)ds}\eta(\xi, \lambda).$$

This function was introduced by Evans [13] for equations arising in neurophysiology; the above definition was given in [1], where the following properties of $D(\lambda)$ were proved:

1. $D(\lambda)$ is analytic in λ for $\lambda \in \Omega$;
2. $D(\lambda)$ is independent of ξ; and
3. the order of the root of $D(\lambda)$ at $\lambda = \lambda_0$ equals the algebraic multiplicity of λ_0 as an eigenvalue of L.

Property 3 was proved in [1] by a complicated perturbation procedure. A simpler, more direct argument can be found in [25].

B.3. The Augmented Unstable Bundle

The previous construction has a geometric flavor in that it specifies how the eigenvalues of L are determined by the intersection of certain distinguished subspaces of solutions of (25.78), namely

$$Y_-(\xi, \lambda) = \text{span}\{Y_1(\xi, \lambda), \ldots, Y_k(\xi, \lambda)\},$$

$$Y_+(\xi, \lambda) = \text{span}\{Y_{k+1}(\xi, \lambda), \ldots, Y_N(\xi, \lambda)\}.$$

This connection with geometry can be carried substantially further, as we shall now see.

The set $Y_-(\xi, \lambda)$ is a family of k-dimensional subspaces of $\mathbf{C}^N$ which depends continuously on the parameters (ξ, λ); in other words, it has the structure of a complex k-plane bundle over the base space $\mathbf{R} \times \Omega$. It will be convenient to compactify the base space by reparametrizing ξ with a parameter $\tau \in [-1, 1]$, namely

$$\xi = \frac{1}{2k} \ln\left(\frac{1+\tau}{1-\tau}\right).$$

We can then express (25.78) as an equivalent, augmented, autonomous system of the form

$$\begin{aligned} Y' &= A(\tau, \lambda)Y, \\ \tau' &= k(1 - \tau^2), \end{aligned} \tag{25.79}$$

where $A(\tau, \lambda)$ is obtained from $A(\xi, \lambda)$ by reparametrization by τ. If $k > 0$ is not too large and if the wave decays exponentially to $V_\pm$ at $\xi = \pm\infty$, then $A(\xi, \lambda)$ is a C^1 function.

The "compactified" system (25.79) is very useful for constructing subspaces of solutions of (25.78) having specified behavior at $\xi = \pm\infty$. The idea is to apply the stable/unstable manifold theorem to the rest points of (25.79), namely $(\vec{0}, \pm 1)$. More precisely, since $A_\pm(\lambda)$ have consistent splitting for $\lambda \in \Omega$, there exist k-dimensional (resp. $(N - k)$-dimensional) subspaces $U_\pm(\lambda)$ (resp. $S_\pm(\lambda)$) associated with the portion of the spectra of $A_\pm(\lambda)$ with positive (resp. negative) real part. The rest point $(\vec{0}, -1)$ of (25.79) is therefore hyperbolic with a k (complex) + 1 (real) dimensional unstable subspace (the real unstable direction is the τ direction). The stable manifold theorem (cf. Chapter 12, §C) provides a $(2k + 1)$-(real)-dimensional manifold W^u of solutions of (25.79) which tend to $(\vec{0}, -1)$ in backward time. By the linear structure of the Y-components, $W^u \cap \mathbf{C}^N \times \{\tau\}$ is a k-dimensional vector space. Define Y_- by

$$Y_-(\tau, \lambda) = W^u \cap \mathbf{C}^N \times \{\tau\};$$

then $Y_-(\tau, \lambda)$ tends to the unstable subspace $U_-(\lambda)$ of $A_-(\lambda)$ as $\tau \to -1$ in the topology of $G_k(\mathbf{C}^N)$. We have therefore constructed a k-plane bundle over the base space $[-1, 1) \times \Omega$.

The key idea is to compactify the base space by somehow tracking the behavior of $Y_-(\tau, \lambda)$ as $\tau \to +1$. In general, it is not always possible to characterize this limit; however, we have the following result [1]:

Lemma IV.3. *If $\lambda \in \Omega \backslash \sigma_p(L)$, then $Y_-(\tau, \lambda)$ tends to $U_+(\lambda)$ in the Grassmannian $G_k(\mathbf{C}^N)$ as $\tau \to +1$.*

The situation is depicted in Figure 25.1. This shows that for λ, as in the lemma, $Y_-(\tau, \lambda)$ forms a k-plane bundle over $[-1, 1] \times \{\lambda\}$.

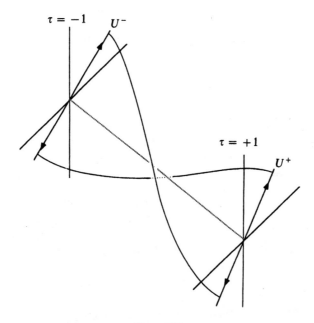

Figure 25.1

Let K be a simple closed curve in Ω with $K \cap \sigma_p(L) = \varnothing$. We assert that the "twisting" of the fibers $Y_-(\tau, \lambda)$ for $(\tau, \lambda) \in [-1, 1] \times K$ determines the number of eigenvalues of L interior to K. More precisely, let $\mathscr{E}(K)$ be the bundle $\pi: \mathscr{E}(K) \to B$ over the base space

$$B = \{-1\} \times K^0 \cup [-1, 1] \times K \cup \{+1\} \times K^0,$$

where K^0 is the interior of K, and the total space $\mathscr{E}(K)$ is defined by

$$\pi^{-1}(\tau, \lambda) = \begin{cases} U_-(\lambda) & \text{for } \tau = -1, \lambda \in K^0, \\ Y_-(\tau, \lambda) & \text{for } \tau \in [-1, 1], \lambda \in K, \\ U_+(\lambda) & \text{for } \tau = +1, \lambda \in K^0. \end{cases}$$

By the defining property of Y_- at $\tau = -1$, together with the above lemma
concerning the behavior of Y_- at $\{1\} \times K$, it follows that $\mathscr{E}(K)$ is a bundle. In
particular, the fibers match up continuously where the caps $\{\pm 1\} \times K^0$ are
glued onto the sides $[-1, 1] \times K$ of the base space. The resulting bundle is
called the *argumented unstable bundle*.

The above construction provides us with a bundle over a 2-sphere B. If K^0
does not contain eigenvalues of L this bundle is easily seen to be trivial; in
that case, $Y_-(1, \lambda)$ is defined and continuous for all λ on, and inside, K, so that
$\mathscr{E}(K)$ is a retract of a bundle over the (contradictible) solid ball. Thus the
eigenvalue count of L inside K is related to the extent to which $\mathscr{E}(K)$ differs
from a trivial bundle. This is measured by a topological invariant, an integer
$c_1(\mathscr{E}(K))$ associated with the bundle, called the *first Chern number of* $\mathscr{E}(K)$.
The following theorem, whose proof can be found in [1], makes this state-
ment precise:

Theorem IV.4. *Suppose that* $K \subset \Omega \backslash \sigma_p(L)$; *then the following three quantities
are equal*:

1. $c_1(\varepsilon(K))$.
2. The winding number of the curve $D(K)$ with respect to the origin.
3. The number of eigenvalues of L interior to K counting algebraic
   multiplicities.

We shall give only a brief account of how this theorem is proved, since the
topological details do not figure prominently in the applications. The base
space B is expressed as a union of hemisphere $B_- \cup B_+$, where B_- is the left
cap together with the sides of B, and B_+ is the right cap. Trivializations for
the restriction of $\mathscr{E}(K)$ to each hemisphere are constructed from Y_- and U_-
(for B_-) and U_+ (for B_+). The "clutching function" of $\mathscr{E}(K)$ is the map f_E from
the equator, $B_- \cap B_+$, into the invertible $k \times k$ matrices which relates two
bases for the fibers determined by the two trivializations. This map represents
a class in $\pi_1(\mathrm{Gl}(k, \mathbf{C}))$. This group is isomorphic to the integers for every k,
and it turns out that if a suitable generator is selected, $[f_E]$ can be equated
with the first Chern number of $\mathscr{E}(K)$ (see Husemoller [31]). An analysis of the
behavior of the section Y_- as $\tau \to +1$ reveals that the Evans function $D(\lambda) =
g(\lambda) \det f_E$, where $g(\lambda)$ is a nonvanishing analytic function. The proof can be
completed because $\det_*$ is an isomorphism π_1 of the invertible $k \times k$ matrices
to that of the nonzero complex numbers.

We finally remark that if K contains $\lambda = 0$ in its interior and if K contains
a sufficiently large portion of $\mathrm{Re}\, \lambda \geq 0$, then $c_1(\mathscr{E}(K)) = 1$ implies that the
hypotheses of Theorem IV.2 are satisfied, so that the underlying wave is
stable. It therefore is appropriate to call $c_1(\mathscr{E}(K))$ the stability index.

B.4. Projectivized Flows and Grassmannians

In order to determine $\mathscr{E}(K)$ and to compute $c_1(E(K))$ we require some appa-
ratus for studying the global behavior of the vector space $Y_-(\xi, \lambda)$. A useful

tool in this regard is the Grassmannian $G_k(\mathbf{C}^N)$ of k-planes in $\mathbf{C}^N$, topologized so that two elements in $G_k(\mathbf{C}^N)$ are close whenever they have bases which are close to the Euclidean topology. Given a linear system $Y' = AY$, with $Y \in \mathbf{C}^N$, there is an associated flow on $G_k(\mathbf{C}^N)$ which we denote by $\hat{Y}' = A_k(\hat{Y})$, where $\hat{Y}$ is the generic point in $G_k(\mathbf{C}^N)$ and A_k is a vector field in the tangent space to the Grassmannian which is induced by A. The latter system is called the *projectivized flow*; its solutions are obtained from the linear flow by forming the span of k independent solutions. The special case when $k = 1$ is of particular importance; here $G_1(\mathbf{C}^N) = \mathbf{C}P^{N-1}$ is complex projective space, the space of complex lines in $\mathbf{C}^N$.

An important observation that will be used repeatedly is that a k-dimensional eigenspace E of A is invariant under A, and thus E is a critical point of A_k. Furthermore, if A is hyperbolic with k (resp. $N - k$) eigenvalues with positive (resp. negative) real part, and if E is the eigenspace associated with the unstable eigenvalues, then E is an attracting rest point for the projectivized flow on $G_k(\mathbf{C}^N)$. This is most easily visualized in the case of a saddle in $\mathbf{R}^2$, as depicted in Figure 25.2. Rays through the origin behave under the linear flow as indicated in the middle figure after being identified to points on S^1; identification of antipodal points in S^1 yields the projectivized flow on $\mathbf{R}P^1$. In regard to the asymptotic matrices $A_{\pm}(\lambda)$ of $A(\xi, \lambda)$, the unstable subspaces $U^{\pm}(\lambda)$ are attractors for the flow $A_{\pm}(\lambda)$ induced on $G_k(\mathbf{C}^N)$ and the stable subspaces $S^{\pm}(\lambda)$ are repellers for the induced flow on $G_{N-k}(\mathbf{C}^N)$. If we now projectivize the augmented system (25.79), the flow on $G_k(\mathbf{C}^N) \times [-1, 1]$,

$$\hat{Y}' = A_k(\tau, \lambda, \hat{Y}),$$

$$\tau' = k(1 - \tau^2),$$

$$(25.80)$$

is qualitatively as depicted in Figure 25.3; i.e., $\hat{Y}_-$ is an orbit of (25.80) which connects the rest point $(U^-(\lambda), -1)$ to $(U^+(\lambda), +1)$, provided, of course, that $\lambda \notin \sigma_p(L)$.

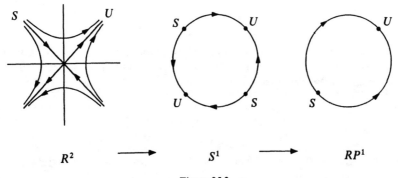

Figure 25.2

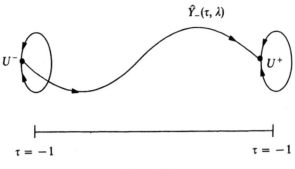

$$\hat{Y}_-(\tau, \lambda)$$

U^- U^+

$\tau = -1$ $\tau = -1$

Figure 25.3

In this manner one obtains a continuous map $\Phi: B \to G_k(\mathbf{C}^N)$, called the *classifying map* of $\mathscr{E}(K)$. The specification of the map Φ is equivalent to the specification of the bundle, and, in particular, $c_1(\mathscr{E}(K))$ is determined by the homotopy type of Φ.

§C. Applications to Fast–Slow Systems

C.1. General Remarks

The stability index is well suited to singular perturbations problems, wherein the underlying equations have a small parameter ε such as a diffusion coefficient tending to zero or, in the study of weak viscous shock profiles, the shock strength. In such situations, the phase space for the underlying wave can be decomposed into manifolds of slowly varying solutions away from which solutions behave on a rapid time scale. This structure is inherited by the linearized equations. In terms of the augmented unstable bundle, it is manifested as a Whitney sum decomposition,

$$\mathscr{E}(K, \varepsilon) = \bigoplus_{i=1}^{l} \mathscr{E}_i(K, \varepsilon)$$

of lower-dimensional subbundles whose fibers are determined by the spans of certain distinguished solutions which exhibit behavior on one of the characteristic time scales. A crucial property of c_1 is that it is additive on the summands in a Whitney sum,

$$c_1(\mathscr{E}(K, \varepsilon)) = \sum_{i=1}^{l} c_1(\mathscr{E}_i(K, \varepsilon)). \tag{25.81}$$

A second crucial property of c_1 is that it is topological so that the stability index is invariant under homotopies of the equations which preserve the bundle. This suggests that, in order to compute c_1 of the summands, we allow the small parameter ε to tend to zero and compute the limits $\mathscr{E}_{iR}(K)$ of each of the perturbed summands. In practice, the reduced summands occur as the augmented unstable bundles for certain (formal) reduced eigenvalue problems obtained directly from the equations in which ε has been set equal to zero. (In a singular perturbation problem, there can be several distinct reduced systems, depending on the scaling of the equations that is being used.) The reduced eigenvalue problems are typically much simpler than the perturbed problem; in particular, it is frequently possible to perform an eigenvalue count directly via Sturm–Liouville theory. By Theorem IV.1, this can be used to compute the first Chern number of the reduced summands, and also by homotopy invariance, this gives the Chern numbers of the perturbed summands. Finally, from (25.81) this information translates into a statement about $c_1(\mathscr{E}(K, \varepsilon))$.

C.2. Geometric Singular Perturbation Theory

Problems with fast–slow dynamics have a convenient expression in terms of invariant manifolds. In an appropriate scaling involving the small parameter ε, the (nonlinear) travelling wave equations can be expressed in the form

$$x' = G(x, \varepsilon), \tag{25.82}$$

where $G(x, 0)$ admits a manifold $\mathscr{M}_0$ of rest points. Since (25.82) is the first-order reduction of a second-order system (25.73), it is usually the case that the dimension of this manifold is even, say $\dim \mathscr{M}_0 = 2r$, with $1 \le r \le n$. Under certain conditions, Fenichel [15] proved the existence of an invariant manifold $\mathscr{M}_\varepsilon$ of (25.82) for small ε near $\mathscr{M}_0$. The main condition for determining when $\mathscr{M}_0$ perturbs smoothly is called *normal hyperbolicity*, which requires that the eigenvalues μ associated with eigenvectors of $dG(x, 0)$, for $x \in \mathscr{M}_0$, which are transverse to the tangent space $T_x\mathscr{M}_0$ satisfy $|\mathrm{Re}\,\mu| > \beta$, for some $\beta > 0$, and all x in some compact subset of $\mathscr{M}_0$. Typically, $\mathscr{M}_0$ is a graph over the show subspace; in other words, the state vector $x = (x_S, x_F)$ splits into a $2r$-dimensional slow subspace x_S and the remaining components x_F are given as a function $x_F = \gamma(x_S)$ on the slow components. Fenichel's theorem also provides equations for the perturbed flow on $\mathscr{M}_\varepsilon$

$$\dot{x}_S = P_S G(x_S, \gamma(x_S), 0) + \mathcal{O}(\varepsilon), \tag{25.83}$$

where P_S is projection on the slow subspace and the independent variable has been rescaled to $s = \varepsilon^{-2}\xi$. It is frequently the case that the reduced problem (25.83) at $\varepsilon = 0$ can be explicitly analyzed, and that the travelling waves occur as a transverse intersection of invariant manifolds of rest points (in the old scaling) inside $\mathscr{M}_0$. The persistence of such connections for $\varepsilon > 0$ follows immediately from Fenichel's theorem and transversality. This provides an efficient machinery for producing travelling wave solutions of the perturbed problem.

This viewpoint is also useful in studying the linearized problem,

$$Y' = A(\tau, \lambda, \varepsilon) Y,$$
$$\tau' = \varepsilon k(1 - \tau^2). \tag{25.84}$$

In most situations, the domain Ω in which $A_{\pm}(\lambda, \varepsilon)$ have consistent splitting contains the set $\{\operatorname{Re} \lambda \geq 0\}$, in which case $k = n = N/2$. Note that $A(\tau, 0, \varepsilon) = dG(x(\tau), \varepsilon)$ so that, by normal hyperbolicity, $A(\tau, 0, \varepsilon)$ has $2r$ slow eigenvalues (i.e., $\mathcal{O}(\varepsilon)$), and $N - 2r$ fast eigenvalues (i.e., $\mathcal{O}(1)$). It follows that the n eigenvalues of $A_{\pm}(\tau, 0, \varepsilon)$ with positive real part split into r slow eigenvalues and $n - r$ fast eigenvalues. It turns out that this splitting frequently persists for all λ in the unstable half-plane. This suggests that Whitney sum decompositions of $\mathscr{E}(K, \varepsilon)$ can be found of the form

$$\mathscr{E}(K, \varepsilon) = \mathscr{E}_S(K, \varepsilon) \oplus \mathscr{E}_F(K, \varepsilon),$$

where $\mathscr{E}_S$ is r-dimensional and $\mathscr{E}_F$ is $(n - r)$-dimensional. The fast summand is determined by considering the projectivized flow on $G_{n-r}(\mathbf{C}_N)$,

$$\hat{Y}' = A_{n-r}(\hat{Y}, \lambda, \tau, \varepsilon),$$
$$\tau' = \varepsilon k(1 - \tau^2). \tag{25.85}$$

Note that at $\varepsilon = 0$, τ appears as a parameter in the equations. Since $A(\tau, \lambda, 0)$ has $n - r$ fast eigenvalues of positive real part, it follows from our remarks in the previous section that this matrix admits an $(n - r)$-dimensional subspace $U_F(\tau, \lambda, 0)$, which is an attracting rest point of (25.85) when $\varepsilon = 0$ for each fixed τ or, equivalently, $(U_F(\tau, \lambda, 0), \tau)$, $-1 \leq \tau \leq 1$, is an attractor for the parametrized flow. Since attractors perturb to nearby flows, it follows that (25.85) admits an attractor $(\hat{Y}_F(\tau, \lambda, 0), \tau)$, $-1 \leq \tau \leq 1$, for small $\varepsilon > 0$. By construction, $\hat{Y}_F(\tau, \lambda, \varepsilon)$ remains near the fast unstable subspace of $A(\tau, \lambda, \varepsilon)$; in particular, this is true at $\tau = \pm 1$. Thus, no solution in $\hat{Y}_F(\tau, \lambda, \varepsilon)$ forms an eigenfunction; in other words, $\hat{Y}_F$ is an orbit of (25.85) connecting $\hat{U}_F(-1, \lambda, \varepsilon)$ to $\hat{U}_F(+1, \lambda, \varepsilon)$, and it can therefore be used to construct a fast subbundle $\mathscr{E}_F$ of $\mathscr{E}$ by the usual capping procedure:

$$\pi^{-1}(\tau, \lambda) = \begin{cases} \hat{U}_F(\tau, \lambda, \varepsilon) & \text{for} \quad \tau = -1, \lambda \in K_0, \\ \hat{Y}_F(\tau, \lambda, \varepsilon) & \text{for } -1 \leq \tau \leq 1, \lambda \in K, \\ \hat{U}_F(\tau, \lambda, \varepsilon) & \text{for} \quad \tau = +1, \lambda \in K. \end{cases}$$

Since the fast–slow splitting is valid for all λ in compact subsets of Ω, it follows that the connection $\hat{Y}_F$ from $\hat{U}_F^-$ to $\hat{U}_F^+$ exists for all $\lambda \in K \cup K^0$; it therefore follows that $\mathscr{E}_F(K, \varepsilon)$ is trivial, so that

$$c_1(\mathscr{E}_F(K, \varepsilon)) = 0.$$

Evidently, all the action in this setting is in the slow summand $\mathscr{E}_S(K, \varepsilon)$. In order to define $\mathscr{E}_S$ we decompose $\mathbf{C}^N$ into fast and slow components, $Y = (Y_S, Y_F)$ where Y_S lies in the $2r$-dimensional slow subspace, and $Y_F = \Gamma(\tau, \lambda, \varepsilon) Y_S$ for an appropriate matrix Γ. When $\varepsilon = 0$, the set of vectors $(Y_S, \Gamma Y_S)$ forms a $2r$-dimensional subspace $S_0(\tau, \lambda)$ for each $\tau \in [-1, 1]$. If we consider the projectivized flow induced by (25.84) on $CP^{n-1} \times [-1, 1]$,

$$\begin{aligned} \hat{Y}' &= A_1(\hat{Y}, \tau, \lambda, \varepsilon), \\ \tau' &= \varepsilon k(1 - \tau^2), \end{aligned} \tag{25.86}$$

it follows that $(\hat{S}_0(\tau, \lambda), \tau)$, $\tau \in [-1, 1]$, forms a normally hyperbolic manifold of rest points of (25.86) at $\varepsilon = 0$. Fenichel's theorem [15] then implies that (25.86) admits a smooth invariant manifold $(\hat{S}_\varepsilon(\tau, \lambda), \tau)$, for $\tau \in [-1, 1]$ and small $\varepsilon > 0$. The flow on this manifold is given (with a rescaled independent variable) by the projectivization on CP^{r-1} of

$$\begin{aligned} \dot{Y}_S &= A^S(\tau, \lambda, \varepsilon) Y_S, \\ \tau' &= k(1 - \tau^2), \\ A^S &= (\tau, \lambda, \varepsilon) = P_S A(\tau, \lambda, \varepsilon)(Y_S, \Gamma(\tau, \lambda, \varepsilon) Y_S). \end{aligned} \tag{25.87}$$

In many applications, $r = 2$ so that (25.87) can be analyzed by Sturm–Liouville theory. Assuming that the spectrum Σ_S of (25.87) is known at $\varepsilon = 0$, the curve K is chosen in the usual manner so that $K \subset \Omega \backslash \Sigma_S$. It then follows, that for $\lambda \in K$ and small $\varepsilon > 0$, an r-dimensional subbundle $\mathscr{E}_S(K, \varepsilon)$ of $\mathscr{E}(K, \varepsilon)$ can be defined by the usual capping procedure, so that

$$\mathscr{E}(K, \varepsilon) = \mathscr{E}_F(K, \varepsilon) \oplus \mathscr{E}_S(K, \varepsilon);$$

by previous remarks,

$$c_1(\mathscr{E}(K, \varepsilon)) = c_1(\mathscr{E}_S(K, \varepsilon))$$

$$= \text{number of eigenvalues of } \Sigma_S \text{ inside } K.$$

C.3. Application to a Phase-Field Model

We briefly sketch the implementation of the methods of the previous section to the stability analysis of travelling wave solutions of the equation

$$u_t = \varepsilon^4 \, \partial^6 u / \partial x^6 + A\varepsilon^3 \, \partial^4 u / \partial x^4 + B \, \partial^2 u / \partial x^2 + f(u),$$

where $f(u)$ is a qualitative bistable cubic with stable roots at $u = 0, 1$; (see [4] for a discussion of the physical origins of the model). Although the stability index was described in the context of second-order systems, the constructions and proofs of the main theorems in [1] easily generalize to this type of higher-order equation; see [25].

The travelling wave equations can be written as:

$$u_1' = \varepsilon u_2,$$
$$u_2' = \varepsilon u_3,$$
$$u_3' = u_4, \qquad\qquad (25.88)$$
$$u_4' = u_5,$$
$$u_5' = u_6,$$
$$u_6' = -Au_5 - Bu_3 - \theta u_2 = f(u_1),$$

where $\xi = (x - \theta t)/\varepsilon$ and "prime" $= d/d\xi$. More compactly, we write (25.88) as

$$x' = G(x, \varepsilon),$$

where $x = (u_1, \ldots, u_6)$. Clearly x splits into fast components $x_F = (u_3, u_4, u_5, u_6)$ and slow components $x_S = (u_1, u_2)$. The slow manifold $\mathcal{M}_0$ at $\varepsilon = 0$ is defined by

$$x_F = \gamma(x_S) = (-\theta B^{-1}u_2 - B^{-1}f(u_1), 0, 0, 0),$$

which is easily seen to be normally hyperbolic. The slow flow (25.83) (with $s = \xi/\varepsilon$) is

$$\dot{u}_1 = u_2,$$
$$B\dot{u}_2 = -\theta u_2 - f(u_1) + \mathcal{O}(\varepsilon), \qquad\qquad (25.89)$$

which is just the travelling wave system for the scalar, bistable reaction–diffusion equation. Existence of the wave connecting $u = 0$ to $u = 1$ follows by appending $\theta' = 0$ in the usual manner to obtain transversality.

In order to implement the linearized stability analysis we need to check (in the notation of the previous section) that $A_\pm(\lambda, \varepsilon)$ have consistent splitting in $\Omega = \{\operatorname{Re} \lambda > \beta\}$ for some $\beta < 0$. It is easily seen that there is a $3/-3$ splitting for $\operatorname{Re} \lambda > \max\{f'(0), f'(1)\} = \beta$ for all sufficiently small ε, so that $\mathscr{E}(K, \varepsilon)$ is three-dimensional, for all appropriate $K \subset \Omega$. Thus, if $\mu_i^\pm(\lambda, \varepsilon)$ are the eigenvalues of $A_\pm(\lambda, \varepsilon)$, then

$$\operatorname{Re} \mu_i^\pm > 0 \quad (i = 1, 2, 3) \quad \text{and} \quad \operatorname{Re} \mu_i^\pm < 0 \quad \text{for } i = 4, 5, 6.$$

Furthermore, for λ in compact subsets of Ω the fast-slow splitting of $Y =$

(Y_S, Y_F) is $Y_S = (Y_1, Y_2)$ and $Y_F = (Y_3, \ldots, Y_6)$ so that $r = 2$. Thus, the eigenvalues of $A_\pm$ satisfy

$$\text{Re } \mu_i^\pm = \mathcal{O}(1) \quad \text{for } i = 1, 2, 5, 6,$$

$$\text{Re } \mu_i^\pm = \mathcal{O}(\varepsilon) \quad \text{for } i = 3, 4.$$

We can therefore expect a fast–slow decomposition $\mathscr{E}_S \oplus \mathscr{E}_F$ of $\mathscr{E}$ with $\dim \mathscr{E}_S = 1$ and $\dim \mathscr{E}_F = 2$. The general analysis of the fast summand via the projectivized flow (25.85) on $G_2(\mathbf{C}^6)$ is applicable here, showing that $\mathscr{E}_F(K, \varepsilon)$ exists and that $c_1(\mathscr{E}_F(K, \varepsilon)) = 0$ for all sufficiently small ε.

We next need to define the slow summand $\mathscr{E}_S$. The general prescription in Section C.2 for passing from the full system (25.86) to the reduced linearized system (25.87) is applicable here. In fact, a simple computation shows that (25.87), at $\varepsilon = 0$, is just

$$\begin{aligned}
\dot{Y}_1 &= Y_2, \\
B\dot{Y}_2 &= -\theta Y_2 - f'(u_R(\xi))Y_1,
\end{aligned} \tag{25.90}$$

where u_R is the connecting solution of (25.89) at $\varepsilon = 0$. This problem is just the linearization of (25.89). The stability of u_R was proved in [FM]; it follows from this that for $\text{Re } \lambda > \beta$, the only eigenvalue of (25.89) in the unstable half-plane is at $\lambda = 0$ and that this eigenvalue is simple. Hence, if K is any simple closed curve in Ω containing $\lambda = 0$ in its interior, $\mathscr{E}_S(K, \varepsilon)$, and hence $\mathscr{E}(K, \varepsilon)$, can be defined, and

$$c_1(\mathscr{E}(K, \varepsilon)) = c_1(\mathscr{E}_S(K, \varepsilon)) = 1$$

for all sufficiently small ε. This proves the (nonlinear) stability of the wave.

We remark that related work on stability of travelling waves, from a different point of view, for systems of two equations having small diffusion in one equation, can be found in [44], and the references therein.

References

[1] Alexander, J., R. Gardner, and C.K.R.T. Jones
   A topological invariant arising in the stability analysis of travelling waves. *J. Reine Angew. Math.*, **410** (1990), 167–212.
[2] Bates P. and C.K.R.T. Jones
   Invariant manifolds for semilinear partial differential equations. *Dynamics Reported*, **2** (1989), 1–38.
[3] Benci, V.
   A new approach to the Morse–Conley theory. In: *Recent Advances in Hamiltonian Systems*, edited by G.F. Dell'Antonio and B. D'Onofrio. World Scientific: Singapore, 1985, pp. 1–52.
[4] Caginalp, G. and P. Fife
   Higher-order phase field models and detailed anisotropy. *Phys. Rev. B.*, **34** (1986), 4940–4943.

[5] Conley, C. and E. Zehnder
The Birkhoff–Lewis fixed point theorem and a conjecture of V.I. Arnold. *Invent. Math.*, **73** (1983), 33–49.

[6] Conley, C. and J. Smoller
Bifurcation and stability of stationary solutions of the Fitz-Hugh–Nagumo equations. *J. Diff. Eqs.*, **63** (1986), 389–405.

[7] Conway, E., R. Gardner, and J. Smoller
Stability and bifurcation of steady-state solutions for predator–prey equations. *Adv. in Appl. Math.*, **3** (1982), 288–334.

[8] Dancer, E.N.
On non-radially symmetric bifurcation. *J. London Math. Soc.*, **20** (1979), 287–292.

[9] Ding Xiaxi, Chen Guiqiang, and Luo Peizhu
Convergence of the Lax–Friedrichs scheme for isentropic gas dynamics, I, II, III. *Acta Math. Sinica* (1985), 415–432, 433–472; (1986), 75–120.

[10] Ding Xiaxi, Chen Guiqiang, and Luo Peizhu
Convergence of the generalized Lax–Friedrichs scheme and Godunov's scheme for isentropic gas dynamics. *Comm. Math. Phys.*, **121** (1989), 63–84.

[11] DiPerna, R.
Convergence of approximate solutions of conservation laws. *Arch. Rat. Mech. Anal.*, **82** (1983), 27–70.

[12] DiPerna, R.
Convergence of the viscosity method for isentropic gas dynamics. *Comm. Math. Phys.*, **91** (1983), 1–30.

[13] Evans, J.W.
Nerve axon equations, I, II, III. *Ind. U. Math. J.*, **21** (1972), 877–895; **22** (1972), 75–90, 577–594.

[14] Evans, J.W.
Nerve axon equations, IV. *Ind. U. Math. J.*, **24** (1975), 1169–1190.

[15] Fenichel, N.
Persistence and smoothness of invariant manifolds for flows. *Ind. U. Math. J.*, **22** (1972), 577–594.

[16] Floer, A. and E. Zehnder
Fixed point results for sympletic maps related to the Arnold conjecture. *Proc. of Workshop in Dynamical Systems and Bifurcation*, Groningen (April 1984), pp. 16–19.

[17] Floer, A.
A refinement of the Conley index and an application to the stability of hyperbolic invariant sets. *Ergodic Theory Dynamical Systems*, **7** (1987), 93–103.

[18] Floer, A.
Proof of the Arnold conjecture for surfaces and generalizations for certain Kähler manifolds. *Duke Math. J.*, **53** (1986), 1–32.

[19] Franzosa, R.
Index filtrations and connection metrics for partially ordered Morse decompositions. *Trans. Amer. Math. Soc.*, **298** (1986), 193–213.

[20] Franzosa, R.
The connection matrix theory for Morse decompositions. *Trans. Amer. Math. Soc.*, **311** (1989), 561–592.

[21] Gardner, R.
Existence of travelling wave solutions of predator–prey systems via the connection index. *SIAM J. Appl. Math.*, **44** (1984), 56–79.

[22] Gardner, R.
On the detonation of a combustible gas. *Trans. Amer. Math. Soc.*, **277** (1983), 431–468.

[23] Gardner, R. and C.K.R.T. Jones
Travelling waves of a perturbed diffusion equation arising in a phase field model. *Ind. U. Math. J.*, **38** (1989), 1197–1222.

[24] Gardner, R. and C.K.R.T. Jones
A stability index for steady-state solutions of boundary-value problems for parabolic systems, *J. Diff. Eqns.*, **91** (1991), 181–203.

[25] Gardner, R. and C.K.R.T. Jones
Stability of travelling wave solutions of diffusive-predator–prey systems. *Trans. Amer. Math. Soc.*, **237** (1991), 465–524.

[26] Gidas, B., W.M. Ni, and L. Nirenberg
Symmetry of positive solutions of nonlinear elliptic equations in $\mathbf{R}^n$. *Comm. Math. Phys.*, **68** (1979), 202–243.

[27] Goodman, J.
Nonlinear asymptotic stability of viscous shock profiles for conservation laws. *Arch. Rat. Mech. Anal.*, **95** (1986), 325–344.

[28] Hastings, S.
On the existence of heteroclinic and periodic orbits for the Fitz-Hugh–Nagumo equations. *Quart. J. Appl. Math., Oxford Ser.* **27** (1976), 123–134.

[29] Helgason, S.
Topics in Harmonic Analysis on Homogeneous Spaces. Boston: Birkhauser, 1981.

[30] Hoff, D. and J. Smoller
Solutions in the large for certain nonlinear parabolic systems. *Ann. Inst. H. Poincaré*, **2** (1985), 213–235.

[31] Husemoller, D.
Fiber Bundles. Springer-Verlag: Berlin, 1966.

[32] Ilin, A.M. and O.A. Oleinik
Behavior of the solution of the Cauchy problem for certain quasilinear equations for unbounded increase of time. *Amer. Math. Soc. Transl., Ser. 2*, **42** (1964), 19–23.

[33] Jones, C.
Stability of travelling wave solutions for the Fitz-Hugh–Nagumo system. *Trans. Amer. Math. Soc.*, **286** (1984), 431–469.

[34] Jones, C.
Some ideas in the proof that the Fitz-Hugh–Nagumo pulse is stable. In: *Nonlinear Partial Differential Equations*, edited by J. Smoller, Contemp. Math., No. 17. Amer. Math. Soc.: Providence, 1983, pp. 287–292.

[35] Kawashima, S. and A. Matzumura
Asymptotic stability of travelling wave solutions of systems of one-dimensional gas motion. *Comm. Math. Phys.*, **101** (1985), 97–127.

[36] Keyfitz, B.L. and H.C. Kranzer (eds.)
Nonstrictly Hyperbolic Conservation Laws. Contemp. Math., No. 60, Amer. Math. Soc.: Providence, 1987.

[37] Langer, R.
Existence of homoclinic travelling wave solutions to the Fitz-Hugh–Nagumo equations. Ph.D. thesis, Northeastern University.

[38] Liu, T.P.
Nonlinear stability of shock waves for viscous conservation laws. Amer. Math. Soc. Memoir, No. 328. Amer. Math. Soc.: Providence, 1985.

[39] Liu, T.P.
Shock waves for compressible Navier–Stokes equations are stable. *Comm. Pure Appl. Math.*, **39** (1986), 565–594.

[40] Liu, T.P. and Z. Xin
Nonlinear stability of rarefaction waves for compressible Navier–Stokes equations. *Comm. Math. Phys.*, **118** (1986), 451–465.

[41] Mischaikow, K.
Conley's connection matrix. In: *Dynamics of Infinite Dimensional Dynamical Systems*, edited by S.-N. Chow and J. Hale, NATO ASI, Series F: Computers and Systems Sciences, #37. Springer-Verlag: New York, 1987.

[42] Mischaikow, K.
Classification of travelling wave solutions of reaction-diffusion systems. Lefschetz Center for Dynamical Systems, Report #86-5, (1985).

[43] Murat, F.
Capacité par compensation. *Ann. Scuola Norm. Pisa Sci. Fis. Mat.*, **5** (1978). 489–507.

[44] Nishiura, Y. and M. Mimura
Layer oscillations in reaction–diffusion systems. *SIAM J. Appl. Math.*, **49** (1989), 481–514.

[45] Reineck, J.
Connecting orbits in one-parameter families of flows. *J. Ergodic Theory Dynamical Systems*, **8** (1988), 359–374.

[46] Rybakowski, K.
On the homotopy index for infinite-dimensional semiflows. *Trans. Amer. Math. Soc.*, **269** (1982), 351–382.

[47] Rybakowski, K.
The Morse index, repellor–attractor pairs, and the connection index for semiflows on noncompact spaces. *J. Diff. Eqns.*, **47** (1983), 66–98.

[48] Rybakowski, K.
Trajectories joining critical points of nonlinear parabolic and hyperbolic partial differential equations. *J. Diff. Eqns.*, **51** (1984), 182–212.

[49] Rybakowski, K. and E. Zehnder
A Morse equation in Conley's index theory for semiflows on metric spaces. *Ergodic Theory Dynamical Systems*, **5** (1985), 123–143.

[50] Salamon, D.
Connected simple systems and the Conley index of isolated invariant sets. *Trans. Amer. Math. Soc.*, **291** (1985), 1–41.

[51] Smoller, J. and A. Wasserman
On the monotonicity of the time map. *J. Diff. Eqns.*, **77** (1989), 287–303.

[52] Smoller, J. and A. Wasserman
Symmetry, degeneracy, and universality in semilinear elliptic equations. *J. Funct. Anal.*, **89** (1990), 364–409.

[53] Smoller, J. and A. Wasserman
Bifurcation and symmetry-breaking. *Invent. Math.*, **100** (1990), 63–95.

[54] Tartar, L.
The compensated compactness method applied to systems of conservation laws. In: *Systems of Non-Linear Partial Differential Equations*, edited by J. Ball. Reidel: Dordrecht, 1983, pp. 263–288.

[55] Tartar, L.
Compensated compactness and applications to partial differential equations. In: *Nonlinear Analysis and Mechanics: Heriot-Watt Symposium*, IV, edited by R.J. Knops. Pitman: San Francisco, 1979, pp. 136–212.

[56] Temple, B.
Global solution of the Cauchy problem for a class of 2×2 nonstrictly hyperbolic conservation laws. *Adv. in Appl. Math.*, **3** (1982), 335–375.

[57] Temple, B.
Nonlinear conservation laws with invariant submanifolds. *Trans. Amer. Math. Soc.*, **280** (1983), 781–795.

[58] Temple, B.
Degenerate systems of conservation laws. In: *Nonstrictly Hyperbolic Conservation Laws*, edited by B. Keyfitz and H. Kranzer. Contemp. Math., No. 60, Amer. Math. Soc.: Providence, 1987, pp. 125–133.

[59] Temple, B.
Stability of Godunov's method for a class of 2×2 systems of conservation laws. *Trans. Amer. Math. Soc.*, **288** (1985), 115–123.

[60] Temple, B.
Systems of conservation laws with coinciding shock and rarefaction waves. *Contemporary Math.*, **17** (1983), 143–151.

[61] Temple, B.
Decay with a rate for noncompactly supported solutions of conservation laws. *Trans. Amer. Math. Soc.*, **298** (1986), 43–82.

[62] Young, L.C.
Lectures on the Calculus of Variations and Optimal Control Theory, W.B. Saunders, Philadelphia, PA (1969).

Bibliography

[Ag] Agmon, S.
  Lectures on Elliptic Boundary Value Problems. Van Nostrand Math. Studies, Vol. 2, Princeton, NJ, 1965.
[ADN] Agmon, S., A. Douglis, and L. Nirenberg
  Estimates near the boundary for solutions of elliptic partial differential equations satisfying general boundary conditions, I. *Comm. Pure Appl. Math.*, **12** (1959), 623–727.
[Al] Alikakos, N.
  1. An application of the invariance principle to reaction–diffusion equations. *J. Diff. Eqns.*, **33** (1979), 201–225.
  2. Remarks on invariance in reaction–diffusion equations. *Nonlinear Anal.*, **5** (1981), 593–614.
[Am] Amann, H.
  1. On the existence of positive solutions of nonlinear elliptic boundary value problems. *Ind. U. Math. J.*, **21** (1971), 125–146.
  2. Invariant sets and existence theorems for semilinear parabolic and elliptic systems. *J. Math. Anal. Appl.*, **65** (1978), 432–467.
[An] Anderson, J. E.
  MHD Shock Waves. MIT Press: Cambridge, MA, 1963.
[Ar] Aris, R.
  The Mathematical Theory of Diffusion and Reaction in Permeable Catalyst. Clarendon Press: Oxford, 1975.
[AW] Aronson, D. and H. Weinberger
  Nonlinear diffusion in population genetics, combustion, and nerve propagation. *Proc. Tulane Program in P.D.E.'s.* Springer Lecture Notes in Math., **446** (1975), 5–49.
[AN] Auchmuty, J. and G. Nicolis
  Dissipative structures, catastrophes, and pattern formation. *Proc. Nat. Acad. Sci., U.S.A.*, **71** (1974), 2748–2751.
[Ba] Ballou, D.
  1. Solutions to nonlinear hyperbolic Cauchy problems without convexity conditions. *Trans. Amer. Math. Soc.*, **152** (1970), 441–460.
  2. Weak solutions with a dense set of discontinuities. *J. Diff. Eqns.*, **10** (1971), 270–280.
  3. The structure and asymptotic behavior of compression waves. *Arch. Rat. Mech. Anal.*, **56** (1974), 170–182.
[BS] Bardos, C. and J. Smoller
  Instabilité des solutions stationaires pour des systèmes de reáction–diffusion. *C. R. Acad. Sci.*, (A), **285** (1977), 249–253.
[BMS] Bardos, C., H. Matano, and J. Smoller
  Some results on the instability of solutions of systems of reaction–diffusion equations. *1er Colloque AFCET-SMF de Math. Appl., Ecole Polytechnique*, t. **II**, (1979), 297–304.
[BCF] Bebernes, J., K. Chueh, and W. Fulks
  Some applications of invariance for parabolic systems. *Ind. U. Math. J.*, **28** (1979), 269–277.

[BB] Berger, M. and M. Berger
      Perspectives in Nonlinearity. Benjamin: New York, 1968.
[BJS] Bers, L., F. John, and M. Schecter
      Partial Differential Equations. Wiley-Interscience: New York, 1964.
[Be] Bethe, H.
      1. Office of Scientific Research and Development, Division B. Report No. 545, 1942.
      2. Report on the Theory of Shock Waves for an Arbitrary Equation of State, Report
         No. PB-32189. Clearinghouse for Federal Scientific and Technical Information,
         U.S. Dept. of Commerce: Washington, DC, 1942.
[Bo] Bochner, S.
      Vorlesungen über Fouriersche Integrale. Akademische Verlag: Leipzig, 1932.
[Bv] Borovikov, V.
      On the problem of discontinuity decay for a system of two quasilinear equations.
      Dokl. Akad. Nauk., SSSR, **185** (1969), 250–252; English transl. in *Sov. Math.,
      Dokl.,* **10** (1969), 321–323.
[BP] Brown, A. and A. Page
      Elements of Functional Analysis. Van Nostrand: London, 1970.
[BDG] Brown, K., P. Donne, and R. Gardner
      A semilinear parabolic system arising in the theory of superconductivity. *J. Diff.
      Eqns.,* **40** (1981), 232–252.
[Bu] Burgers, J.
      Application of a model system to illustrate some points of the statistical theory of
      free turbulence. *Nederl. Akad. Wefensh. Proc.,* **43** (1940), 2–12.
[By] Byhovskii, E.
      Artificial viscosity and automodel solutions of equations for isothermal gas flow.
      Vest. Leningrad. Univ., **23** (1968), 5–23; English transl. in *Vest. Leningrad Math.,* **1**
      (1974), 257–277.
[Ca] Carpenter, G.
      1. A geometric approach to singular perturbation problems with applications to nerve
         impulse equations. *J. Diff. Eqns.,* **23** (1977), 335–367.
      2. Periodic solutions of nerve impulse equations, *J. Math. Anal. Appl.,* **58** (1977),
         152–173.
[CaH] Casten, R. and C. Holland
      Instability results for reaction–diffusion equations with Neumann boundary
      conditions, *J. Diff. Eqns.,* **27** (1978), 266–273.
[Cf] Chafee, N.
      Asymptotic behavior for solutions of a one-dimensional parabolic equation
      with homogeneous Neumann boundary conditions. *J. Diff. Eqns.,* **18** (1975), 111–
      135.
[CI] Chafee, N. and E. Infante
      A bifurcation problem for a nonlinear parabolic equation, *Appl. Anal.,* **4** (1974),
      17–37.
[CW] Chow, P. and S. Williams
      Nonlinear reaction–diffusion models. *J. Math. Anal. Appl.,* **62** (1978), 151–169.
[Ch] Chueh, K.
      A compact positively invariant set of solutions of the Nagumo equation. *J. Diff.
      Eqns.,* **28** (1978), 35–42.
[CCS] Chueh, K., C. Conley, and J. Smoller
      Positively invariant regions for systems of nonlinear diffusion equations. *Ind. U.
      Math. J.,* **26** (1977), 373–392.
[Chr] Churchill, R.
      Isolated invariant sets in compact metric spaces. *J. Diff. Eqns.,* **12** (1970), 330–352.
[CL] Coddington, E. and N. Levinson
      Theory of Ordinary Differential Equations. McGraw-Hill: New York, 1955.
[Co] Cole, J.
      On a quasilinear parabolic equation occurring in aerodynamics. *Q. Appl. Math.,* **9**
      (1951), 225–236.
[Cy] Conley, C.
      1. On travelling wave solutions of nonlinear diffusion equations. *Dynamical Sys-
         tems. Theory and Applications.* Springer Lecture Notes in Physics, **38** (1975), 498–
         510.

  2. *Isolated Invariant Sets and the Morse Index.* Conf. Board Math. Sci., No. 38. Amer. Math. Soc.: Providence, 1978.
[CE] Conley C. and R. Easton
     Isolated invariant sets and isolating blocks. *Trans. Amer. Math. Soc.*, **158** (1971), 35–61.
[CG] Conley, C. and R. Gardner
     An application of the generalized Morse index to travelling wave solutions of a competitive reaction–diffusion model. *Ind. U. Math. J.*, **33** (1984).
[CS] Conley, C. and J. Smoller
  1. Viscosity matrices for two-dimensional nonlinear hyperbolic systems. *Comm. Pure Appl. Math.*, **23** (1970), 867–884.
  2. Shock waves as limits of progressive wave solutions of higher-order equations. *Comm. Pure Appl. Math.*, **24** (1971), 459–472.
  3. Shock waves as limits of progressive wave solutions of higher-order equations, II. *Comm. Pure Appl. Math.*, **25** (1972), 131–146.
  4. Viscosity matrices for two-dimensional nonlinear hyperbolic systems, II. *Amer. J. Math.*, **94** (1972), 631–650.
  5. Topological methods in the theory of shock waves. *Proc. Sympos. Pure Math.*, **23** (1973), 293–302. Amer. Math. Soc.: Providence.
  6. On the structure of magnetohydrodynamic shock waves. *Comm. Pure Appl. Math.*, **28** (1974), 367–375.
  7. The mhd version of a theorem of H. Weyl. *Proc. Amer. Math. Soc.*, **42** (1974), 248–250.
  8. On the structure of magnetohydrodynamic shock waves, II. *J. Math. Pures et Appl.*, **54** (1975), 429–444.
  9. The existence of heteroclinic orbits and applications. *Dynamical Systems, Theory, and Applications. Proc. Battelle Rencontres, 1974.* Springer Lecture Notes in Physics, **38** (1974), 498–510.
 10. Remarks on travelling wave solutions of nonlinear diffusion equations. *Proc. Battelle Sympos. Catastrophe Theory.* Springer Lecture Notes in Math., **525** (1975), 77–89.
 11. Isolated invariant sets of parametrised systems of differential equations. *The Structure of Attractors in Dynamical Systems.* Springer Lecture Notes in Math., **668** (1978), 30–47.
 12. Remarks on the stability of steady-state solutions of reaction–diffusion equations. In: *Bifurcation Phenomena in Mathematical Physics and Related Phenomena*, edited by C. Bardos and D. Bessis. Reidel: Dodrecht, 1980, pp. 47–56.
 13. Topological techniques in reaction–diffusion equations. *Biological Growth and Spread.* Springer Lecture Notes in Biomath., **38** (1980) 473–483.
 14. Bifurcation and stability of stationary solutions of the Fitz-Hugh–Nagumo equations. *J. Diff. Eqns.*, **63** (1986), 389–405.
[CZ] Conley, C. and E. Zehnder
     Morse type index theory for flows and periodic solutions for Hamiltonian equations. *Comm. Pure Appl. Math.*, **37** (1984), 207–254.
[Cn] Conlon, J.
     A theorem in ordinary differential equations with an application to hyperbolic conservation laws. *Adv. in Maths.*, **35** (1980), 1–18.
[Coy] Conway, E.
     The formation and decay of shocks of a conservation law in several dimensions. *Arch. Rat. Mech. Anal.*, **64** (1977), 47–57.
[CoH] Conway, E. and E. Hopf
     Hamilton's theory and generalized solutions of the Hamilton–Jacobi equation. *J. Math. Mech.*, **13** (1964), 939–986.
[CHS] Conway, E., D. Hoff, and J. Smoller
     Large time behavior of solutions of systems of nonlinear reaction–diffusion equations. *SIAM J. Appl. Math.*, **35** (1978), 1–16.
[CR] Conway, E. and S. Rosencrans
     The Riemann problem in gas dynamics, (1969). Unpublished.
[CoS] Conway, E. and J. Smoller
  1. Global solutions of the Cauchy problem for quasilinear first-order equations in several space variables. *Comm. Pure Appl. Math.*, **19** (1966), 95–105.

2. Shocks violating Lax' conditions are unstable. *Proc. Amer. Math. Soc.*, **39** (1973), 253–256.
3. A comparison theorem for systems of reaction–diffusion equations. *Comm. Partial Diff. Eqns.*, **2** (1977), 679–697.
4. Diffusion and the predator–prey interaction. *SIAM J. Appl. Math.*, **33** (1977), 673–686.
5. Diffusion and the classical ecological interactions. In: *Nonlinear Diffusion*, edited by W. Fitzgibbon and H. Walker. Research Notes in Math., No. 14. Pitman: London, 1977.

[CF] Courant, R. and K. Friedrichs
Supersonic Flow and Shock Waves, Wiley-Interscience: New York, 1948.

[CH] Courant, R. and D. Hilbert
Methods of Mathematical Physics, Vols. 1, 2. Wiley-Interscience: New York, 1962.

[Cr] Crandall, M.
1. The semigroup approach to first-order quasilinear equations in several space variables. *Israel J. Math.*, **12** (1972), 108–132.
2. An introduction to constructive aspects of bifurcation theory and the implicit function theorem. In: *Applications of Bifurcation Theory*, edited by P. Rabinowitz. Academic Press: New York, 1977, pp. 1–35.

[CM] Crandall, M. and A. Majda
Monotone difference approximations to scalar conservation laws. *Math. Comp.*, **34** (1980), 1–21.

[CRa] Crandall, M. and P. Rabinowitz
1. Bifurcation, perturbation of simple eigenvalues and linearized stability. *Arch. Rat. Mech. Anal.*, **52** (1973), 161–181.
2. Mathematical theory of bifurcation. In: *Bifurcation Phenomena in Mathematical Physics and Related Topics*, edited by C. Bardos and D. Bessis. Reidel: Dodrecht, 1980.

[Df] Dafermos, C.
1. The entropy rate admissibility criterion for solutions of hyperbolic conservation laws. *J. Diff. Eqns.*, **14** (1973), 202–212.
2. Solution of the Riemann problem for a class of hyperbolic systems of conservation laws by the viscosity method. *Arch. Rat. Mech. Anal.*, **52** (1973), 1–9.
3. Structure of solutions of the Riemann problem for hyperbolic systems of conservation laws. *Arch. Rat. Mech. Anal.*, **53** (1974), 203–217.
4. Generalized characteristics and the structure of solutions of hyperbolic conservation laws. *Ind. U. Math. J.*, **26** (1977), 1097–1119.

[Da] Dancer, E.
Global solution branches for positive mappings. *Arch. Rat. Mech. Anal.*, **53** (1973), 181–192.

[Di] Dieudonné, J.
Foundations of Modern Analysis. Academic Press: New York, 1960.

[DL] Ding Shia-Shi, Chang-Tung, Wang Ching-hua, Hsiao-Ling, and Li Tsai-Chung
A study of the global solutions for quasilinear hyperbolic systems of conservation laws. *Scientica Sinica*, **16** (1973), 317–335.

[Du] Dinu, L.
Teoria Undelor de Şoc în Plasmă. Editura ştiinţifică şi enciclopedică, Bucureşti, 1976, (in Roumanian).

[Dp] DiPerna, R.
1. Existence in the large for nonlinear hyperbolic conservation laws. *Arch. Rat. Mech. Anal.*, **52** (1973), 244–257.
2. Decay and asymptotic behavior of solutions to nonlinear hyperbolic systems of conservation laws. *Ind. U. Math. J.*, **24** (1975), 1047–1071.
3. Singularities of solutions of nonlinear hyperbolic systems of conservation laws. *Arch. Rat. Mech. Anal.*, **60** (1975), 75–100.
4. Decay of solutions of hyperbolic systems of conservation laws with a convex extension. *Arch. Rat. Mech. Anal.*, **64** (1977), 1–46.
5. Uniqueness of solutions to hyperbolic conservation laws. *Ind. U. Math. J.*, **28** (1979), 244–257.
6. Convergence of approximate solutions of conservation laws. *Arch. Rat. Mech. Anal.*, **82** (1983), 27–70.

[Do] Douglis, A.
1. An ordering principle and generalized solutions of certain quasilinear partial differential equations. *Comm. Pure Appl. Math.*, **12** (1959), 87–112.
2. Layering methods for nonlinear partial differential equations of first order. *Ann. Inst. Fourier, Grenoble*, **22** (1972), 141–227.

[Ep] Ehrenpreis, L.
Solutions of some problems of division, I. *Amer. J. Math.*, **76** (1954), 883–903.

[FPU] Fermi, E., J. Pasta, and S. Ulam
Los Alamos Scientific Laboratory Report No. LA-1940, May 1955.

[Fi] Fife, P.
Mathematical Aspects of Reacting and Diffusing Systems. Springer Lecture Notes in Biomath., **28** (1979)

[FM] Fife, P. and J. McLeod
The approach of solutions of nonlinear diffusion equations to travelling front solutions. *Arch. Rat. Mech. Anal.*, **65** (1977), 335–361.

[FT] Fife, P. and M. Tang
Comparison principles for reaction–diffusion systems: irregular comparison functions and applications to questions of stability and speed of propagation of disturbances. *J. Diff. Eqns.*, **40** (1981), 168–185.

[Fl] Flanders, H.
Differential Forms. Academic Press: New York, 1963.

[Fk] Flashka, H.
Nonlinear semi-groups and a hyperbolic conservation law. Carnegie Mellon University. 1972. Unpublished.

[Fo] Foy, R.
Steady-state solutions of hyperbolic systems of conservation laws with viscosity terms. *Comm. Pure Appl. Math.*, **17** (1964), 177–188.

[Fn] Friedman, A.
1. Remarks on the maximum principle for parabolic equations and its applications. *Pacific J. Math.*, **8** (1958), 201–211.
2. A strong maximum principle for subparabolic equations. *Pacific J. Math.*, **11** (1961), 175–184.
3. *Partial Differential Equations of Parabolic Type.* Prentice-Hall: Englewood Cliffs, NJ, 1964.

[Fr] Friedrichs, K.
1. On differential operators in Hilbert space, *Amer. J. Math.*, **61** (1939), 523–544.
2. The identity of weak and strong extensions of differential operators, *Trans. Amer. Math. Soc.*, **55** (1944), 132–151.

[FL] Friedrichs, K. and P. Lax
Systems of conservation laws with a convex extension, *Proc. Nat. Acad. Sci., U.S.A.*, **68** (1971), 1686–1688.

[Ga] Garabedian, P.
Partial Differential Equations. Wiley: New York, 1964.

[Gå] Gårding, L.
Linear hyperbolic partial differential equations with constant coefficients. *Acta Math.*, **85** (1950), 1–62.

[Gd] Gardner, R.
1. Asymptotic behavior of semilinear reaction–diffusion systems with Dirichlet boundary conditions. *Ind. U. Math. J.*, **29** (1980), 161–190.
2. Global stability of stationary solutions of reaction–diffusion systems. *J. Diff. Eqns.*, **37** (1980), 60–69.
3. Comparison and stability theorems for reaction–diffusion systems. *SIAM J. Math. Anal.*, **12** (1980), 603–616.
4. Existence and stability of travelling wave solutions of competition models, a degree theoretic approach. *J. Diff. Eqns.*, **44** (1982), 343–364.
5. Existence of travelling wave solutions of predator–prey systems via the connection index. *SIAM J. Appl. Math.*, **44** (1984), 56–79.

[GS] Gardner, R. and J. Smoller
The existence of periodic travelling waves for singularly perturbed predator–prey equations via the Conley index. *J. Diff. Eqns.* **47** (1983), 133–161.

[Ge] Gelfand, I.
Some problems in the theory of quasilinear equations. *Usp. Mat. Nauk.*, **14** (1959), 87–158; English transl. in *Amer. Math. Soc. Transl.* ser 2, (1963), 295–381.

[Gr] Germain, P.
Contribution à la théorie des ondes de choc en magnétodynamique des fluides. ONERA Publ. No. 97, (1959).

[GB] Germain, P. and R. Bader
Unicité des écoulements avec choc dans la mechanique de Burgers. Office Nationale d'Etude et de Recherches Aeronautiques: Paris, 1953, pp. 1–13.

[Gi] Gilbarg, D.
The existence and limit behavior of the one-dimensional shock layer. *Amer. J. Math.*, **73** (1951), 256–274.

[GT] Gilbarg, D. and N. Trudinger
Elliptic Partial Differential Equations of Second Order. Springer-Verlag: Berlin, 1977.

[Gl] Glimm, J.
Solutions in the large for nonlinear hyperbolic systems of equations. *Comm. Pure Appl. Math.*, **18** (1965), 697–715.

[GL] Glimm, J. and P. Lax
Decay of solutions of systems of nonlinear hyperbolic conservation laws. Amer. Math. Soc. Memoir, No. 101. Amer. Math. Soc.: Providence, 1970.

[Go] Godunov, S.
An interesting class of quasilinear systems. *Dokl. Akad. Nauk, SSSR*, **139** (1961), 521–523. English transl. in *Sov. Math*, **2** (1961), 947–949.

[Gf] Goffman, C.
Two remarks on linearly continuous functions. *J. Math. Mech.*, **16** (1967), 1227–1228.

[GP] Goffman, C. and G. Pedrick
A First Course in Functional Analysis. Prentice Hall: Englewood Cliffs, NJ, 1955.

[GoS] Golubitsky, M. and D. Schaeffer
Imperfect bifurcation in the presence of symmetry. *Comm. Math. Phys.*, **67** (1979), 205–232.

[Go] Gorin, E.
Asymptotic properties of polynomials and algebraic functions of several variables. *Usp. Mat. Nauk. (N.S.)*, **16** (1961), 93–118; English transl. in *Russian Math. Surveys*, **16** (1961), 93–119.

[Gb] Greenberg, J.
1. On the interaction of shocks and simple waves of the same family. *Arch. Rat. Mech. Anal.*, **37** (1970) 136–160.
2. On the elementary interactions for the quasilinear ware equation. *Arch. Rat. Mech. Anal.*, **43** (1971), 325–349.
3. Asymptotic behavior of solutions to the quasilinear wave equations. *Partial Differential Equations and Related Topics.* Springer Lecture Notes in Math., **446** (1975), 198–246.
4. Periodic solutions to reaction–diffusion equations. *SIAM J. Appl. Math*, **30** (1976), 199–205.

[HV] Hale, J. and J. Vegas
A nonlinear parabolic equation with varying domain. LCDS Report 81–1, Brown Univ. (1980). *Arch. Rat. Mech. Anal.*, **86** (1984), 99–123.

[HKW] Hassard, B., N. Kazarinoff, and Y. Wan
Theory and Applications of Hopf Bifurcation, Cambridge University Press: Cambridge, 1981.

[Ha] Hastings, S.
1. The existence of periodic solutions to Nagumo's equation. *Quart. J. Appl. Math., Oxford,* Ser. 3, **25** (1974), 369–378.
2. On travelling wave solutions of the Hodgkin–Huxley equations. *Arch. Rat. Mech. Anal.*, **60** (1976), 229–257.

[HM] Hastings, S. and J. Murray
The existence of oscillatory solutions in the Field–Noyes model for the Belousov–Zhabotinskii reaction. *SIAM J. Appl. Math.*, **28** (1975), 678–688.

[Hy] Hayes, W.
    Gasdynamic Discontinuities. Princeton University Press: Princeton, 1960.

[Hn] Heinz, E.
    An elementary analytic theory of degree in n-dimensional space. *J. Math. Mech.*, **8**
    (1959), 231–247.

[Hg] Heisenberg, W.
    Nonlinear problems in physics. In: *"Topics in Nonlinear Physics,"* edited by N. J.
    Zabusky. Springer-Verlag: Berlin–Heidelberg–New York, 1968.

[He] Henry, D.
    Geometric Theory of Semilinear Parabolic Equations. Springer-Verlag:
    Berlin–Heidelberg–New York, 1981.

[Hz] Hernandez J.
    1. Some existence and stability results for solutions of nonlinear reaction–diffusion
       systems with nonlinear boundary conditions. In: *Nonlinear Differential Equations:*
       Invariance, Stability, and Birfurcation, edited by P. Mottoni and L. Salvadori.
       Academic Press: Toronto, 1980.
    2. Positive solutions of reaction–diffusion systems with nonlinear boundary con-
       ditions and the fixed-point index. In: *Nonlinear Phenomena in Mathematical Sciences,*
       edited by V. Lakshmikantham. Academic Press: New York, 1982, pp. 525–535.

[Hki] Hesaraaki, M.
    Magnetohydrodynamic shock waves. Amer. Math. Soc. Memoir, No. 302, Amer.
    Math. Soc.: Providence, 1984.

[HP] Hille, E. and R. Phillips
    Functional Analysis and Semigroups. Amer. Math. Soc. Colloq. Publ. **31**. Amer.
    Math. Soc.: Providence, (1957).

[Hi] Hilton, P.
    General Cohomology Theory and K-Theory. London Math. Soc. Lecture Notes, Vol.
    1. Cambridge University Press: Cambridge, 1971.

[Hof] Hoff, D.
    1. A characterization of the blow-up time for the solution of a conservation law in
       several space variables. *Comm. Partial Diff. Eqns.,* 7 (1982), 141–151.
    2. Stability and convergence of finite difference methods for systems of nonlinear
       reaction–diffusion equations. *SIAM J. Numer. Anal.,* **15** (1978), 1161–1177.

[HS] Hoff, D. and J. Smoller
    1. Solutions in the large for certain nonlinear parabolic systems. *Analyse Non Linéaire,*
       2 (1985), 213–235.
    2. Global existence for systems of parabolic conservation laws in several space vari-
       ables. *J. Diff. Eqns.,* **68** (1987), 210–220.

[Hf] Hopf, E.
    1. Elementare bemerkungen über die lösungen partieller differentialgleichungen
       zweiter ordnung vom elliptischen typus. *Sitber. preuss. Akad. Wiss., Berlin,* **19**
       (1927), 141–152.
    2. The partial differential equation $u_t + uu_x = \mu\mu_{xx}$. *Comm. Pure Appl. Math.,* 3 (1950),
       201–230.

[Ho] Hörmander, L.
    1. Local and global properties of fundamental solutions. *Math. Scand.,* **5** (1957),
       27–39.
    2. *Linear Partial Differential Operators.* Academic Press, Springer-Verlag· New York,
       1963.
    3. On the division of distributions by polynomials. *Ark. Mat.,* 3 (1958), 555–568.

[HK] Howard., L. and N. Kopell
    Plane wave solutions to reaction–diffusion equations. *Stud. Appl. Math.,* **52** (1973),
    291–328.

[Hs] Hsiao, L.
    The entropy rate admissibility criterion in gas dynamics. *J. Diff. Eqns.,* **38** (1980),
    226–238.

[Hu] Hu, S.
    Homotopy Theory. Academic Press: New York, 1959.

[IKO] Ilin, A., A. Kalashnikov, and O. Oleinik
    Second-order equations of parabolic type. *Usp. Mat. Nauk.,* **27** (1962), 3–146;
    English transl. in *Russ. Math. Surveys,* **17** (1962), 1–144.

[It] Itaya, N.
On the Cauchy problem for the system of fundamental equations describing the movement of a compressible fluid. *Kodai Math. Sem. Rep.*, **23** (1971), 60–120.

[Is] Isaacson, E.
Global solution of the Riemann problem for nonstrictly hyperbolic systems of conservation laws arising in enhanced oil recovery. Rockefeller University. *J. Comp. Phys.*, to appear.

[J] John, F.
1. Partial Differential Equations. New York Univ. Lecture Notes, 1952–1953.
2. Formation of singularities in one-dimensional nonlinear wave propagation. *Comm. Pure Appl. Math.*, **27** (1974), 337–405.

[K] Kanel', Ya.
1. On some systems of quasilinear parabolic equations. *Zh., Vychislit., Matem. i Matem. Fiziki*, **6** (1966), 466–477; English transl. in *USSR Comp. Math. and Math. Physics*, **6** (1966), 74–88.
2. On a model system of equations of one-dimensional gas motion. *Diff. Ural.*, **4** (1968), 721–734; English transl. in *Diff. Eqns.*, **4** (1968), 374–380.

[KN] Kawashima, S. and T. Nishida
The initial-value problems for the equations of viscous compressible and perfect compressible fluids. RIMS, Kokyuroku 428, Kyoto University. Nonlinear Functional Analysis, June 1981, pp. 34–59.

[KS] Kazhikhov, A. and V. Shelukhin
Unique global solution in time of initial-boundary-value problems for one-dimensional equations of a viscous gas. *Prikl. Mat. Mech.*, **41** (1977), 282–291; English transl. in *Prikl. Mat. Mech., J. Appl. Meth. Mech.*, **41** (1977), 273–282.

[KC] Keller, H. and D. Cohen
Some positone problems suggested by nonlinear heat generation. *J. Math. Mech.*, **16** (1967), 1361–1376.

[KK] Keyfitz, B. and H. Kranzer
Existence and uniqueness of entropy solutions to the Riemann problem for hyperbolic systems of two nonlinear conservation laws. *J. Diff. Eqns.*, **27** (1978), 444–476.

[Ko] Kolmogroff, A.
Sulla teoria di Volterra della lotta per l'esistenza. *Gioru. Institute Ital. Attuari*, **7** (1936), 74–80.

[KPP] Kolmogoroff, A., I. Petrovskii, and N. Piscunov
A study of the equation of diffusion with increase in the quantity of matter, and its application to a biological problem. *Bull. Univ. Moscow, Ser. Internat., Sec. A.*, **1**, No. 6 (1937), 1–25 (in Russian).

[Kop] Kopell, N.
Waves, shocks, and target patterns in an oscillating chemical reagent. In: *Nonlinear Diffusion*, edited by W. Fitzgibbon and H. Walker. Research Notes in Math., No. 14, Pitman: London, 1977, pp. 129–154.

[Kw] Kotlow, D.
Quasilinear parabolic equations and first-order quasilinear conservation laws with bad Cauchy data. *J. Math. Anal. Appl.*, **35** (1971), 563–576.

[Kr] Kransnosel'skii, M.
Topological Methods in the Theory of Nonlinear Intergral Equations. Macmillan: New York, 1964.

[KR] Krein, M. and M. Rutman
Linear operators leaving invariant a cone in a Banach space. *Usp. Mat. Nauk, (N.S.)*, **3**, (1948), 59–118. English transl in *Amer. Math. Soc. Transl., Ser.* 1, **10** (1962), 199–325.

[Kv] Kruskov, S.
1. A priori estimates for the derivative of a solution to a parabolic equation and some of its applications. *Dokl. Akad. Nauk., SSSR,* **170** (1966), 501–504; English transl. in *Sov. Math. Dokl.*, **7** (1966), 1215–1218.
2. First-order quasilinear equations in several independent variables, *Mat. Sb.*, **123** (1970). 228–255; English transl. in *Math. USSR Sb.*, **10** (1970), 217–273.

[Kp] Kuiper, H.
   1. Existence and comparison theorems for nonlinear diffusion systems. *J. Math. Anal. Appl.*, **60** (1977), 166–181.
   2. Invariant sets for nonlinear elliptic and parabolic systems. *SIAM J. Math. Anal.*, **11** (1980), 1075–1103.
[Ku N] Kuipen, L. and H. Niederreiter
   Uniform Distribution of Sequences. Wiley: New York, 1974.
[KL] Kulikovskiy, A. and G. Lyubimov
   Magnetohydrodynamics. Addison-Wesley: Reading. MA, 1965.
[Ku] Kuznetsov, N.
   Weak solutions of the Cauchy problem for a multi-dimensional quasilinear equation. *Mat. Zam.*, **2** (1967), 401–410; English transl. in *Math. Notes of Acad. USSR*, **2** (1967), 733–739.
[L] Ladyzenskaya, O.
   On the construction of discontinuous solutions of quasilinear hyperbolic equations as a limit of solutions of the corresponding parabolic equations when the "viscosity coefficient" tends to zero. *Dokl. Akad. Nauk, SSSR*, **111** (1956), 291–294 (in Russian).
[LL] Landau, L. and E. Lifshitz
   Fluid Mechanics. Pergamon Press: Oxford, 1959.
[Lx] Lax, P.
   1. Weak solutions of nonlinear hyperbolic equations and their numerical computation. *Comm. Pure Appl. Math.*, **7** (1954), 159–193.
   2. Hyperbolic systems of conservation laws, II. *Comm. Pure Appl. Math.*, **10** (1957), 537–566.
   3. Development of singularities of solutions of nonlinear hyperbolic partial differential equations. *J. Math. Phys.*, **5** (1964), 611–613.
   4. Shock waves and entropy. In: *Contributions to Nonlinear Functional Analysis*, edited by E. Zarantonello. Academic Press: New York, 1971, pp. 603–634.
   5. The formation and decay of shock waves. *Amer. Math. Monthly*, **79** (1972), 227–241.
   6. Hyperbolic systems of conservation laws and the mathematical theory of shock waves. *Conf. Board Math. Sci.*, **11**, SIAM, 1973.
[Le] Lewy, H.
   An example of a smooth linear partial differential equation without solution. *Ann. Math.*, **66** (1957), 155–158.
[Lu] Liu, T.
   1. Shock waves in the nonisentropic gas flow. *J. Diff. Eqns.*, **22** (1976), 442–452.
   2. The deterministic version of the Glimm scheme. *Comm. Math. Phys.*, **57** (1977), 135–148.
   3. Large time behavior of initial and initial-boundary-value problems of general systems of hyperbolic conservation laws. *Comm. Math. Phys.*, **55** (1977), 163–177.
   4. Linear and nonlinear large time behavior of general systems of hyperbolic conservation laws. *Comm. Pure. Appl. Math.*, **30** (1977), 767–796.
   5. Solutions in the large for the equations of nonisentropic gas dynamics. *Ind. U. Math. J.*, **26** (1977), 147–177.
   6. Initial-boundary-value problems for gas dynamics, *Arch. Rat. Mech. Anal.*, **64** (1977), 137–168.
   7. Decay to *N*-waves of solutions of general systems of nonlinear hyperbolic conservation laws. *Comm. Pure Appl. Math.*, **30** (1977), 585–610.
   8. Development of singularities in the nonlinear waves for quasilinear hyperbolic partial differential equations. *J. Diff. Eqns.*, **33** (1979), 92–111.
   9. Admissible solutions of hyperbolic conservation laws. Amer. Math. Soc. Memoir, No. 240. Amer. Math. Soc.: Providence, 1981.
   10. Uniqueness of weak solutions of the Cauchy problem for general 2 × 2 conservation laws. *J. Diff. Eqns.*, **20** (1976), 369–388.
[LS] Liu, T. and J. Smoller
   The vacuum state in isentropic gas dynamics. *Adv. Appl. Math.*, **1** (1980), 345–359.
[LuS] Liusternik, L. and V. Sobolev
   Elements of Functional Analysis. Wiley: New York, 1974.

[Mg] Maginu, K.
  1. Stability of periodic travelling wave solutions of a nerve conduction equation. *J. Math. Biol.*, **6** (1978), 49–57.
  2. Stability of spatially homogeneous periodic solutions of reaction–diffusion equations. *J. Diff. Eqns.*, **31** (1979), 130–138.
  3. Existence and stability of periodic travelling wave solutions to Nagumo's nerve equation. *J. Math. Biol.*, **10** (1980), 133–153.
  4. Stability of periodic travelling wave solutions with large spatial periods in reaction–diffusion systems. *J. Diff. Eqns.*, **39** (1981), 73–99.

[Ma] Malgrange, B.
  Existence et approximation des solutions des équations aux dérivées partielles et des équations de convolution. *Ann. Inst. Fourier Grenoble*, **6** (1955/56), 271–355.

[MM] Marsden, J. and M. McCracken
  The Hopf Bifurcation and Its Applications. Springer-Verlag: New York, 1976.

[Ms] Markus, L.
  Asymptotically autonomous differential systems. *Contribution to the Theory of Nonlinear Oscillations*, Vol. 3. Annals of Math. Studies, No. 36. Princeton University Press: Princeton, 1956, pp. 17–29.

[Mo] Matano, H.
  Asymptotic behavior and stability of solutions of semilinear diffusion equations. *Publ. RIMS, Kyoto Univ.*, **15** (1979), 401–454.

[MN] Matsumura, A. and T. Nishida
  1. The initial-value problem for the equations of motion of compressible viscous and heat conducting fluids. *Proc. Japan Acad., Ser. A.*, **55** (1979), 337–342.
  2. The initial-value problem for the equations of motion of viscous and heat conductive gases. *J. Math. Kyoto Univ.*, **20** (1980), 67–104.

[My] May, R.
  Stability and Complexity in Model Ecosystems, 2nd ed. Monographs in Population Biology. Princeton University Press: Princeton, 1974.

[Mr] Milnor, J.
  1. On manifolds homeomorphic to the 7-sphere, *Ann. Math.*, **64** (1956), 399–405.
  2. *Morse Theory.* Princeton University Press: Princeton, 1963.

[MiN] Mimura, M. and Y. Nishiura
  Spatial patterns for an interaction–diffusion equation in morphogenesis. *J. Math. Biol.*, **7** (1979), 243–263.

[MNY] Mimura, M., Y. Nishiura, and M. Yamaguti
  Some diffusive predator–prey systems and their bifurcation problems. *Ann. New York Acad. Sci.*, **316** (1979), 490–510.

[Mk] Mock, M.
  A topological degree for orbits connecting critical points of autonomous systems. *J. Diff. Eqns.*, **38** (1980) 176–191.

[MS] Moler, C. and J. Smoller
  Elementary interactions in quasilinear hyperbolic systems. *Arch. Rat. Mech. Anal.*, **37** (1970), 309–322.

[Mt] Montgomery, J.
  Cohomology of invariant sets under perturbation. *J. Diff. Eqns.*, **13** (1973), 257–299.

[Mx] Mora, X.
  Semilinear parabolic problems define semiflows on C^k-spaces. *Trans. Amer. Math. Soc.*, **238** (1983), 21–55.

[Mc] Morrey, C.
  Multiple Integrals in the Calculus of Variations. Springer-Verlag: New York, 1966.

[Mu] Murray, J.
  Lectures on Nonlinear Differential Equation Models in Biology. Clarendon Press: Oxford, 1977.

[Na] Nash, J.
  Le problème de Cauchy pour des équations différentielles d'un fluide générale. *Bull. Soc. Math. France*, **90** (1962), 487–497.

[Nt] Natanson, I.
  Theory of Functions of a Real Variable Vol. 1. Ungar: New York, 1955.

[NeS] Nemytskii, V. and V. Stepanov
    Qualitative Theory of Differential Equations. Princeton University Press: Princeton, 1960.

[Nn] von Neumann, J.
    1. Theory of shock waves. *Collected Works,* Vol. 6. Pergamon Press: Oxford, 1963.
    2. *Ibid.,* Vol. 5.

[Ni] Nirenberg, L.
    1. A strong maximum principle for parabolic equations. *Comm. Pure Appl. Math.,* **6** (1956), 167–177.
    2. Topics in nonlinear functional analysis. *Courant Inst. Math. Sci. Notes* (1972/73).
    3. Variational and topological methods in nonlinear problems. *Bull. Amer. Math. Soc. (NS),* **4** (1981), 267–302.

[N] Nishida, T.
    Global solution for an initial-boundary-value problem of a quasilinear hyperbolic system. *Proc. Jap. Acad.,* **44** (1968), 642–646.

[NS] Nishida, T. and J. Smoller
    1. Solutions in the large for some nonlinear hyperbolic conservation laws. *Comm. Pure Appl. Math.,* **26** (1973), 183–200.
    2. Mixed problems for nonlinear conservation laws. *J. Diff. Eqns.,* **23** (1977), 244–269.
    3. A class of convergent finite-difference schemes for certain nonlinear parabolic systems. *Comm. Pure Appl. Math.,* **36** (1983), 785–808.

[O] Oleinik, O.
    1. Discontinuous solutions of nonlinear differential equations. *Usp. Mat. Nauk. (N.S.),* **12**, (1957) 3–73; English transl. in *Amer. Math. Soc. Transl. Ser. 2,* **26**, 95–172.
    2. On the uniqueness of the generalized solution of the Cauchy problem for a nonlinear system of equations occurring in mechanics. *Usp. Mat. Nauk. (N.S.),* **12** (1957), 169–176 (in Russian).
    3. Uniqueness and stability of the generalized solution of the Cauchy problem for a quasilinear equation. *Usp. Mat. Nauk.,* **14** (1959), 165–170; English transl. in *Amer. Math. Soc. Transl., Ser. 2,* **33** (1964), 285–290.

[Ot] Othmer, H.
    Nonlinear wave propagation in reacting systems. *J. Math. Biol.,* **2** (1975), 133–163.

[P] Petrovsky, I.
    Lectures on Partial Differential Equations. Wiley–Interscience: New York, 1954.

[PW] Protter, M. and H. Weinberger
    Maximum Principles in Differential Equations. Prentice Hall: Englewood Cliffs, NJ, 1967.

[Q] Quinn, B
    Solutions with shocks: an example of an L_1-contraction semi-group. *Comm. Pure Appl. Math.,* **24** (1971), 125–132.

[Ra] Rabinowitz, P.
    Some global results for nonlinear eigenvalue problems. *J. Funct. Anal.,* **7** (1971), 487–513.

[R] Rauch, J.
    Stability of motion for semilinear equations. *Boundary Value Problems for Linear Evolution Partial Differential Equations,* edited by H. Garnir. Reidel: Dordrecht, 1977, pp. 319–349.

[RS] Rauch, J. and J. Smoller
    Qualitative theory of the Fitz-Hugh–Nagumo equations. *Adv. in Math.,* **27** (1978), 12–44.

[Re] Reed, M.
    Abstract Nonlinear Wave Equations. Springer Lecture Notes, **507** (1976).

[Ri] Riemann, B.
    Gesammelte Werke, 1896, pp. 149ff.

[Ro] Rosenthal, A.
    On functions with infinitely-many derivatives. *Proc. Amer. Math. Soc.,* **4** (1953), 600–602.

[Rs] Ross, S.
    A First Course in Probability. Macmillan: New York, 1976.

[Rz] Rozdestvenskii, B.
Discontinuous solutions of hyperbolic systems of quasilinear equations. *Usp. Mat. Nauk.* (N.S.), **15** (1960) 59–118; English transl. in *Russian Math. Surveys*, **15** (1960), 53–111.

[RY] Rozdestvenskii, B. and N. Yanenko
Systems of Quasi-Linear Equations, Nauka: Moscow, 1968 (in Russian). 2nd edition: English translation, Translations of Mathematical Monographs, Vol 55, Amer. Math. Soc., Providence, 1983.

[Ru] Rudin, W.
1. *Real and Complex Analysis*. McGraw-Hill: New York, 1976.
2. *Principles of Mathematical Analysis*, 3rd ed. McGraw-Hill: New York, 1976.
3. *Functional Analysis*. McGraw-Hill: New York, 1973.

[Sa] Sattinger, D.
1. Stability of bifurcating solutions by Leray–Schauder degree. *Arch. Rat. Mech. Anal.*, **43** (1971), 154–166.
2. Monotone methods in nonlinear elliptic and parabolic equations. *Ind. U. Math. J.*, **21** (1972) 979–1000.

[Sf] Schaeffer, D.
A regularity theorem for conservation laws. *Adv. Math.*, **11** (1973), 368–386.

[Sc] Schauder, J.
Der fixpunktsatz in Funktionalraüme. *Studia Math.*, **2** (1936), 171–180.

[Sr] Schecter, M.
Principles of Functional Analysis. Academic Press: New York, 1971.

[Sk] Schonbeck, M.
Boundary-value problems for the Fitz-Hugh–Nagumo equations. *J. Diff. Eqns.*, **30** (1978), 119–147.

[Sw] Schwartz, J.
Nonlinear functional analysis. Courant Inst. Math. Sci. Notes (1963/64).

[Sz] Schwartz, L.
Théorie Des Distributions, tomes 1 et 2. Herman: Paris, 1957.

[SZ] Scudo, F. and R. Ziegler
The Golden Age of Theoretical Ecology: 1923–1940. Springer Lecture Notes in Biomath., **22** (1978).

[Sm] Smith, J.
Models in Ecology, Cambridge University Press: Cambridge, 1974.

[St] Smith, R.
The Riemann problem in gas dynamics. *Trans. Amer. Math. Soc.*, **249** (1979), 1–50.

[Smo] Smoller, J.
On the solution of the Riemann problem with general step data for an extended class of hyperbolic systems. *Mich. Math. J.*, **16** (1969), 201–210.

[SJ] Smoller, J. and J. Johnson
Global solutions for an extended class of hyperbolic systems of conservation laws. *Arch. Rat. Mech. Anal.*, **32** (1969), 169–189.

[ST] Smoller, J. and M. Taylor
Wave front sets and the viscosity method. *Bull. Amer. Math. Soc.*, **79** (1973), 431–436.

[STW] Smoller, J., A. Tromba, and A. Wasserman
Nondegenerate solutions of boundary-value problems. *Nonlinear Anal.*, **4** (1980), 207–215.

[SW] Smoller, J. and A. Wasserman
Global bifurcation of steady-state solutions. *J. Diff. Eqns.*, **39** (1981), 269–290.

[Sp] Spanier, E.
Algebraic Topology. McGraw-Hill: New York, 1966.

[So] Sobolev, S.
Sur un théorème de l'analyse functionelle. *Mat. Sb.*, **46** (1938), 471–496.

[Sp] Spanier, E.
Algebraic Topology. McGraw-Hill: New York, 1966.

[Te] Temple, B.
1. Solutions in the large for some nonlinear hyperbolic conservation laws of gas dynamics. *J. Diff. Eqns.*, **41** (1981), 96–161.

2. No L_1-contractive metrics for systems of conservation laws, *Trans. Amer. Math. Soc.*, **288** (1985), 471–480.

[Tr] Treves, F.
1. Relations de domination entre opérateurs differentiels. *Acta. Math.*, **101** (1959), 1–139.
2. *Basic Linear Partial Differential Equations.* Academic Press: New York, 1975.

[Ty] Troy, W.
A threshold phenomenon in the Field–Noyes model of the Belousov–Zhabotinsky reaction. *J. Math. Anal. Appl.*, **58** (1977), 233–248.

[Tu] Turner, R.
Transversality and cone maps. *Arch. Rat. Mech. Anal.*, **58** (1973), 151–179.

[Tv] Tychanov, A.
Théorèmes d'unicité pour l'équation de la chaleur. *Mat. Sb.*, **42** (1935), 199–215.

[V] Volpert, A.
The spaces BV and quasilinear equations. *Mat. Sb.*, **73** (1967), 255–302; English transl. in *Math. USSR, Sb.*, **2** (1967), 225–267.

[Wi] Weinberger, H.
Invariant sets for weakly coupled parabolic and elliptic systems. *Rend. Mat.*, **8** (1975), 295–310.

[We] Wendroff, B.
1. The Riemann problem for materials with nonconvex equations of state, II. General flow. *J. Math. Anal. Appl.*, **38** (1972), 640–658.
2. *Ibid.* I. Isentropic flow.

[Wh] Whyburn, G. T., *Analytic Topology*, AMS Colloq. Publ., Vol. 28, 1942.

[Wy] Weyl, H.
Shock waves in arbitrary fluids. *Comm. Pure Appl. Math.*, **2** (1949), 103–122.

[Wi] Wilansky, A.
Functional Analysis, Blaisdell: New York, 1964.

[Wu] Wu Zhuo-qun
The ordinary differential equations with discontinuous right-hand members and the discontinuous solutions of the quasilinear partial differential equations. *Acta. Math. Sinica*, **13** (1963), 515–530; English transl. in *Scientica Sinica*, **13** (1964), 1901–1907.

[ZR] Zel'dovich, Ya. and Yu. Razier
Elements of Gas Dynamics and the Classical Theory of Shock Waves. Academic Press: New York, 1968.

[ZG] Zhang, Tong and Guo Yu-Fa
A class of initial-value problems for systems of aerodynamic equations. *Acta. Math. Sinica*, **15** (1965), 386–396; English transl. in *Chinese Math.*, **7** (1965), 90–101.

Author Index

Subject Index

Grundlehren der mathematischen Wissenschaften

Continued from page ii

Grundlehren der mathematischen Wissenschaften